Lecture Notes in Computer Science 12901

More information about this subseries at http://www.springer.com/series/7412

Marleen de Bruijne · Philippe C. Cattin ·
Stéphane Cotin · Nicolas Padoy ·
Stefanie Speidel · Yefeng Zheng ·
Caroline Essert (Eds.)

Medical Image Computing and Computer Assisted Intervention – MICCAI 2021

24th International Conference
Strasbourg, France, September 27 – October 1, 2021
Proceedings, Part I

 Springer

Editors
Marleen de Bruijne (iD)
Erasmus MC - University Medical Center
Rotterdam
Rotterdam, The Netherlands

University of Copenhagen
Copenhagen, Denmark

Stéphane Cotin (iD)
Inria Nancy Grand Est
Villers-lès-Nancy, France

Stefanie Speidel (iD)
National Center for Tumor Diseases
(NCT/UCC)
Dresden, Germany

Caroline Essert (iD)
ICube, Université de Strasbourg, CNRS
Strasbourg, France

Philippe C. Cattin (iD)
University of Basel
Allschwil, Switzerland

Nicolas Padoy (iD)
ICube, Université de Strasbourg, CNRS
Strasbourg, France

Yefeng Zheng (iD)
Tencent Jarvis Lab
Shenzhen, China

ISSN 0302-9743   ISSN 1611-3349   (electronic)
Lecture Notes in Computer Science
ISBN 978-3-030-87192-5   ISBN 978-3-030-87193-2   (eBook)
https://doi.org/10.1007/978-3-030-87193-2

LNCS Sublibrary: SL6 – Image Processing, Computer Vision, Pattern Recognition, and Graphics

This Springer imprint is published by the registered company Springer Nature Switzerland AG
The registered company address is: Gewerbestrasse 11, 6330 Cham, Switzerland

Preface

The 24th edition of the International Conference on Medical Image Computing and Computer Assisted Intervention (MICCAI 2021) has for the second time been placed under the shadow of COVID-19. Complicated situations due to the pandemic and multiple lockdowns have affected our lives during the past year, sometimes perturbing the researchers work, but also motivating an extraordinary dedication from many of our colleagues, and significant scientific advances in the fight against the virus. After another difficult year, most of us were hoping to be able to travel and finally meet in person at MICCAI 2021, which was supposed to be held in Strasbourg, France. Unfortunately, due to the uncertainty of the global situation, MICCAI 2021 had to be moved again to a virtual event that was held over five days from September 27 to October 1, 2021. Taking advantage of the experience gained last year and of the fast-evolving platforms, the organizers of MICCAI 2021 redesigned the schedule and the format. To offer the attendees both a strong scientific content and an engaging experience, two virtual platforms were used: Pathable for the oral and plenary sessions and SpatialChat for lively poster sessions, industrial booths, and networking events in the form of interactive group video chats.

These proceedings of MICCAI 2021 showcase all 531 papers that were presented at the main conference, organized into eight volumes in the Lecture Notes in Computer Science (LNCS) series as follows:

- Part I, LNCS Volume 12901: Image Segmentation
- Part II, LNCS Volume 12902: Machine Learning 1
- Part III, LNCS Volume 12903: Machine Learning 2
- Part IV, LNCS Volume 12904: Image Registration and Computer Assisted Intervention
- Part V, LNCS Volume 12905: Computer Aided Diagnosis
- Part VI, LNCS Volume 12906: Image Reconstruction and Cardiovascular Imaging
- Part VII, LNCS Volume 12907: Clinical Applications
- Part VIII, LNCS Volume 12908: Microscopic, Ophthalmic, and Ultrasound Imaging

These papers were selected after a thorough double-blind peer review process. We followed the example set by past MICCAI meetings, using Microsoft's Conference Managing Toolkit (CMT) for paper submission and peer reviews, with support from the Toronto Paper Matching System (TPMS), to partially automate paper assignment to area chairs and reviewers, and from iThenticate to detect possible cases of plagiarism.

Following a broad call to the community we received 270 applications to become an area chair for MICCAI 2021. From this group, the program chairs selected a total of 96 area chairs, aiming for diversity — MIC versus CAI, gender, geographical region, and a mix of experienced and new area chairs. Reviewers were recruited also via an open call for volunteers from the community (288 applications, of which 149 were selected by the program chairs) as well as by re-inviting past reviewers, leading to a total of 1340 registered reviewers.

We received 1630 full paper submissions after an original 2667 intentions to submit. Four papers were rejected without review because of concerns of (self-)plagiarism and dual submission and one additional paper was rejected for not adhering to the MICCAI page restrictions; two further cases of dual submission were discovered and rejected during the review process. Five papers were withdrawn by the authors during review and after acceptance.

The review process kicked off with a reviewer tutorial and an area chair meeting to discuss the review process, criteria for MICCAI acceptance, how to write a good (meta-)review, and expectations for reviewers and area chairs. Each area chair was assigned 16–18 manuscripts for which they suggested potential reviewers using TPMS scores, self-declared research area(s), and the area chair's knowledge of the reviewers' expertise in relation to the paper, while conflicts of interest were automatically avoided by CMT. Reviewers were invited to bid for the papers for which they had been suggested by an area chair or which were close to their expertise according to TPMS. Final reviewer allocations via CMT took account of reviewer bidding, prioritization of area chairs, and TPMS scores, leading to on average four reviews performed per person by a total of 1217 reviewers.

Following the initial double-blind review phase, area chairs provided a meta-review summarizing key points of reviews and a recommendation for each paper. The program chairs then evaluated the reviews and their scores, along with the recommendation from the area chairs, to directly accept 208 papers (13%) and reject 793 papers (49%); the remainder of the papers were sent for rebuttal by the authors. During the rebuttal phase, two additional area chairs were assigned to each paper. The three area chairs then independently ranked their papers, wrote meta-reviews, and voted to accept or reject the paper, based on the reviews, rebuttal, and manuscript. The program chairs checked all meta-reviews, and in some cases where the difference between rankings was high or comments were conflicting, they also assessed the original reviews, rebuttal, and submission. In all other cases a majority voting scheme was used to make the final decision. This process resulted in the acceptance of a further 325 papers for an overall acceptance rate of 33%.

Acceptance rates were the same between medical image computing (MIC) and computer assisted interventions (CAI) papers, and slightly lower where authors classified their paper as both MIC and CAI. Distribution of the geographical region of the first author as indicated in the optional demographic survey was similar among submitted and accepted papers.

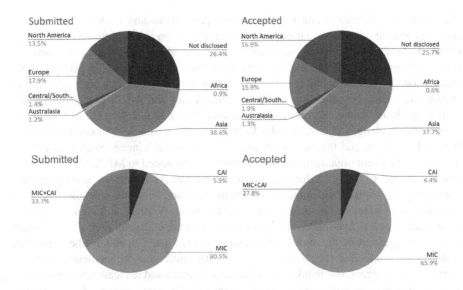

New this year, was the requirement to fill out a reproducibility checklist when submitting an intention to submit to MICCAI, in order to stimulate authors to think about what aspects of their method and experiments they should include to allow others to reproduce their results. Papers that included an anonymous code repository and/or indicated that the code would be made available were more likely to be accepted. From all accepted papers, 273 (51%) included a link to a code repository with the camera-ready submission.

Another novelty this year is that we decided to make the reviews, meta-reviews, and author responses for accepted papers available on the website. We hope the community will find this a useful resource.

The outstanding program of MICCAI 2021 was enriched by four exceptional keynote talks given by Alyson McGregor, Richard Satava, Fei-Fei Li, and Pierre Jannin, on hot topics such as gender bias in medical research, clinical translation to industry, intelligent medicine, and sustainable research. This year, as in previous years, high-quality satellite events completed the program of the main conference: 28 workshops, 23 challenges, and 14 tutorials; without forgetting the increasingly successful plenary events, such as the Women in MICCAI (WiM) meeting, the MICCAI Student Board (MSB) events, the 2nd Startup Village, the MICCAI-RSNA panel, and the first "Reinforcing Inclusiveness & diverSity and Empowering MICCAI" (or RISE-MICCAI) event.

MICCAI 2021 has also seen the first edition of CLINICCAI, the clinical day of MICCAI. Organized by Nicolas Padoy and Lee Swanstrom, this new event will hopefully help bring the scientific and clinical communities closer together, and foster collaborations and interaction. A common keynote connected the two events. We hope this effort will be pursued in the next editions.

We would like to thank everyone who has contributed to making MICCAI 2021 a success. First of all, we sincerely thank the authors, area chairs, reviewers, and session chairs for their dedication and for offering the participants and readers of these proceedings content of exceptional quality. Special thanks go to our fantastic submission platform manager Kitty Wong, who has been a tremendous help in the entire process from reviewer and area chair selection, paper submission, and the review process to the preparation of these proceedings. We also thank our very efficient team of satellite events chairs and coordinators, led by Cristian Linte and Matthieu Chabanas: the workshop chairs, Amber Simpson, Denis Fortun, Marta Kersten-Oertel, and Sandrine Voros; the challenges chairs, Annika Reinke, Spyridon Bakas, Nicolas Passat, and Ingerid Reinersten; and the tutorial chairs, Sonia Pujol and Vincent Noblet, as well as all the satellite event organizers for the valuable content added to MICCAI. Our special thanks also go to John Baxter and his team who worked hard on setting up and populating the virtual platforms, to Alejandro Granados for his valuable help and efficient communication on social media, and to Shelley Wallace and Anna Van Vliet for marketing and communication. We are also very grateful to Anirban Mukhopadhay for his management of the sponsorship, and of course many thanks to the numerous sponsors who supported the conference, often with continuous engagement over many years. This year again, our thanks go to Marius Linguraru and his team who supervised a range of actions to help, and promote, career development, among which were the mentorship program and the Startup Village. And last but not least, our wholehearted thanks go to Mehmet and the wonderful team at Dekon Congress and Tourism for their great professionalism and reactivity in the management of all logistical aspects of the event.

Finally, we thank the MICCAI society and the Board of Directors for their support throughout the years, starting with the first discussions about bringing MICCAI to Strasbourg in 2017.

We look forward to seeing you at MICCAI 2022.

September 2021

Marleen de Bruijne
Philippe Cattin
Stéphane Cotin
Nicolas Padoy
Stefanie Speidel
Yefeng Zheng
Caroline Essert

Organization

General Chair

Caroline Essert — Université de Strasbourg, CNRS, ICube, France

Program Chairs

Marleen de Bruijne — Erasmus MC Rotterdam, The Netherlands, and University of Copenhagen, Denmark
Philippe C. Cattin — University of Basel, Switzerland
Stéphane Cotin — Inria, France
Nicolas Padoy — Université de Strasbourg, CNRS, ICube, IHU, France
Stefanie Speidel — National Center for Tumor Diseases, Dresden, Germany
Yefeng Zheng — Tencent Jarvis Lab, China

Satellite Events Coordinators

Cristian Linte — Rochester Institute of Technology, USA
Matthieu Chabanas — Université Grenoble Alpes, France

Workshop Team

Amber Simpson — Queen's University, Canada
Denis Fortun — Université de Strasbourg, CNRS, ICube, France
Marta Kersten-Oertel — Concordia University, Canada
Sandrine Voros — TIMC-IMAG, INSERM, France

Challenges Team

Annika Reinke — German Cancer Research Center, Germany
Spyridon Bakas — University of Pennsylvania, USA
Nicolas Passat — Université de Reims Champagne-Ardenne, France
Ingerid Reinersten — SINTEF, NTNU, Norway

Tutorial Team

Vincent Noblet — Université de Strasbourg, CNRS, ICube, France
Sonia Pujol — Harvard Medical School, Brigham and Women's Hospital, USA

Clinical Day Chairs

Nicolas Padoy   Université de Strasbourg, CNRS, ICube, IHU, France
Lee Swanström   IHU Strasbourg, France

Sponsorship Chairs

Anirban Mukhopadhyay   Technische Universität Darmstadt, Germany
Yanwu Xu   Baidu Inc., China

Young Investigators and Early Career Development Program Chairs

Marius Linguraru   Children's National Institute, USA
Antonio Porras   Children's National Institute, USA
Daniel Racoceanu   Sorbonne Université/Brain Institute, France
Nicola Rieke   NVIDIA, Germany
Renee Yao   NVIDIA, USA

Social Media Chairs

Alejandro Granados   King's College London, UK
  Martinez
Shuwei Xing   Robarts Research Institute, Canada
Maxence Boels   King's College London, UK

Green Team

Pierre Jannin   INSERM, Université de Rennes 1, France
Étienne Baudrier   Université de Strasbourg, CNRS, ICube, France

Student Board Liaison

Éléonore Dufresne   Université de Strasbourg, CNRS, ICube, France
Étienne Le Quentrec   Université de Strasbourg, CNRS, ICube, France
Vinkle Srivastav   Université de Strasbourg, CNRS, ICube, France

Submission Platform Manager

Kitty Wong   The MICCAI Society, Canada

Virtual Platform Manager

John Baxter   INSERM, Université de Rennes 1, France

Program Committee

Ehsan Adeli	Stanford University, USA
Iman Aganj	Massachusetts General Hospital, Harvard Medical School, USA
Pablo Arbelaez	Universidad de los Andes, Colombia
John Ashburner	University College London, UK
Meritxell Bach Cuadra	University of Lausanne, Switzerland
Sophia Bano	University College London, UK
Adrien Bartoli	Université Clermont Auvergne, France
Christian Baumgartner	ETH Zürich, Switzerland
Hrvoje Bogunovic	Medical University of Vienna, Austria
Weidong Cai	University of Sydney, Australia
Gustavo Carneiro	University of Adelaide, Australia
Chao Chen	Stony Brook University, USA
Elvis Chen	Robarts Research Institute, Canada
Hao Chen	Hong Kong University of Science and Technology, Hong Kong SAR
Albert Chung	Hong Kong University of Science and Technology, Hong Kong SAR
Adrian Dalca	Massachusetts Institute of Technology, USA
Adrien Depeursinge	HES-SO Valais-Wallis, Switzerland
Jose Dolz	ÉTS Montréal, Canada
Ruogu Fang	University of Florida, USA
Dagan Feng	University of Sydney, Australia
Huazhu Fu	Inception Institute of Artificial Intelligence, United Arab Emirates
Mingchen Gao	University at Buffalo, The State University of New York, USA
Guido Gerig	New York University, USA
Orcun Goksel	Uppsala University, Sweden
Alberto Gomez	King's College London, UK
Ilker Hacihaliloglu	Rutgers University, USA
Adam Harrison	PAII Inc., USA
Mattias Heinrich	University of Lübeck, Germany
Yi Hong	Shanghai Jiao Tong University, China
Yipeng Hu	University College London, UK
Junzhou Huang	University of Texas at Arlington, USA
Xiaolei Huang	The Pennsylvania State University, USA
Jana Hutter	King's College London, UK
Madhura Ingalhalikar	Symbiosis Center for Medical Image Analysis, India
Shantanu Joshi	University of California, Los Angeles, USA
Samuel Kadoury	Polytechnique Montréal, Canada
Fahmi Khalifa	Mansoura University, Egypt
Hosung Kim	University of Southern California, USA
Minjeong Kim	University of North Carolina at Greensboro, USA

Zhong Xue   Shanghai United Imaging Intelligence, China
Xin Yang   Huazhong University of Science and Technology,
                               China
Jianhua Yao   National Institutes of Health, USA
Zhaozheng Yin   Stony Brook University, USA
Yixuan Yuan   City University of Hong Kong, Hong Kong SAR
Liang Zhan   University of Pittsburgh, USA
Tuo Zhang   Northwestern Polytechnical University, China
Yitian Zhao   Chinese Academy of Sciences, China
Luping Zhou   University of Sydney, Australia
S. Kevin Zhou   Chinese Academy of Sciences, China
Dajiang Zhu   University of Texas at Arlington, USA
Xiahai Zhuang   Fudan University, China
Maria A. Zuluaga   EURECOM, France

Reviewers

Alaa Eldin Abdelaal   Chloé Audigier
Khalid Abdul Jabbar   Kamran Avanaki
Purang Abolmaesumi   Angelica Aviles-Rivero
Mazdak Abulnaga   Suyash Awate
Maryam Afzali   Dogu Baran Aydogan
Priya Aggarwal   Qinle Ba
Ola Ahmad   Morteza Babaie
Sahar Ahmad   Hyeon-Min Bae
Euijoon Ahn   Woong Bae
Alireza Akhondi-Asl   Junjie Bai
Saad Ullah Akram   Wenjia Bai
Dawood Al Chanti   Ujjwal Baid
Daniel Alexander   Spyridon Bakas
Sharib Ali   Yaël Balbastre
Lejla Alic   Marcin Balicki
Omar Al-Kadi   Fabian Balsiger
Maximilian Allan   Abhirup Banerjee
Pierre Ambrosini   Sreya Banerjee
Sameer Antani   Shunxing Bao
Michela Antonelli   Adrian Barbu
Jacob Antunes   Sumana Basu
Syed Anwar   Mathilde Bateson
Ignacio Arganda-Carreras   Deepti Bathula
Mohammad Ali Armin   John Baxter
Md Ashikuzzaman   Bahareh Behboodi
Mehdi Astaraki   Delaram Behnami
Angélica Atehortúa   Mikhail Belyaev
Gowtham Atluri   Aicha BenTaieb

Camilo Bermudez
Gabriel Bernardino
Hadrien Bertrand
Alaa Bessadok
Michael Beyeler
Indrani Bhattacharya
Chetan Bhole
Lei Bi
Gui-Bin Bian
Ryoma Bise
Stefano B. Blumberg
Ester Bonmati
Bhushan Borotikar
Jiri Borovec
Ilaria Boscolo Galazzo
Alexandre Bousse
Nicolas Boutry
Behzad Bozorgtabar
Nathaniel Braman
Nadia Brancati
Katharina Breininger
Christopher Bridge
Esther Bron
Rupert Brooks
Qirong Bu
Duc Toan Bui
Ninon Burgos
Nikolay Burlutskiy
Hendrik Burwinkel
Russell Butler
Michał Byra
Ryan Cabeen
Mariano Cabezas
Hongmin Cai
Jinzheng Cai
Yunliang Cai
Sema Candemir
Bing Cao
Qing Cao
Shilei Cao
Tian Cao
Weiguo Cao
Aaron Carass
M. Jorge Cardoso
Adrià Casamitjana
Matthieu Chabanas

Ahmad Chaddad
Jayasree Chakraborty
Sylvie Chambon
Yi Hao Chan
Ming-Ching Chang
Peng Chang
Violeta Chang
Sudhanya Chatterjee
Christos Chatzichristos
Antong Chen
Chang Chen
Cheng Chen
Dongdong Chen
Geng Chen
Hanbo Chen
Jianan Chen
Jianxu Chen
Jie Chen
Junxiang Chen
Lei Chen
Li Chen
Liangjun Chen
Min Chen
Pingjun Chen
Qiang Chen
Shuai Chen
Tianhua Chen
Tingting Chen
Xi Chen
Xiaoran Chen
Xin Chen
Xuejin Chen
Yuhua Chen
Yukun Chen
Zhaolin Chen
Zhineng Chen
Zhixiang Chen
Erkang Cheng
Jun Cheng
Li Cheng
Yuan Cheng
Farida Cheriet
Minqi Chong
Jaegul Choo
Aritra Chowdhury
Gary Christensen

Daan Christiaens
Stergios Christodoulidis
Ai Wern Chung
Pietro Antonio Cicalese
Özgün Çiçek
Celia Cintas
Matthew Clarkson
Jaume Coll-Font
Toby Collins
Olivier Commowick
Pierre-Henri Conze
Timothy Cootes
Luca Corinzia
Teresa Correia
Hadrien Courtecuisse
Jeffrey Craley
Hui Cui
Jianan Cui
Zhiming Cui
Kathleen Curran
Claire Cury
Tobias Czempiel
Vedrana Dahl
Haixing Dai
Rafat Damseh
Bilel Daoud
Neda Davoudi
Laura Daza
Sandro De Zanet
Charles Delahunt
Yang Deng
Cem Deniz
Felix Denzinger
Hrishikesh Deshpande
Christian Desrosiers
Blake Dewey
Neel Dey
Raunak Dey
Jwala Dhamala
Yashin Dicente Cid
Li Ding
Xinghao Ding
Zhipeng Ding
Konstantin Dmitriev
Ines Domingues
Liang Dong

Mengjin Dong
Nanqing Dong
Reuben Dorent
Sven Dorkenwald
Qi Dou
Simon Drouin
Niharika D'Souza
Lei Du
Hongyi Duanmu
Nicolas Duchateau
James Duncan
Luc Duong
Nicha Dvornek
Dmitry V. Dylov
Oleh Dzyubachyk
Roy Eagleson
Mehran Ebrahimi
Jan Egger
Alma Eguizabal
Gudmundur Einarsson
Ahmed Elazab
Mohammed S. M. Elbaz
Shireen Elhabian
Mohammed Elmogy
Amr Elsawy
Ahmed Eltanboly
Sandy Engelhardt
Ertunc Erdil
Marius Erdt
Floris Ernst
Boris Escalante-Ramírez
Maria Escobar
Mohammad Eslami
Nazila Esmaeili
Marco Esposito
Oscar Esteban
Théo Estienne
Ivan Ezhov
Deng-Ping Fan
Jingfan Fan
Xin Fan
Yonghui Fan
Xi Fang
Zhenghan Fang
Aly Farag
Mohsen Farzi

Liang Han
Xiaoguang Han
Xu Han
Zhi Han
Zhongyi Han
Jonny Hancox
Xiaoke Hao
Nandinee Haq
Ali Hatamizadeh
Charles Hatt
Andreas Hauptmann
Mohammad Havaei
Kelei He
Nanjun He
Tiancheng He
Xuming He
Yuting He
Nicholas Heller
Alessa Hering
Monica Hernandez
Carlos Hernandez-Matas
Kilian Hett
Jacob Hinkle
David Ho
Nico Hoffmann
Matthew Holden
Sungmin Hong
Yoonmi Hong
Antal Horváth
Md Belayat Hossain
Benjamin Hou
William Hsu
Tai-Chiu Hsung
Kai Hu
Shi Hu
Shunbo Hu
Wenxing Hu
Xiaoling Hu
Xiaowei Hu
Yan Hu
Zhenhong Hu
Heng Huang
Qiaoying Huang
Yi-Jie Huang
Yixing Huang
Yongxiang Huang

Yue Huang
Yufang Huang
Arnaud Huaulmé
Henkjan Huisman
Yuankai Huo
Andreas Husch
Mohammad Hussain
Raabid Hussain
Sarfaraz Hussein
Khoi Huynh
Seong Jae Hwang
Emmanuel Iarussi
Kay Igwe
Abdullah-Al-Zubaer Imran
Ismail Irmakci
Mobarakol Islam
Mohammad Shafkat Islam
Vamsi Ithapu
Koichi Ito
Hayato Itoh
Oleksandra Ivashchenko
Yuji Iwahori
Shruti Jadon
Mohammad Jafari
Mostafa Jahanifar
Amir Jamaludin
Mirek Janatka
Won-Dong Jang
Uditha Jarayathne
Ronnachai Jaroensri
Golara Javadi
Rohit Jena
Rachid Jennane
Todd Jensen
Won-Ki Jeong
Yuanfeng Ji
Zhanghexuan Ji
Haozhe Jia
Jue Jiang
Tingting Jiang
Xiang Jiang
Jianbo Jiao
Zhicheng Jiao
Amelia Jiménez-Sánchez
Dakai Jin
Yueming Jin

Bin Jing
Anand Joshi
Yohan Jun
Kyu-Hwan Jung
Alain Jungo
Manjunath K N
Ali Kafaei Zad Tehrani
Bernhard Kainz
John Kalafut
Michael C. Kampffmeyer
Qingbo Kang
Po-Yu Kao
Neerav Karani
Turkay Kart
Satyananda Kashyap
Amin Katouzian
Alexander Katzmann
Prabhjot Kaur
Erwan Kerrien
Hoel Kervadec
Ashkan Khakzar
Nadieh Khalili
Siavash Khallaghi
Farzad Khalvati
Bishesh Khanal
Pulkit Khandelwal
Maksim Kholiavchenko
Naji Khosravan
Seyed Mostafa Kia
Daeseung Kim
Hak Gu Kim
Hyo-Eun Kim
Jae-Hun Kim
Jaeil Kim
Jinman Kim
Mansu Kim
Namkug Kim
Seong Tae Kim
Won Hwa Kim
Andrew King
Atilla Kiraly
Yoshiro Kitamura
Tobias Klinder
Bin Kong
Jun Kong
Tomasz Konopczynski

Bongjin Koo
Ivica Kopriva
Kivanc Kose
Mateusz Kozinski
Anna Kreshuk
Anithapriya Krishnan
Pavitra Krishnaswamy
Egor Krivov
Frithjof Kruggel
Alexander Krull
Elizabeth Krupinski
Serife Kucur
David Kügler
Hugo Kuijf
Abhay Kumar
Ashnil Kumar
Kuldeep Kumar
Nitin Kumar
Holger Kunze
Tahsin Kurc
Anvar Kurmukov
Yoshihiro Kuroda
Jin Tae Kwak
Yongchan Kwon
Francesco La Rosa
Aymen Laadhari
Dmitrii Lachinov
Alain Lalande
Tryphon Lambrou
Carole Lartizien
Bianca Lassen-Schmidt
Ngan Le
Leo Lebrat
Christian Ledig
Eung-Joo Lee
Hyekyoung Lee
Jong-Hwan Lee
Matthew Lee
Sangmin Lee
Soochahn Lee
Étienne Léger
Stefan Leger
Andreas Leibetseder
Rogers Jeffrey Leo John
Juan Leon
Bo Li

Chongyi Li
Fuhai Li
Hongming Li
Hongwei Li
Jian Li
Jianning Li
Jiayun Li
Junhua Li
Kang Li
Mengzhang Li
Ming Li
Qing Li
Shaohua Li
Shuyu Li
Weijian Li
Weikai Li
Wenqi Li
Wenyuan Li
Xiang Li
Xiaomeng Li
Xiaoxiao Li
Xin Li
Xiuli Li
Yang Li
Yi Li
Yuexiang Li
Zeju Li
Zhang Li
Zhiyuan Li
Zhjin Li
Gongbo Liang
Jianming Liang
Libin Liang
Yuan Liang
Haofu Liao
Ruizhi Liao
Wei Liao
Xiangyun Liao
Roxane Licandro
Gilbert Lim
Baihan Lin
Hongxiang Lin
Jianyu Lin
Yi Lin
Claudia Lindner
Geert Litjens

Bin Liu
Chi Liu
Daochang Liu
Dong Liu
Dongnan Liu
Feng Liu
Hangfan Liu
Hong Liu
Huafeng Liu
Jianfei Liu
Jingya Liu
Kai Liu
Kefei Liu
Lihao Liu
Mengting Liu
Peng Liu
Qin Liu
Quande Liu
Shengfeng Liu
Shenghua Liu
Shuangjun Liu
Sidong Liu
Siqi Liu
Tianrui Liu
Xiao Liu
Xinyang Liu
Xinyu Liu
Yan Liu
Yikang Liu
Yong Liu
Yuan Liu
Yue Liu
Yuhang Liu
Andrea Loddo
Nicolas Loménie
Daniel Lopes
Bin Lou
Jian Lou
Nicolas Loy Rodas
Donghuan Lu
Huanxiang Lu
Weijia Lu
Xiankai Lu
Yongyi Lu
Yueh-Hsun Lu
Yuhang Lu

Imanol Luengo
Jie Luo
Jiebo Luo
Luyang Luo
Ma Luo
Bin Lv
Jinglei Lv
Junyan Lyu
Qing Lyu
Yuanyuan Lyu
Andy J. Ma
Chunwei Ma
Da Ma
Hua Ma
Kai Ma
Lei Ma
Anderson Maciel
Amirreza Mahbod
S. Sara Mahdavi
Mohammed Mahmoud
Saïd Mahmoudi
Klaus H. Maier-Hein
Bilal Malik
Ilja Manakov
Matteo Mancini
Tommaso Mansi
Yunxiang Mao
Brett Marinelli
Pablo Márquez Neila
Carsten Marr
Yassine Marrakchi
Fabio Martinez
Andre Mastmeyer
Tejas Sudharshan Mathai
Dimitrios Mavroeidis
Jamie McClelland
Pau Medrano-Gracia
Raghav Mehta
Sachin Mehta
Raphael Meier
Qier Meng
Qingjie Meng
Yanda Meng
Martin Menten
Odyssée Merveille
Islem Mhiri

Liang Mi
Stijn Michielse
Abhishek Midya
Fausto Milletari
Hyun-Seok Min
Zhe Min
Tadashi Miyamoto
Sara Moccia
Hassan Mohy-ud-Din
Tony C. W. Mok
Rafael Molina
Mehdi Moradi
Rodrigo Moreno
Kensaku Mori
Lia Morra
Linda Moy
Mohammad Hamed Mozaffari
Sovanlal Mukherjee
Anirban Mukhopadhyay
Henning Müller
Balamurali Murugesan
Cosmas Mwikirize
Andriy Myronenko
Saad Nadeem
Vishwesh Nath
Rodrigo Nava
Fernando Navarro
Amin Nejatbakhsh
Dong Ni
Hannes Nickisch
Dong Nie
Jingxin Nie
Aditya Nigam
Lipeng Ning
Xia Ning
Tianye Niu
Jack Noble
Vincent Noblet
Alexey Novikov
Jorge Novo
Mohammad Obeid
Masahiro Oda
Benjamin Odry
Steffen Oeltze-Jafra
Hugo Oliveira
Sara Oliveira

Arnau Oliver
Emanuele Olivetti
Jimena Olveres
John Onofrey
Felipe Orihuela-Espina
José Orlando
Marcos Ortega
Yoshito Otake
Sebastian Otálora
Cheng Ouyang
Jiahong Ouyang
Xi Ouyang
Michal Ozery-Flato
Danielle Pace
Krittin Pachtrachai
J. Blas Pagador
Akshay Pai
Viswanath Pamulakanty Sudarshan
Jin Pan
Yongsheng Pan
Pankaj Pandey
Prashant Pandey
Egor Panfilov
Shumao Pang
Joao Papa
Constantin Pape
Bartlomiej Papiez
Hyunjin Park
Jongchan Park
Sanghyun Park
Seung-Jong Park
Seyoun Park
Magdalini Paschali
Diego Patiño Cortés
Angshuman Paul
Christian Payer
Yuru Pei
Chengtao Peng
Yige Peng
Antonio Pepe
Oscar Perdomo
Sérgio Pereira
Jose-Antonio Pérez-Carrasco
Fernando Pérez-García
Jorge Perez-Gonzalez
Skand Peri

Matthias Perkonigg
Mehran Pesteie
Jorg Peters
Jens Petersen
Kersten Petersen
Renzo Phellan Aro
Ashish Phophalia
Tomasz Pieciak
Antonio Pinheiro
Pramod Pisharady
Kilian Pohl
Sebastian Pölsterl
Iulia A. Popescu
Alison Pouch
Prateek Prasanna
Raphael Prevost
Juan Prieto
Sergi Pujades
Elodie Puybareau
Esther Puyol-Antón
Haikun Qi
Huan Qi
Buyue Qian
Yan Qiang
Yuchuan Qiao
Chen Qin
Wenjian Qin
Yulei Qin
Wu Qiu
Hui Qu
Liangqiong Qu
Kha Gia Quach
Prashanth R.
Pradeep Reddy Raamana
Mehdi Rahim
Jagath Rajapakse
Kashif Rajpoot
Jhonata Ramos
Lingyan Ran
Hatem Rashwan
Daniele Ravì
Keerthi Sravan Ravi
Nishant Ravikumar
Harish RaviPrakash
Samuel Remedios
Yinhao Ren

Yudan Ren
Mauricio Reyes
Constantino Reyes-Aldasoro
Jonas Richiardi
David Richmond
Anne-Marie Rickmann
Leticia Rittner
Dominik Rivoir
Emma Robinson
Jessica Rodgers
Rafael Rodrigues
Robert Rohling
Michal Rosen-Zvi
Lukasz Roszkowiak
Karsten Roth
José Rouco
Daniel Rueckert
Jaime S. Cardoso
Mohammad Sabokrou
Ario Sadafi
Monjoy Saha
Pramit Saha
Dushyant Sahoo
Pranjal Sahu
Maria Sainz de Cea
Olivier Salvado
Robin Sandkuehler
Gianmarco Santini
Duygu Sarikaya
Imari Sato
Olivier Saut
Dustin Scheinost
Nico Scherf
Markus Schirmer
Alexander Schlaefer
Jerome Schmid
Julia Schnabel
Klaus Schoeffmann
Andreas Schuh
Ernst Schwartz
Christina Schwarz-Gsaxner
Michaël Sdika
Suman Sedai
Anjany Sekuboyina
Raghavendra Selvan
Sourya Sengupta

Youngho Seo
Lama Seoud
Ana Sequeira
Maxime Sermesant
Carmen Serrano
Muhammad Shaban
Ahmed Shaffie
Sobhan Shafiei
Mohammad Abuzar Shaikh
Reuben Shamir
Shayan Shams
Hongming Shan
Harshita Sharma
Gregory Sharp
Mohamed Shehata
Haocheng Shen
Li Shen
Liyue Shen
Mali Shen
Yiqing Shen
Yiqiu Shen
Zhengyang Shen
Kuangyu Shi
Luyao Shi
Xiaoshuang Shi
Xueying Shi
Yemin Shi
Yiyu Shi
Yonghong Shi
Jitae Shin
Boris Shirokikh
Suprosanna Shit
Suzanne Shontz
Yucheng Shu
Alberto Signoroni
Wilson Silva
Margarida Silveira
Matthew Sinclair
Rohit Singla
Sumedha Singla
Ayushi Sinha
Kevin Smith
Rajath Soans
Ahmed Soliman
Stefan Sommer
Yang Song

Youyi Song
Aristeidis Sotiras
Arcot Sowmya
Rachel Sparks
William Speier
Ziga Spiclin
Dominik Spinczyk
Jon Sporring
Chetan Srinidhi
Anuroop Sriram
Vinkle Srivastav
Lawrence Staib
Marius Staring
Johannes Stegmaier
Joshua Stough
Robin Strand
Martin Styner
Hai Su
Yun-Hsuan Su
Vaishnavi Subramanian
Gérard Subsol
Yao Sui
Avan Suinesiaputra
Jeremias Sulam
Shipra Suman
Li Sun
Wenqing Sun
Chiranjib Sur
Yannick Suter
Tanveer Syeda-Mahmood
Fatemeh Taheri Dezaki
Roger Tam
José Tamez-Peña
Chaowei Tan
Hao Tang
Thomas Tang
Yucheng Tang
Zihao Tang
Mickael Tardy
Giacomo Tarroni
Jonas Teuwen
Paul Thienphrapa
Stephen Thompson
Jiang Tian
Yu Tian
Yun Tian

Aleksei Tiulpin
Hamid Tizhoosh
Matthew Toews
Oguzhan Topsakal
Antonio Torteya
Sylvie Treuillet
Jocelyne Troccaz
Roger Trullo
Chialing Tsai
Sudhakar Tummala
Verena Uslar
Hristina Uzunova
Régis Vaillant
Maria Vakalopoulou
Jeya Maria Jose Valanarasu
Tom van Sonsbeek
Gijs van Tulder
Marta Varela
Thomas Varsavsky
Francisco Vasconcelos
Liset Vazquez Romaguera
S. Swaroop Vedula
Sanketh Vedula
Harini Veeraraghavan
Miguel Vega
Gonzalo Vegas Sanchez-Ferrero
Anant Vemuri
Gopalkrishna Veni
Mitko Veta
Thomas Vetter
Pedro Vieira
Juan Pedro Vigueras Guillén
Barbara Villarini
Satish Viswanath
Athanasios Vlontzos
Wolf-Dieter Vogl
Bo Wang
Cheng Wang
Chengjia Wang
Chunliang Wang
Clinton Wang
Congcong Wang
Dadong Wang
Dongang Wang
Haifeng Wang
Hongyu Wang

Changchun Yang
Chao-Han Huck Yang
Dong Yang
Erkun Yang
Fan Yang
Ge Yang
Guang Yang
Guanyu Yang
Heran Yang
Hongxu Yang
Huijuan Yang
Jiancheng Yang
Jie Yang
Junlin Yang
Lin Yang
Peng Yang
Xin Yang
Yan Yang
Yujiu Yang
Dongren Yao
Jiawen Yao
Li Yao
Qingsong Yao
Chuyang Ye
Dong Hye Ye
Menglong Ye
Xujiong Ye
Jingru Yi
Jirong Yi
Xin Yi
Youngjin Yoo
Chenyu You
Haichao Yu
Hanchao Yu
Lequan Yu
Qi Yu
Yang Yu
Pengyu Yuan
Fatemeh Zabihollahy
Ghada Zamzmi
Marco Zenati
Guodong Zeng
Rui Zeng
Oliver Zettinig
Zhiwei Zhai
Chaoyi Zhang

Daoqiang Zhang
Fan Zhang
Guangming Zhang
Hang Zhang
Huahong Zhang
Jianpeng Zhang
Jiong Zhang
Jun Zhang
Lei Zhang
Lichi Zhang
Lin Zhang
Ling Zhang
Lu Zhang
Miaomiao Zhang
Ning Zhang
Qiang Zhang
Rongzhao Zhang
Ru-Yuan Zhang
Shihao Zhang
Shu Zhang
Tong Zhang
Wei Zhang
Weiwei Zhang
Wen Zhang
Wenlu Zhang
Xin Zhang
Ya Zhang
Yanbo Zhang
Yanfu Zhang
Yi Zhang
Yishuo Zhang
Yong Zhang
Yongqin Zhang
You Zhang
Youshan Zhang
Yu Zhang
Yue Zhang
Yueyi Zhang
Yulun Zhang
Yunyan Zhang
Yuyao Zhang
Can Zhao
Changchen Zhao
Chongyue Zhao
Fenqiang Zhao
Gangming Zhao

Outstanding Reviewers

Honorable Mentions (Reviewers)

Mazdak Abulnaga	Massachusetts Institute of Technology, USA
Pierre Ambrosini	Erasmus University Medical Center, The Netherlands
Hyeon-Min Bae	Korea Advanced Institute of Science and Technology, South Korea
Mikhail Belyaev	Skolkovo Institute of Science and Technology, Russia
Bhushan Borotikar	Symbiosis International University, India
Katharina Breininger	Friedrich-Alexander-Universität Erlangen-Nürnberg, Germany
Ninon Burgos	CNRS, Paris Brain Institute, France
Mariano Cabezas	The University of Sydney, Australia
Aaron Carass	Johns Hopkins University, USA
Pierre-Henri Conze	IMT Atlantique, France
Christian Desrosiers	École de technologie supérieure, Canada
Reuben Dorent	King's College London, UK
Nicha Dvornek	Yale University, USA
Dmitry V. Dylov	Skolkovo Institute of Science and Technology, Russia
Marius Erdt	Fraunhofer Singapore, Singapore
Ruibin Feng	Stanford University, USA
Enzo Ferrante	CONICET/Universidad Nacional del Litoral, Argentina
Antonio Foncubierta-Rodríguez	IBM Research, Switzerland
Isabel Funke	National Center for Tumor Diseases Dresden, Germany
Adrian Galdran	University of Bournemouth, UK
Ben Glocker	Imperial College London, UK
Cristina González	Universidad de los Andes, Colombia
Maged Goubran	Sunnybrook Research Institute, Canada
Sobhan Goudarzi	Concordia University, Canada
Vicente Grau	University of Oxford, UK
Andreas Hauptmann	University of Oulu, Finland
Nico Hoffmann	Technische Universität Dresden, Germany
Sungmin Hong	Massachusetts General Hospital, Harvard Medical School, USA
Won-Dong Jang	Harvard University, USA
Zhanghexuan Ji	University at Buffalo, SUNY, USA
Neerav Karani	ETH Zurich, Switzerland
Alexander Katzmann	Siemens Healthineers, Germany
Erwan Kerrien	Inria, France
Anitha Priya Krishnan	Genentech, USA
Tahsin Kurc	Stony Brook University, USA
Francesco La Rosa	École polytechnique fédérale de Lausanne, Switzerland
Dmitrii Lachinov	Medical University of Vienna, Austria
Mengzhang Li	Peking University, China
Gilbert Lim	National University of Singapore, Singapore
Dongnan Liu	University of Sydney, Australia

Bin Lou	Siemens Healthineers, USA
Kai Ma	Tencent, China
Klaus H. Maier-Hein	German Cancer Research Center (DKFZ), Germany
Raphael Meier	University Hospital Bern, Switzerland
Tony C. W. Mok	Hong Kong University of Science and Technology, Hong Kong SAR
Lia Morra	Politecnico di Torino, Italy
Cosmas Mwikirize	Rutgers University, USA
Felipe Orihuela-Espina	Instituto Nacional de Astrofísica, Óptica y Electrónica, Mexico
Egor Panfilov	University of Oulu, Finland
Christian Payer	Graz University of Technology, Austria
Sebastian Pölsterl	Ludwig-Maximilians Universität, Germany
José Rouco	University of A Coruña, Spain
Daniel Rueckert	Imperial College London, UK
Julia Schnabel	King's College London, UK
Christina Schwarz-Gsaxner	Graz University of Technology, Austria
Boris Shirokikh	Skolkovo Institute of Science and Technology, Russia
Yang Song	University of New South Wales, Australia
Gérard Subsol	Université de Montpellier, France
Tanveer Syeda-Mahmood	IBM Research, USA
Mickael Tardy	Hera-MI, France
Paul Thienphrapa	Atlas5D, USA
Gijs van Tulder	Radboud University, The Netherlands
Tongxin Wang	Indiana University, USA
Yirui Wang	PAII Inc., USA
Jelmer Wolterink	University of Twente, The Netherlands
Lei Xiang	Subtle Medical Inc., USA
Fatemeh Zabihollahy	Johns Hopkins University, USA
Wei Zhang	University of Georgia, USA
Ya Zhang	Shanghai Jiao Tong University, China
Qingyu Zhao	Stanford University, China
Yushan Zheng	Beihang University, China

Mentorship Program (Mentors)

Shadi Albarqouni	Helmholtz AI, Helmholtz Center Munich, Germany
Hao Chen	Hong Kong University of Science and Technology, Hong Kong SAR
Nadim Daher	NVIDIA, France
Marleen de Bruijne	Erasmus MC/University of Copenhagen, The Netherlands
Qi Dou	The Chinese University of Hong Kong, Hong Kong SAR
Gabor Fichtinger	Queen's University, Canada
Jonny Hancox	NVIDIA, UK

Nobuhiko Hata   Harvard Medical School, USA
Sharon Xiaolei Huang   Pennsylvania State University, USA
Jana Hutter   King's College London, UK
Dakai Jin   PAII Inc., China
Samuel Kadoury   Polytechnique Montréal, Canada
Minjeong Kim   University of North Carolina at Greensboro, USA
Hans Lamecker   1000shapes GmbH, Germany
Andrea Lara   Galileo University, Guatemala
Ngan Le   University of Arkansas, USA
Baiying Lei   Shenzhen University, China
Karim Lekadir   Universitat de Barcelona, Spain
Marius George Linguraru   Children's National Health System/George
                             Washington University, USA
Herve Lombaert   ETS Montreal, Canada
Marco Lorenzi   Inria, France
Le Lu   PAII Inc., China
Xiongbiao Luo   Xiamen University, China
Dzung Pham   Henry M. Jackson Foundation/Uniformed Services
                             University/National Institutes of Health/Johns
                             Hopkins University, USA
Josien Pluim   Eindhoven University of Technology/University
                             Medical Center Utrecht, The Netherlands
Antonio Porras   University of Colorado Anschutz Medical
                             Campus/Children's Hospital Colorado, USA
Islem Rekik   Istanbul Technical University, Turkey
Nicola Rieke   NVIDIA, Germany
Julia Schnabel   TU Munich/Helmholtz Center Munich, Germany,
                             and King's College London, UK
Debdoot Sheet   Indian Institute of Technology Kharagpur, India
Pallavi Tiwari   Case Western Reserve University, USA
Jocelyne Troccaz   CNRS, TIMC, Grenoble Alpes University, France
Sandrine Voros   TIMC-IMAG, INSERM, France
Linwei Wang   Rochester Institute of Technology, USA
Yalin Wang   Arizona State University, USA
Zhong Xue   United Imaging Intelligence Co. Ltd, USA
Renee Yao   NVIDIA, USA
Mohammad Yaqub   Mohamed Bin Zayed University of Artificial
                             Intelligence, United Arab Emirates, and University
                             of Oxford, UK
S. Kevin Zhou   University of Science and Technology of China, China
Lilla Zollei   Massachusetts General Hospital, Harvard Medical
                             School, USA
Maria A. Zuluaga   EURECOM, France

Contents – Part I

Image Segmentation

Noisy Labels are Treasure: Mean-Teacher-Assisted Confident Learning for Hepatic Vessel Segmentation

Zhe Xu^{1,3}, Donghuan Lu$^{2(\boxtimes)}$, Yixin Wang^4, Jie Luo^3, Jagadeesan Jayender^3, Kai Ma^2, Yefeng Zheng^2, and Xiu Li$^{1(\boxtimes)}$

^{1} Shenzhen International Graduate School, Tsinghua University, Shenzhen, China
li.xiu@sz.tsinghua.edu.cn
^{2} Tencent Jarvis Lab, Shenzhen, China
caleblu@tencent.com
^{3} Brigham and Women's Hospital, Harvard Medical School, Boston, USA
^{4} Institute of Computing Technology, Chinese Academy of Sciences, Beijing, China

Abstract. Manually segmenting the hepatic vessels from Computer Tomography (CT) is far more expertise-demanding and laborious than other structures due to the low-contrast and complex morphology of vessels, resulting in the extreme lack of high-quality labeled data. Without sufficient high-quality annotations, the usual data-driven learning-based approaches struggle with deficient training. On the other hand, directly introducing additional data with low-quality annotations may confuse the network, leading to undesirable performance degradation. To address this issue, we propose a novel mean-teacher-assisted confident learning framework to robustly exploit the noisy labeled data for the challenging hepatic vessel segmentation task. Specifically, with the adapted confident learning assisted by a third party, i.e., the weight-averaged teacher model, the noisy labels in the additional low-quality dataset can be transformed from 'encumbrance' to 'treasure' via progressive pixel-wise soft-correction, thus providing productive guidance. Extensive experiments using two public datasets demonstrate the superiority of the proposed framework as well as the effectiveness of each component.

Keywords: Hepatic vessel · Noisy label · Confident learning

1 Introduction

Segmenting hepatic vessels from Computer Tomography (CT) is essential to many hepatic surgeries such as liver resection and transplantation. Benefiting from a large amount of high-quality (HQ) pixel-wise labeled data, deep learning has greatly advanced in automatic abdominal segmentation for various structures, such as liver, kidney and spleen [5,9,13,16]. Unfortunately, due

---

Z. Xu—This work was done as a research intern at Tencent Jarvis Lab.

M. de Bruijne et al. (Eds.): MICCAI 2021, LNCS 12901, pp. 3–13, 2021.
https://doi.org/10.1007/978-3-030-87193-2_1

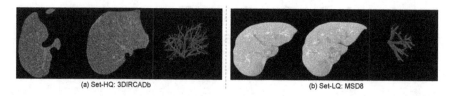

(a) Set-HQ: 3DIRCADb   (b) Set-LQ: MSD8

Fig. 1. 2D and 3D visualization of the processed example cases of (a) 3DIRCADb dataset [1] with high-quality annotations (Set-HQ), and (b) MSD8 dataset [21] with numerous mislabeled and unlabeled pixels (Set-LQ). Red represents the labeled vessels, while the yellow arrows at (b) point at some unlabeled pixels. (Color figure online)

to the noises in CT images, pathological variations, poor-contrast and complex morphology of vessels, manually delineating the hepatic vessels is far more expertise-demanding, laborious and error-prone than other structures. Thus, limited amount of data with HQ pixel-wise hepatic vessel annotations, as exampled in Fig. 1(a), is available. Most data, as exampled in Fig. 1(b), have considerable unlabeled or mislabeled pixels, also known as "noises".

For a typical fully-supervised segmentation method, training with tiny HQ labeled dataset often results in overfitting and inferior performance. However, additionally introducing data with low-quality (LQ) annotation may provide undesirable guidance, and offset the efficacy of the HQ labeled data. Experimentally, the considerable noises become the *'encumbrance'* for training, leading to substantial performance degradation, as shown in Fig. 3 and Table 1. Therefore, how to robustly exploit the additional information in the abundant LQ noisy labeled data remains an open challenge.

Related Work. Due to the lack of HQ labeled data and the complex morphology, few efforts have been made on hepatic vessel segmentation. Huang et al. applied the U-Net with a new variant Dice loss to balance the foreground (vessel) and background (liver) classes [8]. Kitrungrotsakul et al. used three deep networks to extract the vessel features from different planes of hepatic CT images [11]. Neglecting the data with LQ annotation because of their potential misleading guidance, only 10 and 5 HQ labeled volumes were used for training in [8] and [11], resulting in unsatisfactory performance. To introduce auxiliary image information from additional dataset, Semi-Supervised Learning (SSL) technique [4,22,24] is a promising method. However, the standard SSL-based methods fail to exploit the potential useful information of the noisy label. To make full use of the LQ labeled data, several efforts have been made to alleviate the negative effects brought by the noisy labels, such as assigning lower weights to the noisy labeled samples [18,28], modeling the label corrupting process [7] and confident learning [17]. However, these studies focused on image-level noise identification, while the localization of pixel-wise noises is necessary for the segmentation task.

In this paper, we propose a novel Mean-Teacher-assisted Confident Learning (MTCL) framework for hepatic vessel segmentation to leverage the additional *'cumbrous'* noisy labels in LQ labeled data. Specifically, our framework shares the same architecture as the mean-teacher model [22]. By encouraging consistent segmentation under different perturbations for the same input, the network can additionally exploit the image information of the LQ labeled data. Then, assisted by the weight-averaged teacher model, we adapt the Confident Learning (CL) technique [17], which was initially proposed for removing noisy labels in image-level classification, to characterize the pixel-wise label noises based on the Classification Noise Process (CNP) assumption [3]. With the guidance of the identified noise map, the proposed Smoothly Self-Denoising Module (SSDM) progressively transforms the LQ labels from *'encumbrance'* to *'treasure'*, allowing the network to robustly leverage the additional noisy labels towards superior segmentation performance. We conduct extensive experiments on two public datasets with hepatic vessel annotations [1,21]. The results demonstrate the superiority of the proposed framework as well as the effectiveness of each component.

2   Methods

The detailed explanation of the experimental materials, the hepatic CT preprocessing approach, and the proposed Mean-Teacher-assisted Confident Learning (MTCL) framework are presented in the following three sections, respectively.

2.1   Materials

Two public datasets, 3DIRCADb [1] and MSD8 [21], with obviously different qualities of annotation (shown in Fig. 1) are used in this study, tersely referred as Set-HQ (i.e., *high quality*) and Set-LQ (i.e., *low quality*), respectively.

1) **Set-HQ: 3DIRCADb** [1]. The first dataset, 3DIRCADb, maintained by the French Institute of Digestive Cancer Treatment, serves as Set-HQ. It only consists of 20 contrast-enhanced CT hepatic scans with high-quality liver and vessel annotation. In this dataset, different volumes share the same axial slice size (512×512 pixels), while the pixel spacing varies from 0.57 to 0.87 mm, the slice thickness varies from 1 to 4 mm, and the slice number is between 74 and 260.

2) **Set-LQ: MSD8** [21]. The second dataset MSD8 provides 443 CT hepatic scans collected from Memorial Sloan Kettering Cancer Center, serving as the Set-LQ. The properties of the CT scans are similar to that of the 3DIRCADb dataset but with low-quality annotations. According to the statistics [15], around 65.5% of the vessel pixels are unlabeled and approximately 8.5% are mislabeled as vessels for this dataset, resulting in the necessity of laborious manual refinement in previous work [15].

In our experiments, the images in Set-HQ are randomly divided into two groups: 10 cases for training, and the remaining 10 cases for testing, while all the samples in Set-LQ are only used for training since their original low-quality noisy labels are not appropriate for unbiased evaluation [15].

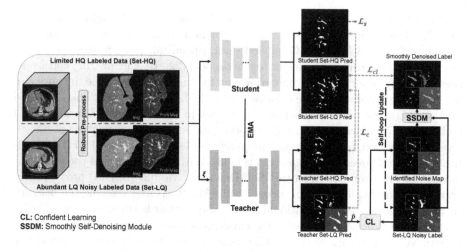

Fig. 2. Illustration of the proposed Mean-Teacher-assisted Confident Learning (MTCL) framework for hepatic vessel segmentation.

2.2 Hepatic CT Preprocessing

A standard preprocessing strategy is firstly applied to all the CT images: (1) the images are masked and cropped to the liver region based on the liver segmentation masks. Note that for the MSD8 dataset, the liver masks are obtained with the trained H-DenseUNet model [13] because no manual annotation of the liver is provided. All the cropped images are adjusted to $320 \times 320 \times D$, where D denotes the slice number. Since the slice thickness varies greatly, we do not perform any resampling to avoid the potential artifacts [23] caused by the interpolation; (2) The intensity of each pixel is truncated to the range of $[-100, 250]$ HU, followed by Min-Max normalization.

However, we observe that many cases have different intensity ranges (shown in Fig. 1) and intrinsic image noises [6], which could drive the model to be oversensitive to the high-intensity regions, as demonstrated in Table 1 and Fig. 3. Therefore, the vessel probability map based on the Sato tubeness filter [20] is introduced to provide auxiliary information. By calculating the Hessian matrix's eigenvectors, the similarity of the image to tubes can be obtained, so that the potential vessel regions can be enhanced with high probability (illustrated in Fig. 2). Following the input-level fusion strategy used in other multimodal segmentation tasks [27], we regard the vessel probability map as an auxiliary modality and directly concatenate it with the processed CT images in the original input space. By jointly considering the information in both the images and the probability maps, the network could perceive more robust vessel signals towards better segmentation performance (demonstrated in Table 1 and Fig. 3).

2.3   Mean-Teacher-assisted Confident Learning Framework

Learn from Images of Set-LQ. To additionally exploit the image information of Set-LQ, the mean-teacher model (MT) [22] is adopted as our basic architecture with the backbone network U-Net [19], as shown in Fig. 2. Denoting the weights of the student model at training step t as θ_t, Exponential Moving Average (EMA) is applied to update the teacher model's weights θ'_t, formulated as $\theta'_t = \alpha\theta'_{t-1} + (1 - \alpha)\theta_t$, where α is the EMA decay rate and set to 0.99 as recommended by [22]. By encouraging the teacher model's temporal ensemble prediction to be consistent with that of the student model under different perturbations (e.g., adding random noise ξ to the input samples) for the same inputs, superior prediction performance can be achieved as demonstrated in previous studies [22,24,25]. As shown in Fig. 2, the student model is optimized by minimizing the supervised loss $\mathcal{L}_s$ on Set-HQ, along with the (unsupervised) consistency loss $\mathcal{L}_c$ between predictions of the student model and the teacher model on both datasets.

Learn from Progressively Self-Denoised Soft Labels of Set-LQ. The above MT model can only leverage the image information, while the potential useful information of the noisy labels is still unexploited. To further leverage the LQ annotation without being affected by the label noises, we propose a progressive self-denoising process to alleviate the potential misleading guidance.

Inspired by the arbitration based manual annotation procedure where a third party, e.g., the radiologists, is consulted for disputed cases, the teacher model serves as the *'third party'* here to provide guidance for identifying label noises. With its assistance, we adapt the Confident Learning [17], which was initially proposed for pruning mislabeled samples in image-level classification, to characterize the pixel-wise label noises based on the Classification Noise Process (CNP) assumption [3]. The self-denoising process can be formulated as follows:

(1) *Characterize the pixel-wise label errors via adapted CL.* First, we estimate the joint distribution $Q_{\tilde{y},y^*}$ between the noisy (observed) labels $\tilde{y}$ and the true (latent) labels y^*. Given a dataset $\mathbf{X} := (\mathbf{x}, \tilde{y})^n$ consisting of n samples of $\mathbf{x}$ with m-class noisy label $\tilde{y}$, the out-of-sample predicted probabilities $\hat{\boldsymbol{P}}$ can be obtained via the *'third party'*, i.e., our teacher model. Ideally, such a *third party* is also jointly enhanced during training. If the sample $\mathbf{x}$ with label $\tilde{y} = i$ has *large enough* $\hat{p}_j(\mathbf{x}) \geq t_j$, the true latent label y^* of $\mathbf{x}$ can be suspected to be j instead of i. Here, the threshold t_j is obtained by calculating the average (expected) predicted probabilities $\hat{p}_j(\mathbf{x})$ of the samples labeled with $\tilde{y} = j$, which can be formulated as $t_j := \frac{1}{|\mathbf{X}_{\tilde{y}=j}|}\sum_{\mathbf{x}\in\mathbf{X}_{\tilde{y}=j}} \hat{p}_j(\mathbf{x})$. Based on the predicted label, we further introduce the confusion matrix $\boldsymbol{C}_{\tilde{y},y^*}$, where $\boldsymbol{C}_{\tilde{y},y^*}[i][j]$ is the number of $\mathbf{x}$ labeled as i ($\tilde{y} = i$), yet the true latent label may be j ($y^* = j$). Formally, $\boldsymbol{C}_{\tilde{y},y^*}$ can be defined as:

$$\mathbf{C}_{\tilde{y},y^*}[i][j] := \left|\hat{\mathbf{X}}_{\tilde{y}=i,y^*=j}\right|, \text{ where}$$

$$\hat{\mathbf{X}}_{\tilde{y}=i,y^*=j} := \left\{ \mathbf{x} \in \mathbf{X}_{\tilde{y}=i} : \hat{p}_j(\mathbf{x}) \geq t_j, j = \arg\max_{l\in M : \hat{p}_l(\mathbf{x})\geq t_l} \hat{p}_l(\mathbf{x}) \right\}. \tag{1}$$

With the constructed confusion matrix $\boldsymbol{C}_{\tilde{y},y^*}$, we can further estimate the $m \times m$ joint distribution matrix $\mathbf{Q}_{\tilde{y},y^*}$ for $p(\tilde{y}, y^*)$:

$$\mathbf{Q}_{\tilde{y},y^*}[i][j] = \frac{\frac{\mathbf{C}_{\tilde{y},y^*}[i][j]}{\sum_{j \in M} \mathbf{C}_{\tilde{y},y^*}[i][j]} \cdot |\mathbf{X}_{\tilde{y}=i}|}{\sum_{i \in M, j \in M} \left(\frac{\mathbf{C}_{\tilde{y},y^*}[i][j]}{\sum_{j \in M} \mathbf{C}_{\tilde{y},y^*}[i][j]} \cdot |\mathbf{X}_{\tilde{y}=i}| \right)}. \tag{2}$$

Then, we utilize the *Prune by Class (PBC)* [17] method recommended by [26] to identify the label noises. Specifically, for each class $i \in M$, PBC selects the $n \cdot \sum_{j \in M : j \neq i}(\mathbf{Q}_{\tilde{y},y^*}[i][j])$ samples with the lowest self-confidence $\hat{p}(\tilde{y} = i; \boldsymbol{x} \in \boldsymbol{X}_i)$ as the wrong-labeled samples, thereby obtaining the binary noise identification map $\mathbf{X}_n$, where "1" denotes that the pixel has a wrong label and vice versa. It is worth noting that the adapted CL module is computationally efficient and does not require any extra hyper-parameters.

(2) *Smoothly refine the noisy labels of Set-LQ to provide rewarding supervision.* Experimentally, the CL still has uncertainties in distinguishing the label noises. Therefore, instead of directly imposing the hard-correction, we introduce the Smoothly Self-Denoising Module (SSDM) to impose a soft correction [2] on the given noisy segmentation masks $\tilde{y}$. Based on the binary noise identification map $\mathbf{X}_n$, the smoothly self-denoising operation can be formulated as follows:

$$\dot{y}(\mathbf{x}) = \tilde{y}(\mathbf{x}) + \mathbb{I}(\mathbf{x} \in \mathbf{X}_n) \cdot (-1)^{\tilde{y}} \cdot \tau, \tag{3}$$

where $\mathbb{I}(\cdot)$ is the indicator function, and $\tau \in [0, 1]$ is the smooth factor, which is empirically set as 0.8. After that, the updated soft-corrected LQ labels of Set-LQ are used as the auxiliary CL guidance $\mathcal{L}_{cl}$ to the student model.

(3) *Self-loop updating.* With the proposed SSDM, we construct a self-loop updating process that substitutes the noisy labels of Set-LQ with the updated denoised ones for the next training epoch, so that the framework can progressively refine the noisy vessel labels during training.

2.4   Loss Function

The total loss is a weighted combination of the supervised loss $\mathcal{L}_s$ on Set-HQ, the perturbation consistency loss $\mathcal{L}_c$ on both datasets and the auxiliary self-denoised CL loss $\mathcal{L}_{cl}$ on Set-LQ, calculated by:

$$\mathcal{L} = \mathcal{L}_s + \lambda_c \mathcal{L}_c + \lambda_{cl} \mathcal{L}_{cl}, \tag{4}$$

where λ_c and λ_{cl} are the trade-off weights for $\mathcal{L}_c$ and $\mathcal{L}_{cl}$, respectively. We adopt the time-dependent Gaussian function [4] to schedule the ramp-up weight $\mathcal{L}_c$. Meanwhile, the teacher model needs to be "warmed up" to provide reliable out-of-sample predicted probabilities. Therefore, λ_{cl} is set as 0 in the first 4,000 iterations, and adjusted to 0.5 during the rest training iterations. Note that the supervised loss $\mathcal{L}_s$ is a combination of cross-entropy loss, Dice loss, focal loss [14] and boundary loss [10] with weights of 0.5, 0.5, 1 and 0.5, respectively, as such a combination can provide better performance in our exploratory experiments with the fully supervised baseline method. The consistency loss $\mathcal{L}_c$ is calculated by the voxel-wise Mean Squared Error (MSE), and the CL loss $\mathcal{L}_{cl}$ is composed of cross-entropy loss and focal loss with equal weights.

Table 1. Quantitative results of different methods. Best results are shown in bold.

Method	Dice ↑	PRE ↑	ASD ↓	HD ↓
Huang et al. [8]	0.5991	0.6352	2.5477	10.5088
U-Net(i)	0.6093	0.5601	2.7209	10.3103
U-Net(p)	0.6082	0.5553	2.3574	10.2864
U-Net(c)	0.6685	0.6699	2.0463	9.2078
U-Net(c, Mix)	0.6338	0.6322	1.6040	9.2038
MT(c)	0.6963	0.6931	1.4860	7.5912
MT(c)+NL w/o CL	0.6807	0.7270	1.3205	8.0893
MTCL(c) w/o SSDM	0.7046	0.7472	1.2252	8.3667
MTCL(c)	**0.7245**	**0.7570**	**1.1718**	**7.2111**

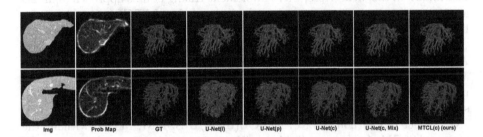

Fig. 3. Visualization of the fused segmentation results of different methods. The red voxels represent the ground truth, while the green voxels denote the difference between the ground truth and the segmented vessel of different methods. (Color figure online)

3 Experiments and Results

Evaluation Metrics and Implementation. For inference, the student model segments each volume slice-by-slice and the segmentation of each slice is concatenated back into 3D volume. Then, a post-processing step that removes very small regions (less than 0.1% of the volume size) is performed. We adopt four metrics for a comprehensive evaluation, including Dice score, Precision (PRE), Average Surface Distance (ASD) and Hausdorff Distance (HD). The framework is based on the PyTorch implementation of [25] using an NVIDIA Titan X GPU. SGD optimizer is also adopted and the batch size is set to 4. Standard data augmentation, including randomly flipping and rotating, is applied. Our implementation is publicly available at https://github.com/lemoshu/MTCL.

Comparison Study. A comprehensive qualitative and quantitative comparison study is performed on the hold-out test set of Set-HQ, as shown in Fig. 3 and Table 1. Succinctly, "*i*", "*p*" and "*c*" represent different input types: processed image, the vessel probability map and the concatenated one, respectively.

Surprisingly, the performance of 3D networks is far worse than the 2D ones in our experiments, which may result from inadequate training data or the thickness variation [23]. Therefore, all the rest experiments are performed in 2D. The exploratory fully supervised experiments are performed on the Set-HQ. We can observe that using the concatenated slices as input (U-Net(c)) achieves superior performance. Next, we additionally introduce the Set-LQ to train the model in the fully supervised manner, denoted as **U-Net(c, Mix)**. As predicted, the noisy labels of Set-LQ cause unavoidable performance degradation. Compared with U-Net(c) with only Set-HQ, U-Net(c, Mix)'s Dice score and PRE drop from 0.6685 to 0.6338, and from 0.6699 to 0.6322, respectively. Note that the previous learning-based studies [8,11,12] on hepatic vessel segmentation performed the evaluation on manually refined annotation without making the improved 'ground truth' or their implementation publicly available, resulting in excessive lack of benchmark in this field. Here, we re-implement Huang et al.'s approach [8] in 2D as another baseline. The proposed method, denoted as **MTCL(c)**, achieves the best performance in terms of all four metrics and more appealing visual results.

Ablation Study. To verify the effectiveness of each component, we perform an ablation study with the following variants: a) **MT(c)**: a typical mean-teacher model that additionally uses the image information of Set-LQ, i.e., the SSL setting; b) **MT(c)+NL w/o CL**: extended MT(c) by leveraging the noisy labels (NL) of Set-LQ without CL; c) **MTCL(c) w/o SSDM**: MTCL without the proposed SSDM. As shown in Table 1, with the assistance of image information of Set-LQ, adding the perturbation consistency loss can improve the segmentation performance, as well as alleviate the performance degradation caused by noisy labels. Superior performance can be achieved through the self-denoising process via the adapted CL, and further improved by the SSDM.

Effectiveness of Label Self-denoising. The visualization of two example slices from MSD8 is shown in Fig. 4 to further illustrate the label self-denoising process. Some noticeable noises can be identified with the proposed framework. Moreover, an additional experiment, which uses the denoised label of Set-LQ along with Set-HQ to train a U-Net (same setting as U-Net(c, Mix)), is performed and obtains 7.67%, 8.46%, 0.61% and 3.91% improvement in terms of Dice, PRE, ASD and HD, respectively, compared to the one using the original label of Set-LQ. The extended experiment further demonstrates the capability of the proposed framework in correcting the label errors, indicating a potential application of our framework to explicably refine the label quality of large datasets by taking advantage of limited HQ labeled data for many other tasks.

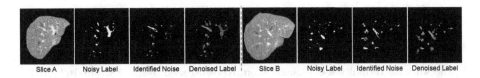

Fig. 4. Illustration of the self-denoising performance for the MSD8 dataset (Set-LQ).

4 Conclusion

In this work, we proposed a novel Mean-Teacher-assisted Confident Learning (MTCL) framework for the challenging hepatic vessel segmentation task with a limited amount of high-quality labeled data and abundant low-quality noisy labeled data. The superior performance we achieved using two public datasets demonstrated the effectiveness of the proposed framework. Furthermore, the additional experiment with refined annotation showed that the proposed framework could improve the annotation quality of noisy labeled data with only a small amount of high-quality labeled data.

Acknowledgements. This project was partly supported by the National Natural Science Foundation of China (Grant No. 41876098), the National Key R&D Program of China (Grant No. 2020AAA0108303), Shenzhen Science and Technology Project (Grant No. JCYJ20200109143041798), Key-Area Research and Development Program of Guangdong Province, China (No. 2018B010111001), and the Scientific and Technical Innovation 2030- "New Generation Artificial Intelligence" Project (No. 2020AAA0104100).

References

1. 3DIRCADb Dataset. https://www.ircad.fr/research/3d-ircadb-01/
2. Ainam, J.P., Qin, K., Liu, G., Luo, G.: Sparse label smoothing regularization for person re-identification. IEEE Access **7**, 27899–27910 (2019)
3. Angluin, D., Laird, P.: Learning from noisy examples. Mach. Learn. **2**(4), 343–370 (1988)
4. Cui, W., et al.: Semi-supervised brain lesion segmentation with an adapted mean teacher model. In: Chung, A.C.S., Gee, J.C., Yushkevich, P.A., Bao, S. (eds.) IPMI 2019. LNCS, vol. 11492, pp. 554–565. Springer, Cham (2019). https://doi.org/10.1007/978-3-030-20351-1_43
5. Dou, Q., Chen, H., Jin, Y., Yu, L., Qin, J., Heng, P.-A.: 3D deeply supervised network for automatic liver segmentation from CT volumes. In: Ourselin, S., Joskowicz, L., Sabuncu, M.R., Unal, G., Wells, W. (eds.) MICCAI 2016. LNCS, vol. 9901, pp. 149–157. Springer, Cham (2016). https://doi.org/10.1007/978-3-319-46723-8_18
6. Duan, X., Wang, J., Leng, S., Schmidt, B., Allmendinger, T., Grant, K., Flohr, T., McCollough, C.H.: Electronic noise in CT detectors: impact on image noise and artifacts. Am. J. Roentgenology **201**(4), W626–W632 (2013)
7. Goldberger, J., Ben-Reuven, E.: Training deep neural-networks using a noise adaptation layer. In: International Conference on Learning Representations (2016)

8. Huang, Q., Sun, J., Ding, H., Wang, X., Wang, G.: Robust liver vessel extraction using 3D U-Net with variant Dice loss function. Comput. Biology Med. **101**, 153–162 (2018)
9. Jin, Q., Meng, Z., Pham, T.D., Chen, Q., Wei, L., Su, R.: DUNet: a deformable network for retinal vessel segmentation. Knowl.-Based Syst. **178**, 149–162 (2019)
10. Kervadec, H., Bouchtiba, J., Desrosiers, C., Granger, E., Dolz, J., Ayed, I.B.: Boundary loss for highly unbalanced segmentation. In: International Conference on Medical Imaging with Deep Learning, pp. 285–296. PMLR (2019)
11. Kitrungrotsakul, T., Han, X.H., Iwamoto, Y., Foruzan, A.H., Lin, L., Chen, Y.W.: Robust hepatic vessel segmentation using multi deep convolution network. In: Medical Imaging 2017: Biomedical Applications in Molecular, Structural, and Functional Imaging, vol. 10137, p. 1013711. International Society for Optics and Photonics (2017)
12. Kitrungrotsakul, T., Han, X.H., Iwamoto, Y., Lin, L., Foruzan, A.H., Xiong, W., Chen, Y.W.: Vesselnet: a deep convolutional neural network with multi pathways for robust hepatic vessel segmentation. Computerized Med. Imaging Graph. **75**, 74–83 (2019)
13. Li, X., Chen, H., Qi, X., Dou, Q., Fu, C.W., Heng, P.A.: H-DenseUNet: hybrid densely connected UNet for liver and tumor segmentation from CT volumes. IEEE Trans. Med. Imaging **37**(12), 2663–2674 (2018)
14. Lin, T.Y., Goyal, P., Girshick, R., He, K., Dollár, P.: Focal loss for dense object detection. In: Proceedings of the IEEE International Conference on Computer Vision, pp. 2980–2988 (2017)
15. Liu, L., Tian, J., Zhong, C., Shi, Z., Xu, F.: Robust hepatic vessels segmentation model based on noisy dataset. In: Medical Imaging 2020: Computer-Aided Diagnosis, vol. 11314, p. 113140L. International Society for Optics and Photonics (2020)
16. Livne, M., Rieger, J., Aydin, O.U., Taha, A.A., Akay, E.M., Kossen, T., Sobesky, J., Kelleher, J.D., Hildebrand, K., Frey, D., et al.: A U-Net deep learning framework for high performance vessel segmentation in patients with cerebrovascular disease. Frontiers Neurosci. **13**, 97 (2019)
17. Northcutt, C.G., Jiang, L., Chuang, I.L.: Confident learning: estimating uncertainty in dataset labels. arXiv preprint arXiv:1911.00068 (2019)
18. Ren, M., Zeng, W., Yang, B., Urtasun, R.: Learning to reweight examples for robust deep learning. In: International Conference on Machine Learning, pp. 4334–4343. PMLR (2018)
19. Ronneberger, O., Fischer, P., Brox, T.: U-Net: convolutional networks for biomedical image segmentation. In: Navab, N., Hornegger, J., Wells, W.M., Frangi, A.F. (eds.) MICCAI 2015. LNCS, vol. 9351, pp. 234–241. Springer, Cham (2015). https://doi.org/10.1007/978-3-319-24574-4_28
20. Sato, Y., Nakajima, S., Shiraga, N., Atsumi, H., Yoshida, S., Koller, T., Gerig, G., Kikinis, R.: Three-dimensional multi-scale line filter for segmentation and visualization of curvilinear structures in medical images. Med. Image Anal. **2**(2), 143–168 (1998)
21. Simpson, A.L., et al.: A large annotated medical image dataset for the development and evaluation of segmentation algorithms. arXiv preprint arXiv:1902.09063 (2019)
22. Tarvainen, A., Valpola, H.: Mean teachers are better role models: weight-averaged consistency targets improve semi-supervised deep learning results. In: Advances in Neural Information Processing Systems, pp. 1195–1204 (2017)

23. Wang, S., Cao, S., Chai, Z., Wei, D., Ma, K., Wang, L., Zheng, Y.: Conquering data variations in resolution: a slice-aware multi-branch decoder network. IEEE Trans. Med. Imaging **39**(12), 4174–4185 (2020)
24. Wang, Y., Zhang, Y., Tian, J., Zhong, C., Shi, Z., Zhang, Y., He, Z.: Double-uncertainty weighted method for semi-supervised learning. In: International Conference on Medical Image Computing and Computer Assisted Intervention, pp. 542–551. Springer (2020)
25. Yu, L., Wang, S., Li, X., Fu, C.-W., Heng, P.-A.: Uncertainty-aware self-ensembling model for semi-supervised 3D left atrium segmentation. In: Shen, D., Liu, T., Peters, T.M., Staib, L.H., Essert, C., Zhou, S., Yap, P.-T., Khan, A. (eds.) MICCAI 2019. LNCS, vol. 11765, pp. 605–613. Springer, Cham (2019). https://doi.org/10.1007/978-3-030-32245-8_67
26. Zhang, M., Gao, J., Lyu, Z., Zhao, W., Wang, Q., Ding, W., Wang, S., Li, Z., Cui, S.: Characterizing label errors: confident learning for noisy-labeled image segmentation. In: Martel, A.L., Abolmaesumi, P., Stoyanov, D., Mateus, D., Zuluaga, M.A., Zhou, S.K., Racoceanu, D., Joskowicz, L. (eds.) MICCAI 2020. LNCS, vol. 12261, pp. 721–730. Springer, Cham (2020). https://doi.org/10.1007/978-3-030-59710-8_70
27. Zhou, T., Ruan, S., Canu, S.: A review: Deep learning for medical image segmentation using multi-modality fusion. Array **3**, 100004 (2019)
28. Zhu, H., Shi, J., Wu, J.: Pick-and-learn: automatic quality evaluation for noisy-labeled image segmentation. In: Shen, D., Liu, T., Peters, T.M., Staib, L.H., Essert, C., Zhou, S., Yap, P.-T., Khan, A. (eds.) MICCAI 2019. LNCS, vol. 11769, pp. 576–584. Springer, Cham (2019). https://doi.org/10.1007/978-3-030-32226-7_64

TransFuse: Fusing Transformers and CNNs for Medical Image Segmentation

Yundong Zhang[1], Huiye Liu[1,2(✉)], and Qiang Hu[1]

[1] Rayicer, Suzhou, China
huiyeliu@rayicer.com
[2] Georgia Institute of Technology, Atlanta, GA, USA

Abstract. Medical image segmentation - the prerequisite of numerous clinical needs - has been significantly prospered by recent advances in convolutional neural networks (CNNs). However, it exhibits general limitations on modeling explicit long-range relation, and existing cures, resorting to building deep encoders along with aggressive downsampling operations, leads to redundant deepened networks and loss of localized details. Hence, the segmentation task awaits a better solution to improve the efficiency of modeling global contexts while maintaining a strong grasp of low-level details. In this paper, we propose a novel parallel-in-branch architecture, TransFuse, to address this challenge. TransFuse combines Transformers and CNNs in a parallel style, where both global dependency and low-level spatial details can be efficiently captured in a much shallower manner. Besides, a novel fusion technique - BiFusion module is created to efficiently fuse the multi-level features from both branches. Extensive experiments demonstrate that TransFuse achieves the newest state-of-the-art results on both 2D and 3D medical image sets including polyp, skin lesion, hip, and prostate segmentation, with significant parameter decrease and inference speed improvement.

Keywords: Medical image segmentation · Transformers · Convolutional neural networks · Fusion

1 Introduction

Convolutional neural networks (CNNs) have attained unparalleled performance in numerous medical image segmentation tasks [9,12], such as multi-organ segmentation, liver lesion segmentation, brain 3D MRI, etc., as it is proved to be powerful at building hierarchical task-specific feature representation by training the networks end-to-end. Despite the immense success of CNN-based methodologies, its lack of efficiency in capturing global context information remains a challenge. The chance of sensing global information is equaled by the risk of efficiency, because existing works obtain global information by generating very large receptive fields, which requires consecutively down-sampling and stacking

---

Y. Zhang and H. Liu—These authors contributed equally to this work.

© Springer Nature Switzerland AG 2021
M. de Bruijne et al. (Eds.): MICCAI 2021, LNCS 12901, pp. 14–24, 2021.
https://doi.org/10.1007/978-3-030-87193-2_2

convolutional layers until deep enough. This brings several drawbacks: 1) training of very deep nets is affected by the diminishing feature reuse problem [23], where low-level features are washed out by consecutive multiplications; 2) local information crucial to dense prediction tasks, e.g., pixel-wise segmentation, is discarded, as the spatial resolution is reduced gradually; 3) training parameter-heavy deep nets with small medical image datasets tends to be unstable and easily overfitting. Some studies [29] use the non-local self-attention mechanism to model global context; however, the computational complexity of these modules typically grows quadratically with respect to spatial size, thus they may only be appropriately applied to low-resolution maps.

Transformer, originally used to model sequence-to-sequence predictions in NLP tasks [26], has recently attracted tremendous interests in the computer vision community. The first purely self-attention based vision transformers (ViT) for image recognition is proposed in [7], which obtained competitive results on ImageNet [6] with the prerequisite of being pretrained on a large external dataset. SETR [32] replaces the encoders with transformers in the conventional encoder-decoder based networks to successfully achieve state-of-the-art (SOTA) results on the natural image segmentation task. While Transformer is good at modeling global context, it shows limitations in capturing fine-grained details, especially for medical images. We independently find that SETR-like pure transformer-based segmentation network produces unsatisfactory performance, due to lack of spatial inductive-bias in modelling local information (also reported in [4]).

To enjoy the benefit of both, efforts have been made on combining CNNs with Transformers, e.g., TransUnet [4], which first utilizes CNNs to extract low-level features and then passed through transformers to model global interaction. With skip-connection incorporated, TransUnet sets new records in the CT multi-organ segmentation task. However, past works mainly focus on replacing convolution with transformer layers or stacking the two in a sequential manner. To further unleash the power of CNNs plus Transformers in medical image segmentation, in this paper, we propose a different architecture—*TransFuse*, which runs shallow CNN-based encoder and transformer-based segmentation network in parallel, followed by our proposed *BiFusion* module where features from the two branches are fused together to jointly make predictions. TransFuse possesses several advantages: 1) both low-level spatial features and high-level semantic context can be effectively captured; 2) it does not require very deep nets, which alleviates gradient vanishing and feature diminishing reuse problems; 3) it largely improves efficiency on model sizes and inference speed, enabling the deployment at not only cloud but also edge. To the best of our knowledge, TransFuse is the first parallel-in-branch model synthesizing CNN and Transformer. Experiments demonstrate the superior performance against other competing SOTA works.

2   Proposed Method

As shown in Fig. 1, TransFuse consists of two parallel branches processing information differently: 1) CNN branch, which gradually increases the receptive field and encodes features from local to global; 2) Transformer branch, where it starts with global self-attention and recovers the local details at the end. Features with same resolution extracted from both branches are fed into our proposed BiFusion Module, where self-attention and bilinear Hadamard product are applied to selectively fuse the information. Then, the multi-level fused feature maps are combined to generate the segmentation using gated skip-connection [20]. There are two main benefits of the proposed branch-in-parallel approach: firstly, by leveraging the merits of CNNs and Transformers, we argue that TransFuse can capture global information without building very deep nets while preserving sensitivity on low-level context; secondly, our proposed BiFusion module may simultaneously exploit different characteristics of CNNs and Transformers during feature extraction, thus making the fused representation powerful and compact.

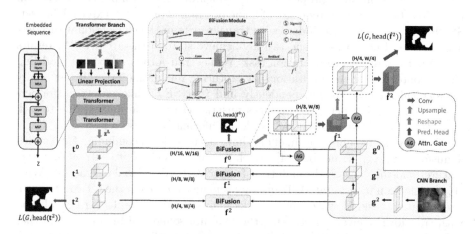

Fig. 1. Overview of TransFuse (best viewed in color): two parallel branches - CNN (bottom right) and transformer (left) fused by our proposed BiFusion module.

Transformer Branch. The design of Transformer branch follows the typical encoder-decoder architecture. Specifically, the input image $\mathbf{x} \in \mathbb{R}^{H \times W \times 3}$ is first evenly divided into $N = \frac{H}{S} \times \frac{W}{S}$ patches, where S is typically set to 16. The patches are then flattened and passed into a linear embedding layer with output dimension D_0, obtaining the raw embedding sequence $\mathbf{e} \in \mathbb{R}^{N \times D_0}$. To utilize the spatial prior, a learnable positional embeddings of the same demension is added to $\mathbf{e}$. The resulting embeddings $\mathbf{z}^0 \in \mathbb{R}^{N \times D_0}$ is the input to Transformer encoder, which contains L layers of multiheaded self-attention (MSA) and Multilayer Perceptron (MLP). We highlight that the self-attention (SA) mechanism, which is the core principal of Transformer, updates the states of each embedded patch by aggregating information globally in every layer:

$$\mathrm{SA}(\mathbf{z}_i) = \mathrm{softmax}\left(\frac{\mathbf{q}_i \mathbf{k}^T}{\sqrt{D_h}}\right) \mathbf{v}, \tag{1}$$

where $[\mathbf{q}, \mathbf{k}, \mathbf{v}] = \mathbf{z}\mathbf{W}_{qkv}$, $\mathbf{W}_{qkv} \in \mathbb{R}^{D_0 \times 3D_h}$ is the projection matrix and vector $\mathbf{z}_i \in \mathbb{R}^{1 \times D_0}$, $\mathbf{q_i} \in \mathbb{R}^{1 \times D_h}$ are the ith row of $\mathbf{z}$ and $\mathbf{q}$, respectively. MSA is an extension of SA that concatenates multiple SAs and projects the latent dimension back to $\mathbb{R}^{D_0}$, and MLP is a stack of dense layers (refer to [7] for details of MSA and MLP). Layer normalization is applied to the output of the last transformer layer to obtain the encoded sequence $\mathbf{z}^L \in \mathbb{R}^{N \times D_0}$. For the decoder part, we use progressive upsampling (PUP) method, as in SETR [32]. Specifically, we first reshape $\mathbf{z}^L$ back to $\mathbf{t}^0 \in \mathbb{R}^{\frac{H}{16} \times \frac{W}{16} \times D_0}$, which could be viewed as a 2D feature map with D_0 channels. We then use two consecutive standard upsampling-convolution layers to recover the spatial resolution, where we obtain $\mathbf{t}^1 \in \mathbb{R}^{\frac{H}{8} \times \frac{W}{8} \times D_1}$ and $\mathbf{t}^2 \in \mathbb{R}^{\frac{H}{4} \times \frac{W}{4} \times D_2}$, respectively. The feature maps of different scales $\mathbf{t}^0$, $\mathbf{t}^1$ and $\mathbf{t}^2$ are saved for late fusion with corresponding feature maps of the CNN branch.

CNN Branch. Traditionally, features are progressively downsampled to $\frac{H}{32} \times \frac{W}{32}$ and hundreds of layers are employed in deep CNNs to obtain global context of features, which results in very deep models draining out resources. Considering the benefits brought by Transformers, we remove the last block from the original CNNs pipeline and take advantage of the Transformer branch to obtain global context information instead. This gives us not only a shallower model but also retaining richer local information. For example, ResNet-based models typically have five blocks, each of which downsamples the feature maps by a factor of two. We take the outputs from the 4th ($\mathbf{g}^0 \in \mathbb{R}^{\frac{H}{16} \times \frac{W}{16} \times C_0}$), 3rd ($\mathbf{g}^1 \in \mathbb{R}^{\frac{H}{8} \times \frac{W}{8} \times C_1}$) and 2nd ($\mathbf{g}^2 \in \mathbb{R}^{\frac{H}{4} \times \frac{W}{4} \times C_2}$) blocks to fuse with the results from Transformer (Fig. 1). Moreover, our CNN branch is flexible that any off-the-shelf convolutional network can be applied.

BiFusion Module. To effectively combine the encoded features from CNNs and Transformers, we propose a new BiFusion module (refer to Fig. 1) that incorporates both self-attention and multi-modal fusion mechanisms. Specifically, we obtain the fused feature representation $\mathbf{f}^i, i = 0, 1, 2$ by the following operations:

$$\hat{\mathbf{t}}^i = \text{ChannelAttn}(\mathbf{t}^i) \qquad \hat{\mathbf{g}}^i = \text{SpatialAttn}(\mathbf{g}^i)$$
$$\hat{\mathbf{b}}^i = \text{Conv}(\mathbf{t}^i\mathbf{W}_1^i \odot \mathbf{g}^i\mathbf{W}_2^i) \qquad \mathbf{f}^i = \text{Residual}([\hat{\mathbf{b}}^i, \hat{\mathbf{t}}^i, \hat{\mathbf{g}}^i]) \tag{2}$$

where $W_1^i \in \mathbb{R}^{D_i \times L_i}$, $W_2^i \in \mathbb{R}^{C_i \times L_i}$, $| \odot |$ is the Hadamard product and Conv is a 3×3 convolution layer. The channel attention is implemented as SE-Block proposed in [10] to promote global information from the Transformer branch. The spatial attention is adopted from CBAM [30] block as spatial filters to enhance local details and suppress irrelevant regions, as low-level CNN features could be noisy. The Hadamard product then models the fine-grained interaction between features from the two branches. Finally, the interaction features $\hat{\mathbf{b}}^i$ and attended features $\hat{\mathbf{t}}^i$, $\hat{\mathbf{g}}^i$ are concatenated and passed through a Residual block. The resulting feature $\mathbf{f}^i$ effectively captures both the global and local context for the current spatial resolution. To generate final segmentation, $\mathbf{f}^i$s are combined using the attention-gated (AG) skip-connection [20], where we have $\hat{\mathbf{f}}^{i+1} = \text{Conv}([\text{Up}(\hat{\mathbf{f}}^i), \text{AG}(\mathbf{f}^{i+1}, \text{Up}(\hat{\mathbf{f}}^i))])$ and $\hat{\mathbf{f}}^0 = \mathbf{f}^0$, as in Fig. 1.

Loss Function. The full network is trained end-to-end with the weighted IoU loss and binary cross entropy loss $L = L_{IoU}^w + L_{bce}^w$, where boundary pixels receive larger weights [17]. Segmentation prediction is generated by a simple head, which directly resizes the input feature maps to the original resolution and applies convolution layers to generate M maps, where M is the number of classes. Following [8], We use deep supervision to improve the gradient flow by additionally supervising the transformer branch and the first fusion branch. The final training loss is given by $\mathcal{L} = \alpha L\left(G, \text{head}(\hat{\mathbf{f}}^2)\right) + \gamma L\left(G, \text{head}(\mathbf{t}^2)\right) + \beta L\left(G, \text{head}(\mathbf{f}^0)\right)$, where α, γ, β are tunnable hyperparameters and G is groundtruth.

3   Experiments and Results

Data Acquisition. To better evaluate the effectiveness of TransFuse, four segmentation tasks with different imaging modalities, disease types, target objects, target sizes, etc. are considered: 1) *Polyp Segmentation*, where five public polyp datasets are used: Kvasir [14], CVC-ClinicDB [2], CVC-ColonDB [24], EndoScene [27] and ETIS [21]. We adopt the same split and training setting as in [8,11], i.e. 1450 training images are solely selected from Kvasir and CVC-ClinicDB while 798 testing images are from all five datasets. Before processing, the resolution of each image is resized into 352×352 as [8,11]. 2) *Skin Lesion Segmentation*, where the publicly available 2017 International Skin Imaging Collaboration skin lesion segmentation dataset (ISIC2017) [5] is used[1]. ISIC2017 provides 2000 images for training, 150 images for validation and 600 images for testing. Following the setting in [1], we resize all images to 192×256. 3) *Hip Segmentation*, where a total of 641 cases are collected from a hospital with average size of 2942×2449 and pixel spacing as $0.143\,\text{mm}^2$. Each image is annotated by a clinical expert and double-blind reviewed by two specialists. We resized all images into 352×352, and randomly split images with a ratio of 7:1:2 for training, validation and testing. 4) *Prostate Segmentation*, where volumetric Prostate Multi-modality MRIs from the Medical Segmentation Decathlon [22] are used. The dataset contains multi-modal MRIs from 32 patients, with a median volume shape of $20 \times 320 \times 319$. Following the setting in [12], we reshape all MRI slices to 320×320, and independently normalize each volume using z-score normalization.

Implementation Details. TransFuse was built in PyTorch framework [16] and trained using a single NVIDIA-A100 GPU. The values of α, β and γ were set to 0.5, 0.3, 0.2 empirically. Adam optimizer with learning rate of 1e−4 was

---

[1] Another similar dataset ISIC2018 was not used because of the missing test set annotation, which makes fair comparison between existing works can be hardly achieved.

[2] All data are from different patients and with ethics approval, which consists of 267 patients of Avascular Necrosis, 182 patients of Osteoarthritis, 71 patients of Femur Neck Fracture, 33 patients of Pelvis Fracture, 26 patients of Developmental Dysplasia of the Hip and 62 patients of other dieases.

Table 1. Quantitative results on polyp segmentation datasets compared to previous SOTAs. The results of [4] is obtained by running the released code and we implement SETR-PUP. '−' means results not available.

Methods	Kvasir		ClinicDB		ColonDB		EndoScene		ETIS	
	mDice	mIoU	mDice	mIoU	mDice	mIoU	mDice	mIoU	mDice	mIoU
U-Net [18]	0.818	0.746	0.823	0.750	0.512	0.444	0.710	0.627	0.398	0.335
U-Net++ [33]	0.821	0.743	0.794	0.729	0.483	0.410	0.707	0.624	0.401	0.344
ResUNet++ [13]	0.813	0.793	0.796	0.796	−	−	−	−	−	−
PraNet [8]	0.898	0.840	0.899	0.849	0.709	0.640	0.871	0.797	0.628	0.567
HarDNet-MSEG [11]	0.912	0.857	0.932	0.882	0.731	0.660	0.887	0.821	0.677	0.613
TransFuse-S	**0.918**	**0.868**	0.918	0.868	**0.773**	**0.696**	0.902	0.833	0.733	0.659
TransFuse-L	**0.918**	**0.868**	0.934	0.886	0.744	0.676	**0.904**	**0.838**	0.737	**0.661**
SETR-PUP [32]	0.911	0.854	0.934	0.885	0.773	0.690	0.889	0.814	0.726	0.646
TransUnet [4]	0.913	0.857	0.935	0.887	0.781	0.699	0.893	0.824	0.731	0.660
TransFuse-L*	**0.920**	**0.870**	0.942	0.897	0.781	0.706	0.894	0.826	0.737	0.663

adopted and all models were trained for 30 epochs as well as batch size of 16, unless otherwise specified.

In polyp segmentation experiments, no data augmentation was used except for multi-scale training, as in [8,11]. For skin lesion and hip segmentation, data augmentation including random rotation, horizontal flip, color jittering, etc. were applied during training. A smaller learning rate of 7e−5 was found useful for skin lesion segmentation. Finally, we follow the nnU-Net framework [12] to train and evaluate our model on Prostate Segmentation, using the same data augmentation and post-processing scheme. As selected pretrained datasets and branch backbones may affect the performance differently, three variants of TransFuse are provided to 1) better demonstrate the effectiveness as well as flexibility of our approach; 2) conduct fair comparisons with other methods. *TransFuse-S* is implemented with ResNet-34 (R34) and 8-layer DeiT-Small (DeiT-S) [25] as backbones of the CNN branch and Transformer branch respectively. Similarly, *TransFuse-L* is built based on Res2Net-50 and 10-layer DeiT-Base (DeiT-B), while *TransFuse-L** uses ResNetV2-50 and ViT-B [7]. Note that ViTs and DeiTs have the same backbone architecture and they mainly differ in the pre-trained strategy and dataset: the former is trained on ImageNet21k while the latter is trained on ImageNet1k with heavier data augmentation.

Evaluation Results. TransFuse is evaluated on both 2D and 3D datasets to demonstrate the effectiveness. As different medical image segmentation tasks serve different diagnosis or operative purposes, we follow the commonly used evaluation metrics for each of the segmentation tasks to quantitatively analyze the results. Selected visualization results of *TransFuse-S* are shown in Fig. 2.

Results of Polyp Segmentation. We first evaluate the performance of our proposed method on polyp segmentation against a variety of SOTA methods, in terms of mean Dice (mDice) and mean Intersection-Over-Union (mIoU). As in Table 1, our *TransFuse-S/L* outperform CNN-based SOTA methods by a

Table 2. Quantitative results on ISIC 2017 test set. Results with backbones use weights pretrained on ImageNet.

Methods	Backbones	Epochs	Jaccard	Dice	Accuracy
CDNN [31]	-	-	0.765	0.849	0.934
DDN [15]	ResNet-18	600	0.765	0.866	0.939
FrCN [1]	VGG16	200	0.771	0.871	0.940
DCL-PSI [3]	ResNet-101	150	0.777	0.857	0.941
SLSDeep [19]	ResNet-50	100	0.782	**0.878**	0.936
Unet++ [33]	ResNet-34	30	0.775	0.858	0.938
TransFuse-S	R34+DeiT-S	30	**0.795**	0.872	**0.944**

Table 3. Results on in-house hip dataset. All models use pretrained backbones from ImageNet and are of similar size ($\sim 26\,M$). HD and ASD are measured in mm.

Methods	Pelvis		L-Femur		R-Femur	
	HD	ASD	HD	ASD	HD	ASD
Unet++ [33]	14.4	1.21	9.33	0.932	5.04	0.813
HRNetV2 [28]	14.2	1.13	6.36	0.769	5.98	0.762
TransFuse-S	**9.81**	**1.09**	**4.44**	**0.767**	**4.19**	**0.676**

large margin. Specifically, TransFuse-S achieves 5.2% average mDice improvement on the *unseen* datasets (ColonDB, EndoSene and ETIS). Comparing to other transformer-based methods, *TransFuse-L** also shows superior learning ability on Kvasir and ClinicDB, observing an increase of 1.3% in mIoU compared to TransUnet. Besides, the efficiency in terms of the number of parameters as well as inference speed is evaluated on an RTX2080Ti with Xeon(R) Gold 5218 CPU. Comparing to prior CNN-based arts, *TransFuse-S* achieves the best performance while using only 26.3 M parameters, about 20% reduction with respect to HarDNet-MSEG (33.3 M) and PraNet (32.5 M). Moreover, *TransFuse-S* is able to run at 98.7 FPS, much faster than HarDNet-MSEG (85.3 FPS) and PraNet (63.4 FPS), thanks to our proposed parallel-in-branch design. Similarly, *TransFuse-L** not only achieves the best results compared to other Transformer-based methods, but also runs at 45.3 FPS, about 12% faster than TransUnet.

Results of Skin Lesion Segmentation. The ISBI 2017 challenge ranked methods according to Jaccard Index [5] on the ISIC 2017 test set. Here, we use Jaccard Index, Dice score and pixel-wise accuracy as evaluation metrics. The comparison results against leading methods are presented in Table 2. *TransFuse-S* is about 1.7% better than the previous SOTA SLSDeep [19] in Jaccard score, without any pre- or post-processing and converges in less than 1/3 epochs. Besides, our results outperform Unet++ [33] that employs pretrained R34 as backbone and has comparable number of parameters with TransFuse-S (26.1M vs 26.3M). Again, the results prove the superiority of our proposed architecture.

Results of Hip Segmentation. Tab. 3 shows our results on hip segmentation task, which involves three human body parts: Pelvis, Left Femur (L-Femur) and Right Femur (R-Femur). Since the contour is more important in dianosis and THA preoperative planning, we use Hausdorff Distance (HD) and Average Surface Distance (ASD) to evaluate the prediction quality. Compared to the two advanced segmentation methods [28,33], *TransFuse-S* performs the best on both metrics and reduces HD significantly (30% compared to HRNetV2 as well as 34% compared to Unet++ on average), indicating that our proposed method is able to capture finer structure and generates more precise contour.

Table 4. Quantitative results on prostate MRI segmentation. PZ, TZ stand for the two labeled classes (peripheral and transition zone) and performance (PZ, TZ and mean) is measure by dice score.

Methods	PZ	TZ	Mean	Params	Throughput
nnUnet-2d [12]	0.6285	0.8380	0.7333	29.97M	0.209s/vol
nnUnet-3d_full [12]	0.6663	0.8410	0.7537	44.80M	0.381s/vol
TransFuse-S	**0.6738**	**0.8539**	**0.7639**	**26.30M**	**0.192s/vol**

Table 5. Ablation study on parallel-in-branch design. Res: Residual.

Index	Backbones	Composition	Fusion	Kvasir	ColonDB
E.1	R34	Sequential	–	0.890	0.645
E.2	DeiT-S	Sequential	–	0.889	0.727
E.3	R34+DeiT-S	Sequential	–	0.908	0.749
E.4	R34+VGG16	Parallel	BiFusion	0.896	0.651
E.5	R34+DeiT-S	Parallel	Concat+Res	0.912	0.764
E.6	R34+DeiT-S	Parallel	BiFusion	0.918	0.773

Table 6. Ablation study on BiFusion module. Res: Residual; TFM: Transformer; Attn: Attention.

Fusion	Jaccard	Dice	Accuracy
Concat+Res	0.778	0.857	0.939
+CNN Spatial Attn	0.782	0.861	0.941
+TFM Channel Attn	0.787	0.865	0.942
+Dot Product	0.795	0.872	0.944

Results of Prostate Segmentation. We compare TransFuse-S with nnU-Net [12], which ranked 1st in the prostate segmentation challenge [22]. We follow the same preprocessing, training as well as evaluation schemes of the publicly available nnU-Net framework[3] and report the 5-fold cross validation results in Table 4. We can find that TransFuse-S surpasses nnUNet-2d by a large margin (+4.2%) in terms of the mean dice score. Compared to nnUNet-3d, TransFuse-S not only achieves better performance, but also reduces the number of parameters by ~41% and increases the throughput by ~50% (on GTX1080).

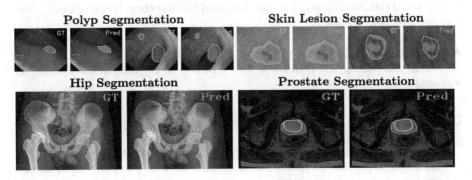

Polyp Segmentation   **Skin Lesion Segmentation**

Hip Segmentation   **Prostate Segmentation**

Fig. 2. Results visualization on all four tasks (best viewed in color). Each row follows the repeating sequence of ground truth (GT) and predictions (Pred).

---

[3] https://github.com/MIC-DKFZ/nnUNet.

Ablation Study. An ablation study is conducted to evaluate the effectiveness of the parallel-in-branch design as well as BiFusion module by varying design choices of different backbones, compositions and fusion schemes. A *seen* (Kvasir) and an *unseen* (ColonDB) datasets from polyp are used, and results are recorded in mean Dice. In Table 5, by comparing E.3 against E.1 and E.2, we can see that combining CNN and Transformer leads to better performance. Further, by comparing E.3 against E.5, E.6, we observe that the parallel models perform better than the sequential counterpart. Moreover, we evaluate the performance of a double branch CNN model (E.4) using the same parallel structure and fusion settings with our proposed E.6. We observe that E.6 outperforms E.4 by 2.2% in Kvasir and 18.7% in ColonDB, suggesting that the CNN branch and transformer branch are complementary to each other, leading to better fusion results. Lastly, performance comparison is conducted between another fusion module comprising concatenation followed by a residual block and our proposed BiFusion module (E.5 and E.6). Given the same backbone and composition setting, E.6 with BiFusion achieves better results. Additional experiments conducted on ISIC2017 are presented in Table 6 to verify the design choice of BiFusion module, from which we find that each component shows its unique benefit.

4   Conclusion

In this paper, we present a novel strategy to combine Transformers and CNNs with late fusion for medical image segmentation. The resulting architecture, TransFuse, leverages the inductive bias of CNNs on modeling spatial correlation and the powerful capability of Transformers on modelling global relationship. TransFuse achieves SOTA performance on a variety of segmentation tasks whilst being highly efficient on both the parameters and inference speed. We hope that this work can bring a new perspective on using transformer-based architecture. In the future, we plan to improve the efficiency of the vanilla transformer layer as well as test TransFuse on other medical-related tasks such as landmark detection and disease classification.

Acknowledgement. We gratefully thank Weijun Wang, MD, Zhefeng Chen, MD, Chuan He, MD, Zhengyu Xu, Huaikun Xu for serving as our medical advisors on hip segmentation project.

References

1. Al-Masni, M.A., Al-Antari, M.A., et al.: Skin lesion segmentation in dermoscopy images via deep full resolution convolutional networks. Computer methods and programs in biomedicine (2018)
2. Bernal, J., Sánchez, F.J., et al.: Wm-dova maps for accurate polyp highlighting in colonoscopy: Validation vs. saliency maps from physicians. Computerized Medical Imaging and Graphics (2015)
3. Bi, L., Kim, J., et al.: Step-wise integration of deep class-specific learning for dermoscopic image segmentation. Pattern recognition (2019)

4. Chen, J., Lu, Y., et al.: Transunet: transformers make strong encoders for medical image segmentation. arXiv preprint arXiv:2102.04306 (2021)
5. Codella, N.C., Gutman, D., et al.: Skin lesion analysis toward melanoma detection: a challenge at the 2017 international symposium on biomedical imaging (isbi), hosted by the international skin imaging collaboration (isic). In: 2018 IEEE 15th International Symposium on Biomedical Imaging (ISBI 2018) (2018)
6. Deng, J., Dong, W., et al.: Imagenet: a large-scale hierarchical image database. In: 2009 IEEE Conference on Computer Vision and Pattern Recognition (2009)
7. Dosovitskiy, A., Beyer, L., et al.: An image is worth 16x16 words: transformers for image recognition at scale. arXiv preprint arXiv:2010.11929 (2020)
8. Fan, D.P., Ji, G.P., Zhou, T., Chen, G., Fu, H., Shen, J., Shao, L.: Pranet: parallel reverse attention network for polyp segmentation. In: International Conference on Medical Image Computing and Computer-Assisted Intervention (2020)
9. Hesamian, M.H., Jia, W., He, X., Kennedy, P.: Deep learning techniques for medical image segmentation: achievements and challenges. Journal of digital imaging (2019)
10. Hu, J., Shen, L., Sun, G.: Squeeze-and-excitation networks. In: Proceedings of the IEEE Conference on Computer Vision and Pattern Recognition (2018)
11. Huang, C.H., Wu, H.Y., Lin, Y.L.: Hardnet-mseg: a simple encoder-decoder polyp segmentation neural network that achieves over 0.9 mean dice and 86 fps. arXiv preprint arXiv:2101.07172 (2021)
12. Isensee, F., Jäger, P.F., et al.: Automated design of deep learning methods for biomedical image segmentation. arXiv preprint arXiv:1904.08128 (2019)
13. Jha, D., Smedsrud, P.H., Riegler, M.A., Johansen, D., De Lange, T., Halvorsen, P., Johansen, H.D.: Resunet++: an advanced architecture for medical image segmentation. In: 2019 IEEE International Symposium on Multimedia (ISM) (2019)
14. Jha, D., Smedsrud, P.H., et al.: Kvasir-seg: a segmented polyp dataset. In: International Conference on Multimedia Modeling (2020)
15. Li, H., He, X., et al.: Dense deconvolutional network for skin lesion segmentation. IEEE J. Biomed. Health Inform. **23**, 527–537 (2018)
16. Paszke, A., Gross, S., Massa, F., Lerer, A., Bradbury, J., Chanan, G., Killeen, T., Lin, Z., Gimelshein, N., Antiga, L., et al.: Pytorch: an imperative style, high-performance deep learning library. Adv. Neural. Inf. Process. Syst. **32**, 8026–8037 (2019)
17. Qin, X., Zhang, Z., Huang, C., Gao, C., Dehghan, M., Jagersand, M.: Basnet: boundary-aware salient object detection. In: Proceedings of the IEEE/CVF Conference on Computer Vision and Pattern Recognition (2019)
18. Ronneberger, O., Fischer, P., Brox, T.: U-net: convolutional networks for biomedical image segmentation. In: International Conference on Medical Image Computing and Computer-Assisted Intervention (2015)
19. Sarker, M.M.K., Rashwan, H.A., et al.: Slsdeep: skin lesion segmentation based on dilated residual and pyramid pooling networks. In: International Conference on Medical Image Computing and Computer-Assisted Intervention (2018)
20. Schlemper, J., Oktay, O., et al.: Attention gated networks: learning to leverage salient regions in medical images. Medical image analysis (2019)
21. Silva, J., Histace, A., Romain, O., Dray, X., Granado, B.: Toward embedded detection of polyps in wce images for early diagnosis of colorectal cancer. Int. J. Comput. Assisted Radiol. Surg. **9**, 283–293(2014)
22. Simpson, A.L., Antonelli, M., et al.: A large annotated medical image dataset for the development and evaluation of segmentation algorithms. arXiv preprint arXiv:1902.09063 (2019)

23. Srivastava, R.K., Greff, K., Schmidhuber, J.: Highway networks. arXiv preprint arXiv:1505.00387 (2015)
24. Tajbakhsh, N., et al.: Automated polyp detection in colonoscopy videos using shape and context information. IEEE Trans. Med. Imaging **35**, 630–644 (2015)
25. Touvron, H., Cord, M., et al.: Training data-efficient image transformers & distillation through attention. arXiv preprint arXiv:2012.12877 (2020)
26. Vaswani, A., Shazeer, N., Parmar, N., Uszkoreit, J., Jones, L., Gomez, A.N., Kaiser, L., Polosukhin, I.: Attention is all you need. arXiv preprint arXiv:1706.03762 (2017)
27. Vázquez, D., Bernal, J., et al.: A benchmark for endoluminal scene segmentation of colonoscopy images. J. Healthcare Eng. (2017)
28. Wang, J., Sun, K., et al.: Deep high-resolution representation learning for visual recognition. IEEE Trans. Pattern Anal. Mach. Intell. **43**, 3349–3364 (2020)
29. Wang, X., Girshick, R., Gupta, A., He, K.: Non-local neural networks. In: Proceedings of the IEEE Conference on Computer Vision and Pattern Recognition (2018)
30. Woo, S., Park, J., et al.: Cbam: convolutional block attention module. In: Proceedings of the European Conference on Computer Vision (ECCV) (2018)
31. Yuan, Y., Lo, Y.C.: Improving dermoscopic image segmentation with enhanced convolutional-deconvolutional networks. IEEE J. Biomed. Health Inf. **23**, 519–526 (2017)
32. Zheng, S., Lu, J., et al.: Rethinking semantic segmentation from a sequence-to-sequence perspective with transformers. arXiv preprint arXiv:2012.15840 (2020)
33. Zhou, Z., et al.: Unet++: redesigning skip connections to exploit multiscale features in image segmentation. IEEE Trans. Med. Imaging **39**, 1856–1867 (2019)

Pancreas CT Segmentation by Predictive Phenotyping

Yucheng Tang[1](✉), Riqiang Gao[1], Hohin Lee[1], Qi Yang[1], Xin Yu[1], Yuyin Zhou[2], Shunxing Bao[1], Yuankai Huo[1], Jeffrey Spraggins[1,5], Jack Virostko[3], Zhoubing Xu[4], and Bennett A. Landman[1,5]

[1] Vanderbilt University, Nashville, TN 37203, USA
yucheng.tang@vanderbilt.edu
[2] Stanford University, Stanford, CA 94305, USA
[3] University of Texas at Austin, Austin, TX 78705, USA
[4] Siemens Healthineers, Princeton, NJ 08540, USA
[5] Vanderbilt University Medical Center, Nashville, TN 37235, USA

Abstract. Pancreas CT segmentation offers promise at understanding the structural manifestation of metabolic conditions. To date, the medical primary record of conditions that impact the pancreas is in the electronic health record (EHR) in terms of diagnostic phenotype data (*e.g.*, ICD-10 codes). We posit that similar structural phenotypes could be revealed by studying subjects with similar medical outcomes. Segmentation is mainly driven by imaging data, but this direct approach may not consider differing canonical appearances with different underlying conditions (*e.g.*, pancreatic atrophy versus pancreatic cysts). To this end, we exploit clinical features from EHR data to complement image features for enhancing the pancreas segmentation, especially in high-risk outcomes. Specifically, we propose, to the best of our knowledge, the first phenotype embedding model for pancreas segmentation by predicting representatives that share similar comorbidities. Such an embedding strategy can adaptively refine the segmentation outcome based on the discriminative contexts distilled from clinical features. Experiments with 2000 patients' EHR data and 300 CT images with the healthy pancreas, type II diabetes, and pancreatitis subjects show that segmentation by predictive phenotyping significantly improves performance over state-of-the-arts (Dice score 0.775 to 0.791, $p < 0.05$, Wilcoxon signed-rank test). The proposed method additionally achieves superior performance on two public testing datasets, BTCV MICCAI Challenge 2015 and TCIA pancreas CT. Our approach provides a promising direction of advancing segmentation with phenotype features while without requiring EHR data as input during testing.

Keywords: Pancreas segmentation · Patient phenotype · Predictive clustering

1 Introduction

Patient care data, such as CT scans and electronic health records (EHR) with the pancreatic disease, are heterogeneous in nature. Disease progression and treatment delivery are associated with different care trajectories, which in turn lead to varying pancreas patterns. In anticipation of diabetes, patients are observed with atrophic pancreas

© Springer Nature Switzerland AG 2021
M. de Bruijne et al. (Eds.): MICCAI 2021, LNCS 12901, pp. 25–35, 2021.
https://doi.org/10.1007/978-3-030-87193-2_3

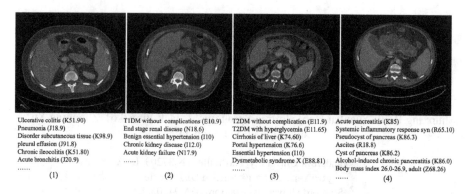

Ulcerative colitis (K51.90)	T1DM without complications (E10.9)	T2DM without complication (E11.9)	Acute pancreatitis (K85)
Pneumonia (J18.9)	End stage renal disease (N18.6)	T2DM with hyperglycemia (E11.65)	Systemic inflammatory response syn (R65.10)
Disorder subcutaneous tissue (K98.9)	Benign essential hypertension (I10)	Cirrhosis of liver (K74.60)	Pseudocyst of pancreas (K86.3)
pleural effusion (J91.8)	Chronic kidney disease (I12.0)	Portal hypertension (K76.6)	Ascites (R18.8)
Chronic ileocolitis (K51.80)	Acute kidney failure (N17.9)	Essential hypertension (I10)	Cyst of pancreas (K86.2)
Acute bronchitis (J20.9)		Dysmetabolic syndrome X (E88.81)	Alcohol-induced chronic pancreatitis (K86.0)
......	......	......	Body mass index 26.0-26.9, adult (Z68.26)
(1)	(2)	(3)	...... (4)

Fig. 1. Representative images are predicted to associate with comorbidities and ICD-10 codes (phenotype components) identified in each risk category. The red outlines show the pancreas tissue can be different under phenotyping contexts. (1) is from a nominally healthy pancreas group with potential lung infections; (2) is from type I diabetes and other chronic kidney disease patients with atrophic pancreas; (3) is from other metabolic syndromes including type II diabetes; (4) is from patients with weight loss and pancreatitis.

tissues [5], with progression noted in the patients' medical history in terms of International Classification of Diseases (ICD) codes. Current pancreas segmentation methods [15,18,24,25] are typically driven by imaging data, while phenotype covariates [3,6,11,13,22] that indicate underlying patient conditions are not well considered. We observe that different disease types present heterogeneous textures (Fig. 1), and thereby hypothesize that identifying different pancreas patterns can extract the discriminative contexts which can well benefit pancreas segmentation.

Data-driven phenotype clustering has been recently used to group patients sharing close outcomes [1,4,8,9]. Combining imaging biomarkers, Virostko et al. [20] assessed the pancreas size with type I diabetes patients. Tang et al. [19] showed the feasibility of onset type II diabetes prediction using CT scans. However, to date, how to fully exploit the EHR data for guiding medical image segmentation has been rarely studied. A naïve approach is to simply concatenate both image and EHR data as a two-channel input, and then train a standard convolutional neural network for deriving the outcome. However, this fusion strategy is not directly applicable for our task since 1) a patient can have hundreds of phenotype categories; 2) the fusion strategy cannot account for patients' observed outcomes (e.g., onset of comorbidities, chronic progression of metabolic syndrome); and 3) it requires EHR data as input during inference, which does not commonly exist for many real-world pancreas segmentation datasets (e.g., BTCV MICCAI Challenge [7], TCIA pancreas CT [14]).

To address above challenges, we propose the first pancreas segmentation framework to model both the pancreas imaging features and clinical features via predictive phenotyping. The rationale is that the larger scale of EHR data with (e.g., ICD-10 code) which indicates phenotype subgroups can be potentially correlated to the different appearance of the pancreas. Specifically, the proposed approach consists of an encoder, a segmentation decoder, and a predictor with sets of phenotypes candidates' centroids. Our method is designed to meet the following requirements: 1) The subject image should be partitioned into several subgroups sharing similar future outcomes; 2) The assigned discrete representation should retain the patient phenotype context, and 3) The phenotype representation is used as prior knowledge for predicting. Particularly, in our framework, the encoder maps an image into a latent representation; the predictor assigns one or several phenotype categories by taking the latent variable as input; the segmentation decoder estimates the pixel-wise labels conditioned on the assigned centroid. To homogenize future outcomes in each subgroup, we introduce a phenotyping objective given CT images by regularizing Kullback-Leibler (KL) divergence between the learned latent representation and the embedding centroid. Finally, the segmentation model estimates the pancreas segmentation mask given encoding of the image and the risk embedding.

Our contribution is four-folds: we successfully (1) learn a phenotype embedding between CT and EHR; (2) formulate a pancreas segmentation framework that benefits from predicting phenotype subgroups; (3) demonstrate improved pancreas segmentation performance on healthy and disease patient cohorts; (4) design the embedding approach without requiring EHR at the testing phase. The significance of the study is that we use an experimental CT imaging-phenotyping approach for investing clinical underpinnings of pancreas segmentation. The phenotype embedded model enriches segmentation contexts improving the characterization of heterogeneous disease and allows for deeper consideration of patient phenotype in image-based learning.

2 Method

2.1 Problem Formulation

Let $X \in \mathcal{X}$ and $Y \in \mathcal{Y}$ be variables for input images and an output segmentation label. $C \in \mathcal{C}$ is the patient phenotype onset (*i.e.*, one or a combination of future outcomes) where $\mathcal{X}, \mathcal{Y}$ and $\mathcal{C}$ are the image feature, label, and phenotype onset space, respectively. Specifically, CT image X selection is censored from timestamp: date of diagnostic code is later than date of scan at least 1 year. The input of C is the sequence of covariate admissions, the feature of one admission is a multi-hot vector containing the comorbidities or demographics: $C'_{EHR} = [M', M_1, M_2, ..., M_C]$. M' is the set of demographic values, M_C is the set of phenotype admissions constructed by binary vectors of aggregate ICD-10 codes. In the training phase, we are given the dataset $\mathcal{D} = \{x^n, y^n, c^n\}_{n=1}^N$ consists of observations (x, y, c) for N subjects. In the testing phase, we assume the dataset only comprising image volume $\{x^n\}_{n=1}^N$. The goal is first to identify a set of K predictive phenotypes $\mathcal{Z} = \{z_1, z_2, ..., z_K\}$ lying in the latent space. Each phenotype cluster is supposed comprising of homogeneous patients that can be represented by the cluster centroid. This predictive phenotyping updates the encoder to suggest the context to which cluster a patient belongs. Second, we design the segmentation model to estimate the pixel-wise label given the encoding variable and the predictive phenotyping distribution. Let $\bar{z}$ be the random variable that lying in the phenotype onset latent space and s be the image feature. The predictive phenotyping can be fulfilled by optimizing the Kullback-Leibler (KL) divergence between distributions conditioned on the image:

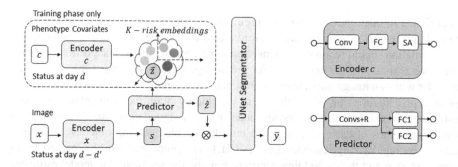

Fig. 2. Phenotype embedded segmentation architecture. The left diagram shows the embedding network combining image features s, predictive phenotyping $\hat{z}$, pre-existing risk conditions $\bar{z}$ lying in the latent space to be fed into the segmentation model. The predictor is trained for predicting phenotype-dependent feature maps and selecting "similar" cluster assignment, where the phenotype information is not required as input in the testing phase. Right top: encoder for processing phenotype covariates. Right bottom: the predictor follows the self-training scheme for image feature. Here, SA denotes soft assignment for risk embeddings, R for ReLU, FC for fully connected layers, and $\otimes$ denotes concatenation.

$p(\hat{z}|s)$ and onset phenotypes $p(\bar{z}|c)$, respectively. Combining the aim of segmentation, we establish our goal as following objective:

$$\underset{Y}{\text{minimize}} \, \mathbb{E}_{Y \sim (x,y,c)} \left[-\log P(y|s,\hat{z}_k) \right] + KL \left(\hat{z}_k|s \, \| \, \bar{z}_k|c \right). \quad (1)$$

2.2 Loss Functions

Loss functions are designed to meet the objective in Eq. 1 and are proposed to iteratively refine the predicted phenotyping from image features. Specifically, our model is trained by matching image distribution to the target distribution defined by future outcomes. To this end, we define the objective as KL divergence between an expectation and the cluster assignment:

$$\mathcal{L}_1(\bar{z},\hat{z}) = \mathbb{E}_{z \sim P(x,c)} \left[-\sum_{k=1}^{K} \bar{z}_k \log \hat{z}_k \right], \quad (2)$$

where $\bar{z}_k$ and $\hat{z}_k$ indicate the k-component of $\bar{z}$, $\hat{z}$, respectively. Note that the KL divergence loss reaches its minimum when two latent distributions are equivalent. Additionally, the segmentation loss penalizes the predicted mask $\bar{y}$ and the ground truth label y by DSC-loss:

$$\mathcal{L}_2(\bar{y},y) = \mathbb{E}_{Y \sim P(s,\hat{z})} \left[1 - \frac{2 \times \sum_i y_i \bar{y}_i}{\sum_i y_i + \sum_i \bar{y}_i} \right], \quad (3)$$

where the form follows [23] to prevent a model from background bias.

2.3 Phenotype Embedding

To encourage homogeneous future outcomes in each phenotyping cluster, we employ embedded mapping [21] as our initialization method. Given an initial estimate of the non-linear mapping c' and cluster centroid μ. We adopt the self-supervision [12] training strategy that iteratively 1) optimizes soft-assignment between embedded points and clustering centroids; 2) updates deep mapping and centroids. The soft-assignment block in Fig. 2 follows [10] using Student's t-distribution as a kernel to estimate between data points and cluster centroid:

$$q_{ik} = \frac{\exp\left(1+\left\|c'_i-\mu_k\right\|^2/\alpha\right)}{\sum_j \exp\left(1+\left\|c'_i-\mu_j\right\|^2/\alpha\right)}, \tag{4}$$

where α denotes the degrees of freedom of Student's t-distribution ($\alpha = 1$ for all experiments), exp is the exponential operation with power $-(\alpha + 1)/2$ and q_{ik} can be interpreted as the probability of assigning sample i to cluster k. More comparisons of clustering benchmarks can be found in [8,9]. After initialization, the embedding learning is iteratively updated during segmentation training.

3 Experiments

3.1 Dataset

The Abnormal Pancreas Segmentation Dataset. We have curated an abnormal pancreas dataset that contains 2000 adult patients (aged 18–50 years) with 14927 recorded visits and de-identified longitudinal CT scans under IRB approval. Each patient is associated with 101 covariates under radiologists' query, including information on demographic and abdomen-related comorbidities that can potentially impact pancreas tissues. CT images are acquired at least one year (range from 1.0 to 2.1 years) earlier than diagnosis codes for each patient, to meet the requirements of the prognostic task with predictive phenotyping. For the segmentation task, 300 patients' CT images are annotated and used for experiments. Each CT scan is $512 \times 512 \times$ Slices, where the number of slices ranges from 72 to 121 under the body part regression process of [17] to acquire relatively same abdomen region of interest (ROI). The slice thickness ranges from 1mm to 2.5mm.

BTCV MICCAI Challenge 2015. We used the MICCAI 2015 Multi-Atlas Abdomen Labeling Challenge [7] as one of the external testing sets. The challenge dataset contains 50 abdominal CT scans. For evaluating the testing phase of the proposed method, the dataset does not include patient phenotype information. Each CT scan is manually labeled with 13 structures including pancreas with a spatial resolution of ($[0.54 \sim 0.54] \times [0.98 \sim 0.98] \times [2.5 \sim 5.0]$ mm^3).

TCIA Pancreas CT. We use the 82 abdominal contrast enhanced CT scans from National Institutes of Health Clinical Center as the second external testing set. The publicly available study cohort contains 17 kidney donor subjects, and 65 patients were selected with no pancreatic cancer lesions and pathology. Each CT scan is in a resolution of 512×512 and slice thickness of $[1.5 \sim 2.5]$ mm.

3.2 Implementation Details

Follow prevailing pancreas segmentation baselines [18, 24, 25], we adopt the coarse-to-fine strategy for 3D pancreas segmentation. The coarse stage takes a highly down-sampled CT volume at an input dimension of $164 \times 164 \times 64$. For the fine stage, we cropped $64 \times 64 \times 64$ sub-volumes constrained to be in the pancreas region of interest (ROI). For experiments, 10% and 20% of subjects are randomly selected as validation and testing sets with the in-house dataset. Note that, the two external datasets are only used for testing, no subjects are used for the training procedure. We used 1) CT window range of [-175, 275] HU; 2) scaled intensities of [0.0,1.0]; 3) training with Nvidia 2080 11GB GPU with Pytorch implementation; 4) Adam optimizer with momentum 0.9. The Learning rate is initialized to 0.001 followed by a factor of 10 every 50 epochs decay.

Metrics. Segmentation performance is evaluated between ground truth and prediction by Dice-Sorensen coefficient (DSC), Averaged Surface Distance (ASD), and symmetric Hausdorff Distance (HD).

3.3 Comparison with State-of-the-Arts

We compare the proposed method with various state-of-the-art methods: 1) 3D-UNet [2]; 2) hierarchical 3D FCN [16] (denoted as "3D FCN"); 3) the fixed-point model [24] (denoted as "C2F Fixed-point"); 4) 3D ResDSN [25] (denoted as "C2F ResDSN"); and 5) the random patches model [18] (denoted as "C2F Random-patches"). Here "C2F" denotes the coarse-to-fine training strategies [18, 24, 25].

3.4 Results

We compare our method against state-of-the-art approaches with respect to the cluster number at 4 (Table 1). Our method significantly improves performance in terms of DSC, ASD and HD, with $p < 0.05$ under Wilcoxon signed-rank test. Importantly, the Hausdorff distance (HD) improvement shows that EHR information provided useful context to reduce outliers. In Table 2, we further investigate the comparison experiment results with external testing sets. The two public challenge data do not include patient EHR, *i.e.*, demographics, ICD codes. Our method implicitly predicts the future outcomes from the image feature and fused to the segmentation task. The method achieves a mean DSC of 0.757 on BTCV data, and 0.827 on TCIA pancreas CT. Predictive phenotyping improves several outlier cases, showing less variance (Fig. 3). Qualitative inspection confirms the numerical results (Fig. 4). First, we inspect the data of a patient with potential lung infections and relatively normal pancreas tissue. In the second case, the patient has type I diabetes, observing a degraded pancreas tissue. Importantly, the improvement with respect to the degraded pancreas is larger than the healthy pancreas, showing the predictive phenotyping can be informative for identifying variant patterns.

Ablative Study. *Efficacy of the predictive phenotyping and network architecture* In Table 1, we compared the backbone model (row 5) and predictive phenotyping (row 6). The EHR improved performance on two datasets by 1.5%, significant Wilcoxon signed-rank test, p<0.001. For the diabetic patients, performance improvement gains

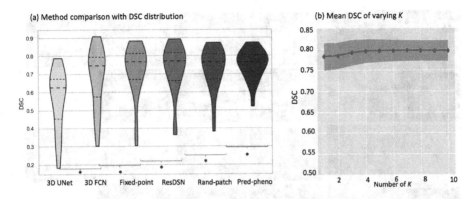

Fig. 3. Testing performance on the in-house dataset. Left: Distribution (median and quartiles) of DSC, the predictive phenotyping shows smaller variance and reduces the number of outliers (DSC < 0.4). Right: The DSC (mean) comparison with varying K. The performance shows higher improvement as K increases from 1 to 4, then becomes marginal after $K = 4$. * denotes statistically significant under Wilcoxon signed-rank test ($p < 0.05$).

Table 1. Performance comparison on the abnormal pancreas segmentation dataset. C2F denotes coarse-to-fine training. * denotes statistically significant against above method with Wilcoxon signed-rank test.

Methods	DSC	ASD	HD
3D-UNet (Cicek *et al.*)	0.697	5.592	27.154
3D FCN (Holger *et al.*)	0.724*	4.042*	25.195*
C2F Fixed-point (Zhou *et al.*)	0.746*	2.981*	22.516*
C2F ResDSN (Zhu *et al.*)	0.767*	2.105	22.017
C2F Random-patches (Tang *et al.*)	0.775*	1.976*	20.591*
Predictive phenotyping (Ours, K = 4)	**0.791***	**1.697***	**19.482***

were larger. Predictive phenotyping with the EHR outperforms naïve approach with feature concatenation by a large margin, from 74.5% to 77.9% (Fig. 3). In Table 1 and Fig. 3, we compared with pancreas segmentation state-of-the-art methods. Predictive phenotyping significantly improved performance in terms of DSC, ASD and HD, with p<0.05, Wilcoxon signed-rank test. Importantly, HD improvement shows that EHR information provided useful context to reduce outliers. In Table 2, we further investigate the comparison experiment results with external testing sets. For external validation, the two public challenge data do not include patient EHR, i.e., demographics, ICD codes. Our method implicitly predicts the future outcomes from the image feature and fused to the segmentation task. The method achieves a mean DSC of 0.757 on BTCV data, and 0.827 on TCIA pancreas CT.

Importance of hyper-parameter K. We further evaluate the performance by varying the number of clusters K from 1 to 10 on the in-house dataset. Figure 3 shows improved

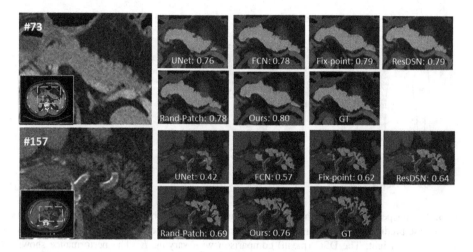

Fig. 4. Two representative cases. The top subject has potential lung infections and relative normal pancreas tissue. The bottom case has type I diabetes with degraded pancreas tissue. The accuracy gain of the diabetes case is larger than the normal case, showing the method's ability for identifying variant morphological pancreas.

Table 2. External testing performance comparison on BTCV MICCAI Challenge 2015 and TCIA pancreas (mean DSC) with our model trained on the internal data. Note that no subject from these two datasets are used for training. C2F denotes coarse-to-fine training strategies. * for statistically significant against above method with Wilcoxon signed-rank test.

Methods	BTCV	TCIA
3D-UNet (Cicek *et al.*)	0.685	0.770
3D FCN (Holger *et al.*)	0.709*	0.776*
C2F Fixed-point (Zhou *et al.*)	0.726*	0.797*
C2F ResDSN (Zhu *et al.*)	0.730*	0.804
C2F Random-patches (Tang *et al.*)	0.742*	0.813*
Predictive phenotyping (Ours, K = 4)	**0.757***	**0.822***

DSC as K increased, the DSC improves from 0.7814 to 0.7956 as K from 1 to 4. The performance is observed no significant improvement after $k = 4$ ($p < 0.1$, Wilcoxon signed-rank test).

4   Discussion and Conclusion

Comparing Table 1 and Table 2, the proposed method shows higher improvement over baseline methods if the cohort has more severe cases of abdominal diseases. Specifically, the performance improvement on the abnormal pancreas segmentation dataset in terms of the average Dice is 1.6%, which is larger than that of the BTCV dataset (1.5%), and the TCIA dataset (1.1%), respectively. We have also demonstrated two qualitative examples in Fig. 4, to show that our method can lead to more performance gain for the atrophic pancreas than the normal pancreas. The larger improvement on diseased cohort can be a potential advantage of the phenotype embedding. In addition to the major segmentation objective, the case-specific feature projected to the phenotype embedding space can be observed in Fig. 1. The comorbidities developed in the next two years with 4 identified clusters, and listed ICD-10 codes are with most frequencies in each grouped phenotype component. The first component shares the most cases with relative normal pancreas, while the second, third and fourth indicate varying phenotype outcomes of the atrophic pancreas, metabolic syndrome, and pancreas with inflammatory fats, respectively. The number of phenotype components K is one of the most important parameters in the study: increasing k can potentially impact the predictive embedding with higher diversity representing data distribution. However, the interpretability will decrease as it shares fewer similar data points. In the future, the interpretability of the predicted phenotyping can be further evaluated with more clinically meaningful investigations.

In this work, we introduce pancreas segmentation by predictive phenotyping, a patient-oriented approach for understanding between EHR and CT data. The experimental imaging-phenotyping approach is used for investigating the phenotype underpinnings of the pancreas. We demonstrate a predictive task to encourage image embedding to the phenotyping cluster with similar patient outcomes. The EHR data is designed as input at the training phase, and only images are required for inferencing phenotyping context and segmentation at the test phase. Throughout experiments on the in-house dataset and two public challenge datasets, we show that the method highlights a significant role over state-of-the-art segmentations. The integrated imaging-phenotyping method could encourage solutions that better respect anatomical variability, especially associated with disease progression or comorbidities. When EHR data is available, the method can be applied for boosting performance.

Acknowledgements. This research is supported by NIH Common Fund and National Institute of Diabetes, Digestive and Kidney Diseases U54DK120058, NSF CAREER 1452485, NIH grants, 2R01EB006136, 1R01EB017230 (Landman), and R01NS09529. The identified datasets used for the analysis described were obtained from the Research Derivative (RD), database of clinical and related data. The imaging dataset(s) used for the analysis described were obtained from ImageVU, a research repository of medical imaging data and image-related metadata. ImageVU and RD are supported by the VICTR CTSA award (ULTR000445 from NCATS/NIH) and Vanderbilt University Medical Center institutional funding. ImageVU pilot work was also funded by PCORI (contract CDRN-1306-04869).

References

1. Baytas, I.M., Xiao, C., Zhang, X., Wang, F., Jain, A.K., Zhou, J.: Patient subtyping via time-aware lstm networks. In: Proceedings of the 23rd ACM SIGKDD International Conference on Knowledge Discovery and Data Mining, pp. 65–74 (2017)
2. Çiçek, Ö., Abdulkadir, A., Lienkamp, S.S., Brox, T., Ronneberger, O.: 3D U-Net: learning dense volumetric segmentation from sparse annotation. In: Ourselin, S., Joskowicz, L., Sabuncu, M.R., Unal, G., Wells, W. (eds.) MICCAI 2016. LNCS, vol. 9901, pp. 424–432. Springer, Cham (2016). https://doi.org/10.1007/978-3-319-46723-8_49

3. Evans, J.A.: Electronic medical records system (Jul 13 1999), uS Patent 5,924,074
4. Giannoula, A., Gutierrez-Sacristán, A., Bravo, Á., Sanz, F., Furlong, L.I.: Identifying temporal patterns in patient disease trajectories using dynamic time warping: a population-based study. Sci. Rep. **8**(1), 1–14 (2018)
5. Goda, K., Sasaki, E., Nagata, K., Fukai, M., Ohsawa, N., Hahafusa, T.: Pancreatic volume in type 1 und type 2 diabetes mellitus. Acta Diabetol. **38**(3), 145–149 (2001)
6. Hales, C.N., Barker, D.J.: Type 2 (non-insulin-dependent) diabetes mellitus: the thrifty phenotype hypothesis. Diabetologia **35**(7), 595–601 (1992)
7. Landman, B., Xu, Z., Igelsias, J., Styner, M., Langerak, T., Klein, A.: Miccai multi-atlas labeling beyond the cranial vault-workshop and challenge. In: Proceedings of MICCAI Multi-Atlas Labeling Beyond Cranial Vault—Workshop Challenge (2015)
8. Lee, C., Van Der Schaar, M.: Temporal phenotyping using deep predictive clustering of disease progression. In: International Conference on Machine Learning, pp. 5767–5777. PMLR (2020)
9. Luong, D.T.A., Chandola, V.: A k-means approach to clustering disease progressions. In: 2017 IEEE International Conference on Healthcare Informatics (ICHI), pp. 268–274. IEEE (2017)
10. Van der Maaten, L., Hinton, G.: Visualizing data using t-sne. J. Mach. Learn. Res. **9**(11) (2008)
11. Mani, S., Chen, Y., Elasy, T., Clayton, W., Denny, J.: Type 2 diabetes risk forecasting from emr data using machine learning. In: AMIA Annual Symposium Proceedings, vol. 2012, p. 606. American Medical Informatics Association (2012)
12. Misra, I., Maaten, L.v.d.: Self-supervised learning of pretext-invariant representations. In: Proceedings of the IEEE/CVF Conference on Computer Vision and Pattern Recognition, pp. 6707–6717 (2020)
13. Quan, H., et al.: Coding algorithms for defining comorbidities in icd-9-cm and icd-10 administrative data. Medical care, pp. 1130–1139 (2005)
14. Roth, H., Farag, A., Turkbey, E., Lu, L., Liu, J., Summers, R.: Data from pancreas-ct. The cancer imaging archive (2016)
15. Roth, H.R., Farag, A., Lu, L., Turkbey, E.B., Summers, R.M.: Deep convolutional networks for pancreas segmentation in CT imaging. In: Medical Imaging 2015: Image Processing, vol. 9413, p. 94131G. International Society for Optics and Photonics (2015)
16. Roth, H.R., et al.: DeepOrgan: multi-level deep convolutional networks for automated pancreas segmentation. In: Navab, N., Hornegger, J., Wells, W.M., Frangi, A.F. (eds.) MICCAI 2015. LNCS, vol. 9349, pp. 556–564. Springer, Cham (2015). https://doi.org/10.1007/978-3-319-24553-9_68
17. Tang, Y., et al.: Body part regression with self-supervision. IEEE Trans. Med. Imaging **40**, 1499–1507 (2021)
18. Tang, Y., et al.: High-resolution 3d abdominal segmentation with random patch network fusion. Med. Image Anal. **69**, 101894 (2021)
19. Tang, Y., et al.: Prediction of Type II diabetes onset with computed tomography and electronic medical records. In: Syeda-Mahmood, T., et al. (eds.) CLIP/ML-CDS -2020. LNCS, vol. 12445, pp. 13–23. Springer, Cham (2020). https://doi.org/10.1007/978-3-030-60946-7_2
20. Virostko, J., Hilmes, M., Eitel, K., Moore, D.J., Powers, A.C.: Use of the electronic medical record to assess pancreas size in type 1 diabetes. PLoS ONE **11**(7), e0158825 (2016)
21. Xie, J., Girshick, R., Farhadi, A.: Unsupervised deep embedding for clustering analysis. In: International Conference on Machine Learning, pp. 478–487. PMLR (2016)
22. Zheng, T., et al.: A machine learning-based framework to identify type 2 diabetes through electronic health records. Int. J. Med. Informatics **97**, 120–127 (2017)

23. Zhou, Y., Xie, L., Fishman, E.K., Yuille, A.L.: Deep supervision for pancreatic cyst segmentation in abdominal CT scans. In: Descoteaux, M., Maier-Hein, L., Franz, A., Jannin, P., Collins, D.L., Duchesne, S. (eds.) MICCAI 2017. LNCS, vol. 10435, pp. 222–230. Springer, Cham (2017). https://doi.org/10.1007/978-3-319-66179-7_26
24. Zhou, Y., Xie, L., Shen, W., Wang, Y., Fishman, E.K., Yuille, A.L.: A fixed-point model for pancreas segmentation in abdominal CT scans. In: MICCAI, pp. 693–701 (2017)
25. Zhu, Z., Xia, Y., Shen, W., Fishman, E., Yuille, A.: A 3d coarse-to-fine framework for volumetric medical image segmentation. In: 2018 International Conference on 3D Vision (3DV), pp. 682–690. IEEE (2018)

Medical Transformer: Gated Axial-Attention for Medical Image Segmentation

Jeya Maria Jose Valanarasu[1]([⊠]), Poojan Oza[1], Ilker Hacihaliloglu[2], and Vishal M. Patel[1]

[1] Johns Hopkins University, Baltimore, MD, USA
[2] Rutgers, The State University of New Jersey, New Brunswick, NJ, USA

Abstract. Over the past decade, deep convolutional neural networks have been widely adopted for medical image segmentation and shown to achieve adequate performance. However, due to inherent inductive biases present in convolutional architectures, they lack understanding of long-range dependencies in the image. Recently proposed transformer-based architectures that leverage self-attention mechanism encode long-range dependencies and learn representations that are highly expressive. This motivates us to explore transformer-based solutions and study the feasibility of using transformer-based network architectures for medical image segmentation tasks. Majority of existing transformer-based network architectures proposed for vision applications require large-scale datasets to train properly. However, compared to the datasets for vision applications, in medical imaging the number of data samples is relatively low, making it difficult to efficiently train transformers for medical imaging applications. To this end, we propose a gated axial-attention model which extends the existing architectures by introducing an additional control mechanism in the self-attention module. Furthermore, to train the model effectively on medical images, we propose a Local-Global training strategy (LoGo) which further improves the performance. Specifically, we operate on the whole image and patches to learn global and local features, respectively. The proposed Medical Transformer (MedT) is evaluated on three different medical image segmentation datasets and it is shown that it achieves better performance than the convolutional and other related transformer-based architectures. Code: https://github.com/jeya-maria-jose/Medical-Transformer

Keywords: Transformers · Medical image segmentation · Self-attention

---

Electronic supplementary material The online version of this chapter (https://doi.org/10.1007/978-3-030-87193-2_4) contains supplementary material, which is available to authorized users.

M. de Bruijne et al. (Eds.): MICCAI 2021, LNCS 12901, pp. 36–46, 2021.
https://doi.org/10.1007/978-3-030-87193-2_4

1 Introduction

Developing automatic, accurate, and robust medical image segmentation methods have been one of the principal problems in medical imaging as it is essential for computer-aided diagnosis and image-guided surgery systems. Segmentation of organs or lesion from a medical scan helps clinicians make an accurate diagnosis, plan the surgical procedure, and propose treatment strategies. Following the popularity of deep convolutional neural networks (ConvNets) in computer vision, ConvNets were quickly adopted for medical image segmentation. Networks like U-Net [15], V-Net [13], 3D U-Net [3], Res-UNet [25], Dense-UNet [11], Y-Net [12], U-Net++ [28], KiU-Net [19,20] and U-Net3+ [7] have been proposed specifically for performing image and volumetric segmentation for various medical imaging modalities. These methods achieve impressive performance on many difficult datasets, proving the effectiveness of ConvNets in learning discriminative features to segment the organ or lesion from a medical scan.

ConvNets are currently the basic building blocks of most methods proposed for image segmentation. However, they lack the ability to model long-range dependencies present in an image. More precisely, in ConvNets each convolutional kernel attends to only a local-subset of pixels in the whole image and forces the network to focus on local patterns rather than the global context. There have been works that have focused on modeling long-range dependencies for ConvNets using image pyramids [26], atrous convolutions [2] and attention mechanisms [8]. However, it can be noted that there is still a scope of improvement for modeling long-range dependencies as the majority of previous methods do not focus on this aspect for medical image segmentation tasks.

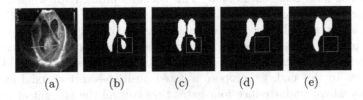

(a)   (b)   (c)   (d)   (e)

Fig. 1. (a) Input Ultrasound of in vivo preterm neonatal brain ventricle. Predictions by (b) U-Net, (c) Res-UNet, (d) MedT, and (e) Ground Truth. The red box highlights the region which are miss-classified by ConvNet based methods due to lack of learned long-range dependencies. The ground truth here was segmented by an expert clinician. Although it shows some bleeding inside the ventricle area, it does not correspond to the segmented area. This information is correctly captured by transformer-based models. (Color figure online)

To first understand why long-range dependencies matter for medical images, we visualize an example ultrasound scan of a preterm neonate and segmentation predictions of brain ventricles from the scan in Fig. 1. For a network to provide an efficient segmentation, it should be able to understand which pixels correspond to the mask and which to the background. As the background of the image is scattered, learning long-range dependencies between the pixels corresponding to the background can help in the network to prevent miss-classifying a pixel as the mask leading to reduction of false positives (considering 0 as background and 1 as segmentation mask). Similarly, whenever the segmentation mask is large, learning long-range dependencies between the pixels corresponding to the mask is also helpful in making efficient predictions. In Fig. 1(b) and (c), we can see that the convolutional networks miss-classify the background as a brain ventricle while the proposed transformer-based method does not make that mistake. This happens as our proposed method learns long-range dependencies of the pixel regions with that of the background.

In many natural language processing (NLP) applications, transformers [4] have shown to be able to encode long-range dependencies. This is due to the self-attention mechanism which finds the dependency between given sequential input. Following their popularity in NLP applications, transformers have been adopted to computer vision applications very recently [5,18]. With regard to transformers for segmentation tasks, Axial-Deeplab [22] utilized the axial attention module [6], which factorizes 2D self-attention into two 1D self-attentions and introduced position-sensitive axial attention design for segmentation. In Segmentation Transformer (SETR) [27], a transformer was used as encoder which inputs a sequence of image patches and a ConvNet was used as decoder resulting in a powerful segmentation model. In medical image segmentation, transformer-based models have not been explored much. The closest works are the ones that use attention mechanisms to boost the performance [14,24]. However, the encoder and decoder of these networks still have convolutional layers as the main building blocks.

It was observed that the transformer-based models work well only when they are trained on large-scale datasets [5]. This becomes problematic while adopting transformers for medical imaging tasks as the number of images, with corresponding labels, available for training in any medical dataset is relatively scarce. Labeling process is also expensive and requires expert knowledge. Specifically, training with fewer images causes difficulty in learning positional encoding for the images. To this end, we propose a gated position-sensitive axial attention mechanism where we introduce four gates that control the amount of information the positional embedding supply to key, query, and value. These gates are learnable parameters which make the proposed mechanism to be applied to any dataset of any size. Depending on the size of the dataset, these gates would learn whether the number of images would be sufficient enough to learn proper position embedding. Based on whether the information learned by the positional embedding is useful or not, the gate parameters either converge to 0 or to some higher value. Furthermore, we propose a Local-Global (LoGo) training strategy, where we use a shallow global branch and a deep local branch that operates on the patches of the medical image. This strategy improves the segmentation performance as we do not only operate on the entire image but focus on finer details present in the local patches. Finally, we propose Medical Transformer (MedT), which uses our gated position-sensitive axial attention as the building blocks and adopts our LoGo training strategy.

In summary, this paper (1) proposes a gated position-sensitive axial attention mechanism that works well even on smaller datasets, (2) introduces Local-Global (LoGo) training methodology for transformers which is effective, (3) proposes Medical-Transformer (MedT) which is built upon the above two concepts proposed specifically for medical image segmentation, and (4) successfully improves the performance for medical image segmentation tasks over convolutional networks and fully attention architectures on three different datasets.

2   Medical Transformer (MedT)

2.1   Self-attention Overview

Let us consider an input feature map $x \in \mathbb{R}^{C_{in} \times H \times W}$ with height H, weight W and channels C_{in}. The output $y \in \mathbb{R}^{C_{out} \times H \times W}$ of a self-attention layer is computed with the help of projected input using the following equation:

$$y_{ij} = \sum_{h=1}^{H} \sum_{w=1}^{W} \text{softmax}\left(q_{ij}^{T} k_{hw}\right) v_{hw}, \tag{1}$$

where queries $q = W_Q x$, keys $k = W_K x$ and values $v = W_V x$ are all projections computed from the input x. Here, q_{ij}, k_{ij}, v_{ij} denote query, key and value at

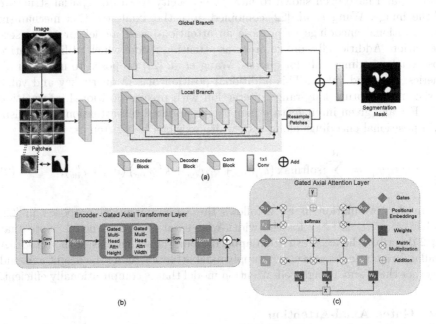

Fig. 2. (a) The main architecture diagram of MedT which uses LoGo strategy for training. (b) The gated axial transformer layer which is used in MedT. (c) Gated Axial Attention layer which is the basic building block of both height and width gated multi-head attention blocks found in the gated axial transformer layer.

any arbitrary location $i \in \{1, \ldots, H\}$ and $j \in \{1, \ldots, W\}$, respectively. The projection matrices $W_Q, W_K, W_V \in \mathbb{R}^{C_{in} \times C_{out}}$ are learnable. As shown in Eq. 1, the values v are pooled based on global affinities calculated using softmax($q^T k$). Hence, unlike convolutions the self-attention mechanism is able to capture non-local information from the entire feature map. However, computing such affinities are computationally very expensive and with increased feature map size it often becomes infeasible to use self-attention for vision model architectures. Moreover, unlike convolutional layer, self-attention layer does not utilize any positional information while computing the non-local context. Positional information is often useful in vision models to capture structure of an object.

Axial-Attention. To overcome the computational complexity of calculating the affinities, self-attention is decomposed into two self-attention modules. The first module performs self-attention on the feature map height axis and the second one operates on the width axis. This is referred to as axial attention [6]. The axial attention consequently applied on height and width axis effectively model original self-attention mechanism with much better computational efficacy. To add positional bias while computing affinities through self-attention mechanism, a position bias term is added to make the affinities sensitive to the positional information [16]. This bias term is often referred to as relative positional encodings. These positional encodings are typically learnable through training and have been shown to have the capacity to encode spatial structure of the image. Wang *et al.* [22] combined both the axial-attention mechanism and positional encodings to propose an attention-based model for image segmentation. Additionally, unlike previous attention model which utilizes relative positional encodings only for queries, Wang *et al.* [22] proposed to use it for all queries, keys and values. This additional position bias in query, key and value is shown to capture long-range interaction with precise positional information [22]. For any given input feature map x, the updated self-attention mechanism with positional encodings along with width axis can be written as:

$$y_{ij} = \sum_{w=1}^{W} \text{softmax} \left(q_{ij}^T k_{iw} + q_{ij}^T r_{iw}^q + k_{iw}^T r_{iw}^k \right) (v_{iw} + r_{iw}^v), \qquad (2)$$

where the formulation in Eq. 2 follows the attention model proposed in [22] and $r^q, r^k, r^v \in \mathbb{R}^{W \times W}$ for the width-wise axial attention model. Note that Eq. 2 describes the axial attention applied along the width axis of the tensor. A similar formulation is also used to apply axial attention along the height axis and together they form a single self-attention model that is computationally efficient.

2.2   Gated Axial-Attention

We discussed the benefits of using the axial-attention mechanism proposed in [22] for visual recognition. Specifically, the axial-attention proposed in [22] is able to compute non-local context with good computational efficiency, able to encode positional bias into the mechanism and enables the ability to encode long-range interaction within an input feature map. However, their model is evaluated on large-scale segmentation datasets and hence it is easier for the axial-attention to learn positional bias at key, query and value. We argue that for experiments with small-scale datasets, which is often the case in medical image segmentation, the positional bias is difficult to learn and hence will not always be accurate in encoding long-range interactions. In the case where the learned relative positional encodings are not accurate enough, adding them to the respective key, query and value tensor would result in reduced performance. Hence, we propose a modified axial-attention block that can control the influence positional bias can exert in the encoding of non-local context. With the proposed modification the self-attention mechanism applied on the width axis can be formally written as:

$$
y_{ij} = \sum_{w=1}^{W} \mathrm{softmax} \left(q_{ij}^{T} k_{iw} + G_Q q_{ij}^{T} r_{iw}^{q} + G_K k_{iw}^{T} r_{iw}^{k} \right) (G_{V1} v_{iw} + G_{V2} r_{iw}^{v}), \quad (3)
$$

where the self-attention formula closely follows Eq. 2 with added gating mechanism. Also, $G_Q, G_K, G_{V1}, G_{V2} \in \mathbb{R}$ are learnable parameters and together they create gating mechanism which control influence of the learned relative positional encodings have on encoding non-local context. Typically, if a relative positional encoding is learned accurately, the gating mechanism will assign it high weight compared to the ones which are not learned accurately. Figure 2(c) illustrates the feed-forward in a typical gated axial attention layer.

2.3   Local-Global Training

It is evident that a transformer on patches is faster but patch-wise training alone is not sufficient for the tasks like medical image segmentation. Patch-wise training restricts the network in learning any information or dependencies for inter-patch pixels. To improve the overall understanding of the image, we propose to use two branches in the network, i.e., a global branch which works on the original resolution of the image, and a local branch which operates on patches of the image. In the global branch, we reduce the number of gated axial transformer layers as we observe that the first few blocks of the proposed transformer model is sufficient to model long range dependencies. In the local branch, we create 16 patches of size $I/4 \times I/4$ of the image where I is the dimensions of the original image. In the local branches, each patch is feed forwarded through the network and the output feature maps are re-sampled based on their location to get the output feature maps. The output feature maps of both of the branches are then added and passed through a 1×1 convolution layer to produce the output segmentation mask. This strategy improves the performance as the global branch focuses on high-level information and the local branch can focus on finer details. The proposed Medical Transformer (MedT) uses gated axial attention layer as the basic building block and uses LoGo strategy for training. It is illustrated in Fig. 2(a). More details on the architecture and an ablation study with regard to the architecture can be found in the supplementary file.

3   Experiments and Results

3.1   Dataset Details

We use Brain anatomy segmentation (ultrasound) [21,23], Gland segmentation (microscopic) [17] and MoNuSeg (microscopic) [9,10] datasets for evaluating our method. More details about the datasets can be found in the supplementary.

3.2   Implementation Details

We use binary cross-entropy (CE) loss between the prediction and the ground truth to train our network and can be written as:

$$\mathcal{L}_{CE(p,\hat{p})} = - \left(\frac{1}{wh} \sum_{x=0}^{w-1} \sum_{y=0}^{h-1} (p(x,y) \log(\hat{p}(x,y))) + (1 - p(x,y)) \log(1 - \hat{p}(x,y)) \right)$$

where w and h are the dimensions of the image, $p(x,y)$ corresponds to the pixel in the image and $\hat{p}(x,y)$ denotes the output prediction at a specific location (x,y). The training details are provided in the supplementary document.

For baseline comparisons, we first run experiments on both convolutional and transformer-based methods. For convolutional baselines, we compare with fully convolutional network (FCN) [1], U-Net [15], U-Net++ [28] and Res-Unet [25]. For transformer-based baselines, we use Axial-Attention U-Net with residual connections inspired from [22]. For our proposed method, we experiment with all the individual contributions. In gated axial attention network, we use axial attention U-Net with all its axial attention layers replaced with the proposed gated axial attention layers. In LoGo, we perform local global training for axial attention U-Net without using the gated axial attention layers. In MedT, we use gated axial attention as the basic building block for global branch and axial attention without positional encoding for local branch.

3.3   Results

Table 1. Quantitative comparison of the proposed methods with convolutional and transformer based baselines in terms of F1 and IoU scores.

Type	Network	Brain US		GlaS		MoNuSeg	
		F1	IoU	F1	IoU	F1	IoU
Convolutional Baselines	FCN [1]	82.79	75.02	66.61	50.84	28.84	28.71
	U-Net [15]	85.37	79.31	77.78	65.34	79.43	65.99
	U-Net++ [28]	86.59	79.95	78.03	65.55	79.49	66.04
	Res-UNet [25]	87.50	79.61	78.83	65.95	79.49	66.07
Fully Attention Baseline	Axial Attention U-Net [22]	87.92	80.14	76.26	63.03	76.83	62.49
Proposed	Gated Axial Attn	88.39	80.7	79.91	67.85	76.44	62.01
	LoGo	88.54	80.84	79.68	67.69	79.56	66.17
	MedT	**88.84**	**81.34**	**81.02**	**69.61**	**79.55**	**66.17**

For quantitative analysis, we use F1 and IoU scores for comparison. The quantitative results are tabulated in Table 1. It can be noted that for datasets with relatively more images like Brain US, fully attention (transformer) based baseline performs better than convolutional baselines. For GlaS and MoNuSeg datasets, convolutional baselines perform better than fully attention baselines as it is difficult to train fully attention models with less data [5]. The proposed method is able to overcome such issue with the help of gated axial attention and LoGo both individually perform better than the other methods. Our final architecture MedT performs better than Gated axial attention, LoGo and all the previous methods. The improvements over fully attention baselines are 0.92 %, 4.76 % and 2.72 % for Brain US, GlaS and MoNuSeg datasets, respectively. Improvements over the best convolutional baseline are 1.32 %, 2.19 % and 0.06 %. All of these values are in terms of F1 scores. For the ablation study, we use the Brain US data for all our experiments. The results for the same has been tabulated in Table 2.

Furthermore, we visualize the predictions from U-Net [15], Res-UNet [25], Axial Attention U-Net [22] and our proposed method MedT in Fig. 3. It can be seen that the predictions of MedT captures the long range dependencies really well. For example, in the second row of Fig. 3, we can observe that the small

Table 2. Ablation study

Network	U-Net [15]	Res-UNet [25]	Axial UNet [22]	Gated Axial UNet	Global only	Local only	LoGo	MedT
F1 Score	85.37	87.5	87.92	88.39	87.67	77.55	88.54	88.84

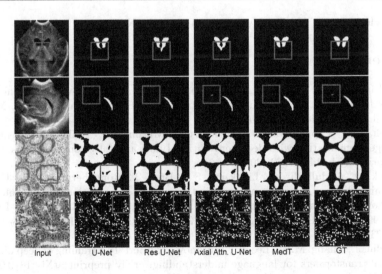

Input   U-Net   Res U-Net   Axial Attn. U-Net   MedT   GT

Fig. 3. Qualitative results on sample test images from Brain US, Glas and MoNuSeg datasets. The red box highlights regions where exactly MedT performs better than the other methods in comparison making better use of long range dependencies. (Color figure online)

segmentation mask highlighted on red box goes undetected in all the convolutional baselines. However, as fully attention model encodes long range dependencies, it learns to segment well thanks to the encoded global context. In the first and fourth row, other methods make false predictions at the highlighted regions as those pixels are in close proximity to the segmentation mask. As our method takes into account pixel-wise dependencies that are encoded with gating mechanism, it is able to learn those dependencies better than the axial attention U-Net. This makes our predictions more precise as they do not miss-classify pixels near the segmentation mask.

4   Conclusion

In this work, we explored the use of transformer-based architectures for medical image segmentation. Specifically, we propose a gated axial attention layer which is used as the building block for multi-head attention models. We also proposed a LoGo training strategy to train the image in both full resolution as well in patches. The global branch helps learn global context features by modeling long-range dependencies, where as the local branch focus on finer features by operating on patches. Using these, we propose MedT (Medical Transformer) which has gated axial attention as its main building block for the encoder and uses LoGo strategy for training. Unlike other transformer-based model the proposed method does not require pre-training on large-scale datasets. Finally, we conduct extensive experiments on three datasets where we achieve a good performance for MedT over ConvNets and other related transformer-based architectures.

Acknowledgment. This work was supported by the NSF grant 1910141.

References

1. Badrinarayanan, V., Kendall, A., Cipolla, R.: Segnet: a deep convolutional encoder-decoder architecture for image segmentation. IEEE Trans. Pattern Anal. Mach. Intell. **39**(12), 2481–2495 (2017)
2. Chen, L.C., Papandreou, G., Kokkinos, I., Murphy, K., Yuille, A.L.: Semantic image segmentation with deep convolutional nets and fully connected crfs. arXiv preprint arXiv:1412.7062 (2014)
3. Çiçek, Ö., Abdulkadir, A., Lienkamp, S.S., Brox, T., Ronneberger, O.: 3D U-Net: learning dense volumetric segmentation from sparse annotation. In: Ourselin, S., Joskowicz, L., Sabuncu, M.R., Unal, G., Wells, W. (eds.) MICCAI 2016. LNCS, vol. 9901, pp. 424–432. Springer, Cham (2016). https://doi.org/10.1007/978-3-319-46723-8_49
4. Devlin, J., Chang, M.W., Lee, K., Toutanova, K.: Bert: Pre-training of deep bidirectional transformers for language understanding. arXiv preprint arXiv:1810.04805 (2018)
5. Dosovitskiy, A., et al.: An image is worth 16x16 words: Transformers for image recognition at scale. arXiv preprint arXiv:2010.11929 (2020)
6. Ho, J., Kalchbrenner, N., Weissenborn, D., Salimans, T.: Axial attention in multi-dimensional transformers. arXiv preprint arXiv:1912.12180 (2019)

7. Huang, H., et al.: Unet 3+: a full-scale connected unet for medical image segmentation. In: ICASSP 2020–2020 IEEE International Conference on Acoustics, Speech and Signal Processing (ICASSP), pp. 1055–1059. IEEE (2020)

8. Huang, Z., Wang, X., Huang, L., Huang, C., Wei, Y., Liu, W.: Ccnet: criss-cross attention for semantic segmentation. In: Proceedings of the IEEE/CVF International Conference on Computer Vision, pp. 603–612 (2019)

9. Kumar, N., et al.: A multi-organ nucleus segmentation challenge. IEEE Trans. Med. Imaging **39**(5), 1380–1391 (2019)

10. Kumar, N., Verma, R., Sharma, S., Bhargava, S., Vahadane, A., Sethi, A.: A dataset and a technique for generalized nuclear segmentation for computational pathology. IEEE Trans. Med. Imaging **36**(7), 1550–1560 (2017)

11. Li, X., Chen, H., Qi, X., Dou, Q., Fu, C.W., Heng, P.A.: H-denseunet: hybrid densely connected unet for liver and tumor segmentation from ct volumes. IEEE Trans. Med. Imaging **37**(12), 2663–2674 (2018)

12. Mehta, S., Mercan, E., Bartlett, J., Weaver, D., Elmore, J.G., Shapiro, L.: Y-net: joint segmentation and classification for diagnosis of breast biopsy images. In: International Conference on Medical Image Computing and Computer-Assisted Intervention, pp. 893–901. Springer (2018)

13. Milletari, F., Navab, N., Ahmadi, S.A.: V-net: fully convolutional neural networks for volumetric medical image segmentation. In: 2016 Fourth International Conference on 3D Vision (3DV), pp. 565–571. IEEE (2016)

14. Oktay, O., et al.: Attention u-net: Learning where to look for the pancreas. arXiv preprint arXiv:1804.03999 (2018)

15. Ronneberger, O., Fischer, P., Brox, T.: U-Net: convolutional networks for biomedical image segmentation. In: Navab, N., Hornegger, J., Wells, W.M., Frangi, A.F. (eds.) MICCAI 2015. LNCS, vol. 9351, pp. 234–241. Springer, Cham (2015). https://doi.org/10.1007/978-3-319-24574-4_28

16. Shaw, P., Uszkoreit, J., Vaswani, A.: Self-attention with relative position representations. In: Proceedings of the 2018 Conference of the North American Chapter of the Association for Computational Linguistics: Human Language Technologies, Volume 2 (Short Papers), pp. 464–468 (2018)

17. Sirinukunwattana, K., et al.: Gland segmentation in colon histology images: the glas challenge contest. Med. Image Anal. **35**, 489–502 (2017)

18. Touvron, H., Cord, M., Douze, M., Massa, F., Sablayrolles, A., Jégou, H.: Training data-efficient image transformers & distillation through attention. arXiv preprint arXiv:2012.12877 (2020)

19. Valanarasu, J.M.J., Sindagi, V.A., Hacihaliloglu, I., Patel, V.M.: Kiu-net: over-complete convolutional architectures for biomedical image and volumetric segmentation. arXiv preprint arXiv:2010.01663 (2020)

20. Valanarasu, J.M.J., Sindagi, V.A., Hacihaliloglu, I., Patel, V.M.: KiU-Net: towards accurate segmentation of biomedical images using over-complete representations. In: Martel, A.L., et al. (eds.) MICCAI 2020. LNCS, vol. 12264, pp. 363–373. Springer, Cham (2020). https://doi.org/10.1007/978-3-030-59719-1_36

21. Valanarasu, J.M.J., Yasarla, R., Wang, P., Hacihaliloglu, I., Patel, V.M.: Learning to segment brain anatomy from 2d ultrasound with less data. IEEE J. Selected Topics Signal Process. **14**(6), 1221–1234 (2020)

22. Wang, H., Zhu, Y., Green, B., Adam, H., Yuille, A., Chen, L.C.: Axial-deeplab: stand-alone axial-attention for panoptic segmentation. arXiv preprint arXiv:2003.07853 (2020)

23. Wang, P., Cuccolo, N.G., Tyagi, R., Hacihaliloglu, I., Patel, V.M.: Automatic real-time cnn-based neonatal brain ventricles segmentation. In: 2018 IEEE 15th International Symposium on Biomedical Imaging (ISBI 2018), pp. 716–719. IEEE (2018)
24. Wang, X., Han, S., Chen, Y., Gao, D., Vasconcelos, N.: Volumetric attention for 3D medical image segmentation and detection. In: Shen, D., et al. (eds.) MICCAI 2019. LNCS, vol. 11769, pp. 175–184. Springer, Cham (2019). https://doi.org/10.1007/978-3-030-32226-7_20
25. Xiao, X., Lian, S., Luo, Z., Li, S.: Weighted res-unet for high-quality retina vessel segmentation. In: 2018 9th International Conference on Information Technology in Medicine and Education (ITME), pp. 327–331. IEEE (2018)
26. Zhao, H., Shi, J., Qi, X., Wang, X., Jia, J.: Pyramid scene parsing network. In: Proceedings of the IEEE Conference on Computer Vision and Pattern Recognition, pp. 2881–2890 (2017)
27. Zheng, S., et al.: Rethinking semantic segmentation from a sequence-to-sequence perspective with transformers. arXiv preprint arXiv:2012.15840 (2020)
28. Zhou, Z., Rahman Siddiquee, M.M., Tajbakhsh, N., Liang, J.: UNet++: a nested u-net architecture for medical image segmentation. In: Stoyanov, D., et al. (eds.) DLMIA/ML-CDS -2018. LNCS, vol. 11045, pp. 3–11. Springer, Cham (2018). https://doi.org/10.1007/978-3-030-00889-5_1

Anatomy-Constrained Contrastive Learning for Synthetic Segmentation Without Ground-Truth

Bo Zhou[1(✉)], Chi Liu[1,2], and James S. Duncan[1,2]

[1] Biomedical Engineering, Yale University, New Haven, CT, USA
bo.zhou@yale.edu
[2] Radiology and Biomedical Imaging, Yale University, New Haven, CT, USA

Abstract. A large amount of manual segmentation is typically required to train a robust segmentation network so that it can segment objects of interest in a new imaging modality. The manual efforts can be alleviated if the manual segmentation in one imaging modality (e.g., CT) can be utilized to train a segmentation network in another imaging modality (e.g., CBCT/MRI/PET). In this work, we developed an anatomy-constrained contrastive synthetic segmentation network (AccSeg-Net) to train a segmentation network for a target imaging modality without using its ground-truth. Specifically, we proposed to use anatomy-constraint and patch contrastive learning to ensure the anatomy fidelity during the unsupervised adaptation, such that the segmentation network can be trained on the adapted image with correct anatomical structure/content. The training data for our AccSeg-Net consists of 1) imaging data paired with segmentation ground-truth in source modality, and 2) unpaired source and target modality imaging data. We demonstrated successful applications on CBCT, MRI, and PET imaging data, and showed superior segmentation performances as compared to previous methods. Our code is available at https://github.com/bbbbbbzhou/AccSeg-Net

Keywords: Contrastive learning · Anatomy-constraint · Synthetic segmentation · Unsupervised learning

1 Introduction

Deep learning based image segmentation has wide applications in various medical imaging modalities. Over the recent years, numerous segmentation networks have been proposed to continuously improve the segmentation performance [1–6]. These segmentation networks require training from large amounts of segmentation ground-truth on their target domain to achieve robust performance.

Electronic supplementary material The online version of this chapter (https://doi.org/10.1007/978-3-030-87193-2_5) contains supplementary material, which is available to authorized users.

M. de Bruijne et al. (Eds.): MICCAI 2021, LNCS 12901, pp. 47–56, 2021.
https://doi.org/10.1007/978-3-030-87193-2_5

However, a large amount of ground-truth data is not always available to several imaging modalities, such as intra-procedural CBCT, gadolinium-enhanced T1 MRI, and PET with different tracers, thus it is infeasible to directly obtain robust segmentation networks for them. In this work, we aim to obtain a robust segmenter on target modality without using target modality's ground-truth by leveraging the large amounts of source domain data (e.g., CT) with segmentation ground-truth.

Previous works on synthetic segmentation can be classified into two categories, including two-stage method [7] and end-to-end methods [8,9]. Zhang et al. [7] developed a two-step strategy, called TD-GAN, for chest x-ray segmentation, where they first use a CycleGAN [10] to adapt the target domain image to the domain with a well-trained segmenter, and then predict the segmentation on the adapted image. However, the segmentation performance relies on the image adaptation performance, thus the two-step process may prone to error aggregation. On the other hand, Kamnitsas et al. [8] developed an end-to-end unsupervised domain adaptation for MRI brain lesion segmentation. However, they only used overlapping MRI modalities (e.g., FLAIR, T2, PD, MPRAGE) in both source and target imaging modalities to ensure performance. Later, Huo et al. [9] proposed to directly concatenate CycleGAN and segmenter as an end-to-end network, called SynSeg-Net, and performed studies on two independent imaging modalities (e.g., MRI and CT). While SynSeg-Net achieved reasonable performance, there are still several issues. First, the training image of the segmentation network relies on high-quality adapted images from the CycleGAN part of SynSeg-Net. Without preserving anatomy structures and contents during the adaptation, the image could be adapted to a target domain image with incorrect structure/content, and negatively impact the subsequent segmentation network's training. Second, SynSeg-Net is a heavy design that relies on training 5 different networks simultaneously which requires careful hyper-parameter tuning, long training time, and high GPU memory consumption.

To tackle these issues, we developed an anatomoy-constrained contrastive learning for synthetic segmentation network (AccSeg-Net), where we proposed to use anatomy-constraint and patch contrastive learning to preserve the structure and content while only using only 3 sub-networks. Given large amounts of CT data with segmentation ground-truth available from public dataset, we used CT as our source domain and validated our method's segmentation performance on target domains of CBCT, MRI, and PET. Our experimental results demonstrated that our AccSeg-Net achieved superior performance over previous methods. We also found our AccSeg-Net achieved better performance as compared to fully supervised methods trained with relatively limited amounts of ground-truth on the target domain.

2   Methods

Our AccSeg-Net includes two parts: 1) an anatomy-constraint contrastive translation network (Acc-Net), and 2) a segmentation network for the target domain

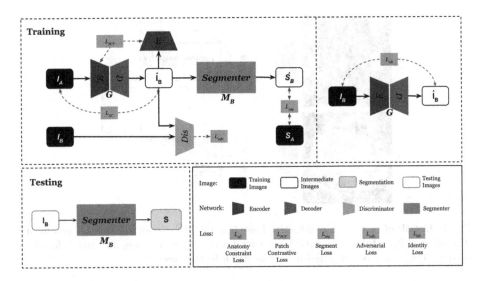

Fig. 1. The architecture of AccSeg-Net. Our AccSeg-Net consisting of a generator, a discriminator, and a segmented, is trained in an end-to-end fashion by 5 loss functions.

segmentation. The architecture and training/test stages are shown in Fig. 1. Acc-Net aims to adapt images from domain $\mathbb{A}$ to domain $\mathbb{B}$. The anatomy-constraint loss and contrastive loss ensure structural information are not lost during the unpaired domain adaptation process, thus critical for training a robust segmenter in domain $\mathbb{B}$. On the other hand, the segmentation loss from the segmentation network is also back-propagated into the Acc-Net, providing addition supervision information for Acc-Net to synthesize image with correct organ delineation. More specifically, our AccSeg-Net contains a generator, a discriminator, and a segmenter. The generator G adapts images from domain $\mathbb{A}$ to domain $\mathbb{B}$, the discriminator D identifies real image from domain $\mathbb{B}$ or the adapted ones from G, and the segmenter M_B predicts the segmentation $\hat{S}_B$ on adapted image from generator G. Training supervision comes from five sources:

(a) adversarial loss $\mathcal{L}_{adv}$ uses discriminator to minimize the perpetual difference between the generative image and the ground truth image in domain $\mathbb{B}$ by:

$$\mathcal{L}_{adv}(G, D, I_B, I_A) = \mathbb{E}_{I_B \sim \mathbb{B}} [\log D(I_B)] + \mathbb{E}_{I_A \sim \mathbb{A}} [\log (1 - D(G(I_A)))] \quad (1)$$

(b) identity loss $\mathcal{L}_{idt}$ ensures the generator does not change target domain appearance when real sample in domain $\mathbb{B}$ is fed:

$$\mathcal{L}_{idt} = \mathbb{E}_{I_B \sim \mathbb{B}} \left[\|G(I_B) - I_B\|_2^2 \right] \quad (2)$$

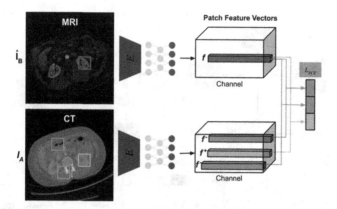

Fig. 2. Details of our patch contrastive loss ($\mathcal{L}_{PCT}$). The encoder (E) of the generator (G) followed by a fully-connected layer encodes the image patches into patch feature vectors. The $\mathcal{L}_{PCT}$ is computed using the patch feature vectors.

(c) segmentation loss $\mathcal{L}_{seg}$ is computed from the segmentation prediction from the adapted image $\hat{I}_B$ and the ground truth segmentation label from domain $\mathbb{A}$:

$$\mathcal{L}_{seg} = \mathbb{E}_{I_A \sim \mathbb{A}} \left[1 - \frac{2|M_B(G(I_A)) \cap S_A|}{|M_B(G(I_A))| + |S_A|} \right] \tag{3}$$

(d) patch contrastive loss $\mathcal{L}_{PCT}$ aims to associate the shared parts between the input image and the output image, while disassociating the non-shared parts. As illustrated in Fig. 2, the kidney in the synthesized image $\hat{I}_B$ should have stronger association with the kidney in the input image I_A than the other parts of the input image. We use the already trained encoder E of generator G (Fig. 1) to generate a feature stack of the image by choosing L layers of interest in E. Each spatial location within the individual level of the stack represents a patch of the input image [11]. In our implementation, each feature level goes through an additional fully-connected layer H_l. Thus, the feature stack can be formulated as $\{f_l\}_L = \{H_l(E_l(I_A))\}_L$, where E_l is the l-th layer in E and H_l is the full-connected layer. Denoting the number of spatial location in each layer as $s \in \{1, ..., S_l\}$ and the number of feature channel in each layer as C_l, we can write the input associated feature as $f_l^s \in \mathbb{R}^{1 \times C_l}$ and the input disassociated feature as $f_l^{S \backslash s} \in \mathbb{R}^{(S_l-1) \times C_l}$. Similarly, the generator output's associated feature stack can be written as $\{\hat{f}_l\}_L = \{H_l(E_l(G(I_A)))\}_L$. Then, we can compute our patch contrastive loss [12] by:

$$\mathcal{L}_{PCT} = \mathbb{E}_{I_A \sim \mathbb{A}} \left[\sum_{l=1}^{L} \sum_{s=1}^{S_l} \mathcal{L}_{CE}(\hat{f}_l^s, f_l^s, f_l^{S \backslash s}) \right] \tag{4}$$

$$\mathcal{L}_{CE}(f, f^+, f^-) = -\log \left[\frac{\exp(f \cdot f^+/\alpha)}{\exp(f \cdot f^+/\alpha) + \sum_{n=1}^{N} \exp(f \cdot f^-/\alpha)} \right] \tag{5}$$

where α is a temperature hyper-parameter for scaling the feature vector distance [13], and is empirically set to $\alpha = 0.07$ here. The patches are randomly cropped during the training.

(e) anatomy-constraint loss $\mathcal{L}_{AC}$ ensures the adaptation only alter the appearance of image while maintaining the anatomical structure, such that segmentation network M_B can be trained correctly to recognize the anatomical content in the adapted image. We use both MIND loss [14] and correlation coefficient (CC) loss to preserve the anatomy in our unpaired adaptation process:

$$\mathcal{L}_{AC} = \lambda_{cc} \mathbb{E}_{I_A \sim \mathbb{A}} \left[\frac{Cov(G(I_A), I_A)}{\sigma_{G(I_A)} \sigma_{I_A}} \right] + \lambda_{mind} \mathbb{E}_{I_A \sim \mathbb{A}} [\|F(G(I_A)) - F(I_A)\|_1] \quad (6)$$

where the first term is the correlation coefficient loss. Cov is the variance operator and σ is the standard deviation operator. The second term is the MIND loss, and F is a modal-independent feature extractor defined as $F_x(I) = \frac{1}{Z} exp \left(-\frac{K_x(I)}{V_x(I)} \right)$. Specifically, $K_x(I)$ is a distance vector of image patches around voxel x with all the neighborhood patches within a non-local region in image I. $V_x(I)$ is the local variance at voxel x in image I. Here, dividing $K_x(I)$ with $V_x(I)$ reduce the influence of image modality and intensity range, and Z is a normalization constant to ensure that the maximum element of F_x equals to 1. We set $\lambda_{cc} = \lambda_{mind} = 1$ here to achieve a balanced training.

Finally, the overall objective is a weighted combination of all loss listed above:

$$\mathcal{L}_{all} = \lambda_1 \mathcal{L}_{adv} + \lambda_2 \mathcal{L}_{idt} + \lambda_3 \mathcal{L}_{seg} + \lambda_4 \mathcal{L}_{PCT} + \lambda_5 \mathcal{L}_{AC} \quad (7)$$

where weighting parameters are set to $\lambda_1 = \lambda_2 = \lambda_3 = \lambda_4 = \lambda_5 = 1$ to achieve a balanced training. We use a decoder-encoder network with 9 residual bottleneck for our generators, and a 3-layer CNN for our discriminators. For segmenter, we use a default setting of a 5-level UNet with concurrent SE module [3] concatenated to each level's output, named DuSEUNet. The segmenter in AccSeg-Net is interchangeable with other segmentation networks, such as R2UNet [4], Attention-UNet [2], and UNet [1].

3   Experimental Results

Dataset Preparation. As liver segmentation is commonly available in CT domain, we chose CT as our source domain (domain $\mathbb{A}$), and aims to obtain CBCT/MRI/PET liver segmenters without their segmentation ground truth. For CT, we collected 131 and 20 CT volumes with liver segmentation from LiTS [15] and CHAOS [16], respectively. In the CBCT/MR domain, we collected 16 TACE patients with both intraprocedural CBCT and pre-operative MRI for our segmentation evaluations. In the PET domain, we collected 100 18F-FDG patients with abdominal scan. All the CBCT were acquired using a Philips C-arm system with a reconstructed image size of $384 \times 384 \times 297$ and voxel size of $0.65 \times 0.65 \times 0.65$ mm^3. The MRI and PET were acquired using different scanners with different spatial resolutions. Thus, we re-sampled all the CBCT, MR, PET and CT to an isotropic spatial resolution of $1 \times 1 \times 1$ mm^3. As a result, we obtained 13, 241 2D CT images with liver segmentation, 3, 792 2D CBCT images, 1, 128 2D MR images, and 6, 150 2D PET images. All the 2D images were resized to 256 × 256. With 16 CBCT and MRI patients in our dataset, we performed four-fold cross-validation with 12 patients used as training and 4 patients used as testing in each validation. For PET, we used 60 patients as training and 40 patients as testing, and qualitatively evaluated the results. Implementation details are summarized in our supplemental materials.

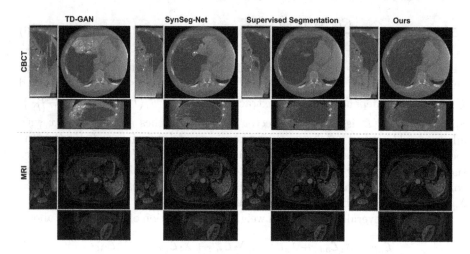

Fig. 3. Comparison of CBCT and MRI segmentation using different methods. Red mask: liver segmentation prediction. Green contour: liver segmentation ground truth. The coronal and saggital view are shown on left and bottom, respectively (Color figure online).

Table 1. Quantitative comparison of CBCT and MRI segmentation results using DSC and ASD (mm). * means supervised training with ground truth segmentation on the target domain (CBCT/MRI domain). -PCT means without patch contrastive learning. Best results are marked in blue, and † means the difference between ours and SynSeg-Net are significant at $p < 0.05$.

CBCT	TD-GAN	SynSeg-Net	Seg_{CBCT}*	Ours-PCT	Ours
Median DSC	0.685	0.870	0.882	0.827	0.885
Mean (Std) DSC	0.695(0.092)	0.862(0.051)	0.874(0.035)	0.831(0.079)	0.893(0.029)†
Median ASD	10.144	7.289	9.190	9.824	5.539
Mean (Std) ASD	10.697(2.079)	7.459(2.769)	10.742(4.998)	10.381(3.739)	5.620(1.352)†
MRI	TD-GAN	SynSeg-Net	Seg_{MRI}*	Ours-PCT	Ours
Median DSC	0.907	0.915	0.907	0.893	0.920
Mean (Std) DSC	0.900(0.044)	0.912(0.029)	0.859(0.102)	0.898(0.037)	0.921(0.018)
Median ASD	1.632	1.681	2.838	1.743	1.468
Mean (Std) ASD	2.328(2.070)	1.916(1.142)	3.660(2.522)	2.478(2.014)	1.769(1.002)

Segmentation Results. After AccSeg-Nets are trained, we can extract the segmenters for liver segmentation on CBCT, MRI, and PET. We used Dice Similarity Coefficient (DSC) and Average Symmetric Surface Distance (ASD) to evaluate the quantitative segmentation performance. First, we compared our segmentation performance with TD-GAN [7], SynSeg-Net [9], and segmenter directly supervised trained on target domain images with limited liver annotations (Seg_{CBCT}/Seg_{MRI}). For a fair comparison, we used DuSEUNet as segmentation network in all the compared methods. The visual comparison is shown in Fig. 3, and the quantitative comparison is summarized in Table 1. As we can observe, the image quality of CBCT is degraded by metal artifacts and low CNR. TD-GAN's results are non-ideal as it requires adapting the input CBCT to CT first, and the segmentation relies on the translated image quality. The unpaired and unconstrained adaption from CBCT to CT is challenging as it consists of metal artifact removal and liver boundary enhancement. The multi-stage inference in TD-GAN thus prone to aggregate prediction errors into the final segmentation. SynSeg-Net with single-stage segmentation help mitigate the prediction error aggregation, but the segmentation is still non-ideal due to lack of structure and content constraint in the unpaired adaptation. On the other hand, our AccSeg-Net with patch contrastive learning and anatomy constraint achieved the best segmentation results. Compared to the segmenters trained on target domains using relatively limited annotation data (12 CBCT/MRI patients), our AccSeg-Net trained from large-scale conventional CT data (151 patients) can also provide slightly better segmenters.

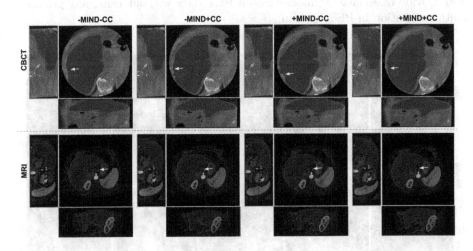

Fig. 4. Comparison of CBCT and MRI segmentation using different anatomy-constraint settings. Red mask: liver segmentation prediction. Green contour: liver segmentation ground truth (Color figure online).

Table 2. Quantitative comparison of CBCT and MRI segmentation results using different anatomy-constraint settings in AccSeg-Net. -MIND and -CC means without MIND loss and without CC loss, respectively. Best results are marked in blue.

CBCT	−MIND-CC	+MIND-CC	−MIND+CC	+MIND+CC
Mean (Std) DSC	0.869(0.059)	0.883(0.041)	0.885(0.039)	0.893(0.029)
Mean (Std) ASD	7.397(2.268)	6.648(2.147)	6.424(2.110)	5.620(1.352)
MRI	−MIND-CC	+MIND-CC	−MIND+CC	+MIND+CC
Mean (Std) DSC	0.913(0.030)	0.916(0.024)	0.918(0.020)	0.921(0.018)
Mean (Std) ASD	1.923(1.153)	1.886(1.138)	1.803(1.109)	1.769(1.002)

Then, we analyzed the effect of using different anatomy-constraint settings in our AccSeg-Net. The results are visualized in Fig. 4, and summarized in Table 2. As we can see, adding either MIND loss or CC loss help improve our segmentation performance, while combining both anatomy constraint losses yields the best segmentation performance of our AccSeg-Net. We also performed ablation studies on using different segmentation networks in our AccSeg-Net, including R2UNet [4], Attention-UNet [2], and UNet [1]. The visualization and quantitative evaluation are summaried in our supplemental materials. We observed that AccSeg-Net can be adapted to different segmentation networks, and yields reasonable segmentation results. Additional PET liver segmentation results from our AccSeg-Net are shown in Fig. 5 for visual evaluation. Our AccSeg-Net can also provide reasonable segmentation on PET data without using any ground-truth annotation in PET domain.

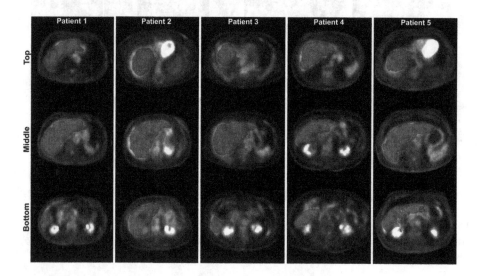

Fig. 5. PET liver segmentation prediction on 5 different patients at three different latitudes.

4   Discussion and Conclusion

In this work, we proposed a novel framework, called AccSeg-Net, for synthetic segmentation without the target domain's ground truth. Specifically, we proposed to use anatomy-constraint and patch contrastive learning in our AccSeg-Net to ensure the anatomy fidelity during the unsupervised adaptation, such that the segmenter can be trained on the adapted image with correct anatomical contents. We demonstrated successful applications on CBCT, MRI, and PET imaging data, and showed superior segmentation performances as compared to previous methods. The presented work also has potential limitations. First, our segmentation performance is far from perfect, and extending our framework to 3D and with enhanced segmentation loss may further improve the performance. Moreover, our method is an open framework, and substituting the segmentation network with a more advanced network may also help improve the performance. In our future work, we will also explore diseases and multi-organ segmentation on other imaging modalities, such as ultrasound and radiography.

References

1. Ronneberger, O., Fischer, P., Brox, T.: U-Net: convolutional networks for biomedical image segmentation. In: Navab, N., Hornegger, J., Wells, W.M., Frangi, A.F. (eds.) MICCAI 2015. LNCS, vol. 9351, pp. 234–241. Springer, Cham (2015). https://doi.org/10.1007/978-3-319-24574-4_28
2. Oktay, O., et al.: Attention u-net: learning where to look for the pancreas. arXiv preprint arXiv:1804.03999 (2018)
3. Roy, A.G., Navab, N., Wachinger, C.: Concurrent spatial and channel 'squeeze & excitation' in fully convolutional networks. In: Frangi, A.F., Schnabel, J.A., Davatzikos, C., Alberola-López, C., Fichtinger, G. (eds.) MICCAI 2018. LNCS, vol. 11070, pp. 421–429. Springer, Cham (2018). https://doi.org/10.1007/978-3-030-00928-1_48
4. Alom, M.Z., Yakopcic, C., Hasan, M., Taha, T.M., Asari, V.K.: Recurrent residual u-net for medical image segmentation. J. Med. Imaging 6(1), 014006 (2019)
5. Isensee, F., Jaeger, P.F., Kohl, S.A., Petersen, J., Maier-Hein, K.H.: nnu-net: a self-configuring method for deep learning-based biomedical image segmentation. Nature Methods, pp. 1–9 (2020)
6. Yu, Q., et al.: C2fnas: coarse-to-fine neural architecture search for 3d medical image segmentation. In: Proceedings of the IEEE/CVF Conference on Computer Vision and Pattern Recognition, pp. 4126–4135 (2020)
7. Zhang, Y., Miao, S., Mansi, T., Liao, R.: Task driven generative modeling for unsupervised domain adaptation: application to x-ray image segmentation. In: Frangi, A.F., Schnabel, J.A., Davatzikos, C., Alberola-López, C., Fichtinger, G. (eds.) MICCAI 2018. LNCS, vol. 11071, pp. 599–607. Springer, Cham (2018). https://doi.org/10.1007/978-3-030-00934-2_67
8. Kamnitsas, K., et al.: Unsupervised domain adaptation in brain lesion segmentation with adversarial networks. In: Niethammer, M., Styner, M., Aylward, S., Zhu, H., Oguz, I., Yap, P.-T., Shen, D. (eds.) IPMI 2017. LNCS, vol. 10265, pp. 597–609. Springer, Cham (2017). https://doi.org/10.1007/978-3-319-59050-9_47

9. Huo, Y., et al.: Synseg-net: synthetic segmentation without target modality ground truth. IEEE Trans. Med. Imaging **38**(4), 1016–1025 (2018)
10. Zhu, J.Y., Park, T., Isola, P., Efros, A.A.: Unpaired image-to-image translation using cycle-consistent adversarial networks. In: Proceedings of the IEEE International Conference on Computer Vision, pp. 2223–2232 (2017)
11. Chen, T., Kornblith, S., Norouzi, M., Hinton, G.: A simple framework for contrastive learning of visual representations. In: International Conference on Machine Learning, PMLR, pp. 1597–1607 (2020)
12. Park, T., Efros, A.A., Zhang, R., Zhu, J.-Y.: Contrastive learning for unpaired image-to-image translation. In: Vedaldi, A., Bischof, H., Brox, T., Frahm, J.-M. (eds.) ECCV 2020. LNCS, vol. 12354, pp. 319–345. Springer, Cham (2020). https://doi.org/10.1007/978-3-030-58545-7_19
13. Wu, Z., Xiong, Y., Yu, S.X., Lin, D.: Unsupervised feature learning via non-parametric instance discrimination. In: Proceedings of the IEEE Conference on Computer Vision and Pattern Recognition, pp. 3733–3742 (2018)
14. Yang, H., et al.: Unsupervised mr-to-ct synthesis using structure-constrained cyclegan. IEEE Trans. Med. Imaging **39**, 4249–4261 (2020)
15. Bilic, P., et al.: The liver tumor segmentation benchmark (lits). arXiv preprint arXiv:1901.04056 (2019)
16. Kavur, A.E., et al.: Chaos challenge-combined (ct-mr) healthy abdominal organ segmentation. arXiv preprint arXiv:2001.06535 (2020)

Study Group Learning: Improving Retinal Vessel Segmentation Trained with Noisy Labels

Yuqian Zhou[1(✉)], Hanchao Yu[1], and Humphrey Shi[1,2]

[1] University of Illinois at Urbana-Champaign, Champaign, USA
[2] University of Oregon, Eugene, Oregon, USA

Abstract. Retinal vessel segmentation from retinal images is an essential task for developing the computer-aided diagnosis system for retinal diseases. Efforts have been made on high-performance deep learning-based approaches to segment the retinal images in an end-to-end manner. However, the acquisition of retinal vessel images and segmentation labels requires onerous work from professional clinicians, which results in smaller training dataset with incomplete labels. As known, data-driven methods suffer from data insufficiency, and the models will easily over-fit the small-scale training data. Such a situation becomes more severe when the training vessel labels are incomplete or incorrect. In this paper, we propose a Study Group Learning (SGL) scheme to improve the robustness of the model trained on noisy labels. Besides, a learned enhancement map provides better visualization than conventional methods as an auxiliary tool for clinicians. Experiments demonstrate that the proposed method further improves the vessel segmentation performance in DRIVE and CHASE_DB1 datasets, especially when the training labels are noisy. Our code is available at https://github.com/SHI-Labs/SGL-Retinal-Vessel-Segmentation.

Keywords: Retinal vessel segmentation · Image enhancement.

1 Introduction

Retinal inspection is an effective approach for the diagnose of multiple retinal diseases including diabetic retinopathy, epiretinal membrane, retinal detachment, retinal tear *etc.*. Among them, retinal vascular disorders which affects retinal blood vessels are usually caused by other medical diseases like atherosclerosis, hypertension, or human blood circulation problems [2]. Those disorders will severely influence human's vision functions and cause obvious symptoms, but can be effectively diagnosed and analyzed by retinal vessel inspection in the collected fundus images. Advanced medical imaging system makes it possible to obtain high-resolution fundus images. However, in practical medical services, visual inspection may still require the involvement and tedious work of neurologists, cardiologists, ophthalmologists, and other experts in retinal vascular diseases. To

© Springer Nature Switzerland AG 2021
M. de Bruijne et al. (Eds.): MICCAI 2021, LNCS 12901, pp. 57–67, 2021.
https://doi.org/10.1007/978-3-030-87193-2_6

release their burden on screening multiple diseased retina from thousands even millions of healthy retinas, an automatic and high-performance Computer-Aided Diagnosis (CAD) system is desirable to conduct pre-screening and other auxiliary works. Specifically for retinal vascular diseases, we expect the system to provide high-quality enhanced images for a better visualization, and reasonable segmentation of the vessel patterns from the complex and noisy images.

Plenty of previous efforts have been made in automatic retinal vessel segmentation. Conventionally, hand-crafted filters [13,14,17,27] like Gabor [14] and Gaussian-based ones [13] are explored to extract features for pixel selection, vessel clustering and segmentation. Recently, data-driven based methods utilize UNet-based model [18] or its variants [11,24,25,28,29] to achieve significant performance compared with traditional methods. Those deep learning methods focus on the design of UNet structures with better feature representation [24,28], or the decouple of structure and textures of retinal images [29]. However, data-driven methods highly suffer from over-fitting issues when the given training data is insufficient. Previously proposed methods cannot overcome the issues of small-scale training data with noisy labels given by the clinicians.

Effectively training networks with noisy labels [21] is a rigid need in industry and an interesting task in academia. Previous research works mostly focus on image classification tasks and develop methodologies like optimizing robust loss functions [4,31], regularizing labels [5,15], or actively selecting samples [7,9] *etc.*. However, noisy pixel-level labels existing in segmentation tasks are not well-studied, especially for medical image processing tasks. Shu *et al.* proposed a LVC-Net [19] to adjust the incorrect pixel-wise labels via a deformable spatial transformation module guided by low-level visual cues. But their method cannot be applied to retinal blood vessel images, because the entire small blood vessels may be mislabeled and cannot be corrected via spatial transformation. Xue *et al.* [26] studied a multi-stage training framework with sample selection for chest X-ray images. They synthesized noisy annotations with image dilation and erosion. However, due to the thickness of blood vessels and less training data compared to other medical image data, neither label synthesis nor sampling procedures are suitable for blood vessel images. Considering the characteristics of blood vessel images, we introduce a novel noisy label synthesis pipeline for retinal vessel images, and propose a Study-Group Learning (SGL) framework to improve the model robustness on noisy labels.

In this paper, we mainly study the two main practical problems for retinal vessel segmentation task. First, we explore the deeply unsupervised learned enhancement of the original retinal images compared with traditional contrast adjustment methods like CLAHE [16]. Second, suppose the ground truth segmentation labels given by the clinicians are incomplete and noisy, which yields missing annotations of some vessel segments, we study the effective learning scheme to improve the robustness of the model while training on noisy labels. Therefore, the contributions can be summarized in the following aspects.

- First, to better visualize the retinal image and understand the model, we design the network to output the learned enhancement map. Compared with

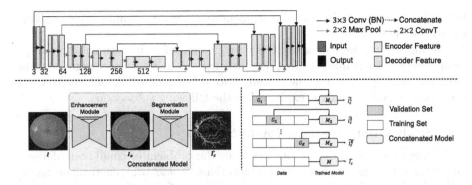

Fig. 1. Baseline model structure and the proposed Study Group Learning (SGL) scheme. The baseline network is a concatenated UNet consisting of an enhancement module and a segmentation module. The three-channel enhancement map I_e is obtained from the bottleneck structure. The SGL is inspired by the cross-validation training scheme and knowledge distillation idea. We first split the whole training set G into K subsets $\{G_k\}$ and feed the model M_k with G_k. The obtained estimation $\tilde{I}_c^k$ of G_k is utilized as the pseudo labels for joint optimization.

other baselines, the high-contrast learned map are better visually plausible, and provides an additional auxiliary tool to aid clinicians for visual inspection or manual segmentation.

- Second, we design a pipeline to manipulate vessel segmentation labels. Given a complete annotations, the proposed approach simulates to automatically erase some vessel segments.
- Third, we propose a Study Group Learning (SGL) framework to improve the generalization ability of the learned model, and better address the missing annotation problems in the training set. The model achieves a more robust performance even some of the vessel pixels are mislabeled.

2 Methodology

2.1 Model Structure

The proposed baseline model structure is illustrated in Fig. 1. We utilize the concatenated UNet consisting of the enhancement and segmentation modules to learn both the enhancement and the segmentation map. Different from previous works, we do not pre-process the retinal images, but directly utilize the raw captured images to preserve the entire information. Specifically, given a three-channel raw retinal image I as the input, we aim to process the image by enhancing its contrast and highlighting the vessel structures in I_e, and estimate the segmentation map $\hat{I}_c$ of the vessels matched with the ground truth segmentation map I_c given by professional clinicians. Notice that I_e preserves the maximum image contents including the vessel structures and retinal textures. It helps the clinicians to inspect the segmentation results $\hat{I}_c$ and better explains the learned model.

During testing, we enhance the raw image and infer the segmentation maps from unseen retinal images. To cope with noisy labels in vessel segmentation datasets, we propose and follow a Study Group Learning (SGL) scheme to train the baseline model, which is explained in the following subsections.

2.2   Study Group Learning (SGL) Scheme

Tasks like retinal vessel segmentation faces the problem of small-scale dataset and incorrect or incomplete vessel annotations. These two practical problems will make the deeply trained model very easily over-fitting the training set. The severe over-fitting problem does harm to the generalization ability and robustness of the model to unseen testing data. In addition to conventional data augmentation approaches like image transformation and random warping, we propose to alternate the training scheme. Inspired by K-fold cross-validation scheme and knowledge distillation [8] approaches, we propose the K-fold Study Group Learning (SGL) to better cope with noisy labels in small datasets, especially for the retinal vessel segmentation task.

The pipeline is illustrated in Fig. 1. Specifically, we first randomly and averagely split the whole training set (G, I_c) into K subsets $\{(G_k, I_{ck})\}$, where $k \in [1, K]$. Like the cross-validation scheme, we train totally K models $\{M_k\}$. For each M_k, we train it on $(G \backslash G_k, I_c \backslash I_{ck})$ using pixel-wise binary cross entropy loss. After optimizing the model set $\{M_k\}$ as $\{M_k^*\}$, we infer the estimated segmentation label $\tilde{I}_{ck}$ of G_k as the pseudo label, where

$$\tilde{I}_{ck} = M_k^*(G_k), k \in [1, K]. \tag{1}$$

Finally, we train a model M from scratch by jointly optimizing the ground truth vessel labels I_c and the obtained pseudo label set $\tilde{I}_c = \bigcup_{k=1}^{K} \tilde{I}_{ck}$ as,

$$\mathcal{L}_{SGL} = CE(\hat{I}_c, I_c) + \lambda CE(\hat{I}_c, \tilde{I}_c), \tag{2}$$

where $\hat{I}_c = M(G)$, CE is the cross entropy loss, and $\lambda = 1$.

Intuitively, we name the proposed learning scheme as Study Group Learning (SGL) because each model M_k trained with partial training set can be regarded as a 'Study Group'. The final model M is a process of group discussion by merging and fusing the knowledge from different study groups. Theoretically, the second term in Eq. 2 can be regarded as a regularization to avoid over-fitting the ground-truth labels, especially when the given labels contain some noises and possibly incorrect annotations. Figure 2 shows one example of such cases. In this example, we set $K = 2$, and the inference of the training samples I from M_1 and M_2 are shown in the right two columns, where I is in the training set of M_1 but not in the training partition of M_2. M_1 highly overfits the given labels of I by ignoring multiple thin and ambiguous vessels. However, if not trained on I, M_2 can infer some of the ignored vessels as the pseudo labels. While combining these two labels, the final model M can intuitively learn better representation and become more generalized.

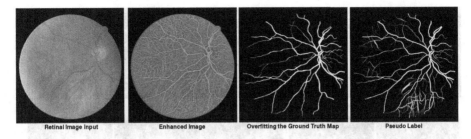

| Retinal Image Input | Enhanced Image | Overfitting the Ground Truth Map | Pseudo Label |

Fig. 2. Inference results on one training example. Clinicians may only label some salient vessels while ignoring the ambiguous ones. The model trained on the entire set will overfit the ground truth annotations. However, inference results from the model trained on the subset can serve as the pseudo label for model regularization purpose.

2.3   Vessel Label Erasing

Annotating the retinal vessels requires the involvement of professional clinicians, and the process of manual labeling is onerous, which reveals one of the reasons why public retinal vessel databases are always small-scale or partially-labeled. It is also common that some labels of thinner vessels are missing due to the annotators' errors. To resemble this practical situation in industry, we propose to synthesize an incomplete map I_c^r by erasing some labeled vessel segments I_r from the ground truth segmentation map $I_c = I_c^r \cup I_r$, where r is the removal ratio. The process is illustrated in Fig. 3.

To generate I_c^r, we first compute and approximate the skeleton of I_c using the method proposed by Zhang *et al.*[30] followed by a skeleton tracing approach [1]. This algorithm converts the binary segmentation map into a set of polylines $L_c = \{l_i\}_c$, where $i \in [1, M]$ and each l_i is stored as an array of coordinates *i.e.* $l_i = \{p_d\}$. The geometric polylines and their spatial relationship represent the topological skeleton of the annotated vessels. We utilize it to locate the vessel center lines, and roughly compute the thickness of the vessels.

Second, to compute and rank the thickness of each vessel segment, we redraw the polylines with larger width t on a black canvas to form the complete mask M_c^t. Notice that t is the least value which makes M_c^t cover the entire vessel regions in I_c. For each l_i, the corresponding pixel regions covered by the drawing lines of width t is denoted by l_i^t, then the thickness s_i of the vessel segment is measured by $s_i = \frac{\|I_r \cap l_i^t\|}{\|l_i^t\|}$. We rank the polylines according to their thickness s_i, and include the thickest vessels in order in the partial set $L_c^r = \{l_j\}$, where $j \in [1, N]$ and $N \leq M$. In L_c^r, we include the top-r thick vessels. Finally, we form the selected mask M_r^t from L_c^r, and the synthetic segmentation mask ablating some of the thin vessels with ratio r can be computed by $I_c^r = M_r^t \odot I_c$. Figure 3 shows one example of I_c^r where r is from 1 to 0.3. Adding a small portions of thin vessels helps the model maintain the ability of segmenting smaller objects. Therefore, we also randomly select 50% of thin vessels in the set and add them to L_c^r.

---

[1] skeleton-tracing: https://github.com/LingDong-/skeleton-tracing.

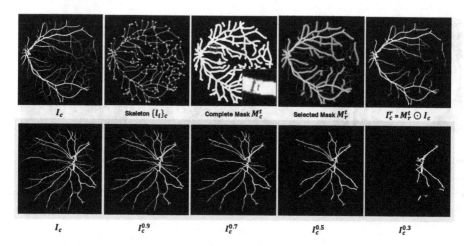

Fig. 3. Vessel label erasing process. Given the complete label map I_c, we compute the skeleton of the vessels, and dilate the vessel skeleton to mask by drawing the polylines with width t. We then rank the vessel segments by their approximated thickness, and include the thick vessels according to ratio r. The second row shows one instance of label erasing with r from 1 to 0.3.

3 Experiments

3.1 Dataset and Implementation

DRIVE: The Digital Retinal Images for Vessel Extraction (DRIVE) [23] dataset consists of 40 images of size 565×584. The training and testing set are fixed, and the ground truth manual annotations are given. We do not resize the image to alternate the resolution or change the Aspect-Ratio.

CHASE_DB1: [3] It consists of 28 retinal images of size 999×960. For a fair comparison with previous methods, we use the first 20 images for training, and the remaining 8 images for testing.

While training the model, we randomly crop the images into 256×256 patches, and apply data augmentation including horizontal and vertical flip, rotation, transpose, and random elastic warping [20]. We train the model using Adam (0.9, 0.999) with learning rate 10^{-4} and batch size 16. We train it for 30 epochs separately for each M_k to generate pseudo label sets, and train for another 20 epochs using the pseudo and ground truth labels. While testing, suppose the image size is $W \times H$, we zero-pad the input image to size $(2^4 \times m, 2^4 \times m)$, where $2^4 \times m > \max(W, H) > 2^4 \times (m - 1)$, and crop the estimated map accordingly. Specifically, m is 37 for DRIVE and 63 for CHASE_DB1 dataset. It makes the arbitrary-sized inputs adaptive to the UNet structure with four down-sampling layers, and retain the original resolution and aspect-ratio of the image. We mask out the non-retinal area in DRIVE, and compute the metrics on the entire image. More details can be found in our code.

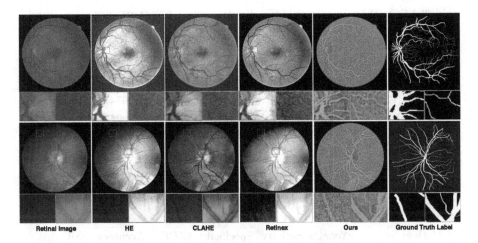

Fig. 4. The learned enhancement map I_e compared with other baseline methods including Histogram Equalization (HE), Contrast Limited Adaptive Histogram Equalization (CLAHE) [16], and Single-scale Retinex [32]. The learned map demonstrates a better contrast and intensity, enhancing the vessel information for a better identifiable visualization for clinicians. Top row is from DRIVE dataset and the bottom row is from CHASE_DB1 dataset. Zooming in for better visualization.

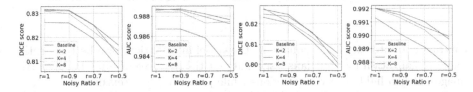

Fig. 5. The performance of the model in the simulated training set with label noise ratio $r = [1, 0.9, 0.7, 0.5]$. The proposed SGL learning scheme overall improves the robustness in all K. Left two columns: DICE and AUC scores on DRIVE. Right two columns: on CHASE_DB1 dataset.

3.2   Learned Retinal Image Enhancement

We learn the enhanced map of the original input images by supervising the model on the manual segmentation labels. We extract the bottleneck 3-channel output which can be visualized in a better contrast level. We compare the learned enhancement of the retinal images with other baseline methods including Histogram Equalization (HE), Contrast Limited Adaptive Histogram Equalization (CLAHE) [16], and Retinex [32].

Figure 4 shows one example of the visual comparison. Traditional methods like HE, CLAHE and Retinex cannot achieve a uniform contrast level both locally and globally. The cropped patches are from either a brighter or darker regions of the image, making the vessel pixels hard to parse accurately by the inspector. However, the learned map in the fifth column demonstrates a better

Table 1. Comparison with other baseline methods on DRIVE dataset.

Method	Year	Sensitivity	Specificity	DICE	Accuracy	AUC
R2U-Net [1]	2018	0.7792	0.9813	0.8171	0.9556	0.9784
LadderNet [33]	2018	0.7856	0.9810	0.8202	0.9561	0.9793
IterNet [12]	2020	0.7791	0.9831	0.8218	0.9574	0.9813
SA-UNet [6]	2020	0.8212	0.9840	0.8263	0.9698	0.9864
BEFD-UNet [28]	2020	0.8215	**0.9845**	0.8267	0.9701	0.9867
Our Baseline	2021	0.8341	0.9827	0.8262	0.9695	0.9867
Our SGL (K=8)	2021	**0.8380**	0.9834	**0.8316**	**0.9705**	**0.9886**

Table 2. Comparison with other baseline methods on CHASE_DB1 dataset.

Method	Year	Sensitivity	Specificity	DICE	Accuracy	AUC
UNet [12]	2018	0.7840	0.9880	0.7994	0.9752	0.9870
DUNet [10]	2019	0.7858	0.9880	0.8000	0.9752	0.9887
IterNet [12]	2020	0.7969	0.9881	0.8072	0.9760	0.9899
SA-UNet [6]	2020	0.8573	0.9835	0.8153	0.9755	0.9905
Our Baseline	2021	0.8502	**0.9854**	0.8232	0.9769	0.9913
Our SGL (K=8)	2021	**0.8690**	0.9843	**0.8271**	**0.9771**	**0.9920**

contrast and intensity level, enhancing the vessel information for a better identifiable visualization for clinicians, especially for the dark regions in CHASE_DB1 images. It highlights the vessel regions while preserving the textures.

In summary, compared with CLAHE, the learned map has three advantages: first, the region of interests can be better highlighted. It aggregates the information from RGB channels and extracts the features relevant to the vessels, while CLAHE only uses grayscale inputs. Second, no saturation artifacts exist like CLAHE. Third, it helps us interpret the segmentation errors to debug the system. In real applications, the obtained enhancement images can be combined with other enhancement methods for a better visual inspection or labelling.

3.3 Study Group Learning

Table 1 and 2 illustrate the effectiveness of the proposed SGL scheme. Compared with previous works, the proposed learning scheme can boost the DICE score [22] and other evaluation metrics by a large margin. Figure 5 shows the DICE and Area Under the Receiver Operating Characteristic (ROC) Curve (AUC) of the model in the simulated training set with label noise ratio $r = [1, 0.9, 0.7, 0.5]$, where $r = 1$ represents the original training set. As shown in the figure, erasing some vessel labels in the training set will drastically degrade the system performance, while the SGL learning scheme overall improves the robustness on both datasets. Besides, a better sensitivity indicates the model is able to extract more thin vessels and boundary pixels. More results can be found in the supplementary material (click here to open).

4 Conclusions

In this paper, we studied the learning-based retinal vessel segmentation model trained with noisy labels. Specifically, we designed the pipeline of synthesizing noisy labels, and proposed a Study Group Learning (SGL) scheme boosting the performance of model trained with imperfect labels. Besides, the learned enhanced images as a side product made the model explainable and helped the clinicians for visual inspection. We still discovered the gap between models trained with different levels of noisy labels, leaving for future work to further improve the model sensitivity.

Acknowledgement. This project has been funded by the Jump ARCHES endowment through the Health Care Engineering Systems Center. This work also utilizes resources supported by the National Science Foundation's Major Research Instrumentation program, grant number 1725729, as well as the University of Illinois at Urbana-Champaign.

References

1. Alom, M.Z., Yakopcic, C., Hasan, M., Taha, T.M., Asari, V.K.: Recurrent residual u-net for medical image segmentation. J. Med. Imaging **6**(1), 014006 (2019)
2. Brand, C.S.: Management of retinal vascular diseases: a patient-centric approach. Eye **26**(2), S1–S16 (2012)
3. Fraz, M.M., et al.: An ensemble classification-based approach applied to retinal blood vessel segmentation. IEEE Trans. Biomed. Eng. **59**(9), 2538–2548 (2012)
4. Ghosh, A., Kumar, H., Sastry, P.: Robust loss functions under label noise for deep neural networks. In: Proceedings of the AAAI Conference on Artificial Intelligence, vol. 31 (2017)
5. Goodfellow, I.J., Shlens, J., Szegedy, C.: Explaining and harnessing adversarial examples. arXiv preprint arXiv:1412.6572 (2014)
6. Guo, C., Szemenyei, M., Yi, Y., Wang, W., Chen, B., Fan, C.: Sa-unet: Spatial attention u-net for retinal vessel segmentation. arXiv preprint arXiv:2004.03696 (2020)
7. Han, B., et al.: Co-teaching: Robust training of deep neural networks with extremely noisy labels. arXiv preprint arXiv:1804.06872 (2018)
8. Hinton, G., Vinyals, O., Dean, J.: Distilling the knowledge in a neural network. arXiv preprint arXiv:1503.02531 (2015)
9. Jiang, L., Zhou, Z., Leung, T., Li, L.J., Fei-Fei, L.: Mentornet: learning data-driven curriculum for very deep neural networks on corrupted labels. In: International Conference on Machine Learning, pp. 2304–2313. PMLR (2018)
10. Jin, Q., Meng, Z., Pham, T.D., Chen, Q., Wei, L., Su, R.: Dunet: a deformable network for retinal vessel segmentation. Knowl.-Based Syst. **178**, 149–162 (2019)
11. Lan, Y., Xiang, Y., Zhang, L.: An elastic interaction-based loss function for medical image segmentation. In: Martel, A.L., et al. (eds.) MICCAI 2020. LNCS, vol. 12265, pp. 755–764. Springer, Cham (2020). https://doi.org/10.1007/978-3-030-59722-1_73

12. Li, L., Verma, M., Nakashima, Y., Nagahara, H., Kawasaki, R.: Iternet: retinal image segmentation utilizing structural redundancy in vessel networks. In: The IEEE Winter Conference on Applications of Computer Vision, pp. 3656–3665 (2020)
13. Niemeijer, M., Staal, J., van Ginneken, B., Loog, M., Abramoff, M.D.: Comparative study of retinal vessel segmentation methods on a new publicly available database. In: Medical imaging 2004: Image Processing, vol. 5370, pp. 648–656. International Society for Optics and Photonics (2004)
14. Oloumi, F., Rangayyan, R.M., Oloumi, F., Eshghzadeh-Zanjani, P., Ayres, F.J.: Detection of blood vessels in fundus images of the retina using gabor wavelets. In: 2007 29th Annual International Conference of the IEEE Engineering in Medicine and Biology Society, pp. 6451–6454. IEEE (2007)
15. Pereyra, G., Tucker, G., Chorowski, J., Kaiser, Ł., Hinton, G.: Regularizing neural networks by penalizing confident output distributions. arXiv preprint arXiv:1701.06548 (2017)
16. Pizer, S.M., et al.: Adaptive histogram equalization and its variations. Comput. Vis. Graph. Image Process. **39**(3), 355–368 (1987)
17. Ricci, E., Perfetti, R.: Retinal blood vessel segmentation using line operators and support vector classification. IEEE Trans. Med. Imaging **26**(10), 1357–1365 (2007)
18. Ronneberger, O., Fischer, P., Brox, T.: U-Net: convolutional networks for biomedical image segmentation. In: Navab, N., Hornegger, J., Wells, W.M., Frangi, A.F. (eds.) MICCAI 2015. LNCS, vol. 9351, pp. 234–241. Springer, Cham (2015). https://doi.org/10.1007/978-3-319-24574-4_28
19. Shu, Y., Wu, X., Li, W.: LVC-Net: medical image segmentation with noisy label based on local visual cues. In: Shen, D., et al. (eds.) MICCAI 2019. LNCS, vol. 11769, pp. 558–566. Springer, Cham (2019). https://doi.org/10.1007/978-3-030-32226-7_62
20. Simard, P.Y., Steinkraus, D., Platt, J.C., et al.: Best practices for convolutional neural networks applied to visual document analysis. In: Icdar, vol. 3 (2003)
21. Song, H., Kim, M., Park, D., Lee, J.G.: Learning from noisy labels with deep neural networks: A survey. arXiv preprint arXiv:2007.08199 (2020)
22. Sorensen, T.A.: A method of establishing groups of equal amplitude in plant sociology based on similarity of species content and its application to analyses of the vegetation on danish commons. Biol. Skar. **5**, 1–34 (1948)
23. Staal, J., Abràmoff, M.D., Niemeijer, M., Viergever, M.A., Van Ginneken, B.: Ridge-based vessel segmentation in color images of the retina. IEEE Trans. Med. Imaging **23**(4), 501–509 (2004)
24. Wang, W., Zhong, J., Wu, H., Wen, Z., Qin, J.: RVSeg-Net: an efficient feature pyramid cascade network for retinal vessel segmentation. In: Martel, A.L., et al. (eds.) MICCAI 2020. LNCS, vol. 12265, pp. 796–805. Springer, Cham (2020). https://doi.org/10.1007/978-3-030-59722-1_77
25. Xu, R., Liu, T., Ye, X., Lin, L., Chen, Y.-W.: Boosting connectivity in retinal vessel segmentation via a recursive semantics-guided network. In: Martel, A.L., et al. (eds.) MICCAI 2020. LNCS, vol. 12265, pp. 786–795. Springer, Cham (2020). https://doi.org/10.1007/978-3-030-59722-1_76
26. Xue, C., Deng, Q., Li, X., Dou, Q., Heng, P.-A.: Cascaded robust learning at imperfect labels for chest x-ray segmentation. In: Martel, A.L., et al. (eds.) MICCAI 2020. LNCS, vol. 12266, pp. 579–588. Springer, Cham (2020). https://doi.org/10.1007/978-3-030-59725-2_56

27. You, X., Peng, Q., Yuan, Y., Cheung, Y.M., Lei, J.: Segmentation of retinal blood vessels using the radial projection and semi-supervised approach. Pattern Recognit. **44**(10–11), 2314–2324 (2011)
28. Zhang, M., Yu, F., Zhao, J., Zhang, L., Li, Q.: BEFD: boundary enhancement and feature denoising for vessel segmentation. In: Martel, A.L., et al. (eds.) MICCAI 2020. LNCS, vol. 12265, pp. 775–785. Springer, Cham (2020). https://doi.org/10.1007/978-3-030-59722-1_75
29. Zhang, S., Fu, H., Xu, Y., Liu, Y., Tan, M.: Retinal image segmentation with a structure-texture demixing network. In: Martel, A.L., et al. (eds.) MICCAI 2020. LNCS, vol. 12265, pp. 765–774. Springer, Cham (2020). https://doi.org/10.1007/978-3-030-59722-1_74
30. Zhang, T., Suen, C.Y.: A fast parallel algorithm for thinning digital patterns. Commun. ACM **27**(3), 236–239 (1984)
31. Zhang, Z., Sabuncu, M.R.: Generalized cross entropy loss for training deep neural networks with noisy labels. arXiv preprint arXiv:1805.07836 (2018)
32. Zhao, Y., Liu, Y., Wu, X., Harding, S.P., Zheng, Y.: Retinal vessel segmentation: an efficient graph cut approach with retinex and local phase. PLoS ONE **10**(4), e0122332 (2015)
33. Zhuang, J.: Laddernet: Multi-path networks based on u-net for medical image segmentation. arXiv preprint arXiv:1810.07810 (2018)

Multi-phase Liver Tumor Segmentation with Spatial Aggregation and Uncertain Region Inpainting

Yue Zhang[1]([⊠]), Chengtao Peng[2], Liying Peng[1], Huimin Huang[1],
Ruofeng Tong[1,3], Lanfen Lin[1], Jingsong Li[3], Yen-Wei Chen[4], Qingqing Chen[5],
Hongjie Hu[5], and Zhiyi Peng[6]

[1] College of Computer Science and Technology, Zhejiang University,
Hangzhou, China
yuezhang95@zju.edu.cn
[2] Department of Electronic Engineering and Information Science,
University of Science and Technology of China, Hefei, China
[3] Research Center for Healthcare Data Science, Zhejiang Lab, Hangzhou, China
[4] College of Information Science and Engineering, Ritsumeikan University,
Kusatsu, Japan
[5] Department of Radiology, Sir Run Run Shaw Hospital, Hangzhou, China
[6] Department of Radiology, The First Affiliated Hospital, College of Medicine,
Zhejiang University, Hangzhou, China

Abstract. Multi-phase computed tomography (CT) images provide crucial complementary information for accurate liver tumor segmentation (LiTS). State-of-the-art multi-phase LiTS methods usually fused cross-phase features through phase-weighted summation or channel-attention based concatenation. However, these methods ignored the spatial (pixel-wise) relationships between different phases, hence leading to insufficient feature integration. In addition, the performance of existing methods remains subject to the uncertainty in segmentation, which is particularly acute in tumor boundary regions. In this work, we propose a novel LiTS method to adequately aggregate multi-phase information and refine uncertain region segmentation. To this end, we introduce a spatial aggregation module (SAM), which encourages per-pixel interactions between different phases, to make full use of cross-phase information. Moreover, we devise an uncertain region inpainting module (URIM) to refine uncertain pixels using neighboring discriminative features. Experiments on an in-house multi-phase CT dataset of focal liver lesions (MPCT-FLLs) demonstrate that our method achieves promising liver tumor segmentation and outperforms state-of-the-arts.

Keywords: Multi-phase segmentation · Liver tumor segmentation · Bi-directional feature fusion

© Springer Nature Switzerland AG 2021
M. de Bruijne et al. (Eds.): MICCAI 2021, LNCS 12901, pp. 68–77, 2021.
https://doi.org/10.1007/978-3-030-87193-2_7

1   Introduction

Liver cancer is one of the leading causes of cancer-induced death, which poses a serious risk to human health [2]. Accurate liver tumor segmentation (LiTS) is a vital prerequisite for liver cancer diagnosis and treatment, which helps to increase the five-year survival rate. Most existing LiTS solutions [1,4,7,15,22] tended to use single-phase computed tomography (CT) images to segment liver tumors. However, these methods usually produce unsatisfactory results due to inherent challenges in medical images (e.g., low contrast and fuzzy tumor boundaries). Alternatively, segmentation relying on multi-phase images could assimilate complementary information from different phases, which is helpful to probe the complete morphology of tumors.

In clinical practice, contrast enhanced CT (CECT) images of different phases present distinct liver tumor morphology and gray scales. Generally, tumors may be not salient in one phase but show clear outlines in another phase. Therefore, making good use of the complementary inter-phase information could effectively improve the segmentation results. In view of this, several methods were proposed to explore this issue, which can be classified into three categories according to multi-phase feature fusion strategies: input-level fusion (ILF) methods [11], decision-level fusion (DLF) methods [11,13,16] and feature-level fusion (FLF) methods [18,20]. Among these methods, FLF methods were demonstrated to achieve the best performance since they exploited multi-level cross-phase features. For instance, Wu et al. [18] proposed an MW-UNet, which integrated different phases by weighting their features from hidden layers of U-Net [14] using trainable coefficients. Xu et al. [20] proposed a ResNet [5] based PA-ResSeg to re-weight features of different phases using channel-attention mechanism [3,6]. However, known FLF methods merely focused on phase-wise or channel-wise inter-phase relationships, but neglected the pixel-wise correspondence between different phases, thus leading to redundancy and low efficiency in information aggregation. The insufficient feature fusion may even bring in interference factors at spatial positions. Moreover, like other segmentation tasks, the performance of existing multi-phase LiTS methods suffer from uncertain region segmentation. That is, the segmentation results usually present some blurry or ambiguous regions (especially in tumor boundaries). This problem is mainly caused by (1) high-frequency information loss during down- and up-sampling operations and (2)low contrast between tumors and surroundings.

Motivated by these observations, in this work, we propose a novel method to segment liver tumors from multi-phase CECT images. In overall, our method exploits complementary information from arterial (ART) phase images to facilitate LiTS in portal venous (PV) phase images. We boost the segmentation performance by introducing pixel-wise inter-phase feature fusion and uncertain region refinement. Specifically, to ensure sufficient multi-phase information aggregation, we devise a spatial aggregation module (SAM). The proposed SAM module mines macro and local inter-phase relationships and yields a pixel-wise response map for each phase. Afterwards, multi-phase features are modulated and fused pixel-by-pixel according to the response maps. Besides, we devise an uncertain region inpainting module (URIM) to refine uncertain regions and

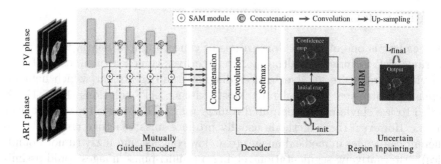

Fig. 1. Illustration of our proposed multi-phase liver tumor segmentation network.

obtain fine segmentation. The key idea of the URIM module is to employ confident pixels (with high confidence in segmentation scores) to inpaint surrounding uncertain pixels. To that end, a local-confidence convolution (LC-Conv) operation is introduced to make uncertain pixels absorb neighboring discriminative features. After several LC-Conv operations, the adjusted features are adopted to do the final prediction. Comprehensive experiments on an in-house multi-phase CT dataset of focal liver lesions (MPCT-FLLs) demonstrate that our method achieves accurate liver tumor segmentation and is superior to state-of-the-arts.

Our main contributions are: (1) we devise a spatial aggregation module to ensure sufficient inter-phase interactions. The module extracts macro and local inter-phase relationships, thereby modulating each pixel with a response value; (2) we devise an uncertain region inpainting module to refine uncertain and blurry regions, which particularly helps to obtain fine-grained tumor boundary segmentation; (3) we validate our method on the multi-phase MPCT-FLLs dataset. Our codes will be available at https://github.com/yzhang-zju/multi_phase_LiTS.

2   Method

In overall, our network takes both PV- and ART-slices as inputs, and produces tumor segmentation of the primary PV phase. Figure 1 illustrates the overview of the proposed network, which mainly comprises three parts.

The mutually guided encoder part takes ResNeXt-50 [19] as the backbone. It uses two Siamese streams, i.e., PV-stream and ART-stream, to extract phase-specific features. Convolution blocks of the two streams are denoted as $B_{PV}^{(i)}$ and $B_{ART}^{(i)}$ ($i \in \{1,2,3,4,5\}$). To integrate cross-phase information, features from $B_{PV}^{(i)}$ and $B_{ART}^{(i)}$ ($i \in \{2,3,4,5\}$) are aggregated through SAMs in a bi-directional manner. By doing so, the two streams provide information to assist each other, thus mutually guiding their feature extraction.

The decoder part takes four-level aggregated features from the encoder as input, and yields initial probability maps. To incorporate multi-level features, all inputs are up-sampled using bilinear interpolation, and fused through concatenation and convolutions.

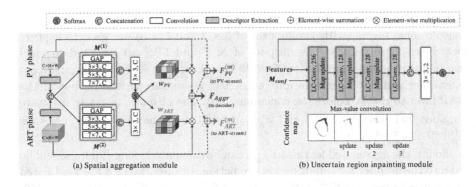

Fig. 2. Detailed structures of the proposed SAM module (a) and URIM module (b).

The uncertain region inpainting part on top of the decoder aims to refine the uncertain regions in the initial maps. Intuitively, it employs confident pixels to inpaint neighboring uncertain pixels. To achieve this, uncertain pixels are allowed to absorb surrounding discriminative features using a proposed local-confidence convolution (LC-Conv) operation. The refined features are adopted for the final prediction.

2.1 Spatial Aggregation Module

State-of-the-art multi-phase LiTS methods ignored feature fusion at spatial positions. This may lead to redundancy and low efficiency in information integration. Hence, we propose the spatial aggregation module (SAM) to ensure sufficient cross-phase feature fusion by weighting each pixel.

Figure 2 (a) shows the detailed structure of the proposed SAM. Having two input feature maps $F_{PV} \in \mathbb{R}^{C \times H \times W}$ (from PV-stream) and $F_{ART} \in \mathbb{R}^{C \times H \times W}$ (from ART-stream), the SAM module calculates two pixel-wise response maps, which are denoted as $w_{PV} \in \mathbb{R}^{C \times H \times W}$ and $w_{ART} \in \mathbb{R}^{C \times H \times W}$, to modulate F_{PV} and F_{ART}, respectively. Accordingly, the overall cross-phase feature aggregation can be formulated as:

$$F_{Aggr} = w_{PV} \otimes F_{PV} + w_{ART} \otimes F_{ART} \tag{1}$$

where $F_{Aggr} \in \mathbb{R}^{C \times H \times W}$ is aggregated features; $\otimes$ is element-wise multiplication.

How to obtain appropriate response maps is the key point of the SAM module. Concretely, the SAM module first extracts efficient descriptors of input features to reduce dimensions and preserve informative characteristics. To do so, we apply average-pooling and max-pooling operations to inputs along the channel direction [17]. The obtained descriptors are denoted as $F'_{PV} \in \mathbb{R}^{2 \times H \times W}$ and $F'_{ART} \in \mathbb{R}^{2 \times H \times W}$, respectively. Then, SAM module learns two mapping functions $M^{(1)}$ and $M^{(2)}$ to model local and global inter-phase complementary relationships from the feature descriptors. Specifically, $M^{(1)}$ and $M^{(2)}$ are built on a pyramid convolution structure (see Fig. 2(a)), i.e., a global average pooling (GAP) layer and a 7×7 convolutional layer are applied to distill global correspondence; two convolutional layers (with kernel sizes of 3×3 and 5×5) are used to capture local inter-phase details. The outputs of $M^{(1)}$ and $M^{(2)}$ are adopted to yield two initial response maps $w_{PV}^{(0)} \in \mathbb{R}^{C \times H \times W}$ and $w_{ART}^{(0)} \in \mathbb{R}^{C \times H \times W}$ through concatenation and 3×3 convolution (note that we up-sample the output of the GAP layer to $H \times W$ before the concatenation). The ultimate response maps are obtained by normalizing $w_{PV}^{(0)}$ and $w_{ART}^{(0)}$ through a softmax layer, which ensures $w_{PV}^{(c,h,w)} + w_{ART}^{(c,h,w)} = 1$.

So far, the aggregated features F_{Aggr} can be calculated using Eq. 1 and will be fed into the decoder for tumor region prediction. Besides, we feed the modulated phase-specific features $F_{PV}^{(m)} \in \mathbb{R}^{C \times H \times W}$ and $F_{ART}^{(m)} \in \mathbb{R}^{C \times H \times W}$ to PV- and ART-stream to mutually guide their feature extraction, where $F_{PV}^{(m)}$ and $F_{ART}^{(m)}$ are obtained by:

$$F_{PV}^{(m)} = (F_{PV} + F_{Aggr})/2, \ F_{ART}^{(m)} = (F_{ART} + F_{Aggr})/2 \tag{2}$$

2.2   Uncertain Region Inpainting Module

The decoding stage took four-level aggregated features from $B_{PV}^{(i)}$ and $B_{ART}^{(i)}$ ($i \in \{2,3,4,5\}$) to predict preliminary probability maps. However, the initial results usually present some blurry and uncertain regions. Accordingly, we propose an uncertain region inpainting module (URIM) to refine the ambiguous regions (especially the tumor boundaries). The core idea of our URIM is leveraging pixels with confident classification scores to inpaint neighboring uncertain pixels.

Confidence Map Calculation. Inspired by Liang's work [8], we derive the concept of the confidence map. Let $S_i \in \mathbb{R}^{1 \times H \times W} (i \in [1,2])$ denote the initial segmentation maps. S_i is the probability of each pixel p belonging to class i (liver tumor or background), where $\sum_{i=1}^{2} S_i(p) = 1$. Thus, the classification confidence of each pixel can be represented by the confidence map $M_{conf} \in \mathbb{R}^{1 \times H \times W}$:

$$M_{conf} = 1 - \exp(1 - S^{max}/S^{min}) \tag{3}$$

where S^{max} represents the largest score of each pixel in initial maps and S^{min} represents the smallest score of each pixel. M_{conf} ranges in $[0,1)$, and a larger value in M_{conf} means the higher confidence.

Local-Confidence Convolutions. Uncertain pixels usually have indistinguishable features, thus it is hard to identify their classes. Intuitively, if we can let uncertain pixels assimilate discriminative features from neighboring confident pixels, the classification of these uncertain pixels may become easier. To achieve this, we propose the local-confidence convolution (LC-Conv) operation, which is formulated as:

$$x' = (W^T(X \otimes M_{conf}))/\text{sum}(M_{conf}) + b \tag{4}$$

where X denotes the input features in the current sliding window; x' denotes the refined features; M_{conf} denotes the pixel-wise confidence map; W denotes the weights of the convolution filter and b denotes the bias. The scaling factor $1/\text{sum}(M_{conf})$ is used to regularize the effect of the confidence map within different sliding windows. During each convolution operation, LC-Conv emphasizes discriminative features and suppresses uncertain features. With this mechanism, pixels with higher confidence in the neighboring window contribute more to the filtering result, thereby making uncertain pixels receive surrounding distinguishable features. After each LC-Conv operation, M_{conf} is updated through a 3×3 max-value convolutional layer.

Figure 2(b) presents the detailed structure of the URIM module, which consists of four LC-Conv layers with 3×3 kernel. The URIM module takes M_{conf} and decision features from the decoder (feature maps before the softmax layer) as inputs, and yields the refined prediction. During the refinement stage, uncertain pixels gradually absorb more distant confident features while the uncertain regions shrink, which works like the image inpainting [9,21]. At last, we concatenate the refined features and input features to predict the final result.

2.3   Loss Function and Training Strategy

Our loss function comprises two cross-entropy losses (see Fig. 1), i.e., the L_{init} between initial segmentation and ground truths, and L_{final} between final predictions and ground truths. L_{init} and L_{final} contribute equally to the total loss. Our method is implemented based on PyTorch 1.5.0 [12] and trained on a NVIDIA GTX 2080 ti GPU (12 GB). We use the SGD optimizer to train our network with an initial learning rate of 5×10^{-4}, which is divided by 10 every 50 epoches.

3   Materials and Experiments

Dataset. We evaluate our method on an in-house multi-phase CT dataset of focal liver lesions (MPCT-FLLs). The dataset contains 121 multi-phase CT cases of five typical liver tumor types with liver and tumor delineations (including 36 cases of cysts, 20 cases of focal nodular hyperplasia (FNH), 25 cases of hemangiomas (HEM), 26 cases of hepatocellular carcinoma (HCC) and 14 cases of metastasis (METS)). The image size is 512×512, the slice thickness is 0.5 mm or 0.7 mm and the inter-plane resolution varies from 0.52×0.52 mm^2 to 0.86×0.86 mm^2. To validate our models, we adopt the five-fold cross-validation technique. Accordingly, 121 cases are randomly divided into five mutually-exclusive subsets. All the quantitative results are averaged on five testing sets.

Pre-processing. All the images are truncated into the range of [−70, 180] hounsfield unit to eliminate unrelated tissues. Besides, to avoid false-positives outside liver regions, we train a simple ResUNet [4] to segment livers. Each input of our network contains three adjacent slices masked by corresponding liver masks. Besides, we follow Xu's method [20] to register multi-phase inputs,

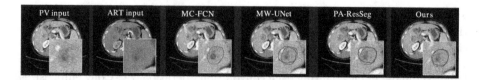

Fig. 3. Visual comparison between four different segmentation methods. Ground truths are delineated in red and the corresponding predictions are delineated in blue. (Color figure online)

Table 1. Quantitative comparison between our method and state-of-the-arts.

Methods	DPC (%)	DG (%)	VOE (%)	RVD (%)	ASSD (mm)	RMSD (mm)
MC-FCN [16]	51.80	71.63	59.30	−5.92	27.26	30.45
MW-UNet [18]	73.37	86.34	37.10	10.11	10.73	18.16
PA-ResSeg [20]	77.26	86.21	33.46	**4.46**	3.71	7.82
Ours	**80.12**	**86.51**	**30.08**	−5.28	**2.81**	**5.47**
ILF	73.19	83.13	37.6	−17.69	14.2	18.15
DLF [16]	75.87	85.36	35.14	−5.73	4.45	9.51
MW [18]	77.39	85.52	33.07	5.43	6.31	8.89
PA [20]	78.24	86.29	32.46	6.01	3.39	6.90
SAM (Ours)	**80.12**	**86.51**	**30.08**	−5.28	**2.81**	**5.47**

which simply aligned multi-phase tumor volumes according to the tumor center voxels. In the training phase, we augment the data by randomly shifting, rotating and scaling to prevent potential over-fitting problem.

Evaluation Metrics. To quantitatively measure the performance, six common metrics are used, i.e., dice per case (DPC), dice global (DG), volumetric overlap error (VOE), relative volume difference (RVD), average symmetric surface distance (ASSD) and root mean square symmetric surface distance (RMSD). The higher DPC and DG scores mean the better segmentation results. For the rest four evaluation metrics, the smaller the absolute value is, the better the results.

Comparison with State-of-the-arts. To validate the effectiveness of the proposed method, we compare it with three state-of-the-art multi-phase LiTS methods: (1) MC-FCN [16], which simply concatenated multi-phase features before the classification layer of FCN [10]; (2) MW-UNet [18], which trained a specific weight value for each phase at multiple layers of the U-Net; (3) PA-ResSeg [20], which incorporated channel-attention mechanism to re-weight each channels of multi-phase features at specific layers of the ResNet.

Figure 3 depicts the visual example of the comparison experiments. It is observed that MC-FCN produces poor results as it rudely concatenated decision-level features from different phases. These raw multi-phase features may bring in conflicts or interference factors, thus leading to the bad performance. MW-UNet and PA-ResSeg produces better results and are able to capture the rough tumor shape. However, the pixel-wise segmentation, especially on the tumor

Table 2. Quantitative comparison (in DPC (%)) regarding to different tumor types.

Methods	Cyst	FNH	HCC	HEM	METS	Average
MC-FCN	67.82	21.58	46.98	51.59	61.07	51.80
MW-UNet	83.07	64.18	64.80	71.42	79.35	73.37
PA-ResSeg	85.70	**73.13**	71.89	74.63	79.64	77.26
Ours	**90.87**	66.08	**79.11**	**75.42**	**83.27**	**80.12**

Table 3. Quantitative results for ablation studies.

SP	MP	SAM	URIM	DPC (%)	DG (%)	VOE (%)	RVD (%)	ASSD (mm)	RMSD (mm)
✓				68.85	83.31	40.07	−19.22	21.53	27.73
✓	✓			75.31	85.13	35.97	−5.89	5.00	10.05
✓	✓	✓		79.10	86.12	31.22	−5.51	3.44	6.46
✓	✓	✓	✓	**80.12**	**86.51**	**30.08**	**−5.28**	**2.81**	**5.47**

boundaries, is not satisfactory. The reason is that both of the methods neglected the spatial-wise feature fusion, thereby aggregating multi-phase features insufficiently. Besides, they did not provide any strategies to handle the uncertainty problem, which made it hard to do boundary pixel classification. In contrast, our method encourages per-pixel inter-phase interactions and incorporates uncertain region inpainting mechanism, which is demonstrated to achieve the best value in DPC, DG, VOE, ASSD and RMSD scores (see Table 1).

To analyze the performance of our method on specific tumor types, we divide the testing sets into five subsets according to tumor categories. As seen in Table 2, our method boosts prominent performance gains on most types of tumors compared to other methods, and achieves the top performance on the average.

Further, to validate our proposed cross-phase fusion strategy (the SAM module), we compare it with other four fusion methods: (1) Input-level fusion (ILF) strategy, which concatenates multi-phase images in the inputs; (2) Decision-level fusion (DLF) strategy used in MC-FCN [16]; (3) Modality weighting (MW) strategy adopted in MW-UNet [18] and (4) Phase Attention (PA) strategy adopted in PA-ResSeg [20]. To ensure the fairness of comparison, all the fusion modules are plugged into the proposed network except that the feature aggregation module is replaced. As shown in Table 1, our SAM strategy makes better use of multi-phase information and outperforms other fusion strategies.

Ablation Study. To validate the effectiveness of each component in our method, we start from a single-phase (PV phase) ResNeXt-50 network and gradually add the modules. Table 3 summaries the quantitative results, in which SP means single-phase segmentation and MP means multi-phase segmentation by simply adding PV- and ART-features at four convolution blocks (we denote this fusion strategy as FLF-add). It is seen that adding multi-phase information donates a performance boost of +6.46% in DPC; employing SAM modules to fuse features improves the performance by +3.79% in DPC; refining uncertain regions via the URIM module contributes a performance gain of +1.02% in DPC.

Robustness Validation. Our SAM and URIM modules could be plugged into various multi-phase segmentation networks. To validate their robustness on different backbones, we replace our backbone with U-Net [14] and ResNet-50 [5]. The DPC gains of SAM and URIM modules on multi-phase U-Net, ResNet-50 and ResNeXt-50 (adopting FLF-add strategy) are 5.28%, 4.50% and 4.81%, respectively. It is demonstrated that the SAM and URIM modules can improve the segmentation performance stably in different networks.

4   Discussion and Conclusions

In this work, we propose a novel multi-phase network to segment liver tumors from PV- and ART-phase images. In our network, we devise a spatial aggregation module to take full advantage of multi-phase information by modulating each pixel. We also devise an uncertain region inpainting module to handle uncertainty in tumor boundary segmentation. Experiments on the MPCT-FLLs dataset demonstrates the superiority of our method against state-of-the-arts. In the future, we may focus on multi-phase segmentation problem on more than two phases. In this situation, we may use the SAM module (with more parameters) to generate the response map for each phase. This issue will be more challenging since we need to consider the memory consumption of the network.

References

1. Christ, P.F., et al.: Automatic liver and lesion segmentation in CT using cascaded fully convolutional neural networks and 3d conditional random fields. In: Ourselin, S., Joskowicz, L., Sabuncu, M.R., Unal, G., Wells, W. (eds.) MICCAI 2016. LNCS, vol. 9901, pp. 415–423. Springer, Cham (2016). https://doi.org/10.1007/978-3-319-46723-8_48
2. El-Serag, H.B.: Epidemiology of hepatocellular carcinoma. Liver: Biol. Pathobiology **59**(1), 758–772 (2020)
3. Fu, J., et al.: Dual attention network for scene segmentation. In: Proceedings of the IEEE/CVF Conference on Computer Vision and Pattern Recognition, pp. 3146–3154 (2019)
4. Han, X.: Automatic liver lesion segmentation using a deep convolutional neural network method. arXiv preprint arXiv:1704.07239 (2017)
5. He, K., Zhang, X., Ren, S., Sun, J.: Deep residual learning for image recognition. In: Proceedings of the IEEE Conference on Computer Vision and Pattern Recognition, pp. 770–778 (2016)
6. Hu, J., Shen, L., Sun, G.: Squeeze-and-excitation networks. In: Proceedings of the IEEE Conference on Computer Vision and Pattern Recognition, pp. 7132–7141 (2018)
7. Li, X., Chen, H., Qi, X., Dou, Q., Fu, C.W., Heng, P.A.: H-denseunet: hybrid densely connected UNet for liver and tumor segmentation from CT volumes. IEEE Trans. Med. Imaging **37**(12), 2663–2674 (2018)

8. Liang, Y., Li, X., Jafari, N., Chen, Q.: Video object segmentation with adaptive feature bank and uncertain-region refinement. In: Advances in neural information processing systems (NeurIPS) (2020)
9. Liu, G., Reda, F.A., Shih, K.J., Wang, T.C., Tao, A., Catanzaro, B.: Image inpainting for irregular holes using partial convolutions (2018)
10. Long, J., Shelhamer, E., Darrell, T.: Fully convolutional networks for semantic segmentation. In: Proceedings of the IEEE Conference on Computer Vision and Pattern Recognition, pp. 3431–3440 (2015)
11. Ouhmich, F., Agnus, V., Noblet, V., Heitz, F., Pessaux, P.: Liver tissue segmentation in multiphase CT scans using cascaded convolutional neural networks. Int. J. Comput. Assist. Radiol. Surg. **14**(8), 1275–1284 (2019)
12. Paszke, A., et al.: Automatic differentiation in pytorch (2017)
13. Raju, A., et al.: Co-heterogeneous and adaptive segmentation from multi-source and multi-phase CT imaging data: a study on pathological liver and lesion segmentation. In: Vedaldi, A., Bischof, H., Brox, T., Frahm, J.-M. (eds.) ECCV 2020. LNCS, vol. 12368, pp. 448–465. Springer, Cham (2020). https://doi.org/10.1007/978-3-030-58592-1_27
14. Ronneberger, O., Fischer, P., Brox, T.: U-Net: convolutional networks for biomedical image segmentation. In: Navab, N., Hornegger, J., Wells, W.M., Frangi, A.F. (eds.) MICCAI 2015. LNCS, vol. 9351, pp. 234–241. Springer, Cham (2015). https://doi.org/10.1007/978-3-319-24574-4_28
15. Seo, H., Huang, C., Bassenne, M., Xiao, R., Xing, L.: Modified u-net (mu-net) with incorporation of object-dependent high level features for improved liver and liver-tumor segmentation in CT images. IEEE Trans. Med. Imaging **39**(5), 1316–1325 (2019)
16. Sun, C., et al.: Automatic segmentation of liver tumors from multiphase contrast-enhanced CT images based on FCNs. Artif. Intell. Med. **83**, 58–66 (2017)
17. Woo, S., Park, J., Lee, J.Y., Kweon, I.S.: Cbam: Convolutional block attention module (2018)
18. Wu, Y., Zhou, Q., Hu, H., Rong, G., Li, Y., Wang, S.: Hepatic lesion segmentation by combining plain and contrast-enhanced CT images with modality weighted u-net. In: 2019 IEEE International Conference on Image Processing (ICIP), pp. 255–259. IEEE (2019)
19. Xie, S., Girshick, R., Dollár, P., Tu, Z., He, K.: Aggregated residual transformations for deep neural networks. In: Proceedings of the IEEE Conference on Computer Vision and Pattern Recognition, pp. 1492–1500 (2017)
20. Xu, Y., et al.: Pa-resseg: a phase attention residual network for liver tumor segmentation from multiphase CT images. Med. Phys. **48**(7), 3752–3766 (2021)
21. Yu, J., Lin, Z., Yang, J., Shen, X., Lu, X., Huang, T.S.: Free-form image inpainting with gated convolution. In: Proceedings of the IEEE/CVF International Conference on Computer Vision, pp. 4471–4480 (2019)
22. Zhang, J., Xie, Y., Zhang, P., Chen, H., Xia, Y., Shen, C.: Light-weight hybrid convolutional network for liver tumor segmentation. In: IJCAI,. pp. 4271–4277 (2019)

Convolution-Free Medical Image Segmentation Using Transformers

Davood Karimi[(✉)], Serge Didenko Vasylechko, and Ali Gholipour

Computational Radiology Laboratory (CRL), Department of Radiology, Boston
Children's Hospital, and Harvard Medical School, Boston, USA
davood.karimi@childrens.harvard.edu

Abstract. Like other applications in computer vision, medical image
segmentation and his email address have been most successfully
addressed using deep learning models that rely on the convolution operation as their main building block. Convolutions enjoy important properties such as sparse interactions, weight sharing, and translation equivariance. These properties give convolutional neural networks (CNNs) a
strong and useful inductive bias for vision tasks. However, the convolution operation also has important shortcomings: it performs a fixed
operation on every test image regardless of the content and it cannot
efficiently model long-range interactions. In this work we show that a
network based on self-attention between neighboring patches and without any convolution operations can achieve better results. Given a 3D
image block, our network divides it into n^3 3D patches, where $n = 3$ or 5
and computes a 1D embedding for each patch. The network predicts the
segmentation map for the center patch of the block based on the self-
attention between these patch embeddings. We show that the proposed
model can achieve higher segmentation accuracies than a state of the
art CNN. For scenarios with very few labeled images, we propose methods for pre-training the network on large corpora of unlabeled images.
Our experiments show that with pre-training the advantage of our proposed network over CNNs can be significant when labeled training data
is small.

Keywords: Segmentation · Deep learning · Transformers · Attention

1 Introduction

Medical image segmentation is routinely used for quantifying the size and shape
of the volume/organ of interest, population studies, disease quantification, treatment planning, and intervention. Classical methods in medical image segmentation range from region growing [11] and deformable models [33] to atlas-
based methods [29], and Bayesian approaches [26]. In recent years, deep learning
(DL) methods have emerged as a highly competitive alternative and they have
achieved remarkable levels of accuracy [3,5,17]. It appears that DL methods
have largely replaced the classical methods for medical image segmentation.

© Springer Nature Switzerland AG 2021
M. de Bruijne et al. (Eds.): MICCAI 2021, LNCS 12901, pp. 78–88, 2021.
https://doi.org/10.1007/978-3-030-87193-2_8

Due to their success, DL methods for medical image segmentation have attracted much attention from the research community. Recent surveys of the published research on this topic can be found in [13,28]. Some of the main directions of these studies include improving network architectures, loss functions, and training strategies. Surprisingly, the one common feature in all of these works is the use of the convolution operation as the main building block of the networks. The proposed network architectures differ in terms of the way the convolutional operations are arranged, but they all build upon the same convolution operation. There have been several attempts to use recurrent neural networks [2,10] and attention mechanisms [6] for medical image segmentation. However, all of those models still rely on the convolution operation. Some recent studies have gone so far as to suggest that a rather simple encoder-decoder-type CNN can be as accurate as more elaborate network architectures [16].

Similar to semantic segmentation, other vision tasks have also been successfully addressed with convolutional neural networks (CNNs) [20,27]. These observations attest to the importance of the convolution operation to image modeling. The effectiveness of the convolution operation can be attributed to a number of key properties, including: 1) local (sparse) connections, 2) parameter (weight) sharing, and 3) translation equivariance [21,22]. These properties are in part inspired by the neuroscience of vision [25], and give CNNs a strong inductive bias. A convolutional layer can be interpreted as a fully connected layer with an "infinitely strong prior" over its weights [12]. As a result, modern deep CNNs have been able to tackle a variety of vision tasks with amazing success.

In other applications, most prominently in natural language processing (NLP), the dominant architectures were those based on recurrent neural networks (RNNs) [7,15]. The dominance of RNNs ended with the introduction of the transformer model, which proposed to replace the recurrence with an attention mechanism that could learn the relationship between different parts of a sequence [31].

The attention mechanism has had a profound impact on the field of NLP. In vision applications, on the other hand, its impact has been much more limited ([9,14,32]). A survey of applications of transformer networks in computer vision can be found in [18]. The number of pixels in a typical image is orders of magnitude larger than the number of units of data (e.g., words) in NLP. This makes it impossible to directly apply standard attention models to image pixels. The recently-proposed vision transformer (ViT) appears to be a major step towards addressing this problem [8]. The main insight in ViT is to consider image patches, rather than pixels, to be the units of information in images. ViT embeds image patches into a shared space and learns the relation between these embeddings using self-attention modules. Given massive amounts of training data and computational resources, ViT surpassed CNNs in image classification. Subsequent work has shown that by using knowledge distillation from a CNN teacher and using standard training data and computational resources, transformer networks can achieve image classification accuracy levels on par with CNNs [30].

The goal of this work is to explore the potential of self-attention-based deep neural networks for 3D medical image segmentation. We propose a network architecture that is based on self-attention between linear embeddings of 3D image patches, without any convolution operations. We train the network on three medical image segmentation datasets and compare its performance with a state of the art CNN. The contributions of this work are as follows:

1. We propose the first convolution-free deep neural network for 3D medical image segmentation.
2. We show that our proposed network can achieve segmentation accuracies that are better than or at least on par with a state of the art CNN on three different medical image segmentation datasets. We show that, unlike recent works on image classification ([8,30]), our network can be effectively trained for 3D medical image segmentation with datasets of $\sim 20-200$ labeled images.
3. For scenarios where very few labeled training images are available, we propose methods to pre-train the network on large corpora of unlabeled images. We show that in such scenarios our network performs better than a state of the art CNN with pre-training. This demonstrates the potential advantage of our network over CNNs in applications where obtaining expert labels is costly, which is very common in medical imaging.

2   Materials and Methods

2.1   Proposed Network

Figure 1 shows our proposed network for convolution-free medical image segmentation. The input to the network is a 3D block $B \in \mathbb{R}^{W \times W \times W \times c}$, where W denotes the extent of the block (in voxels) in each dimension and c is the number of image channels. The block B is partitioned into n^3 contiguous non-overlapping 3D patches $\{p_i \in \mathbb{R}^{w \times w \times w \times c}\}_{i=1}^N$, where $w = W/n$ is the side length of each patch and $N = n^3$ denotes the number of patches in the block. We choose n to be an odd number. In the experiments presented in this paper n is either 3 or 5, corresponding to 27 or 125 patches, respectively. The network uses the image information in all N patches in B to predict the segmentation for *the center patch*, as described below. (We refer the interested reader who is not familiar with the transformer and the attention mechanism to [8,24,31].)

Each of the N patches is flattened into a vector of size $\mathbb{R}^{w^3 c}$ and embedded into $\mathbb{R}^D$ using a trainable mapping $E \in \mathbb{R}^{D \times w^3 c}$. Unlike ViT [8], we do not use the extra "class token" that has been inherited from NLP applications, because that did not improve the segmentation performance of our network in any way. The sequence of embedded patches $X^0 = [Ep_1; ...; Ep_N] + E_{\text{pos}}$ constitutes the input to our network. The matrix $E_{\text{pos}} \in \mathbb{R}^{D \times N}$ is intended to learn a positional encoding. Without such positional information, the transformer ignores the ordering of the input sequence because the attention mechanism is permutation-invariant. Therefore, in most NLP applications such embedding has been very important. For image classification, on the other hand, authors of [8] found that positional encoding resulted in relatively small improvements in accuracy and that simple 1D raster encoding was as good as more elaborate 2D positional encodings. For the experiments presented in this paper, we left the positional encoding as a free parameter to be learned along with the network parameters during training because we did not know a priori what form of positional encoding could be useful. We discuss the results of experimental comparisons with different positional encodings below.

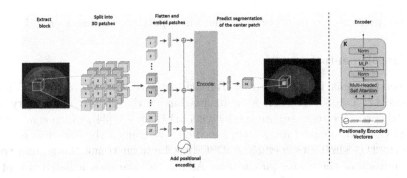

Fig. 1. Proposed convolution-free 3D medical image segmentation method.

The encoder has K stages, each consisting of a multi-head self-attention (MSA) and a subsequent two-layer fully connected feed-forward network (FFN). Both the MSA and FFN modules include residual connections, ReLU activations, and layer normalization. More specifically, starting with the sequence of embedded and position-encoded patches, X^0 described above, the kth stage of the encoder will perform the following operations to map X^k to X^{k+1}:

1. X^k goes through n_h separate heads in MSA. The ith head:
   (a) Computes the query, key, and value sequences from the input sequence:

   $$Q^{k,i} = E_Q^{k,i}\text{LN}(X^k), K^{k,i} = E_K^{k,i}\text{LN}(X^k), V^{k,i} = E_V^{k,i}\text{LN}(X^k)$$

   where $E_Q, E_K, E_v \in \mathbb{R}^{D_h \times D}$ and LN denotes layer normalization.
   (b) Computes the self-attention matrix and then the transformed values:

   $$A^{k,i} = \text{Softmax}(Q^T K)/\sqrt{D_h} \qquad \text{SA}^{k,i} = A^{k,i} V^{k,i}$$

2. Outputs of the n_h heads are stacked and reprojected back onto $\mathbb{R}^D$

   $$\text{MSA}^k = E_{\text{reproj}}^k [\text{SA}^{k,0}; ...; \text{SA}^{k,n_h}]^T \quad \text{where } E_{\text{reproj}} \in \mathbb{R}^{D \times D_h n_h}$$

3. The output of the MSA module is computed as:

   $$X_{\text{MSA}}^k = \text{MSA}^k + X^k$$

4. X_{MSA}^k goes through a two-layer FFN to obtain the output of the kth stage:

$$X^{k+1} = X_{\mathrm{MSA}}^k + E_2^k \Big(\mathrm{ReLU}((E_1^k \mathrm{LN}(X_{\mathrm{MSA}}^k) + b_1^k) \Big) + b_2^k$$

The output of the last encoder stage, X^K, will pass through a final FFN layer that projects it onto the $\mathbb{R}^{N n_{\mathrm{class}}}$ and then reshaped into $\mathbb{R}^{n \times n \times n \times n_{\mathrm{class}}}$, where n_{class} is the number of classes (for binary segmentation, $n_{\mathrm{class}} = 2$):

$$\hat{Y} = \mathrm{Softmax}\Big(E_{\mathrm{out}} X^K + b_{\mathrm{out}} \Big).$$

$\hat{Y}$ is the predicted segmentation for the center patch of the block (Fig. 1).

2.2   Implementation

The network was implemented in TensorFlow 1.16 and run on an NVIDIA GeForce GTX 1080 GPU on a Linux machine with 120 GB of memory and 16 CPU cores. We compare our model with 3D UNet++ [34], which is a state of the art CNN for medical image segmentation. We trained the networks to maximize the Dice similarity coefficient, DSC [23], between $\hat{Y}$ and the ground-truth segmentation of the center patch using Adam [19]. We used a batch size of 10 and a learning rate of 10^{-4}, which was reduced by half if after a training epoch the validation loss did not decrease. For UNet++ a larger initial learning rate of 3×10^{-4} was used because that led to the best results with UNet++.

Pre-training. Manual segmentation of complex structures such as the brain cortical plate can take several hours of a medical expert's time for a single image. Therefore, methods that can achieve high accuracy with fewer labeled training images are highly advantageous. To improve the network's performance when labeled training images are insufficient, we propose to first pre-train the network on unlabeled data for denoising or inpainting tasks. For *denoising* pre-training, we add Gaussian noise with SNR = 10 dB to the input block B, whereas for *inpainting* pre-training we set the center patch of the block to constant 0. In both cases, the target is the noise-free center patch of the block. For pre-training, we use a different output layer (without the softmax operation) and train the network to minimize the ℓ_2 norm between the prediction and the target. To fine-tune the pre-trained network for the segmentation task, we introduce a new output layer with softmax and fine-tune the network on the labeled data as explained in the above paragraph. We fine-tune the entire network, rather than only the output layer, on the labeled images because we have found that fine-tuning the entire network for the segmentation task leads to better results.

Data. Table 1 shows the datasets used in this work. We used ∼1/5 of the training images for validation. The final models were evaluated on the test images. The images were split into train/validation/test at random. The same train/validation/test split was used for both our network and UNet++.

Table 1. Datasets used for experiments in this work.

Target organ	Image modality	$[n_{train}, n_{test}]$
Brain cortical plate	T2 MRI	[18, 9]
Pancreas	CT	[231, 50]
Hippocampus	MRI	[220, 40]

3 Results and Discussion

Table 2 presents test segmentation accuracy of the proposed method and UNet++ in terms of DSC, the 95 percentile of the Hausdorff Distance (HD95) and Average Symmetric Surface Distance (ASSD). In these experiments, we used these parameter settings for our network: $K = 7, W = 24, n = 3, D = 1024, D_h = 256, n_h = 4$. We used the same parameter settings for all experiments reported in the rest of the paper, unless otherwise stated. For each dataset and each of the three criteria, we ran paired t-tests to see if the differences were statistically significant. As shown in the table, segmentation accuracy for the proposed convolution-free method was significantly better than, or at least on part with, UNet++. Figure 2 shows example slices from test images in each dataset and the segmentations predicted by the proposed method and UNet++. We have observed that the proposed method is capable of accurately segmenting fine and complex structures such as the brain cortical plate. In terms of training time, our network converged in approximately 24 h of GPU time. For fairness, we also let UNet++ train for 24 h, but it usually converged in approximately 12 h.

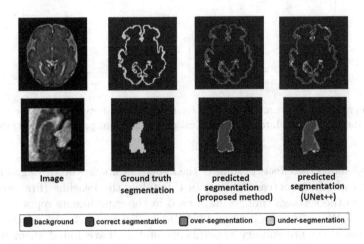

| Image | Ground truth segmentation | predicted segmentation (proposed method) | predicted segmentation (UNet++) |

■ background   ■ correct segmentation   ■ over-segmentation   □ under-segmentation

Fig. 2. Example segmentations predicted by the proposed method and UNet++.

Table 2. Segmentation accuracy of the proposed method and UNet++. Better results for each dataset/criterion are in bold type. Asterisks denote statistically significant difference ($p < 0.01$ in paired t-test.)

Dataset	Method	DSC	HD95 (mm)	ASSD (mm)
Brain cortical plate	Proposed	**0.879 ± 0.026***	0.92 ± 0.04	**0.24 ± 0.03**
	UNet++	0.860 ± 0.024	**0.91 ± 0.04**	0.25 ± 0.04
Hippocampus	Proposed	**0.881 ± 0.017***	**1.10 ± 0.19***	**0.40 ± 0.04**
	UNet++	0.852 ± 0.022	1.33 ± 0.26	0.41 ± 0.07

To assess the segmentation accuracy with reduced number of labeled training images, we trained our method and UNet++ with $n_{train} = 5, 10$, or 15 labeled training images from cortical plate and pancreas datasets. For cortical plate segmentation, we used 500 unlabeled images from the dHCP dataset [4] for pre-training. For pancreas segmentation, we used the remaining training images, i.e., $231 - n_{train}$ (see Table 1) for pre-training. We pre-trained our model as explained in Sect. 2.2. To pre-train UNet++, we used the method proposed in [1]. Figure 3 shows the results of this experiment. With the proposed pre-training, the convolution-free network achieves significantly more accurate segmentations than UNet++ with fewer labeled training images. The differences in terms of DSC were statistically significant ($p < 0.01$) on both datasets for $n_{train} = 5, 10$, and 15.

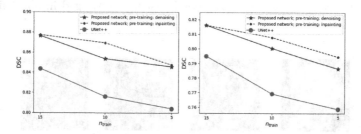

Fig. 3. Segmentation accuracy (in DSC) for the proposed network and UNet++ with reduced labeled training data on the cortical plate (left) and pancreas (right) datasets.

Table 3 shows the effects of some of the hyper-parameters on the segmentation accuracy with the pancreas dataset. In this table, the baseline (first row) corresponds to the settings that we have used in the experiments reported above, i.e., $K = 7, W = 24, n = 3, D = 1024, D_h = 256, n_h = 4$. We selected these settings based on preliminary experiments and we have found them to work well for different datasets. Some of the observations from Table 3 are as follows: 1) Increasing the number/size of the patches or increasing the network depth typically led to only slight improvements in accuracy. Using $n = 5$,

$w = 36$, or $K = 10$ did not significantly improve the segmentation accuracy ($p \approx 0.1$). On the other hand, the reduction in DSC with a shallower network ($K = 4$) was significant ($p < 0.01$). 2) Using a fixed positional encoding or no positional encoding slightly reduced the segmentation accuracy compared with free-parameter/learnable positional encoding. However, the differences were not statistically significant ($p \approx 0.1$). Here, for fixed positional encoding we used the common approach based on trigonometric functions [31], and for no positional encoding we set E_{pos} to zero. 3) Using a single-head attention significantly reduced the segmentation accuracy ($p < 0.01$), which indicates the importance of the multi-head design to enable the network to learn a more complex relation between neighboring patches.

Table 3. Effect of some of the network hyperparameters on the segmentation accuracy with the pancreas dataset. The baseline (first row) corresponds to these settings: $\{K = 7, W = 24, n = 3, D = 1024, D_h = 256, n_h = 4\}$, which are the hyperparameter values used in all experiments reported in this paper other than in this table.

Parameter settings	DSC	HD95 (mm)	ASSD (mm)
baseline	0.826 ± 0.020	5.72 ± 1.61	2.13 ± 0.24
larger number of patches, $n = 5$	0.828 ± 0.023	5.68 ± 1.63	2.01 ± 0.18
larger blocks, $W = 36$	0.829 ± 0.017	5.60 ± 1.56	1.98 ± 0.19
deeper network, $K = 10$	0.827 ± 0.018	5.50 ± 1.48	2.10 ± 0.21
shallower network, $K = 4$	0.810 ± 0.023	6.14 ± 1.80	2.29 ± 0.38
no positional encoding	0.818 ± 0.026	5.84 ± 1.74	2.26 ± 0.27
fixed positional encoding	0.823 ± 0.021	5.81 ± 1.54	2.16 ± 0.23
more heads, $n_h = 8$	0.824 ± 0.017	5.68 ± 1.56	2.14 ± 0.21
single head, $n_h = 1$	0.802 ± 0.026	6.82 ± 1.40	2.31 ± 0.35

4   Conclusions

The convolution operation has strong neuroscientific and machine learning bases and has played a central role in the success of deep learning models in tackling various vision tasks. However, there is no reason to expect that no other model can outperform CNNs on a specific task. Medical image analysis applications, in particular, pose specific challenges such as the 3D nature of the images and typically small number of labeled images. In such applications, other models can be more effective than CNNs. In this work, we proposed a new model for 3D medical image segmentation. Unlike all recent models that use the convolution as their main building block, our model is based on self-attention between neighboring 3D patches. Our results show that the proposed network can outperform a state of the art CNN on three medical image segmentation datasets. With pre-training for denoising and in-painting tasks on unlabeled images, our network also performed better than a CNN when only 5–15 labeled training images were used. This is a very important advantage of our method in many medical imaging applications where obtaining expert labels is expensive and time-consuming.

Acknowledgement. This work was supported in part by the National Institutes of Health (NIH) award numbers R01NS106030, R01EB018988, and R01EB031849; by the Office of the Director of the NIH under number S10OD0250111; and by a Technological Innovations in Neuroscience Award from the McKnight Foundation. The content is solely the responsibility of the authors and does not necessarily represent the official views of the NIH or the McKnight Foundation.

References

1. Bai, W., et al.: Semi-supervised learning for network-based cardiac MR image segmentation. In: Descoteaux, M., Maier-Hein, L., Franz, A., Jannin, P., Collins, D.L., Duchesne, S. (eds.) MICCAI 2017. LNCS, vol. 10434, pp. 253–260. Springer, Cham (2017). https://doi.org/10.1007/978-3-319-66185-8_29
2. Bai, W., et al.: Recurrent neural networks for aortic image sequence segmentation with sparse annotations. In: Frangi, A.F., Schnabel, J.A., Davatzikos, C., Alberola-López, C., Fichtinger, G. (eds.) MICCAI 2018. LNCS, vol. 11073, pp. 586–594. Springer, Cham (2018). https://doi.org/10.1007/978-3-030-00937-3_67
3. Bakas, S., et al.: Identifying the best machine learning algorithms for brain tumor segmentation, progression assessment, and overall survival prediction in the brats challenge. arXiv preprint arXiv:1811.02629 (2018)
4. Bastiani, M., et al.: Automated processing pipeline for neonatal diffusion MRI in the developing human connectome project. NeuroImage **185**, 750–763 (2019)
5. Bernard, O., et al.: Deep learning techniques for automatic MRI cardiac multi-structures segmentation and diagnosis: is the problem solved? IEEE Trans. Med. Imaging **37**(11), 2514–2525 (2018)
6. Chen, J., et al.: Transunet: Transformers make strong encoders for medical image segmentation. arXiv preprint arXiv:2102.04306 (2021)
7. Chung, J., Gulcehre, C., Cho, K., Bengio, Y.: Empirical evaluation of gated recurrent neural networks on sequence modeling. arXiv preprint arXiv:1412.3555 (2014)
8. Dosovitskiy, A., et al.: An image is worth 16x16 words: Transformers for image recognition at scale. arXiv preprint arXiv:2010.11929 (2020)
9. Dou, H., et al.: A deep attentive convolutional neural network for automatic cortical plate segmentation in fetal MRI. arXiv preprint arXiv:2004.12847 (2020)
10. Gao, Y., Phillips, J.M., Zheng, Y., Min, R., Fletcher, P.T., Gerig, G.: Fully convolutional structured lstm networks for joint 4d medical image segmentation. In: 2018 IEEE 15th International Symposium on Biomedical Imaging (ISBI 2018), pp. 1104–1108 (2018). https://doi.org/10.1109/ISBI.2018.8363764
11. Gibbs, P., Buckley, D.L., Blackband, S.J., Horsman, A.: Tumour volume determination from MR images by morphological segmentation. Phys. Med. Biol. **41**(11), 2437 (1996)
12. Goodfellow, I., Bengio, Y., Courville, A., Bengio, Y.: Deep Learning, vol. 1. MIT Press, Cambridge (2016)

13. Hesamian, M.H., Jia, W., He, X., Kennedy, P.: Deep learning techniques for medical image segmentation: achievements and challenges. J. Digit. Imaging **32**(4), 582–596 (2019)
14. Ho, J., Kalchbrenner, N., Weissenborn, D., Salimans, T.: Axial attention in multi-dimensional transformers. arXiv preprint arXiv:1912.12180 (2019)
15. Hochreiter, S., Schmidhuber, J.: Long short-term memory. Neural Comput. **9**(8), 1735–1780 (1997)
16. Isensee, F., Kickingereder, P., Wick, W., Bendszus, M., Maier-Hein, K.H.: No new-net. In: Crimi, A., Bakas, S., Kuijf, H., Keyvan, F., Reyes, M., van Walsum, T. (eds.) BrainLes 2018. LNCS, vol. 11384, pp. 234–244. Springer, Cham (2019). https://doi.org/10.1007/978-3-030-11726-9_21
17. Kamnitsas, K., et al.: Efficient multi-scale 3d CNN with fully connected CRF for accurate brain lesion segmentation. Med. Image Anal. **36**, 61–78 (2017)
18. Khan, S., Naseer, M., Hayat, M., Zamir, S.W., Khan, F.S., Shah, M.: Transformers in vision: A survey. arXiv preprint arXiv:2101.01169 (2021)
19. Kingma, D.P., Ba, J.: Adam: a method for stochastic optimization. In: Proceedings of the 3rd International Conference on Learning Representations (ICLR) (2014)
20. Krizhevsky, A., Sutskever, I., Hinton, G.E.: Imagenet classification with deep convolutional neural networks. In: Advances in Neural Information Processing Systems, pp. 1097–1105 (2012)
21. Le Cun, Y., et al.: Handwritten digit recognition with a back-propagation network. In: Proceedings of the 2nd International Conference on Neural Information Processing Systems, pp. 396–404 (1989)
22. LeCun, Y., Bengio, Y., Hinton, G.: Deep learning. Nature **521**(7553), 436 (2015)
23. Milletari, F., Navab, N., Ahmadi, S.A.: V-net: fully convolutional neural networks for volumetric medical image segmentation. In: 3D Vision (3DV), 2016 Fourth International Conference on, pp. 565–571. IEEE (2016)
24. Murphy, K.P.: Machine learning: a probabilistic perspective (2012)
25. Olshausen, B.A., Field, D.J.: Emergence of simple-cell receptive field properties by learning a sparse code for natural images. Nature **381**(6583), 607–609 (1996)
26. Prince, J.L., Pham, D., Tan, Q.: Optimization of MR pulse sequences for bayesian image segmentation. Med. Phys. **22**(10), 1651–1656 (1995)
27. Ren, S., He, K., Girshick, R., Sun, J.: Faster r-cnn: towards real-time object detection with region proposal networks. In: Advances in Neural Information Processing Systems, pp. 91–99 (2015)
28. Taghanaki, S.A., Abhishek, K., Cohen, J.P., Cohen-Adad, J., Hamarneh, G.: Deep semantic segmentation of natural and medical images: a review. Artif. Intell. Rev. 1–42 (2020)
29. Thompson, P.M., Toga, A.W.: Detection, visualization and animation of abnormal anatomic structure with a deformable probabilistic brain atlas based on random vector field transformations. Med. Image Anal. **1**(4), 271–294 (1997)
30. Touvron, H., Cord, M., Douze, M., Massa, F., Sablayrolles, A., Jégou, H.: Training data-efficient image transformers & distillation through attention. arXiv preprint arXiv:2012.12877 (2020)
31. Vaswani, A., et al.: Attention is all you need. arXiv preprint arXiv:1706.03762 (2017)
32. Wang, H., Zhu, Y., Green, B., Adam, H., Yuille, A., Chen, L.-C.: Axial-DeepLab: stand-alone axial-attention for panoptic segmentation. In: Vedaldi, A., Bischof, H., Brox, T., Frahm, J.-M. (eds.) ECCV 2020. LNCS, vol. 12349, pp. 108–126. Springer, Cham (2020). https://doi.org/10.1007/978-3-030-58548-8_7

33. Wang, Y., Guo, Q., Zhu, Y.: Medical image segmentation based on deformable models and its applications. In: Deformable Models. Topics in Biomedical Engineering. International Book Series, pp. 209–260. Springer, New York (2007). https://doi.org/10.1007/978-0-387-68343-0_7
34. Zhou, Z., Rahman Siddiquee, M.M., Tajbakhsh, N., Liang, J.: UNet++: a nested U-net architecture for medical image segmentation. In: Stoyanov, D., et al. (eds.) DLMIA/ML-CDS -2018. LNCS, vol. 11045, pp. 3–11. Springer, Cham (2018). https://doi.org/10.1007/978-3-030-00889-5_1

Consistent Segmentation of Longitudinal Brain MR Images with Spatio-Temporal Constrained Networks

Jie Wei[1,2,3], Feng Shi[4], Zhiming Cui[3,5], Yongsheng Pan[1,2,3], Yong Xia[1,2(✉)], and Dinggang Shen[3,4(✉)]

[1] National Engineering Laboratory for Integrated Aero-Space-Ground-Ocean Big Data Application Technology, School of Computer Science and Engineering, Northwestern Polytechnical University, Xi'an, Shaanxi 710072, China
yxia@nwpu.edu.cn
[2] Research and Development Institute of Northwestern Polytechnical University in Shenzhen, Shenzhen 518057, China
[3] School of Biomedical Engineering, ShanghaiTech University, Shanghai, China
dgshen@shanghaitech.edu.cn
[4] Shanghai United Imaging Intelligence Co., Ltd., Shanghai, China
[5] Department of Computer Science, The University of Hong Kong, Hong Kong, China

Abstract. Accurate and consistent segmentation of longitudinal brain magnetic resonance (MR) images is of great importance in studying brain morphological and functional changes over time. However, current available brain segmentation methods, especially deep learning methods, are mostly trained with cross-sectional brain images that might generate inconsistent results in longitudinal studies. To overcome this limitation, we present a novel coarse-to-fine spatio-temporal constrained deep learning model for consistent longitudinal segmentation based on limited labeled cross-sectional data with semi-supervised learning. Specifically, both segmentation smoothness and temporal consistency are imposed in the loss function. Moreover, brain structural changes over time are summarized as age constraint, to make the model better reflect the trends of longitudinal aging changes. We validate our proposed method on 53 sets of longitudinal T1-weighted brain MR images from ADNI, with an average of 4.5 time-points per subject. Both quantitative and qualitative comparisons with comparison methods demonstrate the superior performance of our proposed method.

Keywords: Brain MR images · Consistent longitudinal segmentation · Semi-supervised learning

1 Introduction

Longitudinal studies usually acquire brain scans at multiple time points for each subject, to analyze brain morphological changes. These changes are subtle, i.e.,

© Springer Nature Switzerland AG 2021
M. de Bruijne et al. (Eds.): MICCAI 2021, LNCS 12901, pp. 89–98, 2021.
https://doi.org/10.1007/978-3-030-87193-2_9

normal aging brain often has 0.5–1.0% change per year and thus has to be quantified precisely. In this regard, consistent and accurate measurement of brain structure changes from longitudinal magnetic resonance (MR) images is essential in many neuroimaging studies, such as predicting the development or aging of human brain, and quantifying the stages of underlying pathology for Alzheimer's disease (AD). Compared to brain tissues, some Regions of Interest (ROIs) related with AD can reflect more noticeable changes. Therefore, we focus on ROIs instead of whole brain to measure the local longitudinal changes. However, it is a challenging task to compute consistent longitudinal segmentation results for serial brain MR images, due to the inevitable differences in contrast, orientation, noises in images scanned at different time points, where the subtle brain changes are hided behind. In addition, manual annotation is time-consuming and highly subjective for longitudinal data. Specifically, the subjective difference of manual annotations for a same subject at different time points may be greater than actual brain tissue changes. Numerous methods have been proposed for automated segmentation of brain MR images. There are mainly 3 categories, i.e., 1) multi-atlas registration-based methods [8,11,12], 2) machine learning based methods with hand-crafted features [1,3,10], and 3) deep learning based methods with automatically learned features [2,7,9,14,15]. In fact, most of current methods perform segmentation on each time-point image independently, which introduces inconsistent segmentation and thus impair the accuracy of longitudinal quantification. Moreover, it is a formidable and unrealistic task to train a learning based model for longitudinal segmentation from supervised learning. To this end, it is highly desired of a segmentation method with specifically considering brain change patterns and providing consistent quantification.

In this study, we propose a coarse-to-fine deep learning model based on supervised and semi-supervised learning to provide consistent longitudinal segmentation for brain MR images. The method has 3 steps: 1) We adopt supervised learning to train an initial segmentation model based on a dataset of labeled cross-sectional images; 2) We introduce a semi-supervised learning approach on unlabeled longitudinal images to make the segmentation model capable of providing longitudinal consistent segmentation; 3) We present three constraints at the stage of semi-supervised learning, i.e., segmentation consistency, smoothness and age constraints. Here we select four ROIs that are highly related with aging, such as left and right hippocampi (related to memory and brain disorders [4]), lateral ventricle (closely related to changes in cognitive performance [16]), and entorhinal cortex (a network hub for memory, navigation and the perception of time) in this study.

The main contributions of this work are three-fold: (1) We propose a novel method with supervised learning by using labeled cross-sectional images, as well as semi-supervised learning by using unlabeled longitudinal images; (2) We design spatial and temporal constraints in the loss function to ensure more consistent, smoother and meaningful segmentation results with time; (3) We propose to use learned age-specific longitudinal changes to specifically guide longitudinal segmentation, thus measuring accurate longitudinal changes for different subject groups.

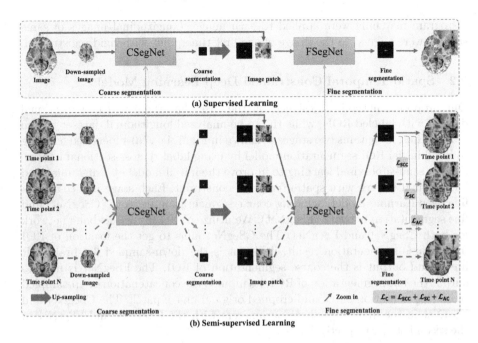

Fig. 1. Framework of our proposed method consisting of two stages (by using right hippocampus segmentation as example), i.e., (a) supervised learning for labeled cross-sectional images and (b) semi-supervised learning for unlabeled longitudinal images. Each stage contains two networks, i.e., coarse segmentation network (CSegNet) and fine segmentation network (FSegNet). $\mathcal{L}_C$ is the total loss with spatio-temporal constraint, and $\mathcal{L}_{SCC}$, $\mathcal{L}_{SC}$, and $\mathcal{L}_{AC}$ represent segmentation consistency, smoothness and age constraint, respectively.

2 Method

2.1 Dataset

There are two datasets used in this study. First, we have 257 in-house images with manual annotations of ROIs from trained raters. This dataset is used for supervised learning of initial deep learning models for ROI segmentation. Second, we use 53 sets of longitudinal T1-weighted MR images with age from 55 to 70 obtained from the Alzheimer's Disease Neuroimaging Initiative (ADNI) database [5]. Each subject has 2 to 11 time-points, with mean of 4.5 and std of 1.9. These images have no ROI annotations and thus serve as unlabeled data for semi-supervised learning. Out of the 53 subjects, there are 8 subjects with Alzheimer's disease (AD), 15 subjects with mild cognitive impairment (MCI), 4 normal controls (NC), 16 subjects with MCI while converted into AD (MCI2AD), 7 subjects with MCI while converted into NC (MCI2NC), 3 subjects with NC while converted into MCI (NC2MCI) using either 1.5T or 3.0T scanners. Each image has a dimension of $256 \times 256 \times 256$ and a voxel size of $1.0 \times 1.0 \times 1.0$ mm^3. Intensity inhomogeneity correction, skull stripping, and histogram matching were applied to each image. Longitudinal images of same subject were aligned to the image of the first time point with rigid registration.

2.2   Spatio-Temporal Constrained Deep Learning Model

Our proposed method is designed on the case that there is a cross-sectional dataset with labeled ROIs, while the to-be-analyzed longitudinal dataset has no labels. Thus, it contains two stages as shown in Fig. 1, i.e., (a) supervised learning to build initial ROI segmentation model by using labeled cross-sectional images, and (b) semi-supervised learning to improve the initial model by using unlabeled longitudinal images with spatio-temporal constraint. Each stage is a coarse-to-fine deep learning model including coarse segmentation network (CSegNet) and fine segmentation network (FSegNet). We utilize V-Net [6] as backbone network for both CSegNet and FSegNet. The CSegNet aims to get the location of ROI from coarse segmentation result, its input is the down-sampled 3D brain MR image and output is the coarse segmentation of ROI. The FSegNet is used to output the fine segmentation of ROI, its input is the concatenation of up-sampled coarse segmentation patch and cropped original image patch. The CSegNet and FSetNet in the first stage share parameters with the CSegNet and FSetNet in the second stage, respectively.

2.3   Initial ROI Segmentation with Supervised Learning

At the stage of supervised learning, we use labeled cross-sectional images to train an initial ROI segmentation model. Since the ROIs we selected only occupy a small proportion of voxels in a brain MR image, large unrelated regions will make the training ineffective. To overcome this, we apply a coarse-to-fine strategy to generate more accurate segmentation results. First, we use the down-sampled image I_{down} of the original brain MR image I as input of CSegNet, at a factor of 2. Since the down-sampling operation results in a smaller area of ROI, the down-sampled ground truth is dilated for 2 voxels. After getting the coarse segmentation result S^{coarse} and 3D bounding box containing the ROI from CSegNet, we apply the up-sampled 3D bounding box to crop the original image I to obtain the image patch I_{P}. Next, we concatenate the up-sampled coarse segmentation result $S^{\mathrm{coarse}}_{\mathrm{up}}$ and the image patch I_{P} as input of FSegNet, and obtain the fine segmentation I^{seg}. We denote the combination of CSegNet and FSegNet as CFSegNet.

2.4   Consistent Segmentation with Semi-supervised Learning

At the stage of semi-supervised learning, we aim to improve the longitudinal consistency of the segmentation model. Let I_t denote the brain MR image at time point t. The CFSegNet in this stage is the same as that in the supervised learning stage, by sharing weights. Considering the relation between the images from same subject at different time points, we input image series $I_t, t \in T = \{t_1, t_2, \cdots, t_N\}$

to the CFSegNet, and obtain segmentation results I_t^{seg}, $t \in T = \{t_1, t_2, \cdots, t_N\}$. To obtain significant segmentation results for longitudinal images, we introduce three constraints as the loss terms.

Segmentation Consistency Constraint (SCC). The segmentation consistency constraint is used to ensure similarity among segmentation results at different time points. The loss function for segmentation consistency constraint consists of Dice loss term $\mathcal{L}_{\text{Dice}}$, binary cross entropy (BCE) loss term $\mathcal{L}_{\text{BCE}}$ and high-level feature consistency (FC) loss term $\mathcal{L}_{\text{FC}}$, defined as follows:

$$\mathcal{L}_{\text{SCC}} = \lambda_{\text{Dice}}\mathcal{L}_{\text{Dice}} + \lambda_{\text{BCE}}\mathcal{L}_{\text{BCE}} + \lambda_{\text{FC}}\mathcal{L}_{\text{FC}}, \tag{1}$$

where λ_{Dice}, λ_{BCE}, and λ_{FC} are the weighting parameters for three loss terms, respectively. $\mathcal{L}_{\text{Dice}}$ and $\mathcal{L}_{\text{BCE}}$ constrain the similarity among segmentation results of ROI, and $\mathcal{L}_{\text{FC}}$ constrains the similarity among high-level features.

The Dice loss term represents the average of Dice between segmentation results of any two images from image series:

$$\mathcal{L}_{\text{Dice}} = mean \sum_{i=1}^{N} \sum_{j=1(j\neq i)}^{N} Dice\left(I_{t_j}^{\text{seg}}, I_{t_i}^{\text{seg}}\right). \tag{2}$$

Similarly, the binary cross entropy loss term represents the average of binary cross entropy between segmentation results of any two images from image series:

$$\mathcal{L}_{\text{BCE}} = mean \sum_{i=1}^{N} \sum_{j=1(j\neq i)}^{N} BCE\left(I_{t_j}^{\text{seg}}, I_{t_i}^{\text{seg}}\right). \tag{3}$$

Specifically, the high-level feature consistency loss term represents the average of mean square error between high-level features of any two images from image series:

$$\mathcal{L}_{\text{FC}} = mean \sum_{i=1}^{N} \sum_{j=1(j\neq i)}^{N} \|F_{t_j} - F_{t_i}\|_2, \tag{4}$$

where F_{t_i} is the high-level feature of the image at time point t_i.

However, it is unlikely to align the images precisely at different time points from the same subject due to the change of ROI with time. Therefore, it is not feasible to directly apply the above segmentation consistency constraint to the segmentation results of these images. To solve the above issue, we add a block matching operation to find the most similar image block from two image patches I_P while calculating the SCC loss. Given two image patches at different time points from a same subject, we set the image patch at the first time point as a reference image patch $I_{P_{\text{ref}}}$ (with its corresponding segmentation $I_{P_{\text{ref}}}^{\text{seg}}$), and set the other image as a target image patch $I_{P_{\text{tar}}}$ (with its corresponding segmentation $I_{P_{\text{tar}}}^{\text{seg}}$). We use a sliding window of size $5 \times 5 \times 5$ with a stride of $5 \times 5 \times 5$ to traverse the location in $I_{P_{\text{ref}}}$. For each image block $I_{P_{\text{ref}}}^{b}, b \in B = \{b_1, b_2, \cdots, b_M\}$ from $I_{P_{\text{ref}}}$, where M is the number of extracted image blocks from $I_{P_{\text{ref}}}$, we search in the eight-neighbor of the center of the image block $I_{P_{\text{ref}}}^{b}$ and find the most similar image block $I_{P_{\text{tar}}}^{*}$ from $I_{P_{\text{tar}}}$. In this study, Pearson correlation coefficient is applied to calculate the similarity between two image

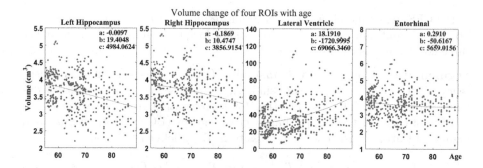

Fig. 2. The functions of volume change curve with age on four ROIs.

blocks. The corresponding segmentation results of these similar image blocks from the two image patches are used to calculate the SCC loss.

Smoothness Constraint (SC). The smoothness constraint is used to impose the smoothness of volume change among segmentation results at different time points. The loss function for smoothness constraint is defined as below:

$$\mathcal{L}_{\mathrm{SC}} = mean \sum\nolimits_{i=1}^{N-2} \sum\nolimits_{j=2(j>i)}^{N-1} \sum\nolimits_{k=3(k>j)}^{N} \|V_{t_j} - (V_{t_i} + V_{t_k})/2\|_2^2, \quad (5)$$

where V_{t_i} is the volume of ROI in the image at time point t_i. When the subject only contains images with two time points, this SC loss term would be ignored.

Age Constraint (AC). The age constraint is used to impose the volume change curve of each ROI, statistically calculated from totally 431 images of ADNI including 53 longitudinal subjects and 209 individual subjects, to the segmentation model. The loss function of age constraint is defined as:

$$\mathcal{L}_{\mathrm{AC}} = 1/N \sum\nolimits_{i=1}^{N} \|V_{A_{t_i}} - \Phi_{ROI}(A_{t_i})\|_2^2, \quad (6)$$

where A_{t_i} is the age of the subject at time point t_i, and $\Phi_{ROI}(A_{t_i})$ represents the function of volume change curve with age on each ROI, as shown in Fig. 2. We define the function is $\Phi_{ROI}(A_{t_i}) = aA_{t_i}^2 + bA_{t_i} + c$. The hyper-parameters are obtained by using second-order polynomial curve fitting shown in Fig. 2.

Loss Function. Thus, the total loss with spatial and temporal constraints for CFSegNet at the stage of semi-supervised learning is the weighted sum of the above three loss terms, shown as follows:

$$\mathcal{L}_{\mathrm{C}} = \lambda_{\mathrm{SCC}} \mathcal{L}_{\mathrm{SCC}} + \lambda_{\mathrm{SC}} \mathcal{L}_{\mathrm{SC}} + \lambda_{\mathrm{AC}} \mathcal{L}_{\mathrm{AC}}, \quad (7)$$

where λ_{SCC}, λ_{SC}, and λ_{AC} are the weighting parameters for three constraint loss terms, respectively.

Table 1. Mean ± standard deviation of DSC (%) on four ROIs.

Method	Left hippocampus	Right hippocampus	Lateral ventricle	Entorhinal
V-Net	83.86 ± 4.05	83.22 ± 5.12	81.59 ± 5.18	77.88 ± 6.65
CSegNet	83.15 ± 3.48	82.83 ± 4.96	82.09 ± 5.01	77.79 ± 4.86
CFSegNet	89.08 ± 2.54	88.57 ± 1.14	87.80 ± 2.37	85.91 ± 3.47
Proposed method	**89.53 ± 2.09**	**89.59 ± 1.11**	**88.05 ± 2.42**	**86.05 ± 2.96**

Table 2. Mean ± standard deviation of TC (%) on four ROIs.

Method	Left hippocampus	Right hippocampus	Lateral ventricle	Entorhinal
V-Net	80.59 ± 5.89	80.35 ± 4.34	82.61 ± 4.50	78.81 ± 4.47
CFSegNet	79.77 ± 3.49	78.10 ± 3.17	82.83 ± 3.55	78.98 ± 5.46
CFSegNet+SCC	92.25 ± 2.76	91.67 ± 2.87	93.92 ± 1.81	91.60 ± 2.40
CFSegNet+SCC+SC	93.02 ± 2.08	92.68 ± 1.91	94.37 ± 1.35	92.05 ± 2.63
Proposed method	**93.43 ± 1.84**	**93.53 ± 2.26**	**95.21 ± 1.46**	**93.01 ± 1.69**

2.5 Implement Details

We trained all networks on a platform with a NVIDIA TITAN Xp GPU (32 GB). For supervised learning, we adopted the Adam optimizer and set batch size to 64 (8 for lateral ventricle), maximum epoch number to 1,000, and initial learning rate to 0.0001, which was divided by 10 after 100 iterations. For semi-supervised learning, we adopted the Adam optimizer and set an adaptive batch size, maximum epoch number to 100, and initial learning rate to 0.0001, which was divided by 10 after 50 iterations. Since the main aim of this study is to provide the longitudinal consistent segmentation, we randomly sampled 231 training and 26 testing from the labeled dataset three times for cross-validation in supervised learning stage. The unlabeled longitudinal images were split for 5-fold cross-validation in semi-supervised learning stage. For other hyper-parameters, we set the loss-weights λ_{SCC}, λ_{SC}, and λ_{AC} for three loss terms in Eq. (7) to 1, 0.5, and 0.1, and the weighting parameters λ_{Dice}, λ_{BCE}, and λ_{FC} in Eq. (1) to 1, 1, and 0.1. We split 20% of the training data as the validation data. These parameters were empirically determined using cross-validation on the validation data. The sizes of bounding box to crop the image are set to 64, 128, and 64 for right and left hippocampi, lateral ventricle, and entorhinal, respectively.

3   Experiments and Results

Segmentation of Labeled Cross-sectional Brain MR Images. We employ Dice Similarity Coefficient (DSC) to evaluate segmentation performance on labeled cross-sectional images. The mean and standard deviation of DSC on four ROIs were presented in Table 1. It shows that the DSC values of CFSegNet was improved, and the unlabeled longitudinal images also contributed to the improvement of segmentation accuracy.

Segmentation of Unlabeled Longitudinal Brain MR Images. To quantitatively evaluate the temporal consistency of the segmentation results on unlabeled longitudinal images, we use the Temporal Consistency (TC) $TC = 1/S(\Omega') \sum_{i \in \Omega'} (1 - L_i/(N-1))$ [13], where L_i is the number of label changes of corresponding voxels across time, Ω' is the voxel set of ROI, $S(\Omega')$ is the number of voxels in Ω', and N is the number of time points. A larger TC indicates a better temporal consistency. Small values indicate relatively less temporal consistent segmentation. Note that consistency value 1 can be ideally obtained only if the brain has not changed at all, which is not the case in these older adults. The mean and standard deviation of TC on four ROIs were shown in Table 2.

For visual comparisons, Fig. 3 shows typical segmentation results of our proposed method and V-Net on four ROIs respectively. For comparative purposes, the images shown are aligned images using rigid transformations. The two images on the right column illustrate the number of label changes of corresponding voxels projected onto the first image space, where red indicates many label changes and black means no changes across time. By comparing the segmentation results, we can see that our proposed method yields longitudinally consistent segmentation results while adapting to the longitudinal changes of anatomical structures.

Finally, it is worth noting that, although our proposed method incorporates temporal smoothness constraint loss, it still maintains longitudinal change information. For example, Fig. 4 shows volume changes of four ROIs on five subjects from different group. We can see from the Fig. 4 that the curves obtained from

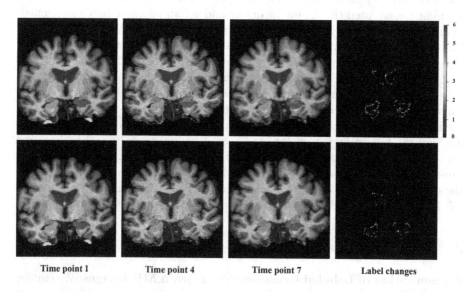

Time point 1   Time point 4   Time point 7   Label changes

Fig. 3. Comparison of typical segmentation results of V-Net (top row) and our proposed method (bottom row) on four ROIs, i.e., left hippocampus (brown), right hippocampus (green), lateral ventricle (blue), and entorhinal (pink). The two images in the last column show the number of label changes L_i. (Color figure online)

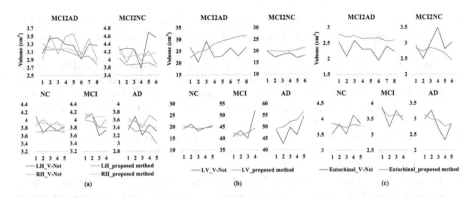

Fig. 4. Volume changes of four ROIs on different subjects from different group. (a) Left hippocampus (LH) and right hippocampus (RH), (b) lateral ventricle (LV) and (c) entorhinal.

our proposed method are quite smooth. Moreover, the left and right hippocampal and entorhinal volumes, calculated from the results of our proposed method on different groups, show steady and different decreases with time. While the lateral ventricular volumes, calculated from the results of our proposed method on different groups, show steady and different increases with time.

4   Conclusion

In this study, we proposed a coarse-to-fine spatio-temporal constrained deep learning model based on supervised and semi-supervised learning to provide consistent longitudinal segmentation for brain MR images. We presented three constraints at the stage of semi-supervised learning, including segmentation consistency constraint, smoothness constraint, and age constraint. The experimental results demonstrated the best precision and robustness of our proposed method than the comparison methods in terms of DSC and TC.

Acknowledgment. This work was supported in part by the National Natural Science Foundation of China under Grants 61771397, and in part by the CAAI-Huawei MindSpore Open Fund under Grants CAAIXSJLJJ-2020-005B and China Postdoctoral Science Foundation under Grants BX2021333

References

1. Chen, A., Yan, H.: An improved fuzzy c-means clustering for brain MR images segmentation. J. Med. Imaging Health Inf. **11**(2), 386–390 (2021)
2. Coupé, P., Mansencal, B., Clément, M., Giraud, R., Manjon, J.V.: AssemblyNet: a large ensemble of CNNs for 3D whole brain MRI segmentation. NeuroImage **219**, 117026 (2020)

3. Dadar, M., Collins, D.L.: BISON: Brain tissue segmentation pipeline using T1-weighted magnetic resonance images and a random forest classifier. Magn. Reson. Med. **85**(4), 1881–1894 (2020)
4. Frisoni, G.B., Fox, N.C., Jack, C.R., Scheltens, P., Thompson, P.M.: The clinical use of structural MRI in Alzheimer disease. Nat. Rev. Neurol. **6**(2), 66–67 (2010)
5. Jack, C.R., Bernstein, M.A., Fox, N.C., Thompson, P., Weiner, M.W.: The Alzheimer's disease neuroimaging initiative (ADNI): MRI methods. J. Magn. Reson. Imaging **27**(4), 685–691 (2010)
6. Milletari, F., Navab, N., Ahmadi, S.: V-Net: fully convolutional neural networks for volumetric medical image segmentation. In: 2016 Fourth International Conference on 3D Vision (3DV), pp. 565–571 (2016)
7. Novosad, P., Fonov, V., Collins, D.L., Initiative†, A.D.N.: Accurate and robust segmentation of neuroanatomy in T1-weighted MRI by combining spatial priors with deep convolutional neural networks. Hum. Brain Mapp. **41**(2), 309–327 (2020)
8. Sun, L., Shao, W., Wang, M., Zhang, D., Liu, M.: High-order feature learning for multi-atlas based label fusion: application to brain segmentation with MRI. IEEE Trans. Image Process. **29**, 2702–2713 (2020)
9. Sun, L., Shao, W., Zhang, D., Liu, M.: Anatomical attention guided deep networks for ROI segmentation of brain MR images. IEEE Trans. Med. Imaging **39**(6), 2000–2012 (2020)
10. Van Leemput, K., Maes, F., Vandermeulen, D., Suetens, P.: Automated model-based tissue classification of MR images of the brain. IEEE Trans. Med. Imaging **18**(10), 897–908 (1999)
11. Wang, H., Suh, J.W., Das, S.R., Pluta, J.B., Craige, C., Yushkevich, P.A.: Multi-atlas segmentation with joint label fusion. IEEE Trans. Pattern Anal. Mach. Intell. **35**(3), 611–623 (2013)
12. Wu, J., Tang, X.: Brain segmentation based on multi-atlas and diffeomorphism guided 3D fully convolutional network ensembles. Pattern Recognit. **115**, 107904 (2021)
13. Xue, Z., Shen, D., Davatzikos, C.: CLASSIC: consistent longitudinal alignment and segmentation for serial image computing. Neuroimage **30**(2), 388–399 (2005)
14. Yu, Q., et al.: C2FNAS: coarse-to-fine neural architecture search for 3D medical image segmentation. In: 2020 IEEE/CVF Conference on Computer Vision and Pattern Recognition (CVPR), pp. 4125–4134 (2020)
15. Zhai, J., Li, H.: An improved full convolutional network combined with conditional random fields for brain MR image segmentation algorithm and its 3D visualization analysis. J. Med. Syst. **43**(9), 1–10 (2019). https://doi.org/10.1007/s10916-019-1424-0
16. Zhang, W., et al.: Morphometric analysis of hippocampus and lateral ventricle reveals regional difference between cognitively stable and declining persons. In: 2016 IEEE 13th International Symposium on Biomedical Imaging (ISBI), pp. 14–18 (2016)

A Multi-branch Hybrid Transformer Network for Corneal Endothelial Cell Segmentation

Yinglin Zhang[1,6], Risa Higashita[1,5(✉)], Huazhu Fu[7], Yanwu Xu[8],
Yang Zhang[1], Haofeng Liu[1], Jian Zhang[6], and Jiang Liu[1,2,3,4(✉)]

[1] Department of Computer Science and Engineering, Southern University of Science
and Technology, Shenzhen 518055, China
liuj@sustech.edu.cn
[2] Cixi Institute of Biomedical Engineering, Chinese Academy of Sciences,
Ningbo, China
[3] Guangdong Provincial Key Laboratory of Brain-inspired Intelligent Computation,
Department of Computer Science and Engineering, Southern University of Science
and Technology, Shenzhen 518055, China
[4] Research Institute of Trustworthy Autonomous Systems,
Southern University of Science and Technology, Shenzhen 518055, China
[5] Tomey Corporation, Nagoya 451-0051, Japan
k-chen@tomey.co.jp
[6] Global Big Data Technologies Centre, University of Technology Sydney,
Ultimo, NSW, Australia
[7] Inception Institute of Artificial Intelligence, Abu Dhabi, UAE
[8] Intelligent Healthcare Unit, Baidu, Beijing 100085, China

Abstract. Corneal endothelial cell segmentation plays a vital role in quantifying clinical indicators such as cell density, coefficient of variation, and hexagonality. However, the corneal endothelium's uneven reflection and the subject's tremor and movement cause blurred cell edges in the image, which is difficult to segment, and need more details and context information to release this problem. Due to the limited receptive field of local convolution and continuous downsampling, the existing deep learning segmentation methods cannot make full use of global context and miss many details. This paper proposes a Multi-Branch hybrid Transformer Network (MBT-Net) based on the transformer and body-edge branch. Firstly, we use the convolutional block to focus on local texture feature extraction and establish long-range dependencies over space, channel, and layer by the transformer and residual connection. Besides, we use the body-edge branch to promote local consistency and to provide edge position information. On the self-collected dataset TM-EM3000 and public Alisarine dataset, compared with other State-Of-The-Art (SOTA) methods, the proposed method achieves an improvement.

Keywords: Corneal endothelial cell segmentation · Deep learning · Transformer · Multi-branch

© Springer Nature Switzerland AG 2021
M. de Bruijne et al. (Eds.): MICCAI 2021, LNCS 12901, pp. 99–108, 2021.
https://doi.org/10.1007/978-3-030-87193-2_10

1 Introduction

Corneal endothelial cell abnormalities may be related to many corneal and systemic diseases. Quantifying corneal endothelial cell density, the coefficient of variation, and hexagonality have essential clinical significance [2]. Cell segmentation is a crucial step to quantify the above parameters. Nevertheless, manual segmentation is time-consuming, laborious, and unstable. Therefore, an accurate and fully automatic corneal endothelial cell segmentation method is essential to improve diagnosis efficiency and accuracy.

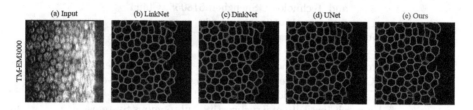

Fig. 1. Segmentation results on TM-EM3000.(a) is the input equalized corneal endothelium cell image. (b), (c), (d), (e) are the segmentation results of LinkNet, DinkNet, UNet, and our method. The red line represents the prediction result, and the green line represents ground truth, orange when the two overlap. (Color figure online)

The main challenge of accurate segmentation is the blurred cell edges, which are difficult to segment, as shown in Fig. 1, and needs more details and context information to release this problem. UNet [12] captures contextual semantic information through the contracting path and combines high-resolution features in the contracted path with upsampled output to achieve precise localization. UNet++ [18] optimizes it through a series of nested, dense skip connections to reduce the semantic gap between the encoder and decoder's feature maps. Fabijánska [7] first applied UNet to the task of corneal endothelial cell segmentation. Vigueras-Guillén et al. [15] applied the complete convolution method based on UNet and the sliding window version to the analysis of cell images obtained by SP-1P Topcon corneal endothelial microscope. Fu et al. [8] proposed a multi-conetxt deep network, by combining prior knowledge of regions of interest and clinical parameters. However, due to the limited receptive field of local convolution and continuous downsampling, they cannot make full use of the global context and still miss many details.

The transformer has been proved to be an effective method for establishing long-range dependencies. Vaswani et al. [14] proposed a transformer structure system for language translation tasks through a complete attention mechanism to establish the global dependence of input and output among time, space, and levels. Prajit et al. [11] explored the use of the transformer mechanism on visual classification tasks, replacing all spatial convolutional layers in ResNet with stand-alone self-attention layers. However, local self-attention will still lose part of the global information. Wang et al. [16] establish a stand-alone attention layer by using two decomposed axial attention blocks, to reduce the number of parameters and calculations, and allow performing attention in a larger or even global range.

Some previous works obtain better segmentation results by taking full advantage of edge information. Chen et al. [1] proposed the deep contour-aware network, using a multi-task learning framework to study the complementary information of gland objects and contours, which improves the discriminative capability of intermediate features. Chen et al. [5] improved the network output by learning the reference edge map of CNN intermediate features. Ding et al. [6] proposed to use boundary as an additional semantic category to introduce boundary layout constraints and promote intra-class consistency through the boundary feature propagation module based on unidirectional acyclic graphs.

We need to preserve more local details and make full use of the global context. In this paper, we propose a Multi-Branch hybrid Transformer Network(MBT-Net). At first, we apply a hybrid residual transformer feature extraction module to give full play to the advantages of convolution block and transformer block in terms of local details and global semantics. Specifically, we use the convolutional block to focus on local texture feature extraction and establish long-range dependencies over space, channel, and layer by the transformer and residual connection. Besides, we define the corneal endothelial cell's segmentation task more entirely from the perspective of edge and body. Body-edge branches provide precise edge location information and promote local consistency. The experimental results show that the proposed method is superior to other state-of-the-art methods and has achieved better performance on two corneal endothelial datasets.

2   Method

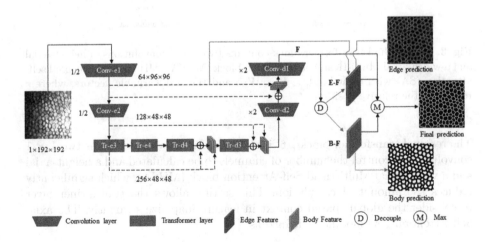

Fig. 2. The pipeline of multi-branch hybrid transformer network. Conv-e1, Conv-e2, Conv-d1 and Conv-d2 represent encoder and decoder layer based on convolution block. Tr-e3, Tr-e4, Tr-d3, and Tr-d4 are based on transformer blocks.

In this paper, we propose the MBT-Net, as shown in Fig. 2. Firstly, the feature F of equalized corneal endothelium cell image is extracted by the hybrid residual transformer encoder-decoder module. Each convolution layer contains two basic residual blocks with a kernel size = 3 × 3. Each transformer layer contains two residual transformer blocks with kernel size = 1 × 48. Then, the feature is decoupled into two parts, body and edge. Also, the edge texture information from Conv-e1 is fused into the edge feature. Finally, we take the maximum response of edge feature E-F, body feature B-F, and feature F to obtain the fused feature to predict the final segmentation result. The training process of these three branches is explicitly supervised.

In this pipeline, the convolutional layer focuses on local texture feature extraction, which retains more details. The residual connection and transformer make full use of the feature map's global context information in a more extensive range of space, different channels, and layers. The edge perspective helps preserve boundary details, and the body perspective promotes local consistency. The low-resolution feature map of d_{x+1} is refined by features from e_x by concatenating and addition operation.

2.1   Residual Transformer Block

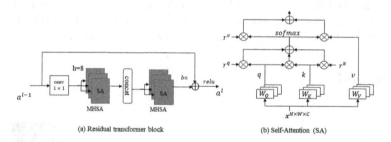

(a) Residual transformer block   (b) Self-Attention (SA)

Fig. 3. The residual transformer block contains two 1 × 1 convolutions, a height-axial and a width-axial Multi-Head Self-Attention block (MHSA). MHSA compute axial self-attention (SA) with eight head. r, W_Q, W_K, and W_V are learnable vectors, where r related to the relative position

The residual transformer block [16] is shown in Fig. 3, which contains two 1 × 1 convolution to control the number of channels to be calculated and a height-axial and a width-axial Multi-Head Self-Attention block (MHSA), which significantly reduces the amount of calculation. This setting allows the transformer layer to consider the global spatial context in feature map size straightly. The axial Self-Attention(SA) module is defined as:

$$y_o = \sum_{p \in N_{1 \times m}(o)} softmax_p(q_o^T k_p + q_o^T r_{p-o}^q + k_p^T r_{p-o}^k)(v_p + r_{p-o}^v) \qquad (1)$$

For a given input feature map x, queries $q = W_Q x$, keys $k = W_K x$, values $v = W_V x$ are linear projections of feature map x, where W_Q, W_K, W_V are learnable parameters. $r^q_{(p-o)}, r^k_{(p-o)}, r^v_{(p-o)}$ measure the compatibility from position p to o in query, key and value. They are also learnable paramters. The $softmax_p$ denotes a softmax function applied to all possible p positions. $N_{1 \times m}(o)$ represents the local $1 \times m$ square region centered around location o, y_o is the output at position o.

$$a^{l_2} = a^{l_1} + \sum_{i=l_1}^{l_2-1} f(a^i) \tag{2}$$

Besides, all the block used in encoder-decoder is in residual form, which can propagate input signal directly from any low layer to the high layer, optimizing information interaction [9,10]. Taking any two layers $l_2 > l_1$ into consideration, the forward information propagation process is formulated as Eq. (2).

2.2   The Body, Edge, and Final Branches

The information from the body and edge perspectives is combined to better define corneal endothelial cell segmentation. The body branch provides general shape and overall consistency information to promote local consistency, while the edge branch provides edge localization information to improve the segmentation accuracy of image details.

We decouple the feature F extracted by the hybrid residual transformer encoder-decoder module into $F_{body} = \phi(F)$ and $F_{edge} = F - F_{body}$, where ϕ is implemented by convolution layer. Also, the low level information from encoder Conv-e1 is fused into the edge feature, $F_{edge} = F_{edge} + \psi(F_{e1})$, where ψ is dimension operation. Finally, the above three feature maps are fused into $F_{final} = \varphi(F, F_{edge}, F_{body})$ for final segmentation prediction, where φ represents the maximum response.

The training process of these three branches is explicitly supervised. The three masks used in training are shown in Fig. 4. The final prediction mask is

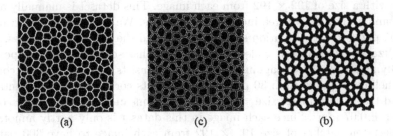

    (a)   (c)   (b)

Fig. 4. Three kinds of masks on TM-EM3000. (a) The final prediction mask from the annotation of an expert, (b) The edge prediction mask extracted from (a) through the canny operator, (c) The body prediction mask by relaxing the edge of invert image of (a) with a Gaussian kernel.

the ground truth annotated by experts. The edge prediction mask is extracted from the final prediction mask by the canny operator. The body prediction mask is obtained by inverting the final prediction and then performing Gaussian blurring at the edges.

2.3   Loss Function

$$Loss = \lambda_1 L_b(\hat{y}_b, y_b) + \lambda_2 L_e(\hat{y}_e, y_e) + \lambda_3 L_f(\hat{y}_f, y_f) \tag{3}$$

In this paper, we jointly optimize the body, edge, and final losses, as shown in Eq. (3), where $\lambda_1, \lambda_2, \lambda_3$ are hyper parameters to adjust the weight of three different losses. As the final prediction is the output we finally use to compare with ground truth, we give it a higher weight than edge and body branch. In our experiment, we set $\lambda_1 = 0.5, \lambda_2 = 0.5, \lambda_3 = 1.2$. y_b, y_e, y_f represent the ground truth of body, edge and final prediction respectively, and $\hat{y}_b, \hat{y}_e, \hat{y}_f$ are corresponding prediction from model. The binary cross entropy loss is used, as shown in Eq. (4).

$$L = \frac{1}{N} \sum_i [y_i \ln \hat{y}_i + (1 - y_i) \ln(1 - \hat{y}_i)] \tag{4}$$

Where N represents the total number of pixels, y_i denotes target label for pixel i, $\hat{y}_i$ is the predicted probability.

3   Experiments

3.1   Datasets and implementation Details

TM-EM3000 contains 184 images of corneal endothelium cell and its corresponding segmentation ground truth, with size = 266× 480, collected by specular microscope EM3000, Tomey, Japan. To reduce the interference of lesions and artifacts and build a data set with almost the same imaging quality, we select a patch with a size of 192 × 192 from each image. This dataset is manually annotated and reviewed by three independent experts. We split it into the training set 155 patches, the validation set 10 patches, and the test set 19 patches.

Alizarine Dataset is collected by inverse phase-contrast microscope (CK 40, Olympus) at 200× magnification [13]. It consists of 30 images of corneal endothelium acquired from 30 porcine eyes and its corresponding segmentation ground truth, with image size = 768 × 576, and mean area assessed per cornea = 0.54 ± 0.07 mm². Since each image in this dataset is only partly annotated, we select ten patches of size 192 × 192 from each image to have 300 patches in total. And then split it into the training set 260 patches, validation set 40 patches.The training set and validation set do not overlap.

Implementation Details. We use the RMSprop optimization strategy during model training. The initial learning rate is 2e−4, epochs = 100, batch size = 1.

The learning rate optimization strategy is ReduceLROnPlateau, and the network input size is 192 × 192. All the models are trained and tested with PyTorch on the platform of NVIDIA GeForce TITAN XP.

3.2   Comparison with SOTA methods

We compare performance of the proposed method with LinkNet [3], DinkNet [17], UNet [12], UNet++ [18] and TransUNet [4] on **TM-EM3000** and **Alisarine** dataset. We use dice coefficient(DICE), F1 score(F1), sensitivity(SE), and specificity(SP) as evaluation indicators, where DICE and F1 are most important because they are related to the overall performance.

Table 1. Quantitative evaluation of different methods. The proposed method achieves the best performance.

Model	TM-EM3000				Alisarine			
	DICE	F1	SE	SP	DICE	F1	SE	SP
LinkNet34 [3]	0.711	0.712	0.719	0.941	0.766	0.801	0.805	0.956
DinkNet34 [17]	0.717	0.718	0.724	0.944	0.767	0.805	0.821	0.953
UNet [12]	0.730	0.743	0.763	0.945	0.775	0.811	0.814	**0.960**
UNet++ [18]	0.728	0.739	**0.775**	0.938	0.773	0.811	0.850	0.947
TransUNet [4]	0.734	0.742	0.769	0.941	0.783	0.821	0.866	0.948
Proposed	**0.747**	**0.747**	0.768	**0.946**	**0.786**	**0.821**	**0.877**	0.944

As shown in Table 1, The proposed method has obtained the best overall performance on both TM-EM3000 and Alisarine data sets. On TM-EM3000, the DICE accuracy and F1 score of our approach are 0.747 and 0.747. On the Alisarine data set, the Dice accuracy and F1 score of our method are 0.786 and 0.821. UNet++ [18] is modified from UNet, through a series of nested, dense skip connections to capture more semantic information. However, in general, there is no noticeable improvement observed in this experiment. TransUNet [4] optimized the UNet by using the transformer layer to capture the global context in the encoder part, but our method has achieved better performance. It is mainly due to the following advantages. 1) Long-range dependencies are established through the transformer in both the encoder and decoder. 2) Performing transformer layer on the whole feature map, further reducing the loss of semantic information. 3) The body-edge branch encourages the network to learn more general features and provide edge localization information.

As shown in Fig. 5, on the left side of the TM-EM3000 image, the cell boundary is clear. The segmentation performance of different methods is not much different. Nevertheless, on the right side with uneven illumination and the blur cell boundary, the proposed method achieves better segmentation results, closer to the ground truth, and in line with the real situation.

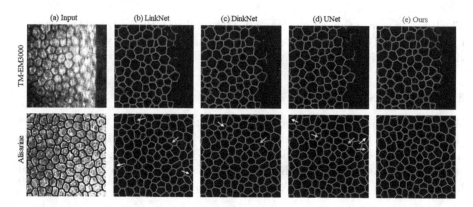

Fig. 5. Qualitative results on TM-EM3000 and the Alisarine Dataset.(a) is the input equalized corneal endothelium cell image. The red line represents the prediction result, and the green line represents ground truth, orange when the two overlap. The white arrow indicates the missing segmentation location, and the yellow arrow indicates the location where the segmentation result does not fit well with the ground truth (Color figure online)

There is no extensive range of fuzzy area in the Alisarine image, and all methods obtained satisfied segmentation results. However, the baselines have varying degrees of loss in details, which lead to discontinuous cell edge segmentation as the white arrow indicated, and the segmentation results do not match the ground truth well as the yellow arrow indicated. The proposed method obtains better segmentation accuracy.

3.3   Ablation Study

The ablation experiment is conducted to explore the transformer's replacement design, as shown in Table 2. In this process, we gradually replace the encoder-decoder structure's convolution layer with the transformer from inside to outside. In the beginning, the model captures more semantic information, and the

Table 2. Ablation study on the replacement design of transformer. 0-0-TR means no transformer is used. 1-1-TR means e4, d4 is transformer layer. 2-2-TR means e3, e4, and d3, d4 is transformer layer. 3-3-TR means e2, e3, e4, and d2, d3, d4 is transformer layer. 4-4-TR means complete transformer structure.

Model	TM-EM3000				Alisarine			
	DICE	F1	SE	SP	DICE	F1	SE	SP
0-0-TR	0.731	0.737	0.774	0.937	0.776	0.813	0.852	0.948
1-1-TR	0.737	0.746	**0.778**	0.941	0.778	0.816	0.857	**0.948**
2-2-TR	**0.747**	**0.747**	0.768	**0.946**	**0.786**	**0.821**	**0.877**	0.944
3-3-TR	0.702	0.714	0.742	0.935	0.777	0.812	0.874	0.940
4-4-TR	0.687	0.707	0.717	0.940	0.769	0.802	0.869	0.936

performance is improved. Then, with the transformer replacing the shallow convolutional layer further, the model starts to lose local information, resulting in the decline of performance. Model 2-2-TR achieves the best balance between local details and global context.

Table 3. Ablation study on transformer and body-edge branch on TM-EM3000.TR means transformer, and B-E means body-edge branch. When transformer is used, it means 2-2-TR.

TR	B-E	Dice	F1	SE	SP
✗	✗	0.720	0.733	0.746	0.945
✗	✓	0.731	0.737	0.774	0.937
✓	✗	0.736	0.741	**0.786**	0.936
✓	✓	**0.747**	**0.747**	0.768	**0.946**

We also study the influence of transformer and body-edge branch on performance on TM-EM3000 dataset, as shown in Table 3. When neither transformer nor body-edge branches are used, the DICE accuracy and F1 score on TM-EM3000 and Alisarine are 0.720 and 0.733, respectively. After adding the body-edge branch, the performance is improved to 0.731 and 0.737. When the transformer is used, the DICE accuracy and F1 score are 0.736 and 0.741. Using both the body-edge branch and transformer, we improve the performance by 2.7% and 1.4% in total to 0.747 and 0.747.

4   Conclusion

This paper proposes a multi-branch hybrid transformer network for corneal endothelial cell segmentation, which combines the convolution and transformer block's advantage and uses the body-edge branch to promote local consistency and provide edge localization information. Our method achieves superior performance to various state-of-the-art methods, especially in the fuzzy region. The ablation study shows that both the well-designed transformer replacement and body-edge branches contribute to improved performance.

References

1. C.H., Qi, X., Yu, L., Heng, P.A.: Dcan: deep contour-aware networks for accurate gland segmentation. In: CVPR, pp. 2487–2496 (2016)
2. Al-Fahdawi, S., et al.: A fully automated cell segmentation and morphometric parameter system for quantifying corneal endothelial cell morphology. Comput. Methods Programs Biomed. **160**, 11–23 (2018)
3. Chaurasia, A., Culurciello, E.A.: Linknet: exploiting encoder representations for efficient semantic segmentation. In: VCIP, pp. 1–4 (2017)

4. Chen, J., et al.: Transunet: Transformers make strong encoders for medical image segmentation (2021)
5. Chen, L.C., Barron, J.T., Papandreou, G., Murphy, K., Yuille, A.L.: Semantic image segmentation with task-specific edge detection using CNNs and a discriminatively trained domain transform. In: Computer Vision and Pattern Recognition (2016)
6. Ding, H., Jiang, X., Liu, Q.A., Magnenat-Thalmann, N., Wang, G.: Boundary-aware feature propagation for scene segmentation. In: ICCV, pp. 6819–6829 (2019)
7. Fabijanska, A.: Segmentation of corneal endothelium images using a u-net-based convolutional neural network. Artif. Intell. Med. **88**, 1–13 (2018)
8. Fu, H., et al.: Multi-context deep network for angle-closure glaucoma screening in anterior segment OCT. In: Frangi, A.F., Schnabel, J.A., Davatzikos, C., Alberola-López, C., Fichtinger, G. (eds.) MICCAI 2018. LNCS, vol. 11071, pp. 356–363. Springer, Cham (2018). https://doi.org/10.1007/978-3-030-00934-2_40
9. He, K., Zhang, X., Ren, S., Sun, J.: Deep residual learning for image recognition. In: CVPR, pp. 770–778 (2016)
10. He, K., Zhang, X., Ren, S., Sun, J.: Identity mappings in deep residual networks. In: Leibe, B., Matas, J., Sebe, N., Welling, M. (eds.) ECCV 2016. LNCS, vol. 9908, pp. 630–645. Springer, Cham (2016). https://doi.org/10.1007/978-3-319-46493-0_38
11. Ramachandran, P., Parmar, N., Vaswani, A., Bello, I., Levskaya, A., Shlens, J.: Stand-alone self-attention in vision models. In: NIPS (2019)
12. Ronneberger, O., Fischer, P., Brox, T.: U-net: convolutional networks for biomedical image segmentation. In: MICCAI (2015)
13. Ruggeri, A., Scarpa, F., Luca, D.M., Meltendorf, C., Schroeter, J.: A system for the automatic estimation of morphometric parameters of corneal endothelium in alizarine red-stained images. Br. J. Ophthalmol. **94**, 643–647 (2010)
14. Vaswani, A., et al.: Attention is all you need. In: NIPS, pp. 5998–6008 (2017)
15. Vigueras-Guillén, J.P., et al.: Fully convolutional architecture vs sliding-window CNN for corneal endothelium cell segmentation. BMC Biomed. Eng. **1**(1), 1–16 (2019)
16. Wang, H., Zhu, Y., Green, B., Adam, H., Yuille, A., Chen, L.-C.: Axial-DeepLab: stand-alone axial-attention for panoptic segmentation. In: Vedaldi, A., Bischof, H., Brox, T., Frahm, J.-M. (eds.) ECCV 2020. LNCS, vol. 12349, pp. 108–126. Springer, Cham (2020). https://doi.org/10.1007/978-3-030-58548-8_7
17. Zhou, L., Zhang, C., Wu, M.: D-linknet: linknet with pretrained encoder and dilated convolution for high resolution satellite imagery road extraction. In: CVPR Workshops, pp. 182–186 (2018)
18. Zhou, Z., Siddiquee, M.R.M., Tajbakhsh, N., Liang, J.: Unet++: a nested u-net architecture for medical image segmentation. In: DLMIA/ML-CDS@MICCAI, pp. 3–11 (2018)

TransBTS: Multimodal Brain Tumor Segmentation Using Transformer

Wenxuan Wang[1], Chen Chen[2], Meng Ding[3], Hong Yu[1], Sen Zha[1], and Jiangyun Li[1(✉)]

[1] School of Automation and Electrical Engineering, University of Science and Technology, Beijing, China
{s20200579,g20198754,g20198675}@xs.ustb.edu.cn, leejy@ustb.edu.cn
[2] Center for Research in Computer Vision, University of Central Florida, Orlando, USA
chen.chen@crcv.ucf.edu
[3] Scoop Medical, Houston, TX, USA
meng.ding@okstate.edu

Abstract. Transformer, which can benefit from global (long-range) information modeling using self-attention mechanisms, has been successful in natural language processing and 2D image classification recently. However, both local and global features are crucial for dense prediction tasks, especially for 3D medical image segmentation. In this paper, we for the first time exploit Transformer in 3D CNN for MRI Brain Tumor Segmentation and propose a novel network named TransBTS based on the encoder-decoder structure. To capture the local 3D context information, the encoder first utilizes 3D CNN to extract the volumetric spatial feature maps. Meanwhile, the feature maps are reformed elaborately for tokens that are fed into Transformer for global feature modeling. The decoder leverages the features embedded by Transformer and performs progressive upsampling to predict the detailed segmentation map. Extensive experimental results on both BraTS 2019 and 2020 datasets show that TransBTS achieves comparable or higher results than previous state-of-the-art 3D methods for brain tumor segmentation on 3D MRI scans. The source code is available at https://github.com/Wenxuan-1119/TransBTS.

Keywords: Segmentation · Brain tumor · MRI · Transformer · 3D CNN

1 Introduction

Gliomas are the most common malignant brain tumors with different levels of aggressiveness. Automated and accurate segmentation of these malignancies on magnetic resonance imaging (MRI) is of vital importance for clinical diagnosis.

Electronic supplementary material The online version of this chapter (https://doi.org/10.1007/978-3-030-87193-2_11) contains supplementary material, which is available to authorized users.

© Springer Nature Switzerland AG 2021
M. de Bruijne et al. (Eds.): MICCAI 2021, LNCS 12901, pp. 109–119, 2021.
https://doi.org/10.1007/978-3-030-87193-2_11

Convolutional Neural Networks (CNN) have achieved great success in various vision tasks such as classification, segmentation and object detection. Fully Convolutional Networks (FCN) [10] realize end-to-end semantic segmentation for the first time with impressive results. U-Net [15] uses a symmetric encoder-decoder structure with skip-connections to improve detail retention, becoming the mainstream architecture for medical image segmentation. Many U-Net variants such as U-Net++ [24] and Res-UNet [23] further improve the performance for image segmentation. Although CNN-based methods have excellent representation ability, it is difficult to build an explicit **long-distance** dependence due to limited receptive fields of convolution kernels. This limitation of convolution operation raises challenges to learn global semantic information which is critical for dense prediction tasks like segmentation.

Inspired by the attention mechanism [1] in natural language processing, existing research overcomes this limitation by fusing the attention mechanism with CNN models. Non-local neural networks [21] design a plug-and-play non-local operator based on the self-attention mechanism, which can capture the long-distance dependence in the feature map but suffers from the high memory and computation cost. Schlemper et al. [16] propose an attention gate model, which can be integrated into standard CNN models with minimal computational overhead while increasing the model sensitivity and prediction accuracy. On the other hand, Transformer [19] is designed to model long-range dependencies in sequence-to-sequence tasks and capture the relations between arbitrary positions in the sequence. This architecture is proposed based *solely on self-attention*, dispensing with convolutions entirely. Unlike previous CNN-based methods, Transformer is not only powerful in modeling global context, but also can achieve excellent results on downstream tasks in the case of large-scale pre-training.

Recently, Transformer-based frameworks have also reached state-of-the-art performance on various computer vision tasks. Vision Transformer (ViT) [7] splits the image into patches and models the correlation between these patches as sequences with Transformer, achieving satisfactory results on image classification. DeiT [17] further introduces a knowledge distillation method for training Transformer. DETR [4] treats object detection as a set prediction task with the help of Transformer. TransUNet [5] is a concurrent work which employs ViT for medical image segmentation. We will elaborate the differences between our approach and TransUNet in Sec. 2.3.

Research Motivation. The success of Transformer has been witnessed mostly on image classification. For dense prediction tasks such as segmentation, both local and global (or long-range) information is important. However, as pointed out by [22], local structures are ignored when directly splitting images into patches as tokens for Transformer. Moreover, for medical volumetric data (e.g. 3D MRI scans) which is **beyond 2D**, local feature modeling among continuous slices (i.e. depth dimension) is also critical for volumetric segmentation. We are therefore inspired to ask: *How to design a neural network that can effectively model local and global features in spatial and depth dimensions of volumetric data by leveraging the highly expressive Transformer?*

In this paper, we present the *first attempt* to exploit **Trans**former in 3D CNN for 3D MRI **B**rain **T**umor **S**egmentation (TransBTS). The proposed TransBTS builds upon the encoder-decoder structure. The network encoder first utilizes 3D CNN to extract the volumetric spatial features and downsample the input 3D images at the same time, resulting in compact volumetric feature maps that effectively captures the local 3D context information. Then each volume is reshaped into a vector (i.e. token) and fed into Transformer for global feature modeling. The 3D CNN decoder takes the feature embedding from Transformer and performs progressive upsampling to predict the full resolution segmentation map. Experiments on BraTS 2019 and 2020 datasets show that TransBTS achieves comparable or higher results than previous state-of-the-art 3D methods for brain tumor segmentation on 3D MRI scans. We also conduct comprehensive ablation study to shed light on architecture engineering of incorporating Transformer in 3D CNN to unleash the power of both architectures.

2   Method

2.1   Overall Architecture of TransBTS

An overview of the proposed TransBTS is presented in Fig. 1. Given an input MRI scan $X \in \mathbb{R}^{C \times H \times W \times D}$ with a spatial resolution of $H \times W$, depth dimension of D (# of slices) and C channels (# of modalities), we first utilize 3D CNN to generate compact feature maps capturing spatial and depth information, and then leverage the Transformer encoder to model the long-distance dependency in a global space. After that, we repeatedly stack the upsampling and convolutional layers to gradually produce a high-resolution segmentation result.

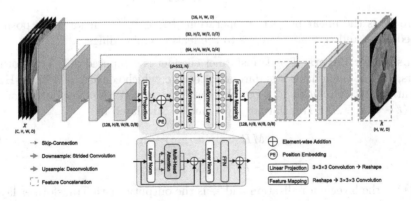

Fig. 1. Overall architecture of the proposed TransBTS.

2.2   Network Encoder

As the computational complexity of Transformer is quadratic with respect to the number of tokens (i.e. sequence length), directly flattening the input image to a sequence as the Transformer input is impractical. Therefore, ViT [7] splits an image into fixed-size (16×16) patches and then reshapes each patch into a token, reducing the sequence length to 16^2. For 3D volumetric data, the straightforward tokenization, following ViT, would be splitting the data into 3D patches. However, this simple strategy makes Transformer unable to model the image *local context information across spatial and depth dimensions* for volumetric segmentation. To address this challenge, our solution is to stack the $3 \times 3 \times 3$ convolution blocks with downsamping (strided convolution with stride=2) to gradually encode input images into low-resolution/high-level feature representation $F \in \mathbb{R}^{K \times \frac{H}{8} \times \frac{W}{8} \times \frac{D}{8}}$ ($K = 128$), which is 1/8 of input dimensions of H, W and D (overall stride (OS)=8). In this way, rich local 3D context features are effectively embedded in F. Then, F is fed into the Transformer encoder to further learn long-range correlations with a global receptive field.

Feature Embedding of Transformer Encoder. Given the feature map F, to ensure a comprehensive representation of each volume, a linear projection (a $3 \times 3 \times 3$ convolutional layer) is used to increase the channel dimension from $K = 128$ to $d = 512$. The Transformer layer expects a sequence as input. Therefore, we collapse the spatial and depth dimensions into one dimension, resulting in a $d \times N$ ($N = \frac{H}{8} \times \frac{W}{8} \times \frac{D}{8}$) feature map f, which can be also regarded as N d-dimensional tokens. To encode the location information which is vital in segmentation task, we introduce the learnable position embeddings and fuse them with the feature map f by direct addition, creating the feature embeddings as follows:

$$z_0 = f + PE = W \times F + PE \tag{1}$$

where W is the linear projection operation, $PE \in \mathbb{R}^{d \times N}$ denotes the position embeddings, and $z_0 \in \mathbb{R}^{d \times N}$ refers to the feature embeddings.

Transformer Layers. The Transformer encoder is composed of L Transformer layers, each of them has a standard architecture, which consists of a Multi-Head Attention (MHA) block and a Feed Forward Network (FFN). The output of the ℓ-th ($\ell \in [1, 2, ..., L]$) Transformer layer can be calculated by:

$$z_\ell' = MHA(LN(z_{\ell-1})) + z_{\ell-1} \tag{2}$$

$$z_\ell = FFN(LN(z_\ell')) + z_\ell' \tag{3}$$

$LN(*)$ is the layer normalization and z_ℓ is the output of ℓ-th Transformer layer.

2.3   Network Decoder

In order to generate the segmentation results in the original 3D image space ($H \times W \times D$), we introduce a 3D CNN decoder to perform feature upsampling and pixel-level segmentation (see the right part of Fig. 1).

Feature Mapping. To fit the input dimension of 3D CNN decoder, we first design a feature mapping module to project the sequence data back to a standard 4D feature map. Specifically, the output sequence of Transformer $z_L \in \mathbb{R}^{d \times N}$ is first reshaped to $d \times \frac{H}{8} \times \frac{W}{8} \times \frac{D}{8}$. In order to reduce the computational complexity of decoder, a convolution block is employed to reduce the channel dimension from d to K. Through these operations, the feature map $Z \in \mathbb{R}^{K \times \frac{H}{8} \times \frac{W}{8} \times \frac{D}{8}}$, which has the same dimension as F in the feature encoding part, is obtained.

Progressive Feature Upsampling. After the feature mapping, cascaded upsampling operations and convolution blocks are applied to Z to gradually recover a full resolution segmentation result $R \in \mathbb{R}^{H \times W \times D}$. Moreover, skip-connections are employed to fuse the encoder features with the decoder counterparts by concatenation for finer segmentation masks with richer spatial details.

Discussion. A recent work TransUNet [5] also employs Transformer for medical image segmentation. We highlight a few key distinctions between our TransBTS and TransUNet. (1) TransUNet is a 2D network that processes each 3D medical image in a **slice-by-slice** manner. However, our TransBTS is based on 3D CNN and processes all the image slices at once, allowing the exploitation of better representations of continuous information between slices. In other words, TransUNet only focuses on the spatial correlation between tokenized image patches, but TransBTS can model the long-range dependencies in both slice/depth dimension and spatial dimension simultaneously for volumetric segmentation. (2) As TransUNet adopts the ViT structure, it relies on pre-trained ViT models on large-scale image datasets. In contrast, TransBTS has a flexible network design and is trained from scratch on task-specific dataset without the dependence on pre-trained weights.

3   Experiments

Data and Evaluation Metric. The first 3D MRI dataset used in the experiments is provided by the Brain Tumor Segmentation (BraTS) 2019 challenge [2,3,11]. It contains 335 cases of patients for training and 125 cases for validation. Each sample is composed of four modalities of brain MRI scans. Each modality has a volume of $240 \times 240 \times 155$ which has been aligned into the same space. The labels contain 4 classes: background (label 0), necrotic and non-enhancing tumor (label 1), peritumoral edema (label 2) and GD-enhancing tumor (label 4). The segmentation accuracy is measured by the Dice score and the Hausdorff distance (95%) metrics for enhancing tumor region (ET, label 1), regions of the tumor core (TC, labels 1 and 4), and the whole tumor region (WT, labels 1,2 and 4). The second 3D MRI dataset is provided by the Brain Tumor Segmentation Challenge (BraTS) 2020 [2,3,11]. It consists of 369 cases for training, 125 cases for validation and 166 cases for testing. Except for the number of samples in the dataset, the other information about these two datasets are the same.

Implementation Details. The proposed TransBTS is implemented in Pytorch and trained with 8 NVIDIA Titan RTX GPUs (each has 24 GB memory) for

8000 epochs from scratch using a batch size of 16. We adopt the Adam optimizer to train the model. The initial learning rate is set to 0.0004 with a poly learning rate strategy, in which the initial rate decays by each iteration with power 0.9. The following data augmentation techniques are applied: (1) random cropping the data from $240 \times 240 \times 155$ to $128 \times 128 \times 128$ voxels; (2) random mirror flipping across the axial, coronal and sagittal planes by a probability of 0.5; (3) random intensity shift between $[-0.1, 0.1]$ and scale between $[0.9, 1.1]$. The softmax Dice loss is employed to train the network and $L2$ Norm is also applied for model regularization with a weight decay rate of 10^{-5}. In the testing phase, we utilize Test Time Augmentation (TTA) to further improve the performance of our proposed TransBTS.

Table 1. Comparison on BraTS 2019 validation set.

Method	Dice Score (%) ↑			Hausdorff Dist. (mm) ↓		
	ET	WT	TC	ET	WT	TC
3D U-Net [6]	70.86	87.38	72.48	5.062	9.432	8.719
V-Net [12]	73.89	88.73	76.56	6.131	6.256	8.705
KiU-Net [18]	73.21	87.60	73.92	6.323	8.942	9.893
Attention U-Net [14]	75.96	88.81	77.20	5.202	7.756	8.258
Wang et al. [20]	73.70	89.40	80.70	5.994	5.677	7.357
Li et al. [9]	77.10	88.60	81.30	6.033	6.232	7.409
Frey et al. [8]	78.7	89.6	80.0	6.005	8.171	8.241
Myronenko et al. [13]	**80.0**	89.4	**83.4**	3.921	5.89	6.562
TransBTS w/o TTA	78.36	88.89	81.41	5.908	7.599	7.584
TransBTS w/ TTA	78.93	**90.00**	81.94	**3.736**	**5.644**	**6.049**

3.1   Main Results

BraTS 2019. We first conduct five-fold cross-validation evaluation on the training set – a conventional setting followed by many existing works. Our TransBTS achieves average Dice scores of 78.69%, 90.98%, 82.85% respectively for ET, WT and TC. We also conduct experiments on the BraTS 2019 **validation** set and compare TransBTS with state-of-the-art (SOTA) 3D approaches. The quantitative results are presented in Table 1. TransBTS achieves the Dice scores of 78.93%, 90.00%, 81.94% on ET, WT, TC, respectively, which are comparable or higher results than previous SOTA 3D methods presented in Table 1. In terms of Hausdorff distance metric, a considerable improvement has also been achieved for segmentation. Compared with 3D U-Net [6], TransBTS shows great superiority in both metrics with significant improvements. This clearly reveals the benefit of leveraging Transformer for modeling the global relationships. For **qualitative**

Table 2. Comparison on BraTS 2020 validation set.

Method	Dice Score (%) ↑			Hausdorff Dist. (mm) ↓		
	ET	WT	TC	ET	WT	TC
3D U-Net [6]	68.76	84.11	79.06	50.983	13.366	13.607
Basic V-Net [12]	61.79	84.63	75.26	47.702	20.407	12.175
Deeper V-Net [12]	68.97	86.11	77.90	43.518	14.499	16.153
Residual 3D U-Net	71.63	82.46	76.47	37.422	12.337	13.105
TransBTS w/o TTA	78.50	89.00	81.36	**16.716**	6.469	10.468
TransBTS w/ TTA	**78.73**	**90.09**	**81.73**	17.947	**4.964**	**9.769**

analysis, we also show a visual comparison of the brain tumor segmentation results of various methods including 3D U-Net [6], V-Net [12], Attention U-Net [14] and our TransBTS in Fig. 2. Since the ground truth for the validation set is not available, we conduct five-fold cross-validation evaluation on the training set for all methods. It is evident from Fig. 2 that TransBTS can describe brain tumors more accurately and generate much better segmentation masks by modeling long-range dependencies between each volume.

BraTS 2020. We also evaluate TransBTS on BraTS 2020 validation set and the results are reported in Table 2. We adopt the hyperparameters on BraTS19 for model training, TransBTS achieves Dice scores of 78.73%, 90.09%, 81.73% and HD of 17.947mm, 4.964mm, 9.769mm on ET, WT, TC. Compared with 3D U-Net [6], V-Net [12] and Residual 3D U-Net, TransBTS shows great superiority in both metrics with significant improvements. This clearly reveals the benefit of leveraging Transformer for modeling the global relationships.

3.2   Model Complexity

TransBTS has 32.99 M parameters and 333G FLOPs which is a moderate size model. Besides, by reducing the number of stacked Transformer layers from 4 to 1 and halving the hidden dimension of the FFN, we reach a lightweight Trans-BTS which only has 15.14 M parameters and 208G FLOPs while achieving Dice scores of 78.94%, 90.36%, 81.76% and HD of 4.552 mm, 6.004 mm, 6.173 mm on ET, WT, TC on BraTS2019 validation set. In other words, by reducing the layers in Transformer as a simple and straightforward way to reduce complexity (54.11% reduction in parameters and 37.54% reduction in FLOPs of our lightweight TransBTS), the performance only drops marginally. Compared with 3D U-Net [6] which has 16.21 M parameters and 1670G FLOPs, our lightweight TransBTS shows great superiority in terms of model complexity. Note that efficient Transformer variants can be used in our framework to replace the vanilla Transformer to further reduce the memory and computation complexity while maintaining the accuracy. But this is beyond the scope of this work.

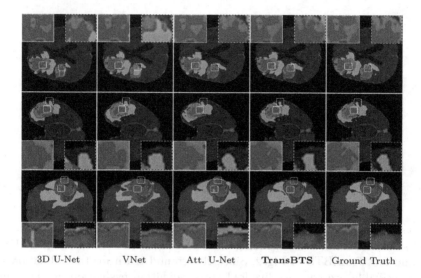

3D U-Net   VNet   Att. U-Net   **TransBTS**   Ground Truth

Fig. 2. The visual comparison of MRI brain tumor segmentation results.

3.3 Ablation Study

We conduct extensive ablation experiments to verify the effectiveness of Trans-BTS and justify the rationale of its design choices based on five-fold cross-validation evaluations on the BraTS 2019 training set. (1) We investigate the impact of the sequence length (N) of tokens for Transformer, which is controlled by the overall stride (OS) of 3D CNN in the network encoder. (2) We explore Transformer at various model scales (i.e. depth (L) and embedding dimension (d)). (3) We also analyze the impact of different positions of skip-connections.

Sequence Length N. Table 3 presents the ablation study of various sequence lengths for Transformer. The first row (OS = 16) and the second row (OS = 8) both reshape each volume of the feature map to a feature vector after down-sampling. It is noticeable that increasing the length of tokens, by adjusting the OS from 16 to 8, leads to a significant improvement on performance. Specifically, 1.66% and 2.41% have been attained for the Dice score of ET and WT respectively. Due to the memory constraint, after setting the OS to 4, we can not directly reshape each volume to a feature vector. So we make a slight modification to keep the sequence length to 4096, which is unfolding each $2 \times 2 \times 2$ patch into a feature vector before passing to the Transformer. We find that although the OS drops from 8 to 4, without the essential increase of sequence length, the performance does not improve or even gets worse.

Transformer Scale. Two hyper-parameters, the feature embedding dimension (d) and the number of Transformer layers (depth L), mainly determines the scale of Transformer. We conduct ablation study to verify the impact of Transformer scale on the segmentation performance. For efficiency, we only train each model configuration for 1000 epochs. As shown in Table 4, the network with $d = 512$

Table 3. Ablation study on sequence length (N).

OS	Sequence Length(N)	Dice Score(%)		
		ET	WT	TC
16	512	73.30	87.59	**81.36**
8	4096	**74.96**	**90.00**	79.96
4	4096	74.86	87.10	77.46

Table 4. Ablation study on transformer.

Depth (L)	Embedding dim (d)	Dice score(%)		
		ET	WT	TC
4	384	68.95	83.31	66.89
4	512	**73.72**	**88.02**	73.14
4	768	69.38	83.54	**74.16**
1	512	70.11	85.84	70.95
8	512	66.48	79.16	67.22

Table 5. Ablation study on the positions of skip-connections (SC).

Number of SC	Position of SC	Dice score(%)		
		ET	WT	TC
3	Transformer layer	74.96	90.00	79.96
3	3D Conv (Fig. 1)	**78.92**	**90.23**	**81.19**

and $L = 4$ achieves the best scores of ET and WT. Increasing the embedding dimension (d) may not necessarily lead to improved performance ($L = 4$, d: 512 vs. 768) yet brings extra computational cost. We also observe that $L = 4$ is a "sweet spot" for the Transformer in terms of performance and complexity.

Positions of Skip-connections (SC). To improve the representation ability of the model, we further investigate the positions for skip-connections (orange dash lines "- →" in Fig. 1). The ablation results are listed in Table 5. If skip-connections are attached to the first three Transformer layers, it is more alike to feature aggregation from adjacent layers without the compensation for loss of spatial details. Following the traditional design of skip-connections from U-Net (i.e. attach to the 3D Conv layers as shown in Fig. 1), considerable gains (3.96% and 1.23%) have been achieved for the important ET and TC, thanks to the recovery of low-level spatial detail information.

4 Conclusion

We present a novel segmentation framework that effectively incorporates Transformer in 3D CNN for multimodal brain tumor segmentation in MRI. The resulting architecture, TransBTS, not only inherits the advantage of 3D CNN for modeling local context information, but also leverages Transformer on learning global semantic correlations. Experimental results on two datasets (BraTS 2019 and 2020) validate the effectiveness of the proposed TransBTS. In future work, we will explore computational and memory efficient attention mechanisms in Transformer to develop efficiency-focused models for volumetric segmentation.

References

1. Bahdanau, D., Cho, K., Bengio, Y.: Neural machine translation by jointly learning to align and translate. arXiv preprint arXiv:1409.0473 (2014)
2. Bakas, S., et al.: Advancing the cancer genome atlas glioma MRI collections with expert segmentation labels and radiomic features. Sci. Data **4**, 170117 (2017)
3. Bakas, S., et al.: Identifying the best machine learning algorithms for brain tumor segmentation, progression assessment, and overall survival prediction in the brats challenge. arXiv preprint arXiv:1811.02629 (2018)
4. Carion, N., Massa, F., Synnaeve, G., Usunier, N., Kirillov, A., Zagoruyko, S.: End-to-end object detection with transformers. In: Vedaldi, A., Bischof, H., Brox, T., Frahm, J.-M. (eds.) ECCV 2020. LNCS, vol. 12346, pp. 213–229. Springer, Cham (2020). https://doi.org/10.1007/978-3-030-58452-8_13
5. Chen, J., et al.: Transunet: Transformers make strong encoders for medical image segmentation. arXiv preprint arXiv:2102.04306 (2021)
6. Çiçek, Ö., Abdulkadir, A., Lienkamp, S.S., Brox, T., Ronneberger, O.: 3D U-net: learning dense volumetric segmentation from sparse annotation. In: Ourselin, S., Joskowicz, L., Sabuncu, M.R., Unal, G., Wells, W. (eds.) MICCAI 2016. LNCS, vol. 9901, pp. 424–432. Springer, Cham (2016). https://doi.org/10.1007/978-3-319-46723-8_49
7. Dosovitskiy, A., et al.: An image is worth 16x16 words: Transformers for image recognition at scale. arXiv preprint arXiv:2010.11929 (2020)
8. Frey, M., Nau, M.: Memory efficient brain tumor segmentation using an autoencoder-regularized u-net. In: Crimi, A., Bakas, S. (eds.) BrainLes 2019. LNCS, vol. 11992, pp. 388–396. Springer, Cham (2020). https://doi.org/10.1007/978-3-030-46640-4_37
9. Li, X., Luo, G., Wang, K.: Multi-step cascaded networks for brain tumor segmentation. In: Crimi, A., Bakas, S. (eds.) BrainLes 2019. LNCS, vol. 11992, pp. 163–173. Springer, Cham (2020). https://doi.org/10.1007/978-3-030-46640-4_16
10. Long, J., Shelhamer, E., Darrell, T.: Fully convolutional networks for semantic segmentation. In: Proceedings of the IEEE Conference on Computer Vision and Pattern Recognition, pp. 3431–3440 (2015)
11. Menze, B.H., et al.: The multimodal brain tumor image segmentation benchmark (brats). IEEE Trans. Med. Imaging **34**(10), 1993–2024 (2014)
12. Milletari, F., Navab, N., Ahmadi, S.A.: V-net: Fully convolutional neural networks for volumetric medical image segmentation. In: 2016 Fourth International Conference on 3D Vision (3DV), pp. 565–571. IEEE (2016)
13. Myronenko, A., Hatamizadeh, A.: Robust semantic segmentation of brain tumor regions from 3D MRIs. In: Crimi, A., Bakas, S. (eds.) BrainLes 2019. LNCS, vol. 11993, pp. 82–89. Springer, Cham (2020). https://doi.org/10.1007/978-3-030-46643-5_8
14. Oktay, O., et al.: Attention u-net: Learning where to look for the pancreas. arXiv preprint arXiv:1804.03999 (2018)
15. Ronneberger, O., Fischer, P., Brox, T.: U-Net: convolutional networks for biomedical image segmentation. In: Navab, N., Hornegger, J., Wells, W.M., Frangi, A.F. (eds.) MICCAI 2015. LNCS, vol. 9351, pp. 234–241. Springer, Cham (2015). https://doi.org/10.1007/978-3-319-24574-4_28
16. Schlemper, J., et al.: Attention gated networks: learning to leverage salient regions in medical images. Med. Image Anal. **53**, 197–207 (2019)

17. Touvron, H., Cord, M., Douze, M., Massa, F., Sablayrolles, A., Jégou, H.: Training data-efficient image transformers & distillation through attention. arXiv preprint arXiv:2012.12877 (2020)
18. Valanarasu, J.M.J., Sindagi, V.A., Hacihaliloglu, I., Patel, V.M.: Kiu-net: Over-complete convolutional architectures for biomedical image and volumetric segmentation. arXiv preprint arXiv:2010.01663 (2020)
19. Vaswani, A., et al.: Attention is all you need. In: Advances in Neural Information Processing Systems, pp. 5998–6008 (2017)
20. Wang, F., Jiang, R., Zheng, L., Meng, C., Biswal, B.: 3D U-net based brain tumor segmentation and survival days prediction. In: Crimi, A., Bakas, S. (eds.) BrainLes 2019. LNCS, vol. 11992, pp. 131–141. Springer, Cham (2020). https://doi.org/10.1007/978-3-030-46640-4_13
21. Wang, X., Girshick, R., Gupta, A., He, K.: Non-local neural networks. In: Proceedings of the IEEE Conference on Computer Vision and Pattern Recognition, pp. 7794–7803 (2018)
22. Yuan, L., et al.: Tokens-to-token vit: Training vision transformers from scratch on imagenet. arXiv preprint arXiv:2101.11986 (2021)
23. Zhang, Z., Liu, Q., Wang, Y.: Road extraction by deep residual u-net. IEEE Geosci. Remote Sens. Lett. 15(5), 749–753 (2018)
24. Zhou, Z., Rahman Siddiquee, M.M., Tajbakhsh, N., Liang, J.: UNet++: a nested u-net architecture for medical image segmentation. In: Stoyanov, D., et al. (eds.) DLMIA/ML-CDS -2018. LNCS, vol. 11045, pp. 3–11. Springer, Cham (2018). https://doi.org/10.1007/978-3-030-00889-5_1

Automatic Polyp Segmentation via Multi-scale Subtraction Network

Xiaoqi Zhao[1], Lihe Zhang[1(⊠)], and Huchuan Lu[1,2]

[1] Dalian University of Technology, Dalian, China
[2] Peng Cheng Laboratory, Shenzhen, China
zxq@mail.dlut.edu.cn, {zhanglihe,lhchuan}@dlut.edu.cn

Abstract. More than 90% of colorectal cancer is gradually transformed from colorectal polyps. In clinical practice, precise polyp segmentation provides important information in the early detection of colorectal cancer. Therefore, automatic polyp segmentation techniques are of great importance for both patients and doctors. Most existing methods are based on U-shape structure and use element-wise addition or concatenation to fuse different level features progressively in decoder. However, both the two operations easily generate plenty of redundant information, which will weaken the complementarity between different level features, resulting in inaccurate localization and blurred edges of polyps. To address this challenge, we propose a multi-scale subtraction network (MSNet) to segment polyp from colonoscopy image. Specifically, we first design a subtraction unit (SU) to produce the difference features between adjacent levels in encoder. Then, we pyramidally equip the SUs at different levels with varying receptive fields, thereby obtaining rich multi-scale difference information. In addition, we build a training-free network "LossNet" to comprehensively supervise the polyp-aware features from bottom layer to top layer, which drives the MSNet to capture the detailed and structural cues simultaneously. Extensive experiments on five benchmark datasets demonstrate that our MSNet performs favorably against most state-of-the-art methods under different evaluation metrics. Furthermore, MSNet runs at a real-time speed of ∼70fps when processing a 352×352 image. The source code will be publicly available at https://github.com/Xiaoqi-Zhao-DLUT/MSNet.

Keywords: Colorectal cancer · Automatic polyp segmentation · Subtraction · LossNet.

1 Introduction

According to GLOBOCAN 2020 data, colorectal cancer is the third most common cancer worldwide and the second most common cause of death. It usually begins as small, noncancerous (benign) clumps of cells called polyps that form on the inside of the colon. Over time some of these polyps can become colon cancers. Therefore, the best way of preventing colon cancer is to identify and

© Springer Nature Switzerland AG 2021
M. de Bruijne et al. (Eds.): MICCAI 2021, LNCS 12901, pp. 120–130, 2021.
https://doi.org/10.1007/978-3-030-87193-2_12

remove polyps before they turn into cancer. At present, colonoscopy is the most commonly used means of examination, but this process involves manual and expensive labor, not to mention its high misdiagnosis rate. Hence, automatic and accurate polyp segmentation is of great practical significance.

The automatic polyp segmentation has gradually evolved from the traditional methods [23] based on manually designed features to the deep learning methods [13, 26, 27]. Although these methods have made progress in clinical, they are limited by box-level prediction results, thus failing to capture the shape and contour of polyps. To address this issue, Brandao et al. [3] utilize the FCN [17] to segment polyps by a pixel-level prediction. Akbari et al. [1] also use FCN-based segmentation network and combine the patch selection mechanism to improve the accuracy of polyp segmentation. However, FCN-based methods rely on low-resolution features to generate the final prediction, resulting in rough segmentation results and fuzzy boundaries.

In recent years, U-shape structures [14, 16, 20] have received considerable attention due to their abilities of utilizing multi-level information to reconstruct high-resolution feature maps. Many polyp segmentation networks [7, 8, 20, 28] adopt the U-shape architecture. In UNet [20], the up-sampled feature maps are concatenated with feature maps skipped from the encoder and convolutions and non-linearities are added between up-sampling steps. UNet++ [28] uses nested and dense skip connections to reduce the semantic gap between the feature maps of encoder and decoder. Later, ResUNet++ [12] combines many advanced techniques such as residual computation [9], squeeze and excitation [10], atrous spatial pyramidal pooling [4], and attention mechanism to further improve performance. Recent works, SFA [8] and PraNet [7], focus on recovering the sharp boundary between a polyp and its surrounding mucosa. The former proposes a selective feature aggregation structure and a boundary-sensitive loss function under a shared encoder and two mutually constrained decoders. The latter utilizes reverse attention module to establish the relationship between region and boundary cues.

Generally speaking, different level features in encoder have different characteristics. High-level ones have more semantic information which helps localize the objects, while low-level ones have more detailed information which can capture the subtle boundaries of objects. The decoder leverages the level-specific and cross-level characteristics to generate the final high-resolution prediction. Nevertheless, the aforementioned methods directly use an element-wise addition or concatenation to fuse any two level features from the encoder and transmit them to the decoder. These simple operations do not pay more attention to differential information between different levels. This drawback not only generates redundant information to dilute the really useful features but also weakens the characteristics of level-specific features, which results in that the network can not balance accurate polyp localization and subtle boundary refinement.

In this paper, we propose a novel multi-scale subtraction network (MSNet) for the polyp segmentation task. We first design a subtraction unit (SU) and apply it to each pair of adjacent level features. To address the scale diversity

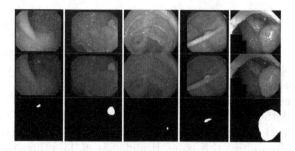

Fig. 1. Visualization of RGB color slices, our prediction and gold standard.

of polyps, we pyramidally concatenate multiple SUs to capture the large-span cross-level information. Then, we aggregate level-specific features and multi-path cross-level differential features and then generate the final prediction in decoder. Moreover, we propose a LossNet to automatically supervise the extracted feature maps from bottom layer to top layer, which can optimize the segmentation from detail to structure with a simple L2-loss function. Our main contributions can be summarized as follows:

- We propose a novel multi-scale subtraction network for automatic polyp segmentation. With multi-level and multi-stage cascaded subtraction operations, the complementary information from lower order to higher order among different levels can be effectively obtained, thereby comprehensively enhancing the perception of polyp areas.
- We build a general training-free loss network to implement the detail-to-structure supervision in the feature levels, which provides important supplement to the loss design based on the prediction itself.
- The proposed MSNet can accurately segment polyps as shown in Fig. 1. Extensive experiments demonstrate that our MSNet advances the state-of-the-art methods by a large margin under different evaluation metrics on five challenging datasets, with a real-time inference speed of ∼70fps.

2   Method

The MSNet architecture is shown in Fig. 2, in which there are five encoder blocks ($\mathbf{E}^i$, $i \in \{1, 2, 3, 4, 5\}$), a multi-scale subtraction module and four decoder blocks ($\mathbf{D}^i$, $i \in \{1, 2, 3, 4\}$). Following the PraNet [7], we adopt the Res2Net-50 as the backbone to extract five levels of features. First, we separately adopt a 3×3 convolution for feature maps of each encoder block to reduce the channel to 64, which can decrease the number of parameters for subsequent operations. Next, these different level features are fed into the multi-scale subtraction module and output five complementarity enhanced features (CE^i, $i \in \{1, 2, 3, 4, 5\}$). Finally, each CE^i progressively participates in the decoder and generate the final prediction. In the training phase, both the prediction and ground truth

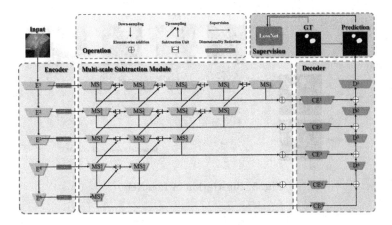

Fig. 2. Overview of the proposed MSNet.

are fed into the LossNet to achieve supervision. We describe the multi-scale subtraction module in Sect. 2.1 and give the details of LossNet in Sect. 2.2.

2.1 Multi-scale Subtraction Module

We use F_A and F_B to represent adjacent level feature maps. They all have been activated by the ReLU operation. We define a basic subtraction unit (SU):

$$SU = Conv(|F_A \ominus F_B|),$$ (1)

where $\ominus$ is the element-wise subtraction operation, $|\cdot|$ calculates the absolute value and $Conv(\cdot)$ denotes the convolution layer. The SU unit can capture the complementary information of F_A and F_B and highlight their differences, thereby providing richer information for the decoder.

To obtain higher-order complementary information across multiple feature levels, we horizontally and vertically concatenate multiple SUs to calculate a series of differential features with different orders and receptive fields. The detail of the multi-scale subtraction module can be found in Fig. 2. We aggregate the level-specific feature (MS_1^i) and cross-level differential features ($MS_{n \neq 1}^i$) between the corresponding level and any other levels to generate complementarity enhanced feature (CE^i). This process can be formulated as follows:

$$CE^i = Conv(\sum_{n=1}^{6-i} MS_n^i) \quad i = 1, 2, 3, 4, 5.$$ (2)

Finally, all CE^i participate in decoding and then the polyp region is segmented.

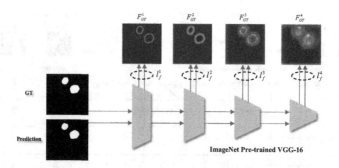

Fig. 3. Illustration of LossNet.

2.2 LossNet

In the proposed model, the total training loss can be written as:

$$\mathcal{L}_{total} = \mathcal{L}_{IoU}^{w} + \mathcal{L}_{BCE}^{w} + \mathcal{L}_f, \tag{3}$$

where $\mathcal{L}_{IoU}^{w}$ and $\mathcal{L}_{BCE}^{w}$ represent the weighted IoU loss and binary cross entropy (BCE) loss which have been widely adopted in segmentation tasks. We use the same definitions as in [7,19,25] and their effectiveness has been validated in theses works. Different from them, we extra use a LossNet to further optimize the segmentation from detail to structure. Specifically, we use an ImageNet pre-trained classification network, such as VGG-16, to extract the multi-scale features of the prediction and ground truth, respectively. Then, their feature difference is computed as loss $\mathcal{L}_f$:

$$\mathcal{L}_f = l_f^1 + l_f^2 + l_f^3 + l_f^4. \tag{4}$$

Let F_P^i and F_G^i separately represent the i-th level feature maps extracted from the prediction and ground truth. The l_f^i is calculated as their Euclidean distance (L2-Loss), which is supervised at the pixel level:

$$l_f^i = ||F_P^i - F_G^i||_2, \quad i = 1, 2, 3, 4. \tag{5}$$

The structure of LossNet is shown in Fig. 3. It can be seen that the low-level feature maps contain rich boundary information and the high-level ones depict location information. Thus, the LossNet can generate comprehensive supervision in the feature levels.

3 Experiments

3.1 Datasets

We evaluate the proposed model on five benchmark datasets: CVC-ColonDB [22], ETIS [21], Kvasir [11], CVC-T [24] and CVC-ClinicDB [2]. We adopt the same training set as the latest polyp segmentation method [7], that is, 900 samples from the Kvasir and 550 samples from the CVC-ClinicDB are used for training. The remaining images and other three datasets are used for testing.

3.2 Evaluation Metrics

We adopt several widely used metrics for quantitative evaluation: mean Dice, mean IoU, the weighted F-measure (F_β^w) [18], mean absolute error (MAE), the recently released S-measure (S_α) [5] and E-measure (E_ϕ^{max}) [6] scores. The lower value is better for the MAE and the higher is better for others.

3.3 Implementation Details

Our model is implemented based on the PyTorch framework and trained on a single 2080Ti GPU for 50 epochs with mini-batch size 16. We resize the inputs to 352×352 and employ a general multi-scale training strategy as the PraNet [7]. Random horizontally flipping and random rotate data augmentation are used to avoid overfitting. For the optimizer, we adopt the stochastic gradient descent (SGD). The momentum and weight decay are set as 0.9 and 0.0005, respectively. Maximum learning rate is set to 0.005 for backbone and 0.05 for other parts. Warm-up and linear decay strategies are used to adjust the learning rate.

3.4 Comparisons with State-of-the-art

We compare our MSNet with U-Net [20], U-Net++ [28], SFA [8] and PraNet [7]. To be fair, the predictions of these competitors are directly provided by their respective authors or computed by their released codes.

Quantitative Evaluation. Table 1 shows performance comparisons in terms of six metrics. It can be seen that our MSNet outperforms other approaches across all datasets. In particular, MSNet achieves a predominant performance on the CVC-ColonDB and ETIS datasets. Compared to the second best method (PraNet), our method achieves an important improvement on the challenging ETIS of 14.1%, 15.3%, 13.0%, 6.2%, 4.8% and 35.5% in terms of mDice, mIoU, F_β^w, S_α, E_ϕ^{max} and MAE, respectively. In addition, Table 2 lists the model average speed of different methods. Our model runs at a real-time speed of ∼70fps that is the fastest one among these state-of-art methods.

Qualitative Evaluation. Figure 4 illustrates visual comparison with other approaches. It can be seen that the proposed method has good detection performance for small, medium, and large scale polyps (see the 1st - 3th rows). Moreover, for the images with multiple polyps, our method can accurately detect them and capture more details (see the 4th rows).

Table 1. Quantitative comparison. ↑ and ↓ indicate that the larger and smaller scores are better, respectively. The best results are shown in red.

	Methods	mDice ↑	mIoU ↑	F_β^w ↑	S_α ↑	E_ϕ^{max} ↑	MAE ↓
ColonDB	U-Net(MICCAI'15) [20]	0.519	0.449	0.498	0.711	0.763	0.061
	U-Net++(TMI'19) [28]	0.490	0.413	0.467	0.691	0.762	0.064
	SFA (MICCAI'19) [8]	0.467	0.351	0.379	0.634	0.648	0.094
	PraNet (MICCAI'20) [7]	0.716	0.645	0.699	0.820	0.847	0.043
	MSNet (Ours)	0.755	0.678	0.737	0.836	0.883	0.041
ETIS	U-Net (MICCAI'15) [20]	0.406	0.343	0.366	0.682	0.645	0.036
	U-Net++ (TMI'19) [28]	0.413	0.342	0.390	0.681	0.704	0.035
	SFA (MICCAI'19) [8]	0.297	0.219	0.231	0.557	0.515	0.109
	PraNet (MICCAI'20) [7]	0.630	0.576	0.600	0.791	0.792	0.031
	MSNet (Ours)	0.719	0.664	0.678	0.840	0.830	0.020
Kvasir	U-Net (MICCAI'15) [20]	0.821	0.756	0.794	0.858	0.901	0.055
	U-Net++ (TMI'19) [28]	0.824	0.753	0.808	0.862	0.907	0.048
	SFA (MICCAI'19) [8]	0.725	0.619	0.670	0.782	0.828	0.075
	PraNet (MICCAI'20) [7]	0.901	0.848	0.885	0.915	0.943	0.030
	MSNet (Ours)	0.907	0.862	0.893	0.922	0.944	0.028
CVC-T	U-Net (MICCAI'15) [20]	0.717	0.639	0.684	0.842	0.867	0.022
	U-Net++ (TMI'19) [28]	0.714	0.636	0.687	0.838	0.884	0.018
	SFA (MICCAI'19) [8]	0.465	0.332	0.341	0.640	0.604	0.065
	PraNet (MICCAI'20) [7]	0.873	0.804	0.843	0.924	0.938	0.010
	MSNet (Ours)	0.869	0.807	0.849	0.925	0.943	0.010
ClinicDB	U-Net (MICCAI'15) [20]	0.824	0.767	0.811	0.889	0.917	0.019
	U-Net++ (TMI'19) [28]	0.797	0.741	0.785	0.872	0.898	0.022
	SFA (MICCAI'19) [8]	0.698	0.615	0.647	0.793	0.816	0.042
	PraNet (MICCAI'20) [7]	0.902	0.858	0.896	0.935	0.958	0.009
	MSNet (Ours)	0.921	0.879	0.914	0.941	0.972	0.008

Table 2. The average speed of different methods.

Methods	U-Net	U-Net++	SFA	PraNet	*MSNet* (Ours)
Average speed	~8fps	~7fps	~40fps	~50fps	~ 70fps

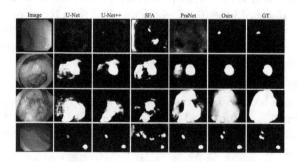

Fig. 4. Visual comparison of different methods.

Table 3. Ablation study on the CVC-ColonDB and ETIS datasets.

Metric	ColonDB				ETIS			
	mDice	mIoU	F_β^w	E_ϕ^{max}	mDice	mIoU	F_β^w	E_ϕ^{max}
baseline (MS_1^i)	0.678	0.607	0.659	0.825	0.588	0.549	0.532	0.707
+ MS_2^i	0.731	0.652	0.703	0.861	0.642	0.579	0.586	0.745
+ MS_3^i	0.733	0.659	0.712	0.861	0.642	0.580	0.581	0.745
+ MS_4^i	0.750	0.676	0.729	0.872	0.643	0.580	0.585	0.757
+ MS_5^i	0.749	0.676	0.729	0.878	0.643	0.582	0.600	0.787
+ $\mathcal{L}_f$	0.755	0.678	0.737	0.883	0.719	0.664	0.678	0.830
Replace MS with MA	0.697	0.630	0.676	0.839	0.680	0.621	0.636	0.820

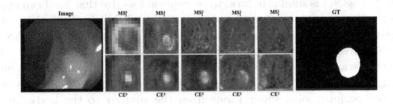

Fig. 5. Visual comparison between encoder features and the cross-level complementarity enhanced features.

3.5 Ablation Study

We take the common FPN network as the baseline to analyze the contribution of each component. The results are shown in Table 3. These defined feature symbols are the same as those in Fig. 2. First, we apply the subtraction module to the baseline to get a series of MS_i^2 features to participate in the feature aggregation calculated by Eq. 2. The gap between the "+ MS_i^2" and the baseline demonstrates the effectiveness of the subtraction unit (SU). It can be seen that the usage of SU has a significant improvement on the CVC-ColonDB dataset compared to the baseline, with the gain of 7.8%, 7.4%, 6.7% and 4.4% in terms of mDice, mIoU, F_β^w, and E_ϕ^{max}, respectively. Next, we gradually add MS_i^3, MS_i^4 and MS_i^5 to achieve multi-scale aggregation. The gap between the "+ MS_i^5" and the " + MS_i^2" quantitatively demonstrates the effectiveness of multi-scale strategy. To more intuitively show its effectiveness, we visualize features of each encoder level (MS_1^i) and the complementarity enhanced features (CE^i) in Fig. 5. We can see that the multi-scale subtraction module can clearly highlight the difference between high-level features and other level features and propagate its localization effect to the low-level ones. Thus, both the global structural information and local boundary information is well depicted in the enhanced features of different levels. Finally, we evaluate the benefit of $\mathcal{L}_f$. Compared to the "+ MS_i^5" model, the "+ $\mathcal{L}_f$" achieves significant performance improvement on the ETIS dataset, with the gain of 11.8%, 14.1%, 13.0% and 5.5% in terms of mDice, mIoU, F_β^w, and E_ϕ^{max}, respectively. Besides, we replace all subtraction units with the element-wise addition units and compare their performance. It can be seen that our subtraction units have significant advantage and no additional parameters are introduced.

4 Discussion

Multi-scale Subtraction Module: Different from previous addition operation, using subtraction in multi-scale module make resulted features input to the decoder have much less redundancy among different levels and their scale-specific properties are significantly enhanced. This mechanism can be explored in more segmentation tasks in the future.

LossNet: LossNet is similar in form to perception loss [15] that has been applied in many tasks, such as style transfer and inpainting. While in those vision tasks, the perception-like loss is mainly used to speed the convergence of GAN and obtain high frequency information and ease checkerboard artifacts, but it does not bring obvious accuracy improvement. In our paper, the inputs are binary segmentation masks, LossNet can directly target the geometric features of the lesion and perform joint supervisions from the contour to the body, thereby improving the overall segmentation accuracy. In the binary segmentation task, our work is the first one.

5 Conclusion

In this paper, we present a novel multi-scale subtraction network (MSNet) to automatically segment polyps from colonoscopy images. We pyramidally concatenate multiple subtraction units to extract lower-order and higher-order cross-level complementary information and combine with level-specific information to enhance multi-scale feature representation. Besides, we design a loss function based on a training-free network to supervise the prediction from different feature levels, which can optimize the segmentation on both structure and details during the backward phase. Extensive experimental results demonstrate that MSNet notably outperforms the state-of-the-art methods under different evaluation metrics. Moreover, the proposed model runs at the fastest speed of ∼70fps among the existing polyp segmentation methods.

Acknowledgements. This work was supported in part by the National Natural Science Foundation of China #61876202, #61725202, #61751212 and #61829102, the Dalian Science and Technology Innovation Foundation #2019J12GX039, and the Fundamental Research Funds for the Central Universities #DUT20ZD212.

References

1. Akbari, M., et al.: Polyp segmentation in colonoscopy images using fully convolutional network. In: IEEE EMBC, pp. 69–72 (2018)

2. Bernal, J., Sánchez, F.J., Fernández-Esparrach, G., Gil, D., Rodríguez, C., Vilariño, F.: Wm-dova maps for accurate polyp highlighting in colonoscopy: Validation vs. saliency maps from physicians. In: CMIG, vol. 43, pp. 99–111 (2015)

3. Brandao, P., et al.: Fully convolutional neural networks for polyp segmentation in colonoscopy. In: MICAD, vol. 10134, p. 101340F (2017)

4. Chen, L.C., Papandreou, G., Kokkinos, I., Murphy, K., Yuille, A.L.: Deeplab: Semantic image segmentation with deep convolutional nets, atrous convolution, and fully connected crfs. IEEE TPAMI 40(4), 834–848 (2017)

5. Fan, D.P., Cheng, M.M., Liu, Y., Li, T., Borji, A.: Structure-measure: a new way to evaluate foreground maps. In: IEEE ICCV, pp. 4548–4557 (2017)

6. Fan, D.P., Gong, C., Cao, Y., Ren, B., Cheng, M.M., Borji, A.: Enhanced-alignment measure for binary foreground map evaluation. In: IJCAI (2018)

7. Fan, D.-P., et al.: PraNet: parallel reverse attention network for polyp segmentation. In: Martel, A.L., et al. (eds.) MICCAI 2020. LNCS, vol. 12266, pp. 263–273. Springer, Cham (2020). https://doi.org/10.1007/978-3-030-59725-2_26

8. Fang, Y., Chen, C., Yuan, Y., Tong, K.: Selective feature aggregation network with area-boundary constraints for polyp segmentation. In: Shen, D., et al. (eds.) MICCAI 2019. LNCS, vol. 11764, pp. 302–310. Springer, Cham (2019). https://doi.org/10.1007/978-3-030-32239-7_34

9. He, K., Zhang, X., Ren, S., Sun, J.: Deep residual learning for image recognition. In: IEEE CVPR, pp. 770–778 (2016)

10. Hu, J., Shen, L., Sun, G.: Squeeze-and-excitation networks. In: IEEE CVPR, pp. 7132–7141 (2018)

11. Jha, D., et al.: Kvasir-SEG: a segmented polyp dataset. In: Ro, Y.M., et al. (eds.) MMM 2020. LNCS, vol. 11962, pp. 451–462. Springer, Cham (2020). https://doi.org/10.1007/978-3-030-37734-2_37

12. Jha, D., et al.: Resunet++: An advanced architecture for medical image segmentation. In: IEEE ISM, pp. 225–2255 (2019)

13. Ji, G.P., et al.: Progressively normalized self-attention network for video polyp segmentation. In: MICCAI (2021)

14. Ji, W., et al.: Learning calibrated medical image segmentation via multi-rater agreement modeling. In: IEEE CVPR, pp. 12341–12351 (2021)

15. Johnson, J., Alahi, A., Fei-Fei, L.: Perceptual losses for real-time style transfer and super-resolution. In: Leibe, B., Matas, J., Sebe, N., Welling, M. (eds.) ECCV 2016. LNCS, vol. 9906, pp. 694–711. Springer, Cham (2016). https://doi.org/10.1007/978-3-319-46475-6_43

16. Lin, T.Y., Dollár, P., Girshick, R., He, K., Hariharan, B., Belongie, S.: Feature pyramid networks for object detection. In: IEEE CVPR, pp. 2117–2125 (2017)

17. Long, J., Shelhamer, E., Darrell, T.: Fully convolutional networks for semantic segmentation. In: IEEE CVPR, pp. 3431–3440 (2015)

18. Margolin, R., Zelnik-Manor, L., Tal, A.: How to evaluate foreground maps? In: IEEE CVPR, pp. 248–255 (2014)

19. Qin, X., Zhang, Z., Huang, C., Gao, C., Dehghan, M., Jagersand, M.: Basnet: boundary-aware salient object detection. In: IEEE CVPR, pp. 7479–7489 (2019)

20. Ronneberger, O., Fischer, P., Brox, T.: U-Net: convolutional networks for biomedical image segmentation. In: Navab, N., Hornegger, J., Wells, W.M., Frangi, A.F. (eds.) MICCAI 2015. LNCS, vol. 9351, pp. 234–241. Springer, Cham (2015). https://doi.org/10.1007/978-3-319-24574-4_28

21. Silva, J., Histace, A., Romain, O., Dray, X., Granado, B.: Toward embedded detection of polyps in wce images for early diagnosis of colorectal cancer. IJCARS 9(2), 283–293 (2014)

22. Tajbakhsh, N., Gurudu, S.R., Liang, J.: Automated polyp detection in colonoscopy videos using shape and context information. IEEE TMI **35**(2), 630–644 (2015)
23. Tajbakhsh, N., Gurudu, S.R., Liang, J.: Automatic polyp detection in colonoscopy videos using an ensemble of convolutional neural networks. In: IEEE ISBI, pp. 79–83 (2015)
24. Vázquez, D., et al.: A benchmark for endoluminal scene segmentation of colonoscopy images. JHE **2017** (2017)
25. Wei, J., Wang, S., Huang, Q.: F3Net: fusion, AAAI, feedback and focus for salient object detection (2020)
26. Yu, L., Chen, H., Dou, Q., Qin, J., Heng, P.A.: Integrating online and offline three-dimensional deep learning for automated polyp detection in colonoscopy videos. IEEE JBHI **21**(1), 65–75 (2016)
27. Zhang, R., Zheng, Y., Poon, C.C., Shen, D., Lau, J.Y.: Polyp detection during colonoscopy using a regression-based convolutional neural network with a tracker. PR **83**, 209–219 (2018)
28. Zhou, Z., Siddiquee, M.M.R., Tajbakhsh, N., Liang, J.: Unet++: a nested u-net architecture for medical image segmentation. In: IEEE TMI, pp. 3–11 (2019)

Patch-Free 3D Medical Image Segmentation Driven by Super-Resolution Technique and Self-Supervised Guidance

Hongyi Wang[1], Lanfen Lin[1(✉)], Hongjie Hu[2], Qingqing Chen[2], Yinhao Li[3,4], Yutaro Iwamoto[4], Xian-Hua Han[5], Yen-Wei Chen[4,3,1], and Ruofeng Tong[1,3]

[1] College of Computer Science and Technology, Zhejiang University, Hangzhou, China
llf@zju.edu.cn
[2] Department of Radiology, Sir Run Run Shaw Hospital, Hangzhou, China
[3] Research Center for Healthcare Data Science, Zhejiang Lab, Hangzhou, China
[4] College of Information Science and Engineering, Ritsumeikan University, Kusatsu, Japan
[5] Artificial Intelligence Research Center, Yamaguchi University, Yamaguchi, Japan

Abstract. 3D medical image segmentation with high resolution is an important issue for accurate diagnosis. The main challenge for this task is its large computational cost and GPU memory restriction. Most of the existing 3D medical image segmentation methods are patch-based methods, which ignore the global context information for accurate segmentation and also reduce the efficiency of inference. To tackle this problem, we propose a patch-free 3D medical image segmentation method, which can realize high-resolution (HR) segmentation with low-resolution (LR) input. It contains a multi-task learning framework (Semantic Segmentation and Super-Resolution (SR)) and a Self-Supervised Guidance Module (SGM). SR is used as an auxiliary task for the main segmentation task to restore the HR details, while the SGM, which uses the original HR image patch as a guidance image, is designed to keep the high-frequency information for accurate segmentation. Besides, we also introduce a Task-Fusion Module (TFM) to exploit the inter connections between the segmentation and SR tasks. Since the SR task and TFM are only used in the training phase, they do not introduce extra computational costs when predicting. We conduct the experiments on two different datasets, and the experimental results show that our framework outperforms current patch-based methods as well as has a 4× higher speed when predicting. Our codes are available at https://github.com/Dootmaan/PFSeg.

Keywords: 3D medical image segmentation · Patch-free · Multi-task learning

© Springer Nature Switzerland AG 2021
M. de Bruijne et al. (Eds.): MICCAI 2021, LNCS 12901, pp. 131–141, 2021.
https://doi.org/10.1007/978-3-030-87193-2_13

1   Introduction

Segmentation is one of the most important tasks in medical image analysis. Recent years, with the help of deep learning, there are many inspiring progresses are made in this field. However, most medical images are of high resolution and cannot be directly processed by mainstream graphics cards. Thus, many previous works are 2D networks, which only focus on the segmentation of one single slice at a time [4,7,8,18,20]. Nevertheless, such methods ignore the valuable information along the z-axis, which limits the improvement of model's performance. To better capture all the information along the three dimensions, many algorithms such as 2D multiple views [23,24] and 2.5D [19,28] are developed, alleviating the problem to some extent. However, these methods still mainly use 2D convolution to extract features and cannot capture the overall 3D spatial information. Therefore, to thoroughly solve this problem, the better way is to use the intuitive 3D convolution [5,10,11,14,17]. Since training 3D segmentation models needs more computational cost, patch-sampling, that each time only crop a small part from the original medical image as the model's input, becomes a necessity [15,21,25].

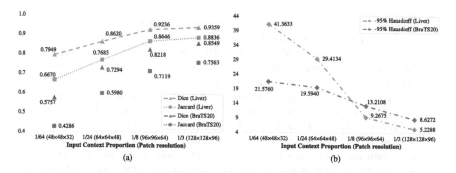

Fig. 1. (a) The Dice Similarity Coefficient, Jaccard Coefficient and (b) 95% Hausdorff Distance of different patch size using 3D ResUNet on two datasets. As is shown, a bigger patch size with more context tends to have a better performance.

Though widely used, patch-sampling also has some flaws. Firstly, the patch-based methods ignore the global context information, which is important for accurate segmentation [27]. As is shown in Fig. 1, our experiments illustrate that the size of a patch can greatly affect the model's performance, and because bigger patches contain more context, they can usually achieve higher accuracy; Secondly, if the network is trained with patches, it also have to use patches (such as sliding window strategy) in inference stage, which may not only severely decrease the efficiency, but also reduce the accuracy due to inconsistencies introduced during the fusion process that takes place in areas where patches overlap [9].

To solve these problems, we need to design a patch-free segmentation method with moderate computational budgets. Motivated by SR technique, which can recover HR images from LR inputs, we concrete our idea by lowering the resolution of the input image. We propose a novel 3D patch-free segmentation method which can realize HR segmentation with LR input (i.e. the down-sampled 3D image). We call this kind of tasks as Upsample-Segmentation (US) tasks. Inspired by [22], we use SR as an auxiliary task for the US task to restore the HR details lost in the down-sampling procedure. In addition, we introduce a Self-Supervised Guidance Module (SGM), which uses a patch of the original HR image as the subsidiary guidance input. High-frequency features can be directly extracted from it, and then be concatenated with the features from the down-sampled input image. To further improve the model's performance, we also propose a novel Task-Fusion Module (TFM) to exploit the connections between the US and SR tasks. It should be noted that TFM as well as the auxiliary SR branch are only used in training phase. They do not introduce extra computational costs when predicting.

Our contributions can be mainly concluded into three points: (1) We propose a patch-free 3D medical image segmentation method, which can realize HR segmentation with LR input. (2) We propose a Self-Supervised Guidance Module (SGM), which uses the original HR image patch as guidance, to keep the high-frequency representations for accurate segmentation. (3) We further design a Task-Fusion Module (TFM) to exploit the inter connections between the US and SR tasks, by which the two tasks can be optimized jointly.

2 Methodology

The proposed method is shown in Fig. 2. For a given image, we first down-sample it by 2× to lower the resolution, and then use it as the framework's main input. The encoder will process it into shared features which can be used for both SR and US tasks. In addition, we also crop a patch from the original HR image as guidance, using the features extracted from it to provide the network with more high-frequency information. In training phase, outputs of the US task and SR task will also be sent into Task-Fusion Module (TFM), where the two tasks are fused together to help each other optimize. Note that in the testing phase, only the main segmentation network is used for 3D segmentation, with no extra computational cost.

2.1 Multi-task Learning

Multi-task Learning is the foundation of our framework, including a US task a SR task. Ground truth for US is the labeled segmentation mask of original high resolution, while that of SR is the HR image itself. Since our goal is about generating an accurate segmentation mask, here we will treat US as the main task and SR only as an auxiliary one, which can be removed in testing phase. The two branches are both designed on the basis of ResUNet3D [26] for better consistency, since they share an encoder and will be fused together afterwards. The details are shown in Fig. 3(a).

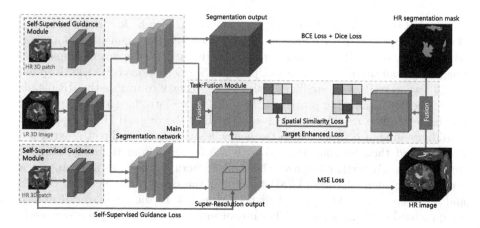

Fig. 2. A schematic view of the proposed framework.

The loss functions for this part can be divided into segmentation loss L_{seg} and SR loss L_{sr} for each task. L_{seg} consists of Binary Cross Entropy (BCE) Loss and Dice Loss, while L_{sr} is a simple Mean Square Error (MSE) Loss.

2.2 Self-Supervised Guidance Module

To make proper use of the original HR images and further improve the framework's performance, we propose the Self-Supervised Guidance Module (SGM). This module uses a typical patch cropped from the original HR image to extract some representative high-frequency features. Through the experiments, we found that simply cropping the central area performs even better than random cropping. Random cropping may cause instability since for every testing case the content of the guidance patch may vary a lot. In our experiment, the size of the guidance patch is set to be 1/64 of the original image.

SGM is designed according to the guidance patch size to make sure the features extracted from it can be correctly concatenated with the shared features. To avoid too much computational cost, SGM is built to be very concise, as is shown in Fig. 3(c). We also introduce a Self-Supervised Guidance Loss (SGL) to evaluate the distance between the guidance and its corresponding part of SR output. The loss function can be described as:

$$L_{sgl} = \frac{1}{N} \sum_{i=1}^{N} ||SIG(i) \cdot SR(X \downarrow)_i - SIG(i) \cdot X_i||^2, \tag{1}$$

where N refers to the total number of all voxels, and $SIG(i)$ denotes the signal function that will output 1 if i-th voxel is in the cropping window and 0 otherwise. X and $X \downarrow$ denote the original medical image and the one after down-sampling, while $SR(\cdot)$ represents the SR output.

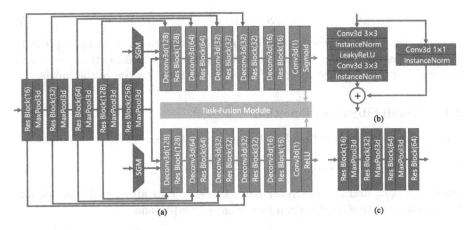

Fig. 3. Detailed introduction of the framework structure. (a) Multi-task learning as the foundation of the framework. (b) Residual Block used in the framework. (c) Network structure of Self-Supervised Guidance Module (SGM).

2.3   Task-Fusion Module

To better utilize the connections between US and SR, we design a Task-Fusion Module (TFM) combining the two tasks together to let them help each other. This module will first calculate the element-wise product of the two tasks' outputs (the estimated HR mask and HR image), and then optimize it by two different streams. For the first stream, we propose a Target-Enhanced Loss (TEL), which calculates the average square Euclidean distance of target area voxels. It can be viewed as adding weight to the loss of segmentation target area. Thus, the US task will tend to segment more precisely, and the SR task will pay more attention on the target part. As to the second one, inspired by Spatial Attention Mechanism in [6], we propose a Spatial Similarity Loss (SSL) to make the internal differences between prediction voxels similar to that of the ground truth. SSL is calculated using Spatial Similarity Matrix, which mainly describes the pairwise relationship between voxels. For a $D \times W \times H \times C$ image I (for medical images C usually equals 1), to compute its Spatial Similarity Matrix, first we need to reshape it into $V \times C$, where $V = D \times W \times H$. After that, by multiplying this matrix with its transpose, we can have the $V \times V$ similarity matrix and calculate the loss of it with ground truth. The loss function for this module can be defined as follows.

$$L_{tfm} = L_{tel} + L_{ssl}, \tag{2}$$

$$L_{tel} = \frac{1}{N} \sum_{i=1}^{N} ||p_i \cdot SR(X \downarrow)_i - y_i \cdot X_i||^2, \tag{3}$$

$$L_{ssl} = \frac{1}{D^2 W^2 H^2} \sum_{i=1}^{D \cdot W \cdot H} \sum_{j=1}^{D \cdot W \cdot H} ||S_{ij}^{predict} - S_{ij}^{gt}||^2, \tag{4}$$

$$S_{ij} = I_i \cdot I_j^{\mathrm{T}}, \tag{5}$$

where p_i denotes the prediction of i-th voxel after binarization, and y_i represents its corresponding ground truth. S_{ij} refers to the correlation between i-th and j-th voxel of fusion image I, while I_i represents the i-th voxel of the image.

2.4   Overall Objective Function

The overall objective function L of the proposed framework is:

$$L = L_{seg} + \omega_{sr}L_{sr} + \omega_{tfm}L_{tfm} + \omega_{sgl}L_{sgl}, \tag{6}$$

where ω_{sr}, ω_{tfm} and ω_{sgl} are hyper-parameters, and are all set to 0.5 by default. The whole objective function can be optimized end-to-end.

3   Experiments

3.1   Datasets

We used BRATS2020 dataset [2,3,16] and a privately-owned liver segmentation dataset in the experiment. BRATS2020 dataset contains a total number of 369 subjects, each with four-modality MRI images (T1, T2, T1ce and FLAIR) of size $240 \times 240 \times 155$ and spacing $1 \times 1 \times 1$ mm^3. The ground truth includes masks of Tumor Core (TC), Enhanced Tumor (ET) and Whole Tumor (WT). For each image, we removed the edges without brain part by 24 voxels and resized the rest part to resolution $192 \times 192 \times 128$. In our experiment, we used the down-sampled T2-weighted images as input, the original T2-weighted images as SR ground truth, and WT masks as US ground truth.

The privately-owned liver segmentation dataset contains 347 subjects. Each one has an MRI image and a segmentation ground truth labeled by experienced doctors. In our experiment, spacing of the images were all regulated to $1.5 \times 1.5 \times 1.5$ mm^3, and we then cropped the central $192 \times 192 \times 128$ area. The cropped MRI image and its segmentation mask are used as ground truth of SR and US, while the input is the cropped image after down-sampling.

3.2   Implementation Details

We compared our framework with different patch-based 3D segmentation models. For those methods, we predict the test image using sliding window strategy with a stride of 48, 48 and 32 for x-axis, y-axis and z-axis, respectively. Besides, we also tested our method with other patch-free segmentation models (i.e., ResUNet3D↑ and HDResUNet). ResUNet3D↑ conducts ordinary segmentation with a down-sampled image, then enlarging the result by tricubic interpolation [13]; HDResUNet uses Holistic Decomposition [27] with ResUNet3D, and the down-shuffling factors of it are all set to 2.

We employed three quantitative evaluation indices in the experiments, which are Dice Similarity Coefficient, 95% Hausdorff Distance and Jaccard Coefficient.

Table 1. Ablation study results on BRATS2020. HD95 refers to 95% Hausdorff Distance.

Method	UNet3D			ResUNet3D		
	Dice (%)	HD95 (mm)	Jaccard (%)	Dice (%)	HD95 (mm)	Jaccard (%)
US	80.58	10.56	70.21	81.83	8.88	71.75
US+SR	81.71	9.64	71.15	82.09	8.41	71.52
US+SR+TEL	82.17	9.31	71.55	82.96	8.16	72.82
US+SR+TEL+SSL	82.53	9.24	72.24	83.58	8.64	73.56
US+SR+TEL+SSL+SGM	**83.16**	**8.00**	**72.83**	**83.82**	**7.83**	**74.01**

Dice and Jaccard mainly focus on the segmentation area. 95% Hausdorff pays more attention to the edges.

All the experiments run on a Nvidia GTX 1080Ti GPU with 11 GB video memory. For fair comparison, the input sizes were all set to $96 \times 96 \times 64$, except HDResUNet, which uses the original HR image. Therefore, the patch size for patch-based methods and the input image size for patch-free methods are the same. For both datasets, we used 80% for training and the rest for testing. Data augmentation includes random cropping (only for patch-based methods), random flip, random rotation, and random shift. All the models are optimized by Adam [12] with the initial learning rate set to $1e^{-4}$. The rate will be divided by 10 if the loss does not continuously reduce over 20 epochs, and the training phase will end when it reaches $1e^{-7}$.

3.3 Ablation Study

We conduct an ablation study on BRATS2020 to investigate how the designed modules affect the framework's performance. The framework is tested with two different backbones (i.e., UNet3D and ResUNet3D).

In Table 1, for both backbones, appending SR as the auxiliary task improves the segmentation performance, indicating that the framework successfully rebuild some high-frequency information with the help of SR. Moreover, the segmentation result after introducing TEL and SSL also proves the effectivity of TFM, showing that the inter connection between US and SR is very useful for joint optimization. At last, the increase of all the metrics after adding SGM demonstrate that the framework benefits from the self-supervised guidance for the high-frequency features it brings.

3.4 Experimental Results

The experimental results are summarized in Table 2. Our framework surpasses traditional 3D patch-based methods and also outperforms the other patch-free methods. Patch-free methods have the most obvious improvements in 95% Hausdorff Distance: with the global context, the model can more easily segment the target area as a whole, hence making the segmentation edges smoother and more

Table 2. Segmentation results on two datasets. HD95 refers to 95% Hausdorff Distance, and Time denotes the average inference time for each case.

Model	Patch-free	BRATS2020			Liver dataset			Time (s)
		Dice (%)	HD95 (mm)	Jaccard (%)	Dice (%)	HD95 (mm)	Jaccard (%)	
V-Net		79.91	13.86	68.29	89.49	14.93	81.83	5.26
UNet3D		81.21	14.63	69.89	90.80	11.64	84.11	4.00
ResUNet3D		82.18	13.21	71.19	92.36	9.27	86.46	4.21
ResUNet3D↑	✓	80.89	8.56	70.02	91.98	6.40	85.84	2.21
HDResUNet	✓	82.45	9.21	72.07	92.88	5.58	87.33	**0.19**
Ours	✓	**83.82**	**7.83**	**74.01**	**93.83**	**4.29**	**88.70**	0.95

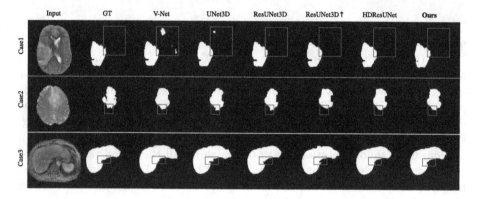

Fig. 4. Typical segmentation results of the experiment. Case1 and Case2 are from BRATS2020, while Case3 is from the liver dataset. For the convenience of visualization, we only select one slice from every case.

accurate. Since our framework can directly output a complete segmentation mask at a time, it also has a faster inference speed than most of the other methods.

Some typical segmentation results are listed in Fig. 4. As is shown, the patch-based results have many obvious flaws (labeled in red): in Case1, there is some segmentation noise. This problem mainly results from the limited context in patches. When conducting segmentation on the upper right corner patch, the model does not have the information of the real tumor area and it will be more likely to misdiagnose normal area as lesion. In Case2 and Case3, there are some failed segmentation in the corner area due to the padding technique. In [1], the authors pointed out that padding may result in artifacts on the edges of feature maps, and these artifacts may confuse the network. The problem of Case2 and Case3 is commonly seen when target area leaps over several patches. Under such circumstances, it is difficult for the patch-based models to correctly estimate the voxels on the edges, hence resulting in inconsistencies during the fusion process. Although patch-free segmentation can solve the above-mentioned problems, it may lead to significant performance degradation due to the loss of high-frequency information during down-scaling. In our method, we build a multi-task learning framework (US and SR) with two well-designed modules (TFM and SGM) to keep the HR representations, thus avoiding this issue. Therefore, our framework outperforms other existing patch-free methods.

4   Conclusion

In this work, we propose a novel framework for fast and accurate patch-free segmentation, which is capable of capturing global context while not introducing too much extra computational cost. We validate the framework's performance on two datasets to demonstrate its effectiveness, and the result shows that it can efficiently generate better segmentation mask than other patch-based and patch-free methods.

Acknowledgement. This work was supported in part by Major Scientific Research Project of Zhejiang Lab under the Grant No. 2020ND8AD01, and in part by the Grant-in Aid for Scientific Research from the Japanese Ministry for Education, Science, Culture and Sports (MEXT) under the Grant No. 20KK0234, No. 21H03470, and No. 20K21821.

References

1. Alsallakh, B., Kokhlikyan, N., Miglani, V., Yuan, J., Reblitz-Richardson, O.: Mind the pad - cnns can develop blind spots. In: International Conference on Learning Representations (2021). https://openreview.net/forum?id=m1CD7tPubNy
2. Bakas, S., Akbari, H., Sotiras, A., Bilello, M., Rozycki, M., Kirby, J.S., Freymann, J.B., Farahani, K., Davatzikos, C.: Advancing the cancer genome atlas glioma MRI collections with expert segmentation labels and radiomic features. Sci. data 4(1), 1–13 (2017)
3. Bakas, S., et al.: Identifying the best machine learning algorithms for brain tumor segmentation, progression assessment, and overall survival prediction in the brats challenge. arXiv preprint arXiv:1811.02629 (2018)
4. Christ, P.F., et al.: Automatic liver and lesion segmentation in CT using cascaded fully convolutional neural networks and 3D conditional random fields. In: Ourselin, S., Joskowicz, L., Sabuncu, M.R., Unal, G., Wells, W. (eds.) MICCAI 2016. LNCS, vol. 9901, pp. 415–423. Springer, Cham (2016). https://doi.org/10.1007/978-3-319-46723-8_48
5. Çiçek, Ö., Abdulkadir, A., Lienkamp, S.S., Brox, T., Ronneberger, O.: 3D U-Net: learning dense volumetric segmentation from sparse annotation. In: Ourselin, S., Joskowicz, L., Sabuncu, M.R., Unal, G., Wells, W. (eds.) MICCAI 2016. LNCS, vol. 9901, pp. 424–432. Springer, Cham (2016). https://doi.org/10.1007/978-3-319-46723-8_49
6. Fu, J., et al.: Dual attention network for scene segmentation. In: Proceedings of the IEEE/CVF Conference on Computer Vision and Pattern Recognition, pp. 3146–3154 (2019)
7. Huang, H., et al.: Unet 3+: a full-scale connected unet for medical image segmentation. In: ICASSP 2020–2020 IEEE International Conference on Acoustics, Speech and Signal Processing (ICASSP), pp. 1055–1059. IEEE (2020)
8. Huang, H., et al.: Medical image segmentation with deep atlas prior. IEEE Trans. Med. Imaging (2021)

9. Huang, Y., Shao, L., Frangi, A.F.: Simultaneous super-resolution and cross-modality synthesis of 3D medical images using weakly-supervised joint convolutional sparse coding. In: Proceedings of the IEEE Conference on Computer Vision and Pattern Recognition, pp. 6070–6079 (2017)

10. Kao, P.Y., et al.: Improving patch-based convolutional neural networks for MRI brain tumor segmentation by leveraging location information. Front. Neurosci. **13**, 1449 (2020)

11. Kim, H., et al.: Abdominal multi-organ auto-segmentation using 3d-patch-based deep convolutional neural network. Sci. Rep. **10**(1), 1–9 (2020)

12. Kingma, D.P., Ba, J.: Adam: a method for stochastic optimization. In: International Conference on Learning Representations (2015)

13. Lekien, F., Marsden, J.: Tricubic interpolation in three dimensions. Int. J. Numer. Methods Eng. **63**(3), 455–471 (2005)

14. Li, Z., Pan, J., Wu, H., Wen, Z., Qin, J.: Memory-efficient automatic kidney and tumor segmentation based on non-local context guided 3D U-Net. In: Martel, A.L., et al. (eds.) MICCAI 2020. LNCS, vol. 12264, pp. 197–206. Springer, Cham (2020). https://doi.org/10.1007/978-3-030-59719-1_20

15. Madesta, F., Schmitz, R., Rösch, T., Werner, R.: Widening the focus: biomedical image segmentation challenges and the underestimated role of patch sampling and inference strategies. In: Martel, A.L., et al. (eds.) MICCAI 2020. LNCS, vol. 12264, pp. 289–298. Springer, Cham (2020). https://doi.org/10.1007/978-3-030-59719-1_29

16. Menze, B.H., et al.: The multimodal brain tumor image segmentation benchmark (brats). IEEE Trans. Med. Imaging **34**(10), 1993–2024 (2014)

17. Milletari, F., Navab, N., Ahmadi, S.A.: V-net: fully convolutional neural networks for volumetric medical image segmentation. In: 2016 Fourth International Conference on 3D Vision (3DV), pp. 565–571. IEEE (2016)

18. Ronneberger, O., Fischer, P., Brox, T.: U-Net: convolutional networks for biomedical image segmentation. In: Navab, N., Hornegger, J., Wells, W.M., Frangi, A.F. (eds.) MICCAI 2015. LNCS, vol. 9351, pp. 234–241. Springer, Cham (2015). https://doi.org/10.1007/978-3-319-24574-4_28

19. Shao, Q., Gong, L., Ma, K., Liu, H., Zheng, Y.: Attentive CT lesion detection using deep pyramid inference with multi-scale booster. In: Shen, D., et al. (eds.) MICCAI 2019. LNCS, vol. 11769, pp. 301–309. Springer, Cham (2019). https://doi.org/10.1007/978-3-030-32226-7_34

20. Tang, Y., Tang, Y., Zhu, Y., Xiao, J., Summers, R.M.: E^2Net: an edge enhanced network for accurate liver and tumor segmentation on CT scans. In: Martel, A.L., et al. (eds.) MICCAI 2020. LNCS, vol. 12264, pp. 512–522. Springer, Cham (2020). https://doi.org/10.1007/978-3-030-59719-1_50

21. Tang, Y., et al.: High-resolution 3D abdominal segmentation with random patch network fusion. Med. Image Anal. **69**, 101894 (2021)

22. Wang, L., Li, D., Zhu, Y., Tian, L., Shan, Y.: Dual super-resolution learning for semantic segmentation. In: Proceedings of the IEEE/CVF Conference on Computer Vision and Pattern Recognition, pp. 3774–3783 (2020)

23. Wang, Y., Zhou, Y., Shen, W., Park, S., Fishman, E.K., Yuille, A.L.: Abdominal multi-organ segmentation with organ-attention networks and statistical fusion. Med. Image Anal. **55**, 88–102 (2019)

24. Xia, Y., et al.: Uncertainty-aware multi-view co-training for semi-supervised medical image segmentation and domain adaptation. Med. Image Anal. **65**, 101766 (2020)

25. Yang, H., Shan, C., Bouwman, A., Kolen, A.F., de With, P.H.: Efficient and robust instrument segmentation in 3D ultrasound using patch-of-interest-fusenet with hybrid loss. Med. Image Anal. **67**, 101842 (2021)
26. Yu, L., Yang, X., Chen, H., Qin, J., Heng, P.A.: Volumetric convnets with mixed residual connections for automated prostate segmentation from 3D MR images. In: Proceedings of the AAAI Conference on Artificial Intelligence, vol. 31 (2017)
27. Zeng, G., Zheng, G.: Holistic decomposition convolution for effective semantic segmentation of medical volume images. Med. Image Anal. **57**, 149–164 (2019)
28. Zlocha, M., Dou, Q., Glocker, B.: Improving RetinaNet for CT lesion detection with dense masks from weak RECIST labels. In: Shen, D., et al. (eds.) MICCAI 2019. LNCS, vol. 11769, pp. 402–410. Springer, Cham (2019). https://doi.org/10.1007/978-3-030-32226-7_45

Progressively Normalized Self-Attention Network for Video Polyp Segmentation

Ge-Peng Ji[1,2], Yu-Cheng Chou[2], Deng-Ping Fan[1(✉)], Geng Chen[1], Huazhu Fu[1], Debesh Jha[3], and Ling Shao[1]

[1] Inception Institute of AI (IIAI), Abu Dhabi, UAE
[2] Wuhan University, Wuhan, China
[3] SimulaMet, Oslo, Norway

Abstract. Existing video polyp segmentation(VPS) models typically employ convolutional neural networks (CNNs) to extract features. However, due to their limited receptive fields, CNNs cannot fully exploit the global temporal and spatial information in successive video frames, resulting in false positive segmentation results. In this paper, we propose the novel **PNS-Net** (Progressively Normalized Self-attention Network), which can efficiently learn representations from polyp videos with real-time speed ($\sim$**140fps**) on a single RTX 2080 GPU and no post-processing. Our *PNS-Net* is based solely on a basic normalized self-attention block, equipping with recurrence and CNNs entirely. Experiments on challenging VPS datasets demonstrate that the proposed *PNS-Net* achieves state-of-the-art performance. We also conduct extensive experiments to study the effectiveness of the channel split, soft-attention, and progressive learning strategy. We find that our *PNS-Net* works well under different settings, making it a promising solution to the VPS task.

Keywords: Normalized self-attention · Polyp segmentation · Colonoscopy

1 Introduction

Early diagnosis of colorectal cancer (CRC) plays a vital role in improving the survival rate of CRC patients. In fact, the survival rate in the first stage of CRC is over 95%, decreasing to below 35% in the fourth and fifth stages [4]. Currently, colonoscopy is widely adopted in clinical practice and has become a standard method for screening CRC. During the colonoscopy, physicians visually inspect

---

G.-P. Ji and Y.-C. Chou—Contributed equally. Code: http://dpfan.net/pnsnet/.

Electronic supplementary material The online version of this chapter (https:// doi.org/10.1007/978-3-030-87193-2_14) contains supplementary material, which is available to authorized users.

© Springer Nature Switzerland AG 2021
M. de Bruijne et al. (Eds.): MICCAI 2021, LNCS 12901, pp. 142–152, 2021.
https://doi.org/10.1007/978-3-030-87193-2_14

the bowel with an endoscope to identify polyps, which can develop into CRC if left untreated. In practice, colonoscopy is highly dependent on the physicians' level of experience and suffers from a high polyp miss rate [19]. These limitations can be resolved with automatic polyp segmentation techniques, which segment polyps from colonoscopy images/videos without intervention from physicians. However, accurate and real-time polyp segmentation is a challenging task due to the low boundary contrast between a polyp and its surroundings and the large shape variation of polyps [9].

Significant efforts have been dedicated to overcoming these challenges. In early studies, learning-based methods turned to handcrafted features [17,21], such as color, shape, texture, appearance, or some combination. These methods train a classifier to separate the polyps from the background. However, they usually suffer from low accuracy due to the limited representation capability of handcrafted features in depicting heterogeneous polyps, as well as the close resemblance between polyps and hard mimics [25]. In more recent studies, deep learning methods have been used for polyp segmentation [25,27,28]. Although these methods have made some progress, they only detect polyps via drawing bounding boxes, and therefore cannot accurately locate the boundaries. To solve this, Brandao et al. [5] adopted a fully convolutional networks (FCN) with a pre-trained model to recognize and segment polyps. Later, Akbari et al. [1] introduced a modified FCN to increase the accuracy of polyp segmentation. Inspired by the success of UNet [20] in biomedical image segmentation, UNet++ [30] and ResUNet [14] are employed for polyp segmentation and achieved good results. Some methods also focus on area-boundary constraints. For instance, Psi-Net [18] makes use of polyp boundary and area information simultaneously. Fang et al. [10] introduced a three-step selective feature aggregation network. ACSNet [26] utilized an adaptive context selection based encoder-decoder framework. Zhong et al. [29] propose a context-aware network based on adaptive scale and global semantic context. Introduced more recently, the current golden standard for image polyp segmentation, PraNet [9], applies area and boundary cues in a reverse attention module, achieving the cutting-edge performance. However, these methods have only been trained and evaluated on still images and focus on static information, ignoring the temporal information in endoscopic videos which can be exploited for better results. To this end, Puyal et al. [19] propose a hybrid 2D/3D CNN architecture. This model aggregates spatial and temporal correlations and achieves better segmentation results. However, the spatial correlation between frames is restricted by the size of the kernel, preventing the accurate segmentation of fast videos.

Recently, the self-attention network [23] has shown superior performance in computer vision tasks such as video object segmentation [11], image super-resolution [24], and others. Inspired by this, in this paper, we propose a novel self-attention framework, called the **P**rogressively **N**ormalized **S**elf-attention **N**etwork (***PNS-Net***), for the video polyp segmentation (VPS) task. Our contributions are as follows:

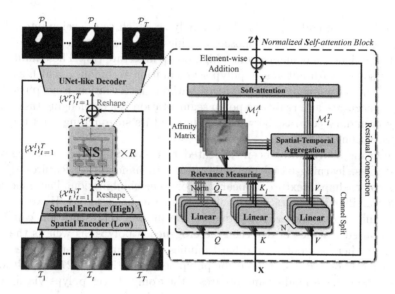

Fig. 1. Pipeline of the proposed *PNS-Net*, including the normalized self-attention block (see Sect. 2.1) with a stacked (×R) learning strategy (see Sect. 2.2).

- Different from existing CNN-based models, the proposed *PNS-Net* framework is a self-attention model for VPS, introducing a new perspective for addressing this task.
- To fully utilize the temporal and spatial cues, we propose a simple normalized self-attention (NS) block. The NS block is flexible (backbone-free) and efficient, enabling it to easily be embedded into current CNN-based encoder-decoder architectures for better performance.
- We evaluate the proposed *PNS-Net* (See Fig. 1) on challenging VPS datasets and compare it with two classical methods (*i.e.*, UNet [20] and UNet++ [30]) and three cutting-edge models (*i.e.*, ResUNet [14], ACSNet [26], and PraNet [9]). Experimental results show that *PNS-Net* achieves state-of-the-art performance with real-time speed (∼140fps). All the training data, models, results, and evaluation tools will be released http://dpfan.net/pnsnet/ to advance the development of this field.

2   Method

2.1   Normalized Self-attention (NS)

Motivation. Recently, the self-attention mechanism [23] has been widely exploited in many popular computer vision tasks. However, in our initial studies, we found that introducing the original self-attention mechanism to the VPS task does not achieve satisfactory results (*i.e.*, high accuracy and speed).

Analysis. For the VPS task, multi-scale polyps move at various speeds. Thus, dynamically updating the receptive field of the network is important. Further, the self-attention, such as the non-local network [23], incurs a high computational and memory cost, which limits the inference speed for our fast and dense prediction task. Motivated by the recent video salient object detection model [11], we utilize the **channel split**, **query-dependent**, and **normalization** rules to reduce the computational cost and improve the accuracy, respectively.

Channel Split Rule. Specifically, given an input feature (*i.e.*, $\mathbf{X} \in \mathbb{R}^{T \times H \times W \times C}$) extracted from T video frames with a size of $H \times W$ and C channels, we first utilize three linear embedding functions $\theta(\cdot)$, $\phi(\cdot)$, and $g(\cdot)$ to generate the corresponding attention features, which are implemented by a $1 \times 1 \times 1$ convolutional layer [23], respectively. This can be expressed as:

$$Q = \theta(\mathbf{X}), K = \phi(\mathbf{X}), V = g(\mathbf{X}). \tag{1}$$

Then we split each attention feature $\{Q, K, V\} \in \mathbb{R}^{T \times H \times W \times C}$ into N groups along the channel dimension and generate query, key, and value features, *i.e.*, $\{Q_i, K_i, V_i\} \in \mathbb{R}^{T \times H \times W \times \frac{C}{N}}$, where $i = \{1, 2, \cdots, N\}$.

Query-Dependent Rule. To extract the spatial-temporal relationship between successive video frames, we need to measure the similarity between query features Q_i and key features K_i. Inspired by [11], we introduce N relevance measuring (*i.e.*, query-dependent rule) blocks to compute the spatial-temporal affinity matrix for the *constrained neighborhood* of the target pixel. Rather than computing the response between a query position and the feature at all positions, as done in [23], the relevance measuring block can capture more relevance regarding the target object within T frames. More specifically, given a sliding window with fixed kernel size k and dilation rate $d_i = 2i - 1$, we get the corresponding constrained neighborhood in K_i for query pixel $\mathbf{X}^q$ of Q_i in position (x, y, z), which can be obtained by a sampling function $\mathcal{F}^S$. This is computed by:

$$\mathcal{F}^S \langle \mathbf{X}^q, K_i \rangle \in \mathbb{R}^{T(2k+1)^2 \times \frac{C}{N}} = \sum_{m=x-kd_i}^{x+kd_i} \sum_{n=y-kd_i}^{y+kd_i} \sum_{t=1}^{T} K_i(m, n, t), \tag{2}$$

where $1 \leq x \leq H$, $1 \leq y \leq W$, and $1 \leq z \leq T$. Thus, the size of the constrained neighborhood depends on the various spatial-temporal receptive fields with different kernel size k, dilation rate d_i, and frame number T, respectively.

Normalization Rule. However, the internal covariate shift problem [12] exists in the feed-forward of input Q_i, incurring that the layer parameters cannot dynamically adapt the next mini-batch. Therefore, we maintain a fixed distribution for Q_i via:

$$\hat{Q}_i = \texttt{Norm}(Q_i), \tag{3}$$

where $\texttt{Norm}$ is implemented by layer normalization [2] *along temporal dimension.*

Relevance Measuring. Finally, the affinity matrix is computed as:

$$\mathcal{M}_i^A \in \mathbb{R}^{THW \times T(2k+1)^2} = \texttt{Softmax}(\frac{\hat{Q}_i \mathcal{F}^S \langle \hat{\mathbf{X}}^q, K_i \rangle^{\mathbf{T}}}{\sqrt{C/N}}), \text{ when } \hat{\mathbf{X}}^q \in \hat{Q}_i, \tag{4}$$

where $\sqrt{C/N}$ is a scaling factor to balance the multi-head attention [22].

Spatial-Temporal Aggregation. Similar to relevance measuring, we also compute the spatial-temporally aggregated features $\mathcal{M}_i^T$ within the constrained neighborhood during temporal aggregation. This can be formulated as:

$$\mathcal{M}_i^T \in \mathbb{R}^{THW \times \frac{C}{N}} = \mathcal{M}_i^A \mathcal{F}^S \langle \mathbf{X}^a, V_i \rangle, \text{ when } \mathbf{X}^a \in \mathcal{M}_i^A, \tag{5}$$

Soft-Attention. We use a soft-attention block to synthesize features from the group of affinity matrices $\mathcal{M}_i^A$ and aggregated features $\mathcal{M}_i^T$. During the synthesis process, relevant spatial-temporal patterns should be enhanced while less relevant ones should be suppressed. We first concatenate a group of affinity matrices $\mathcal{M}_i^A$ along the channel dimension to generate $\mathcal{M}^A$. Thus, the soft-attention map $\mathcal{M}^S$ is computed by:

$$\mathcal{M}^S \in \mathbb{R}^{THW \times 1} \leftarrow \max \mathcal{M}^A, \text{ when } \mathcal{M}^A \in \mathbb{R}^{THW \times T(2k+1)^2 N}, \tag{6}$$

where the max function computes the channel-wise maximum value. Then we concatenate a group of the spatial-temporally aggregated features $\mathcal{M}_i^T$ along the channel dimension to generate $\mathcal{M}^T$.

Normalized Self-attention. Finally, our NS block can be defined as:

$$\mathbf{Z} \in \mathbb{R}^{T \times H \times W \times C} = \mathbf{X} + \mathbf{Y} = \mathbf{X} + (\mathcal{M}^T \mathbf{W}_T) \circledast \mathcal{M}^S, \tag{7}$$

where $\mathbf{W}_T$ is the learnable weight and $\circledast$ is the channel-wise Hadamard product.

2.2   Progressive Learning Strategy

Encoder. For fair comparison, we use the same backbone (*i.e.*, Res2Net-50) as in [7,9]. Given a polyp video clip with T frames as input (*i.e.*, $\{\mathcal{I}\}_{t=1}^T \in \mathbb{R}^{H' \times W' \times 3}$), we first feed them into a spatial encoder to extract two spatial features from the conv3_4 and conv4_6 layers, respectively. To alleviate the computational burden, we adopt an RFB-like [16] module to reduce the feature channel. Thus, we generate two spatial features, including low-level (*i.e.*, $\{\mathcal{X}_t^l\}_{t=1}^T \in \mathbb{R}^{H^l \times W^l \times C^l}$) and high-level (*i.e.*, $\{\mathcal{X}_t^h\}_{t=1}^T \in \mathbb{R}^{H^h \times W^h \times C^h}$)[1].

Progressively Normalized Self-attention (PNS). Most attention strategies aim to refine candidate features, such as first-order [9] and second-order [22,23] functions. As such, the strong semantic information in high-level features might be diffused gradually during the forward pass of the network. To alleviate this, we introduce a progressive residual learning strategy in our NS block. Specifically, we first reshape the corresponding high-level features $\{\mathcal{X}_t^h\}_{t=1}^T$ of consecutive

---

[1] We set $H^l = \frac{H'}{4}$, $W^l = \frac{W'}{4}$, $C^l = 24$, $H^h = \frac{H'}{8}$, $W^h = \frac{W'}{8}$, and $C^h = 32$.

input frames into a temporal feature, which can be viewed as a four-dimensional tensor (*i.e.*, $\widetilde{\mathcal{X}}^h \in \mathbb{R}^{T \times H^h \times W^h \times C^h}$). Then we refine $\widetilde{\mathcal{X}}^h$ via stacked normalized self-attention in a progressive manner:

$$\widetilde{\mathcal{X}}^r \in \mathbb{R}^{T \times H^h \times W^h \times C^h} = \text{NS}^{\times R}(\widetilde{\mathcal{X}}^h) = \text{NS}^{\times R}(\mathcal{F}^R(\{\mathcal{X}_t^h\}_{t=1}^T)), \qquad (8)$$

where $\text{NS}^{\times R}$ means that R normalized self-attention blocks are stacked in the refinement process. $\mathcal{F}^R$ is the reshaping function for the temporal dimension. To allow this block to easily be plugged into pre-trained networks, the commonly adopted solution is to add a residual learning process. Finally, the refined spatial-temporal feature is generated by:

$$\{\mathcal{X}_t^r\}_{t=1}^T \in \mathbb{R}^{H^h \times W^h \times C^h} = \mathcal{F}^R(\widetilde{\mathcal{X}}^h + \widetilde{\mathcal{X}}^r). \qquad (9)$$

Decoder and Learning Strategy. We combine the low-level feature $\{\mathcal{X}_t^l\}_{t=1}^T$ from the spatial encoder and the spatial-temporal feature $\{\mathcal{X}_t^r\}_{t=1}^T$ from the PNS block via a two-stage UNet-like decoder $\mathcal{F}^D$. Thus, the output of our method is computed by $\{\mathcal{P}_t\}_{t=1}^T = \mathcal{F}^D(\{\mathcal{X}_t^l\}_{t=1}^T, \{\mathcal{X}_t^r\}_{t=1}^T)$. We adopt the standard *binary cross-entropy* loss function in the learning process.

3   Experiments

3.1   Implementation Details

Datasets. We adopt four widely used polyp datasets in our experiments, including image-based (*i.e.*, Kvasir [13]) and video-based (*i.e.*, CVC-300 [4], CVC-612 [3], and ASU-Mayo [21]) ones. Kvasir is a large-scale and challenging dataset, which consists of 1,000 polyp images with fully annotated pixel-level ground truths (GTs). The whole Kvasir is used for training. ASU-Mayo contains 10 negative video samples from normal subjects and 10 positive samples from patients. We only adopt the positive part for training. Following the same protocol as [3,4], we split the videos from CVC-300 (12 clips) and CVC-612 (29 clips) into 60% for training, 20% for validation, and 20% for testing.

Training. Due to the limited video training data, we try to fully utilize large-scale polyp image data to capture more appearances of the polyp and scene. Thus, we train our model in two steps: *i) Pre-training phase.* We remove the normalized self-attention (NS) block from *PNS-Net* and pre-train the static backbone using an image-based polyp dataset (*i.e.*, Kvasir [13]) and the training set of video-based polyp datasets (*i.e.*, CVC-300 [4], CVC-612 [3], and ASU-Mayo [21]). The initial learning rate of the Adam algorithm and the weight decay are both 1e-4. The static part of our *PNS-Net* convergences after 100 epochs. *ii) Fine-tuning phase.* We plug the NS block into our *PNS-Net* and fine-tune the whole network using the video polyp datasets, including the ASU-Mayo and the training sets of CVC-300 and CVC-612. We set the number of attention groups

$N = 4$ and the number of stacked normalized self-attention blocks $R = 2$, along with a kernel size of $k = 3$. The initial learning is set to 1e-4, and the whole model is fine-tuned over one epoch. In this way, although the densely labeled VPS data is scarce, our *PNS-Net* still achieves good generalization performance.

Testing and Runtime. To test the performance of our *PNS-Net*, we validate it on challenging datasets, including the test set of CVC-612 (*i.e.*, CVC-612-T), the validation set of CVC-612 (*i.e.*, CVC-612-V), and the test/validation set of CVC-300 (*i.e.*, CVC-300-TV). During inference, we sample $T=5$ frames from a polyp clip and resize them to 256×448 as the input. For final prediction, we use the output $\mathcal{P}_t$ of the network followed by a *sigmoid* function. Our *PNS-Net* achieves a speed of ~140fps on a single RTX 2080 GPU without any post-processing (*e.g.*, CRF [15]). The speeds of the compared methods are listed in Table 1.

3.2 Evaluation on Video Polyp Segmentation

Baselines. We re-train five cutting-edge polyp segmentation baselines (*i.e.*, UNet [20], UNet++ [30], ResUNet [14], ACSNet [26], and PraNet [9]) with the same data used by our *PNS-Net*, under their default settings, for fair comparison.

Metrics. The metrics used included: (1) maximum Dice (maxDice), which measures the similarity between two sets of data; (2) maximum specificity (maxSpe), which refers to the percentage of the samples that are negative and are judged

Table 1. Quantitative results on different datasets.

	Metrics	2018~2019			2020		2021
		UNet	UNet++	ResUNet	ACSNet	PraNet	*PNS-Net*
		MICCAI [20]	TMI [30]	ISM [14]	MICCAI [26]	MICCAI [9]	**(OUR)**
	Speed	108 fps	45 fps	20 fps	35 fps	97 fps	**140 fps**
CVC-300-TV	maxDice↑	0.639	0.649	0.535	0.738	0.739	**0.840**
	maxSpe↑	0.963	0.944	0.852	0.987	0.993	**0.996**
	maxIoU↑	0.525	0.539	0.412	0.632	0.645	**0.745**
	S_α ↑	0.793	0.796	0.703	0.837	0.833	**0.909**
	E_ϕ ↑	0.826	0.831	0.718	0.871	0.852	**0.921**
	M ↓	0.027	0.024	0.052	0.016	0.016	**0.013**
CVC-612-V	maxDice↑	0.725	0.684	0.752	0.804	0.869	**0.873**
	maxSpe↑	0.971	0.952	0.939	0.929	0.983	**0.991**
	maxIoU↑	0.610	0.570	0.648	0.712	0.799	**0.800**
	S_α ↑	0.826	0.805	0.829	0.847	0.915	**0.923**
	E_ϕ ↑	0.855	0.830	0.877	0.887	0.936	**0.944**
	M ↓	0.023	0.025	0.023	0.054	0.013	**0.012**
CVC-612-T	maxDice↑	0.729	0.740	0.617	0.782	0.852	**0.860**
	maxSpe↑	0.971	0.975	0.950	0.975	0.986	**0.992**
	maxIoU↑	0.635	0.635	0.514	0.700	0.786	**0.795**
	S_α ↑	0.810	0.800	0.727	0.838	0.886	**0.903**
	E_ϕ ↑	0.836	0.817	0.758	0.864	**0.904**	0.903
	M ↓	0.058	0.059	0.084	0.053	0.038	**0.038**

as such; (3) maximum IoU (maxIoU), which measures the overlap between two masks; (4) S-measure [6] (S_α), which evaluates region- and object-aware structural similarity; (5) enhanced-alignment measure [8] (E_ϕ), which measures pixel-level matching and image-level statistics; and (6) mean absolute error (M), which measures the pixel-level error between the prediction and GT.

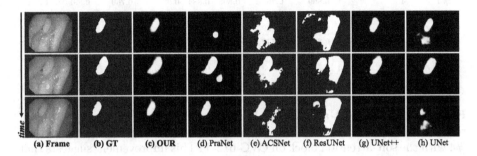

(a) Frame   (b) GT   (c) OUR   (d) PraNet   (e) ACSNet   (f) ResUNet   (g) UNet++   (h) UNet

Fig. 2. Qualitative results on CVC-612-T [3]. For more visualization results please refer to the supplementary material (*i.e.*, video).

Qualitative Comparison. In Fig. 2, we provide the polyp segmentation results of our *PNS-Net* on CVC-612-T. Our model can accurately locate and segment polyps in many difficult situations, such as different sizes, homogeneous areas, different textures, *etc.*

Quantitative Comparison. Quantitative comparison results are summarized in Table 1. We conduct three experiments on test datasets to verify the model's performance. CVC-300-TV consists of both validation set and test set, which include six videos in total. CVC-612-V and CVC-612-T each contain five videos. On CVC-300, where all the baseline methods perform poorly, our *PNS-Net* achieves remarkable performance in all metrics and outperforms all SOTA methods by a large margin (max Dice: ∼10%). On CVC-612-V and CVC-612-T, our *PNS-Net* consistently outperforms other SOTAs.

3.3 Ablation Study

Effectiveness of Channel Split. We investigate the contribution of channel split rule under different scales. The results are listed in rows #2 to #5 in Table 2. We observe that #4 (N=4) outperforms other settings (*i.e.*, #2, #3, and #5) on CVC-300-TV, in all metrics. This improvement shows that an improper receptive field (RF) harms the ability to excavate temporal information, since a large RF will pay more attention to the global environment rather than local motion information. On the other hand, when the split number is too small, the model fails to capture multi-scale polyps moving at various speeds.

Table 2. Ablation studies. See Sect. 3.3 for more details.

No	Variants					CVC-300-TV				CVC-612-T			
	Base	N	Soft	Norm	R	maxDice↑	maxIoU↑	S_α ↑	E_ϕ ↑	maxDice↑	maxIoU↑	S_α ↑	E_ϕ ↑
#1	✓					0.778	0.665	0.850	0.858	0.850	0.778	0.896	0.885
#2	✓	1			1	0.755	0.650	0.865	0.844	0.850	0.779	0.896	0.891
#3	✓	2			1	0.790	0.679	0.876	0.872	0.825	0.746	0.870	0.856
#4	✓	4			1	0.809	0.709	0.893	0.884	0.834	0.760	0.881	0.867
#5	✓	8			1	0.763	0.663	0.867	0.842	0.787	0.702	0.841	0.829
#6	✓	4	✓		1	0.829	0.729	0.896	0.903	0.852	0.784	0.895	0.897
#7	✓	4	✓	✓	1	0.827	0.732	0.897	0.898	0.856	0.792	0.898	0.896
#8	✓	4	✓	✓	2	**0.840**	**0.745**	**0.909**	**0.921**	**0.860**	**0.795**	**0.903**	**0.903**
#9	✓	4	✓	✓	3	0.737	0.609	0.793	0.751	0.732	0.613	0.776	0.728

Effectiveness of Soft-attention. We further investigate the contribution of the soft-attention mechanism. As shown in Table 2, #6 is generally better than #4 with the soft-attention block on CVC-612-T. This improvement suggests that introducing the soft-attention block to synthesize the aggregation feature and affinity matrix is necessary for increasing performance.

Effectiveness of the Number of NS Blocks. To access the number of normalized self-attention blocks under different settings, we derive three variants as #7, #8, and #9. We observe that #8 (*PNS-Net* setting) is significantly better than #7 and #9, with $R = 2$, in all metrics on CVC-300-TV and CVC-612-T. This improvement illustrates that too many iterations of NS blocks may cause overfitting on small datasets (#9). In contrast, the model fails to alleviate the diffusion issue of high-level features with a single residual block. Empirically, we recommend increasing the number of NS blocks when training on larger datasets.

4 Conclusion

We have proposed a self-attention based framework, *PNS-Net*, to accurately segment polyps from colonoscopy videos with super high speed (~140fps). Our basic normalized self-attention blocks can be easily plugged into existing CNN-based architectures. We experimentally show that our *PNS-Net* achieves the best performance on all existing publicly available datasets under six metrics. Further, extensive ablation studies demonstrate that the core components in our *PNS-Net* are all effective. We hope that the proposed *PNS-Net* can serve as a catalyst for progressing both in VPS as well as other closely related video-based medical segmentation tasks. Exploring the performance of *PNS-Net* on a larger VPS dataset will be left to our future work.

References

1. Akbari, M., et al.: Polyp segmentation in colonoscopy images using fully convolutional network. In: IEEE EMBC, pp. 69–72 (2018)

2. Ba, J.L., Kiros, J.R., Hinton, G.E.: Layer normalization. arXiv preprint arXiv:1607.06450 (2016)

3. Bernal, J., Sánchez, F.J., Fernández-Esparrach, G., Gil, D., Rodríguez, C., Vilariño, F.: Wm-dova maps for accurate polyp highlighting in colonoscopy: validation vs. saliency maps from physicians. CMIG **43**, 99–111 (2015)

4. Bernal, J., Sánchez, J., Vilarino, F.: Towards automatic polyp detection with a polyp appearance model. PR **45**(9), 3166–3182 (2012)

5. Brandao, P., et al.: Fully convolutional neural networks for polyp segmentation in colonoscopy. In: MICAD, vol. 10134, p. 101340F (2017)

6. Fan, D.P., Cheng, M.M., Liu, Y., Li, T., Borji, A.: Structure-measure: a new way to evaluate foreground maps. In: IEEE ICCV, pp. 4548–4557 (2017)

7. Fan, D.P., Ji, G.P., Cheng, M.M., Shao, L.: Concealed object detection. IEEE TPAMI **66**, 9909–9917 (2021)

8. Fan, D.P., Ji, G.P., Qin, X., Cheng, M.M.: Cognitive vision inspired object segmentation metric and loss function. SSI (2020)

9. Fan, D.P., et al.: Pranet: parallel reverse attention network for polyp segmentation. In: MICCAI, pp. 263–273 (2020)

10. Fang, Y., Chen, C., Yuan, Y., Tong, K.: Selective feature aggregation network with area-boundary constraints for polyp segmentation. In: Shen, D., et al. (eds.) MICCAI 2019. LNCS, vol. 11764, pp. 302–310. Springer, Cham (2019). https://doi.org/10.1007/978-3-030-32239-7_34

11. Gu, Y., Wang, L., Wang, Z., Liu, Y., Cheng, M.M., Lu, S.P.: Pyramid constrained self-attention network for fast video salient object detection. AAAI **34**, 10869–10876 (2020)

12. Guo, L., Liu, J., Zhu, X., Yao, P., Lu, S., Lu, H.: Normalized and geometry-aware self-attention network for image captioning. In: IEEE CVPR, pp. 10327–10336 (2020)

13. Jha, D., et al.: Kvasir-SEG: a segmented polyp dataset. In: Ro, Y.M., et al. (eds.) MMM 2020. LNCS, vol. 11962, pp. 451–462. Springer, Cham (2020). https://doi.org/10.1007/978-3-030-37734-2_37

14. Jha, D., et al.: Resunet++: an advanced architecture for medical image segmentation. In: IEEE ISM, pp. 225–2255 (2019)

15. Krähenbühl, P., Koltun, V.: Efficient inference in fully connected CRFs with gaussian edge potentials. NIPS **24**, 109–117 (2011)

16. Liu, S., Huang, D., et al.: Receptive field block net for accurate and fast object detection. In: ECCV, pp. 385–400 (2018)

17. Mamonov, A.V., Figueiredo, I.N., Figueiredo, P.N., Tsai, Y.H.R.: Automated polyp detection in colon capsule endoscopy. IEEE TMI **33**(7), 1488–1502 (2014)

18. Murugesan, B., Sarveswaran, K., Shankaranarayana, S.M., Ram, K., Joseph, J., Sivaprakasam, M.: Psi-Net: shape and boundary aware joint multi-task deep network for medical image segmentation. In: IEEE EMBC, pp. 7223–7226 (2019)

19. Puyal, J.G.B., et al.: Endoscopic polyp segmentation using a hybrid 2D/3D CNN. In: Martel, A.L., et al. (eds.) MICCAI 2020. LNCS, vol. 12266, pp. 295–305. Springer, Cham (2020). https://doi.org/10.1007/978-3-030-59725-2_29

20. Ronneberger, O., Fischer, P., Brox, T.: U-Net: convolutional networks for biomedical image segmentation. In: Navab, N., Hornegger, J., Wells, W.M., Frangi, A.F. (eds.) MICCAI 2015. LNCS, vol. 9351, pp. 234–241. Springer, Cham (2015). https://doi.org/10.1007/978-3-319-24574-4_28

21. Tajbakhsh, N., Gurudu, S.R., Liang, J.: Automated polyp detection in colonoscopy videos using shape and context information. IEEE TMI **35**(2), 630–644 (2015)

22. Vaswani, A., et al.: Attention is all you need. In: NIPS (2017)
23. Wang, X., Girshick, R., Gupta, A., He, K.: Non-local neural networks. In: IEEE CVPR, pp. 7794–7803 (2018)
24. Yang, F., Yang, H., Fu, J., Lu, H., Guo, B.: Learning texture transformer network for image super-resolution. In: IEEE CVPR, pp. 5791–5800 (2020)
25. Yu, L., Chen, H., Dou, Q., Qin, J., Heng, P.A.: Integrating online and offline three-dimensional deep learning for automated polyp detection in colonoscopy videos. IEEE JBHI **21**(1), 65–75 (2016)
26. Zhang, R., Li, G., Li, Z., Cui, S., Qian, D., Yu, Y.: Adaptive context selection for polyp segmentation. In: Martel, A.L., et al. (eds.) MICCAI 2020. LNCS, vol. 12266, pp. 253–262. Springer, Cham (2020). https://doi.org/10.1007/978-3-030-59725-2_25
27. Zhang, R., Zheng, Y., Poon, C.C., Shen, D., Lau, J.Y.: Polyp detection during colonoscopy using a regression-based convolutional neural network with a tracker. PR **83**, 209–219 (2018)
28. Zhao, X., Zhang, L., Lu, H.: Automatic polyp segmentation via multi-scale subtraction network. In: MICCAI (2021)
29. Zhong, J., Wang, W., Wu, H., Wen, Z., Qin, J.: PolypSeg: an efficient context-aware network for polyp segmentation from colonoscopy videos. In: Martel, A.L., et al. (eds.) MICCAI 2020. LNCS, vol. 12266, pp. 285–294. Springer, Cham (2020). https://doi.org/10.1007/978-3-030-59725-2_28
30. Zhou, Z., Rahman Siddiquee, M.M., Tajbakhsh, N., Liang, J.: UNet++: A nested U-Net architecture for medical image segmentation. In: Stoyanov, D., et al. (eds.) DLMIA/ML-CDS -2018. LNCS, vol. 11045, pp. 3–11. Springer, Cham (2018). https://doi.org/10.1007/978-3-030-00889-5_1

SGNet: Structure-Aware Graph-Based Network for Airway Semantic Segmentation

Zimeng Tan[1,2], Jianjiang Feng[1,2(✉)], and Jie Zhou[1,2]

[1] Department of Automation, Tsinghua University, Beijing, China
jfeng@tsinghua.edu.cn
[2] Beijing National Research Center for Information Science and Technology,
Beijing, China

Abstract. Airway semantic segmentation, which refers to segmenting airway from background and dividing it into anatomical segments, provides clinically valuable information for lung lobe analysis, pulmonary lesion localization, and comparison between different patients. It is technically challenging due to the complicated tree-like structure, individual variations, and severe class imbalance. We propose a structure-aware graph-based network (SGNet) for airway semantic segmentation directly from chest CT scans. The proposed framework consists of a feature extractor combining a multi-task U-Net with a structure-aware GCN, and an inference module comprised of two convolutional layers. The multi-task U-Net is trained to regress bifurcation landmark heatmaps, binary and semantic segmentation maps simultaneously, providing initial predictions for graph construction. By introducing irregular edges connecting voxels with the sampled points around corresponding bifurcation landmarks, the two-layer GCN incorporates the structural prior explicitly. Experiments on both public and private datasets demonstrate that the SGNet achieves superior and robust performance, even on subjects affected by severe pulmonary diseases.

Keywords: Airway semantic segmentation · Graph convolutional network · Structural prior

1 Introduction

Airway segmentation on chest CT scans is crucial for pulmonary disease diagnosis and surgical navigation [5–7]. Airway semantic segmentation (see Fig. 1(b)), which can be regarded as an extension of airway binary segmentation (see Fig. 1(a)), refers to segmenting airway from background and dividing it into 32 anatomical segments defined in [1,2] according to the anatomical topology.

This work was supported in part by the National Natural Science Foundation of China under Grants 82071921.

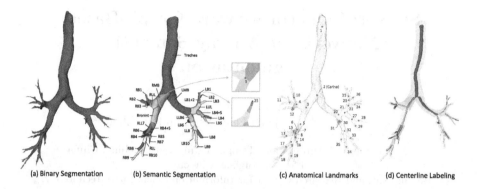

(a) Binary Segmentation   (b) Semantic Segmentation   (c) Anatomical Landmarks   (d) Centerline Labeling

Fig. 1. Illustration of airway binary segmentation, 32 semantic segments, 35 anatomical bifurcation landmarks and centerline labeling.

Airway semantic segmentation can not only capture the whole tree-like structure, but also provide more clinically valuable information for many subsequent automated procedures, such as lung lobe analysis, pulmonary lesion localization, and comparison of the same bronchi from different subjects.

Considering that manual annotation is time-consuming and laborious, several methods based on convolutional neural networks (CNNs) [3,4] have been developed for airway binary segmentation [5–12]. Despite the remarkable progress in binary segmentation, airway semantic segmentation can be much more challenging. Firstly, severe class imbalance occurs naturally due to the sparse distribution of tertiary bronchi, which are very tenuous and tend to be misdetected or undetected. Secondly, there is no obvious interface between adjacent segments (e.g., RMB and RUL in Fig. 1(b)), that is, the voxels around the interface may have similar local appearance and grayscale distribution but different labels. Furthermore, the individual variations, pathological changes, and the large amount of segments contained in an airway tree pose threat to the network training. Another more relevant task is the centerline labeling (see Fig. 1(d)), which takes the extracted centerline as a prerequisite and divides it into different categories for airway [13,14], coronary arteries [15–17], brain arteries [18] and abdominal arteries [19]. An intuitive approach for airway semantic segmentation is to combine binary segmentation with centerline extraction and labeling [13]. Nevertheless, it is not trivial to accomplish these tasks robustly. Completing them independently and integrating results may not achieve the best performance.

Recently, graph convolutional network (GCN) [20] is developed to reason the relationships among sampled points and has already been used in tree-like anatomical structure segmentation [21–25]. Shin et al. [23] combined GCN with a CNN framework to explore global connectivity for 2D vessel segmentation in retinal and coronary artery X-ray images. Juarez et al. [24] replaced the deepest convolutional layer in the U-Net by a series graph convolutions to improve airway binary segmentation. In a more relevant paper, Yao et al. [25] proposed a GCN-based point cloud approach to conduct semantic segmentation of head and neck vessels in CTA images, in which CNN-based binary segmentation is converted to point cloud and a GCN is employed to further label vessels into 13 major segments. However, only the neighbors on regular image grid are taken into consideration in [25] and the tree-like structural prior has not been fully utilized.

In this paper, we propose an end-to-end deep learning framework, called Structure-aware Graph-based Network (SGNet), for airway semantic segmentation directly from chest CT scans. Specifically, as shown in Fig. 2, a multi-task U-Net is combined with a specially designed GCN for feature extraction and two stacked convolution layers serve as interference module. A total of 35 bifurcation landmarks are predefined at the center of interface between adjacent segments (see Fig. 1(c)) for both auxiliary task of multi-task U-Net and sampling strategy of graph vertices. Our main contributions are summarized as follows: (1) The U-Net is modified to accomplish anatomical landmark detection, airway binary and semantic segmentation simultaneously. The synergy among different objectives guides the network to learn more discriminative features. (2) Based on the initial prediction of the U-Net, graph vertices are sampled randomly on the segments and around the detected landmark positions. Besides the regular edges between vertex and its nearest neighbors, we explore another type of edge connecting each vertex with the points sampled around the corresponding landmarks. In this way, we introduce the structural prior knowledge which boosts the final performance. The voxels' coordinates and distances relative to the carina (shown in Fig. 1(c)) are also fed into the GCN to enrich feature representation. (3) To our knowledge, this is the first paper concentrated on airway semantic segmentation. In order to train and evaluate the proposed system, we have made annotations of 100 CT scans, in which the annotations of 60 public CT scans are released to promote further study of airway semantic segmentation[1]. Experiments on both public and private datasets demonstrated that our proposed method achieves superior performance on airway semantic segmentation, even on subjects affected by severe pulmonary diseases such as COVID-19 and interstitial pneumonia.

2 Method

2.1 Multi-task U-Net Module

Given an input 3D volume, we exploit a modified multi-task U-Net [3] to obtain the coarse initial prediction of airway semantic segmentation. Based on the generic pipeline of pixel-wise classification approach, we introduce landmark detection and airway binary segmentation as auxiliary objectives to guide the network to learn more discriminative features. A total of 35 anatomical landmarks are predefined at the interface center of adjacent semantic classes, so that each segment can be expressed by the two points at its ends. Inspired by [26], we convert landmark detection object to a heatmap regression task. Compared with regressing coordinates directly, the voxel-to-voxel heatmap regression

---

[1] Annotations are available at https://cloud.tsinghua.edu.cn/d/43bbc05fb9714f71 a56f/

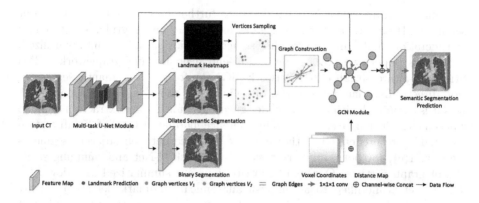

Fig. 2. Overview of the proposed SGNet for airway semantic segmentation.

method focuses on each position and is more suitable intrinsically for landmark localization. The discrete coordinates of each landmark are modeled to a channel heatmap with a Gaussian distribution centered at the position. The heatmap value $G_k(x)$ of voxel x ranging in $[0,1]$ can be regarded as the probability to be the kth landmark, which is determined according to the standard deviation δ and the distance from the position x to the target landmark x_k. During inference, the voxel with maximum predicted probability is chosen to be the corresponding landmark position. Specifically, the formulation of the heatmap is as follows:

$$G_k(x) = e^{-\frac{1}{2\delta^2}(x-x_k)^2}, k = 1, 2, ..., 35. \tag{1}$$

For airway binary segmentation, it should be noted that unlike the original binary segmentation task, only the 32 predefined anatomical segments are taken into consideration and the peripheral bronchioles are omitted. In this way, the auxiliary objectives are highly correlated with the main task, and the synergy among them boosts the individual performance of each task. Furthermore, we introduce the global context information by capturing the whole airway structure and incorporate the spatial relationships among landmarks implicitly.

2.2 Structure-Aware Graph Convolutional Network

Graph Construction. Based on the initial prediction of airway semantic segmentation S_1 and landmark detection L_1 obtained from the multi-task U-Net, we construct a structure-aware graph $G(V,E)$, where $V = \{v_i\}_{i=1}^N$ and E are sets of vertices and edges, respectively. As shown in Fig. 2, we have two types of vertices, one is sampled randomly from the whole segmented airway region and denoted by $V_1 = \{v_i\}_{i=1}^{N_1}$, the other is sampled around the detected landmark positions and denoted by $V_2 = \{v_i\}_{i=1}^{N_2}$, where $N_1 + N_2 = N$. The sampling density of vertices V_1 varies with different segments to pay extra attention to the tenuous tertiary bronchi. Each vertex $v_i \in V$ is linked with its k-nearest neighbors in the Euclidean space. Besides this regular connectivity, we propose an additional edge between per vertex $v_i \in V_1$ and point $v_j \in V_2$ sampled around the corresponding detected landmarks according to its predicted semantic class $S_1(v_i)$. Specifically, if vertex v_i is classified on the lth segment (i.e. $S_1(v_i) = l$), it should be connected with the vertices sampled around the two endpoints x_A and x_B (see Fig. 2). These novel edges introduce the airway structural prior information explicitly, allowing vertices to access faraway but meaningful neighbors. Moreover, connecting with a series points distributed around the landmark prediction brings a degree of error tolerance in landmark detection.

Graph Convolutional Network. A two-layer graph convolutional network (GCN) [20] operating on the constructed graph $G(V, E)$ is adopted for further feature extraction in our framework. Input feature vector of each vertex is sampled from the immediate feature map $F_0 \in \mathbb{R}^{h \times w \times d \times C_1}$, which is generated by the convolutional layer in the multi-task U-Net before the three branch subnetworks. To enrich feature representation, the voxels' coordinates in 3 dimensions and distances relative to the carina prediction, which is always detected accurately in experiments, are also fed into the input. The matrix of all input feature vectors is denoted as $F_1 \in \mathbb{R}^{N \times (C_1+4)}$. The binary adjacency matrix $A \in \{0,1\}^{N \times N}$ can be obtained from the constructed graph, which is largely sparse with non-zero entries corresponding to the predefined edges. Then the degree matrix D is derived from A with diagonal entries $D_{ii} = \sum_j A_{ij}$. The two-layer graph convolutional network with self connections is formulated as:

$$F_2 = \sigma \left(\hat{A} \sigma \left(\hat{A} F_1 W^{(0)} \right) W^{(1)} \right) \tag{2}$$

where $\tilde{A} = A + I_N, \tilde{D}_{ii} = \sum_j \tilde{A}_{ij}$, and $\hat{A} = \tilde{D}^{-\frac{1}{2}} \tilde{A} \tilde{D}^{-\frac{1}{2}}$. $\sigma(\cdot)$ is the rectified linear unit activation function. $W^{(0)}$ and $W^{(1)}$ are trainable filter weights. F_2 is the output feature representation.

Unlike standard convolution operating on a local regular grid, graph convolution intuitively enables irregular information exchange on extracted graph and hence guarantees a much wider receptive field. Benefiting from the specially designed graph, the GCN module can capture more discriminative structural features. To conduct final inference for airway semantic segmentation, we convert the output feature matrix $F_2 \in \mathbb{R}^{N \times C_2}$ to a sparse matrix form $F3 \in \mathbb{R}^{h \times w \times d \times C_2}$ by projecting the feature vector of each vertex to the corresponding voxel position. Then the feature map F_0 from the U-Net and F_3 from the GCN are concatenated channel-wise and fed into the interference module, which consists of two $1 \times 1 \times 1$ convolutional layers followed by non-linear activation function.

2.3 Loss Functions and Training Methodology

We apply weighted L2 loss, cross-entropy loss, and Dice loss function [27] for landmark detection, binary segmentation and semantic segmentation in both the U-Net module and the whole SGNet, respectively. To address the class imbalance issue, the weights are set to be the voxel number ratio of the background class

Table 1. Quantitative results on both public and private test set evaluated using mean Dice coefficient (%) and standard deviation.

Public dataset

Trachea	RMB	RUL	BronInt	RB1	RB2	RB3	RB4+5	RB4
91.3(5.25)	83.4(4.78)	82.5(4.76)	83.6(4.80)	84.2(4.83)	83.3(4.86)	81.8(4.70)	79.9(4.57)	84.5(5.43)
RB5	RB6	RLL7	RB7	RLL	RB8	RB9	RB10	LMB
81.7(4.73)	74.6(4.35)	80.2(4.60)	85.8(4.83)	80.5(4.62)	74.6(4.35)	88.4(5.11)	86.8(5.01)	82.5(4.90)
LUL	LLB6	LB1+2	LB1	LB2	LB3	LB4+5	LB4	LB5
77.8(4.50)	77.7(4.46)	83.2(4.83)	85.8(5.34)	84.8(5.06)	68.8(3.82)	80.7(4.62)	86.4(5.55)	86.1(5.01)
LB6	LLB	LB8	LB9	LB10	Average		Overall	
78.8(4.57)	81.2(4.66)	70.3(3.95)	83.4(4.72)	78.1(4.57)	81.7(4.76)		89.3(1.96)	

Private dataset

Trachea	RMB	RUL	BronInt	RB1	RB2	RB3	RB4+5	RB4
90.9(2.63)	82.8(2.37)	82.0(2.36)	82.1(2.31)	76.9(2.16)	78.5(2.25)	78.5(2.27)	79.6(2.28)	76.3(2.21)
RB5	RB6	RLL7	RB7	RLL	RB8	RB9	RB10	LMB
73.9(2.13)	78.0(2.22)	75.8(2.13)	70.5(2.01)	78.8(2.27)	72.8(2.17)	62.6(1.79)	65.9(1.82)	80.2(2.33)
LUL	LLB6	LB1+2	LB1	LB2	LB3	LB4+5	LB4	LB5
76.0(2.21)	75.3(2.15)	68.1(2.15)	73.2(2.24)	7.14(2.18)	67.7(2.03)	71.3(2.13)	85.0(2.42)	80.8(2.35)
LB6	LLB	LB8	LB9	LB10	Average		Overall	
74.9(2.12)	77.7(2.24)	71.3(2.04)	70.4(2.09)	84.2(2.39)	76.0(2.20)		88.5(1.61)	

to each type of foreground. We adopt a sequential training scheme composed of initial pretraining of the U-Net and joint training of the whole framework. The objective functions of the U-Net module and the SGNet are defined as:

$$\mathcal{L}_{U-Net} = \mathcal{L}_{land} + \alpha\mathcal{L}_{bseg} + \beta\mathcal{L}_{sseg-ini} \tag{3}$$

$$\mathcal{L}_{SGNet} = \mathcal{L}_{U-Net} + \gamma\mathcal{L}_{sseg-final} \tag{4}$$

where the hyper-parameters α, β, and γ are adjusted to make different components having the same scale. Similar to [25], we dilate the ground truth mask of semantic segmentation in the U-Net, ensuring inclusion of all meaningful voxels in the GCN sampling stage and improving the performance especially around the airway wall. During training, the vertices set V_2 is sampled around the ground truth landmark positions to avoid unnecessary disturbance. During inference, the landmark predictions of the U-Net are utilized for graph construction.

3   Experiments and Results

3.1   Datasets and Implementation Details

Datasets. We evaluated our approach on both public and private chest CT datasets. The public dataset [6] is comprised of 40 scans from LIDC-IDRI [28] and 20 scans from EXACT'09 [29], with airway binary segmentation annotation provided. The private dataset contains 40 scans with 10 kinds of severe pulmonary diseases (e.g. COVID-19, pulmonary cryptococcosis, interstitial pneumonia and adenocarcinoma). 35 bifurcation anatomical landmarks and binary

Table 2. Comparison of different networks on both public and private test set measured by DSC (%), TPR (%), and FPR (10^{-4}). The U-Net, mU-Net, and +GCN represent the original U-Net, the proposed multi-task U-Net, and the combination of the multi-task U-Net and a GCN with regular connectivity, respectively.

Public dataset														
Models	Trachea			Primary			Lobar			Tertiary			Average		
	DSC	TPR	FPR	DSC	TPR	FPR	DSC	TPR	FPR	DSC	TPR	FPR	DSC	TPR	FPR
U-Net	88.7	84.1	1.66	79.2	73.0	**0.51**	74.9	71.1	0.20	73.6	73.7	0.07	74.7	73.3	0.18
mU-Net	89.3	85.1	1.66	80.5	74.6	0.55	77.6	76.1	0.24	76.3	77.6	0.07	77.3	77.3	0.19
+GCN	90.5	86.6	1.5	81.5	80.2	0.69	78.7	75.6	0.19	79.9	79.6	0.05	80.0	78.8	0.17
SGNet	**91.3**	**88.0**	**1.48**	**83.0**	**81.5**	0.67	**80.4**	**77.6**	**0.18**	**81.6**	**79.7**	**0.04**	**81.7**	**79.5**	**0.16**
Private dataset														
Models	Trachea			Primary			Lobar			Tertiary			Average		
	DSC	TPR	FPR	DSC	TPR	FPR	DSC	TPR	FPR	DSC	TPR	FPR	DSC	TPR	FPR
U-Net	90.2	88.0	3.37	77.5	70.3	0.56	75.4	73.9	0.30	63.1	70.2	0.11	67.9	71.6	0.28
mU-Net	90.3	88.4	3.29	77.3	70.1	**0.53**	75.3	74.4	0.27	65.5	70.1	0.07	69.4	71.7	0.25
+GCN	90.7	89.2	2.50	80.2	74.4	0.57	77.8	75.8	0.26	71.6	72.4	**0.06**	74.3	73.9	0.22
SGNet	**90.9**	**89.8**	**2.20**	**81.5**	**76.1**	0.56	**78.4**	**76.3**	**0.24**	**73.9**	**73.7**	0.06	**76.0**	**75.0**	**0.20**

segmentation (only for private dataset) were annotated by well-trained experts. Based on the binary segmentation annotation, the airway semantic segmentation ground truth of each CT scan was established automatically according to the landmark distribution, followed by manual correction and delineation. All scans were spatially normalized to the same resolution of $0.7 \times 0.7 \times 1$ mm^3 with average size of $486 \times 486 \times 317$. The datasets were randomly split into training set (40 public and 30 private scans) and test set (20 public and 10 private scans).

Implementation Details. To improve the model's generalizability, we applied data augmentation via random translation and rotation for 3 times totally on the training set. We trained our U-Net module and SGNet using Adam optimizer ($\beta_1 = 0.5, \beta_2 = 0.999$) with a learning rate of 0.0001 for 100 and 50 epochs, respectively. For hyper-parameters, we empirically set $\alpha = 10^{-3}, \beta = 10^{-7}$, and $\gamma = 10^{-5}$. For the constructed graph, we randomly sampled 1024 points for each airway tree to form vertex set V_1 and 8 points around each bifurcation landmark to form vertex set V_2. Each vertex is connected with its 16-nearest neighbors in V_1 and the vertices located around the corresponding two endpoints in V_2. The proposed model was implemented in TensorFlow framework with 2 NVIDIA GeForce RTX 3090 GPUs.

3.2 Results

The performance of the proposed SGNet was evaluated utilizing Dice coefficient, and the quantitative results on both public and private test sets are presented in Table 1. The overall Dice coefficient refers to converting the semantic segmentation prediction to binary form by regarding voxels classified in airway segments

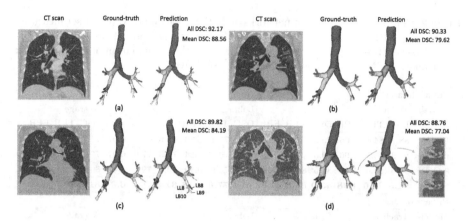

Fig. 3. Qualitative results of (a) a public test case, (b) a private test case, (c) individual variations of airway anatomical topology, and (d) pathological effects of severe pulmonary diseases.

as foreground, and then comparing with the ground truth. The average Dice coefficient indicates the mean of all airway segments. The qualitative results of 4 patients are illustrated in Fig. 3. The proposed SGNet achieves satisfactory performance on most airway segments (see Fig. 3(a) and (b)), while some failure cases may be ascribed to the sparse distribution of complicated airway structure, tenuous tertiary bronchi, and individual variations of airway anatomical topology (e.g. in some subjects, as shown in Fig. 3(a), LB8-LB10 diverge from LLB simultaneously, while in others, as shown in Fig. 3(c), LB8 first diverges from LLB, and then LLB is divided into LB9 and LB10. A similar situation can be seen in RB8-RB10.). There are some ruptures between the adjacent airway segments, which may be improved by taking voxel-wise spatial connectivity into account and will be explored in future work. Meanwhile, the performance on the private dataset shows a slight decrease. It is intuitively more difficult due to the pathological effects of severe pulmonary diseases, such as the indistinct airway wall and changes of the lung grayscale distribution (e.g. see Fig. 3(d) from a patient with interstitial pneumonia). The experimental results demonstrate the effectiveness and robustness of the proposed framework on challenging cases.

Furthermore, we conducted an ablation study to verify each component of our proposed framework. For a more comprehensive comparison, we additionally introduced true positive rate (TPR) and false positive rate (FPR) together with Dice coefficient (DSC) as evaluation metrics. For demonstration purpose, we divided 32 airway anatomical segments into 4 groups and computed the average metrics of each one, i.e., trachea, primary bronchi (including LMB and RMB), lobar bronchi (including BronInt, RUL, RB4+5, RLL7, RLL, LUL, LLB6, LLB), and tertiary bronchi (including RB1-RB10, LB1+2, LB4+5, LB1-LB6, LB8-LB10). Note that in order to cover all the segments, there are slight differences between this division and anatomical definition (e.g. we classified BronInt to the lobar bronchi group). We compared the performance of four different networks: the original U-Net trained for airway semantic segmentation, the proposed multi-task U-Net, combination of the multi-task U-Net and a GCN with regular connectivity, and the proposed SGNet. Results on both public and private test sets are presented in Table 2, where the last 4 columns represent the average measurements on all 32 segments. The proposed SGNet achieved the highest DSC and TPR of each airway segment on both test sets with comparable FPR. Adding auxiliary objectives, combining with GCN, and introducing the structure-aware connectivity successively improves the average DSC by 2.6%, 2.7%, 1.7%, and 1.5%, 4.9%, 1.7% on public and private datasets, respectively. It implies that integrating global context information and introducing structural prior knowledge explicitly contribute to the overall performance improvement, demonstrating the effectiveness of the proposed framework.

4 Conclusion

In this paper, we presented an end-to-end structure-aware graph-based network (SGNet) for airway semantic segmentation directly from chest CT scans, in which a multi-task U-Net and a GCN are combined for more discriminative feature extraction. The structural prior knowledge was incorporated explicitly by introducing irregular graph connectivity. Experiments showed the effectiveness and robustness of the proposed method on both public and private datasets. In the future, we will extend our framework to other tree-like anatomical structures and perform a more comprehensive comparison with related techniques.

References

1. van Ginneken, B., Baggerman, W., van Rikxoort, E.M.: Robust segmentation and anatomical labeling of the airway tree from thoracic CT scans. In: Metaxas, D., Axel, L., Fichtinger, G., Székely, G. (eds.) MICCAI 2008. LNCS, vol. 5241, pp. 219–226. Springer, Heidelberg (2008). https://doi.org/10.1007/978-3-540-85988-8_27

2. Tschirren, J., McLennan, G., Palágyi, K., et al.: Matching and anatomical labeling of human airway tree. IEEE T-MI **24**(12), 1540–1547 (2005)

3. Ronneberger, O., Fischer, P., Brox, T.: U-Net: convolutional networks for biomedical image segmentation. In: Navab, N., Hornegger, J., Wells, W.M., Frangi, A.F. (eds.) MICCAI 2015. LNCS, vol. 9351, pp. 234–241. Springer, Cham (2015). https://doi.org/10.1007/978-3-319-24574-4_28

4. Çiçek, Ö., Abdulkadir, A., Lienkamp, S.S., Brox, T., Ronneberger, O.: 3D U-Net: learning dense volumetric segmentation from sparse annotation. In: Ourselin, S., Joskowicz, L., Sabuncu, M.R., Unal, G., Wells, W. (eds.) MICCAI 2016. LNCS, vol. 9901, pp. 424–432. Springer, Cham (2016). https://doi.org/10.1007/978-3-319-46723-8_49

5. Qin, Y., et al.: AirwayNet: a voxel-connectivity aware approach for accurate airway segmentation using convolutional neural networks. In: Shen, D., et al. (eds.) MICCAI 2019. LNCS, vol. 11769, pp. 212–220. Springer, Cham (2019). https://doi.org/10.1007/978-3-030-32226-7_24

6. Qin, Y., Gu, Y., Zheng, H., et al.: AirwayNet-SE: a simple-yet-effective approach to improve airway segmentation using context scale fusion. In: ISBI (2020)

7. Qin, Y., et al.: Learning bronchiole-sensitive airway segmentation CNNs by feature recalibration and attention distillation. In: Martel, A.L., et al. (eds.) MICCAI 2020. LNCS, vol. 12261, pp. 221–231. Springer, Cham (2020). https://doi.org/10.1007/978-3-030-59710-8_22

8. Charbonnier, J.P., Van Rikxoort, E.M., Setio, A.A., et al.: Improving airway segmentation in computed tomography using leak detection with convolutional networks. MedIA **36**, 52–60 (2017)

9. Meng, Q., Roth, H.R., Kitasaka, T., Oda, M., Ueno, J., Mori, K.: Tracking and segmentation of the airways in chest CT using a fully convolutional network. In: Descoteaux, M., Maier-Hein, L., Franz, A., Jannin, P., Collins, D.L., Duchesne, S. (eds.) MICCAI 2017. LNCS, vol. 10434, pp. 198–207. Springer, Cham (2017). https://doi.org/10.1007/978-3-319-66185-8_23

10. Yun, J., Park, J., Yu, D., et al.: Improvement of fully automated airway segmentation on volumetric computed tomographic images using a 2.5 dimensional convolutional neural net. MedIA **51**, 13–20 (2019)

11. Wang, C., et al.: Tubular structure segmentation using spatial fully connected network with radial distance loss for 3D medical images. In: Shen, D., et al. (eds.) MICCAI 2019. LNCS, vol. 11769, pp. 348–356. Springer, Cham (2019). https://doi.org/10.1007/978-3-030-32226-7_39

12. Zhao, T., Yin, Z., Wang, J., Gao, D., Chen, Y., Mao, Y.: Bronchus segmentation and classification by neural networks and linear programming. In: Shen, D., et al. (eds.) MICCAI 2019. LNCS, vol. 11769, pp. 230–239. Springer, Cham (2019). https://doi.org/10.1007/978-3-030-32226-7_26

13. Gu, S., Wang, Z., Siegfried, J.M., et al.: Automated lobe-based airway labeling. Int. J. Biomed. Imaging **2012** (2012)

14. Feragen, A.: A hierarchical scheme for geodesic anatomical labeling of airway trees. In: Ayache, N., Delingette, H., Golland, P., Mori, K. (eds.) MICCAI 2012. LNCS, vol. 7512, pp. 147–155. Springer, Heidelberg (2012). https://doi.org/10.1007/978-3-642-33454-2_19

15. Wu, D., et al.: Automated anatomical labeling of coronary arteries via bidirectional tree LSTMs. Int. J. Comput. Assist. Radiol. Surg. **14**(2), 271–280 (2018). https://doi.org/10.1007/s11548-018-1884-6

16. Yang, H., Zhen, X., Chi, Y., et al.: CPR-GCN: conditional partial-residual graph convolutional network in automated anatomical labeling of coronary arteries. In: CVPR (2020)

17. Cao, Q., et al.: Automatic identification of coronary tree anatomy in coronary computed tomography angiography. Int. J. Cardiovasc. Imaging **33**(11), 1809–1819 (2017). https://doi.org/10.1007/s10554-017-1169-0

18. Robben, D., Türetken, E., Sunaert, S., et al.: Simultaneous segmentation and anatomical labeling of the cerebral vasculature. MedIA **32**, 201–215 (2016)

19. Matsuzaki, T., Oda, M., Kitasaka, T., et al.: Automated anatomical labeling of abdominal arteries and hepatic portal system extracted from abdominal CT volumes. MedIA **20**(1), 152–161 (2015)

20. Kipf, T.N., Welling, M.: Semi-supervised classification with graph convolutional networks. In: ICLR (2017)

21. Selvan, R., Kipf, T., Welling, M., et al.: Extraction of airways using graph neural networks. In: MIDL (2018)

22. Selvan, R., Kipf, T., Welling, M., et al.: Graph refinement based airway extraction using mean-field networks and graph neural networks. MedIA **64**, 101751 (2020)

23. Shin, S.Y., Lee, S., Yun, I.D., et al.: Deep vessel segmentation by learning graphical connectivity. MedIA **58**, 101556 (2019)
24. Juarez, A.G.U., Selvan, R., Saghir, Z., de Bruijne, M.: A joint 3D unet-graph neural network-based method for airway segmentation from chest CTs. In: MLMI (2019)
25. Yao, L., Jiang, P., Xue, Z., et al.: Graph convolutional network based point cloud for head and neck vessel labeling. In: MLMI (2020)
26. Payer, C., Štern, D., Bischof, H., Urschler, M.: Regressing heatmaps for multiple landmark localization using CNNs. In: Ourselin, S., Joskowicz, L., Sabuncu, M.R., Unal, G., Wells, W. (eds.) MICCAI 2016. LNCS, vol. 9901, pp. 230–238. Springer, Cham (2016). https://doi.org/10.1007/978-3-319-46723-8_27
27. Milletari, F., Navab, N., Ahmadi, S.A.: V-net: fully convolutional neural networks for volumetric medical image segmentation. In: 3DV (2016)
28. Armato, I.I.I., Samuel, G., et al.: The lung image database consortium (LIDC) and image database resource initiative (IDRI): a completed reference database of lung nodules on CT scans. Med. Phys. **38**(2), 915–931 (2011)
29. Lo, P., Ginneken, B.V., Reinhardt, J.M., et al.: Extraction of airways from CT (EXACT'09). IEEE T-MI **31**(11), 2093–2107 (2012)

NucMM Dataset: 3D Neuronal Nuclei Instance Segmentation at Sub-Cubic Millimeter Scale

Zudi Lin[1(✉)], Donglai Wei[1], Mariela D. Petkova[1], Yuelong Wu[1],
Zergham Ahmed[1], Krishna Swaroop K[2], Silin Zou[1], Nils Wendt[3],
Jonathan Boulanger-Weill[1], Xueying Wang[1], Nagaraju Dhanyasi[1],
Ignacio Arganda-Carreras[4,5,6], Florian Engert[1], Jeff Lichtman[1],
and Hanspeter Pfister[1]

[1] Harvard University,Cambridge, USA
{linzudi,donglai}@g.harvard.edu
[2] NIT Karnataka, Mangalore, India
[3] Technical University of Munich, Munich, Germany
[4] Donostia International Physics Center (DIPC), Donostia-San Sebastian, Spain
[5] University of the Basque Country (UPV/EHU), Leioa, Spain
[6] Ikerbasque, Basque Foundation for Science,Bilbao, Spain

Abstract. Segmenting 3D cell nuclei from microscopy image volumes is critical for biological and clinical analysis, enabling the study of cellular expression patterns and cell lineages. However, current datasets for *neuronal* nuclei usually contain volumes smaller than $10^{-3}\,\mathrm{mm}^3$ with fewer than 500 instances per volume, unable to reveal the complexity in large brain regions and restrict the investigation of neuronal structures. In this paper, we have pushed the task forward to the sub-cubic millimeter scale and curated the *NucMM* dataset with two fully annotated volumes: one $0.1\,\mathrm{mm}^3$ electron microscopy (EM) volume containing nearly the entire zebrafish brain with around 170,000 nuclei; and one $0.25\,\mathrm{mm}^3$ micro-CT (uCT) volume containing part of a mouse visual cortex with about 7,000 nuclei. With two imaging modalities and significantly increased volume size and instance numbers, we discover a great diversity of neuronal nuclei in appearance and density, introducing new challenges to the field. We also perform a statistical analysis to illustrate those challenges quantitatively. To tackle the challenges, we propose a novel hybrid-representation learning model that combines the merits of foreground mask, contour map, and signed distance transform to produce high-quality 3D masks. The benchmark comparisons on the NucMM dataset show that our proposed method significantly outperforms state-of-the-art nuclei segmentation approaches. Code and data are available at https://connectomics-bazaar.github.io/proj/nucMM/index.html.

Z. Lin and D. Wei—Equally contributed.

Krishna Swaroop K and N. Wendt—Works were done during internship at Harvard University.

M. de Bruijne et al. (Eds.): MICCAI 2021, LNCS 12901, pp. 164–174, 2021.
https://doi.org/10.1007/978-3-030-87193-2_16

Keywords: 3D Instance Segmentation · Nuclei · Brain · Electron
Microscopy (EM) · Micro-CT (uCT) · Zebrafish · Mouse

1 Introduction

Segmenting cell nuclei from volumetric (3D) microscopy images is an essential
task in studying biological systems, ranging from specific tissues [20] to entire
organs [28] and developing animals [18,25]. There has been a great success in
benchmarking 2D and 3D nuclei segmentation methods using datasets covering
samples from various species [4,27,30,33]. However, existing datasets only have
relatively small samples from brain tissues (*e.g.*, volumes from mouse and rat
brains [23] smaller than 10^{-3} mm^3, with less than 200 instances each), restricting
the investigation of neuronal nuclei in a larger and more diverse scale. Besides,
most of the images are collected with optical microscopy, which can not reflect
the challenges in other imaging modalities widely used in studying brain tissues,
including electron microscopy (EM) [14,24] and micro-CT (uCT) [9].

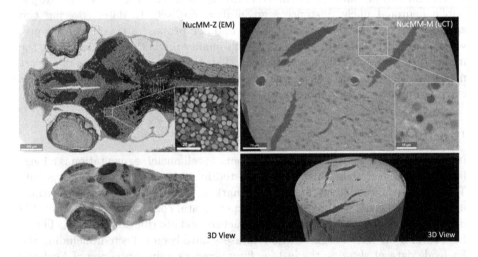

Fig. 1. Overview of the NucMM dataset. NucMM contains two large volumes for
3D nuclei instance segmentation, including **(left)** the NucMM-Z electron microscopy
(EM) volume covering nearly a whole zebrafish brain, and **(right)** the NucMM-M
micro-CT volume from the visual cortex of a mouse.

To address this deficiency in the field, we have curated a large-scale 3D nuclei
instance segmentation dataset, **NucMM**, which is over two magnitudes larger
in terms of volume size and the number of instances than previous neuronal
nuclei datasets [4,23] and widely-used non-neuronal benchmark datasets [1,33].
Our NucMM consists of one EM image volume covering a nearly entire larval
zebrafish brain and a uCT image volume from the visual cortex of an adult

mouse, facilitating large-scale cross-tissue and cross-modality comparison. Challenges in the two volumes for automatic approaches include the high density of closely touching instances (Fig. 1, left) and low contrast between object and non-object regions (Fig. 1, right). We also perform a statistical analysis to provide a quantitative justification of the challenges in these two volumes (Fig. 2).

To tackle the challenges introduced by the large-scale NucMM dataset, we propose a new hybrid-representation model that learns foreground mask, instance contour map, and signed distance transform simultaneously with a 3D U-Net [5] architecture, which is denoted as U3D-BCD. At inference time, we utilize a multi-target watershed segmentation algorithm that combines all three predictions to separate closely touching instances and generate high-quality instance masks. Under the average precision (AP) metric for evaluating instance segmentation approaches, we show that our U3D-BCD model significantly outperforms existing state-of-the-art approaches by **22%** on the NucMM dataset. We also perform ablation studies on the validation data to demonstrate the sensitivity of segmentation parameters.

To summarize, we have three main contributions in this paper. First, we collected and densely annotated the NucMM dataset, which is one of the largest public neuronal nuclei instance segmentation datasets to date, covering two species with two imaging modalities. Second, we propose a novel hybrid representation model, U3D-BC, to produce high-quality predictions by combining the merits of different mask representations. Third, we benchmark state-of-the-art nuclei segmentation approaches and show that our proposed model outperforms existing approaches by a large margin on the NucMM dataset.

1.1   Related Works

Nuclei Segmentation Datasets. Automatic cell nuclei segmentation is a long-lasting problem that is usually the first step in microscopy image analysis [19]. There has been a great success in benchmarking nuclei segmentation algorithms in 2D images across various types and experimental conditions [4]. However, 2D images can not completely display the structure and distribution of nuclei. Therefore, several 3D nuclei segmentation datasets have been collected, including the Parhyale dataset showing the histone fluorescent protein expression of *Parhyale hawaiensis* [1,33], the BBBC050 dataset showing nuclei of mouse embryonic cells [27], and multiple 3D volumes recording human cancer cells and animal embryonic cells [30]. However, there are few datasets covering neuronal nuclei in brain tissues, and the volumes collected from brain tissue usually contain less than 200 instances per volume [4,23]. Besides, most public datasets mentioned above only have fluorescence images obtained with optical microscopy.

The NucMM dataset we curated contains two volumes from animal brain tissues and covers two imaging modalities widely used in neuroscience, including EM [14,24] and uCT [9]. In addition, our dataset is over two magnitudes larger than previous neuronal nuclei volumes [23] in terms of size and instance number.

Table 1. NucMM dataset characteristics. We collected and fully annotated a *neuronal nuclei* segmentation dataset at the sub-cubic millimeter scale. The two volumes in the dataset cover two species and two imaging modalities.

Name	Sample	Modality	Volume size	Resolution (μm)	#Instance
NucMM-Z	Zebrafish	SEM	$397 \times 1450 \times 2000$ ($0.14\,\mathrm{mm}^3$)	$0.48 \times 0.51 \times 0.51$	170K
NucMM-M	Mouse	Micro-CT	$700 \times 996 \times 968$ ($0.25\,\mathrm{mm}^3$)	$0.72 \times 0.72 \times 0.72$	7K

Instance Segmentation in Microscopy Images. Instance segmentation requires assigning each pixel (voxel) not only a semantic label but also an instance index to differentiate objects that belong to the same category. The permutation-invariance of object indices makes the task challenging. For segmenting instances in microscopy images, recent learning-based approaches first train 2D or 3D convolutional neural network (CNN) models to predict an intermediate representation of the object masks such as boundary [6,22] or affinity maps [16,29]. Then techniques including watershed transform [8,35] and graph partition [15] are applied to convert the representations into object masks. Since one representation can be vulnerable to some specific kinds of errors (*e.g.*, small mispredictions on the affinity or boundary map can cause significant merge errors), some approaches also employ hybrid-representation learning models that learn multiple representations and combine their information in the segmentation step [26,32,33]. However, those approaches only optimize 2D CNNs [26] or learn targets that may not be suitable for nuclei segmentation [32], which leads to unsatisfactory results for down-stream analysis. Thus we propose a 3D hybrid-representation learning model that predicts foreground mask, instance contour, and signed distance to better capture nuclei with different textures and shapes.

2   NucMM Dataset

Dataset Acquisition. The EM dataset was collected from an entire larval zebrafish brain with *serial-section* electron microscopy (SEM) [21]. The original resolution is 30 nm × 4 nm × 4 nm for z-, y-, x-axis. We downsample the images to 480 nm × 512 nm × 512 nm (Table 1) to make the resolution close to other imaging modalities for nuclei analysis. The micro-CT dataset was collected from layer II/III in the primary visual cortex of an adult male mouse using 3D X-ray microscopy. The images have an isotropic voxel size of $720\,\mathrm{nm}^3$ (Table 1).

Dataset Annotation. For annotating both volumes, we use a semi-automatic pipeline that first applies automatic algorithms to generate instance candidates and then asks neuroscientists to proofread the masks. Since EM images have high contrast and the object boundaries are well-defined (Fig. 1), we apply filtering and thresholding to get the binary foreground mask and run a watershed transform to produce the segments. For the uCT data, we iteratively enlarge the labeled set by alternating between manual correction (annotation) and automatic U-Net [22] prediction. Both volumes are finally proofread by experienced neuroscientists using VAST [2]. We also provide a binary mask for each volume to indicate the valid brain region for evaluation.

Dataset Statistics. The EM data has a size of $397 \times 1450 \times 2000$ voxels, equivalent to a physical size of $0.14\,\mathrm{mm}^3$; the uCT data has $700 \times 996 \times 968$ voxels, equivalent to $0.25\,\mathrm{mm}^3$ (Table 1). Although the zebrafish volume is physically smaller, it contains significantly more objects than the mouse volume, showing the distinction between brain structures. We show the distribution of nuclei size (Fig. 2a) and the nearest-neighbor distance between nuclei centers (Fig. 2b). The density plots illustrate that objects in the zebrafish volume are smaller and more densely packed, which poses challenges in separating closely touching nuclei.

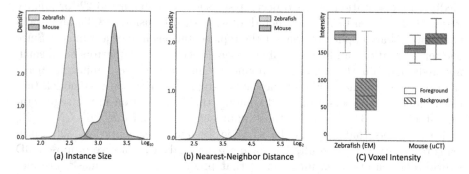

(a) Instance Size   (b) Nearest-Neighbor Distance   (C) Voxel Intensity

Fig. 2. NucMM dataset statistics. We show the distribution of **(a)** instance size (in terms of number of voxels) and **(b)** the distance between adjacent nuclei centers. The density plots are normalized by the total number of instances in each volume. We also show **(c)** the voxel intensity distribution in object (foreground) and non-object (background) regions for both volumes.

We further show the voxel intensity distribution of object and non-object regions (Fig. 2c). The boxplots demonstrate that objects in the EM volume are better separated from the background than the uCT volume. To quantify the separation, we calculate the *Kullback–Leibler* (KL) divergence between the foreground and background intensity distributions in the two volumes. The results are $D_{KL} = 3.43$ and 1.33 for the zebrafish and mouse data, respectively, showing the foreground-background contrast is significantly lower in the uCT data.

Dataset Splits and Evaluation Metric. We split each volume into 5%, 5%, and 90% parts for training, validation, and testing. The limited training data makes the task more challenging, but it is also closer to the realistic annotation budget when neuroscientists handle newly collected data. Besides, to avoid sampling data in a local region without enough diversity, we follow previous practice [13] and sample 27 small chunks of size $64 \times 64 \times 64$ voxels from the zebrafish volume for training. Since the nuclei in the mouse uCT volume are sparser (Fig. 1 and Fig. 2b), we sample 4 chunks of size $192 \times 192 \times 192$ voxels for training.

For evaluation, we use average precision (AP), a standard metric in assessing instance segmentation methods [7,17]. Specifically, we use the code optimized for large-scale 3D image volumes [32] to facilitate efficient evaluation.

3 Method

3.1 Hybrid-Representation Learning

Recent 3D instance segmentation methods for microscopy images, including Cellpose [26], StarDist [33] and U3D-BC [32], all use a single model to learn multiple mask representations simultaneously. Specifically, Cellpose [26] regresses the horizontal and vertical spatial gradients of the instances; StarDist [33] learns object probability and the star-convex distance within the masks; U3D-BC [32] learns the foreground mask with the instance contour map. By analyzing the representations used in those models, we notice that all the targets emphasize the learning of object masks (foreground), but the structure of the *background* of the segmentation map is less utilized. That is, pixels close to and far away from the object masks are treated equitably.

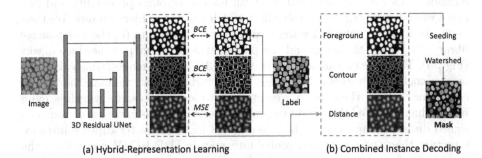

(a) Hybrid-Representation Learning   (b) Combined Instance Decoding

Fig. 3. Hybrid-representation learning model. (a) Our U3D-BCD model learns a set of hybrid representations simultaneously, including foreground mask, instance contour, and signed distance transform map calculated from the segmentation. (b) The representations are combined in seeding and watershed transform to produce high-quality segmentation masks.

Therefore based on the U3D-BC model that predicts both the binary foreground mask (B) and instance contour map (C), we develop a **U3D-BCD** model that in addition predicts a signed Euclidean distance map (Fig. 3a). Let x_i denote a pixel in the image, we have:

$$f(x_i) = \begin{cases} +\text{dist}(x_i, B)/\alpha, & \text{if } x \in F. \\ -\text{dist}(x_i, F)/\beta, & \text{if } x \in B. \end{cases} \tag{1}$$

where F and B denote the foreground and background masks, respectively. The scaling parameters α and β are applied to control the range of the distance. Compared with the U3D-BC baseline, the signed distance map is more informative in capturing the shape information of masks. In comparison with all discussed approaches [26,32,33], the signed distance map also model the landscape of background regions. In implementation, we apply a *tanh* activation to restrict the range to $(-1, 1)$ and directly regress the target with a 3D U-Net [5]. A similar learning target has been used for semantic segmentation of synapses [12], but it has not been integrated into a multi-task learning model nor has it been explored for 3D instance segmentation. Specifically, the loss we optimize is

$$\mathcal{L}_{bcd} = h(\sigma(y_1), y_b) + h(\sigma(y_2), y_c) + g(\phi(y_3), y_d) \qquad (2)$$

where y_i $(i = 1, 2, 3)$ denote the three output channels, while σ and ϕ denote the sigmoid and tanh function. The foreground (y_b) and contour (y_c) maps are learned by optimizing the binary cross-entropy (BCE) loss (h), while the signed distance map (y_d) is learned with the mean-squared-error (MSE) loss (g).

3.2 Instance Decoding

In this part, we describe how to combine all three model predictions in the U3D-BCD model to generate the segmentation masks. We first threshold predictions to locate *seeds* (or markers) with high foreground probability and distance value but low contour probability, which points to object centers. We then apply the marker-controlled watershed transform algorithm (in the *scikit-image* library [31]) using the seeds and the predicted distance map to produce masks (Fig. 3b). There are two advantages over the U3D-BC [32] model, which also uses marker-controlled watershed transform for decoding. First, we make use of the consistency among the three representations to locate the seeds, which is more robust than U3D-BC that uses two predictions. Furthermore, we use the smooth signed distance map in watershed decoding, which can better capture instance structure than the foreground probability map in U3D-BC. We also show the sensitivity of decoding parameters in the experiments.

3.3 Implementation

We use a customized 3D U-Net model that substitutes the convolutional layers at each stage with residual blocks [10]. We also change the concatenation operation to addition to save memory. Since recent work [34] has shown that ADAM-alike adaptive optimization algorithms do not generalize as well as stochastic gradient descent (SGD) [3], we optimize our models with SGD and adopt *cosine learning rate decay* [11]. We set the scaling parameters α and β of the signed distance map to 8 and 50, respectively, without tweaking. We follow the open-source code of U3D-BC[1] and apply data augmentations including brightness, flip, rotation,

---

[1] https://github.com/zudi-lin/pytorch_connectomics.

elastic transform, and missing parts augmentations to U3D-BC [32] and our U3D-BCD. We also use the official implementation of StarDist[2] and Cellpose[3].

4   Experiments

4.1   Methods in Comparison

We benchmark state-of-the-art microscopy image segmentation models including Cellpose [26], StarDist [33] and U3D-BC [32] based on their public implementations. Specifically, the Cellpose model is trained on the 2D xy, yz, and xz planes from the image volumes, and the predicted spatial gradients are averaged to generate a 3D gradient before segmentation. For StarDist, we calculate the optimal number of rays on the training data, which are 96 and 64 for the zebrafish and mouse volumes, respectively. For U3D-BC, we use the default 1.0 weight ratio between the losses of the foreground and contour map. Models are trained on a machine with four Tesla V100 GPUs.

Table 2. Benchmark results on the NucMM dataset. We compare state-of-the-art methods on the NucMM dataset using AP score. **Bold** and underlined numbers denote the 1st and 2nd scores, respectively. Our U3D-BCD model significantly improves the performance of previously state-of-the-art approaches.

Method	NucMM-Z			NucMM-M			Overall
	AP-50	AP-75	Mean	AP-50	AP-75	Mean	
Cellpose [26]	0.796	0.342	0.569	0.463	0.002	0.233	0.401
StarDist [33]	<u>0.912</u>	0.328	0.620	0.306	0.004	0.155	0.388
U3D-BC [32]	0.782	<u>0.556</u>	<u>0.670</u>	**0.645**	<u>0.210</u>	<u>0.428</u>	<u>0.549</u>
U3D-BCD (**Ours**)	**0.978**	**0.809**	**0.894**	<u>0.638</u>	**0.250**	**0.444**	**0.669**

4.2   Benchmark Results on the NucMM Dataset

After choosing hyper-parameter on the validation sets, we run predictions on the 90% test data in each volume and evaluate the performance. Specifically, we show the AP scores at intersection-over-union (IoU) thresholds of both 0.5 (AP-50) and 0.75 (AP-75), as well as their average (Table 2). The overall score is averaged over two NucMM volumes. The results show that our U3D-BCD model significantly outperforms previous state-of-the-art models by relatively **22%** in overall performance. Besides, our method ranks 1st in 6 out of 7 scores, showing its robustness in handling different challenges. We argue that Cellpose [26] trains 2D models to estimate 3D spatial gradient, which can be ineffective for challenging 3D cases. The other two models use 3D models, but the representations

---

[2] https://github.com/stardist/stardist.
[3] https://github.com/MouseLand/cellpose.

StarDist [33] uses overlooks the background in the segmentation mask, while U3D-BC [32] overlooks both foreground and background structures. Although conceptually straightforward, introducing the signed distance on top of the U3D-BC baseline gives a notable performance boost.

4.3   Sensitivity of the Decoding Parameters

We also show the sensitivity of the decoding hyper-parameters of our U3D-BCD model. Specifically, there are 5 thresholds for segmentation: 3 values are the thresholds of foreground probability ($\tau1$), contour probability ($\tau2$), and distance value ($\tau3$) to decide the seeds. The other 2 values are the thresholds of foreground probability ($\tau4$) and distance value ($\tau5$) to decide the valid foreground regions.

The validation results show that the final segmentation performance is not sensitive to the $\tau1$ and $\tau2$ in deciding seeds. When fixing $\tau4 = 0.2$ and $\tau5 = 0.0$ changing the foreground probability $\tau1$ from 0.4 to 0.8 only changes the AP-50 score between 0.943 and 0.946. While changing the contour probability $\tau2$ from 0.05 to 0.30 only changes the scores from 0.937 to 0.946. However, when changing $\tau4$ within the range of 0.1 to 0.4, the score changes from 0.872 to 0.946. Those results suggest that the signed distance transform map contains important information about the object structures and performs an important role in generating high-quality segmentation masks.

5   Conclusion

In this paper, we introduce the large-scale NucMM dataset for 3D neuronal nuclei segmentation in two imaging modalities and analysis the challenges quantitatively. We also propose a simple yet effective model that significantly outperforms existing approaches. We expect the densely annotated dataset can inspire various applications beyond its original task, *e.g.*, feature pre-training, shape analysis, and benchmarking active learning and domain adaptation methods.

Acknowledgments. This work has been partially supported by NSF award IIS-1835231 and NIH award U19NS104653. We thank Daniel Franco-Barranco for setting up the challenge using NucMM. M.D.P. would like to acknowledge the support of Howard Hughes Medical Institute International Predoctoral Student Research Fellowship. I.A-C would like to acknowledge the support of the Beca Leonardo a Investigadores y Creadores Culturales 2020 de la Fundación BBVA.

References

1. Alwes, F., Enjolras, C., Averof, M.: Live imaging reveals the progenitors and cell dynamics of limb regeneration. Elife **5**, e19766 (2016)
2. Berger, D.R., Seung, H.S., Lichtman, J.W.: Vast (volume annotation and segmentation tool): efficient manual and semi-automatic labeling of large 3D image stacks. Front. Neural Circ. **12**, 88 (2018)

3. Bottou, L.: Stochastic gradient learning in neural networks. In: Proceedings of Neuro-Nımes (1991)

4. Caicedo, J.C., et al.: Nucleus segmentation across imaging experiments: the 2018 data science bowl. Nat. Methods **16**, 1247–1253 (2019)

5. Çiçek, Ö., Abdulkadir, A., Lienkamp, S.S., Brox, T., Ronneberger, O.: 3D U-Net: learning dense volumetric segmentation from sparse annotation. In: Ourselin, S., Joskowicz, L., Sabuncu, M.R., Unal, G., Wells, W. (eds.) MICCAI 2016. LNCS, vol. 9901, pp. 424–432. Springer, Cham (2016). https://doi.org/10.1007/978-3-319-46723-8_49

6. Ciresan, D., Giusti, A., Gambardella, L.M., Schmidhuber, J.: Deep neural networks segment neuronal membranes in electron microscopy images. In: NeurIPS (2012)

7. Cordts, M., et al.: The cityscapes dataset for semantic urban scene understanding. In: CVPR (2016)

8. Cousty, J., Bertrand, G., Najman, L., Couprie, M.: Watershed cuts: minimum spanning forests and the drop of water principle. TPAMI **31**, 1362–1374 (2008)

9. Dyer, E.L., et al.: Quantifying mesoscale neuroanatomy using x-ray microtomography. Eneuro (2017)

10. He, K., Zhang, X., Ren, S., Sun, J.: Deep residual learning for image recognition. In: CVPR (2016)

11. He, T., Zhang, Z., Zhang, H., Zhang, Z., Xie, J., Li, M.: Bag of tricks for image classification with convolutional neural networks. In: CVPR (2019)

12. Heinrich, L., Funke, J., Pape, C., Nunez-Iglesias, J., Saalfeld, S.: Synaptic cleft segmentation in non-isotropic volume electron microscopy of the complete *drosophila* brain. In: Frangi, A.F., Schnabel, J.A., Davatzikos, C., Alberola-López, C., Fichtinger, G. (eds.) MICCAI 2018. LNCS, vol. 11071, pp. 317–325. Springer, Cham (2018). https://doi.org/10.1007/978-3-030-00934-2_36

13. Januszewski, M., et al.: High-precision automated reconstruction of neurons with flood-filling networks. Nat. Methods **15**, 605–610 (2018)

14. Kasthuri, N., et al.: Saturated reconstruction of a volume of neocortex. Cell **162**, 648–661 (2015)

15. Krasowski, N., Beier, T., Knott, G., Köthe, U., Hamprecht, F.A., Kreshuk, A.: Neuron segmentation with high-level biological priors. TMI **37**, 829–839 (2017)

16. Lee, K., Zung, J., Li, P., Jain, V., Seung, H.S.: Superhuman accuracy on the SNEMI3D connectomics challenge. arXiv:1706.00120 (2017)

17. Lin, T.Y., et al.: Microsoft COCO: common objects in context. In: Fleet, D., Pajdla, T., Schiele, B., Tuytelaars, T. (eds.) ECCV 2014. LNCS, vol. 8693, pp. 740–755. Springer, Cham (2014). https://doi.org/10.1007/978-3-319-10602-1_48

18. Lou, X., Kang, M., Xenopoulos, P., Munoz-Descalzo, S., Hadjantonakis, A.K.: A rapid and efficient 2d/3d nuclear segmentation method for analysis of early mouse embryo and stem cell image data. Stem Cell Rep. **2**, 382–397 (2014)

19. Meijering, E.: Cell segmentation: 50 years down the road. Signal Process. Mag. **29**, 140–145 (2012)

20. Nhu, H.T.T., Drigo, R.A.E., Berggren, P.O., Boudier, T.: A novel toolbox to investigate tissue spatial organization applied to the study of the islets of langerhans. Sci. Rep. **7**, 1–12 (2017)

21. Petkova, M.: Correlative Light and Electron Microscopy in an Intact Larval Zebrafish. Ph.D. thesis (2020)

22. Ronneberger, O., Fischer, P., Brox, T.: U-Net: convolutional networks for biomedical image segmentation. In: Navab, N., Hornegger, J., Wells, W.M., Frangi, A.F. (eds.) MICCAI 2015. LNCS, vol. 9351, pp. 234–241. Springer, Cham (2015). https://doi.org/10.1007/978-3-319-24574-4_28

23. Ruszczycki, B., et al.: Three-dimensional segmentation and reconstruction of neuronal nuclei in confocal microscopic images. Front. Neuroanatomy **13**, 81 (2019)
24. Shapson-Coe, A., et al.: A connectomic study of a petascale fragment of human cerebral cortex. bioRxiv (2021)
25. Stegmaier, J., et al.: Real-time three-dimensional cell segmentation in large-scale microscopy data of developing embryos. Dev. Cell **36**, 225–240 (2016)
26. Stringer, C., Wang, T., Michaelos, M., Pachitariu, M.: Cellpose: a generalist algorithm for cellular segmentation. Nat. Methods **18**, 100–106 (2021)
27. Tokuoka, Y., et al.: 3D convolutional neural networks-based segmentation to acquire quantitative criteria of the nucleus during mouse embryogenesis. NPJ Syst. Biol. Appl. **6**, 1–12 (2020)
28. Toyoshima, Y., et al.: Accurate automatic detection of densely distributed cell nuclei in 3D space. PLoS Comput. Biol. **12**, e1004970 (2016)
29. Turaga, S.C., Briggman, K.L., Helmstaedter, M., Denk, W., Seung, H.S.: Maximin affinity learning of image segmentation. In: NeurIPS (2009)
30. Ulman, V., et al.: An objective comparison of cell-tracking algorithms. Nat. Methods **14**, 1141–1152 (2017)
31. van der Walt, S., et al.: The scikit-image contributors: scikit-image: image processing in Python. PeerJ (2014)
32. Wei, D., et al.: MitoEM dataset: large-scale 3D mitochondria instance segmentation from EM images. In: Martel, A.L., et al. (eds.) MICCAI 2020. LNCS, vol. 12265, pp. 66–76. Springer, Cham (2020). https://doi.org/10.1007/978-3-030-59722-1_7
33. Weigert, M., Schmidt, U., Haase, R., Sugawara, K., Myers, G.: Star-convex polyhedra for 3d object detection and segmentation in microscopy. In: WACV (2020)
34. Zhou, P., Feng, J., Ma, C., Xiong, C., HOI, S., et al.: Towards theoretically understanding why sgd generalizes better than adam in deep learning. arXiv preprint arXiv:2010.05627 (2020)
35. Zlateski, A., Seung, H.S.: Image segmentation by size-dependent single linkage clustering of a watershed basin graph. arXiv:1505.00249 (2015)

AxonEM Dataset: 3D Axon Instance Segmentation of Brain Cortical Regions

Donglai Wei[1]([✉]), Kisuk Lee[2], Hanyu Li[3], Ran Lu[2], J. Alexander Bae[2],
Zequan Liu[4], Lifu Zhang[5], Márcia dos Santos[6], Zudi Lin[1], Thomas Uram[7],
Xueying Wang[1], Ignacio Arganda-Carreras[8,9,10], Brian Matejek[1],
Narayanan Kasthuri[3,7], Jeff Lichtman[1], and Hanspeter Pfister[1]

[1] Harvard University, Cambridge, USA
donglai@seas.harvard.edu
[2] Princeton University, Princeton, USA
[3] University of Chicago, Chicago, USA
[4] RWTH Aachen University, Aachen, Germany
[5] Boston University, Boston, USA
[6] Universidade do Vale do Rio dos Sinos, Sao Leopoldo, Brazil
[7] Argonne National Laboratory, Lemont, USA
[8] Donostia International Physics Center, Donostia-San Sebastian, Spain
[9] University of the Basque Country (UPV/EHU), Leioa, Spain
[10] Ikerbasque, Basque Foundation for Science, Bilbao, Spain

Abstract. Electron microscopy (EM) enables the reconstruction of neural circuits at the level of individual synapses, which has been transformative for scientific discoveries. However, due to the complex morphology, an accurate reconstruction of cortical axons has become a major challenge. Worse still, there is no publicly available large-scale EM dataset from the cortex that provides dense ground truth segmentation for axons, making it difficult to develop and evaluate large-scale axon reconstruction methods. To address this, we introduce the *AxonEM* dataset, which consists of two $30 \times 30 \times 30$ μm^3 EM image volumes from the human and mouse cortex, respectively. We thoroughly proofread over 18,000 axon instances to provide dense 3D axon instance segmentation, enabling large-scale evaluation of axon reconstruction methods. In addition, we densely annotate nine ground truth subvolumes for training, per each data volume. With this, we reproduce two published state-of-the-art methods and provide their evaluation results as a baseline. We publicly release our code and data at https://connectomics-bazaar.github.io/proj/AxonEM/index.html to foster the development of advanced methods.

Keywords: Axon · Electron microscopy · 3D instance segmentation

---

D. Wei and K. Lee—Equal contribution.
Z. Liu, L. Zhang and M. dos Santos—Works were done during the internship at Harvard University.

© Springer Nature Switzerland AG 2021
M. de Bruijne et al. (Eds.): MICCAI 2021, LNCS 12901, pp. 175–185, 2021.
https://doi.org/10.1007/978-3-030-87193-2_17

1   Introduction

With recent technical advances in high-throughput 3D electron microscopy (EM) [16], it has become feasible to map out neural circuits at the level of individual synapses in large brain volumes [10,29,30]. Reconstruction of a wiring diagram of the whole-brain, or a connectome, has seen successes in invertebrates, such as the complete connectome of the roundworm *C. elegans* [35] and the partial connectome of the fruit fly *D. melanogaster* [9,29]. Mapping the connectome of a whole mouse brain is currently being considered as the next transformative challenge in the field of connectomics [3]. Detailed study of large-scale connectomes may lead to new scientific discoveries with potential clinical implications [3].

At present, automated reconstruction of neural circuits from the cortex faces a significant obstacle: the axon instance segmentation. In the cortex, axons are the most abundant type of neurites by path length [24], with the smallest diameter of ∼50 nm in the case of mouse cortical axons [14]. Given their small caliber, complex morphology, and densely packed nature (Fig. 1), axons are prone to reconstruction errors especially when challenged by various image defects such as imperfect staining and image misalignment [17].

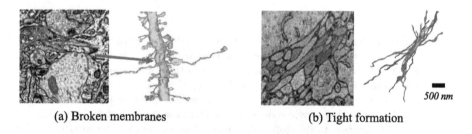

(a) Broken membranes   (b) Tight formation

Fig. 1. Example axons in our AxonEM dataset. **(a)** Due to the unclear cell boundaries resulted from imperfect staining, an axon is falsely merged with an abutting dendrite. **(b)** Axons often form a tight bundle. Due to the similar shape and small size of their cross-sections, axons in such a tight bundle may be prone to errors, especially in the presence of image misalignment.

To the best of our knowledge, there is no existing EM reconstruction from any cortical region that provides *densely and fully* proofread axon instances. Previously the largest work was carried out by Motta *et al.* [24], in which the team spent ∼4,000 human hours in semi-automated proofreading of the initial automated segmentation. However, their finalized reconstruction for axons suffered from a relatively high error rate [24],[1] making their reconstruction less suitable for developing and evaluating large-scale axon reconstruction methods.

To foster new research to address the challenge, we have created a large-scale benchmark for 3D axon instance segmentation, *AxonEM*, which is about 250×

---

[1] 12.8 errors per 1 mm of path length, estimated with 10 randomly chosen axons [24].

larger than the prior art [2] (Fig. 2). AxonEM consists of two $30 \times 30 \times 30\ \mu m^3$ EM image volumes acquired from the mouse and human cortex, respectively. Both image volumes span, at least partially, layer 2 of the cortex, thus allowing for comparative studies on brain structures across different mammalian species.

Contributions. First, we thoroughly proofread over 18,000 axon instances to provide the *largest ever* ground truth for cortical axons, which enables large-scale *evaluation*. Second, we *densely* annotate a handful of dataset subvolumes, which can be used to *train* a broad class of methods for axon reconstruction and allow for an objective comparison. Third, we publicly release our data and baseline code to reproduce two published state-of-the-art neuron instance segmentation methods [15,18], thereby calling for innovative methods to tackle the challenge.

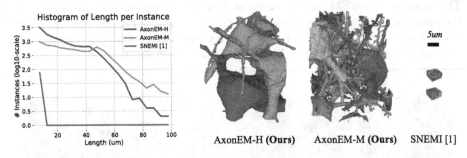

Fig. 2. Comparison of EM datasets with proofread axon segmentation. (Left) Distribution of axon instance length. (Right) 3D rendering of the neurons with somas in our AxonEM dataset and the dense segmentation of the SNEMI3D dataset [2]. For AxonEM-H, two large glial cells (not rendered) occupy the space between the blue and pink neurons, resulting in much fewer long axons compared to AxonEM-M.

1.1   Related Works

Axon Segmentation. In non-cortical regions, myelinated axons are thick and easier to segment. Traditional image processing algorithms have been proposed, including thresholding and morphological operations [8], axon shape-based morphological discrimination [23], watershed [33], region growing [39], active contours without [4] and with discriminant analysis [36]. More recently, deep learning methods have also been used to segment this type of axons [22,25,37].

However, *unmyelinated* axons in the cortex have been a significant source of reconstruction errors, because their thin and intricate branches are vulnerable to various types of image defects [17]. Motta *et al.* [24] took a semi-automated reconstruction approach to densely segment axons in a volume of $\sim$500,000 μm^3 from the mouse cortex. However, their estimate of remaining axon reconstruction errors was relatively high [24].

Neuron Segmentation in EM. There are two mainstream approaches to instance segmentation from 3D EM image volumes. One approach trains convolutional neural networks (CNNs) to predict an intermediate representation for neuronal boundaries [7,27,38] or inter-voxel affinities [12,18,31], which are then often conservatively oversegmented with watershed [40]. For further improvement, adjacent segments are agglomerated by a similarity measure [12,18,26], or with graph partitioning [5]. In the other approach, CNNs are trained to recursively extend a segmentation mask for a single object of interest [15,21]. Recently, recurrent neural networks (RNNs) have also been employed to extend multiple objects simultaneously [13,20].

2   AxonEM Dataset

The proposed AxonEM dataset contains two EM image volumes cropped from the published cortex datasets, one from mouse [11] (AxonEM-M) and the other from human [30,34] (AxonEM-H). We densely annotated axons longer than 5 μm in both volumes (Sect. 2.2), and performed basic analysis (Sect. 2.3).

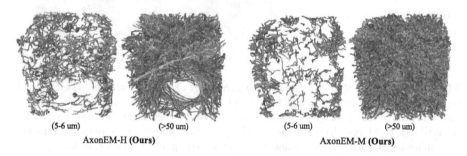

(5-6 um)   (>50 um)   (5-6 um)   (>50 um)

AxonEM-H **(Ours)**   AxonEM-M **(Ours)**

Fig. 3. Visualization of ground truth axon segmentation. We show 3D meshes of axons with length either 5–6 μm or > 50 μm. As expected, the short axons are fragments near the volume boundaries that briefly enter and leave the volume.

2.1   Dataset Description

Two tissue blocks were imaged with serial section EM from layer 2/3 in the primary visual cortex of an adult mouse and from layer 2 in the temporal lobe of an adult human. The image volumes were 2× in-plane-downsampled to the voxel resolutions of $7 \times 7 \times 40$ nm^3 and $8 \times 8 \times 30$ nm^3, respectively. We cropped out two $30 \times 30 \times 30$ μm^3 subvolumes, AxonEM-M and AxonEM-H, intentionally avoiding large blood vessels in order to contain more axon instances. We refer readers to the original papers for more dataset details [11,30].

2.2   Dataset Annotation

For training, we annotated nine $512 \times 512 \times 50$ subvolumes to produce dense ground truth segmentation, respectively, for each volume. For evaluation, we annotated all axon segments that are longer than 5 μm. We choose this threshold because segments shorter than 5 μm mostly concentrate near the borders of the volumes, making it difficult to confidently classify them into axons due to the lack of context. We took a semi-automatic approach for annotation. Starting from the initial segmentation provided by the original datasets [11,30], proofreaders inspected potential error locations suggested by automatic detection [19] in Neuroglancer [1], and manually corrected remaining errors in VAST [6].

Initial Dense Segmentation. For AxonEM-M, we downloaded a publicly available volume segmentation computed from a variant of 3D U-Net model [11,18] and applied the connected components method to obtain the initial segmentation to start with. For AxonEM-H, we manually created small volumes of dense segmentation to train the flood-filling network [15] and ran inference on the entire AxonEM-H volume to generate the initial segmentation to start with.

Axon Identification. We first identified axons from the initial dense segmentation, along with the skeletonization obtained by the TEASER algorithm [28]. We computed for each segment the following statistics: volume, path length, and width (*i.e.*, average radius of all skeleton points). Based on these stats, we used the K-means method to cluster the segments into $K = 10$ groups with similar morphology. Then we visually inspected the 3D meshes of all segments that are longer than 5 μm, and classified them into four categories: glia, dendrite, axon, or uncertain. Instead of going through each individual segments one by one, we pulled a small-sized batch from each of the groups with similar morphology. This enabled us to collectively inspect such a "minibatch" of morphologically similar segments all at once, thus making our classification more efficient and effective.

Axon Proofreading. After identifying axons, we automatically detected the potential error locations based on morphological features and biological prior knowledges [19]. To correct remaining split errors, we measured the direction of the axon skeletons at every end point, and then selected candidate pairs of axons whose end points are close in space and pointing to each other in direction. We examined these candidate pairs, as well as those axon skeletons that terminate in the middle of the volume without reaching the boundaries, which may indicate a premature termination caused by a split error. To fix remaining merge errors, we inspected the junction points in the skeleton graph, because some of them may exhibit a biologically implausible motif such as the "X-shaped" junction [21].

Finalization. Four annotation experts with experience in axon reconstruction were recruited to proofread and cross-check the axon identification and proofreading results until there remains no disagreement between the annotators.

2.3 Dataset Analysis

The physical size of our EM volumes is about 250× larger than the previous SNEMI3D benchmark [2]. AxonEM-H and AxonEM-M have around 11.3K and 6.7K axon instances (longer than 5 μm), respectively, over 200× more than that of SNEMI3D. We compare the distribution of axon instance length in Fig. 2, where our AxonEM dataset contains more and longer axons. Due to the presence of two glial cells, in AxonEM-H there are fewer long, branching axons than in AxonEM-M. In Fig. 3, we show 3D meshes of short (5–6 μm) and long axons (>50 μm) from both volumes. Again, one can see the vacancy of long, branching axons in AxonEM-H where the glial cells (not visualized) occupy the space.

3 Methods

For 3D axon instance segmentation, we describe our choice of evaluation metric (Sect. 3.1) and two state-of-the-art methods adopted for later benchmark (Sect. 4).

3.1 Task and Evaluation Metric

Our task is dense segmentation of 3D EM images of the brain, *e.g.*, as in the SNEMI3D benchmark challenge [2]. We focus, however, our attention only to the accurate reconstruction of axons, which are challenging due to their long, thin morphology and complicated branching patterns. To this end, our evaluation is restricted to axons among all ground truth segments.

To measure the overall accuracy of axon reconstruction, we adopted the expected run length (ERL, [15]) to estimate the average error-free path length of the reconstructed axons. The previous ARAND metric used in SNEMI3D [2] requires dense ground truth segmentation with voxel-level precision, which is impractical for large-scale datasets. In contrast, the original ERL metric focuses on the skeleton-level accuracy with a disproportionately larger penalty on merge errors, in which two or more distinct objects are merged together erroneously. In practice, due to skeletonization artifacts, the original ERL metric may assign a good axon segment *zero* run length if there exists a merge error with just one outlier skeleton point that is mistakenly placed near the ground truth segment boundaries. To mitigate this, we extended the original ERL metric, which is not robust to outlier skeleton nodes, to have a "tolerance" threshold of 50 skeleton nodes (around 2 μm in length) to relax the condition to determine whether the ground truth skeleton encounters a merge error in a segmentation.

3.2 State-of-the-Art Methods

We consider the top two methods for neuron instance segmentation on the SNEMI3D benchmark [2]: U-Net models predicting the affinity graph (affinity U-Net[2] [18]) and flood-filling networks (FFN[3] [15]). For model details, we

---

[2] https://github.com/seung-lab/DeepEM.
[3] https://github.com/google/ffn.

refer readers to the original papers. Note that although their implementations are publicly available, in practice, a substantial amount of engineering effort may still be required in order to reproduce the reported performance.

4   Experiments

4.1   Implementation Details

For the FFN pipeline, we used the default configuration in [15] (convolution module depth = 9, FOV size = [33, 33, 17], movement deltas = [4, 8, 8]) and trained separate models from scratch until accuracy saturates on the training data (94% for AxonEM-M, 97% for AxonEM-H, and 92% for SNEMI3D). We then performed a single pass of inference without the over-segmentation consensus and FFN agglomeration steps [15]. For inference, we ran distributed FFN on $512 \times 512 \times 256$ voxel subvolumes with $64 \times 64 \times 32$ voxel overlap and later reconciliated into one volume [32]. Training and inference were conducted on a multi-GPU cluster and reconciliation was done on a workstation [32]. Overall, for AxonEM-M and AxonEM-H, we used 32 NVidia A100 GPUs, spent 20 h on training and 6.2 and 5.8 h on inference for each dataset. For SNEMI3D, we trained the net for 10 h with 16 NVidia A100 GPUs, then finished inference in 0.28 h.

Table 1. Segmentation results on SNEMI3D dataset. We report the ARAND score for the dense segmentation and ERL score for axon segments among them. The percentage numbers in parentheses represent the ratio of the baseline axon ERL with respect to the upper limit given by the ground truth ERL.

Method		ARAND↓	ERL↑ (μm)
Ground truth		0.000	8.84 (100%)
PNI's Aff. U-Net (mean aff. aggl.) [18]		0.031	8.31 (94%)
Google's FFN [15]		0.029	8.20 (93%)
Our Impl.	Aff. U-Net (mean aff. aggl.)	0.038	8.22 (93%)
	FFN (w/o aggl.)	0.112	5.18 (59%)

For the affinity U-Net pipeline, we first used the Google Cloud Platform (GCP) compute engine to predict the affinity, using 12 NVidia T4 GPUs, each paired with a n1-highmem-8 instance (8 vCPUs and 52 GB memory). Processing the AxonEM-M and AxonEM-H datasets took 2.6 and 3.5 h, respectively. From the predicted affinities, we produced segmentation using the watershed-type algorithm [40] and mean affinity agglomeration [18]. We used 8 e2-standard-32 instances (32 vCPUs and 128 GB memory). The watershed and agglomeration steps together took 0.58 h for AxonEM-M, and 0.75 h for AxonEM-H.

4.2   Benchmark Results on SNEMI3D Dataset

We first show the results of our reproducing experiments on the SNEMI3D benchmark in Table 1, where we obtained the adapted Rand (ARAND) error scores from the challenge evaluation server. Our affinity U-Net error score (0.038) was comparable to that of [18] (0.031), confirming that we are capable of faithfully reproducing the affinity U-Net method. However, our FFN error score (0.112) was severely worse than that of [15] (0.033) by a large margin, suggesting that one should not consider our FFN baseline as state-of-the-art.

Next, we computed the axon ERL for our reproducing results, as well as for the ground truth segmentation provided by the challenge organizer and some of the top submission results [15,18]. The ERL for the ground truth axons was 8.84 μm, which is the average path length of the ground truth axons. As expected from the worst ARAND error score, our FFN result showed the lowest ERL (5.18 μm) among others. The axon ERLs for the top submission results [15,18] and our affinity U-Net result were all above 8 μm, approaching the upper limit given by the ground truth axon ERL (8.84 μm). Interestingly, a lower (better) ARAND error score for dense segmentation does not necessarily result in a higher (better) axon ERL. This is because the ARAND error score is computed for dense segmentation including all segments (axon, dendrite, and glia), whereas we computed the ERL selectively for axon. For example, our affinity U-Net result was better in axon ERL, but worse in ARAND, than [15] because of more problems in dendrite and glia.

Table 2. Axon segmentation results on AxonEM dataset (ERL).

Method		AxonEM-H	AxonEM-M
Ground truth		28.5 μm (100%)	43.6 μm (100%)
Our Impl.	Aff. U-Net (mean aff. aggl.)	18.5 μm (65%)	38.5 μm (88%)
	FFN (w/o aggl.)	9.6 μm (34%)	9.4 μm (22%)

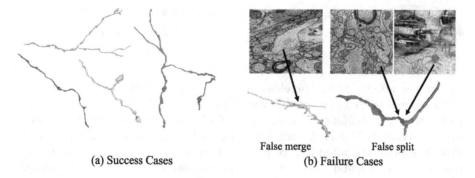

(a) Success Cases   False merge   False split
                                        (b) Failure Cases

Fig. 4. Qualitative results of our affinity U-Net baseline on AxonEM. Failure cases are often caused by invaginating structures, thin parts, and image defects.

4.3 Benchmark Results on AxonEM Dataset

We provide axon segmentation baselines for the AxonEM dataset by reproducing two state-of-the-art methods, affinity U-Net [18] and FFN [15]. One should be cautious that our FFN baseline does not fully reproduce the state-of-the-art accuracy of the original method by [15], due to the reasons detailed below.

Quantitative Results. As shown in Table 2, our affinity U-Net baseline achieved a relatively high axon ERL (88% with respect to the ground truth) on AxonEM-M, whereas its ERL for AxonEM-H was relatively lower (65%). The axon ERL of our FFN baseline was consistently and significantly lower (22% on AxonEM-H and 34% on AxonEM-H) than our affinity U-Net baseline, being consistent with our SNEMI3D reproducing results in the previous section (Table 1). Due to the lack of open-source code for several inference steps, our FFN baseline was not a faithful reproduction of the full reconstruction pipeline proposed by [15]. Our FFN baseline lacks both the multi-scale over-segmentation consensus and FFN agglomeration steps in [15], which are crucial in achieving a very low merge error rate while maximizing ERL. A much better FFN baseline by faithfully reproducing the full reconstruction pipeline remains to be done in the future.

Qualitative Results. We show both successful and unsuccessful axon segmentation results of our affinity U-Net baselines (Fig. 4). Due to the thin parts, the axon reconstruction is more vulnerable to image defects like missing section, staining artifact, knife mark, and misalignment [17] than other structures.

5   Conclusion

In this paper, we have introduced a large-scale 3D EM dataset for axon instance segmentation, and provided baseline results by re-implementing two state-of-the-art methods. We expect that the proposed densely annotated large-scale axon dataset will not only foster development of better axon segmentation methods, but also serve as a test bed for various other applications.

Declaration of Interests. KL, RL, and JAB disclose financial interests in Zetta AI LLC.

Acknowledgment. KL, RL, and JAB were supported by the Intelligence Advanced Research Projects Activity (IARPA) via Department of Interior/ Interior Business Center (DoI/IBC) contract number D16PC0005, NIH/NIMH (U01MH114824, U01MH117072, RF1 MH117815), NIH/NINDS (U19NS104648, R01NS104926), NIH/NEI (R01EY027 036), and ARO (W911NF-12-1-0594), and are also grateful for assistance from Google, Amazon, and Intel. DW, ZL, JL, and HP were partially supported by NSF award IIS-1835231. I. A-C would like to acknowledge the support of the Beca Leonardo a Investigadores y Creadores Culturales 2020 de la Fundación BBVA.

We thank Viren Jain, Michał Januszewski and their team for generating the initial segmentation for AxonEM-H, and Daniel Franco-Barranco for setting up the challenge using AxonEM.

References

1. Neuroglancer. https://github.com/google/neuroglancer
2. SNEMI3D EM segmentation challenge and dataset. http://brainiac2.mit.edu/SNEMI3D/home
3. Abbott, L.F., et al.: The mind of a mouse. Cell **182**, 1372–1376 (2020)
4. Bégin, S., Dupont-Therrien, O., Bélanger, E., Daradich, A., Laffray, S., et al.: Automated method for the segmentation and morphometry of nerve fibers in large-scale cars images of spinal cord tissue. Biomed. Opt. Exp. **5**, 4145–4161 (2014)
5. Beier, T., et al.: Multicut brings automated neurite segmentation closer to human performance. Nat. Methods **14**, 101–102 (2017)
6. Berger, D.R., Seung, H.S., Lichtman, J.W.: Vast (volume annotation and segmentation tool): efficient manual and semi-automatic labeling of large 3D image stacks. Front. Neural Circ. **12**, 88 (2018)
7. Ciresan, D., Giusti, A., Gambardella, L.M., Schmidhuber, J.: Deep neural networks segment neuronal membranes in electron microscopy images. In: NeurIPS (2012)
8. Cuisenaire, O., Romero, E., Veraart, C., Macq, B.M.: Automatic segmentation and measurement of axons in microscopic images. In: Medical Imaging (1999)
9. Dorkenwald, S., McKellar, C., et al.: Flywire: online community for whole-brain connectomics. bioRxiv (2020)
10. Dorkenwald, S., et al.: Automated synaptic connectivity inference for volume electron microscopy. Nat. Methods **14**, 435–442 (2017)
11. Dorkenwald, S., et al.: Binary and analog variation of synapses between cortical pyramidal neurons. BioRxiv (2019)
12. Funke, J., et al.: Large scale image segmentation with structured loss based deep learning for connectome reconstruction. TPAMI **41**, 1669–1680 (2018)
13. Gonda, F., Wei, D., Pfister, H.: Consistent recurrent neural networks for 3d neuron segmentation. In: ISBI (2021)
14. Helmstaedter, M.: Cellular-resolution connectomics: challenges of dense neural circuit reconstruction. Nat. Methods **10**, 501–507 (2013)
15. Januszewski, M., et al.: High-precision automated reconstruction of neurons with flood-filling networks. Nat. Methods **15**, 605–610 (2018)
16. Kornfeld, J., Denk, W.: Progress and remaining challenges in high-throughput volume electron microscopy. Curr. Opin. Neurobiol. **50**, 261–267 (2018)
17. Lee, K., Turner, N., Macrina, T., Wu, J., Lu, R., Seung, H.S.: Convolutional nets for reconstructing neural circuits from brain images acquired by serial section electron microscopy. Curr. Opin. Neurobiol. **55**, 188–198 (2019)
18. Lee, K., Zung, J., Li, P., Jain, V., Seung, H.S.: Superhuman accuracy on the snemi3d connectomics challenge. arXiv:1706.00120 (2017)
19. Matejek, B., Haehn, D., Zhu, H., Wei, D., Parag, T., Pfister, H.: Biologically-constrained graphs for global connectomics reconstruction. In: CVPR (2019)
20. Meirovitch, Y., Mi, L., Saribekyan, H., Matveev, A., Rolnick, D., Shavit, N.: Cross-classification clustering: an efficient multi-object tracking technique for 3-d instance segmentation in connectomics. In: CVPR (2019)
21. Meirovitch, Y., et al.: A multi-pass approach to large-scale connectomics. arXiv preprint arXiv:1612.02120 (2016)

22. Mesbah, R., McCane, B., Mills, S.: Deep convolutional encoder-decoder for myelin and axon segmentation. In: IVCNZ (2016)
23. More, H.L., Chen, J., Gibson, E., Donelan, J.M., Beg, M.F.: A semi-automated method for identifying and measuring myelinated nerve fibers in scanning electron microscope images. J. Neurosci. Methods **201**, 149–158 (2011)
24. Motta, A., et al.: Dense connectomic reconstruction in layer 4 of the somatosensory cortex. Science (2019)
25. Naito, T., Nagashima, Y., Taira, K., Uchio, N., Tsuji, S., Shimizu, J.: Identification and segmentation of myelinated nerve fibers in a cross-sectional optical microscopic image using a deep learning model. J. Neurosci. Methods **291**, 141–149 (2017)
26. Nunez-Iglesias, J., Kennedy, R., Parag, T., Shi, J., Chklovskii, D.B.: Machine learning of hierarchical clustering to segment 2d and 3d images. PloS one **8**, e71715 (2013)
27. Ronneberger, O., Fischer, P., Brox, T.: U-Net: convolutional networks for biomedical image segmentation. In: Navab, N., Hornegger, J., Wells, W.M., Frangi, A.F. (eds.) MICCAI 2015. LNCS, vol. 9351, pp. 234–241. Springer, Cham (2015). https://doi.org/10.1007/978-3-319-24574-4_28
28. Sato, M., Bitter, I., Bender, M., Kaufman, A., Nakajima, M.: TEASAR: tree-structure extraction algorithm for accurate and robust skeletons. In: Pacific Conference on Computer Graphics and Applications (2000)
29. Scheffer, L.K., et al.: A connectome and analysis of the adult drosophila central brain. Elife **9**, e57443 (2020)
30. Shapson-Coe, A., et al.: A connectomic study of a petascale fragment of human cerebral cortex. bioRxiv (2021)
31. Turaga, S.C., et al.: Convolutional networks can learn to generate affinity graphs for image segmentation. Neural Comput. **22**, 511–538 (2010)
32. Vescovi, R., et al.: Toward an automated hpc pipeline for processing large scale electron microscopy data. In: XLOOP (2020)
33. Wang, Y.Y., Sun, Y.N., Lin, C.C.K., Ju, M.S.: Segmentation of nerve fibers using multi-level gradient watershed and fuzzy systems. AI Med. **54**, 189–200 (2012)
34. Wei, D., et al.: MitoEM dataset: large-scale 3D mitochondria instance segmentation from EM images. In: Martel, A.L., et al. (eds.) MICCAI 2020. LNCS, vol. 12265, pp. 66–76. Springer, Cham (2020). https://doi.org/10.1007/978-3-030-59722-1_7
35. White, J.G., Southgate, E., Thomson, J.N., Brenner, S.: The structure of the nervous system of the nematode caenorhabditis elegans. Philos. Trans. R. Soc. B Biol. Sci. **314**, 1–340 (1986)
36. Zaimi, A., Duval, T., Gasecka, A., Côté, D., Stikov, N., Cohen-Adad, J.: AxonSeg: open source software for axon and myelin segmentation and morphometric analysis. Front. Neuroinf. **10**, 37 (2016)
37. Zaimi, A., Wabartha, M., Herman, V., Antonsanti, P.L., Perone, C.S., Cohen-Adad, J.: Axondeepseg: automatic axon and myelin segmentation from microscopy data using convolutional neural networks. Sci. Rep. **8**, 1–11 (2018)
38. Zeng, T., Wu, B., Ji, S.: Deepem3d: approaching human-level performance on 3d anisotropic em image segmentation. Bioinformatics **33**, 2555–2562 (2017)
39. Zhao, X., Pan, Z., Wu, J., Zhou, G., Zeng, Y.: Automatic identification and morphometry of optic nerve fibers in electron microscopy images. Comput. Med. Imaging Graph. **34**, 179–184 (2010)
40. Zlateski, A., Seung, H.S.: Image segmentation by size-dependent single linkage clustering of a watershed basin graph. arXiv:1505.00249 (2015)

Improved Brain Lesion Segmentation with Anatomical Priors from Healthy Subjects

Chenghao Liu[1], Xiangzhu Zeng[2], Kongming Liang[3], Yizhou Yu[4],
and Chuyang Ye[1(✉)]

[1] School of Information and Electronics, Beijing Institute of Technology,
Beijing, China
chuyang.ye@bit.edu.cn
[2] Department of Radiology, Peking University Third Hospital, Beijing, China
[3] School of Artificial Intelligence, Beijing University of Posts and
Telecommunications, Beijing, China
[4] Deepwise AI lab, Beijing, China

Abstract. *Convolutional neural networks* (CNNs) have greatly improved the performance of brain lesion segmentation. However, accurate segmentation of brain lesions can still be challenging when the appearance of lesions is similar to normal brain tissue. To address this problem, in this work we seek to exploit the information in scans of healthy subjects to improve brain lesion segmentation, where anatomical priors about normal brain tissue can be taken into account for better discrimination of lesions. To incorporate such prior knowledge, we propose to register a set of reference scans of healthy subjects to each scan with lesions, and the registered reference scans provide reference intensity samples of normal tissue at each voxel. In this way, the spatially adaptive prior knowledge can indicate the existence of abnormal voxels even when their intensities are similar to normal tissue, because their locations contradict with the prior knowledge about normal tissue. Specifically, with the reference scans, we compute anomaly score maps for the scan with lesions, and these maps are used as auxiliary inputs to the segmentation network to aid brain lesion segmentation. The proposed strategy was evaluated on different brain lesion segmentation tasks, and the results indicate the benefit of incorporating the anatomical priors using our approach.

Keywords: Brain lesion segmentation · Convolutional neural network · Anatomical priors

1 Introduction

Brain lesions may lead to persistent systematic changes in brain functions [4], such as motor impairments [17] and cognitive deficits [11]. Accurate segmentation of brain lesions enables quantitative measurements of the lesion size, shape, and location, and it provides important clues for intervention and prognosis [6,13,17].

© Springer Nature Switzerland AG 2021
M. de Bruijne et al. (Eds.): MICCAI 2021, LNCS 12901, pp. 186–195, 2021.
https://doi.org/10.1007/978-3-030-87193-2_18

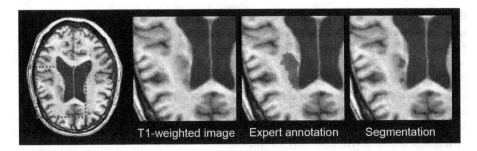

T1-weighted image   Expert annotation   Segmentation

Fig. 1. An example of chronic stroke lesion segmentation on the T1-weighted image. The intensity of the lesion (indicated by the expert annotation) resembles the intensity of normal gray matter, and even a state-of-the-art segmentation pipeline nnU-Net [5] can undersegment the lesion.

The performance of automated brain lesion segmentation has been greatly advanced by *convolutional neural networks* (CNNs) [5,7]. However, accurate segmentation of brain lesions may still be challenging when the appearances of lesions are similar to normal brain tissue. For example, as shown in Fig. 1, the intensity of a chronic stroke lesion in the white matter can resemble that of normal gray matter, and even a state-of-the-art segmentation approach could confuse the lesion with normal tissue and produce low-quality segmentation results.

To address this problem, it is possible to incorporate prior knowledge about normal brain tissue, so that the lesions can be better discriminated even when their intensities are similar to those of normal brain tissue. For example, based on the assumption that normal tissue tends to appear symmetric, an image of interest can be registered to its reflected version, and the difference between them can indicate asymmetric regions that are possibly lesions [10,12]. In [3], based on the assumption that a CNN model trained to reconstruct images of healthy subjects is not capable of truthfully reconstructing the high frequency components of lesions, a scale-space autoencoder is proposed to estimate pixelwise anomalies. The results of these methods can be used as anatomical priors and provide auxiliary information to improve brain lesion segmentation.

In this work, we continue to explore the incorporation of prior knowledge about normal brain tissue for brain lesion segmentation. Although the intensity of brain lesions can be similar to that of the normal tissue elsewhere, the locations of these lesions help to discriminate them. Based on this observation, we propose to use the scans of healthy subjects to provide voxelwise reference intensities of normal tissue, so that abnormal intensity at each voxel can be better identified. Specifically, we register a set of reference images of healthy subjects to a scan of interest with lesions. Then, intuitively, the difference map between the intensities of the scan of interest and the registered reference images can be computed to indicate anomaly. In addition, the intensities of the registered scans at each voxel can be considered samples drawn from a distribution of normal tissue intensities at that voxel. We fit the distribution with these samples and compute the likelihood of the intensities of the scan of interest, which gives a different perspective of anomaly indication. These two types of anomaly score maps are combined with the image with lesions for network training and inference, so that knowledge about normal brain tissue is exploited to improve the segmentation. For evaluation, we integrated the proposed method with the nnU-Net segmentation pipeline [5] and performed experiments for segmenting chronic stroke lesions and ischemic stroke lesions. Results show that the proposed strategy of incorporating anatomical prior information improves the quality of brain lesion segmentation. The code of our method is publicly available at https://github.com/lchdl/NLL_anomaly_detection.

2   Method

2.1   Problem Formulation

Given an image x with brain lesions, we seek to perform CNN-based brain lesion segmentation for x aided by a set $\mathcal{R} = \{r_i\}_{i=1}^K$ of K reference images of healthy subjects without lesions, where r_i ($i \in \{1, \ldots, K\}$) represents the i-th reference image. These reference images can allow the incorporation of anatomical prior information about normal brain tissue for improved discrimination of lesions. We hypothesize that by properly comparing x and $\mathcal{R}$, it is possible to suggest lesion areas that may be confused with normal tissue, and such anomaly information can be used an auxiliary input to the segmentation network. To this end, we design two strategies to compute the anomaly map using image registration, which are based on purely the intensities of the reference images and a distribution fitted from these intensities, respectively. For each of the two strategies, the reference images $\mathcal{R}$ are aligned to x with deformable registration, leading to a new set $\mathcal{R}'$ of registered reference images $\phi_i(r_i)$, where $\phi_i(\cdot)$ represents the spatial transformation for r_i, and then x is compared with $\mathcal{R}'$. The detailed design of these strategies, which for convenience are referred to as the intensity-based strategy and the distribution-based strategy, respectively, is described below.

2.2   Intensity-Based Strategy

The intensity-based strategy directly calculates the intensity difference between the registered reference images and the image with lesions. To avoid noise outside brain tissue, brain masks of the reference images are also used for the computation. We denote the brain mask of the reference image r_i by m_i, and then the anomaly score map α is derived as

$$\alpha = \frac{1}{K} \sum_{i=1}^K \phi_i(m_i) \cdot (x - \phi_i(r_i)), \tag{1}$$

where the average difference between the image with lesions and the registered reference images is computed for each voxel inside the brain. In this way, if at a voxel of x the intensity is very different from the reference intensities of

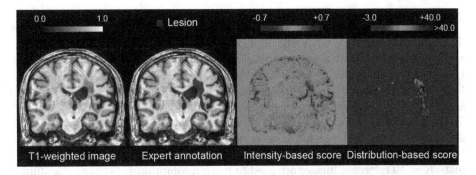

Fig. 2. An example of the anomaly score maps computed with the intensity-based strategy and distribution-based strategy for chronic stroke lesion segmentation on the T1-weighted image. The expert annotation is also shown for reference.

normal tissue, a much higher or lower value is observed on α, indicating the possible existence of brain lesions. Note that the difference may also be caused by registration error, especially at tissue interfaces where the reference intensities can have large variations. Figure 2 provides an example of the anomaly score map computed with the intensity-based strategy for chronic stroke lesion segmentation on the T1-weighted image.

2.3 Distribution-Based Strategy

To reduce the impact of registration error at tissue interfaces, in addition to the intuitive intensity-based strategy, it is also possible to treat the intensities of the registered reference images as samples drawn from a distribution of normal tissue intensities. Then, the distribution can be fitted from these samples, and the likelihood of an intensity being normal can be computed. For simplicity, we assume that the distribution of the normal tissue intensity at each voxel v is a Gaussian distribution $\mathcal{N}(\mu_v, \sigma_v^2)$ parameterized by the mean μ_v and variance σ_v^2. μ_v and σ_v^2 can be computed by fitting the registered reference intensities to the distribution at each voxel.

With the distribution $\mathcal{N}(\mu_v, \sigma_v^2)$, for the intensity x_v at each voxel v of an image with lesions, an anomaly score β_v can then be derived as the negative log-likelihood:

$$\beta_v = \log \sqrt{2\pi}\sigma_v + \frac{(x_v - \mu_v)^2}{2\sigma_v^2}. \tag{2}$$

We denote the anomaly score map comprising β_v by β. When x_v deviates from $\mathcal{N}(\mu_v, \sigma_v^2)$, the negative log-likelihood is higher. In this way, at tissue interfaces where α has a very large or small value, if the variance of the reference intensities is also large due to registration error, then a small anomaly score will be assigned. An example of β is shown in Fig. 2, where regions with higher β_v are consistent with the lesion location. Note that similar to the intensity-based strategy, brain masks are also applied to the anomaly score map.

2.4 Integration with CNNs and Implementation Details

To train a segmentation CNN, the anomaly score maps α and β are concatenated with the image x with lesions as network input. For test scans, the anomaly score maps are also computed with the reference images and concatenated with x for brain lesion segmentation. The proposed method can be integrated with various CNN-based segmentation approaches, such as [5] and [7].

We use ANTs [1] for the registration of reference images, where an affine registration is followed by a deformable registration. The brain masks are extracted from the reference images with BET [14,16] for constraining the anomaly score map. $K = 10$ reference images are used to reach a compromise between the abundance of reference samples and the computational overhead of image registration. Note that since deformable registration is time-consuming, learning-based registration [2] may also be used to accelerate the registration.

3   Results

3.1 Dataset Description

The proposed method was evaluated on two brain lesion segmentation tasks, which are the segmentation of chronic stroke lesions and ischemic stroke lesions. The datasets associated with these tasks are introduced below.

We first used the publicly available ATLAS dataset [8] for evaluation, which comprises T1-weighted MRI scans of 220 subjects with chronic stroke lesions. The lesions in these scans have been annotated by experts. All images were affinely registered to the MNI152 template, intensity normalized, and defaced. The dimension of the preprocessed images is $197 \times 233 \times 189$ with an isotropic $1\,\text{mm}^3$ resolution. Since the ATLAS dataset only contains scans of patients with brain lesions, we further selected 10 T1-weighted scans of healthy subjects from the HCP dataset [15] as the reference images. The registration of each reference image to a scan in the ATLAS dataset took about 20 min. The histograms of the reference images were matched to an image randomly selected from the ATLAS dataset.

A second in-house dataset for ischemic stroke lesion segmentation was also used for evaluation, which comprises 21 normal-appearing *diffusion weighted images* (DWIs) and 219 DWIs with ischemic stroke lesions. Each DWI belongs to a different subject, and the DWIs were annotated in their native space by an experienced radiologist. All images were acquired on a 3T Siemens Verio scanner with a b-value of $1000\,\text{s}/\text{mm}^2$, together with a $b0$ image. The image dimension is $240 \times 240 \times 21$ and the resolution is $0.96\,\text{mm} \times 0.96\,\text{mm} \times 6.5\,\text{mm}$. We randomly picked 10 images from the 21 normal-appearing scans as the reference images. The registration of each reference image to a DWI scan with brain lesions took about 5 min.

3.2 Experimental Settings

For each segmentation task, we selected 50 scans as the test set, and different dataset configurations (numbers of labeled images) were used for training. Specifically, for chronic stroke lesion segmentation, a total number of 170, 80, or 15 labeled images were used for training, and for ischemic stroke lesion segmentation, the numbers were 169, 40, or 10. In addition, for each configuration, 20% of the training images were used as a validation set. Model selection was performed based on the best performance on the validation set.

We evaluated the proposed method based on the state-of-the-art medical image segmentation pipeline nnU-Net [5] with default settings. Given a dataset, nnU-Net automatically determines the optimal U-Net-based structure variant, pre- and post-processing strategy, and performs end-to-end image segmentation. Before training and inference, z-score normalization was applied to all input images. Also, nnU-net automatically performs data augmentation in the training and inference stages. During training, random elastic deformation, rotation, scaling, mirroring, and gamma intensity transform are applied to each training sample. During inference, mirroring is applied to each input sample, and the final prediction is obtained by averaging the predictions from eight differently mirrored inputs. For each test scan, the computation of anomaly score maps and network inference took about one minute in total.

The proposed method was compared with two competing methods. The nnU-Net model trained with the intensity image input only (without using any auxiliary information) was considered the baseline method for comparison. In addition, the proposed method was compared with nnU-Net integrated with the autoencoder-based anomaly detection method in [3], where the anomaly score was also concatenated with the intensity image and fed into nnU-Net. This competing method is referred to as 'baseline+[3]'.

3.3 Segmentation Performance

The proposed method was applied to perform the two brain lesion segmentation tasks. Since in the proposed method the anomaly maps are computed with the intensity-based and distribution-based strategies, our method is referred to as 'baseline+IB+DB'. For reference, the results achieved with the intensity-based or distribution-based strategy alone were also considered (referred to as 'baseline+IB' and 'baseline+DB', respectively). Both qualitative and quantitative evaluation was performed, and we used the *Dice similarity coefficient* (DSC) to quantitatively measure the segmentation quality. The detailed results are described below.

Coronal views of the results of chronic stroke lesion segmentation are shown in Fig. 3, together with the expert annotation. With the auxiliary information computed with both intensity-based and distribution-based strategies, the segmentation results reached a better consensus with the expert annotations than the results of the competing methods. Although using either baseline+IB or baseline+DB could also improve the segmentation performance, when the anomaly score maps were combined, the quality of the results was better.

192   C. Liu et al.

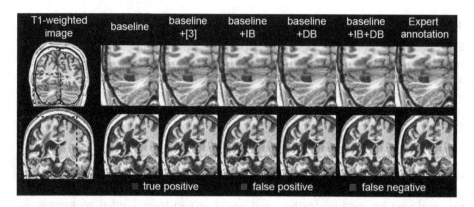

Fig. 3. Coronal views of representative segmentation results for chronic stroke lesions.

Table 1. The DSC results (mean ± std) for the chronic stroke lesion segmentation task. The best results are highlighted in bold. Asterisks indicate that the difference between Baseline+IB+DB and the competing method is significant using a paired student's *t*-test. (*: $p < 0.05$)

#labeled images	170	80	15
baseline	0.5870 ± 0.3254	0.5551 ± 0.3353	$0.4362 \pm 0.3315^*$
baseline+ [3]	$0.5510 \pm 0.3525^*$	0.5403 ± 0.3377	0.4850 ± 0.3242
baseline+IB	0.6075 ± 0.3018	0.5646 ± 0.3231	$\mathbf{0.5268 \pm 0.3308}$
baseline+DB	0.5996 ± 0.3051	0.5442 ± 0.3267	0.4831 ± 0.3288
baseline+IB+DB	$\mathbf{0.6256 \pm 0.2918}$	$\mathbf{0.5722 \pm 0.3076}$	0.5245 ± 0.3274

Table 2. Method ranking for the chronic stroke lesion segmentation task. The best results are highlighted in bold.

#labeled images	170	80	15
baseline	3.22	3.32	3.56
baseline+ [3]	3.08	3.26	3.16
baseline+IB	2.94	2.84	2.58
baseline+DB	3.14	3.06	3.14
baseline+IB+DB	**2.62**	**2.52**	**2.56**

Quantitatively, the means and *standard deviations* (stds) of the DSCs of the segmentation results on the test set for the ATLAS dataset are shown in Table 1 for each data configuration. Baseline+IB+DB has better segmentation accuracy than the two competing methods. Its performance is also better than baseline+IB and baseline+DB for the cases of 170 and 80 training images, and close to the best result of baseline+IB given 15 labeled images for training. We

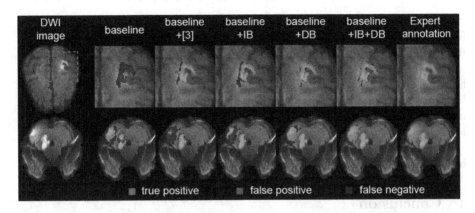

Fig. 4. Axial views of representative segmentation results for ischemic stroke lesions.

Table 3. The DSC results (mean ± std) for the ischemic stroke lesion segmentation task. The best results are highlighted in bold. Asterisks indicate that the difference between Baseline+IB+DB and the competing method is significant using a paired student's t-test. (**: $p < 0.01$, ***: $p < 0.001$)

#labeled images	169	40	10
baseline	0.7788 ± 0.1936***	0.7477 ± 0.1938	0.6481 ± 0.2195
baseline+ [3]	0.7985 ± 0.1904	0.7390 ± 0.2105	0.5981 ± 0.2702**
baseline+IB	0.8038 ± 0.1842	0.7481 ± 0.2099	0.6275 ± 0.2399
baseline+DB	0.8016 ± 0.1899	**0.7625 ± 0.1885**	**0.6897 ± 0.1941**
baseline+IB+DB	**0.8089 ± 0.1826**	0.7538 ± 0.2014	0.6733 ± 0.2048

Table 4. Method ranking for the ischemic stroke lesion segmentation task. The best results are highlighted in bold.

#labeled images	169	40	10
baseline	3.96	3.44	3.08
baseline+ [3]	3.04	3.20	3.48
baseline+IB	2.88	3.26	3.24
baseline+DB	2.60	2.76	**2.46**
baseline+IB+DB	**2.52**	**2.34**	2.74

also computed the average rank of each method for each data configuration using the ranking strategy in [9], where all the methods were ranked together based on the DSC of each test scan. The results are shown in Table 2. In all cases, baseline+IB+DB achieves the best ranking.

The qualitative results for ischemic stroke lesion segmentation are shown in Fig. 4, where the results of baseline+IB+DB better agree with the expert annotations than the competing methods. The quantitative DSC results are shown in Table 3, where baseline+IB+DB outperforms the two competing methods baseline and baseline+[3]. Compared with baseline+IB and baseline+DB, baseline+IB+DB achieves either the best or the second best result. Method ranking was also performed and the results are shown in Table 4, where baseline+IB+DB has better ranking than the two competing methods. Its rank is also better than baseline+IB and baseline+DB except when 10 training images were used, where baseline+IB+DB is the second best. Together with the results for chronic stroke lesion segmentation, these results suggest that baseline+IB+DB outperforms the existing methods, and the combination of the intensity-based and distribution-based strategies is recommended over the use of a single one of them.

4   Conclusion

We have developed a method for improving CNN-based brain lesion segmentation, where prior knowledge about normal brain tissue is incorporated in the segmentation. The prior knowledge is obtained from reference images of healthy subjects via image registration and suggests spatial anomaly maps. The anomaly information is used as auxiliary network inputs to aid the segmentation. The experimental results on two brain lesion segmentation tasks show that our method improves the segmentation performance.

Acknowledgement. This work is supported by Beijing Natural Science Foundation (L192058 & 7192108).

References

1. Avants, B.B., Tustison, N.J., Song, G., Cook, P.A., Klein, A., Gee, J.C.: A reproducible evaluation of ANTs similarity metric performance in brain image registration. NeuroImage **54**(3), 2033–2044 (2011)
2. Balakrishnan, G., Zhao, A., Sabuncu, M.R., Guttag, J., Dalca, A.V.: Voxelmorph: a learning framework for deformable medical image registration. IEEE Trans. Med. Imaging **38**(8), 1788–1800 (2019)
3. Baur, C., Wiestler, B., Albarqouni, S., Navab, N.: Scale-space autoencoders for unsupervised anomaly segmentation in brain MRI. In: Martel, A.L., et al. (eds.) MICCAI 2020. LNCS, vol. 12264, pp. 552–561. Springer, Cham (2020). https://doi.org/10.1007/978-3-030-59719-1_54
4. Churchill, N., et al.: The effects of chronic stroke on brain function while driving. Stroke **47**(Supplement 1), 150 (2016)
5. Isensee, F., Jaeger, P.F., Kohl, S.A., Petersen, J., Maier-Hein, K.H.: nnU-Net: a self-configuring method for deep learning-based biomedical image segmentation. Nat. Methods **18**(2), 203–211 (2021)
6. Jongbloed, L.: Prediction of function after stroke: a critical review. Stroke **17**(4), 765–776 (1986)
7. Kamnitsas, K., et al.: Efficient multi-scale 3D CNN with fully connected CRF for accurate brain lesion segmentation. Med. Image Anal. **36**, 61–78 (2017)
8. Liew, S.L., et al.: A large, open source dataset of stroke anatomical brain images and manual lesion segmentations. Sci. Data **5**, 180011 (2018)

9. Maier, O., et al.: ISLES 2015 - a public evaluation benchmark for ischemic stroke lesion segmentation from multispectral MRI. Med. Image Anal. **35**, 250–269 (2017)
10. Martins, S.B., Telea, A.C., Falcão, A.X.: Investigating the impact of supervoxel segmentation for unsupervised abnormal brain asymmetry detection. Comput. Med. Imaging Graph. **85**, 101770 (2020)
11. Nakling, A.E., et al.: Cognitive deficits in chronic stroke patients: neuropsychological assessment, depression, and self-reports. Dementia Geriatric Cogn. Disord. Extra **7**(2), 283–296 (2017)
12. Raina, K., Yahorau, U., Schmah, T.: Exploiting bilateral symmetry in brain lesion segmentation. arXiv preprint arXiv:1907.08196 (2019)
13. Riley, J.D., et al.: Anatomy of stroke injury predicts gains from therapy. Stroke **42**(2), 421–426 (2011)
14. Smith, S.M.: Fast robust automated brain extraction. Human Brain Mapp. **17**(3), 143–155 (2002)
15. Van Essen, D.C., Smith, S.M., Barch, D.M., Behrens, T.E.J., Yacoub, E., Ugurbil, K.: The WU-Minn human connectome project: an overview. NeuroImage **80**, 62–79 (2013)
16. Woolrich, M.W., et al.: Bayesian analysis of neuroimaging data in FSL. NeuroImage **45**(1), S173–S186 (2009)
17. Zhu, L.L., Lindenberg, R., Alexander, M.P., Schlaug, G.: Lesion load of the corticospinal tract predicts motor impairment in chronic stroke. Stroke **41**(5), 910–915 (2010)

CarveMix: A Simple Data Augmentation Method for Brain Lesion Segmentation

Xinru Zhang[1], Chenghao Liu[1], Ni Ou[2], Xiangzhu Zeng[3], Xiaoliang Xiong[4], Yizhou Yu[4], Zhiwen Liu[1(✉)], and Chuyang Ye[1(✉)]

[1] School of Information and Electronics, Beijing Institute of Technology, Beijing, China
{zwliu,chuyang.ye}@bit.edu.cn
[2] School of Automation, Beijing Institute of Technology, Beijing, China
[3] Department of Radiology, Peking University Third Hospital, Beijing, China
[4] Deepwise AI Lab, Beijing, China

Abstract. Brain lesion segmentation provides a valuable tool for clinical diagnosis, and *convolutional neural networks* (CNNs) have achieved unprecedented success in the task. Data augmentation is a widely used strategy that improves the training of CNNs, and the design of the augmentation method for brain lesion segmentation is still an open problem. In this work, we propose a simple data augmentation approach, dubbed as CarveMix, for CNN-based brain lesion segmentation. Like other "mix"-based methods, such as Mixup and CutMix, CarveMix stochastically combines two existing labeled images to generate new labeled samples. Yet, unlike these augmentation strategies based on image combination, CarveMix is lesion-aware, where the combination is performed with an attention on the lesions and a proper annotation is created for the generated image. Specifically, from one labeled image we carve a *region of interest* (ROI) according to the lesion location and geometry, and the size of the ROI is sampled from a probability distribution. The carved ROI then replaces the corresponding voxels in a second labeled image, and the annotation of the second image is replaced accordingly as well. In this way, we generate new labeled images for network training and the lesion information is preserved. To evaluate the proposed method, experiments were performed on two brain lesion datasets. The results show that our method improves the segmentation accuracy compared with other simple data augmentation approaches.

Keywords: Brain lesion segmentation · Data augmentation · Convolutional neural network

1 Introduction

Quantitative analysis of brain lesions may improve our understanding of brain diseases and treatment planning [1,7]. Automated brain lesion segmentation is

---

X. Zhang and C. Liu—Equal contribution.

© Springer Nature Switzerland AG 2021
M. de Bruijne et al. (Eds.): MICCAI 2021, LNCS 12901, pp. 196–205, 2021.
https://doi.org/10.1007/978-3-030-87193-2_19

desired for reproducible and efficient analysis of brain lesions, and *convolutional neural networks* (CNNs) have achieved state-of-the-art performance of brain lesion segmentation [8,9].

Data augmentation is a widely used strategy for improving the training of CNNs, where additional training data is generated from existing training data. It is shown to reduce the variance of the mapping learned by CNNs [4] and has been effectively applied to brain lesion segmentation [8,10]. Data augmentation can be achieved with basic image transformation, including translation, rotation, flipping, etc. [8], where an existing training image is transformed together with the annotation using hand-crafted rules. Since the diversity of the data generated via basic image transformation can be limited, more advanced approaches based on generative models have also been developed [3,12]. However, the implementation of these methods is usually demanding, and the training of generative models is known to be challenging [14]. The success of generative models for data augmentation may depend on the specific task.

To achieve a compromise between data diversity and implementation difficulty, methods based on combining existing annotated data have been developed for data augmentation, which are easy to implement and allow more variability of the generated data than data augmentation based on basic image transformation. For example, Mixup linearly combines two annotated images and the corresponding annotations [16], and the combination is performed stochastically to create a large number of augmented training images. CutMix is further developed to allow nonlinear combination of two images, where one region in the combined image is from one image and the rest is from the other image [15]. The annotations are still linearly combined according to the contribution of each image [15]. However, these data augmentation approaches based on image combination are mostly applied to image classification problems [2,10], and the development of this type of data augmentation methods for brain lesion segmentation is still an open problem.

In this work, we develop a data augmentation approach that produces diverse training data and is easy to implement for brain lesion segmentation. Similar to Mixup and CutMix, the proposed method combines existing annotated data for the generation of new training data; unlike these methods, the combination in our method is lesion-aware, and thus the proposed method is more appropriate for brain lesion segmentation. Specifically, given a pair of annotated training images, from one image we carve a *region of interest* (ROI) according to the lesion location and geometry, and then the carved region replaces the corresponding voxels in the other labeled image. The size of the ROI is sampled from a probability distribution so that diverse combinations can be achieved. The annotation of the second image in this region is replaced by the corresponding labels in the first image as well. Since the combination is achieved with a carving operation, our method is referred to as CarveMix. To evaluate our method, experiments were performed for two brain lesion segmentation tasks, where CarveMix was integrated with the state-of-the-art segmentation framework nnU-Net [8] and improved the segmentation accuracy. The codes of our method are available at https://github.com/ZhangxinruBIT/CarveMix.git.

2   Method

2.1   Problem Formulation

Suppose we are given a set $\mathcal{X} = \{\mathbf{X}_i\}_{i=1}^N$ of 3D annotated images with brain lesions, where $\mathbf{X}_i$ is the i-th image and N is the total number of images. The annotation of $\mathbf{X}_i$ is denoted by $\mathbf{Y}_i$, and the set of annotations is denoted by $\mathcal{Y} = \{\mathbf{Y}_i\}_{i=1}^N$. In this work, we consider binary brain lesion segmentation, and thus the intensity of $\mathbf{Y}_i$ is either 1 (lesion) or 0 (background).

$\mathcal{X}$ and $\mathcal{Y}$ can be used to train a CNN that automatically segments brain lesions. In addition, it is possible to perform data augmentation to generate new images and annotations from $\mathcal{X}$ and $\mathcal{Y}$, so that more training data can be used to improve the network training. It is shown in classification problems and some segmentation problems that the combination of pairs of existing annotated images is a data augmentation approach that can generate diverse training data and is also easy to implement [15,16]. Specifically, from an image pair $\mathbf{X}_i$ and $\mathbf{X}_j$ as well as the pair of annotations $\mathbf{Y}_i$ and $\mathbf{Y}_j$, a synthetic image $\mathbf{X}$ and its annotation $\mathbf{Y}$ are generated. By repeating the image generation for different image pairs and different sampling of generation parameters, a number of synthetic images and annotations can be created and used together with $\mathcal{X}$ and $\mathcal{Y}$ to improve network training. However, existing methods based on image combination are not necessarily appropriate for brain lesion segmentation problems, because their design is unaware of lesions and the generation of annotations is designed for classification problems. Therefore, it is desirable to develop an effective combination-based data augmentation approach for brain lesion segmentation.

2.2   CarveMix

To develop a data augmentation approach based on image combination for brain lesion segmentation, we propose CarveMix, which is lesion-aware and thus more appropriate for brain lesion segmentation. In CarveMix, given $\{\mathbf{X}_i, \mathbf{Y}_i\}$ and $\{\mathbf{X}_j, \mathbf{Y}_j\}$, we carve a 3D ROI from $\mathbf{X}_i$ according to the lesion location and geometry and mix it with $\mathbf{X}_j$. Specifically, the extracted ROI replaces the corresponding region in $\mathbf{X}_j$, and the replacement is also performed for the annotation using $\mathbf{Y}_i$ and $\mathbf{Y}_j$. Mathematically, the synthetic image $\mathbf{X}$ and annotation $\mathbf{Y}$ are generated as follows

$$\mathbf{X} = \mathbf{X}_i \odot \mathbf{M}_i + \mathbf{X}_j \odot (1 - \mathbf{M}_i), \tag{1}$$
$$\mathbf{Y} = \mathbf{Y}_i \odot \mathbf{M}_i + \mathbf{Y}_j \odot (1 - \mathbf{M}_i). \tag{2}$$

Here, $\mathbf{M}_i$ is the ROI (binary mask) of extraction determined by the lesion location and geometry given by the annotation $\mathbf{Y}_i$, and $\odot$ denotes voxelwise multiplication.

As mentioned above, the ROI $\mathbf{M}_i$ should be lesion-aware. Thus, $\mathbf{M}_i$ is designed to extract the ROI along the lesion contours. In addition, to allow more diversity of the extracted ROI and thus the generated image, the size of $\mathbf{M}_i$ is designed to be randomly sampled from a probability distribution. To this end, we first compute the signed distance function $D(\mathbf{Y}_i)$ for the lesion regions of $\mathbf{X}_i$ using its annotation $\mathbf{Y}_i$, where the intensity $D^v(\mathbf{Y}_i)$ of $D(\mathbf{Y}_i)$ at voxel v is determined as

$$D^v(\mathbf{Y}_i) = \begin{cases} -d(v, \partial\mathbf{Y}_i), & \text{if } \mathbf{Y}_i^v = 1 \\ d(v, \partial\mathbf{Y}_i), & \text{if } \mathbf{Y}_i^v = 0 \end{cases}. \tag{3}$$

Here, $\partial\mathbf{Y}_i$ represents the boundary of the lesions in $\mathbf{Y}_i$, $d(v, \partial\mathbf{Y}_i)$ represents the distance between v and the lesion boundary, and $\mathbf{Y}_i^v$ denotes the intensity of $\mathbf{Y}_i$ at v. Then, we can obtain $\mathbf{M}_i$ that is consistent with the location and shape of the lesions by thresholding $D(\mathbf{Y}_i)$, where the value $\mathbf{M}_i^v$ of $\mathbf{M}_i$ at voxel v is

$$\mathbf{M}_i^v = \begin{cases} 1, & D^v(\mathbf{Y}_i) \le \lambda \\ 0, & \text{otherwise} \end{cases}. \tag{4}$$

Here, λ is a threshold that is sampled from a predetermined distribution. A greater λ leads to a larger carved ROI. To allow the carved region to be larger or smaller than the lesion, λ can be either positive or negative, respectively. Thus, the distribution for sampling λ is defined as a mixture of two uniform distributions

$$\lambda \sim \frac{1}{2}U(\lambda_\mathrm{l}, 0) + \frac{1}{2}U(0, \lambda_\mathrm{u}), \tag{5}$$

where λ_l and λ_u are the lower and upper bounds of the distribution $U(\lambda_\mathrm{l}, 0)$ and $U(0, \lambda_\mathrm{u})$, respectively.

Since the minimum value $D(\mathbf{Y}_i)_\mathrm{min}$ of $D(\mathbf{Y}_i)$ is an indicator of the lesion size, λ_l and λ_u are determined adaptively based on $D(\mathbf{Y}_i)_\mathrm{min}$ as

$$\lambda_\mathrm{l} = -\frac{1}{2}|D(\mathbf{Y}_i)_\mathrm{min}| \text{ and } \lambda_\mathrm{u} = |D(\mathbf{Y}_i)_\mathrm{min}|. \tag{6}$$

In this way, the relative variation of the ROI size with respect to the lesion size is within two. Then, we have

$$\lambda \sim \frac{1}{2}U(-\frac{1}{2}|D(\mathbf{Y}_i)_\mathrm{min}|, 0) + \frac{1}{2}U(0, |D(\mathbf{Y}_i)_\mathrm{min}|). \tag{7}$$

A graphical illustration of the CarveMix procedure described above is shown in Fig. 1, where a synthetic image and its annotation are generated from a pair of annotated images and their annotations. This procedure can be repeated by randomly drawing pairs of annotated images and the annotations as well as the size parameter λ for each image pair, so that a set $\mathcal{X}_\mathrm{s}$ of synthetic images and the corresponding set $\mathcal{Y}_\mathrm{s}$ of annotations can be generated for network training. The complete CarveMix algorithm for generating the sets of synthetic images and annotations is summarized in Algorithm 1. Note that like in Mixup and CutMix, the generated images may not always look realistic. However, existing works have shown that unrealistic synthetic images are also able to improve network training despite the distribution shift, and there exists a tradeoff between the distribution shift and augmentation diversity [6].

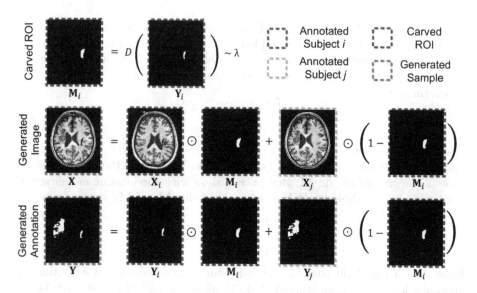

Fig. 1. A graphical illustration of the image generation procedure in CarveMix.

Algorithm 1. CarveMix

Input: Training images $\mathcal{X}$ and annotations $\mathcal{Y}$; the desired number T of synthetic images

Output: Synthetic images $\mathcal{X}_s$ and annotations $\mathcal{Y}_s$

    for $t = 1, 2, ..., T$ **do**:

        Randomly select a pair of training subjects: $\{\mathbf{X}_i, \mathbf{Y}_i\}$ and $\{\mathbf{X}_j, \mathbf{Y}_j\}$

        Compute the signed distance function $D(\mathbf{Y}_i)$ for $\mathbf{Y}_i$ with Eq. (3)

        Sample λ using Eq. (7)

        Threshold $D(\mathbf{Y}_i)$ with λ to obtain a carved ROI $\mathbf{M}_i$ according to Eq. (4)

        Generate a synthetic image and its annotation using Eqs. (1) and (2)

    end for

    return $\mathcal{X}_s$ and $\mathcal{Y}_s$

2.3 Relationship with Mixup and CutMix

The setup in Eqs. (1) and (2) bears similarity with the Mixup [16] and Cut-Mix [15] frameworks, where synthetic samples are also generated from pairs of annotated training subjects. However, Mixup and CutMix may not be suitable for brain lesion segmentation. To see that, we summarize and compare the data generation procedures in Mixup, CutMix, and CarveMix in Table 1. Note that for Mixup and CutMix the generation of annotations is extended to voxelwise combination for 3D image segmentation. Both Mixup and CutMix are unaware of lesions, where the data generation does not pay special attention to the lesions. In addition, in CutMix the synthetic intensity at each voxel originates from one individual image, but the generation of labels simply linearly combines the annotations of the two images at the voxel instead of using the label of the subject that contributes to the voxel. CarveMix addresses these limitations for brain lesion segmentation, where the generation of images is lesion-aware and the generation of annotations is consistent with the image generation.

Table 1. Comparison of Mixup, CutMix, and CarveMix

Method	Image generation	Annotation generation	Notes
Mixup	$\mathbf{X} = \lambda\mathbf{X}_i + (1 - \lambda)\mathbf{X}_j$	$\mathbf{Y} = \lambda\mathbf{Y}_i + (1 - \lambda)\mathbf{Y}_j$	λ is sampled from the beta distribution
CutMix	$\mathbf{X} = \mathbf{X}_i \odot \mathbf{M}_i + \mathbf{X}_j \odot (1 - \mathbf{M}_i)$	$\mathbf{Y} = \lambda\mathbf{Y}_i + (1 - \lambda)\mathbf{Y}_j$	$\mathbf{M}_i$ is a randomly selected cube and λ is determined by the size of $\mathbf{M}_i$
CarveMix	$\mathbf{X} = \mathbf{X}_i \odot \mathbf{M}_i + \mathbf{X}_j \odot (1 - \mathbf{M}_i)$	$\mathbf{Y} = \mathbf{Y}_i \odot \mathbf{M}_i + \mathbf{Y}_j \odot (1 - \mathbf{M}_i)$	$\mathbf{M}_i$ is selected according to the lesion location and shape

2.4 Implementation Details

The proposed method can be used for either online or offline data augmentation. In this work, we choose to perform offline data augmentation, where a desired number T of synthetic samples are generated before network training, and these samples are combined with true annotated data to train the segmentation network. In this way, our method is agnostic to the segmentation approach. For demonstration, we integrate CarveMix with the state-of-the-art nnU-Net method, which has achieved consistent top performance for a variety of medical image segmentation tasks [8] with carefully designed preprocessing and postprocessing[1]. nnU-Net uses the U-net architecture [5,13] and automatically determines the data configuration, including intensity normalization, the selection of 2D or 3D processing, the patch size and batch size, etc. For more details about nnU-Net, the readers are referred to [8]. The default hyperparameters of nnU-Net are used, except for the number of training epochs, because we empirically found that a smaller number was sufficient for training convergence in our experiments (see Sect. 3.2).

3 Experiments

3.1 Data Description

To evaluate the proposed method, we performed experiments on two brain lesion datasets, where chronic and acute ischemic stroke lesions were segmented, respec-

---

[1] Note that CarveMix can also be integrated with other segmentation frameworks if they are shown superior to nnU-net.

tively. The first dataset is the publicly available ATLAS dataset [11] for chronic stroke lesions, which contains 220 annotated T1-weighted images. These images have the same voxel size of 1 mm isotropic. We selected 50 images as the test set and considered several cases for the training set, where different numbers of the remaining images were included in the training set. Specifically, in these cases 170, 85, 43, and 22 annotated training images were used, which corresponded to 100%, 50%, 25%, and 12.5% of the total number of the available annotated images, respectively. For each case, 20% of the images in the training set were further split into a validation set for model selection.

The second dataset is an in-house dataset for acute ischemic stroke lesions, which includes 219 annotated *diffusion weighted images* (DWIs). The DWIs were acquired on a 3T Siemens Verio scanner with a b-value of $1000\,\mathrm{s/mm^2}$. The image resolution is 0.96 mm × 0.96 mm × 6.5 mm. We selected 50 images as the test data and considered four cases of the training set, where 169 (100%), 84 (50%), 42 (25%), and 21 (12.5%) annotated training images were used, respectively. For each case, 20% of the training images were further split into a validation set.

3.2   Evaluation Results

CarveMix was applied to the two datasets separately for each experimental setting. It was compared with the default *traditional data augmentation* (TDA) implemented in nnU-Net [8], including rotation, scaling, mirroring, elastic deformation, intensity perturbation, and simulation of low resolution. CarveMix was also compared with Mixup [16] and CutMix [15] with their default hyperparameters. For CarveMix, Mixup, and CutMix, the synthetic annotated scans were generated so that a total number of 1000 scans (including the true annotated scans) were available for training. CarveMix, Mixup, and CutMix were integrated with nnU-Net offline, and thus TDA was also performed for these synthetic images. Note that since the same number of epochs and the same number of batches per epoch were used for each method during network training, and TDA was performed randomly online, all methods including TDA have used the same number of training samples. Therefore, the comparison with TDA was fair. The maximum number of training epochs was set to 450/200 for the ATLAS/DWI dataset, respectively. The evaluation results are presented below.

Results on the ATLAS Dataset. We first qualitatively evaluated CarveMix. Triplanar views of the segmentation results on a representative test scan are shown in Fig. 2(a) for CarveMix and each competing method, together with the expert annotation. Here, the results were obtained with 100% training data (170 annotated training scans). We can see that CarveMix produced segmentation results that better agree with the annotation than the competing methods.

Next, CarveMix was quantitatively evaluated. For each method and each experimental setting of the training set, we computed the means and standard deviations of the Dice coefficients of the segmentation results on the test set. These results are summarized in Table 2 (the part associated with the ATLAS

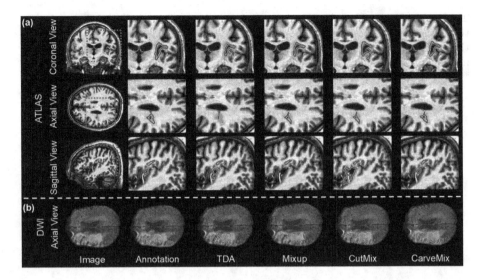

Fig. 2. Cross-sectional views of the segmentation results (red contours) on representative test scans: (a) triplanar views and their zoomed views for the ATLAS dataset and (b) axial views for the DWI dataset. The results of CarveMix and each competing method are overlaid on the T1-weighted image or DWI for the ATLAS or DWI dataset, respectively, and they are shown together with the expert annotation. These results were obtained with 100% annotated training scans. (Color figure online)

dataset). In all cases, CarveMix outperforms the competing methods with higher Dice coefficients. In addition, in most cases the difference between CarveMix and the competing methods is statistically significant using paired Student's t-tests, and this is also indicated in Table 2. Note that Mixup and CutMix are not originally designed for brain lesion segmentation. Compared with TDA they do not necessarily improve the segmentation quality, which is consistent with the previous observations in [3]. The CutMix strategy could even degrade the segmentation performance due to the inappropriate generation of synthetic annotations discussed in Sect. 2.3.

Results on the DWI Dataset. Similar to the evaluation on the ATLAS dataset, qualitative and quantitative evaluation was performed for the DWI dataset. The results are shown in Fig. 2(b) and Table 2 (the part associated with the DWI dataset). From Fig. 2(b) we can see that the result of CarveMix better resembles the expert annotation than those of the competing methods; and Table 2 indicates that CarveMix has better Dice coefficients than the competing methods and its difference with the competing methods is significant in most cases.

Table 2. Means and standard deviations of the Dice coefficients (%) of the segmentation results on the test set for the ATLAS/DWI dataset. The results for each size of the training set are shown. Asterisks indicate that the difference between the proposed method and the competing method is statistically significant (*: $p \leq 0.05$, **: $p \leq 0.01$, ***: $p \leq 0.001$) using a paired Student's t-test. The best results are highlighted in bold.

Dataset	Size	TDA	Mixup	CutMix	CarveMix
ATLAS	100%	59.39 ± 32.45*	59.33 ± 33.06*	56.11 ± 32.44**	**63.91 ± 29.87**
	50%	56.72 ± 30.74	58.40 ± 29.35	54.25 ± 30.24*	**60.57 ± 31.77**
	25%	49.87 ± 32.19***	49.18 ± 32.72***	41.19 ± 33.98***	**55.82 ± 31.58**
	12.5%	41.86 ± 32.87***	42.57 ± 33.54***	24.57 ± 27.01***	**54.77 ± 30.55**
DWI	100%	74.91 ± 25.22*	74.19 ± 25.22	73.33 ± 27.30*	**76.40 ± 25.31**
	50%	73.35 ± 25.91*	71.10 ± 27.50*	69.70 ± 27.36**	**74.99 ± 25.34**
	25%	69.41 ± 27.94*	68.71 ± 28.56*	50.28 ± 32.44***	**72.07 ± 26.64**
	12.5%	64.83 ± 25.23**	57.04 ± 31.86***	07.72 ± 15.31***	**71.32 ± 24.59**

4   Conclusion

We have proposed CarveMix, which is a simple data augmentation approach for brain lesion segmentation. The proposed method combines pairs of annotated training samples to generate synthetic training images, and the combination is lesion-aware. The experimental results on two brain lesion segmentation tasks show that CarveMix improves the segmentation accuracy and compares favorably with competing data augmentation strategies.

Acknowledgement. This work is supported by Beijing Natural Science Foundation (L192058 & 7192108).

References

1. Barber, P.A., Demchuk, A.M., Zhang, J., Buchan, A.M.: Validity and reliability of a quantitative computed tomography score in predicting outcome of hyperacute stroke before thrombolytic therapy. The Lancet **355**(9216), 1670–1674 (2000)
2. Bdair, T., Wiestler, B., Navab, N., Albarqouni, S.: ROAM: random layer mixup for semi-supervised learning in medical imaging. arXiv preprint arXiv:2003.09439 (2020)
3. Chaitanya, K., et al.: Semi-supervised task-driven data augmentation for medical image segmentation. Med. Image Anal. **68**, 101934 (2021)
4. Chen, S., Dobriban, E., Lee, J.H.: A group-theoretic framework for data augmentation. J. Mach. Learn. Res. **21**, 1–71 (2019)
5. Çiçek, Ö., Abdulkadir, A., Lienkamp, S.S., Brox, T., Ronneberger, O.: 3D U-Net: learning dense volumetric segmentation from sparse annotation. In: Ourselin, S., Joskowicz, L., Sabuncu, M.R., Unal, G., Wells, W. (eds.) MICCAI 2016. LNCS, vol. 9901, pp. 424–432. Springer, Cham (2016). https://doi.org/10.1007/978-3-319-46723-8_49

6. Gontijo-Lopes, R., Smullin, S., Cubuk, E.D., Dyer, E.: Tradeoffs in data augmentation: an empirical study. In: International Conference on Learning Representations (2021)
7. Ikram, M.A., et al.: Brain tissue volumes in relation to cognitive function and risk of dementia. Neurobiol. Aging **31**(3), 378–386 (2010)
8. Isensee, F., Jaeger, P.F., Kohl, S.A., Petersen, J., Maier-Hein, K.H.: nnU-Net: a self-configuring method for deep learning-based biomedical image segmentation. Nat. Methods **18**(2), 203–211 (2021)
9. Kamnitsas, K., et al.: Efficient multi-scale 3D CNN with fully connected CRF for accurate brain lesion segmentation. Med. Image Anal. **36**, 61–78 (2017)
10. Li, Z., Kamnitsas, K., Glocker, B.: Analyzing overfitting under class imbalance in neural networks for image segmentation. IEEE Trans. Med. Imaging **40**, 1065–1077 (2020)
11. Liew, S.L., et al.: A large, open source dataset of stroke anatomical brain images and manual lesion segmentations. Sci. Data **5**, 180011 (2018)
12. Pesteie, M., Abolmaesumi, P., Rohling, R.N.: Adaptive augmentation of medical data using independently conditional variational auto-encoders. IEEE Trans. Med. Imaging **38**(12), 2807–2820 (2019)
13. Ronneberger, O., Fischer, P., Brox, T.: U-Net: convolutional networks for biomedical image segmentation. In: Navab, N., Hornegger, J., Wells, W.M., Frangi, A.F. (eds.) MICCAI 2015. LNCS, vol. 9351, pp. 234–241. Springer, Cham (2015). https://doi.org/10.1007/978-3-319-24574-4_28
14. Tajbakhsh, N., Jeyaseelan, L., Li, Q., Chiang, J.N., Wu, Z., Ding, X.: Embracing imperfect datasets: a review of deep learning solutions for medical image segmentation. Med. Image Anal. **63**, 10169 (2020)
15. Yun, S., Han, D., Oh, S.J., Chun, S., Choe, J., Yoo, Y.: CutMix: regularization strategy to train strong classifiers with localizable features. In: Proceedings of the IEEE/CVF International Conference on Computer Vision, pp. 6023–6032 (2019)
16. Zhang, H., Cisse, M., Dauphin, Y.N., Lopez-Paz, D.: mixup: Beyond empirical risk minimization. In: International Conference on Learning Representations (2018)

Boundary-Aware Transformers for Skin Lesion Segmentation

Jiacheng Wang[1], Lan Wei[2], Liansheng Wang[1(✉)], Qichao Zhou[3(✉)], Lei Zhu[4], and Jing Qin[5]

[1] Department of Computer Science at School of Informatics, Xiamen University, Xiamen, China
jiachengw@stu.xmu.edu.cn, lswang@xmu.edu.cn
[2] School of Electrical and Computer Engineering, Xiamen University Malaysia, Bandar Sunsuria, Malaysia
[3] Manteia Technologies Co., Ltd., Xiamen, China
zhouqc@manteiatech.com
[4] Department of Computer Science and Engineering, The Chinese University of Hong Kong, Hong Kong, China
lzhu@cse.cuhk.edu.hk
[5] Center for Smart Health, School of Nursing, The Hong Kong Polytechnic University, Hong Kong, China
harry.qin@polyu.edu.hk

Abstract. Skin lesion segmentation from dermoscopy images is of great importance for improving the quantitative analysis of skin cancer. However, the automatic segmentation of melanoma is a very challenging task owing to the large variation of melanoma and ambiguous boundaries of lesion areas. While convolutional neutral networks (CNNs) have achieved remarkable progress in this task, most of existing solutions are still incapable of effectively capturing global dependencies to counteract the inductive bias caused by limited receptive fields. Recently, transformers have been proposed as a promising tool for global context modeling by employing a powerful global attention mechanism, but one of their main shortcomings when applied to segmentation tasks is that they cannot effectively extract sufficient local details to tackle ambiguous boundaries. We propose a novel boundary-aware transformer (BAT) to comprehensively address the challenges of automatic skin lesion segmentation. Specifically, we integrate a new boundary-wise attention gate (BAG) into transformers to enable the whole network to not only effectively model global long-range dependencies via transformers but also, simultaneously, capture more local details by making full use of boundary-wise prior knowledge. Particularly, the auxiliary supervision of BAG is capable of assisting transformers to learn position embedding as it provides much spatial information. We conducted extensive experiments to evaluate the proposed BAT and experiments corroborate its effectiveness, consistently outperforming state-of-the-art methods in two famous datasets (Code is available at https://github.com/jcwang123/BA-Transformer).

J. Wang and L. Wei—Contributed equally.

© Springer Nature Switzerland AG 2021
M. de Bruijne et al. (Eds.): MICCAI 2021, LNCS 12901, pp. 206–216, 2021.
https://doi.org/10.1007/978-3-030-87193-2_20

Keywords: Transformer · Medical image segmentation · Deep learning

1 Introduction

Melanoma is one of the most rapidly increasing cancers all over the world. According to the American Cancer Society's estimation, there are about 100,350 new cases and over 65,00 deaths in 2020 [13]. Segmenting skin lesions from dermoscopy images is a key step in skin cancer diagnosis and treatment planning. In current clinical practice, dermatologists usually need to manually delineate skin lesions for further analysis. However, manual delineation is usually tedious, time-consuming, and error-prone. To the end, automated segmentation methods are highly demanded in clinical practice to improve the segmentation efficiency and accuracy.

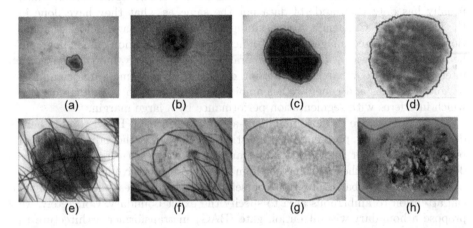

Fig. 1. The challenges of automatic skin lesion segmentation from dermoscopy images: (a)–(d) large skin lesion variations in size, shape, and color, (f)(g) partial occlusion by hair, and (f)–(h) ambiguous boundaries.

It remains, however, a very challenging task because (1) skin lesions have large variations in size, shape, and color (see Fig. 1 (a–d)), (2) present of hair will partially cover the lesions destroying local context, (3) the contrast between some lesions to normal skin are relatively low, resulting in ambiguous boundaries (see Fig. 1 (e–h)), and (4) the limited training data make the task even harder. A lot of effort has been dedicated to overcoming these challenges. Traditional methods based on various hand-crafted features are usually not stable and robust, leading to poor segmentation performance when facing lesions with large variations [15]. The main reason is that these hand-crafted features are incapable of capturing distinctive representations of skin lesions. To solve the problem, deep learning models based on convolutional neural networks (CNN) have been proposed and achieved remarkable performance gains compared with traditional methods such as some advanced version of the fully convolutional network (FCN) [20,21]. However, these models are still insufficient to tackle the challenges of skin lesion segmentation due to the inductive bias caused by the lack of global context. With regard to this, researchers propose various approaches to enlarging the receptive fields inspired by the advancement of dilated convolution [18,19]. Lee *et al.* [10] extensively incorporate the dilated attention module with boundary prior so that the network predict boundary key-points maps to guide the attention module.

Nevertheless, most existing solutions are still incapable of effectively capturing sufficient global context to deal with above mentioned challenges. Recently, *transformers* have been proposed to regard an image as a sequence of patches and aggregate feature in global context by self-attention mechanisms [4,14]. For example, TransUNet [5], a hybrid architecture of CNN and transformer, performs well on Synapse multi-organ segmentation. Yet, it is difficult for transformer based framework to achieve the same success on skin lesion segmentation, which usually has only thousands of data not the same as what they have done in the COCO 2017 Challenge [4] containing 118k training images and 5k validation images. Limited images make it difficult to encode position embedding, and hence will not always be able to accurately and effectively model long-range interactions. Moreover, regions of lesion cover a relatively small area compared to normal tissues and generally has ambiguous boundary not as human organs, which interferes with segmentation performance by a large margin.

In this paper, we propose boundary-aware transformer (BAT) to ably handle aforementioned problems, by holistically leveraging the advancement of boundary-wise prior knowledge and transformer-based network. In fact, this design is based on the intuitions for human beings to perceive lesions in vision, i.e. considering global context to coarsely locate lesion area and paying special attention to ambiguous area to specify the exact boundary. Concretely, we propose a boundary-wise attention gate (BAG) in transformer architecture to make full use of boundary-wise prior knowledge. Firstly, BAG would learn which patches in the sequence belong to ambiguous boundary, thus providing a patch-wise attention map to guide this attention gate. Secondly, a novel key-patch map generation algorithm is introduced for adeptly giving the ground-truth label that can best represent the ambiguous boundary of target lesion. Thirdly, the auxiliary supervision of BAG provides feedback to train transformers that can let it efficiently learn position embedding on a relatively small dataset. We evaluate our model on different publicly available databases. One is the ISBI 2016 and PH2 dataset following the experimental setting in most recent work [10], the other one is the latest ISIC 2018 dataset consisting of 2594 labeled images in total. All the experiment results demonstrate the significant performance gains of our proposed framework.

2   Method

An overview of our proposed model is illustrated in Fig. 2. We will first introduce our basic transformer network that leverages the intrinsic local locality of CNN and innate long-range dependency to complement the skin lesion segmentation in Sect. 2.1. Then, the overall framework of our boundary-aware transformer for segmenting regions of ambiguous boundary will be elaborated in Sect. 2.2. In the end, Sect. 2.3 introduces the hybrid objective function to efficiently train our boundary-aware transformer.

2.1   Basic Transformer for Segmentation

Our basic transformer (BT) blends typical CNN's and transformer's architectures to perform segmentation task, by three procedures: image sequentialization, sequence transformation, and atrous prediction. Image sequentialization is typically the primary step at which an image is converted into 1D sequential embedding fitting the input format of subsequent sequence transformation. Atrous prediction is designed to effectively remedy the local feature representation which has been ignored in sequence transformation and robustly segment lesions at multiple scales.

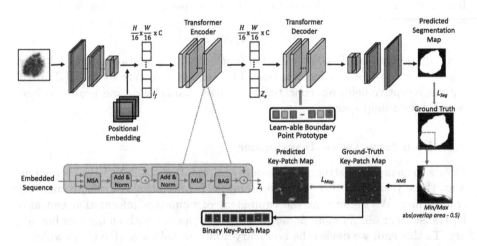

Fig. 2. An overview of the proposed Boundary-Aware Transformer framework.

Image Sequentialization. Current literature processing sequentialization by typical CNN encoder [4,11] or linear projection [7,22]. Despite its successful advancement achieved on some natural vision tasks, linear projection remains the shortcoming that is the high dependency on the amount of dataset, leading to inferior segmentation performance in medical domain [5]. To this end, we perform sequentialization through the former. Given an input image $I \in \mathbb{R}^{H \times W \times 3}$ with width W and height H, the CNN backbone (ResNet50 as default in this work) produces the corresponding image feature map $I_f \in \mathbb{R}^{\frac{H}{16} \times \frac{W}{16} \times C}$, that the size of each patch is 16×16. The feature map is then flattened into 1D patch embedding and added by a learnable positial embedding [8] which is randomly initialized to compensate spatial information destroyed by sequentialization, resulting in final sequential embedding as $E \in \mathbb{R}^{L \times C}, L = \frac{HW}{256}$.

Sequence Transformation. Transformer encoder composed of n stacked encoder layers is applied to capture long-range context in a whole dermoscopic image. Each layer in the encoder consists of a multi-head self-attention module (MSA) and a Multilayer Perceptron (MLP) following typical design [17]. Assumed that the input of i-th layer is Z^{i-1} (specially, $Z^0 \leftarrow E$), the output can be written as follows:

$$Z^i = MSA(Z^{i-1}) \oplus MLP(MSA(Z^{i-1})), \tag{1}$$

Eventually, transformed feature of the last layer Z^n will be reshaped to 2D format as $Z \in \mathbb{R}^{\frac{H}{16} \times \frac{W}{16} \times C}$, for dense prediction in the next.

Atrous Prediction. For segmenting lesions at multiple scales, this module takes transformed feature Z after self-attention mechanism as input and aims to produce a dense prediction. Aiming to enhance the local feature representation and handle the multi-scale lesion context, an atrous prediction module is designed as follows:

$$\hat{S}_{pred} = \delta(d_1^1([d_1^3(Z), d_3^3(Z), d_6^3(Z)])). \tag{2}$$

Here, $d_r^s(\cdot)$ denotes dilated convolution function with a dilation rate r and filter size of $s \times s$. δ is a sigmoid function. The enhanced feature maps $d_r^s(Z)$ with various receptive fields are concatenated across channel-wise and projected into segmentation map space.

2.2 Boundary-Aware Transformer

Efforts to incorporate structural boundary information to CNNs have been made a lot these years, but there is little literature investigating the effectiveness on transformer. We argue that the equipment of boundary information can also let transformer obtain more power in addressing lesions with ambiguous boundary. To this end, we devise the boundary-aware transformer (BAT), in which a boundary-wise attention gate (BAG) is added at end of each transformer encoder layer to refine transformed feature. BAG's architecture is similar to conventional spatial attention gate including (1) a key-patch map generator which takes the transformed feature as input and output a binary patch-wise attention map $\hat{M}_{pred} = \delta(d_1^1(Z)) \in \mathbb{R}^{L \times 1}$, where value 1 indicates that the corresponding patch is at ambiguous boundary. (2) and a residual attention scheme for preserving boundary-wise information. Hence, the boundary-aware transformed feature can be re-written as:

$$V^{i-1} = MSA(Z^{i-1}) \oplus MLP(MSA(Z^{i-1})),$$
$$Z^i = V^{i-1} \oplus (V^{i-1} \otimes \hat{M}^{i-1}), \tag{3}$$

where $\oplus$ and $\otimes$ denote element-wise addition and channel-wise multiplication, respectively.

In addition to BAGs in transformer encoder layers, a query embedding based BAG is applied after encoder to refine the feature Z^n. It plays the same role in boundary-wise attention but comes true by a totally different way. Here, instead of learning the linear projection as classifier, we refer a learnable embedding Q_b as context prototype for regions among ambiguous boundary. It will be compared with all patch embedding (Z) after aforementioned blocks, to produce a similarity map M^n. Those patches with high similarity will be the regions of ambiguous boundary. Similar to other BAGs, a residual attention scheme is also applied here as: $Z^{n+1} = Z^n \oplus (Z^n \otimes \hat{M}^n)$.

By this design, BAT learns robust feature representation of ambiguous boundary in a variety of ways, which is of great significance to handle segmentation of lesions with ambiguous boundary. Following our basic design, feature Z^{n+1} is fed into atrous prediction module to produce the segmentation map $\hat{S}_{pred}$.

Boundary-Supervised Generator. As the generator doesn't necessarily know on its own which patches can best represent structural boundary of target lesion, we introduce a novel algorithm to produce ground-truth key-patch map to train the generator with full supervision. Besides the enhancement of boundary features, this design can also help in accelerating training transformer thanks to the auxiliary constraints.

Specifically, boundary points set is produced using conventional edge detection algorithm at first. For each point in this set, we draw a circle of radius r (set to 10 as default) and calculate the proportion p of lesion area in this circle. Larger or smaller proportion indicates that boundary is not smooth in this cricle. Thus we score each point as $|p - 0.5|$, representing the assistance in segmenting ambiguous parts. Non-maximum suppression is then utilized to filter points with larger proportion than neighbour k (set to 30 as default) points. Next, filtered points' 2D location (x, y) is mapped into 1D location as $\lfloor x/16 \rfloor * 16 + \lfloor y/16 \rfloor$, and patch labels at these location are set to 1 and others are set to 0, leading to final ground-truth M_{GT}.

2.3   Objective Function

To train the segmentation network including the proposed BAGs, we emply two types of loss functions. The first one is a Dice loss function to minimize the difference between the groud-truth segmentation map and the predicted segmentation map as L_{Seg}. The second one is a Cross-Entropy loss to reduce the predicted key-patch map and its ground-truth as L_{Map}. Total loss is defined as:

$$L_{Total} = L_{Seg} + \sum_{i=1}^{n+1} L_{Map}^i, \tag{4}$$

$$L_{Seg} = \phi_{DICE}(S_{GT}, \hat{S}_{Pred}), L_{Map}^i = \phi_{CE}(M_{GT}, \hat{M}_{Pred}^i),$$

where M_{Pred}^i denotes the predicted key-patch map at i-th transformer encoder layer. ϕ_{DICE}, ϕ_{CE} denote Dice loss function and Cross-Entropy loss function, respectively. n denotes the number of transformer encoder layers and is set to 4 as default.

3   Experimental Results

3.1   Datasets

We conduct extensive experiments on the skin lesion segmentation datasets from International Symposium on Biomedical Imaging (ISBI) of the years 2016 and 2018. The datasets are collected from a variety of different treatment centers, archived by the International Skin Imaging Collaboration (ISIC), which hosted a challenge named skin lesion analysis toward melanoma detection to boost the performance of melanoma diagnosis. ISIC 2016 contains a total number of 900 samples for training and a total number of 379 dermoscopy images for testing. We follow up the same experimental protocols in the most recent work [10], in which we train our model on the training set of ISIC 2016 and extensively evaluate it on PH2 dataset. ISIC 2018 contains 2594 training samples in total and annotation of its public test set is missing, therefore we perform five-fold cross-validation on its training set for fair comparison.

3.2   Implementation Details

Our network is implemented on a single NVIDIA RTX 3090Ti. All images are empirically resized to (512×512) considering the efficiency, and we do data augmentation including vertical flip, horizontal flip and random scale change (limited 0.9–1.1). Each mini-batch includes 24 images and we utilize Adam with an initial learning rate of 0.001 to optimize the network. Learning rate decrease in half when loss on the validation set has not dropped by 10 epochs. The encoder of each network has been pre-trained on ImageNet and all parameters are then fine-tuned for 500 epochs in total.

Table 1. Experimental results on different datasets.

Model	ISIC 2016 + PH2		Model	ISIC 2018	
	$Dice \uparrow$	$IoU \uparrow$		$Dice \uparrow$	$IoU \uparrow$
SSLS[1]	0.783	0.681	DeepLabv3 [6]	0.884	0.806
MSCA[3]	0.815	0.723	U-Net++ [23]	0.879	0.805
FCN [12]	0.894	0.821	CE-Net [9]	0.891	0.816
Bi et al. [2]	0.906	0.839	MedT [16]	0.859	0.778
Lee et al. [10]	0.918	0.843	TransUNet [5]	0.894	0.822
BAT	0.921	0.858	BAT	0.912	0.843

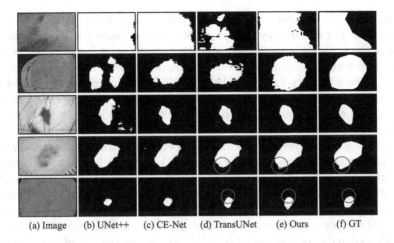

(a) Image   (b) UNet++   (c) CE-Net   (d) TransUNet   (e) Ours   (f) GT

Fig. 3. Visual comparison of lesion segmentation results produced by different methods.

3.3   Comparison with State-of-the-Arts

For baseline comparisons, we run experiments on both convolutional and transformer-based methods. With regard to the evaluation metrics, we employ a Dice coefficient (Dice), and a Intersection over Union (IoU). Table 1 displayed the comparative study of our proposed boundary-aware transformer (BAT) with other methods on different datasets. It's obviously shown that our model achieves the best segmentation performance.

On the *ISIC 2016 + PH2* dataset, we compare our method with five state-of-the-art methods. Among them, Lee *et al.* [10] is a 2D attention-based model with use of boundary-prior knowledge, achieving best segmentation performance on skin lesion segmentation recently. As seen in the Table 1, our BAT achieves 0.920 in Dice and 0.858 in IoU, outperform Lee *et al.* by 0.2% and 1.5% in Dice and IoU, respectively. We extensively conduct experiments with other SOTA segmentation networks on the *ISIC 2018* dataset, including three famous convolutional models for segmentation (DeepLabv3 [6], UNet++ [23], CE-Net [9]) and two transformer-based network to address medical image segmentation (TransUNet [5], MedT [16]). Even compared with other state-of-the-art segmentation

Table 2. Experimental results on different datasets.

Trans.	BAG	ISIC 2016 + PH2		ISIC 2018	
		Dice ↑	*IoU* ↑	*Dice* ↑	*IoU* ↑
		0.884	0.805	0.879	0.810
✓		0.900	0.827	0.890	0.821
✓	✓	0.921	0.858	0.912	0.843

models, our BAT still achieves the consistent and significant improvement on both metrics. It's noteworthy that transformer-based network has superior performance than conventional CNNs, indicating the effectiveness of utilizing global context to detect skin lesion. In addition, compared with TransUNet, our method leveraging the boundary-prior knowledge significantly improves the segmentation performance (1.8% on Dice and 2.1% on IoU), proving that the combination of boundary information and transformer architecture is indeed helpful to segment target lesion.

Figure 3 visualizes five typical challenging cases of lesion segmentation results. It is observed that our results are closest to the ground truth, when compared with our competitors. The first three rows represent cases with various color, size and shape, and our BAT outperforms others with most stable segmentation performance, indicating the robust advancement of global context. The last two rows highlight some small regions of ambiguous boundary and it's shown that our BAT is capable of tacking such problems, due to the use of boundary-wise prior knowledge.

3.4 Ablation Study

We further conduct ablation studies to demonstrate the effectiveness of three major components in BAT: (1) the transformer-based self-attention mechanism (Trans.), (2) boundary-wise attention gate (BAG). As shown in the Table 2, by the incorporation of self-attention mechanism, the IoU increases by a large margin on both datasets. This result indicate that it's essential to integrate global context to improve the skin lesion detection. On the other hand, applying BAGs to guide transformer further improves the performance significantly, confirming the effectiveness of boundary-wise prior knowledge to tackling challenging cases, such as lesions with ambiguous boundary.

4 Conclusion

We present a novel and efficient context-aware network, namely boundary-aware transformer (BAT) network, for accurate segmentation of skin lesion from dermoscopy images. Extensive experiments on two public datasets confirm the effectiveness of our proposed BAT, to help yield much better segmentation results for skin lesions. Our full model outperforms state-of-the-art models by a large margin in segmentation accuracy and the intuitive visualization shows that our BAT has most satisfactory performance on skin lesions with ambiguous boundary.

References

1. Ahn, E., et al.: Automated saliency-based lesion segmentation in dermoscopic images. In: 2015 37th Annual International Conference of the IEEE Engineering in Medicine and Biology Society (EMBC), pp. 3009–3012. IEEE (2015)

2. Bi, L., Kim, J., Ahn, E., Kumar, A., Fulham, M., Feng, D.: Dermoscopic image segmentation via multistage fully convolutional networks. IEEE Trans. Biomed. Eng. **64**(9), 2065–2074 (2017). https://doi.org/10.1109/TBME.2017.2712771

3. Bi, L., Kim, J., Ahn, E., Feng, D., Fulham, M.: Automated skin lesion segmentation via image-wise supervised learning and multi-scale superpixel based cellular automata. In: 2016 IEEE 13th International Symposium on Biomedical Imaging (ISBI), pp. 1059–1062. IEEE (2016)

4. Carion, N., Massa, F., Synnaeve, G., Usunier, N., Kirillov, A., Zagoruyko, S.: End-to-end object detection with transformers. In: Vedaldi, A., Bischof, H., Brox, T., Frahm, J.M. (eds.) Computer Vision, ECCV 2020. Lecture Notes in Computer Science, vol. 12346. Springer, Cham (2020). https://doi.org/10.1007/978-3-030-58452-8_13

5. Chen, J., et al.: TransUNet: transformers make strong encoders for medical image segmentation (2021)

6. Chen, L.C., Papandreou, G., Schroff, F., Adam, H.: Rethinking atrous convolution for semantic image segmentation. arXiv preprint arXiv:1706.05587 (2017)

7. Dosovitskiy, A., et al.: An image is worth 16x16 words: transformers for image recognition at scale. arXiv preprint arXiv:2010.11929 (2020)

8. Gehring, J., Auli, M., Grangier, D., Yarats, D., Dauphin, Y.N.: Convolutional sequence to sequence learning. In: International Conference on Machine Learning, pp. 1243–1252. PMLR (2017)

9. Gu, Z.: Ce-net: Context encoder network for 2d medical image segmentation. IEEE Trans. Med. Imaging **38**(10), 2281–2292 (2019)

10. Lee, H.J., Kim, J.U., Lee, S., Kim, H.G., Ro, Y.M.: Structure boundary preserving segmentation for medical image with ambiguous boundary. In: 2020 IEEE/CVF Conference on Computer Vision and Pattern Recognition (CVPR), pp. 4816–4825 (2020). https://doi.org/10.1109/CVPR42600.2020.00487

11. Li, Z., Liu, X., Creighton, F.X., Taylor, R.H., Unberath, M.: Revisiting stereo depth estimation from a sequence-to-sequence perspective with transformers. arXiv preprint arXiv:2011.02910 (2020)

12. Long, J., Shelhamer, E., Darrell, T.: Fully convolutional networks for semantic segmentation. In: Proceedings of the IEEE Conference on Computer Vision and Pattern Recognition, pp. 3431–3440 (2015)

13. Mathur, P., et al.: Cancer statistics, 2020: report from national cancer registry programme, India. JCO Glob. Oncol. **6**, 1063–1075 (2020)

14. Prangemeier, T., Reich, C., Koeppl, H.: Attention-based transformers for instance segmentation of cells in microstructures. In: 2020 IEEE International Conference on Bioinformatics and Biomedicine (BIBM), pp. 700–707. IEEE (2020)

15. Tu, Z., Bai, X.: Auto-context and its application to high-level vision tasks and 3d brain image segmentation. IEEE Trans. Pattern Anal. Mach. Intell. **32**(10), 1744–1757 (2010). https://doi.org/10.1109/TPAMI.2009.186

16. Valanarasu, J.M.J., Oza, P., Hacihaliloglu, I., Patel, V.M.: Medical transformer: gated axial-attention for medical image segmentation. arXiv preprint arXiv:2102.10662 (2021)

17. Vaswani, A., et al.: Attention is all you need (2017)

18. Yu, F., Koltun, V.: Multi-scale context aggregation by dilated convolutions. arxiv 2015. arXiv preprint arXiv:1511.07122 615 (2019)

19. Yu, F., Koltun, V., Funkhouser, T.: Dilated residual networks. In: Proceedings of the IEEE Conference on Computer Vision and Pattern Recognition, pp. 472–480 (2017)

20. Yu, L., Chen, H., Dou, Q., Qin, J., Heng, P.: Automated melanoma recognition in dermoscopy images via very deep residual networks. IEEE Trans. Med. Imaging **36**(4), 994–1004 (2017). https://doi.org/10.1109/TMI.2016.2642839
21. Yuan, Y., Chao, M., Lo, Y.C.: Automatic skin lesion segmentation using deep fully convolutional networks with Jaccard distance. IEEE Trans. Med. Imaging **36**(9), 1876–1886 (2017)
22. Zheng, S., et al.: Rethinking semantic segmentation from a sequence-to-sequence perspective with transformers. arXiv preprint arXiv:2012.15840 (2020)
23. Zhou, Z., Rahman Siddiquee, M.M., Tajbakhsh, N., Liang, J.: UNet++: a nested U-Net architecture for medical image segmentation. In: Stoyanov, D., et al. (eds.) DLMIA/ML-CDS -2018. LNCS, vol. 11045, pp. 3–11. Springer, Cham (2018). https://doi.org/10.1007/978-3-030-00889-5_1

A Topological-Attention ConvLSTM Network and Its Application to EM Images

Jiaqi Yang[1(✉)], Xiaoling Hu[2], Chao Chen[2], and Chialing Tsai[1]

[1] Graduate Center, CUNY, New York, USA
jyang2@gradcenter.cuny.edu
[2] Stony Brook University, New York, USA

Abstract. Structural accuracy of segmentation is important for fine-scale structures in biomedical images. We propose a novel Topological-Attention ConvLSTM Network (TACLNet) for 3D anisotropic image segmentation with high structural accuracy. We adopt ConvLSTM to leverage contextual information from adjacent slices while achieving high efficiency. We propose a Spatial Topological-Attention (STA) module to effectively transfer topologically critical information across slices. Furthermore, we propose an Iterative Topological-Attention (ITA) module that provides a more stable topologically critical map for segmentation. Quantitative and qualitative results show that our proposed method outperforms various baselines in terms of topology-aware evaluation metrics.

Keywords: Topological-attention · Spatial · Iterative · ConvLSTM

1 Introduction

Deep learning methods have achieved state-of-the-art performance for image segmentation. However, most existing methods focus on per-pixel accuracy (e.g., minimizing the cross-entropy loss) and are prone to structural errors, e.g., missing connected components and broken connections. These structural errors can be fatal in downstream analysis, affecting the functionality of the extracted fine-scale structures such as neuron membranes, vessels and cells.

To address this issue, differentiable topological losses [6,12,13,27] have been proposed to enforce the network to learn to segment with correct topology. However, these methods have their limitations when applied to 3D images, due to the high computational cost of topological information. Furthermore, we often

---

J. Yang, X. Hu—The two authors contributed equally to this paper.

Electronic supplementary material The online version of this chapter (https://doi.org/10.1007/978-3-030-87193-2_21) contains supplementary material, which is available to authorized users.

© Springer Nature Switzerland AG 2021
M. de Bruijne et al. (Eds.): MICCAI 2021, LNCS 12901, pp. 217–228, 2021.
https://doi.org/10.1007/978-3-030-87193-2_21

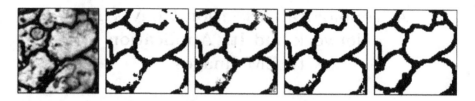

Fig. 1. An illustration of our method. From left to right: original image, ConvLSTM segmentation result, topological attention map (overlaid with segmentation), result of our method, and ground truth.

encounter anisotropic images, i.e., images with low resolution in z-dimension. The topological loss cannot be directly applied to 3D anisotropic images. For example, a tube in 3D may manifest as a series of rings across different slices rather than a seamless tube. Directly enforcing a 3D tube topology cannot work.

In this paper, a novel 3D topology-preserving segmentation method is proposed to address the aforementioned issues. Inspired by existing approaches for anisotropic images [19,28], we propose to first segment individual slices, and then stack the results together as the 3D output. We use *convolutional LSTM (ConvLSTM)* [4] as our backbone. Specifically designed for 3D anisotropic images, ConvLSTM uses 2D convolution and exploits inter-slice correlation to achieve high quality results while being more efficient than 3D CNNs. We incorporate topological loss into each of the 2D slices. This way, the topological computation is restricted within each 2D slice, and thus is very efficient.

However, simply enforcing topological loss at each slice is insufficient. A successful method should account for the fact that the topology of consecutive slices share some similarity, but are not the same. When segmenting one slice, the topology of other slices should help recalibrate the prediction, but in a soft manner. To effectively propagate topological information across slices, we propose a *Spatial Topological-Attention module*, which redirects the convolutional network's attention toward topologically critical locations of each slice, based on the topology of itself and its adjacent slices. These critical locations are locations at which the model is prone to topological mistakes. Redirecting the attention to these locations will enforce the model to make topologically correct predictions. See Fig. 1 for an illustration of our method.

Another challenge is that the topologically critical map can be inconsistent across different slices and unstable through training epochs. During the training process, the predicted probability maps will change slightly, while the corresponding Topological-Attention maps can be quite different, leading to instability of the training process. To this end, we propose an *Iterative Topological-Attention module* that iteratively refines the topologically critical map through epochs.

Our method, called *Topological-Attention ConvLSTM Network (TACLNet)*, fully utilizes topological information from adjacent slices for 3D images without much additional computational cost. Empirically, our method outperforms baselines in terms of topology-aware metrics. In summary, our main contribution is threefold:

1. A novel Spatial Topological-Attention (STA) module to propagate spatial contextual topological information across adjacent slices.
2. An Iterative Topological-Attention (ITA) module to improve the stability of the topologically critical maps, and consequently the quality of the final results.
3. Combining Topological-Attention with ConvLSTM to achieve high performance on 3D image segmentation benchmarks.

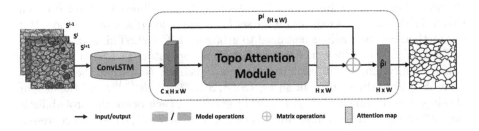

Fig. 2. Overview of the proposed framework

2   Related Works

Standard 3D medical image segmentation methods directly apply the networks to 3D images [5,11,16,17]. These methods could be computationally expensive. Alternatively, one may first segment each 2D slice, and then link the 2D segmentation results to generate 3D results [19,27]. Note that this segment-and-link approach ignores the contextual information shared among adjacent slices at the segmentation step. To address this, one may introduce pooling techniques across adjacent slices [9]. But these methods are not explicitly modeling the topology as our method does.

Persistent Homology. Our topological approach is based on the theory of persistent homology [7,8], which has attracted a great amount of attention both from theory [3,10] and from applications [24,25]. In image segmentation, persistent-homology-based topological loss functions [6,12] have been proposed to train a neural network to preserve the topology of the segmentation. The key insight of these methods is to identify critical locations for topological correctness, and improve the neural network's prediction at these locations. These critical locations are computed using the theory of persistent homology, and correspond to critical points (local maxima/minima and saddles) of the likelihood function.

Attention Mechanism. Attention modules model relationships between pixels/channels/feature maps and have been widely applied in both vision and natural language processing tasks [14,15,21]. Specifically, self-attention mechanism [22] is proposed to draw global dependencies of inputs and has been used in machine translation tasks. [29] tries to learn a better image generator via self-attention mechanism. [23] mainly explores effectiveness of non-local operation, which is similar to self-attention mechanism. [30] learns an attention map to aggregate contextual information for each individual point for scene parsing.

3   Method

The overview of the proposed architecture is illustrated in Fig. 2. To capture the inter-slice information, l consecutive slices along Z-dimension are fed into a ConvLSTM. For ease of exposition, we set $l = 3$ when describing our method. But our method can be easily generalized to arbitrary l. ConvLSTM is an extension of FC-LSTM [26], which has the convolutional operators in LSTM gates and is particularly efficient in exploiting image sequences. Note that the inputs to ConvLSTM are three adjacent slices, $\{S^{i-1}, S^i, S^{i+1}\} \in R^{H \times W}$, and the output also has three channels, $\{P^{i-1}, P^i, P^{i+1}\} \in R^{H \times W}$, each being the probabilistic map P^i of the corresponding input slice S^i. We use $i-1, i, i+1$ to represent input slice indices in this paper.

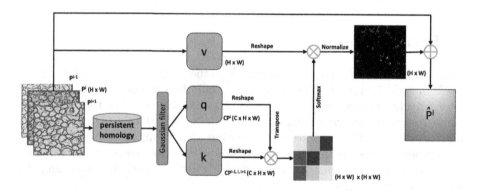

Fig. 3. Illustration of the Spatial Topological-Attention (STA) module

The three probabilistic maps $\{P^{i-1}, P^i, P^{i+1}\}$ are then fed into the Topological-Attention module. In this module, each pixel in the feature maps gathers rich structural information from both the current and adjacent slices, without introducing extra parameters. We propose a Spatial Topological-Attention module to model the correlation between the topologically critical information of adjacent slices. A Topological-Attention map is generated to highlight the locations which are structurally critical. See Sect. 3.1 for details. In Sect. 3.2, we introduce the Iterative Topological-Attention module to stabilize the critical map.

3.1   Spatial Topological-Attention (STA) Module

Continuation in contextual information across slices is essential for 3D image understanding, which can be obtained by taking adjacent slices into consideration. In order to collect contextual information in the Z-dimension to enhance the prediction quality, we introduce a STA module which encodes the inter-slices contextual information into the focused slice.

As illustrated in Fig. 2, we can obtain three predicted probabilistic maps, $\{P^{i-1}, P^i, P^{i+1}\} \in R^{H \times W}$ which corresponds to the input slices after ConvL-STM. Figure 3 shows the complete process that passes the probabilistic maps to the STA module, and yields the final probabilistic map $\hat{P}^i$ in the end. Next, we elaborate the process of aggregating the topological context of adjacent slices.

Persistent Homology and Critical Points. Given a 2D image likelihood map, we obtain the binary segmentation by thresholding at $\alpha = 0.5$. The 2D likelihood map can be represented as a 2D continuous-valued function f. We consider thresholding the continuous function f with all possible thresholds. Denote by Ω the image domain. For a specific threshold α, we define the thresholded results $f^\alpha := \{x \in \Omega | f(x) \ge \alpha\}$. By decreasing α, we obtain a monotonically growing sequence $\varnothing \subseteq f^{\alpha_1} \subseteq f^{\alpha_2} \subseteq \dots \subseteq f^{\alpha_n} = \Omega$, where $\alpha_1 \ge \alpha_2 \ge \dots \ge \alpha_n$. As α changes, the topology of f^α changes. New topological structures are born while existing ones are killed. The theory of persistent homology captures all the birth time and death time of these topological structures and summarize them as a *persistence diagram*. One can define a topological loss as the matching distance between the persistence diagrams of the likelihood function and the ground truth. When the loss is minimized, the two diagrams are the same and the likelihood map will generate a segmentation with the correct topology.

As shown in [12], the topological loss can be written as a polynomial function of the likelihood function at different critical pixels. These critical pixels correspond to critical points of the likelihood function (e.g., saddles, local minima and local maxima), and these critical points are crucial locations at which the current model is prone to make topological mistakes. The loss essentially forces the network to improve its prediction at these *topologically critical* locations.

Aggregating Topologically Critical Maps via Topological Attention. The likelihood maps at different slices generate different critical point maps. Here we propose an attention mechanism to aggregate these critical point maps across different slices to generate topological attention map for the current slice (third column in Fig. 1). For $\{P^{i-1}, P^i, P^{i+1}\}$, using persistent homology algorithm, we identify the critical points. Here we use a Gaussian operation to expand the isolated critical points to blobs because the surrounding regions will likely also be vital for structures. This way we obtain a soft version of critical point map: $CP^i = Gaussian(PH(P^i))$. Here, $PH(\cdot)$ is the operation to generate isolated critical points and $Gaussian(\cdot)$ is a Gaussian operation. CP^i has the same dimension as $P^i \in R^{H \times W}$.

As shown in Fig. 3, these critical maps are used as the query and the key for the attention mechanism. To improve the computational efficiency, we combine the critical maps CP^{i-1}, CP^i, CP^{i+1} into one single $k \in R^{C \times H \times W}$ ($C = 3$) and expand the CP^i into same size as $q \in R^{C \times H \times W}$. To obtain the correlation between target map (q) and consecutive slices (k), we reshape them to $R^{C \times N}$, where $N = H \times W$, and perform a matrix multiplication between the transpose of q and k. The similarity map $SM \in R^{N \times N}$ is generated after a softmax. SM_{nm} measures the correlation between two pixels, m and n.

Next, we reshape probabilistic map P^i and perform a matrix multiplication between P^i and SM. For pixel n, we obtain normalized $o_n^i = \sum_{m=1}^{N}(P^i{}_m SM_{nm})$ and perform an element-wise sum operation with the probabilistic map P^i to get the final output:

$$\hat{P}_n^i = \alpha o_n^i + P_n{}^i \tag{1}$$

where α is initialized as 0 to capture stable probabilistic maps first. As training continues, we assign more weight on attention map so that the $\hat{P}^i$ at each position is a weighted sum across all positions and original probabilistic map P^i.

3.2 Iterative Topological-Attention (ITA) Module

As mentioned above, the critical points generated by persistent homology are sensitive, and consequently the attention map is also relatively unstable.

By exploiting the correlation of probabilistic maps between different epochs, we can further improve the robustness of the obtained attention maps, which can lead to a better representation of the predicted probabilistic maps. Therefore, we introduce an ITA module to explore the relationships between the attention maps of different epochs. The iterative method not only helps with stability, but also enforces faster convergence. Formally, the ITA is calculated as $o_t = \beta o_{t-1} + (1 - \beta)o_t$. Here, β is a parameter to deal with the sensitiveness of the critical points, and t denotes different training epochs. o is the output of attention map which was described in Eq. (1). More details of ITA module is illustrated in Supplementary Fig. 1. During the training process, the final output $\hat{P}^i$ is generated by the sum of iterative attention map o_T and the original probabilistic map P^i. Therefore, it has a global contextual view and selectively aggregates contexts according to the spatial attention map.

4   Experiments

We use three EM datasets with rich structure information to demonstrate the effectiveness of the proposed method. In this section, we will introduce the implementation details, datasets, and the experiment results on both datasets.

Datasets. We demonstrate the effectiveness of our proposed method with three different 3D Electron Microscopic Images datasets: **ISBI12** [2], **ISBI13** [1] and **CREMI**. The size of **ISBI12**, **ISBI13** and **CREMI** are $30 \times 512 \times 512$, $100 \times 1024 \times 1024$ and $125 \times 1250 \times 1250$, respectively.

Train Settings. We adopt ConvLSTM as our backbone architecture. Also, we apply simple data augmentation, Contrast-Limited Adaptive Histogram Equalization (CLAHE) and random flipping (for ISBI12 only to enlarge training size).

For the training parameters, we initialize learning rate (lr) as 0.001 and multiply by 0.5 every 50 epochs. We train our model with batch size of 15 for CREMI and ISBI13, and 30 for ISBI12. The number of training epochs are 35, 900 and 150 for CREMI, ISBI13 and ISBI12, respectively (without attention module). We use cross entropy loss as the optimization metric.

Attention Module Details. As described in Sect. 3.1, Topological-Attention module comes after ConvLSTM. We train the TACLNet for another 15 epochs, with $lr = 0.00001$. Specifically, the patch size is 39×39 for critical points extraction. The iterative rate β is set as 0.5 for ITA module.

Quantitative and Qualitative Results. In this paper we use similar topology-aware metrics as of [12] for structural accuracy, Adapted Rand Index (ARI), Variation of Information (VOI) and Betti number error. We also report dice scores for pixel accuracy for all the baselines and the proposed method for completeness. The details of the evaluation metrics can be found in the Sect. 3 of [12]. For all the experiments, we use three-fold cross-validation to report the average performance and standard deviation over the validation set. Table 1 shows the quantitative results for the three different datasets. Note that we remove small connected components as a post-processing step to obtain final segmentation results. Our method generally outperforms existing methods [9,18,20] in terms of topology-aware metrics. Figure 4 shows qualitative results. Our method achieves better consistency/connection compared with other baselines.

Table 1. Experiment results for different models on CREMI dataset

Datasets	Models	DICE	ARI	VOI	Betti error
CREMI	DIVE	0.9542 ± 0.0037	0.6532 ± 0.0247	2.513 ± 0.047	4.378 ± 0.152
	U-Net	0.9523 ± 0.0049	0.6723 ± 0.0312	2.346 ± 0.105	3.016 ± 0.253
	Mosin	0.9489 ± 0.0053	0.7853 ± 0.0281	1.623 ± 0.083	1.973 ± 0.310
	TopoLoss	0.9596 ± 0.0029	0.8083 ± 0.0104	1.462 ± 0.028	1.113 ± 0.224
	TACLNet	**0.9665 ± 0.0008**	**0.8126 ± 0.0153**	**1.317 ± 0.165**	**0.853 ± 0.183**
ISBI12	DIVE	0.9709 ± 0.0029	0.9434 ± 0.0087	1.235 ± 0.025	3.187 ± 0.307
	U-Net	0.9699 ± 0.0048	0.9338 ± 0.0072	1.367 ± 0.031	2.785 ± 0.269
	Mosin	0.9716 ± 0.0022	0.9312 ± 0.0052	0.983 ± 0.035	1.238 ± 0.251
	TopoLoss	**0.9755 ± 0.0041**	**0.9444 ± 0.0076**	0.782 ± 0.019	0.429 ± 0.104
	TACLNet	0.9576 ± 0.0047	0.9417 ± 0.0045	**0.771 ± 0.027**	**0.417 ± 0.117**
ISBI13	DIVE	0.9658 ± 0.0020	0.6923 ± 0.0134	2.790 ± 0.025	3.875 ± 0.326
	U-Net	0.9649 ± 0.0057	0.7031 ± 0.0256	2.583 ± 0.078	3.463 ± 0.435
	Mosin	0.9623 ± 0.0047	0.7483 ± 0.0367	1.534 ± 0.063	2.952 ± 0.379
	TopoLoss	**0.9689 ± 0.0026**	**0.8064 ± 0.0112**	1.436 ± 0.008	1.253 ± 0.172
	TACLNet	0.9510 ± 0.0022	0.7943 ± 0.0127	**1.305 ± 0.016**	**1.175 ± 0.108**

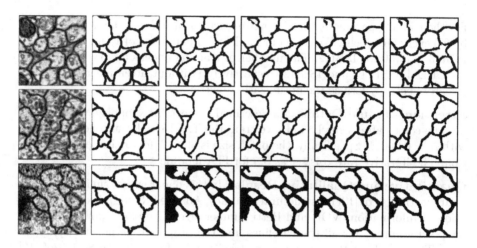

Fig. 4. An illustration of structural accuracy. From left to right: a sample patch, the ground truth, results of UNet, TopoLoss, ConvLSTM and the proposed TACLNet.

In Table 1, the Betti Error of our TACLNet brings 23.3% improvement compared to the best baseline, TopoLoss [12], for CREMI dataset. Meanwhile, TACLNet achieves best performances in terms of Betti Error and VOI on both ISBI12 and ISBI13 datasets. In Fig. 4, the results from TACLNet possess better structures with less broken boundaries comparing to other baseline methods. Results show that our attention module strengthens the structural performance overall. Also, the proposed method computes topological information on a stack of 2D images rather than directly on a 3D image, and this significantly reduces the computational expense. Specifically, for CREMI dataset, our method takes ≈1.2 h per epoch to train, whereas topoloss (3D version) takes ≈2.8h per epoch.

Ablation Study for TACLNet. Table 2 shows the ablation study of the proposed method, which demonstrates the individual contributions of the two proposed modules, Spatial Topological-Attention and Iterative Topological-Attention. As shown in Table 2, compared with the backbone ConvLSTM model (Betti Error = 1.785), the STA improves the performance remarkably to 0.873. After applying the ITA module, the network further improves the performance by 2.3%, 11.9%, 2.1% in Betti Error, VOI, and ARI, respectively. In addition, the combination of STA and ITA also improves the speed of convergence. For our ablation study, STA was trained with 50 epochs, but STA + ITA (our TACLNet) was trained with fewer than 15 epochs for a better performance, which demonstrates that the ITA module can stabilize the training procedure.

Table 2. Ablation study results for TACLNet on CREMI dataset

Models	DICE	ARI	VOI	Betti Error
ConvLSTM	0.9667 ± 0.0007	0.7627 ± 0.0132	1.753 ± 0.212	1.785 ± 0.254
ConvLSTM + STA	0.9663 ± 0.0004	0.7957 ± 0.0144	1.496 ± 0.156	0.873 ± 0.212
Our TACLNet	0.9665 ± 0.0008	$\mathbf{0.8126 \pm 0.0153}$	$\mathbf{1.317 \pm 0.165}$	$\mathbf{0.853 \pm 0.183}$

Table 3. Ablation study results for number of input slices on CREMI

Number	ARI	VOI	Betti Error	Time
1 s	0.7813 ± 0.0141	1.672 ± 0.191	1.386 ± 0.117	$0.99\,\mathrm{h/epoch}$
3 s	$\mathbf{0.8126 \pm 0.0153}$	$\mathbf{1.317 \pm 0.165}$	$\mathbf{0.853 \pm 0.183}$	$1.20\,\mathrm{h/epoch}$
5 s	0.8076 ± 0.0107	1.461 ± 0.125	0.967 ± 0.098	$2.78\,\mathrm{h/epoch}$

Ablation Study for Number of Input Slices. Table 3 is an illustration for the number of input slices. As shown in Table 3, compared with 1 slice (Betti Error = 1.386) or 5 slices (Betti Error = 0.967), the adopted setting of 3 slices achieves the best results (Betti Error = 0.853). It's not surprised that the 3 slices achieves better performance than 1 slice, as it makes use of inter-slice information. On the other hand, the dataset is anisotropic, and the slices further away are increasingly different from the center slice, which degrades the performance for 5 slices setting.

Illustration of the Attention Module. We select two images to show the effectiveness of the attention module in Fig. 5. Compared with the second column showing only the critical points detected in the current slice, the attention map in the third column captures more information with the structural similarity from adjacent slices. The attention areas on the final probabilistic map (last column) are highlighted with red color. We observe that the responses of most broken connections are high with attention module enhancement. In summary, Fig. 5 demonstrates that our TACLNet successfully captures the structure information and further improves responses on those essential areas.

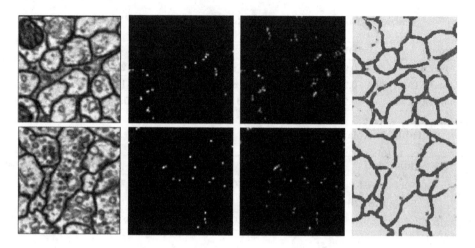

Fig. 5. Illustration of the proposed Topological-Attention. From left to right: original images S^i, the smooth critical points of CP^i, final attention map o^i, and the final probability map $\hat{P}^i$ with o^i superimposed in red (Zoom in and best viewed in color). (Color figure online)

5   Conclusion

In this paper, we proposed a novel Topological-Attention Module with ConvLSTM, named TACLNet, for 3D EM image segmentation. Validated with three EM anisotropic datasets, our method outperforms baselines in terms of topology-aware metrics. We expect the performance to further improve on isotropic datasets, because slices are closer (due to higher sampling rate in the z-dimension) with more consistent topologies across slices. For the future work, we will apply TACLNet to datasets of other medical structures, such as cardiac and vascular images, to prove its efficacy in a broader medical domain.

Acknowledgements. This work was partially supported by grants NSF IIS-1909038, CCF-1855760, NCI 1R01CA253368-01 and PSC-CUNY Research Award 64450-00-52.

References

1. Arganda-Carreras, I., Seung, H., Vishwanathan, A., Berger, D.: 3D segmentation of neurites in EM images challenge. In: ISBI 2013 (2013)
2. Arganda-Carreras, I., et al.: Crowdsourcing the creation of image segmentation algorithms for connectomics. Front. Neuroanat. **9**, 142 (2015)
3. Bubenik, P.: Statistical topological data analysis using persistence landscapes. JMLR **16**(1), 77–102 (2015)
4. Chen, J., Yang, L., Zhang, Y., Alber, M., Chen, D.Z.: Combining fully convolutional and recurrent neural networks for 3d biomedical image segmentation. In: NeurIPS, pp. 3036–3044 (2016)

5. Çiçek, Ö., Abdulkadir, A., Lienkamp, S.S., Brox, T., Ronneberger, O.: 3D U-Net: learning dense volumetric segmentation from sparse annotation. In: Ourselin, S., Joskowicz, L., Sabuncu, M.R., Unal, G., Wells, W. (eds.) MICCAI 2016. LNCS, vol. 9901, pp. 424–432. Springer, Cham (2016). https://doi.org/10.1007/978-3-319-46723-8_49

6. Clough, J.R., Oksuz, I., Byrne, N., Schnabel, J.A., King, A.P.: Explicit topological priors for deep-learning based image segmentation using persistent homology. In: Chung, A.C.S., Gee, J.C., Yushkevich, P.A., Bao, S. (eds.) IPMI 2019. LNCS, vol. 11492, pp. 16–28. Springer, Cham (2019). https://doi.org/10.1007/978-3-030-20351-1_2

7. Edelsbrunner, H., Harer, J.: Computational Topology: An Introduction. American Mathematical Society (2010)

8. Edelsbrunner, H., Letscher, D., Zomorodian, A.: Topological persistence and simplification. In: Proceedings 41st Annual Symposium on Foundations of Computer Science, pp. 454–463. IEEE (2000)

9. Fakhry, A., Peng, H., Ji, S.: Deep models for brain EM image segmentation: novel insights and improved performance. Bioinformatics **32**(15), 2352–2358 (2016)

10. Fasy, B.T., Lecci, F., Rinaldo, A., Wasserman, L., Balakrishnan, S., Singh, A., et al.: Confidence sets for persistence diagrams. Ann. Stat. **42**(6), 2301–2339 (2014)

11. Funke, J., et al.: Large scale image segmentation with structured loss based deep learning for connectome reconstruction. TPAMI **41**(7), 1669–1680 (2018)

12. Hu, X., Li, F., Samaras, D., Chen, C.: Topology-preserving deep image segmentation. In: NeurIPS, pp. 5658–5669 (2019)

13. Hu, X., Wang, Y., Fuxin, L., Samaras, D., Chen, C.: Topology-aware segmentation using discrete Morse theory. In: ICLR (2021). https://openreview.net/forum?id=LGgdb4TS4Z

14. Lin, G., Shen, C., Van Den Hengel, A., Reid, I.: Efficient piecewise training of deep structured models for semantic segmentation. In: CVPR, pp. 3194–3203 (2016)

15. Lin, Z., et al.: A structured self-attentive sentence embedding. arXiv preprint arXiv:1703.03130 (2017)

16. Meirovitch, Y., Mi, L., Saribekyan, H., Matveev, A., Rolnick, D., Shavit, N.: Cross-classification clustering: an efficient multi-object tracking technique for 3-D instance segmentation in connectomics. In: CVPR, pp. 8425–8435 (2019)

17. Milletari, F., Navab, N., Ahmadi, S.A.: V-Net: fully convolutional neural networks for volumetric medical image segmentation. In: 3DV, pp. 565–571. IEEE (2016)

18. Mosinska, A., Marquez-Neila, P., Koziński, M., Fua, P.: Beyond the pixel-wise loss for topology-aware delineation. In: CVPR, pp. 3136–3145 (2018)

19. Nunez-Iglesias, J., Ryan Kennedy, T.P., Shi, J., Chklovskii, D.B.: Machine learning of hierarchical clustering to segment 2d and 3d images. PLOS ONE **8**(8), e71715 (2013)

20. Ronneberger, O., Fischer, P., Brox, T.: U-Net: convolutional networks for biomedical image segmentation. In: Navab, N., Hornegger, J., Wells, W.M., Frangi, A.F. (eds.) MICCAI 2015. LNCS, vol. 9351, pp. 234–241. Springer, Cham (2015). https://doi.org/10.1007/978-3-319-24574-4_28

21. Shen, T., Zhou, T., Long, G., Jiang, J., Pan, S., Zhang, C.: DiSAN: directional self-attention network for RNN/CNN-free language understanding. In: AAAI (2018)

22. Vaswani, A., et al.: Attention is all you need. In: NeurIPS, pp. 5998–6008 (2017)

23. Wang, X., Girshick, R., Gupta, A., He, K.: Non-local neural networks. In: CVPR, pp. 7794–7803 (2018)

24. Wong, E., Palande, S., Wang, B., Zielinski, B., Anderson, J., Fletcher, P.T.: Kernel partial least squares regression for relating functional brain network topology to clinical measures of behavior. In: ISBI, pp. 1303–1306. IEEE (2016)
25. Wu, P., et al.: Optimal topological cycles and their application in cardiac trabeculae restoration. In: Niethammer, M., et al. (eds.) IPMI 2017. LNCS, vol. 10265, pp. 80–92. Springer, Cham (2017). https://doi.org/10.1007/978-3-319-59050-9_7
26. Xingjian, S., Chen, Z., Wang, H., Yeung, D.Y., Wong, W.K., Woo, W.: Convolutional LSTM network: a machine learning approach for precipitation nowcasting. In: NeurIPS, pp. 802–810 (2015)
27. Yang, J., Hu, X., Chen, C., Tsai, C.: 3D topology-preserving segmentation with compound multi-slice representation. In: ISBI, pp. 1297–1301. IEEE (2021)
28. Ye, Z., Chen, C., Yuan, C., Chen, C.: Diverse multiple prediction on neuron image reconstruction. In: Shen, D., et al. (eds.) MICCAI 2019. LNCS, vol. 11764, pp. 460–468. Springer, Cham (2019). https://doi.org/10.1007/978-3-030-32239-7_51
29. Zhang, H., Goodfellow, I., Metaxas, D., Odena, A.: Self-attention generative adversarial networks. In: ICML, pp. 7354–7363. PMLR (2019)
30. Zhao, H., et al.: PSANet: point-wise spatial attention network for scene parsing. In: Ferrari, V., Hebert, M., Sminchisescu, C., Weiss, Y. (eds.) ECCV 2018. LNCS, vol. 11213, pp. 270–286. Springer, Cham (2018). https://doi.org/10.1007/978-3-030-01240-3_17

BiX-NAS: Searching Efficient Bi-directional Architecture for Medical Image Segmentation

Xinyi Wang[1], Tiange Xiang[1], Chaoyi Zhang[1], Yang Song[2], Dongnan Liu[1], Heng Huang[3,4], and Weidong Cai[1(✉)]

[1] School of Computer Science, University of Sydney, Camperdown, Australia
{xwan2191,txia7609}@uni.sydney.edu.au,
{chaoyi.zhang,dongnan.liu,tom.cai}@sydney.edu.au
[2] School of Computer Science and Engineering, University of New South Wales, Sydney, Australia
yang.song1@unsw.edu.au
[3] Electrical and Computer Engineering, University of Pittsburgh, Pittsburgh, USA
[4] JD Finance America Corporation, Mountain View, CA, USA

Abstract. The recurrent mechanism has recently been introduced into U-Net in various medical image segmentation tasks. Existing studies have focused on promoting network recursion via reusing building blocks. Although network parameters could be greatly saved, computational costs still increase inevitably in accordance with the pre-set iteration time. In this work, we study a multi-scale upgrade of a bi-directional skip connected network and then automatically discover an efficient architecture by a novel two-phase Neural Architecture Search (NAS) algorithm, namely BiX-NAS. Our proposed method reduces the network computational cost by sifting out ineffective multi-scale features at different levels and iterations. We evaluate BiX-NAS on two segmentation tasks using three different medical image datasets, and the experimental results show that our BiX-NAS searched architecture achieves the state-of-the-art performance with significantly lower computational cost. Our project page is available at: https://bionets.github.io.

Keywords: Semantic segmentation · Recursive neural networks · Neural architecture search

1 Introduction

Deep learning based methods have prevailed in medical image analysis. U-Net [14], a widely used segmentation network, constructs forward skip connections

---

X. Wang and T. Xiang—Equal contributions.

Electronic supplementary material The online version of this chapter (https://doi.org/10.1007/978-3-030-87193-2_22) contains supplementary material, which is available to authorized users.

© Springer Nature Switzerland AG 2021
M. de Bruijne et al. (Eds.): MICCAI 2021, LNCS 12901, pp. 229–238, 2021.
https://doi.org/10.1007/978-3-030-87193-2_22

(skips) to aggregate encoded features in encoders with the decoded ones. Recent progress has been made on the iterative inference of such architecture by exploiting the reusable building blocks. [15] proposed a recurrent U-Net that recurses a subset of paired encoding and decoding blocks at each iteration. BiO-Net [17] introduces backward skips passing semantics in decoder to the encoder at the same level. Although this recurrent design could greatly slim the network size, the computational costs still increase inevitably in accordance with its pre-set iteration time. Meanwhile, the success of multi-scale approaches [4,20] suggests the usage of multi-scale skips that fuse both fine-grained traits and coarse-grained semantics. To this end, introduction of forward/backward skips from multiple semantic scales and searching efficient aggregation of multi-scale features with low computational cost is of high interest.

Neural Architecture Search (NAS) methods automatically perceive the optimal architecture towards effective and economic performance gain. Classic evolutionary NAS methods [12,13] evolve by randomly drawing samples during searching and validating each sampled model individually. Differentiable NAS algorithms [3,10] relax the discrete search space to be continuous and delegate backpropagation to search for the best candidate. Auto-DeepLab [9] applied differentiable NAS into image segmentation tasks to determine the best operators and topologies for each building block. Concurrently, NAS-Unet [16] applied an automatic gradient-based search of cell structures to construct an U-Net like architecture. Despite the success on feature fusions at same levels, [18] introduced a multi-scale search space to endow their proposed MS-NAS with the capability of arranging multi-level feature aggregations. However, architectures searched by the above NAS algorithms only bring marginal improvements over hand-designed ones and their searching process is empirically inefficient.

In this paper, we present an efficient **multi-scale** (abstracted as 'X') NAS method, namely BiX-NAS, which searches for the optimal **bi-directional** architecture (BiX-Net) by recurrently skipping multi-scale features while discarding insignificant ones at the same time.

Our contributions are three-fold: **(1)** We study the multi-scale fusion scheme in a recurrent bi-directional manner, and present an effective two-phase Neural Architecture Search (NAS) strategy, namely BiX-NAS, that automatically searches for the optimal bi-directional architecture. **(2)** We analyze the bottleneck of the searching deficiency in classic evolution-based search algorithms, and propose a novel progressive evolution algorithm to further discover a subset of searched candidate skips and accelerate the searching process. **(3)** Our method is benchmarked on three medical image segmentation datasets, and surpasses other state-of-the-art counterparts by a considerable margin.

2   Methods

We first discuss the effectiveness of introducing multi-scale skip connections to BiO-Net as an intuitive upgrade (Sect. 2.1), then we demonstrate the details of each phase of BiX-NAS (Sect. 2.2), and eventually, we present the *skip fairness* principle which ensures the search fairness and efficiency (Sect. 2.3).

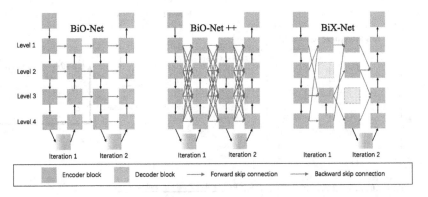

Fig. 1. Overview of BiO-Net, BiO-Net++, and BiX-Net with 4 levels and 2 iterations. Encoder and decoder blocks at the same level are reused [17].

2.1   BiO-Net++: A Multi-scale Upgrade of BiO-Net

BiO-Net [17] triggers multiple encoding and decoding phases that concatenate skipped features at the same level (Fig. 1 left). We optimize the feature fusion scheme as element-wise average for an initial reduction on the total complexity. For better clarity, we denote (1) A sequential encoding or decoding process as an *extraction stage*. There are four extraction stages in a BiO-Net like network with two iterations. (2) Any blocks in a non-first extraction stage as *searching blocks*. To fuse multi-scale features, precedent encoded/decoded features at all levels are densely connected to every decedent decoding/encoding level through bi-directional skips. We align inconsistent spatial dimensions across different levels via bilinear resizing. The suggested BiO-Net++ is outlined in Fig. 1 middle.

Although the above design promotes multi-scale feature fusion and shrinks network size, empirically we found that such *dense* connections bring a mere marginal improvement in terms of the overall performance but with an increase of computational costs (Table 1). We are interested in seeking a *sparser* connected sub-architectures of BiO-Net++ which could not only benefit from multi-scale fusions but also ease computation burdens to the greatest extent.

2.2   BiX-NAS: Hierarchical Search for Efficient BiO-Net++

To this end, we present BiX-NAS, a two-phase search algorithm, to find a sparsely connected sub-architecture of BiO-Net++, where a trainable selection matrix is adopted to narrow down the search space for differentiable NAS in Phase1, and evolutionary NAS is introduced to progressively discover the optimal sub-architecture in Phase2. To spot an adequate candidate, dense skip connections between every pair of extraction stages are further sifted for better efficiency. Suppose there are N incoming feature streams in a desired searching block ($N = 5$ for BiO-Net++), we anticipate that in a sparser connected architecture, only $k \in [1, N-2]$ candidate(s) of them could be accepted to such

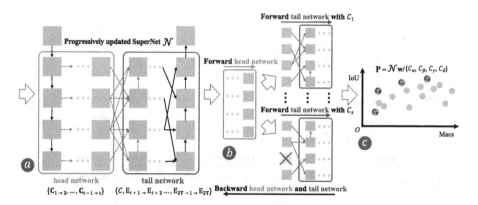

Fig. 2. Overview of progressive evolutionary search. (a) Phase1 searched Super-Net $\mathcal{N}$ can be divided into *head network* and *tail network*. **(b)** Proposed forward and backward schemes. **(c)** Only the searched skips at the Pareto front of **P** are retained.

blocks, which results in a search space of $\approx \sum_{k=1}^{(N-2)} \binom{N}{k}^{\mathbf{L}(2\mathbf{T}-1)}$ in the Super-Net BiO-Net++ with **L** levels and **T** iterations. When $\mathbf{L} = 4$ and $\mathbf{T} = 3$ [17], the search space expands to 5^{40}, escalating the difficulty to find the optimal sub-architectures.

Phase1: Narrowing Down Search Space via Selection Matrix. To alleviate such difficulty, we determine k *candidate skips* from N incoming skips in each searching block by reducing the easy-to-spot ineffective ones. Intuitively, one-to-one relaxation parameters α [10] for each skip connection x could be registered and optimized along with the SuperNet. The skip with the highest relaxation score α is then picked as the output of $\Phi(\cdot)$, such that $\Phi(\cdot) = x_{\arg\max \alpha}$, where $\mathbf{x} = \{x_1, \cdots, x_N\}$ and $\boldsymbol{\alpha} = \{\alpha_1, \cdots, \alpha_N\}$ denote the full set of incoming skips, and their corresponding relaxation scores, respectively.

However, the above formulation outputs a fixed number of skips rather than flexible ones. We continuously relax skips by constructing a learnable *selection matrix* $\mathbf{M} \in \mathbb{R}^{N \times (N-2)}$ that models the mappings between the N incoming skips and k candidates, and formulate $\Phi(\cdot)$ as a fully differentiable equation below:

$$\Phi(\mathbf{x}, \mathbf{M}) = Matmul(\mathbf{x}, Gumbel\_Softmax(\mathbf{M})), \qquad (1)$$

where the gumbel-softmax [5] forces each of the $(N-2)$ columns of $\mathbf{M}$ to be an one-hot vector that votes for one of the N incoming skips. Our formulation generates $(N-2)$ selected skips with repetition allowed, achieving a dynamic selection of candidate skips, where the *unique ones* are further averaged out and fed into subsequent blocks. Differing from [9,10], we design $\Phi(\cdot)$ to unify the forward behaviour during both network training and inference stages.

Phase2: Progressive Evolutionary Search. To further reduce potential redundancies among the candidate skips obtained in Phase1, we perform an

---

Algorithm 1. Progressive evolutionary search

Input: Iteration $\mathbf{T}$, sampling number $\mathbf{s}$, randomly initialized SuperNet weights $\mathcal{W}$,
    Phase1 searched candidate skips $\{\mathbf{C}_{1\to 2},\cdots,\mathbf{C}_{2\mathbf{T}-1\to 2\mathbf{T}}\}$, criterion $\mathcal{L}$.
Output: BiX-Net with evolved skips $\{\mathbf{E}_{1\to 2},\cdots,\mathbf{E}_{2\mathbf{T}-1\to 2\mathbf{T}}\}$

  1: **for** $t = 2\mathbf{T}-1,\cdots,1$ **do**
  2:    **for** $i = 1,\cdots,\mathbf{s}$ **do**
  3:       **for** each searching block b **do**
  4:          Randomly sample n skips from $\mathbf{C}_{t\to t+1}^{b}$: $C_i^b, 1 \leq n \leq |\mathbf{C}_{t\to t+1}^{b}|$.
  5:       **end for**
  6:    **end for**
  7:    **for** data batch X, target Y **do**
  8:       Forward head network with candidate skips $\mathbf{C}_{1\to 2},\cdots,\mathbf{C}_{t-2\to t-1}$.
  9:       **for** $i = 1,\cdots,\mathbf{s}$ **do**
10:          Forward tail network with sampled skips $C_i, \mathbf{E}_{t+1\to t+2},\cdots,\mathbf{E}_{2\mathbf{T}-1\to 2\mathbf{T}}$.
11:          Calculate loss $l_i = \mathcal{L}(\mathrm{X}, \mathrm{Y})$
12:       **end for**
13:       Optimize $\mathcal{W}$ with the average loss $\frac{1}{\mathbf{s}}\sum_{i=1}^{\mathbf{s}} l_i$.
14:    **end for**
15:    Get Pareto front from $\{C_1,\cdots,C_{\mathbf{s}}\}$ and determine $\mathbf{E}_{t\to t+1}$.
16: **end for**

---

additional evolutionary search to find an evolved subset of skips for better network compactness and performance. Specifically, we search the candidate skips for all levels between a certain pair of adjacent extraction stages at the same time, and then progressively move to the next pair once the current search is concluded. As the connectivity of adjacent extraction stages depends on the connectivity of precedent ones, we initiate the search from the last extraction stage pair and progressively move to the first one.

The most straightforward strategies [12,13] optimize the SuperNet with each sampling skip set in a population $\mathbf{P}$ individually, and then update $\mathbf{P}$ when all training finish. However, there are two major flaws of such strategies: first, optimizing SuperNet with sampling skips individually may result in unfair outcomes; second, the searching process is empirically slow. Assuming the forward and backward of each extraction stage takes $\mathbf{I_F}$ and $\mathbf{I_B}$ time, training all $|\mathbf{P}|$ sampling architectures individually for each step takes $2\mathbf{T}|\mathbf{P}|(\mathbf{I_F} + \mathbf{I_B})$ in total.

2.3 Analysis of Searching Fairness and Deficiency

To overcome the first flaw above, we concept the *skip fairness* and claim that all skip search algorithms need to meet such principle. Note that each sampled architecture A^i (with a subset of skips C_i) is randomly drawn from the progressively updated SuperNet $\mathcal{N}$, which makes up the population $\mathbf{P}$ at each iteration.

Definition 1. *Skip Fairness.* Let $\mathbf{F} = \{\mathbf{f}^1,\cdots,\mathbf{f}^m\}$ *be the skipped features to any searching blocks in each sampled architecture* $\mathcal{A}^i$ *within a population* $\mathbf{P}$. *The skip fairness requires* $\mathbf{f}_{\mathcal{A}^1}^1 \equiv \cdots \equiv \mathbf{f}_{\mathcal{A}^{|\mathbf{P}|}}^1,\cdots,\mathbf{f}_{\mathcal{A}^1}^m \equiv \cdots \equiv \mathbf{f}_{\mathcal{A}^{|\mathbf{P}|}}^m \ \forall \mathbf{f}^i \in \mathbf{F}, \forall \mathcal{A}^i \in \mathbf{P}$.

The above principle yields that, when searching between the same extraction stage pair, any corresponding level-to-level skips (e.g., $\mathbf{f}^1_{\mathcal{A}^1}$, $\mathbf{f}^1_{\mathcal{A}^2}$, $\mathbf{f}^1_{\mathcal{A}^3}$) across different sampled architectures (e.g., $\mathcal{A}^1$, $\mathcal{A}^2$, $\mathcal{A}^3$) are required to carry identical features. Otherwise, the inconsistent incoming features would impact the search decision on the skipping topology, hence causing unexpected search unfairness. Gradient-based search algorithms (e.g. Phase1 search algorithm) meet this principle by its definition, as the same forwarded features are distributed to all candidate skips equally. However, the aforementioned straightforward strategy violates such principle due to the inconsistent incoming features produced by the individually trained architectures.

Our proposed Phase2 search algorithm meets the skip fairness by synchronizing partial forwarded features in all sampling networks. Specifically, suppose we are searching skips between the tth and $t+1^{th}$ extraction stages ($t \in [1, 2\mathbf{T}-1]$): network topology from the 1st to $t-1^{th}$ stages is fixed and the forward process between such stages can be shared. We denote such stages as *head network*. On the contrary, network topology from the tth to $2\mathbf{T}^{th}$ stages varies as the changes of different sampled skips. We then denote such unfixed stages as *tail networks*, which share the same SuperNet weights but with distinct topologies. The forwarded features of head network are fed to all sampling tail networks individually, as shown in Fig. 2. We average the losses of all tail networks, and backward the gradients through the SuperNet weights only once. Besides, our Phase2 searching process is empirically efficient and overcomes the second flaw above, as one-step training only requires $\mathbf{I_B} + \sum_{t=1}^{2\mathbf{T}-1}(t\mathbf{I_F} + (2\mathbf{T}-t)\mathbf{I_F} \cdot |\mathbf{P}|)$.

After search between each extraction stage pair completes, we follow a multi-objective selection criterion that retains the architectures at the Pareto front [19] based on both validation accuracy (IoU) and computational complexity (MACs). The proposed progressive evolutionary search details are presented in Algorithm 1 and the searched BiX-Net is shown in supplementary material Fig. 1.

3   Experiments

3.1   Datasets and Implementation Details

Two segmentation tasks across three different medical image datasets were adopted for validating our proposed method including MoNuSeg [8], TNBC [11], and CHAOS [6]. **(1)** The MoNuSeg dataset contains 30 training images and 14 testing images of size 1000×1000 cropped from whole slide images of different organs. Following [17], we extracted 512×512 patches from the corners of each image. **(2)** The TNBC dataset was used as an *extra validation dataset* [2,17], consisting of 512×512 sub-images obtained from 50 histopathology images of different tissue regions. **(3)** We also conducted 5-fold cross-validation on MRI scans from the CHAOS dataset to evaluate the generalization ability of BiX-Net on the abdominal organ segmentation task, which contains 120 DICOM image sequences from T1-DUAL (both in-phase and out-phase) and T2-SPIR with a spatial resolution of 256×256. We pre-processed the raw sequences by min-max normalization and auto-contrast enhancement.

Table 1. Comparison on MoNuSeg and TNBC with three independent runs.

Methods	MoNuSeg		TNBC		Params(M)	MACs(G)
	mIoU (%)	DICE (%)	mIoU (%)	DICE (%)		
U-Net [14]	68.2±0.3	80.7±0.3	46.7±0.6	62.3±0.6	8.64	65.83
R2U-Net [1]	69.1±0.3	81.2±0.3	60.1±0.5	71.3±0.6	9.78	197.16
BiO-Net [17]	69.9±0.2	82.0±0.2	62.2±0.4	75.8±0.5	14.99	115.67
NAS-UNet [16]	68.4±0.3	80.7±0.3	54.5±0.6	69.6±0.5	2.42	67.31
AutoDeepLab [9]	68.5±0.2	81.0±0.3	57.2±0.5	70.8±0.5	27.13	60.33
MS-NAS [18]	68.8±0.4	80.9±0.3	58.8±0.6	71.1±0.5	14.08	72.71
BiO-Net++	70.0±0.3	82.2±0.3	67.5±0.4	80.4±0.5	0.43	34.36
Phase1 searched	69.8±0.2	82.1±0.2	66.8±0.6	80.1±0.4	0.43	31.41
BiX-Net	**69.9±0.3**	**82.2±0.2**	**68.0±0.4**	**80.8±0.3**	**0.38**	**28.00**

Searching Implementation. We utilized MoNuSeg only for searching the optimal BiX-Net, and the same architecture is then transferred to all other tasks. We followed the same data augmentation strategies as in [17]. In Phase1, the BiO-Net++ SuperNet was trained with a total of 300 epochs with a base learning rate of 0.001 and a decay rate of $3e^{-3}$. The network weights and the selection matrices were optimized altogether with the same optimizer, rather than optimized separately [10]. Our Phase1 searching procedure took only 0.09 GPU-Day. In Phase2, there were total 5 searching iterations when $\mathbf{T} = 3$. At each iteration, for each retained architecture from the preceding generation, we sampled $s = 15$ different skip sets to form the new population $\mathbf{P}$ that were trained 40 epochs starting from a learning rate of 0.001 and then decayed by 10 times every 10 epochs. Due to GPU memory limitation, we only retained $|\mathbf{P}| < 3$ architectures with highest IoU at the Pareto front. Our Phase2 searching process consumed 0.37 GPU-Day.

Retraining Implementation. Adam optimizer [7] was used across all experiments to minimize cross entropy loss in both network searching and retraining. Batch size was set to be 2 for MoNuSeg and TNBC, and 16 for CHAOS. We retrained the constructed BiX-Net and all competing models with the same implementation as Phase1, which was identical across all experiments for fair comparisons. Mean intersection of Union (mIoU) and Dice Coefficient (DICE) were reported to evaluate accuracy while Multiplier Accumulator (MACs) was reported to measure computational complexity.

3.2 Experimental Results

Nuclei Segmentation. Our method was compared to the vanilla U-Net [14], state-of-the-art recurrent U-Net variants [1,17], and homogeneous state-of-the-art NAS searched networks [9,16,18]. All models were trained from scratch with final results reported as the average of three independent runs. Table 1 shows that our plain BiX-Net outperforms the state-of-the-art NAS counterparts con-

Table 2. Comparison on CHAOS (MRI) with 5-fold cross validation.

Methods	Liver		Left Kidney		Right Kidney		Spleen	
	mIoU (%)	DICE (%)	mIoU (%)	DICE (%)	mIoU (%)	DICE (%)	mIoU (%)	DICE (%)
U-Net	78.1±2.0	86.8±1.8	61.3±1.1	73.8±1.2	63.5±1.1	76.2±1.1	62.2±2.1	74.4±2.3
BiO-Net	85.8±2.0	91.7±1.8	75.7±1.1	85.1±1.2	78.2±1.0	87.2±1.1	73.2±2.3	82.8±2.3
NAS-UNet	79.1±1.8	87.2±1.8	65.5±1.5	75.0±1.3	66.2±1.2	77.7±1.0	64.1±1.3	75.8±1.6
AutoDeepLab	79.8±1.9	88.1±1.8	66.7±1.6	75.0±1.7	61.9±0.9	75.7±1.1	63.9±1.2	75.5±1.4
MS-NAS	72.6±2.3	82.6±2.1	71.0±1.3	81.9±1.3	70.1±1.9	81.1±1.8	62.5±2.1	74.0±2.3
BiX-Net	**82.6±1.5**	**89.8±1.5**	**71.0±1.0**	**82.1±1.1**	**71.9±0.8**	**82.7±1.0**	**66.0±1.7**	**76.5±2.0**

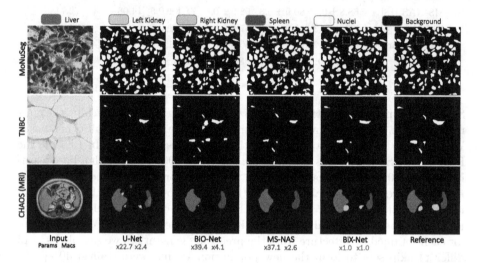

Fig. 3. Qualitative comparison of different segmentation results. Relative parameter (left) and computation overheads (right) compared to BiX-Net are also shown.

siderably, and achieves on par results to BiO-Net [17] with significantly lower network complexity. Noteworthy, our results are much higher than all others on TNBC, which indicates superior generalization ability. Qualitative comparisons on all datasets are shown in Fig. 3.

Multi-class Organ Segmentation. CHAOS challenge aims at the precise segmentation of four abdominal organs separately: liver, left kidney, right kidney, and spleen in a CT or MRI slice. Instead of training networks on each class as several independent binary segmentation tasks [18], we reproduced all models to output the logits for all classes directly. Similar to the nuclei segmentation, Table 2 indicates that BiX-Net achieves the best performance of all classes among state-of-the-art NAS searched networks [9,16,18] with much lower computational complexity. Although the hand-crafted BiO-Net outperforms all comparison methods, it suffers from the computation burdens, which are a 4.1 times of computational complexity, and a 39.4 times of trainable parameter number.

Additionally, BiX-Net produces much better segmentation mask when all organs are presented in a single slice (Fig. 3).

Ablation Studies. In addition, we conducted two ablation studies by training the presented SuperNet BiO-Net++ (the multi-scale upgrade of BiO-Net) and the Phase1 searched architecture (has not been searched by Phase2) directly on nuclei datasets. As shown in Table 1, our Phase1 search algorithm provides a 8.6% MACs reduction compared to BiO-Net++, and our BiX-Net eventually achieves a 18.5% MACs reduction, which validates the necessity of Phase1 and Phase2 search of BiX-NAS. Unlike prior NAS works, our finally searched BiX-Net follows the recurrent bi-directional paradigm with repeated use of the same building blocks at different iterations. Note that there is one building block (encoder at level 4) that has been skipped at all iterations (supplementary material Fig. 1), resulting in a further reduction in the total network parameters.

For all metrics, BiX-Net obtains higher scores than competing NAS counterparts on all datasets and achieves on-par results with our proposed BiO-Net++ with fewer computations. Additionally, we perform two tail paired t-test to analyze the statistical significance between our method and other competing NAS methods. BiX-Net achieves p-values <0.05 on the nuclei datasets and <0.1 on the CHAOS dataset, validating the significance of our method.

4   Conclusion

In this work, we proposed an efficient two-phase NAS algorithm that searches for bi-directional multi-scale skip connections between encoder and decoder, namely BiX-NAS. We first follow differentiable NAS with a novel selection matrix to narrow down the search space. An efficient progressive evolutionary search is then proposed to further reduce skip redundancies. Experimental results on various segmentation tasks show that the searched BiX-Net surpasses state-of-the-art NAS counterparts with considerably fewer parameters and computational costs.

References

1. Alom, M.Z., Yakopcic, C., Taha, T.M., Asari, V.K.: Nuclei segmentation with recurrent residual convolutional neural networks based u-net (R2U-Net). In: IEEE National Aerospace and Electronics Conference, pp. 228–233. IEEE (2018)
2. Graham, S., et al.: Hover-Net: simultaneous segmentation and classification of nuclei in multi-tissue histology images. Medical Image Anal. **58**, 101563 (2019)
3. Guo, Z., et al.: Single path one-shot neural architecture search with uniform sampling. In: Vedaldi, A., Bischof, H., Brox, T., Frahm, J.-M. (eds.) ECCV 2020. LNCS, vol. 12361, pp. 544–560. Springer, Cham (2020). https://doi.org/10.1007/978-3-030-58517-4_32
4. Huang, H., et al.: UNet 3+: a full-scale connected unet for medical image segmentation. In: 2020 IEEE International Conference on Acoustics, Speech and Signal Processing (ICASSP), ICASSP 2020, pp. 1055–1059. IEEE (2020)
5. Jang, E., Gu, S., Poole, B.: Categorical reparameterization with Gumbel-Softmax. In: International Conference on Learning Representations (ICLR) (2017)

6. Kavur, A.E., et al.: CHAOS challenge-combined (CT-MR) healthy abdominal organ segmentation. Med. Image Anal. **69**, 101950 (2021)
7. Kingma, D.P., Ba, J.: Adam: a method for stochastic optimization. In: International Conference on Learning Representations (ICLR) (2015)
8. Kumar, N., Verma, R., Sharma, S., Bhargava, S., Vahadane, A., Sethi, A.: A dataset and a technique for generalized nuclear segmentation for computational pathology. IEEE Trans. Med. Imaging **36**(7), 1550–1560 (2017)
9. Liu, C., et al.: Auto-DeepLab: hierarchical neural architecture search for semantic image segmentation. In: Proceedings of the IEEE/CVF Conference on Computer Vision and Pattern Recognition (CVPR), pp. 82–92 (2019)
10. Liu, H., Simonyan, K., Yang, Y.: DARTS: differentiable architecture search. In: International Conference on Learning Representations (ICLR) (2019)
11. Naylor, P., Laé, M., Reyal, F., Walter, T.: Segmentation of nuclei in histopathology images by deep regression of the distance map. IEEE Trans. Med. Imaging **38**(2), 448–459 (2018)
12. Real, E., Aggarwal, A., Huang, Y., Le, Q.V.: Regularized evolution for image classifier architecture search. Proc. AAAI Conf. Artif. Intell. **33**, 4780–4789 (2019)
13. Real, E., et al.: Large-scale evolution of image classifiers. In: International Conference on Machine Learning, pp. 2902–2911. PMLR (2017)
14. Ronneberger, O., Fischer, P., Brox, T.: U-Net: convolutional networks for biomedical image segmentation. In: Navab, N., Hornegger, J., Wells, W.M., Frangi, A.F. (eds.) MICCAI 2015. LNCS, vol. 9351, pp. 234–241. Springer, Cham (2015). https://doi.org/10.1007/978-3-319-24574-4_28
15. Wang, W., Yu, K., Hugonot, J., Fua, P., Salzmann, M.: Recurrent U-Net for resource-constrained segmentation. In: The IEEE International Conference on Computer Vision (ICCV) (2019)
16. Weng, Y., Zhou, T., Li, Y., Qiu, X.: NAS-Unet: neural architecture search for medical image segmentation. IEEE Access **7**, 44247–44257 (2019)
17. Xiang, T., Zhang, C., Liu, D., Song, Y., Huang, H., Cai, W.: BiO-Net: learning recurrent bi-directional connections for encoder-decoder architecture. In: Martel, A.L., et al. (eds.) MICCAI 2020. LNCS, vol. 12261, pp. 74–84. Springer, Cham (2020). https://doi.org/10.1007/978-3-030-59710-8_8
18. Yan, X., Jiang, W., Shi, Y., Zhuo, C.: MS-NAS: multi-scale neural architecture search for medical image segmentation. In: Martel, A.L., et al. (eds.) MICCAI 2020. LNCS, vol. 12261, pp. 388–397. Springer, Cham (2020). https://doi.org/10.1007/978-3-030-59710-8_38
19. Yang, Z., et al.: CARS: continuous evolution for efficient neural architecture search. In: Proceedings of the IEEE/CVF Conference on Computer Vision and Pattern Recognition (CVPR), pp. 1829–1838 (2020)
20. Zhou, Z., Rahman Siddiquee, M.M., Tajbakhsh, N., Liang, J.: UNet++: a nested U-Net architecture for medical image segmentation. In: Stoyanov, D., et al. (eds.) DLMIA/ML-CDS -2018. LNCS, vol. 11045, pp. 3–11. Springer, Cham (2018). https://doi.org/10.1007/978-3-030-00889-5_1

Multi-task, Multi-domain Deep Segmentation with Shared Representations and Contrastive Regularization for Sparse Pediatric Datasets

Arnaud Boutillon[1,2(✉)], Pierre-Henri Conze[1,2], Christelle Pons[2,3,4], Valérie Burdin[1,2], and Bhushan Borotikar[2,3,5]

[1] IMT Atlantique, Brest, France
arnaud.boutillon@imt-atlantique.fr
[2] LaTIM UMR 1101, Inserm, Brest, France
[3] Centre Hospitalier Régional et Universitaire (CHRU) Brest, Brest, France
[4] Fondation ILDYS, Brest, France
[5] SCMIA, Symbiosis International University, Pune, India

Abstract. Automatic segmentation of magnetic resonance (MR) images is crucial for morphological evaluation of the pediatric musculoskeletal system in clinical practice. However, the accuracy and generalization performance of individual segmentation models are limited due to the restricted amount of annotated pediatric data. Hence, we propose to train a segmentation model on multiple datasets, arising from different parts of the anatomy, in a multi-task and multi-domain learning framework. This approach allows to overcome the inherent scarcity of pediatric data while benefiting from a more robust shared representation. The proposed segmentation network comprises shared convolutional filters, domain-specific batch normalization parameters that compute the respective dataset statistics and a domain-specific segmentation layer. Furthermore, a supervised contrastive regularization is integrated to further improve generalization capabilities, by promoting intra-domain similarity and impose inter-domain margins in embedded space. We evaluate our contributions on two pediatric imaging datasets of the ankle and shoulder joints for bone segmentation. Results demonstrate that the proposed model outperforms state-of-the-art approaches.

Keywords: Multi-task learning · Domain adaptation · Supervised contrastive regularization · Musculoskeletal system

1 Introduction

Segmentation of the pediatric musculoskeletal system serves as an essential pre-processing step to guide clinical decisions, as the generated 3D models of muscles

© Springer Nature Switzerland AG 2021
M. de Bruijne et al. (Eds.): MICCAI 2021, LNCS 12901, pp. 239–249, 2021.
https://doi.org/10.1007/978-3-030-87193-2_23

and bones help clinicians evaluate pathology progression and optimally plan therapeutic interventions [1,9,17]. As manual segmentation is the current standard for delineating pediatric magnetic resonance (MR) images, the implementation of automatic and robust segmentation techniques could reduce analysis time and enhance the reliability of morphological evaluation [1,9,17]. Recently, convolutional neural networks have demonstrated promising results for automatic segmentation tasks [13,15]. However, while the development of supervised deep learning models typically requires large amount of annotated data [13,15], the conception of pediatric imaging datasets is a slow and onerous process [9]. Hence, the inherent scarcity of pediatric imaging resources can induce limited generalization capabilities in deep learning models [13,15].

To address the similar problem of limited annotated data for semantic scene labeling in natural images, Fourure et al. proposed to train a single network over the union of multiple datasets, in order to leverage a greater amount of training data [7]. Their model is optimized in a multi-task and multi-domain framework, in which each dataset is defined by its own task (segmentation label-set) and domain (data distribution). Hence, this approach is more generic than traditional domain adaptation techniques which usually focus on domains containing the same set of objects [7]. Following this, Rebuffi et al. [19,20] proposed to employ a model with agnostic filters, as visual primitives may be shared across tasks and domains, and dataset-specific layers which allow task and domain specialization. These approaches, based on shared representations, have been reported to perform at par or superior to traditional independent models [6,11,14,22].

Even though these models integrate domain-specific information, this prior knowledge could be further exploited to improve the generalizability of the shared representation. For instance, Zhu et al. imposed a Gaussian mixture distribution on the shared representation of their image translation model [23], however, such a hypothesis may be too restrictive and lead to a decrease in performance. In representation learning, a good representation can be characterized by the presence of natural clusters corresponding to the classes of the problem [2]. Hence, a number of self-supervised representation learning approaches focus on pulling together data-points from the same class and pushing apart negative samples in embedded space using a contrastive metric [5,8]. Khosla et al. extended this idea to fully-supervised classification setting by leveraging the label information and considering many positives simultaneously [12]. Thus, the contrastive regularization maximizes the performance of the classifier by imposing intra-class cohesion and inter-class separation in latent space.

In this study, we propose to develop and optimize a single segmentation network over the union of pediatric imaging datasets acquired on separate anatomical regions. The contributions of this study are threefold: 1) We formalize a segmentation model which incorporates shared representations, domain-specific batch normalization [3,4,6,11,14], and a domain-specific output layer. 2) We extend the multi-task, multi-domain segmentation learning framework by integrating a contrastive regularization during optimization. As opposed to classical contrastive approaches that operate on image classes [5,8,12], we leverage dataset label information to enhance intra-domain similarity and impose inter-domain margins, and 3) we illustrate the effectiveness of our approach for multi-task, multi-domain segmentation on two sparse, unpaired (from different patient cohorts), and heterogeneous pediatric musculoskeletal MR imaging datasets.

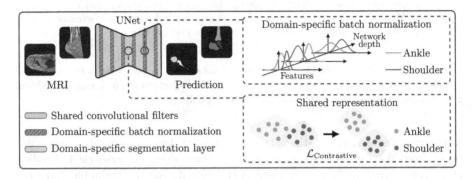

Fig. 1. The proposed multi-task, multi-domain segmentation method is based on UNet [21] with shared convolutional filters as well as domain-specific batch normalization and a domain-specific segmentation layer. The training procedure incorporates a contrastive loss $\mathcal{L}_{\text{Contrastive}}$ to promote inter-domain separation in the shared representation.

2 Method

2.1 Deep Segmentation Model with Domain-Specific Layers (DSL)

Let $\mathcal{D}_1, \ldots, \mathcal{D}_K$ be K different datasets organized such that the k-th dataset $\mathcal{D}_k = \{x_i^k, y_i^k\}_{i=1}^{n_k}$ contains n_k pairs of greyscale images x_i^k and corresponding class label images y_i^k in label space $\mathscr{C}_k$. We considered a network $S : x_i^k \mapsto S(x_i^k; \Theta, \Lambda_k, W_k)$ with shared parameters Θ and domain-specific weights $\{\Lambda_k, W_k\}_{k=1}^{K}$. The shared parameters Θ comprised classical and transposed convolutional filters, while Λ_k represented domain-specific batch normalization weights and W_k corresponded to the weights of the domain-specific segmentation layer (Fig. 1).

Batch normalization aims at improving convergence speed and generalization abilities of the model by normalizing the internal activations of the network [10]. However, as the individual statistics of the K domains can be very different from each other, a domain-agnostic batch normalization layer could lead to defective features. This could result in zero mean activation over domains at certain layers which is meaningless [3,4,6,11,14]. Thus, to more carefully calibrate the internal activations, we employed domain-specific batch normalization (DSBN) functions:

$$\text{DSBN}(v_{l,m}^k; \beta_{l,m}^k, \gamma_{l,m}^k) = \gamma_{l,m}^k \frac{v_{l,m}^k - \mathbb{E}[v_{l,m}^k]}{\sqrt{\mathbb{V}[v_{l,m}^k] + \epsilon}} + \beta_{l,m}^k \qquad (1)$$

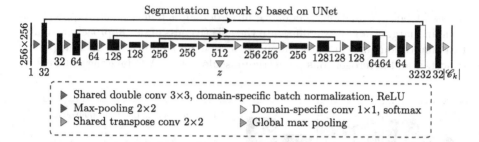

Fig. 2. The architecture of the segmentation network S is based on a convolutional encoder-decoder and a final segmentation layers with $|\mathscr{C}_k|$ classes. The image embedding z is obtained via UNet encoder and global max pooling.

where $v_{l,m}^k$ denoted the m-th features at the l-th layer produced by an input batch from the k-th dataset and $\epsilon = 1e{-}5$ was added for numerical stability. Hence, $\Lambda_k = \{\beta_{l,m}^k, \gamma_{l,m}^k\}_{l,m}$ comprised the domain-specific trainable shift and scale of each features at each layers.

Moreover, it is essential to employ a dedicated output layer, as a domain-agnostic segmentation layer may predict classes from all K datasets, which is counterproductive [7]. Hence, the network was designed such that $S = \psi \circ \phi$ where ψ was the segmentation layer while ϕ contained all the previous layers of the network. Specifically, if $u_i^k = \phi(x_i^k; \Theta, \Lambda_k)$ denotes the output of the penultimate layer then $\psi(u_i^k; W_k) = \mathrm{softmax}(u_i^k * W_k)$ was a domain-specific segmentation layer with a 1×1 convolutional filter W_k which produced a segmentation mask with $|\mathscr{C}_k|$ classes (Fig. 2). $|\mathscr{C}_k|$ denoted the cardinality of the k-th label space.

During training, we used the stochastic gradient descent algorithm to optimize the cross-entropy loss defined in a multi-task and multi-domain setting:

$$\mathcal{L}_{\mathrm{CE}} = -\frac{1}{K} \sum_{k=1}^{K} \frac{1}{n_k |\mathscr{C}_k|} \sum_{i=1}^{n_k} \sum_{c \in \mathscr{C}_k} y_{i,c}^k \log(\hat{y}_{i,c}^k) \qquad (2)$$

where $\hat{y}_i^k = S(x_i^k; \Theta, \Lambda_k, W_k)$ was the predicted segmentation. The shared parameters and the domain-specific weights were thus learnt through this novel optimization scheme. The domain-specific weights were a minimal supplementary parameterization with regards to the number of shared convolutional filters.

2.2 Supervised Contrastive Regularization

The encoder of the segmentation network mapped the greyscale images to a shared representation (Fig. 1). We assumed that learning a shared representation with domain-specific clusters would enhance the generalizability of the decoder and improve the accuracy of the segmentation predictions. More precisely, we assumed that a local variation in the shared representation should preserve the category of the domain. Hence, we designed a novel regularization term aimed at conserving intra-domain cohesion and inter-domain separation in the shared representation. We adapted the supervised contrastive loss [12] to multi-domain segmentation using the known datasets labels. In the following, x_i denotes the i-th image of the joint dataset with n samples.

Let z_i be the embedding of x_i to which we applied global max pooling to obtain a spatially invariant representation (Fig. 2), and k_i the dataset index of x_i. We note $P(i) = \{p \in [\![1, \ldots, n]\!] : k_p = k_i, p \neq i\}$ the set of indexes of all images from the same dataset as x_i. As opposed to classical self-supervised contrastive approaches [5,8,12], it was not necessary to generate an augmented view of x_i as more than one sample was known to be from the same dataset. The contrastive loss was defined as follows:

$$\mathcal{L}_{\text{Contrastive}} = -\frac{1}{n} \sum_{i=1}^{n} \frac{1}{|P(i)|} \sum_{p \in P(i)} \log \left(\frac{\exp(\text{sim}(z_i, z_p)/\tau)}{\sum_{j=1}^{n} \mathbb{1}_{[j \neq i]} \exp(\text{sim}(z_i, z_j)/\tau)} \right) \quad (3)$$

where $\text{sim}(z_i, z_j) = z_i \cdot z_j / \|z_i\| \|z_j\|$ was the cosine similarity between two representations and τ was the temperature hyper-parameter which controlled the smoothness of the loss as well as imposed hard negative/positive predictions [5,12]. During optimization, only the weights of the encoder were penalized by the proposed regularization. Thus, the contrastive regularization gathered the representation from the same domain in embedded space, while simultaneously separating clusters from different domains (Fig. 1). The deep segmentation model with domain-specific layers (DSL) was trained using the proposed regularized loss $\mathcal{L} = \mathcal{L}_{\text{CE}} + \lambda \mathcal{L}_{\text{Contrastive}}$ with weighting hyper-parameter λ.

3   Experiments

3.1   Imaging Datasets

Experiments were conducted on MR images retrospectively extracted from previous clinical studies. These image datasets were acquired using a 3T Achieva scanner (Philips Healthcare, Best, Netherlands) from two pediatric musculoskeletal joints: ankle and shoulder. The ankle dataset was acquired from 17 pediatric patients aged from 7 to 13 years (10 ± 2 years) while MR images of 15 shoulder were acquired from pediatric patients aged from 5 to 17 years old (12 ± 4 years). The ankle dataset comprised 10 healthy and 7 pathological cases and the shoulder dataset comprised 8 healthy and 7 pathological cases. A T1 weighted 3D gradient echo sequence was used (TR: 7.9 ms, TE: 2.8 ms, voxel: $0.4 \times 0.4 \times 1.2$ mm^3, FOV: 140×161 mm^2) to acquire ankle images. An eTHRIVE (enhanced T1-weighted High-Resolution Isotropic Volume Examination) sequence was employed (TR: 8.4 ms, TE: 4.2 ms, voxel: $0.4 \times 0.4 \times 1.2$ mm^3, FOV: 2600×210 mm^2) for shoulder image acquisition. Magnetic resonance images were acquired during separate clinical studies where parents provided informed consent to use the imaging data for research purpose. A medically trained expert (12 years of experience) annotated the images to obtain ground truth segmentation masks of calcaneus, talus and tibia bones for ankle and scapula and humerus bones for shoulder. All axial slices were downsampled to 256×256 pixels and were normalized to have zero-mean and unit variance.

3.2   Implementation Details

The proposed method based on DSL and contrastive regularization was implemented using UNet architecture [21] and its performance was assessed in comparison with other state-of-the-art models. The compared methods included UNet [21] and UNet with attention gates (Att-UNet) [18], which were implemented using baseline (trained on individual dataset), joint (trained on all datasets at once) and DSL schemes. All networks were initialized with randomly distributed weights and trained from scratch without any transfer learning scheme.

All methods were based on the same backbone UNet architecture (Fig. 2). The segmentation networks were trained for 30 epochs with batch size set to 32, using the Adam optimizer with a $1e-4$ learning rate. We explored different values for the hyper-parameters of the supervised contrastive regularization and found $\tau = 1e-1$ and $\lambda = 1e-1$ to be optimal. We used extensive on the fly 2D data augmentation including random flip, translation ($\pm25\%$) and rotation ($\pm45°$) in both directions due to limited available training data. Deep learning networks were implemented on PyTorch and a Nvidia RTX 2080 Ti GPU with 12 GB of RAM was used during optimization.

The same post-processing was employed after each method: first, the obtained 2D segmentation masks were stacked to form a 3D binary volume, then we selected the 3D largest connected set as output prediction and we finally applied 3D morphological closing ($5 \times 5 \times 5$ spherical kernel) to smooth the resulting boundaries.

3.3   Evaluation of Predicted Segmentation

We assessed the performance of each method based on the similarity between 3D predicted and ground truth masks. For each dataset, Dice coefficient, average symmetric surface distance (ASSD) and maximum symmetric surface distance (MSSD) metrics were computed for each bone and we reported the average scores. Due to the limited amount of examinations, the metrics were determined in a leave-one-out manner such that, for each dataset, one examination was retained for validation, one for test and the remaining data were used to train the model. We iterated through the datasets simultaneously and each examination was used at maximum once for test. We did not test all combinations between datasets, as this would have drastically increased computation time and would have introduced redundant observations in the results. Moreover, due to the scarce amount of 3D examinations, we performed statistical analysis between the methods on the 2D MR images from both datasets. We employed the Kolmogorov-Smirnov non-parametric test using Dice scores obtained from the 6215 ankle and shoulder 2D slices corresponding to 32 3D MR images.

4   Results and Discussion

4.1   Segmentation Results

From the quantitative results (Table 1), the proposed method ranked first on all metrics on both datasets. Results obtained on the ankle dataset corresponded to a marginal increase in performance compared to the second best method ($+0.1\%$ Dice and -0.5 mm MSSD) while shoulder performance were substantially increased ($+1.5\%$ Dice, -0.6 mm ASSD and -0.1 mm MSSD). We observed that for a fixed architecture, the DSL scheme outperformed the joint approach, which in turn outranked the baseline scheme. Hence, the shared representation and layer specialization allowed performance improvements over independent models and promoted more precise bone extraction. We also reported a higher variability in shoulder results, which was due to the presence of examinations with a higher level of noise due to patient movements during acquisition. Moreover, the statistical analysis performed on 2D slices using Dice (Table 2) indicated that the proposed model produced significant improvements (p-values <0.05). The improvements in segmentation quality of the proposed approach over state-of-the-art methods were further supported by the visual comparisons on both datasets (Fig. 3).

Table 1. Quantitative assessment of UNet [21] and Att-UNet [18] using baseline, joint and DSL schemes, and the proposed UNet with DSL and contrastive regularization on ankle and shoulder datasets. Metrics include Dice (%), ASSD (mm) and MSSD (mm). Best results are in bold.

Metric		Ankle			Shoulder		
		Dice ↑	ASSD ↓	MSSD ↓	Dice ↑	ASSD ↓	MSSD ↓
UNet	Base	92.2 ± 3.2	1.0 ± 1.1	9.5 ± 9.0	81.8 ± 13.8	2.6 ± 3.8	22.3 ± 15.1
	Joint	93.0 ± 3.3	$\mathbf{0.8 \pm 0.6}$	8.5 ± 8.0	83.6 ± 11.0	2.1 ± 2.6	21.9 ± 19.3
	DSL	93.4 ± 1.5	$\mathbf{0.8 \pm 0.5}$	8.3 ± 7.8	84.3 ± 10.7	2.2 ± 2.4	25.4 ± 26.4
Att	Base	92.4 ± 2.0	1.0 ± 1.1	8.9 ± 7.4	83.0 ± 13.1	2.0 ± 3.5	19.5 ± 16.7
	Joint	92.7 ± 1.6	0.9 ± 0.7	9.4 ± 7.5	83.2 ± 13.0	2.3 ± 3.9	20.2 ± 18.3
	DSL	93.4 ± 1.6	$\mathbf{0.8 \pm 0.4}$	8.8 ± 7.9	83.8 ± 12.1	2.2 ± 4.1	18.4 ± 15.6
Proposed		$\mathbf{93.5 \pm 1.2}$	$\mathbf{0.8 \pm 0.5}$	$\mathbf{7.8 \pm 7.8}$	$\mathbf{85.8 \pm 8.6}$	$\mathbf{1.4 \pm 1.4}$	$\mathbf{18.3 \pm 11.1}$

Table 2. Statistical analysis between the proposed model with UNet [21] and Att-UNet [18] using baseline, joint and DSL schemes, through Kolmogorov-Smirnov non-parametric test using Dice computed on 2D slices from ankle and shoulder datasets. Bold p-values (<0.05) highlight statistically significant results.

Method	UNet			Att-UNet			Proposed
	Base	Joint	DSL	Base	Joint	DSL	
Dice 2D	86.9 ± 18.7	87.6 ± 18.1	88.3 ± 17.6	87.8 ± 18.5	87.5 ± 18.3	88.5 ± 16.6	88.9 ± 17.0
p-value	$\mathbf{1.5e{-}15}$	$\mathbf{3.5e{-}6}$	$\mathbf{0.04}$	$\mathbf{7.2e{-}3}$	$\mathbf{2.6e{-}8}$	$\mathbf{1.0e{-}4}$	–

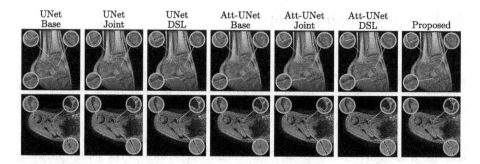

Fig. 3. Visual comparison of UNet [21] and Att-UNet [18] using baseline, joint and DSL schemes, and the proposed model on ankle and shoulder datasets. Ground truth delineations are in red (–). Predicted calcaneus, talus, tibia, humerus and scapula bones respectively appear in green ( ), blue (–), yellow ( ), magenta (–) and cyan (–). (Color figure online)

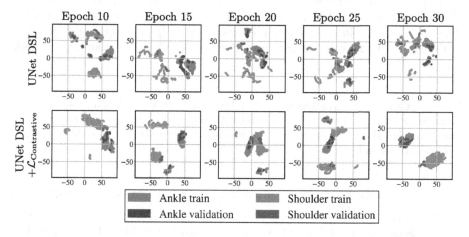

Fig. 4. Visualization of the contrastive regularization which promotes intra-domain cohesion and inter-domain margins in embedded space during optimization. This visualization was obtained using the t-SNE algorithm in which each colored dot represented a 2D MR slice from the training or validation set of the ankle or shoulder datasets. (Color figure online)

4.2   Supervised Contrastive Regularization Visualization

We performed an ablation study and compared the shared representation learnt by the proposed UNet DSL with and without contrastive regularization during optimization. The embedding of ankle and shoulder MR images from the training and validation set were visualized at epochs 10, 15, 20, 25 and 30. In order to visualize the 512 dimensional feature vectors, we applied a two-step dimensionality reduction as recommended in [16]. We first employed principal component analysis which reduced the representations to 50 dimensional feature vectors, then the t-SNE algorithm embedded the data into a 2D space. The perplexity and learning rate of the t-SNE algorithm were set to 30 and 200 respectively.

Visualization of the shared representation provided a qualitative validation of the benefits of the additional supervised contrastive regularization on intra-domain cohesion and inter-domain separation. During optimization, the shared representation learnt by UNet DSL did not present margins between domains, while the addition of the contrastive regularization led to distinctive clusters. Hence, the shared representation of our proposed network was invariant to local variations and preserved the category of the domain. Moreover, the generalization capabilities of the network were visually attested as validation data-points were located inner their respective clusters (Fig. 4).

5   Conclusion

We presented a multi-task, multi-domain segmentation model with promising results for the task of pediatric bone segmentation in MR images. The proposed framework based on shared convolutional filters, domain-specific batch normalization and a domain-specific output segmentation layer, incorporated a contrastive regularization to enforce clustering in the shared representation and enhance generalization capabilities. An important perspective from this study is that this strategy can improve the segmentation performance on other musculoskeletal pediatric imaging datasets. Future studies aim at extending the segmentation model to other tissues such as muscles and cartilages, in order to provide a more complete description of the musculoskeletal system.

Compliance with Ethical Standards. MRI data acquisition on the pediatric cohort used in this study was performed in line with the principles of the Declaration of Helsinki. Ethical approval was granted by the Ethics Committee (Comité Protection de Personnes Ouest VI) of CHRU Brest (2015-A01409-40).

Acknowledgments. This work was funded by IMT, Fondation Mines-Télécom and Institut Carnot TSN (Futur & Ruptures). Data were acquired with the support of Fondation motrice (2015/7), Fondation de l'Avenir (AP-RM-16-041), PHRC 2015 (POPB 282) and Innoveo (CHRU Brest).

References

1. Balassy, C., Hörmann, M.: Role of MRI in paediatric musculoskeletal conditions. Eur. J. Radiol. **68**(2), 245–258 (2008). https://doi.org/10.1016/j.ejrad.2008.07.018
2. Bengio, Y., Courville, A., Vincent, P.: Representation learning: a review and new perspectives. IEEE Trans. Pattern Anal. Mach. Intell. **35**(8), 1798–1828 (2013). https://doi.org/10.1109/TPAMI.2013.50
3. Bilen, H., Vedaldi, A.: Universal representations: the missing link between faces, text, planktons, and cat breeds. arXiv arXiv:1701.07275 [cs, stat] (January 2017). http://arxiv.org/abs/1701.07275

4. Chang, W.G., You, T., Seo, S., Kwak, S., Han, B.: Domain-specific batch normalization for unsupervised domain adaptation. In: Proceedings of the IEEE/CVF Conference on Computer Vision and Pattern Recognition, pp. 7354–7362 (2019)
5. Chen, T., Kornblith, S., Norouzi, M., Hinton, G.: A simple framework for contrastive learning of visual representations. In: International Conference on Machine Learning, pp. 1597–1607. PMLR (November 2020)
6. Dou, Q., Liu, Q., Heng, P.A., Glocker, B.: Unpaired multi-modal segmentation via knowledge distillation. IEEE Trans. Med. Imaging **39**(7), 2415–2425 (2020). https://doi.org/10.1109/tmi.2019.2963882
7. Fourure, D., Emonet, R., Fromont, E., Muselet, D., Trémeau, A., Wolf, C.: Semantic segmentation via multi-task, multi-domain learning. In: S+SSPR 2016 the Joint IAPR International Workshops on Structural and Syntactic Pattern Recognition (SSPR 2016) and Statistical Techniques in Pattern Recognition (SPR 2016) (November 2016)
8. Hadsell, R., Chopra, S., LeCun, Y.: Dimensionality reduction by learning an invariant mapping. In: 2006 IEEE Computer Society Conference on Computer Vision and Pattern Recognition, CVPR 2006, vol. 2, pp. 1735–1742 (June 2006). https://doi.org/10.1109/CVPR.2006.100
9. Hirschmann, A., Cyriac, J., Stieltjes, B., Kober, T., Richiardi, J., Omoumi, P.: Artificial intelligence in musculoskeletal imaging: review of current literature, challenges, and trends. Semin. Musculoskelet. Radiol. **23**(3), 304–311 (2019). https://doi.org/10.1055/s-0039-1684024
10. Ioffe, S., Szegedy, C.: Batch normalization: accelerating deep network training by reducing internal covariate shift. In: International Conference on Machine Learning, pp. 448–456. PMLR (June 2015)
11. Karani, N., Chaitanya, K., Baumgartner, C., Konukoglu, E.: A lifelong learning approach to brain MR segmentation across scanners and protocols. In: Frangi, A.F., Schnabel, J.A., Davatzikos, C., Alberola-López, C., Fichtinger, G. (eds.) MICCAI 2018. LNCS, vol. 11070, pp. 476–484. Springer, Cham (2018). https://doi.org/10.1007/978-3-030-00928-1_54
12. Khosla, P., et al.: Supervised contrastive learning. In: Larochelle, H., Ranzato, M., Hadsell, R., Balcan, M.F., Lin, H. (eds.) Advances in Neural Information Processing Systems, vol. 33, pp. 18661–18673. Curran Associates, Inc. (2020)
13. Litjens, G., et al.: A survey on deep learning in medical image analysis. Med. Image Anal. **42**, 60–88 (2017). https://doi.org/10.1016/j.media.2017.07.005
14. Liu, Q., Dou, Q., Yu, L., Heng, P.A.: MS-Net: multi-site network for improving prostate segmentation with heterogeneous MRI data. IEEE Trans. Med. Imaging **39**(9), 2713–2724 (2020)
15. Lundervold, A.S., Lundervold, A.: An overview of deep learning in medical imaging focusing on MRI. Z. Med. Phys. **29**(2), 102–127 (2019). https://doi.org/10.1016/j.zemedi.2018.11.002
16. Maaten, L., Hinton, G.: Visualizing data using t-SNE. J. Mach. Learn. Res. **9**, 2579–2605 (2008)
17. Meyer, J.S., Jaramillo, D.: Musculoskeletal MR imaging at 3T. Magn. Reson. Imaging Clin. N. Am. **16**(3), 533–545, vi (2008). https://doi.org/10.1016/j.mric.2008.04.004
18. Oktay, O., et al.: Attention U-Net: learning where to look for the pancreas. In: 2018 Conference on Medical Imaging with Deep Learning (MIDL) (April 2018)

19. Rebuffi, S., Vedaldi, A., Bilen, H.: Efficient parametrization of multi-domain deep neural networks. In: 2018 IEEE/CVF Conference on Computer Vision and Pattern Recognition, pp. 8119–8127 (June 2018). ISSN 2575-7075. https://doi.org/10.1109/CVPR.2018.00847

20. Rebuffi, S.A., Bilen, H., Vedaldi, A.: Learning multiple visual domains with residual adapters. Adv. Neural. Inf. Process. Syst. **30**, 506–516 (2017)

21. Ronneberger, O., Fischer, P., Brox, T.: U-Net: convolutional networks for biomedical image segmentation. In: Navab, N., Hornegger, J., Wells, W.M., Frangi, A.F. (eds.) MICCAI 2015. LNCS, vol. 9351, pp. 234–241. Springer, Cham (2015). https://doi.org/10.1007/978-3-319-24574-4_28

22. Valindria, V.V., et al.: Multi-modal learning from unpaired images: application to multi-organ segmentation in CT and MRI. In: 2018 IEEE Winter Conference on Applications of Computer Vision (WACV), pp. 547–556 (2018). https://doi.org/10.1109/WACV.2018.00066

23. Zhu, Y., et al.: Cross-domain medical image translation by shared latent Gaussian mixture model. In: Martel, A.L., et al. (eds.) MICCAI 2020. LNCS, vol. 12262, pp. 379–389. Springer, Cham (2020). https://doi.org/10.1007/978-3-030-59713-9_37

TEDS-Net: Enforcing Diffeomorphisms in Spatial Transformers to Guarantee Topology Preservation in Segmentations

Madeleine K. Wyburd[1]([✉]), Nicola K. Dinsdale[2], Ana I. L. Namburete[1], and Mark Jenkinson[2,3,4]

[1] Department of Engineering Science, Institute of Biomedical Engineering, University of Oxford, Oxford, UK
madeleine.wyburd@hertford.ox.ac.uk
[2] Wellcome Centre for Integrative Neuroimaging, FMRIB, University of Oxford, Oxford, UK
[3] Department of Computer Science, Australian Institute for Machine Learning (AIML), University of Adelaide, Adelaide, Australia
[4] South Australian Health and Medical Research Institute (SAHMRI), North Terrace, Adelaide, Australia

Abstract. Accurate topology is key when performing meaningful anatomical segmentations, however, it is often overlooked in traditional deep learning methods. In this work we propose TEDS-Net: a novel segmentation method that guarantees accurate topology. Our method is built upon a continuous diffeomorphic framework, which enforces topology preservation. However, in practice, diffeomorphic fields are represented using a finite number of parameters and sampled using methods such as linear interpolation, violating the theoretical guarantees. We therefore introduce additional modifications to more strictly enforce it. Our network learns how to warp a binary prior, with the desired topological characteristics, to complete the segmentation task. We tested our method on myocardium segmentation from an open-source 2D heart dataset. TEDS-Net preserved topology in 100% of the cases, compared to 90% from the U-Net, without sacrificing on Hausdorff Distance or Dice performance. Code will be made available at: www.github.com/mwyburd/TEDS-Net.

Keywords: Diffeomorphic · Topology · Segmentation · Spatial transformers

1 Introduction

Anatomical segmentations with incorrect topology can be clinically problematic and highly impractical. For example, when segmenting the heart's myocardium, topological mistakes may present as isolated segmented regions, small holes

A.I.L. Namburete and M. Jenkinson—Equal contribution.

© Springer Nature Switzerland AG 2021
M. de Bruijne et al. (Eds.): MICCAI 2021, LNCS 12901, pp. 250–260, 2021.
https://doi.org/10.1007/978-3-030-87193-2_24

within the wall or a disconnected perimeter. These inaccuracies can all lead to incorrect measurements of circumference and wall thickness, which are often required when diagnosing heart conditions such as hypertrophic cardiomyopathy [11]. Post-processing morphological operations, such as binary closing, can be used to correct some small gaps and disconnected regions. However, they are unaware of the structure's overall topology and need to be customised accordingly, therefore, an unified method is preferred.

Semantic segmentation of medical images is regularly performed using deep convolutional neural networks (CNNs), such as the U-Net [15]. One alternative segmentation method uses a combination of CNNs and spatial transformers to learn the spatial warping required to transform a set of prior shapes into the desired class labels [8,10,17–19]. Such methods have outperformed conventional encoder-decoder and state-of-the-art architectures, however, they have no topological guarantees.

Moreover, both these methods are commonly trained and evaluated using loss functions such as Dice or binary cross-entropy, which evaluate each pixel individually and not the higher level structure, often resulting in large topological errors. Recent work has recognised the importance of accurate whole-structure segmentation, with the development of loss functions that encourage the preservation of topology [5,9,14]. Although these techniques report an improved topology accuracy compared to standalone pixel-wise loss functions, they only encourage topology preservation as opposed to enforcing it, with the best performing method achieving 96.7% topology accuracy in myocardium segmentation [5].

In theory, topology will always be preserved by deforming a prior, with the correct topological characteristic, with a continuous *diffeomorphic field* [2]. Diffeomorphic fields are continuous deformation fields that result in a one-to-one mapping. Their derivatives are invertible, resulting in positive Jacobian determinants, giving an unambiguous mapping between coordinates and therefore preserving topology [2].

Resampling one diffeomorphic field by another, also known as a *composition*, always results in a new diffeomorphic field [2]. In VoxelMorph's methods [6,7], this elegant property has been utilised for brain registration, by initialising with a small diffeomorphic field and amplifying it through a series of novel integration layers, based on the squaring and scaling approach [1,13], to generate a topology-preserving large-deformation field. However, when diffeomorphisms are applied in discrete settings, such as images stored as discrete voxels, their topological guarantees can begin to break down. This is often indicated by the emergence of non-positive Jacobian determinants in the deformation fields, which correspond to folding voxels in the warped space [7]. In brain registration a small fraction of violations may be manageable, conversely, for segmentation tasks these can lead to topological errors in the structure of interest, defeating the purpose. Here, we adapt VoxelMorph's methods to be suitable for segmentation tasks by introducing a novel topology-preserving activation function and additional smoothing terms to counteract the negative effects of discrete sampling. These modifications enforce true diffeomorphisms, shown to have only positive Jacobian determinants, which results in 100% topology preservation.

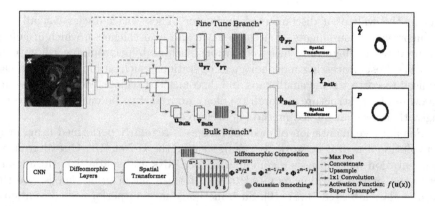

Fig. 1. Schematic of TEDS-Net architecture. A CNN learns two initial fields, **u**, at different resolutions, from an input image **X**. The fields are enforced to be diffeomorphic using an activation function, $\mathbf{v} = f(\mathbf{u})$, amplified through composition layers and "super" upsampled to 2× the resolution of the input. The bulk displacement, Φ_{Bulk}, samples a binary prior, **P**, generating $\mathbf{Y}_{Bulk}$, which is then sampled by the fine tuning field, Φ_{FT}. The asterisks show the elements removed during ablation studies.

In this work, we present the novel Topology Enforcing Diffeomorphic Segmentation Network (TEDS-Net), which to the best of our knowledge is the first deep learning technique to achieve 100% topology accuracy, and to combine spatial transformer networks (STN) and diffeomorphic displacement fields to complete a segmentation as the primary task. Our method uses CNNs to learn a warp field that maps a binary prior on to the input image, with the warp field being diffeomorphic by construction. We show that in the discrete setting, continuous properties of diffeomorphisms are no longer guaranteed, therefore, we introduce model components that enforce topology preservation in all cases.

2   Method

The aim of the proposed TEDS-Net is to predict a segmentation label, $\hat{\mathbf{Y}}$, using learnt diffeomorphic transformations, Φ, applied to a binary prior image, **P**. The network is comprised of three main parts, shown in the bottom left corner of Fig. 1. Firstly, an encoder-decoder style network is used to learn the initial velocity fields. These fields are then enforced to be diffeomorphic and amplified using the squaring and scaling approach [1,13]. Finally, the diffeomorphic fields are used to sample a prior binary shape to complete the segmentation task.

CNN. An encoder-decoder network was used to extract the relevant features from an input image, $\mathbf{X} \in \mathbb{R}^{h \times w}$ with dimensions $[h \times w]$, in order to predict two initial velocity fields: a *bulk field* and a *fine-tuning field*. We used an architecture similar to the U-Net but with two decoder branches [15, 16]. Each branch consisted of a series of convolutions followed by an instance normalisation and ReLU activation function and then repeated, referred to as a convolutional block.

Branching off at the bottleneck were two decoder streams, made up of convolutional blocks and a final 1×1 convolution, used to generate a two-dimensional field, $\mathbf{u}$. The first stream, with one upsampling convolutional block, predicts a low resolution bulk-velocity, $\mathbf{u}_{\text{Bulk}} \in \mathbb{R}^{2 \times (h \times w)/8}$, which focuses on warping the prior shape into the correct region. The second, with three upsampling blocks, predicts a higher resolution fine-tuning field, used to fine-tune the warped shape, $\mathbf{u}_{\text{FT}} \in \mathbb{R}^{2 \times (h \times w)/2}$.

Diffeomorphic Layers. To enforce that the initial fields, $\mathbf{u}$, were diffeomorphic, and therefore suitable for the scaling and squaring approach, a customised tanh activation function, $f(\mathbf{u})$, was applied to the output of each decoder branch to enforce that the displacements were between -0.5 voxel and $+0.5$ voxel and therefore topology preserving:

$$\mathbf{v}(\mathbf{x}) = f(\mathbf{u}(\mathbf{x})) = 0.5 \left(\frac{e^{\mathbf{u}(\mathbf{x})} - e^{-\mathbf{u}(\mathbf{x})}}{e^{\mathbf{u}(\mathbf{x})} + e^{-\mathbf{u}(\mathbf{x})}} \right). \tag{1}$$

A series of composition layers were then used to amplify the initial diffeomorphic fields, $\mathbf{v}$, adapted from the VoxelMorph's implementation [6, 7].

Gaussian Smoothing: As the diffeomorphic fields are represented using a finite number of parameters and sampled using linear interpolation, the theoretical guarantees are violated, risking key properties such as topological preservation. Moreover, these inaccuracies and imperfections can be amplified with the number of compositions performed [2]. To reduce these violations, we introduced Gaussian smoothing between each integration layer. For this work we used a 5 by 5 kernel and $\sigma = 2$ and the impact of this will be investigated in future work.

Super Upsample: To further smooth the fields and minimise any abnormalities, the amplified diffeomorphic flow fields were both "super upsampled" to double the resolution of the input image, $\mathbf{X}$, using bilinear interpolation.

Spatial Transformers. The diffeomorphic field, $\mathbf{\Phi}_{\text{Bulk}} \in \mathbb{R}^{2 \times 2h \times 2w}$, generated from the first decoder branch sampled the prior shape, $\mathbf{P} \in \mathbb{R}^{h \times w}$, using bilinear interpolation: $\mathbf{Y}_{\text{Bulk}} = \mathbf{\Phi}_{\text{Bulk}}(\mathbf{P})$, as shown in Fig. 2. The resulting warped image was then sampled by the fine-tuning field generated by the second decoder branch, $\hat{\mathbf{Y}} = \mathbf{\Phi}_{\text{FT}}(\mathbf{Y}_{\text{Bulk}})$, where $\mathbf{\Phi}_{\text{FT}} \in \mathbb{R}^{2 \times 2h \times 2w}$. The prediction, $\hat{\mathbf{Y}}$, was then downsampled to the same resolution as the label, $\mathbf{Y} \in \mathbb{R}^{h \times w}$, using Max Pooling.

Fig. 2. A prior is sampled with a displacement field at double the resolution, $\mathbf{Y}_{\text{Bulk}} = \mathbf{\Phi}_{\text{Bulk}}(\mathbf{P})$, before the second super upsampled displacement is applied to the shifted image $\hat{\mathbf{Y}} = \mathbf{\Phi}_{\text{FT}}(\mathbf{Y}_{\text{Bulk}})$. The displacement panels show the shift in both directions.

Field Regularisation: The network outputs the final label prediction, $\hat{\mathbf{Y}}$, and the two diffeomorphic transformation fields: $\mathbf{\Phi}_{\text{Bulk}}, \mathbf{\Phi}_{\text{FT}}$, which were all used for the training loss. A combination of Dice Loss ($\mathcal{L}_{\text{Dice}}$) and a field regularisation ($\mathcal{L}_{\text{Grad}}$) function were used for training:

$$\mathcal{L} = \mathcal{L}_{\text{Dice}}(\mathbf{Y}, \hat{\mathbf{Y}}) + \beta \mathcal{L}_{\text{Grad}}(\mathbf{\Phi}_{\text{FT}}) + \beta \mathcal{L}_{\text{Grad}}(\mathbf{\Phi}_{\text{Bulk}}), \qquad (2)$$

where $\mathcal{L}_{\text{Grad}} = \sum_{i,j=1}^{2h,2w} \|\nabla \mathbf{\Phi}(i,j)\|^2$, which encourages smooth flow fields by penalising large spatial gradients between neighbouring voxels, adapted from [3]. A weighting parameter, β, was used to balance the contributions between $\mathcal{L}_{\text{Grad}}$ and $\mathcal{L}_{\text{Dice}}$, in this work we used a constant β of $10,000$, in order to give all the loss function components similar magnitudes, as measured empirically.

2.1   Experimental Setup

TEDS-Net was used to segment the myocardium from an open source dataset of MRI heart slices in the short axis[1] [4]. This dataset consisted of 2 scans from 100 patients, across five different pathologies. Five myocardium-containing slices were taken from each scan, cropped to 144 by 208 pixels and augmented using rotations, shifts, zooming and a combination of the three, resulting in 8,000 images split 75%, 15% and 10% for training, validation and testing, respectively. It was assured that slices from the same patient were assigned to the same subset.

The topology of the myocardium is equivalent to a hollow circle, which we used as the prior shape, $\mathbf{P}$, shown in Fig. 2. It should be noted that the patients' scans were not aligned and therefore the location of the myocardium was inconsistent. Additionally, different pathologies have varying myocardium thickness. Despite this, the same arbitrary prior shape was used throughout, illustrating the robustness of the method.

Two main experiments were completed and evaluated on 100 unseen heart slices without augmentations. Firstly, ablation studies were performed to show the contribution of each additional element used in TEDS-Net to tackle the limitations of diffeomorphic sampling, shown with the asterisks in Fig. 1, using a

---

[1] The ACDC database: www.creatis.insa-lyon.fr/Challenge/acdc.

Table 1. Comparison of myocardium segmentation performed by the baselines and TEDS-Net, accompanied by ablation studies (A1–6). The mean and standard deviation are given for Dice, Hausdorff Distance (HD) and percentage of non-positive Jacobian determinants from each field. The best result from each measure is shown in bold.

Network	Dice	HD	$\%\,\lvert J_{\Phi_{\mathrm{Bulk}}}\rvert \leq 0$	$\%\,\lvert J_{\Phi_{\mathrm{FT}}}\rvert \leq 0$	Incorrect topology
U-Net (baseline)	$\mathbf{0.87 \pm 0.16}$	5.25 ± 13.11	N/A	N/A	10/100
VoxelMorph (baseline)	0.79 ± 0.18	3.0 ± 2.4	N/A	$6.7e^{-05} \pm 0.7e^{-03}$	6/100
(A1) TEDS-Net: No Gaussian, $\mathcal{L}_{\mathrm{Grad}}$ or Super Upsample	0.82 ± 0.19	4.26 ± 5.76	56.32 ± 1.97	43.8 ± 3.46	17/100
(A2) TEDS-Net: No Gaussian Smoothing	0.86 ± 0.13	3.73 ± 8.34	67.10 ± 7.07	35.81 ± 4.41	3/100
(A3) TEDS-Net: No $\mathcal{L}_{\mathrm{Grad}}$	0.85 ± 0.14	2.77 ± 2.01	5.94 ± 2.57	22.15 ± 3.67	35/100
(A4) TEDS-Net: No Super Upsample	0.85 ± 0.11	2.89 ± 1.61	$\mathbf{0.0 \pm 0.0}$	$\mathbf{0.0 \pm 0.0}$	0/100
(A5) TEDS-Net: Only Bulk Branch	0.77 ± 0.15	3.86 ± 3.50	0.03 ± 0.17	N/A	0/100
(A6) TEDS-Net: Only Fine-Tune Branch	0.85 ± 0.14	3.01 ± 2.16	N/A	$\mathbf{0.0 \pm 0.0}$	0/100
TEDS-Net (ours)	0.86 ± 0.12	$\mathbf{2.76 \pm 2.13}$	$\mathbf{0.0 \pm 0.0}$	$\mathbf{0.0 \pm 0.0}$	0/100

prior shape of radius 30 and 8 integration layers, which were empirically set. As a comparison, we used a U-Net [15] and VoxelMorph diffeomorphic network [6,7], using the shape prior **P** as the second input channel. Secondly, we investigated the effect of the number of integration layers and radius of the binary prior.

The networks were all trained end-to-end on a NVIDIA GeForce GTX 1080 GPU and implemented in Pytorch using Python 3.6. Training was performed over 200 epochs, using the Adam optimiser with a learning rate of 0.0001 and a batch size of 5. TEDS-Net and the U-Net were both made up of 5 layers, with 12 initial feature maps, and a 20% dropout applied to each map.

3 Results and Discussion

To evaluate segmentation performance, Dice, Hausdorff Distance (HD), and a count of incorrect topologies were used, as seen in Table 1. A predicted segmentation's topology was classed as incorrect if their Betti numbers [12] varied from the known topological properties of the myocardium. To assess the discrete fields, the Jacobian determinants were also measured, as perfect diffeomorphic fields always have strictly positive determinants.

Large topological errors were found when segmenting the myocardium with the U-Net, as seen from Table 1. These errors were mainly expressed as large gaps separating the myocardium, shown in Fig. 3, making clinical measures of circumference extremely challenging. VoxelMorph was designed to maximise registration performance using diffeomorphisms, however, when applied to segmentation a small fraction of topology errors are observed, shown in Table 1: VoxelMorph.

Unlike the U-Net predictions, these topology violations are seen at the boundaries of the myocardium, forming holes or disconnected regions when defined using 4-pixel connectivity, as done here, but not when using 8-pixel connectivity. Conversely, our TEDS-Net enforced diffeomorphisms, preserving topology for all cases, which to the best of our knowledge, is the first deep learning segmentation technique to achieve 100% topological accuracy. Therefore, TEDS-Net additional modifications are required to prioritise topology preservation.

Paired t-tests were computed between TEDS-Net and U-Net and found that TEDS-Net significantly outperformed the U-Net in Hausdorff Distance, with a p-value of 0.01. Although the Dice accuracies were competitive, the U-Net performed significantly better (p = 0.04). Figure 3c–d shows particularly challenging examples, where the U-Net fails to return the correct topology. However, although TEDS-Net returns labels with accurate topology, the myocardial wall appears too thick in some parts. This is likely due to the smoothing modification, whose parameter will be further investigated in future work.

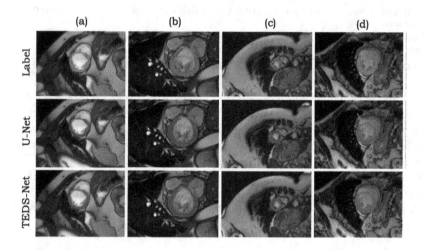

Fig. 3. Examples of myocardium manual annotations (green), compared to U-Net and TEDS-Net segmentations (pink) from the ACDC dataset [4]. (Color figure online)

Ablation studies were performed to show the effect of the additional smoothing modifications used to encourage perfect diffeomorphisms, as shown in Table 1 (A1–6). The continuous guarantees of diffeomorphic sampling, such as topology preservation, break down when applied in the discrete setting, shown by the emergence of non-positive Jacobian determinants that can then lead to topological errors. There are two sources of such violations in TEDS-Net: the composition layers and the image sampling.

To limit the numerical inaccuracies brought about by the discrete composition of two diffeomorphic fields, we introduced Gaussian smoothing between each layer. Without this addition, a large fraction of the Jacobian determinants no longer remained positive, corresponding to a number of topological defects, as shown in Table 1 (A2).

Sampling a discrete binary image with a finite warp field often results in disconnects, due to interpolations and the use of thresholds on the resultant image. To reduce this effect, we regularised the smoothness of the final deformation fields with $\mathcal{L}_{\mathrm{Grad}}$. Without this term, 35 out of 100 images were found to have topological defects and therefore, it played a vital role in mimicking perfect diffeomorphisms, as shown in Table 1 (A3).

Removing either the bulk or fine-tuning sampling branch was found to reduce the performance of both Dice and HD, as shown Table 1 (A5–6). As the images are unaligned, with varying sizes of myocardium, shown in Fig. 3, both branches are required to first align the prior before fine-tuning the warped shape. TEDS-Net performs the worst without the fine-tuning branch, which is likely due to the lack of flexibility in the bulk transformation, as the deformation fields are generated at a much lower resolution before being upsampled.

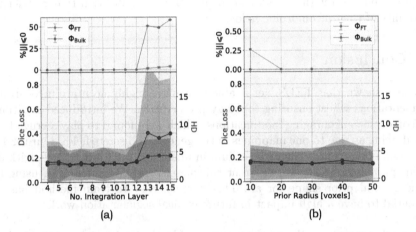

Fig. 4. The effect that the number of integration layers (a) and the radius of the binary prior (b) had on segmentation performance and the diffeomorphic nature of the generated fields. Due to the image dimensions, 50 was the maximum radius used.

Number of Integration Layers: Including Gaussian smoothing between the integration layers played a key role in enforcing topology preservation, as shown in Table 1 (A2). Although theoretically accuracy should increase with the number of composition layers, this requires performing more compositions that can each bring about small violations, due to the discrete nature of the fields [2]. However, limiting the number of composition layers, limits the size of deformations. To investigate this further, we varied the number of integration layers used in TEDS-Net and measured the segmentation performance and Jacobian determinants of the resulting fields, shown by Fig. 4a. When using 12 or fewer integration layers, the segmentation performance is stable whilst the Jacobian determinants all remain positive and all topology is preserved. However, when the number of layers is increased beyond 12, non-positive Jacobian determinants emerge. Therefore, whilst Gaussian smoothing has been shown to be essential in the integration layers, there is a limit to the number of compositions that can be used whilst enforcing diffeomorphic sampling.

Prior Radius: In the test set, the myocardium's radius ranged between approximately 10 to 60 voxels, so we investigated the impact of varying the radius on the final segmentation performance, as shown in Fig. 4b. The segmentation performance remained consistent as the radius was increased between 20 and 50 voxels. However, when the radius was set to 10 voxels, a small percentage of the Jacobian determinants of the fine-tuning fields were non-positive, corresponding to 2/100 incorrect topologies. This is likely due to the decreased number of voxels in the prior representing the myocardium and the area enclosed by it, in combination with the additional smoothing modification that restrict the flexibility of the diffeomorphic fields.

4   Conclusion

We have shown that TEDS-Net is able to achieve highly accurate myocardium segmentations whilst ensuring topology preservation. We introduced additional diffeomorphic-encouraging modifications, which were found to play a crucial role in enforcing an one-to-one mappings in the generated discrete fields. Our method successfully segmented the myocardium in unaligned MRI heart slices, with different pathologies that had different thicknesses and circumferences, using the same general prior shape for all. This flexible, easy to train method has the potential to have a high impact in future clinical segmentation work.

Acknowledgments. MW and ND is supported by the Engineering and Physical Sciences Research Council (EPSRC) and Medical Research Council (MRC) [grant number EP/L016052/1]. MJ is supported by the National Institute for Health Research (NIHR) Oxford Biomedical Research Centre (BRC), and this research was funded by the Wellcome Trust [215573/Z/19/Z]. The Wellcome Centre for Integrative Neuroimaging is supported by core funding from the Wellcome Trust [203139/Z/16/Z]. AN is grateful for support from the UK Royal Academy of Engineering under the Engineering for Development Research Fellowships scheme.

References

1. Arsigny, V., Commowick, O., Pennec, X., Ayache, N.: A log-Euclidean polyaffine framework for locally rigid or affine registration. In: Pluim, J.P.W., Likar, B., Gerritsen, F.A. (eds.) WBIR 2006. LNCS, vol. 4057, pp. 120–127. Springer, Heidelberg (2006). https://doi.org/10.1007/11784012_15

2. Ashburner, J.: A fast diffeomorphic image registration algorithm. Neuroimage **38**(1), 95–113 (2007)

3. Balakrishnan, G., Zhao, A., Sabuncu, M.R., Guttag, J., Dalca, A.V.: VoxelMorph: a learning framework for deformable medical image registration. IEEE Trans. Med. Imaging **38**(8), 1788–1800 (2019)

4. Bernard, O., et al.: Deep learning techniques for automatic MRI cardiac multi-structures segmentation and diagnosis: is the problem solved? IEEE Trans. Med. Imaging **37**(11), 2514–2525 (2018)

5. Clough, J., Byrne, N., Oksuz, I., Zimmer, V.A., Schnabel, J.A., King, A.: A topological loss function for deep-learning based image segmentation using persistent homology. IEEE Trans. Pattern Anal. Mach. Intell. (2020)

6. Dalca, A.V., Balakrishnan, G., Guttag, J., Sabuncu, M.R.: Unsupervised learning for fast probabilistic diffeomorphic registration. In: Frangi, A.F., Schnabel, J.A., Davatzikos, C., Alberola-López, C., Fichtinger, G. (eds.) MICCAI 2018. LNCS, vol. 11070, pp. 729–738. Springer, Cham (2018). https://doi.org/10.1007/978-3-030-00928-1_82

7. Dalca, A.V., Balakrishnan, G., Guttag, J., Sabuncu, M.R.: Unsupervised learning of probabilistic diffeomorphic registration for images and surfaces. Med. Image Anal. **57**, 226–236 (2019)

8. Dinsdale, N.K., Jenkinson, M., Namburete, A.I.L.: Spatial warping network for 3D segmentation of the hippocampus in MR images. In: Shen, D., et al. (eds.) MICCAI 2019. LNCS, vol. 11766, pp. 284–291. Springer, Cham (2019). https://doi.org/10.1007/978-3-030-32248-9_32

9. Hu, X., Fuxin, L., Samaras, D., Chen, C.: Topology-preserving deep image segmentation. arXiv preprint arXiv:1906.05404 (2019)

10. Jaderberg, M., Simonyan, K., Zisserman, A., et al.: Spatial transformer networks. Adv. Neural. Inf. Process. Syst. **28**, 2017–2025 (2015)

11. Marian, A.J., Braunwald, E.: Hypertrophic cardiomyopathy: genetics, pathogenesis, clinical manifestations, diagnosis, and therapy. Circ. Res. **121**(7), 749–770 (2017)

12. Millson, J.J.: On the first Betti number of a constant negatively curved manifold. Ann. Math. **104**, 235–247 (1976)

13. Moler, C., Van Loan, C.: Nineteen Dubious ways to compute the exponential of a matrix, twenty-five years later. SIAM Rev. **45**(1), 3–49 (2003)

14. Mosinska, A., Marquez-Neila, P., Koziński, M., Fua, P.: Beyond the pixel-wise loss for topology-aware delineation. In: Proceedings of the IEEE Conference on Computer Vision and Pattern Recognition, pp. 3136–3145 (2018)

15. Ronneberger, O., Fischer, P., Brox, T.: U-Net: convolutional networks for biomedical image segmentation. In: Navab, N., Hornegger, J., Wells, W.M., Frangi, A.F. (eds.) MICCAI 2015. LNCS, vol. 9351, pp. 234–241. Springer, Cham (2015). https://doi.org/10.1007/978-3-319-24574-4_28

16. Stergios, C., et al.: Linear and deformable image registration with 3d convolutional neural networks. In: Stoyanov, D., et al. (eds.) RAMBO/BIA/TIA -2018. LNCS, vol. 11040, pp. 13–22. Springer, Cham (2018). https://doi.org/10.1007/978-3-030-00946-5_2

17. Vigneault, D.M., Xie, W., Ho, C.Y., Bluemke, D.A., Noble, J.A.: ω-net (omega-net): fully automatic, multi-view cardiac MR detection, orientation, and segmentation with deep neural networks. Med. Image Anal. **48**, 95–106 (2018)
18. Wickramasinghe, U., Knott, G., Fua, P.: Probabilistic atlases to enforce topological constraints. In: Shen, D., et al. (eds.) MICCAI 2019. LNCS, vol. 11764, pp. 218–226. Springer, Cham (2019). https://doi.org/10.1007/978-3-030-32239-7_25
19. Zeng, Q., et al.: Liver segmentation in magnetic resonance imaging via mean shape fitting with fully convolutional neural networks. In: Shen, D., et al. (eds.) MICCAI 2019. LNCS, vol. 11765, pp. 246–254. Springer, Cham (2019). https://doi.org/10.1007/978-3-030-32245-8_28

Learning Consistency- and Discrepancy-Context for 2D Organ Segmentation

Lei Li[1], Sheng Lian[2,3,4], Zhiming Luo[2(✉)], Shaozi Li[2(✉)], Beizhan Wang[1], and Shuo Li[3,4]

[1] Department of Software Engineering, Xiamen University, Xiamen, Fujian, China
[2] Department of Artificial Intelligence, Xiamen University, Xiamen, Fujian, China
{zhiming.luo,szlig}@xmu.edu.cn
[3] Digital Image Group (DIG), London, ON, Canada
[4] School of Biomedical Engineering, Western University, London, ON, Canada

Abstract. Recently, CNN-based methods lead tremendous progress in segmenting abdominal organs (*e.g.*, kidney, liver, and pancreas) and anomaly tumors in CT scans. Although 3D CNN-based methods can significantly improve accuracy by using 3D volume as input, they need more computational cost and may not satisfy the efficiency requirement for many practical applications. In this study, we mainly aim at improving the 2D segmentation by leveraging the consistency- and- discrepancy-context information from adjacent slices. Specifically, the consistency context mainly considers that the prediction variance of two adjacent slices needs to follow the variance in the ground truth. The discrepancy-context assumes the label difference of adjacent slices usually occurs in the edge area of organs. To fully utilize the above context information, we further devise a two-stage 2.5D segmentation framework based on the U-Net that takes three adjacent slices as input. In the first stage, we encourage the predictions of the three slices following the consistency context. In the second stage, we refine the segmentation result by adopting the prediction discrepancy area of adjacent slices as an extra input. Experimental results on several challenging datasets demonstrate the effectiveness of our proposed methods. Moreover, the adjacent-slice context information considered in this study can be effortlessly incorporated into other segmentation frameworks without extra testing overhead.

Keywords: Organ segmentation · Consistency context · Discrepancy context

1 Introduction

Segmenting abdominal organs (e.g., kidney, liver, and pancreas) and anomaly tumors in CT scans is a critical prerequisite for many clinical scenarios, including

---

L. Li and S. Lian—Equal contribution.

© Springer Nature Switzerland AG 2021
M. de Bruijne et al. (Eds.): MICCAI 2021, LNCS 12901, pp. 261–270, 2021.
https://doi.org/10.1007/978-3-030-87193-2_25

computer-aided diagnosis, radiotherapy planning, and computer-assisted surgery overtreatment [8,11,12,15,17,20,22]. Apart from artifact and low contrast in the original CT scan, this segmentation task still needs to deal with challenges from different aspects, such as complex contextual associations, class imbalance, and intricate tissue shapes/sizes. Due to these challenges, a vanilla U-Net [16] model usually suffers the issue of discontinuous prediction, missing the small target, as shown in Fig. 1.

The development of CNN-based methods has significantly boosted the performance of this task. Existing CNN-based methods can be categorized as 2D-based [4,19,23] and 3D-based [5,6,10]. The 2D-based methods are more efficient since they process each slice independently while not utilizing the context information from the adjacent slice, resulting in relatively low performance. On the other hand, the 3D-based approaches are more accurate by using 3D volume as input. Meanwhile, they need more computational cost and may not satisfy the efficiency requirements for many practical applications. Therefore, it raises a question: *Is it possible to utilize the contextual information for improving segmentation performance while still maintaining the efficiency as in 2D-based models?* The work in [7] has tried to tackle this problem by training the segmentation model in a 2.5D manner, in which several adjacent slices are used as input. However, [7] simply uses more input slices and fails to exploit the inter-slice knowledge.

In Fig. 1, we illustrate the predictions from a vanilla U-Net. Moreover, we also compute the mis-segmented regions and the inter-slice difference (inter-diff) regions of the ground truth. Notice that, the *inter-diff* in the ground truth of two adjacent slices (formulated in Eq. 1) is roughly located in the boundary regions of the target tissues. By observing the mis-segmented regions, we can find that a large variance occurs between the predictions of adjacent slices, but the inter-diff of two adjacent slices are very similar. Besides, there are also some errors that happen around the boundary of tissues. Therefore, we could improve the segmentation accuracy by reducing the prediction variance in adjacent slices and leveraging the information of inter-diff regions.

In this paper, for accurate segmentation of abdominal organs in CT scans, we propose to leverage the inter-diff information from two aspects: (1) **Consistency-context** mainly considers that the prediction variance of two adjacent slices needs to follow the variance in the ground truth (GT). (2) **Discrepancy-context** assumes the label difference of adjacent slices usually occurs in the organ's boundary area with more segment difficulty. To fully utilize the context information, we further devise a two-stage 2.5D segmentation framework that takes three adjacent slices as input. In the first stage, we encourage the predictions of the three slices to follow the consistency-context by leveraging it as an additional supervision signal. In the second stage, we further refine the segmentation results by adopting the prediction discrepancy area of adjacent slices as an extra input. In this way, our model can better capture the inter-slice context information of the CT scans, especially the organ/tumor edge regions with more segment difficulty, thus improving segmentation accuracy and maintaining the high efficiency. Besides, the proposed context information can be easily

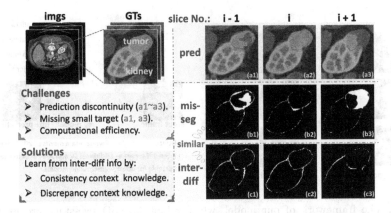

Fig. 1. Schematic diagram of the results of adjacent slices predicted by U-Net model. *Mis-seg*: Regions of inconsistency between predictions and GTs. *Inter-diff*: The inter-slice differences between adjacent GT slices. The inter-diff regions have a high degree of overlapping with the mis-seg regions, which are basically located in tissues' edge regions. These regions are difficult for segmentation models and can act as critical information for model learning. (Best viewed in color) (Color figure online)

incorporated into other promising segmentation models (e.g., DeepLabv3 [3], PSPNet [24]). Experiments on three different organ datasets demonstrate our model's: (a) Effectiveness of different components. (b) Generalization ability on different organ datasets. (c) Portability to different segmentation models.

2 Methodology

For robust segmentation of organ and anomaly tumor in 3D CT scan, our model adopts a two-stage 2.5D framework, which benefits from inter-slice difference (inter-diff) knowledge from two aspects: (1) Consistency context similarity between the predictions and GTs (Sect. 2.1), and (2) Discrepancy context (Sect. 2.2). The overall pipeline of our model is illustrated in Fig. 2.

Firstly, we will introduce the inter-diff, consistency context, and discrepancy context used in our model. We denote a 3D organ volume, the corresponding one-hot prediction and GT as $\mathbf{X} \in \mathbf{R}^{N \times H \times W}$, $\mathbf{Y}_{seg} \in \mathbf{R}^{N \times C \times H \times W}$, $\mathbf{Y}_{GT} \in \mathbf{R}^{N \times C \times H \times W}$, where N is the number of slices, C is the number of categories, H and W are height and width of each slice, respectively.

Inter-diff: The inter-diff in prediction for a slice t is the prediction difference between itself and the adjacent slice. In the same manner, we can get the inter-diff of the GT.

$$\mathcal{R}^t_{seg} = \mathbf{Y}^t_{seg} - \mathbf{Y}^{t-1}_{seg}, \quad \mathcal{R}^t_{GT} = \mathbf{Y}^t_{GT} - \mathbf{Y}^{t-1}_{GT}. \tag{1}$$

Consistency Context. We assume that the prediction's inter-diff of adjacent slices needs to follow the one in GT, and this knowledge act as an additional

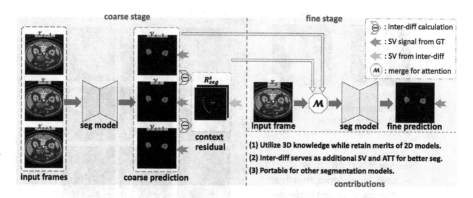

Fig. 2. The framework of our model, which adopts a 2.5D two-stage architecture. Abbreviations: SV = supervision, ATT = attention. Best viewed in color. (Color figure online)

supervision signal in our model. This context constraint can be measured by the L1 distance between the L2-norm of $\mathcal{R}_{\text{seg}}^t$ and $\mathcal{R}_{\text{GT}}^t$, which goes as:

$$\mathcal{D}^t = \left| ||\mathcal{R}_{seg}^t||^2 - ||\mathcal{R}_{GT}^t||^2 \right|. \tag{2}$$

Discrepancy Context. Our model leverages the fact that inter-diff usually occurs in tissues' edge regions which are hard to segment. We denote such knowledge as discrepancy context, and utilize it as the attention guidance for refinement.

2.1 Consistency Context-Based Organ Segmentation

As illustrated in Fig. 2, in the coarse segmentation stage, our model takes the target slice (x_t) and its adjacent slices $(x_{t-1}$ & $x_{t+1})$ as input, and output the predicted segmentation masks of the three slices. For each slice, we calculate DiceCE loss $(\mathcal{L}_{DCE})$ to measure the prediction accuracy, which goes as:

$$\mathcal{L}_{DCE} = \mathcal{L}_{Dice} + \mathcal{L}_{CE}. \tag{3}$$

Specifically, $\mathcal{L}_{Dice} = -\sum_{i=1}^{n}(1 - 2 \cdot \frac{t_i p_i}{t_i + p_i})$ and $\mathcal{L}_{CE} = -\sum_{i=1}^{n} t_i \log(p_i)$, where t_i is the GT label, and p_i is the predicted probability for the ith class.

Apart from the DiceCE loss for the three input slices, we further use a Contextual Residual loss $(\mathcal{L}_{CR})$ to encourage the consistency context of the (x_{t-1}, x_t) and (x_t, x_{t+1}) to be minimum. The Contextual Residual loss $(\mathcal{L}_{CR})$ goes as:

$$\mathcal{L}_{CR} = \mathcal{D}^t + \mathcal{D}^{t+1} \tag{4}$$

Finally, the overall loss function for the first coarse stage is

$$\mathcal{L}_{coarse} = \mathcal{L}_{DCE}(x_{t-1}) + \mathcal{L}_{DCE}(x_t) + \mathcal{L}_{DCE}(x_{t+1}) + \lambda_{CR}\mathcal{L}_{CR}, \tag{5}$$

where λ_{CR} is a hyper-parameter and is set to 0.05 in this study.

2.2   Segmentation Refinement with Discrepancy Context Knowledge

After the coarse stage, we further propose to utilize the discrepancy context to further rectify the segmentation results. As shown in Fig. 1, the edge regions of organ & tumor are with higher difficulty, and the inter-diff mask (M_t) is with a high degree of overlap with the edge regions. Therefore, we leverage the discrepancy context as extra attention guidance. The discrepancy context used in this stage is formulated as the prediction difference regions in the adjacent slices from the coarse stage's outputs, which goes as:

$$M^t = \mathbb{1}(L_{seg}^t \: != L_{seg}^{t-1}) + \mathbb{1}(L_{seg}^t \: != L_{seg}^{t+1}) \tag{6}$$

where L_{seg}^{t-1}, L_{seg}^t and L_{seg}^{t-1} are the predicted labels.

Then, we leverage M^t as the attention information for refining the result L_{seg}^t of the slice X_t. In this study, we directly concatenate the X_t with M^t and the coarse segmentation result Y_{seg}^t as an augment input for a 2D U-Net model, and compute the refined segmentation results. In this stage, we train the model by the DiceCE loss.

3   Datasets and Implementation Details

Datasets. We conduct the experiments to evaluate the performance of our method on three different organ CT datasets: KiTS ($\mathcal{D}_K$, for kidney) [9], LiTS ($\mathcal{D}_L$, for liver) [1], and Pancreas ($\mathcal{D}_P$) [18]. The statistical details of the datasets are summarized in Table 1.

Table 1. The statistics of datasets adopted in our study.

Datasets	Labeled organs	#Volumes		
		Training	Testing	Total
KiTS ($\mathcal{D}_K$)	Kidney	168	42	210
LiTS ($\mathcal{D}_L$)	Liver	104	26	130
Panc ($\mathcal{D}_P$)	Pancreas	225	57	282

Implementation Details. We implement our model with the PyTorch library [14] in a device with an NVIDIA 2080TI GPU. The image intensity is windowed by $[-160, 240]$ for $\mathcal{D}_K$, $[-100, 400]$ for $\mathcal{D}_L$ and $[-100, 240]$ for $\mathcal{D}_P$. Each slice is resized to 512×512 and then randomly cropped to 256×256 for training. We adopt random flip for data augmentation. For the U-Net model, we adopt the ResNeXT50_32x4d [21] as the encoder's backbone, which could be replaced by other segmentation backbones such as VGG-16 and ResNet-50. For both stages, we set the number of training epochs and batch-size to 12 and 8, respectively. The Adam [13] is used for optimization with an initial learning rate of 1e−4.

Evaluation Metrics: We adopt Dice-Sørensen Coefficient (DSC) as the evaluation metrics, which goes as $DSC(\mathcal{P}, \mathcal{G}) = \frac{2 \times |\mathcal{P} \cap \mathcal{G}|}{|\mathcal{P}| + |\mathcal{G}|}$, where $\mathcal{P}$ is the binary prediction and $\mathcal{G}$ is the ground-truth.

Table 2. The segmentation accuracy of our model and the baseline models on three organ segmentation datasets (in DSC (%)). Kid., Liv., and Panc. are short for kidney, liver, and pancreas, respectively.

Model	$\mathcal{D}_K$			$\mathcal{D}_L$			$\mathcal{D}_P$		
	Mean	Kid.	Tumor	Mean	Liv.	Tumor	Mean	Panc.	Tumor
2D U-Net	78.64	94.90	62.37	72.38	94.39	50.37	58.86	76.27	41.45
2.5D U-Net-1	78.71	94.47	62.95	74.64	95.07	54.21	59.54	76.56	42.51
2.5D U-Net-3	79.46	95.06	63.85	75.29	95.07	55.50	59.76	76.72	42.80
Ours (coarse stage)	81.06	95.26	66.86	75.80	95.44	56.16	60.05	77.03	43.07
Ours (fine stage)	**81.78**	**96.57**	**66.98**	**79.63**	**95.62**	**63.63**	**60.49**	**77.45**	**43.53**

4   Experimental Results and Analysis

4.1   The Effectiveness of Our Proposed Method

We first evaluate the effectiveness of the proposed model by comparing it with three baseline models. Specifically, we choose U-Net in vanilla version and two 2.5D variants as baselines. (1) *2D U-Net* is a vanilla 2D U-Net that only uses one slice as input. (2) *2.5D U-Net-1* is a 2.5D U-Net that takes three adjacent slices as input, while only predict the segmentation result of the middle slice for training and testing. (3) *2.5D U-Net-3* takes three adjacent slices as input, while computing the segmentation results for all three slices for training and testing. The DSC results on three different organ datasets ($\mathcal{D}_K$, $\mathcal{D}_L$, $\mathcal{D}_P$) are reported in Table 2. From the table, we can make the following two conclusions.

(1) **More input slices and labels are beneficial for the accuracy.** In Table 2, we observe that 2.5D-based model can consistently improve the performance on three datasets. For example, the mean DSC on $\mathcal{D}_L$ is improved from 72.38% to 74.64% when switching the model from 2D U-Net to 2.5D U-Net-1. Besides, with the supervision of more adjacent slices, the 2.5D U-Net-3 can further increase the mean DSC from 74.64% to 75.29% compared with 2.5D U-Net-1. These results demonstrate that the 2.5D architecture with more input slices and labels can jointly improve the segmentation performance.

(2) **The Consistency- and Discrepancy- Knowledge are Effective for the Segmentation.** From the last two rows in Table 2, we can find that our model outperforms the above baseline models consistently in both coarse stage and fine stage. For example, with consistency context, the result in

the coarse stage outperforms the best baseline model (2.5D U-Net-3) by 3.13% and 2.32% in the tumor DSC and mean DSC on $\mathcal{D}_K$, respectively. In addition, the discrepancy context leveraged in the fine stage further boosts the accuracy by 0.72%, 3.83%, and 0.44% in mean DSC of $\mathcal{D}_K$, $\mathcal{D}_L$, $\mathcal{D}_P$, respectively. Note that the tumor DSC on $\mathcal{D}_L$ witnesses a remarkable gain from 56.16% to 63.63%. These results verify that incorporating consistency- and discrepancy- knowledge can improve model's segmentation performance, especially for relatively smaller targets like tumors.

4.2   Portability with Different Segmentation Models

In this section, we evaluate the portability of our proposed context information with several different promising models, including U-Net [16], PSPNet [24], DeepLabv3 [2]. Notice that all the comparing models are implemented in the same 2.5D two-stage training manner. In Table 3, we report the accuracy without and after using the proposed context, and can have following observations.

(1) **Both contexts can increase the accuracy of different models.** As shown in the left part of Table 3, our model can bring consistent performance gains for the coarse stage of all the three segmentation models (1.61/12.87/2.57 points gains in mean DSC by using U-Net, PSPNet, DeepLabV3, respectively). In the right part of the table, we can observe similar results for all the three models at the fine stage (2.00/11.75/2.68 points gains in mean DSC over U-Net, PSPNet, DeepLabV3, respectively). These experiments demonstrate the portability of the proposed consistency- and discrepancy-context that can be effortlessly incorporated into different models.

(2) **The context information has a more positive impact on small targets.** In Table 3, we can find that tumor DSCs achieve relatively more improvement, e.g., 64.84% to 68.74% for DeepLabv3's coarse stage and 64.90% to 69.31% for DeepLabv3's fine stage. These results suggest that our proposed method has a more positive impact on smaller targets, which can be critical for clinical scenarios since accurate recognition of the small tumor is essential.

Table 3. Experimental results (in DSC) of different segmentation models without and after using the proposed context on the KiTS dataset.

Method	Coarse stage			Fine stage		
	Mean (%)	Kidney (%)	Tumor (%)	Mean (%)	Kidney (%)	Tumor (%)
U-Net [16]	79.46	95.06	63.85	79.78	95.06	64.50
U-Net+ours	81.06	95.26	66.86	81.78	96.57	66.98
PSPNet [24]	52.69	72.78	32.59	54.30	72.55	36.04
PSPNet+ours	65.55	86.48	44.62	66.04	87.12	44.96
DeepLabv3 [2]	79.00	93.16	64.84	79.37	93.84	64.90
DeepLabv3+ours	81.57	94.39	68.74	82.05	94.78	69.31

4.3   Qualitative Results

Figure 3 depicts the qualitative results of our model against baselines on examples from $\mathcal{D}_K$. Although the original adjacent slices are similar, the baseline models give inconsistent predictions, especially for false tumor regions. The 2D UNet is prone to over-segment the tumor regions (a3, b3) and missing kidney regions (d3, e3, f3). The 2.5 D U-Net-1 has better prediction in (a4, b4), but over-segment tumor in (c4), and can not recognize tumors in (d4, e4, f4). The 2.5 D U-Net-3 could alleviate the over-segmentation in (a5, b5, c5), but errors still exist. In addition, 2.5D U-Net-3 locate tumor regions of (d, e, f) better, but all have false-positive regions of kidney. In (b6), our coarse model also suffers the same issue but is corrected by our fine model (b7). Comparing the above results with (d6, e6, f6) and (d7, e7, f7), our coarse model can also help locate the tumor regions and eliminate the false positive region of kidney. Besides, our fine model further recovers the missing tumor areas ((d6) to (d7), (e6) to (e7)).

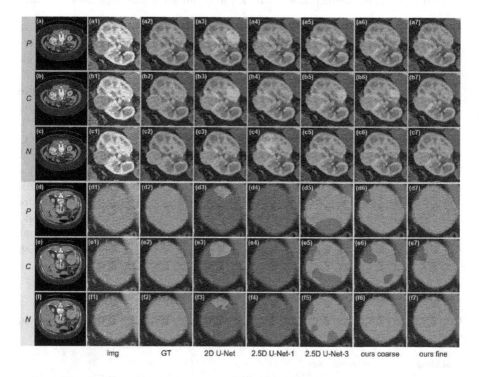

Fig. 3. The qualitative segmentation results of two representative adjacent slices. Blue and Green colors show the predictions of organs and tumors. P = previous, C = current, N = Next. Best viewed in color. (Color figure online)

5   Conclusion

In this study, we aim to boost the 2D segmentation model's accuracy by leveraging the inter-slice context information while maintaining the efficiency of 2D models. Specifically, our model adopts a 2.5D coarse-to-fine architecture, which benefits from the inter-slice context knowledge from two aspects: (1) Consistency context similarity between the predictions and GTs for additional supervision, and (2) Discrepancy context for attention guidance. The experiments demonstrate that our method achieves considerable improvement in different datasets and can be incorporated into other promising segmentation models. For future work, we will further investigate the incorporation of a more extensive range of contextual information and the organ's anatomical priors to further improve accuracy.

Acknowledgements. This work is supported by the National Nature Science Foundation of China (No. 61876159, 61806172, 62076116 & U1705286), the China Postdoctoral Science Foundation Grant (No. 2019M652257), the Guiding Project of Science and Technology Department of Fujian Province (No. 2019Y0018), and China Scholarship Council.

References

1. Bilic, P., et al.: The liver tumor segmentation benchmark (LiTS). arXiv (2019)
2. Chen, L.C., Papandreou, G., Schroff, F., Adam, H.: Rethinking atrous convolution for semantic image segmentation. arXiv (2017)
3. Chen, L.-C., Zhu, Y., Papandreou, G., Schroff, F., Adam, H.: Encoder-decoder with atrous separable convolution for semantic image segmentation. In: Ferrari, V., Hebert, M., Sminchisescu, C., Weiss, Y. (eds.) ECCV 2018. LNCS, vol. 11211, pp. 833–851. Springer, Cham (2018). https://doi.org/10.1007/978-3-030-01234-2_49
4. da Cruz, L.B., et al.: Kidney segmentation from computed tomography images using deep neural network. Comput. Biol. Med. **123**, 103906 (2020)
5. Dou, Q., Chen, H., Jin, Y., Yu, L., Qin, J., Heng, P.-A.: 3D deeply supervised network for automatic liver segmentation from CT volumes. In: Ourselin, S., Joskowicz, L., Sabuncu, M.R., Unal, G., Wells, W. (eds.) MICCAI 2016. LNCS, vol. 9901, pp. 149–157. Springer, Cham (2016). https://doi.org/10.1007/978-3-319-46723-8_18
6. Haghighi, M., Warfield, S.K., Kurugol, S.: Automatic renal segmentation in DCE-MRI using convolutional neural networks. In: ISBI (2018)
7. Han, X.: Automatic liver lesion segmentation using a deep convolutional neural network method. arXiv (2017)
8. Heller, N., et al.: The state of the art in kidney and kidney tumor segmentation in contrast-enhanced CT imaging: results of the KiTS19 challenge. Med. Image Anal. **67**, 101821 (2021)
9. Heller, N., et al.: The KiTS19 challenge data: 300 kidney tumor cases with clinical context, CT semantic segmentations, and surgical outcomes. arXiv (2019)
10. Hou, X., et al.: A triple-stage self-guided network for kidney tumor segmentation. In: ISBI (2020)

11. Howe, R.D., Matsuoka, Y.: Robotics for surgery. Annu. Rev. Biomed. Eng. **1**, 211–240 (1999)
12. Khalifa, F., et al.: 3D kidney segmentation from CT images using a level set approach guided by a novel stochastic speed function. In: Fichtinger, G., Martel, A., Peters, T. (eds.) MICCAI 2011. LNCS, vol. 6893, pp. 587–594. Springer, Heidelberg (2011). https://doi.org/10.1007/978-3-642-23626-6_72
13. Kingma, D.P., Ba, J.: Adam: a method for stochastic optimization. arXiv (2014)
14. Paszke, A., et al.: PyTorch: an imperative style, high-performance deep learning library. In: NeurIPS (2019)
15. Pekar, V., McNutt, T.R., Kaus, M.R.: Automated model-based organ delineation for radiotherapy planning in prostatic region. IJROBP **60**, 973–980 (2004)
16. Ronneberger, O., Fischer, P., Brox, T.: U-Net: convolutional networks for biomedical image segmentation. In: Navab, N., Hornegger, J., Wells, W.M., Frangi, A.F. (eds.) MICCAI 2015. LNCS, vol. 9351, pp. 234–241. Springer, Cham (2015). https://doi.org/10.1007/978-3-319-24574-4_28
17. Roth, H.R., et al.: DeepOrgan: multi-level deep convolutional networks for automated pancreas segmentation. In: Navab, N., Hornegger, J., Wells, W.M., Frangi, A.F. (eds.) MICCAI 2015. LNCS, vol. 9349, pp. 556–564. Springer, Cham (2015). https://doi.org/10.1007/978-3-319-24553-9_68
18. Simpson, A.L., et al.: A large annotated medical image dataset for the development and evaluation of segmentation algorithms. arXiv (2019)
19. Thong, W., Kadoury, S., Piché, N., Pal, C.J.: Convolutional networks for kidney segmentation in contrast-enhanced CT scans. Comput. Meth. Biomech. Biomed. Eng. Imaging Vis. **6**, 277–282 (2018)
20. Van Ginneken, B., Schaefer-Prokop, C.M., Prokop, M.: Computer-aided diagnosis: how to move from the laboratory to the clinic. Radiology **261**, 719–732 (2011)
21. Xie, S., Girshick, R., Dollár, P., Tu, Z., He, K.: Aggregated residual transformations for deep neural networks. In: CVPR (2017)
22. Yang, D., et al.: Automatic liver segmentation using an adversarial image-to-image network. In: Descoteaux, M., et al. (eds.) MICCAI 2017. LNCS, vol. 10435, pp. 507–515. Springer, Cham (2017). https://doi.org/10.1007/978-3-319-66179-7_58
23. Yu, Z., Pang, S., Du, A., Orgun, M.A., Wang, Y., Lin, H.: Fine-grained tumor segmentation on computed tomography slices by leveraging bottom-up and top-down strategies. In: Medical Imaging 2020: Image Processing (2020)
24. Zhao, H., Shi, J., Qi, X., Wang, X., Jia, J.: Pyramid scene parsing network. In: CVPR (2017)

Partially-Supervised Learning for Vessel Segmentation in Ocular Images

Yanyu Xu[1], Xinxing Xu[1(✉)], Lei Jin[2], Shenghua Gao[2], Rick Siow Mong Goh[1],
Daniel S. W. Ting[3,4], and Yong Liu[1]

[1] Institute of High Performance Computing, A*STAR, Singapore, Singapore
{xu_yanyu,xuxinx,gohsm,liuyong}@ihpc.a-star.edu.sg
[2] ShanghaiTech University, Pudong, China
{jinlei,gaoshh}@shanghaitech.edu.cn
[3] Singapore National Eye Centre, Singapore Eye Research Institute,
Singapore, Singapore
daniel.ting.s.w@singhealth.com.sg
[4] Duke-NUS Medical School, Singapore, Singapore

Abstract. The vessel segmentation in ocular images is a fundamental and important step in the diagnosis of eye-related diseases. Existing vessel segmentation methods require a large-scale ocular images with pixel-level annotations. However, manually annotating masks is a laborious and tedious process. Compared with the traditional pipelines which either annotate the complete training set or several images in full, in this paper, we propose a novel supervision manner, named Partially-Supervised Learning (PSL), which only relies on partial annotations in the form of one patch from each of the few images. Targeting it, we propose an active learning framework with latent MixUp. The active learning strategy is employed to select the most informative patch for further annotation, while the latent MixUp is proposed to learn a proper visual representation of both the annotated and unannotated patches. The experimental results on two types of vessel segmentation datasets (Rose-1 (SVC) dataset for OCTA image, and DRIVE dataset for fundus image) validate the effectiveness of our model. With only 5% annotations on Rose-1 (SVC) and DRIVE dataset, our performance is comparable with the previous methods trained on the whole fully annotated dataset.

Keywords: Partially-supervised learning · Vessel segmentation

1 Introduction

The retinal blood vessel segmentation from color fundus images [20] or OCTA images [12] is a fundamental and important step in the diagnosis of eye-related

Electronic supplementary material The online version of this chapter (https://doi.org/10.1007/978-3-030-87193-2_26) contains supplementary material, which is available to authorized users.

M. de Bruijne et al. (Eds.): MICCAI 2021, LNCS 12901, pp. 271–281, 2021.
https://doi.org/10.1007/978-3-030-87193-2_26

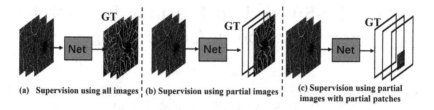

(a) Supervision using all images | (b) Supervision using partial images | (c) Supervision using partial images with partial patches

Fig. 1. An illustration of segmentation with different supervised way.

diseases, including systemic, metabolic, and hematologic diseases [13]. In the past decades, researchers have proposed many automatic segmentation methods, such as tracking [3] or filtering [2] based models, as well as thresholding based models [5,23]. Recently, deep learning based methods [8,12,16] have achieved promising results for the retinal blood vessel segmentation task. As shown in Fig. 1 (a), learning such segmentation models usually requires enough fully annotated pixel-wise masks, which is a laborious and tedious process. Besides, the annotations of retina blood vessels also require professional medical knowledge and proper training.

Naturally, a key question arises. Can we design a model that could still produce satisfactory performance but use as few annotations as possible? Along this direction, some researchers use a part of the dataset with fully annotated images in Fig. 1 (b). For example, some research works propose to use semi-supervised learning for cell segmentation [27], tumor segmentation [4], 3D abdominal CT [22], left atrium segmentation [10] or active learning for skin lesion segmentation [25]. Others propose an interactive annotation process for brain segmentation [21] or heart, aorta segmentation [9] or a full supervision directly on patches [6]. For such a labeling and training strategy, there are limited scenes and image conditions, which might reduce the generalization ability of the trained model. Based on the fact that the thick or thin vessels are usually consistent or the same in one image, it might be redundant to annotate all the vessels.

Therefore, different from the above attempts that fully annotate a few training images, we propose a novel supervision manner, named Partially-Supervised Learning, which only requires one patch from each of the few images. In particular, each annotated image consists of both one annotated patch, *e.g.* 10%, and the unannotated patches, *e.g.* 90% in Fig. 1 (c). In this way, the model could preserve diversity in the scenes and reduce the costs of manual annotation. Besides, only annotating one patch in each of the few images could also reduce the redundancy of annotating similar texture or repeated patterns. In [15], they propose a method using partial point annotation to segment the cell regions. However, their model is specific for the cell data, since they use the prior knowledge, such as the locations of the cell nucleus, while our setting is more general for several kinds of vessel images, such as fundus and OCTA.

There are two challenges of our partially-supervised learning setting. The first one is how to select the most useful patch to be annotated, since different patches include different level of information. The second one is how to leverage

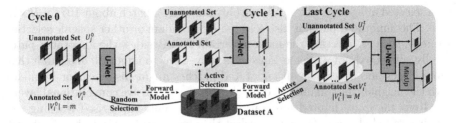

Fig. 2. The pipeline of our proposed model.

those unannotated patches and alleviate the need for labeled data, given the quite limited annotations. Targeting it, we propose an active learning framework with a latent MixUp for vessel segmentation in the partially-supervised learning setting. Given a labeling budget, we employ an active labeling strategy to iteratively annotate the most informative and uncertain patch in the unlabeled set and learn the segmentation model on them. Besides, all the rest unannotated patches in the annotated images and the rest unannotated images are used to train the network. Inspired by MixUp [1, 24], we propose a latent MixUp to learn a proper visual representation of both the annotated and unannotated patches. Different from the combination between the inputs in image level in the existing MixUp, our latent MixUp combines the latent representations in patch level, such as the annotated patches and unannotated patches.

The contributions of this work are summarized as follows: To reduce the annotation cost and produce competitive performance, we study the segmentation task under a new partially-supervised learning setting, where we only annotate one patch from each of the few images. Targeting it, we propose an active learning framework with a latent MixUp. The active learning strategy is used to select the most uncertain patch based on the probability predicted by the segmentation model and the latent MixUp is designed to learn a proper visual representation of both the annotated and unannotated patches. The experimental results on two types of vessel segmentation datasets show the effectiveness of our proposed model. With only 5% annotated data, our proposed model could achieve comparable performance with those methods based on the fully annotated training set on the ROSE-1 (SVC) and DRIVE dataset.

2   Method

2.1   Partially-Supervised Learning

In this work, we propose a partially-supervised learning setting for the ocular image vessel segmentation. We only annotate one patch from each of the few images. All annotated and unannotated patches are used in the training process.

Firstly, we select M images from dataset A ($M < |A|$), using a random or other specific selection strategy, such as selecting the most uncertain one. For each selected image $v_i \in \mathbb{R}^{3 \times H \times W}$, we only annotate a patch about 10% of $H \times W$ area. The annotated patch is also selected by a random or other specific selection strategy, such as selecting the most uncertain one. Besides, the annotated patch is fully connected not disperse, since it is convenient to annotate one patch than several ones and several small annotated patches might disappear or be mixed with the neighboring unannotated patches after a successive convolutional layers.

Given the quite limited annotations, this partially-supervised learning setting has two challenges: (1) how to select the most useful patch to be annotated; (2) how to leverage those unannotated patches and alleviate the need for annotated data. Targeting at it, we propose an active learning framework with a latent MixUp, as illustrated in Fig. 2. The active learning framework is designed to iteratively select the most informative and uncertain patch to be annotated. Inspired by MixUp [1,24], the proposed latent MixUp is to learn a proper visual representation of both the annotated and unannotated patches. In our implementation, we use UNet [16] as backbone. The U-Net includes the contracting path (8 convolution layers following 4 max pooling layers) and expansive path (8 convolution layers and 4 up-sample layers). In the following, we will describe the proposed active learning framework with the latent MixUp in details.

2.2 Active Learning Framework

To select and annotate the most informative patch, we employ an active learning framework. Suppose there is a dataset A with labeling budget M, the maximum number of annotated images [11]. In initial cycle, we randomly select m images from the dataset A, randomly annotate one 10% patch in each selected images, then use them to train the segmentation model R^0. The annotated and unannotated set are represented as $V^0 = \{v_i, p_i\}$ and $U^0 = \{u_j\}$. After the training, we propose an active sample selection to choose m images with the most informative and uncertain patch from the unannotated set U^0, based on the predicted probabilities by model R^0. In the following cycle t, the updated annotated set V^t are used to train the segmentation model R^t. The process goes on until meeting the budget M. Only in the last cycle T ($T = \lceil M/m \rceil$), the unannotated set U is used to train the segmentation model R, using our latent MixUp.

Active Sample Selection. At the ending of cycle t, we want to select and annotate the most informative and uncertain patch and add them into the annotated set $V^t + 1$ to train the segmentation network R^{t+1}. These most uncertain patch usually is too complex or confused to make the right classification decision for the segmentation model with high confidence. The uncertainty of samples is directly indicated by the predicted probabilities of the learned segmentation model R. In particular, we forward the learned model R^t on the unannotated set U^t and get the probability map p_j for each images in U^t. In [19], Simple Margin learns an SVM on the existing labeled data and chooses as the next instance to query the instance that comes closest to the hyperplane. Similarly, a model is trained on labeled data and chooses the next most uncertain patches, which are closest to the decision boundary. In particular, it computes the distance between the predicted probability and the median probability ($w = 0.5$). We calculate the smallest sum values through the following:

$$r_j = \arg \min_j \sum_l^L \sum_k^K (p_{j,lk} - w)^2, \tag{1}$$

where $[l : L, k : K]$ are the coordinate ranges of the unannotated candidate patches. When $p_{j,lk}$ equals w, the sum value is near 0, while $p_{j,lk}$ equals 0 or 1, the sum value is much larger. Thus, the smallest sum value represents this patch has the most uncertainty.

2.3 Latent MixUp

To learn a proper visual representation of the annotated and unannotated patches, we propose a latent MixUp, inspired by MixUp [1,24]. Mixup randomly combine the inputs and their corresponding labels to encourage model to behave linearly between training samples to reduce the oscillations during inference.

Different from the various existing MixUp versions in image level, our latent MixUp are used in patch level. Suppose we randomly select one partial annotated image I_1 and one unannotated image I_2. The proposed latent MixUp includes two kinds of combinations: 1) the combination between the partial annotated patch I_1^v in I_1 and the same coordinate unannotated patch I_2^v in I_2, as well as their corresponding ground truth label p_1^v and guessing label $\hat{p}_2^v$; 2) the combination between the rest unannotated patches I_1^u in I_1 and the same coordinate unannotated patches I_2^u in I_2, as well as their corresponding guessing label $\hat{p}_1^u$ and guessing label $\hat{p}_2^u$. For the guessing labels $\hat{p}$ of the unannotated patches, we use the probability predicted by the segmentation model R.

For a pair of one partial annotated image I_1 and one unannotated image I_2, we compute the new augmented latent representations from their last convolutional layer $\phi(I_1)$ and $\phi(I_2)$. The latent MixUp with a weight λ' is:

$$I^{v'} = \lambda'\phi(I_1^v) + (1 - \lambda')\phi(I_2^v), p^{v'} = p_1^v \tag{2}$$

$$I^{u'} = \lambda'\phi(I_1^u) + (1 - \lambda')\phi(I_2^u), p^{u'} = \lambda'\hat{p}_1^u + (1 - \lambda')\hat{p}_2^u. \tag{3}$$

Following [1], λ' is generated by $\lambda \sim (\alpha, \alpha)$, $\lambda' = \max(\lambda, 1 - \lambda)$, where α is a hyper-parameter. The latent MixUp enriches the distribution in-between training samples. To note that, the combination of existing MixUp occurs at inputs in image level, while ours occurs at latent representations in patch level.

2.4 Loss Function

For vessel segmentation task, we adopt the commonly used pixel-wise cross entropy loss $L_{seg} = - \sum_i p^v \log(R(I^v))$, where $R(I^v)$ is the segmentation mask prediction and p^v is the partial annotated ground truth of image i. Given an image pair $< (I^{v'}, p^{v'}), (I^{u'}, p^{u'}) >$, the latent MixUp loss function is computed on their mixed representations. Following [1], the loss item for the annotated regions is a cross entropy loss $L_V = -\sum p^{v'} \log(R(I^{v'}))$, and the loss item for unannotated regions is MSE loss $L_U = \sum \|p^{u'} - R(I^{u'})\|_2^2$. In the last cycle, the total loss function is $L = L_{seg} + L_V + \lambda_U L_U$. λ_U is a hyper-parameter.

3 Experiment

3.1 Experiment Setting

Implementation Details. We use the PyTorch [14] platform to implement our model with the parameter settings: mini-batch size (4), learning rate (1.0e−4), momentum (0.95), weight decay (0.0005), and 100 epochs. We employ the default initialization to initialize the model. In the latent MixUp, the hyper-parameter α, λ_U are 0.75, 0.01. The labeling budget M is 10 and 15, and m equals 5.

Table 1. The performance comparison **Left** on the ROSE-1 (SVC) dataset and **Right** on the DRIVE dataset. FSL: Fully-Supervised Learning. SSL: Semi-Supervised Learning. PSL: Partially-Supervised Learning. LO: Label-Only. In our model, $m = 5$.

Method	Type	Ratio	ACC ↑	G-mean ↑	Kappa ↑	Dice ↑	FDR ↓	Method	Type	Ratio	ACC ↑	SPE ↑	SEN ↑	Dice ↑
U-Net [16]	FSL	100%	0.8955	0.8068	0.6476	0.7116	0.2627	UNet [16]	FSL	100%	0.9686	0.9861	0.7887	0.8140
CE-Net [8]	FSL	100%	0.9121	0.8256	0.6978	0.7511	0.1997	DEU-Net [20]	FSL	100%	0.9567	0.9816	0.7940	0.8270
OCTA-Net [12]	FSL	100%	0.9182	0.8361	0.7201	0.7697	0.1775	BEFD-UNet [26]	FSL	100%	0.9701	0.9845	0.8215	0.8267
MT [18]	SSL	3.3%	0.8827	0.7696	0.5951	0.6658	0.2896	MT [18]	SSL	5.0%	0.9519	0.9768	0.6403	0.6988
MixMatch [1]	SSL	3.3%	0.9079	0.8046	0.6750	0.7299	0.2313	MixMatch [1]	SSL	5.0%	0.9568	0.9741	0.6519	0.7245
LO (M = 10)	PSL	3.3%	0.9086	0.8263	0.6883	0.7425	0.2260	LO (M = 10)	PSL	5.0%	0.9605	0.9799	0.7584	0.7708
LO (M = 15)	PSL	5.0%	0.9102	0.8291	0.6935	0.7480	0.2228	LO (M = 15)	PSL	7.5%	0.9609	0.9794	0.7688	0.7752
Ours (M = 10)	PSL	3.3%	0.9119	0.8260	0.6948	0.7483	0.2008	Ours (M = 10)	PSL	5.0%	0.9618	0.9768	0.8020	0.7875
Ours (M = 15)	PSL	5.0%	0.9134	0.8273	0.6985	0.7510	0.1794	Ours (M = 15)	PSL	7.5%	0.9630	0.9788	0.8088	0.7919

Dataset. We use two types of eye images datasets: *The DRIVE dataset* [17] consists of 20 training and 20 testing fundus images centered on the macula. Each image includes a circular field-of-view mask; *The ROSE-1 dataset* [12] consists of 117 OCTA images. In this work, we use pixel-level annotation.

Metrics. Following the existing works [8,12,17], we use the following metrics: Dice Coefficient (Dice), Accuracy (ACC), Specificity (SPE), Sensitivity (SEN), Kappa score, False Discovery Rate (FDR) and G-mean score.

3.2 Performance Comparison

We evaluate our model with the following methods on the datasets on the commonly used metrics. Table 1 and Fig. 3 show the quantitative and qualitative results on the ROSE-1 (SVC) and DRIVE datasets.

Baselines. Since this is the first work to study partially-supervised learning setting ('PSL'), we compare our model with the following recent methods, divided into three groups. The first group is related to the fully-supervised learning methods, denoted as 'FSL'. We list some recent state-of-the-art methods using all samples as training, such as CE-Net [8], OCTA-Net [12] on the ROSE-1

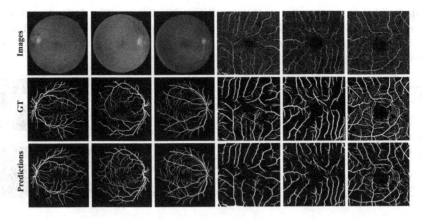

Fig. 3. Some examples of our model on the DRIVE and ROSE-1 (SVC) dataset.

dataset as well as DEU-Net [20] and BEFD-Net [26] on the DRIVE dataset. Secondly, we compare our model with the following related semi-supervised learning methods, denoted as 'SSL'. Mean teacher (MT) [18] is a classic consistency-based semi-supervised learning approach for classification. We modify it to segmentation task, by replacing the classifier by the segmentation mask generation and using the color augmentation in [7]. MixMatch [1] is a recently work for a semi-supervised learning, based on the mixup data augmentation [24] but performed on unlabeled data. The similar modification is applied for segmentation task. Finally, we also design a simple baseline, under partially-supervised learning 'PSL'. Label only ('Label-Only') uses the partial annotation (10% regions) with M images to train the U-Net model using masks on the loss function. To note that, we run five times the SSL and PSL baselines and our models and report the average values in the Table 1 and 2.

The 'Ratio' column represents how many annotated regions percentage the method uses as training samples. The PSL methods use 100% annotated data. The SSL methods use the 5% fully annotated images as training samples. For the PSL methods on the ROSE-1 (SVC) dataset, 5% means they use 15 ($15/30 = 50\%$) partial annotated images, and each of them has 10% annotated region.

The ROSE-1 (SVC) Dataset. We conduct experiments on the OCTA images. The experimental results are shown on the Table 1 Left. With a small amount of annotated regions in the training process, our network can still converge well in this dataset. Under the same ratio of annotated regions, such as 5%, the PSL methods in the third block always perform better than the SSL methods in the second block on the all metrics, which indicates our proposed novel supervision could capture more variance and reduce the redundancy of the dataset. Even using only 5% annotated data, our method improves slightly better performance than fully-supervised method CE-Net [8] in all metrics. We also show some quantitative examples predicted by our model ($M = 15, m = 5$). For the thick vessels, our model could estimate well, while might fall to predict these

Table 2. The ablation studies on the ROSE-1 (SVC) dataset. **The Left**: The effect of Active Sample Selection. RS: Random Selection. AC: Active Selection. **The Middle**: The effect of the Latent MixUp (LM). **The Right**: The visualization of the distribution of the selected patches by random selection and active selection.

Method	Type	Dice ↑	FDR ↓
M=5, m=5	AC/RS	0.7335	0.2343
M=10, m=5	RS	0.7444	0.2318
M=10, m=5	AC	**0.7471**	**0.2145**
M=15, m=5	RS	0.7468	0.2266
M=15, m=5	AC	**0.7494**	**0.2172**
M=20, m=5	RS	0.7521	0.2156
M=20, m=5	AC	**0.7525**	**0.1962**

Method	LM	Dice ↑	FDR ↓
M=5, m=5, AC	×	0.7335	0.2343
M=5, m=5, AC	✓	**0.7378**	**0.2189**
M=10, m=5, AC	×	0.7471	0.2145
M=10, m=5, AC	✓	**0.7483**	**0.2008**
M=15, m=5, AC	×	0.7494	0.2172
M=15, m=5, AC	✓	**0.7510**	**0.1794**
M=20, m=5, AC	×	0.7525	0.1962
M=20, m=5, AC	✓	**0.7546**	**0.1943**

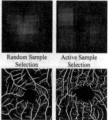

Random Sample Selection   Active Sample Selection

Ground Truth

thin vessels, owning to the lack of enough annotated data. More quantitative comparison with different M could be found in supplementary materials.

The DRIVE Dataset. We also conduct experiments on the fundus images. Table 1 Right is the experimental results. With less 10% annotated data, our network still can perform well on this dataset. Similarly, compared with the SSL methods, the PSL baseline and our PSL model could always achieve higher performance on the all metrics under the same ratio of the annotated regions, such as 5%. Besides, compared with the fully-supervised learning method DEU-Net [20], our method improves comparable performance, even slightly better performance on two metrics. Figure 3 shows some prediction of our model ($M = 15, m = 5$). We can see that our model still could perform well on these thick vessels, while might fall on the very thin vessels. If using more annotated thin regions as training samples, the performance could be further improved.

3.3   Ablation Studies

We conduct ablation studies on the ROSE-1 (SVC) dataset.

The Effect of the Budget M. We use different budgets M to annotated the training samples, such as $M = 5, 10, 15, 20$. The results are shown on the Table 2. We can see that with the increasing M, the model could achieve the better performance. It is obvious that using more annotated regions as training samples, the segmentation model could perform much better. We also train the models using less or more annotations, such as 5%, 25% and 50% in the selected images, to investigate how the annotated ratios effect the performance. Similarly, we find that the bigger the annotated ratio is, the higher performance the model could achieve. More comparisons could be found in supplementary materials.

The Effect of Active Sample Selection. To validate the effect of the active sample selection (AC), we train the baselines, replacing it with random selection (RS). The results are shown in the Table 2 Left part. We can see that our proposed active sample selection always outperforms the random selection baselines on the metrics. Besides, we also show the distribution of the annotated patches, selected by two strategies, in the Fig. 2 Right. The proposed active sample selection could avoid selecting the central regions, where there almost exist no vessels and it is much easier to predict, as shown in the Fig. 2 Right. Both of them could indicate that our proposed active sample selection could select the most informative and uncertain patch to be annotated. Besides, we also use other sample selection ways, such as using $p = 0.25$, 0.75, entropy using output probability $p\log(1 - p)$, or Monte-Carlo Dropout. The results on the Dice metric are 0.750 (MC Dropout), 0.750 ($p\log(1-p)$), 0.747 ($p = 0.25$), 0.750 ($p = 0.75$) and 0.751 (Ours).

The Effect of the Latent MixUp. To evaluate the effect of the latent MixUp, we train the baselines, removing the proposed latent MixUp with different M. The experimental results are shown on the Table 2 Middle part. We can see that our proposed latent MixUp could always outperform the baselines on the metrics, which indicates that our proposed latent MixUp could make use of these large unannotated images to learn more useful visual features.

4 Conclusion

We propose a novel partially-supervised learning setting to estimate the vessel segmentation using as few annotations as possible and producing competitive performance. Compared with semi-supervised learning, it could bring the various challenging scenes using the same even less annotation costs. Besides, we also design an active learning framework with latent MixUp. Further, our model only uses 5% annotated data and the results also indicate there is a further improving space in the vessel segmentation under the partially-supervised learning setting.

Acknowledgements. This research is supported by A*STAR under its PROGRAM-MATIC FUND (Grant No. A20H4g2141).

References

1. Berthelot, D., Carlini, N., Goodfellow, I., Papernot, N., Oliver, A., Raffel, C.A.: MixMatch: a holistic approach to semi-supervised learning. In: Advances in Neural Information Processing Systems, pp. 5049–5059 (2019)
2. Chaudhuri, S., Chatterjee, S., Katz, N., Nelson, M., Goldbaum, M.: Detection of blood vessels in retinal images using two-dimensional matched filters. IEEE Trans. Med. Imaging **8**(3), 263–269 (1989)
3. Chutatape, O., Zheng, L., Krishnan, S.M.: Retinal blood vessel detection and tracking by matched Gaussian and Kalman filters. In: Proceedings of the 20th Annual International Conference of the IEEE Engineering in Medicine and Biology Society, vol. 20 Biomedical Engineering Towards the Year 2000 and Beyond (Cat. No. 98CH36286), vol. 6, pp. 3144–3149. IEEE (1998)
4. Fang, K., Li, W.-J.: DMNet: difference minimization network for semi-supervised segmentation in medical images. In: Martel, A.L., et al. (eds.) MICCAI 2020. LNCS, vol. 12261, pp. 532–541. Springer, Cham (2020). https://doi.org/10.1007/978-3-030-59710-8_52

5. Gao, S.S., et al.: Compensation for reflectance variation in vessel density quantification by optical coherence tomography angiography. Invest. Ophthalmol. Vis. Sci. **57**(10), 4485–4492 (2016)
6. Giarratano, Y., et al.: Automated segmentation of optical coherence tomography angiography images: benchmark data and clinically relevant metrics. Transl. Vis. Sci. Technol. **9**(13), 5 (2020)
7. Godard, C., Mac Aodha, O., Firman, M., Brostow, G.J.: Digging into self-supervised monocular depth prediction (October 2019)
8. Gu, Z., et al.: CE-Net: context encoder network for 2D medical image segmentation. IEEE Trans. Med. Imaging **38**(10), 2281–2292 (2019)
9. Khan, S., Shahin, A.H., Villafruela, J., Shen, J., Shao, L.: Extreme points derived confidence map as a cue for class-agnostic interactive segmentation using deep neural network. In: Shen, D., et al. (eds.) MICCAI 2019. LNCS, vol. 11765, pp. 66–73. Springer, Cham (2019). https://doi.org/10.1007/978-3-030-32245-8_8
10. Li, S., Zhang, C., He, X.: Shape-aware semi-supervised 3D semantic segmentation for medical images. In: Martel, A.L., et al. (eds.) MICCAI 2020. LNCS, vol. 12261, pp. 552–561. Springer, Cham (2020). https://doi.org/10.1007/978-3-030-59710-8_54
11. Liu, X., Van De Weijer, J., Bagdanov, A.D.: Exploiting unlabeled data in CNNs by self-supervised learning to rank. IEEE Trans. Pattern Anal. Mach. Intell. **41**(8), 1862–1878 (2019)
12. Ma, Y., et al.: ROSE: a retinal OCT-angiography vessel segmentation dataset and new model. arXiv preprint arXiv:2007.05201 (2020)
13. Mou, L., et al.: CS-Net: channel and spatial attention network for curvilinear structure segmentation. In: Shen, D., et al. (eds.) MICCAI 2019. LNCS, vol. 11764, pp. 721–730. Springer, Cham (2019). https://doi.org/10.1007/978-3-030-32239-7_80
14. Paszke, A., et al.: PyTorch: an imperative style, high-performance deep learning library. In: Wallach, H., Larochelle, H., Beygelzimer, A., d'Alché-Buc, F., Fox, E., Garnett, R. (eds.) Advances in Neural Information Processing Systems 32, pp. 8024–8035. Curran Associates, Inc. (2019)
15. Qu, H., et al.: Weakly supervised deep nuclei segmentation using partial points annotation in histopathology images. IEEE Trans. Med. Imaging **39**(11), 3655–3666 (2020)
16. Ronneberger, O., Fischer, P., Brox, T.: U-Net: convolutional networks for biomedical image segmentation. In: Navab, N., Hornegger, J., Wells, W.M., Frangi, A.F. (eds.) MICCAI 2015. LNCS, vol. 9351, pp. 234–241. Springer, Cham (2015). https://doi.org/10.1007/978-3-319-24574-4_28
17. Staal, J., Abràmoff, M.D., Niemeijer, M., Viergever, M.A., Van Ginneken, B.: Ridge-based vessel segmentation in color images of the retina. IEEE Trans. Med. Imaging **23**(4), 501–509 (2004)
18. Tarvainen, A., Valpola, H.: Mean teachers are better role models: weight-averaged consistency targets improve semi-supervised deep learning results. In: Advances in Neural Information Processing Systems, pp. 1195–1204 (2017)
19. Tong, S., Koller, D.: Support vector machine active learning with applications to text classification. J. Mach. Learn. Res. **2**, 45–66 (2001)
20. Wang, B., Qiu, S., He, H.: Dual encoding U-Net for retinal vessel segmentation. In: Shen, D., et al. (eds.) MICCAI 2019. LNCS, vol. 11764, pp. 84–92. Springer, Cham (2019). https://doi.org/10.1007/978-3-030-32239-7_10

21. Wang, G., Aertsen, M., Deprest, J., Ourselin, S., Vercauteren, T., Zhang, S.: Uncertainty-guided efficient interactive refinement of fetal brain segmentation from stacks of MRI slices. In: Martel, A.L., et al. (eds.) MICCAI 2020. LNCS, vol. 12264, pp. 279–288. Springer, Cham (2020). https://doi.org/10.1007/978-3-030-59719-1_28

22. Wang, Y., et al.: Double-uncertainty weighted method for semi-supervised learning. In: Martel, A.L., et al. (eds.) MICCAI 2020. LNCS, vol. 12261, pp. 542–551. Springer, Cham (2020). https://doi.org/10.1007/978-3-030-59710-8_53

23. Yousefi, S., Liu, T., Wang, R.K.: Segmentation and quantification of blood vessels for oct-based micro-angiograms using hybrid shape/intensity compounding. Microvasc. Res. **97**, 37–46 (2015)

24. Zhang, H., Cisse, M., Dauphin, Y.N., Lopez-Paz, D.: mixup: beyond empirical risk minimization. arXiv preprint arXiv:1710.09412 (2017)

25. Zhang, M., Dong, B., Li, Q.: Deep active contour network for medical image segmentation. In: Martel, A.L., et al. (eds.) MICCAI 2020. LNCS, vol. 12264, pp. 321–331. Springer, Cham (2020). https://doi.org/10.1007/978-3-030-59719-1_32

26. Zhang, M., Yu, F., Zhao, J., Zhang, L., Li, Q.: BEFD: boundary enhancement and feature denoising for vessel segmentation. In: Martel, A.L., et al. (eds.) MICCAI 2020. LNCS, vol. 12265, pp. 775–785. Springer, Cham (2020). https://doi.org/10.1007/978-3-030-59722-1_75

27. Zhou, Y., Chen, H., Lin, H., Heng, P.-A.: Deep semi-supervised knowledge distillation for overlapping cervical cell instance segmentation. In: Martel, A.L., et al. (eds.) MICCAI 2020. LNCS, vol. 12261, pp. 521–531. Springer, Cham (2020). https://doi.org/10.1007/978-3-030-59710-8_51

Unsupervised Network Learning for Cell Segmentation

Liang Han and Zhaozheng Yin$^{(\boxtimes)}$

Stony Brook University, New York, USA
{liahan,zyin}@cs.stonybrook.edu

Abstract. Cell segmentation is a fundamental and critical step in numerous biomedical image studies. For the fully-supervised cell segmentation algorithms, although highly effective, a large quantity of high-quality training data is required, which is usually labor-intensive to produce. In this work, we formulate the unsupervised cell segmentation as a slightly under-constrained problem, and present the Unsupervised Segmentation network learning by Adversarial Reconstruction (USAR), a novel model able to train cell segmentation networks without any annotation. The key idea is to leverage adversarial learning paradigm to train the segmentation network by adversarially reconstructing the input images based on their segmentation results generated by the segmentation network. The USAR model demonstrates its promising application on training segmentation networks in an unsupervised manner, on two benchmark datasets. The implementation of this project can be found at https://github.com/LiangHann/USAR.

Keywords: Unsupervised learning · Cell segmentation · Adversarial image reconstruction

1 Introduction

Cell segmentation, a fundamental task in many biological and medical discoveries, enables accurate and quantitative analysis of cells. Among all segmentation algorithms to date, deep Convolutional Neural Network (CNN) based approaches give the most promising results in cell segmentation problem, such as Fully Convolutional Network (FCN) [1], U-Net [2], and UNet++ [3], to name a few. Unfortunately, to successfully train these CNN-based models for cell segmentation, we have to label a large number of training images with cell mask annotations, which is rather laborious and expensive.

Motivated by the demand of alleviating or even totally avoiding the workload on collecting training masks, in this work, we propose a fully unsupervised model to train cell segmentation networks, named Unsupervised Segmentation network learning by Adversarial Reconstruction (USAR). It relies on the idea of adversarial learning paradigm [4], learning the segmentation network by adversarially reconstructing the input images based on their segmentation results generated

© Springer Nature Switzerland AG 2021
M. de Bruijne et al. (Eds.): MICCAI 2021, LNCS 12901, pp. 282–292, 2021.
https://doi.org/10.1007/978-3-030-87193-2_27

by the segmentation network. Specifically, a generator, which consists of a segmentation network to generate foreground cell masks and a decoder to generate the background image, is designed to reconstruct the cell image, which is compared to the original input image by a discriminator. The contribution of this work is mainly threefold:

- We formulate the unsupervised cell image segmentation as a slight underconstrained problem, i.e., solving $N + 1$ unknowns (N cell mask images + 1 background image) from N equations (N input images), compared to solving $3N$ unknowns (N mask images, N foreground images, N background images) from N input images.
- We present an unsupervised network learning scheme with adversarial reconstruction to train cell segmentation networks without any annotation.
- Experiments demonstrate the effectiveness of the proposed model, which outperforms the state-of-the-arts unsupervised segmentation by a large margin.

2   Related Work

Cell segmentation in microscopy images are to segment cell pixels from the background pixels and group the cell pixels into detected cells. Recently, deep learning has been firmly established as a robust tool in biomedical image segmentation [2,3,5–7]. In [2], U-Net is proposed to segment the neuronal structures in 2D electron microscopy images. [3] further improves the U-Net by introducing a nested structure. In [5], V-Net is derived for 3D medical image segmentation based on a volumetric neural network. In [6], dense V-Net is designed for multi-organ segmentation on abdominal CT images. In [7], a pyramid-based fully convolutional network is proposed to segment cells in a coarse-to-fine manner. Though impressive performance has achieved with these methods, an extensive amount of annotated data are required to train these deep learning models.

To reduce the label acquisition cost, researchers mainly focus on two directions: (1) training with fewer labeled images [8–16]; (2) training with weaker supervision [17–20]. In the first direction, data augmentation and transfer learning are two popular and effective strategies which either increasing labeled images for training by rotating, cropping and flipping the original labeled training images [8–10] or leveraging other available labeled datasets [11–13]. Besides, semi-supervised methods [14–16] reduce the demand of labeled training data by training the networks with a few pixel-level labeled images and a large amount of images without any annotation information. In the second direction, instead of using the cell mask annotations, weaker annotations (e.g., lesion location information without accurate boundary in [17], user interaction for point clicks in [18], etc.) are collected to train the networks.

To completely avoid the annotation workload, unsupervised segmentation is investigated [21–26]. In [23,24], the generative adversarial network is leveraged to automatically synthesize some labelled data for training segmentation networks. Unfortunately, prior domain knowledge needed for training (e.g., edge diagrams [23] or cell shape [24]) heavily limits the generalization of the proposed methods.

CNN is adopted in [21,22] to perform the unsupervised segmentation by exploiting the intensity similarity and spatial continuity. These two works, however, fail to model the statistical properties (e.g., cell size, shape, etc.) which are critical for meaningful segmentation. In [25,26], a segmentation network is trained in an adversarial manner, relying on the idea that it should be possible to change the textures or colors of the objects without changing the overall distribution of the dataset. The formulated problem in [25,26] is quite under-constrained, which can hardly guarantee accurate segmentation. To overcome these challenges, in this work, we mathematically formulate the unsupervised segmentation as a less under-constrained system, and propose an unsupervised segmentation network learning scheme to train cell segmentation networks without any annotation.

3   Methodology

In this section, first we derive the mathematical formulation of the unsupervised segmentation problem in Subsect. 3.1. Then, based on the derived formulation, we describe our designed USAR model for the unsupervised segmentation network learning in Subsect. 3.2. After that, we introduce how to avoid trivial solutions on the network learning in Subsect. 3.3, followed by the objective function used to train the network in Subsect. 3.4. Finally, pseudo label is introduced to further refine the segmentation network in Subsect. 3.5.

3.1   Formulation of the Unsupervised Segmentation Problem

Given an input cell image set $\{I_i\}_{i=1}^N$ to be segmented, where N is the number of images in this set, each image I_i can be decomposed into foreground F_i (i.e., cells) and background B_i with a foreground mask M_i:

$$I_i = M_i * F_i + (1 - M_i) * B_i, \tag{1}$$

where $*$ denotes the element-wise multiplication. For cell segmentation, the goal is to obtain the foreground mask M_i for image I_i. But, in Eq. 1, all the three variables (M_i, F_i, and B_i) on the right-hand side are unknown. Thus, solving the cell mask M_i from Eq. 1 is severely under-constrained. To alleviate the under-constrained challenge, we replace the foreground F_i with the original image I_i as the foreground part of I_i should be identical to F_i. Now, Eq. 1 becomes:

$$I_i = M_i * I_i + (1 - M_i) * B_i. \tag{2}$$

Equation 2 has two unknown variables (M_i and B_i), which is less than the three unknowns in Eq. 1, but it is still under-constrained.

When performing a specific biological experiment to analyze cells' properties in still images or analyze their dynamics in image sequences, we know that the background information (i.e., the cell culture medium) is relatively stable, thus, we can make an assumption that the cell images in a specific biological experiment share similar background. With this assumption, Eq. 2 can be re-written as:

$$\begin{cases} I_1 = M_1 * I_1 + (1 - M_1) * B \\ \quad \cdots \\ I_N = M_N * I_N + (1 - M_N) * B. \end{cases} \qquad (3)$$

Now we have N equations to solve $N + 1$ unknown variables, and the under-constrained dilemma is greatly alleviated. Though some trivial solutions (e.g., $M_i = 1$) still exist in Eq. 3, we will discuss how to avoid them in Sect. 3.3.

3.2 USAR for Unsupervised Segmentation Network Learning

After mathematically formulating the unsupervised problem (i.e., Eq. 3), we pro-pose the Unsupervised Segmentation network learning by Adversarial Recon-struction (USAR) (Fig. 1) to train the segmentation network without annota-tion. Given an input cell image set $\{I_i\}_{i=1}^{N}$, a segmentation network (e.g., U-Net) is adopted to generate the corresponding cell masks $\{M_i\}_{i=1}^{N}$. Meanwhile, a ran-domly sampled noise vector $\mathbf{z}$ is input into a decoder to synthesize only one background B for all the input images $\{I_i\}_{i=1}^{N}$. The foreground (cell) regions are calculated by element-wisely multiplying the original image set $\{I_i\}_{i=1}^{N}$ and the non-cell background regions are obtained by element-wisely multiplying the synthesized background B and the inverse masks (i.e., $\{1 - M_i\}_{i=1}^{N}$). Finally, the reconstructed images $\{\hat{I}_i\}_{i=1}^{N}$ are obtained by combining the cell regions and the non-cell background regions with an element-wise summation. The segmen-tation network and the background decoder constitute the generator for image reconstruction, while a discriminator is built to regulate the generator such that the distribution of the reconstructed images is aligned to the original ones.

In a nutshell, our proposed USAR model in Fig. 1 does not need any ground truth segmentation mask to train the segmentation network. The USAR model can be applied to train various segmentation networks (e.g., U-Net, FCN, Pyramid-FCN). Without loss of generality, we use U-Net as an example of seg-mentation network in this paper. During testing, only the trained segmentation network is used to perform the cell segmentation.

3.3 Avoid Trivial Solutions in Unsupervised Network Learning

For Eq. 3, there are two possible trivial segmentation: (1) a **"full"** segmenta-tion. The U-Net segments the whole cell image as foreground (i.e., $M_i = 1$). In this case, any synthesized background image B can satisfy the reconstruction; (2) an **"empty"** segmentation. The U-Net segments the whole cell image as background (i.e., $M_i = 0$). In this case, any synthesized background image B which aligns with the data distribution of $\{I_i\}_{i=1}^{N}$, can satisfy the unsupervised training, since a default discriminator can only gauge the distribution-wise dif-ference, but has no constraint on the content of the reconstructed images $\{\hat{I}_i\}_{i=1}^{N}$. To avoid these two trivial solutions, we propose two constraints accordingly:

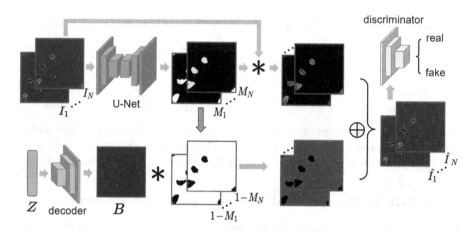

Fig. 1. Overview of the USAR model for unsupervised cell segmentation network learning. $*$ denotes the element-wise multiplication, and $\oplus$ is the element-wise summation.

(1) **Sparsity for "non-full" segmentation:** To prevent the U-Net from generating a "full" segmentation, we add a sparse constraint on the generated cell mask, $\|M_i\|_1$, which forces that only some parts of the generated mask are non-zero (i.e., foreground), while the major parts are zero (i.e., background).

(2) **Content loss for "non-empty" segmentation:** To avoid the "empty" segmentation, we add a content loss between the reconstructed images $\{\hat{I}_i\}_{i=1}^N$ and the original ones $\{I_i\}_{i=1}^N$, which is defined as the difference between the extracted image features $f(\hat{I}_i)$ and $f(I_i)$ (where f is a pre-trained CNN for feature extraction, e.g., VGG16 pre-trained on the ImageNet used in this paper). This forces that each reconstructed image $\hat{I}_i$ should have the same content information with the original I_i. Under this constraint, the generated masks $\{M_i\}_{i=1}^N$ can not be "empty" (i.e., $M_i = \mathbf{0}$), otherwise we would have $\hat{I}_1 = ... = \hat{I}_N = B = I_1 = ... = I_N$ in Eq. 3.

3.4   Objective Function for Learning the Segmentation Network

To this end, the segmentation network can be learnt by alternatively optimizing the following objective functions:

$$\min_G \left\{ -\frac{1}{N}\sum_{i=1}^N [D(G(I_i, \mathbf{z}))] + \lambda_s \cdot \frac{1}{N}\sum_{i=1}^N \|M_i\|_1 + \lambda_p \cdot \frac{1}{N}\sum_{i=1}^N \|f(G(I_i, \mathbf{z})) - f(I_i)\|_2^2 \right\},$$

$$\tag{4}$$

$$\max_D \left\{ \frac{1}{N}\sum_{i=1}^N [D(G(I_i, \mathbf{z}))] - \frac{1}{N}\sum_{i=1}^N [D(I_i)] \right\}, \tag{5}$$

where G and D stand for the generator (U-Net and background decoder in Fig. 1) and the discriminator, respectively. $\mathbf{z}$ is a noise vector to generate the background image B in Fig. 1. $G(I_i, \mathbf{z})$ is the reconstructed image $\hat{I}_i$. λ_s and λ_p are hyper-parameters to balance the loss terms (λ_s and λ_p are determined by validation set).

3.5 Pseudo Labels to Continually Refine the Segmentation Network

To further improve the segmentation network, we can leverage the self-training strategy based on pseudo labels. Specifically, after obtaining the segmentation mask M_i by alternatively optimizing Eq. 4 and Eq. 5, we define the pseudo label ($\tilde{M}_i$) of image i as those high confident cell regions in M_i (e.g., big blobs in M_i). Then, we use the pseudo labels to refine the segmentation network. Mathematically, the generator's objective function for the refinement is:

$$
\min_{G} \Big\{ -\frac{1}{N} \sum_{i=1}^{N} [D(G(I_i, \mathbf{z}))] + \lambda_s \cdot \frac{1}{N} \sum_{i=1}^{N} \|M_i\|_1
$$
$$
+ \lambda_p \cdot \frac{1}{N} \sum_{i=1}^{N} \|f(G(I_i, \mathbf{z})) - f(I_i)\|_2^2 + \lambda_{CE} \cdot \frac{1}{N} \sum_{i=1}^{N} CE(M_i, \tilde{M}_i) \Big\},
$$

$$(6)$$

where CE stands for the cross-entropy loss, and λ_{CE} is a hyper-parameter to control the segmentation loss (λ_{CE} is determined by validation set).

4 Experimental Results

In this section, we first introduce the data and evaluation metric we used in our experiments. Then, we design some experiments to evaluate our USAR model.

4.1 Data and Evaluation Metric

Two public segmentation data sets[1] are used to evaluate the proposed USAR model. The first data set "PhC-U373" contains Glioblastoma-astrocytoma U373 cells on a polyacrylimide substrate recorded by phase contrast microscopy. It contains 196 unlabeled images which are used for training, and 34 annotated images, among which 10 are used for validation and 24 for testing. The second data set "DIC-HeLa" are HeLa cells on a flat glass recorded by differential interference contrast (DIC) microscopy. The training/validation/test set splitting is 230/48/120. The image size in our experiments is $512 * 512$.

The segmentation metric $Dice$ is used for quantitative evaluation. For unsupervised segmentation, we have no mapping information of segmentation mask to cell, so we do not compute cell-level $Dice$, but pixel-level $Dice$ for each image. $Dice$ falls in $[0, 1]$, and a larger value indicates a better segmentation.

---

[1] http://celltrackingchallenge.net/2d-datasets/.

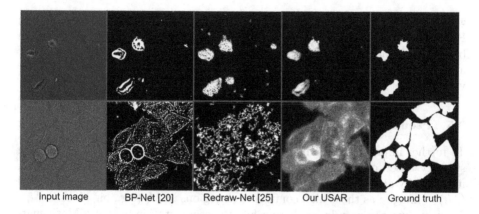

Fig. 2. Visual comparison between the proposed USAR and state-of-the-art unsupervised segmentation models. First row: "Phc-U373", second row: "DIC-HeLa".

4.2   Evaluation Results

Comparison with Unsupervised Algorithms: We first try a traditional unsupervised segmentation method, Otsu [32], on the datasets, but get unsatisfactory segmentation results. This is because in these two datasets, some cells have quite similar pixel intensities with the background, especially in the "DIC-HeLa" dataset. Then we perform the comparison between the proposed USAR model and two state-of-the-art unsupervised methods: BP-Net [21] and Redraw-Net [26]. As shown in Table 1, our proposed USAR model achieves the best segmentation results on both "Phc-U373" and "DIC-HeLa" data sets, with large improvement over BP-Net and Redraw-Net, especially on the more challenging "DIC-HeLa". For fair comparison, we use the same U-Net structure in these two models and our proposed USAR. BP-Net trains the U-Net by back-propagating self-constrained feature similarity and spatial continuity, while ignores the cell's statistical property (e.g., cell shape, size, etc.). Thus, BP-Net faces challenges for cell segmentation, especially when the cell-background contrast is low. Redraw-Net trains the U-Net by redrawing the foreground content without changing the image distribution. It can model the cell's statistical property better, however, its segmentation problem formulation is severely ill-posed, and without proper constraints, the segmentation training is not stable or accurate. Figure 2 presents some visual comparison between these three models, which also illustrates the superiority of the proposed USAR model.

Ablation Studies: First, we evaluate the effectiveness of constraint terms for avoiding trivial solutions. We denote "Ours w/o Sparsity" as the variant that is without the sparsity constraint, and "Ours w/o Content Loss" as the variant without the content constraint. Table 1 shows that the trained segmentation network without constraints performs poorly. For "Ours w/o Content Loss", we

Table 1. Quantitative evaluation of different segmentation methods with *Dice* (%).

Settings	Method	Phc-U373	DIC-HeLa
Unsupervised	Otsu thresholding [32]	46.95	31.24
	BP-Net [21]	68.41	52.10
	Redraw-Net [26]	77.95	66.14
	Ours w/o Sparsity	20.56	63.03
	Ours w/o Content Loss	0.0	0.0
	Ours w/o Pseudo Label	83.12	79.61
	Ours	**85.86**	**83.01**
Fully supervised	U-Net	95.83	87.64

get the "empty segmentation" (trivial solution), and the *Dice* is 0. Figure 3 shows that without the training constraint, trivial segmentation results are obtained.

Secondly, we evaluate the effectiveness of the pseudo labels. The segmentation results without pseudo labels ("Ours w/o Pseudo Label" in Table 1,) has ∼ 3% lower Dice, compared to the refined segmentation network by pseudo labels.

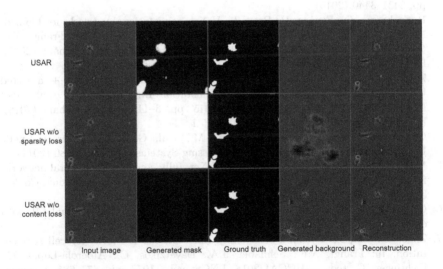

Fig. 3. Ablation study on training constraints.

Future Work - Close the Gap with Fully-Supervised Training: Our unsupervised learning has the big advantage to reduce the annotation effort to zero, compared to the costly annotation in fully-supervised training. The accuracy of our unsupervised learning is close to that of fully-supervised training, as shown in Table 1. To bridge the gap, we can consider semi-supervised learning [27–29] and active learning with human in the loop [30, 31] as future work to improve the accuracy with a little human effort.

5 Conclusion

Cell segmentation is a fundamental task in many biomedical image analyses. Motivated by alleviating the annotation workload, in this work, we formulate the unsupervised cell segmentation as a slightly under-constrained system, and present the Unsupervised Segmentation network learning by Adversarial Reconstruction (USAR), a novel model able to train cell segmentation networks without any annotation. Validated on two public cell segmentation datasets, the USAR model shows its promising results to train high-quality segmentation networks in an unsupervised manner.

Acknowledgement. This project was supported by Stony Brook University - Brookhaven National Laboratory (SBU-BNL) seed grant on annotation-efficient deep learning.

References

1. Long, J., Shelhamer, E., Darrell, T.: Fully convolutional networks for semantic segmentation. In: IEEE Conference on Computer Vision and Pattern Recognition, pp. 3431–3440 (2015)
2. Ronneberger, O., Fischer, P., Brox, T.: U-Net: convolutional networks for biomedical image segmentation. In: Navab, N., Hornegger, J., Wells, W.M., Frangi, A.F. (eds.) MICCAI 2015. LNCS, vol. 9351, pp. 234–241. Springer, Cham (2015). https://doi.org/10.1007/978-3-319-24574-4_28
3. Zhou, Z., Rahman Siddiquee, M.M., Tajbakhsh, N., Liang, J.: UNet++: a nested U-Net architecture for medical image segmentation. In: Stoyanov, D., et al. (eds.) DLMIA/ML-CDS -2018. LNCS, vol. 11045, pp. 3–11. Springer, Cham (2018). https://doi.org/10.1007/978-3-030-00889-5_1
4. Goodfellow, I., Pouget-Abadie, J., Mirza, M., et al.: Generative adversarial nets. In: Advances in Neural Information Processing Systems, pp. 2672–2680 (2014)
5. Milletari, F., Navab, N., Ahmadi, S.A.: V-Net: fully convolutional neural networks for volumetric medical image segmentation. In IEEE Fourth International Conference on 3D Vision (3DV), pp. 565–571 (2016)
6. Gibson, E., et al.: Automatic multi-organ segmentation on abdominal CT with dense v-networks. IEEE Trans. Med. Imaging **37**, 1822–1834 (2018)
7. Zhao, T., Yin, Z.: Pyramid-based fully convolutional networks for cell segmentation. In: Frangi, A.F., Schnabel, J.A., Davatzikos, C., Alberola-López, C., Fichtinger, G. (eds.) MICCAI 2018. LNCS, vol. 11073, pp. 677–685. Springer, Cham (2018). https://doi.org/10.1007/978-3-030-00937-3_77
8. Zhong, Z., Zheng, L., Kang, G., et al.: Random erasing data augmentation. In: AAAI Conference on Artificial Intelligence, pp. 1–8 (2020)
9. Cubuk, E., Zoph, B., Mane, D., et al.: AutoAugment: learning augmentation strategies from data. In: IEEE International Conference on Computer Vision, pp. 113–123 (2019)
10. Zhao, A., Balakrishnan, G., Durand, F., et al.: Data augmentation using learned transformations for one-shot medical image segmentation. In: IEEE Conference on Computer Vision and Pattern Recognition, pp. 8543–8553 (2019)

11. Van Opbroek, A., Achterberg, H., Vernooij, M., et al.: Transfer learning for image segmentation by combining image weighting and kernel learning. IEEE Trans. Med. Imaging **38**(1), 213–224 (2018)

12. Sun, R., Zhu, X., Wu, C., et al.: Not all areas are equal: transfer learning for semantic segmentation via hierarchical region selection. In: IEEE International Conference on Computer Vision, pp. 4360–4369 (2019)

13. Majurski, M., Manescu, P., Padi, S., et al.: Cell image segmentation using generative adversarial networks, transfer learning, and augmentations. In: IEEE Conference on Computer Vision and Pattern Recognition Workshops (2019)

14. Wang, Y., et al.: Double-uncertainty weighted method for semi-supervised learning. In: Martel, A.L., et al. (eds.) MICCAI 2020. LNCS, vol. 12261, pp. 542–551. Springer, Cham (2020). https://doi.org/10.1007/978-3-030-59710-8_53

15. Fang, K., Li, W.-J.: DMNet: difference minimization network for semi-supervised segmentation in medical images. In: Martel, A.L., et al. (eds.) MICCAI 2020. LNCS, vol. 12261, pp. 532–541. Springer, Cham (2020). https://doi.org/10.1007/978-3-030-59710-8_52

16. Cao, X., Chen, H., Li, Y., et al.: Uncertainty aware temporal-ensembling model for semi-supervised ABUS mass segmentation. IEEE Trans. Med. Imaging **40**(1), 431–443 (2020)

17. Zheng, H., Zhuang, Z., Qin, Y., Gu, Y., Yang, J., Yang, G.-Z.: Weakly supervised deep learning for breast cancer segmentation with coarse annotations. In: Martel, A.L., et al. (eds.) MICCAI 2020. LNCS, vol. 12264, pp. 450–459. Springer, Cham (2020). https://doi.org/10.1007/978-3-030-59719-1_44

18. Roth, H.R., Yang, D., Xu, Z., et al.: Going to extremes: weakly supervised medical image segmentation. arXiv preprint arXiv:2009.11988 (2020)

19. Kervadec, H., Dolz, J., Tang, M., et al.: Constrained-CNN losses for weakly supervised segmentation. Med. Image Anal. **54**, 88–99 (2019)

20. Cai, J., et al.: Accurate weakly-supervised deep lesion segmentation using large-scale clinical annotations: slice-propagated 3D mask generation from 2D RECIST. In: Frangi, A.F., Schnabel, J.A., Davatzikos, C., Alberola-López, C., Fichtinger, G. (eds.) MICCAI 2018. LNCS, vol. 11073, pp. 396–404. Springer, Cham (2018). https://doi.org/10.1007/978-3-030-00937-3_46

21. Kanezaki, A.: Unsupervised image segmentation by backpropagation. In: 2018 IEEE International Conference on Acoustics, Speech and Signal Processing (ICASSP), pp. 1543–1547 (2018)

22. Kim, W., Kanezaki, A., Tanaka, M.: Unsupervised learning of image segmentation based on differentiable feature clustering. IEEE Trans. Image Process. **29**, 8055–8068 (2020)

23. Sivanesan, U., et al.: Unsupervised medical image segmentation with adversarial networks: from edge diagrams to segmentation maps. arXiv:1911.05140 (2019)

24. Hou, L., et al.: Robust histopathology image analysis: to label or to synthesize? In: IEEE Conference on Computer Vision and Pattern Recognition, pp. 8533–8542 (2019)

25. Chen, M., Artières, T., Denoyer, L.: Unsupervised object segmentation by redrawing. arXiv preprint arXiv:1905.13539 (2019)

26. Song, Y., Zhou, T., Teoh, J.Y.-C., Zhang, J., Qin, J.: Unsupervised learning for CT image segmentation via adversarial redrawing. In: Martel, A.L., et al. (eds.) MICCAI 2020. LNCS, vol. 12264, pp. 309–320. Springer, Cham (2020). https://doi.org/10.1007/978-3-030-59719-1_31

27. Bortsova, G., Dubost, F., Hogeweg, L., Katramados, I., de Bruijne, M.: Semi-supervised medical image segmentation via learning consistency under transformations. In: Shen, D., et al. (eds.) MICCAI 2019. LNCS, vol. 11769, pp. 810–818. Springer, Cham (2019). https://doi.org/10.1007/978-3-030-32226-7_90

28. Xie, Y., Zhang, J., Liao, Z., Verjans, J., Shen, C., Xia, Y.: Pairwise relation learning for semi-supervised gland segmentation. In: Martel, A.L., et al. (eds.) MICCAI 2020. LNCS, vol. 12265, pp. 417–427. Springer, Cham (2020). https://doi.org/10.1007/978-3-030-59722-1_40

29. Li, S., Zhang, C., He, X.: Shape-aware semi-supervised 3D semantic segmentation for medical images. In: Martel, A.L., et al. (eds.) MICCAI 2020. LNCS, vol. 12261, pp. 552–561. Springer, Cham (2020). https://doi.org/10.1007/978-3-030-59710-8_54

30. Li, H., Yin, Z.: Attention, suggestion and annotation: a deep active learning framework for biomedical image segmentation. In: Martel, A.L., et al. (eds.) MICCAI 2020. LNCS, vol. 12261, pp. 3–13. Springer, Cham (2020). https://doi.org/10.1007/978-3-030-59710-8_1

31. Ravanbakhsh, M., Tschernezki, V., Last, F., et al.: Human-machine collaboration for medical image segmentation. In: ICASSP 2020–2020 IEEE International Conference on Acoustics, Speech and Signal Processing (ICASSP), pp. 1040–1044 (2020)

32. Otsu, N.: A threshold selection method from gray-level histograms. IEEE Trans. Syst. Man Cybern. **9**(1), 62–66 (1979)

MT-UDA: Towards Unsupervised Cross-modality Medical Image Segmentation with Limited Source Labels

Ziyuan Zhao[1,2], Kaixin Xu[2], Shumeng Li[1,2], Zeng Zeng[2(✉)], and Cuntai Guan[1]

[1] Nanyang Technological University, Singapore, Singapore
[2] Institute for Infocomm Research, A*STAR, Singapore, Singapore
zengz@i2r.a-star.edu.sg

Abstract. The success of deep convolutional neural networks (DCNNs) benefits from high volumes of annotated data. However, annotating medical images is laborious, expensive, and requires human expertise, which induces the label scarcity problem. Especially When encountering the domain shift, the problem becomes more serious. Although deep unsupervised domain adaptation (UDA) can leverage well-established source domain annotations and abundant target domain data to facilitate cross-modality image segmentation and also mitigate the label paucity problem on the target domain, the conventional UDA methods suffer from severe performance degradation when source domain annotations are scarce. In this paper, we explore a challenging UDA setting - limited source domain annotations. We aim to investigate how to efficiently leverage unlabeled data from the source and target domains with limited source annotations for cross-modality image segmentation. To achieve this, we propose a new label-efficient UDA framework, termed MT-UDA, in which the student model trained with limited source labels learns from unlabeled data of both domains by two teacher models respectively in a semi-supervised manner. More specifically, the student model not only distills the intra-domain semantic knowledge by encouraging prediction consistency but also exploits the inter-domain anatomical information by enforcing structural consistency. Consequently, the student model can effectively integrate the underlying knowledge beneath available data resources to mitigate the impact of source label scarcity and yield improved cross-modality segmentation performance. We evaluate our method on MM-WHS 2017 dataset and demonstrate that our approach outperforms the state-of-the-art methods by a large margin under the source-label scarcity scenario.

Keywords: Segmentation · Unsupervised domain adaptation · Semi-supervised learning · Self-ensembling

1 Introduction

Deep convolutional neural networks (DCNNs) have obtained promising performance on medical image segmentation tasks [17,18], which further promotes the

© Springer Nature Switzerland AG 2021
M. de Bruijne et al. (Eds.): MICCAI 2021, LNCS 12901, pp. 293–303, 2021.
https://doi.org/10.1007/978-3-030-87193-2_28

development of automated medical image analysis. DCNNs are data-hungry and require large amounts of well-annotated data, however, in real-world clinical settings, medical image annotations are pricey and labor-intensive, which require extensive domain knowledge from biomedical experts. This leads to that scarce annotations are available for training DCNNs, *i.e.*, label scarcity.

To alleviate the burden on human annotation, plenty of methods beyond supervised learning have been proposed for improving label efficiency on medical imaging [19], including self-supervised learning [27], semi-supervised learning [2,29] and disentangled representation learning [3]. In recent years, semi-supervised learning (SSL) methods based on the self-ensembling strategy [7,14,20] have received much attention in medical image analysis, achieving state-of-the-art results in many SSL benchmarks. For instance, Laine and Aila [14] propose Temporal Ensembling to enforce the consistent outputs of the network-in-training across different epochs. Tarvainen and Valpola [20] build the mean teacher (MT) model based on the exponential moving average (EMA) of the weights of the student network, forcing the prediction consistency and further boosting the model performance. Subsequently, many studies have been devoted to leveraging abundant unlabeled data based on MT to mitigate the paucity-of-label problem in biomedical image segmentation [8,15,26]. These methods, however, are presented for label-efficient learning on a single partially labeled dataset, failing to use cross-domain information well when using multi-domain datasets.

On the other hand, given the various imaging modalities with different physical principles, such as CT and MR, the domain shift problem is severe in cross-modality image segmentation, resulting in significantly reduced performance when applying well-trained DCNNs on one domain (*e.g.*, MR) to another domain (*e.g.*, CT), especially in the absence of target labels. To tackle this serious issue, much research has been devoted to investigating unsupervised domain adaptation (UDA) for minimizing the discrepancy between the source and target domains, consequently boosting the generalization ability on the target domain for cross-modality medical image segmentation [11,25]. Inspired by the great success of generative adversarial networks (GANs) on image-to-image translation [1,12,13], many approaches have been developed with adversarial learning from different perspectives for domain alignment, including image-level adaptation [4,28], feature-level adaptation [9,10,22] and their mixtures [5,6]. For example, Chen *et al.* design a synergistic image and feature adaptation model [6], which achieves the state-of-the-art performance in UDA for cross-modality medical image segmentation. Despite the success of adversarial learning in UDA, these methods heavily rely on abundant source labels, which become sub-optimal when only limited source labels are available in clinical deployment.

These motivate us to advocate studying a practical, challenging, and different UDA setting from the past, where only limited source labels are accessible. In this paper, we investigate the feasibility of integrating SSL into UDA under source label scarcity and propose a novel label-efficient UDA framework for cross-modality medical image segmentation. We first present a dual cycle alignment

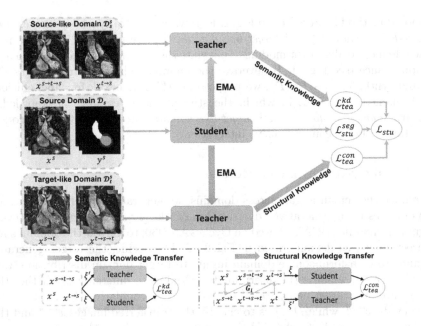

Fig. 1. Overall framework of our proposed MT-UDA. The student model learns from labeled source samples $\mathcal{D}_s^l$ by the $\mathcal{L}_{stu}^{seg}$ loss, and distills the intra-domain semantic knowledge and inter-domain anatomical information from *source-like domain* and *target-like domain* by $\mathcal{L}_{tea}^{kd}$ and $\mathcal{L}_{tea}^{con}$, simultaneously.

module (DCAM) to bridge the appearance gap across domains, synthesizing *source-like domain* images and *target-like domain* images via adversarial learning [12]. We further develop an MT framework [20] for UDA, named MT-UDA, to exploit the knowledge from both intermediate domains. In MT-UDA, the student model distills the intra-domain semantic knowledge by encouraging the prediction consistency of the source domain and exploits the inter-domain anatomical information by enforcing the structural consistency across domains. We evaluate the proposed MT-UDA on a public multi-modality cardiac image segmentation dataset, MM-WHS 2017, and demonstrate that our method outperforms the state-of-the-art methods by a lot under the challenging UDA scenario.

2   Methodology

Let $\mathcal{D}_s^l = \{(\mathbf{x}_i^s, y_i^s)\}_{i=1}^N$ and $\mathcal{D}_s^u = \{(\mathbf{x}_i^s)\}_{i=N+1}^M$ denote the labeled samples and unlabeled samples from source domain (*e.g.*, MR), respectively. In conventional UDA setting, abundant labeled source data $\mathcal{D}_s^l$ is given, *i.e.*, $N = M$. Differently, in our setting, only limited labeled source data $\mathcal{D}_s^l$ is used for UDA, *i.e.*, $N << M$, which is more practical and challenging. We aim to exploit $\mathcal{D}_s^l$, $\mathcal{D}_s^u$ and unlabeled samples $\mathcal{D}_t = \{(\mathbf{x}_i^t)\}_{i=1}^P$ from target domain (*e.g.*, CT) for UDA to improve the model performance on the target domain. The overview of the proposed method is presented in Fig. 1. Firstly, two sets of synthetic images, *i.e.*, *source-like domain* $\mathcal{D}_s^s$ and *target-like domain,* $\mathcal{D}_t^s$ are generated with the proposed dual cycle alignment module to alleviate the notorious domain discrepancy in appearance (see Fig. 2). To leverage the knowledge beneath real images $\mathcal{D}_s$, $\mathcal{D}_t$ and synthetic ones $\mathcal{D}_s^s$, $\mathcal{D}_t^s$, we propose an MT framework for label-efficient UDA, named MT-UDA, in which, the student model explore the knowledge beneath *source-like domain* and *target-like domain* through two teacher models simultaneously for comprehensive integration.

2.1 Dual Cycle Alignment Module

To reduce the semantic gap across domains, we generate synthetic samples for two domains using generative adversarial networks [12]. We design a dual cycle alignment module (DCAM) based on CycleGANs [30] to narrow the domain shift bidirectionally, as demonstrated in Fig. 2. To be specific, the target generator G_t aims to transform source domain inputs to target domain distribution, *i.e.*, $G_t(x^s) = x^{s \to t}$, whereas the discriminator D_t aims to differentiate whether the images are fake target images $x^{s \to t}$ or real ones x^t. Similarly, with x^t, G_s aims to generate $x^{t \to s}$, while D_s aims to classify the transferred images $x^{t \to s}$ and the original images x^s. In CycleGAN, a reverse generator is employed to impose a cycle consistency between source domain images x^s and reconstructed images $x^{s \to t \to s}$. It is noted that both the reverse generator and the source generator G_s aim to generate source-like images, therefore, we share the weights between them. In similar fashion, we refactor G_t to generate $x^{t \to s \to t}$. Different from CycleGAN, we further force the discriminator D_s to differentiate source images x^s, synthetic source images $x^{t \to s}$ or reconstructed source images $x^{s \to t \to s}$ in order to bridge the domain gap better, Similarly, we construct a powerful discriminator D_t. Finally, we can obtain two newly-augmented intermediate domains, *i.e.*, *source-like domain* $\mathcal{D}_s^s = \{x^{t \to s}, x^{s \to t \to s}\}$ and *target-like domain* $\mathcal{D}_t^s = \{x^{s \to t}, x^{t \to s \to t}\}$.

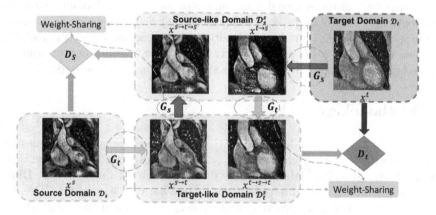

Fig. 2. Overall framework of Dual Cycle Alignment Module (DCAM).

2.2 Semantic Knowledge Transfer

Following image-level adaptation by DCAM, *source-like domain* images $\mathcal{D}_s^s$ and source domain images $\mathcal{D}_s$ maintain a similar visual appearance, allowing us to leverage the knowledge beneath $\mathcal{D}_s^s$ to improve the segmentation performance on $\mathcal{D}_s$ under label scarcity. As shown in Fig. 1, we follow the mean teacher (MT) paradigm and adopt the same architecture for the student and teacher models based on self-ensembling [21]. Specifically, the teacher model $f_{\theta'}$ at training step t is updated with the exponential moving average (EMA) weights of the student model f_θ, *i.e.*, $\theta'_t = \alpha\theta'_{t-1} + (1 - \alpha)\theta_t$, where α is the EMA decay rate that reflects the influence level of the current student model parameters. Given different perturbations (*e.g.*, noises ξ and ξ') to the inputs of teacher and student models, we expect their predictions to be consistent by minimizing the difference between them with a mean square error (MSE) loss $\mathcal{L}_{tea}^{kd}$ as

$$\mathcal{L}_{tea}^{kd} = \frac{1}{N} \sum_{i=1}^{N} \left\| f\left(x_i; \theta'_t, \xi'\right) - f\left(x_i; \theta_t, \xi\right) \right\|^2, \tag{1}$$

where $f(\cdot)$ is the segmentation network. $f\left(x_i; \theta_t, \xi\right)$ and $f\left(x_i; \theta_t, \xi'\right)$ represent the outputs of the student model and the teacher model, respectively.

2.3 Structural Knowledge Transfer

Despite distinct differences like image appearance across domains, the transformed images obtained from generators should have the same structural information as the original ones. In other words, source domain image x_i^s and its synthesis target-like image $x_i^{s \to t}$ should have the same segmentation masks, *i.e.*, $y^s = y_i^{s \to t}$. In this regard, We propose a teacher model for keeping structural consistency between predictions of source images and corresponding synergistic target images, *i.e.*, $f(x; \theta, \xi) = f(G_i[x]; \theta', \xi')$, where x are source (-like) domain images, and G_i is generator G_t or reverse generator G_s. Transferring structural knowledge across domains not only regularizes the student model for semi-supervised learning, but also helps increase adaptation performance at the feature level. Instead of the conventional consistency loss, *e.g.*, MSE loss [16], we exploit the structural information based on weighted self-information [23,24], and calculate the structural consistency loss $\mathcal{L}_{tea}^{con}$ between the teacher and student networks as

$$\mathcal{L}_{tea}^{con} = \frac{1}{N} \sum_{i=1}^{N} \frac{1}{H \times W} \sum_{v=1}^{V} \left\| \mathbf{I}_{i,v}^s - \mathbf{I}_{i,v}^t \right\|^2 \tag{2}$$

where $V = \{1, 2, \ldots, H \times W\}$, $\mathbf{I}_{i,v}^s = -\mathbf{p}_{i,v}^s \circ \log \mathbf{p}_{i,v}^s$ is the weighted self-information of the predicted label at v-th pixel of i-th input from the student network, and similarly $\mathbf{I}_{i,v}^t$ is that from the teacher network. The notation $\circ$ is Hadamard product and log is the logarithmic expression using base 2.

2.4   MT-UDA Framework

With the supervision of corresponding labels y^s, the student model is trained by the supervised loss $\mathcal{L}_{stu}^{seg}$ as

$$\mathcal{L}_{stu}^{se.g.} = \frac{1}{2} \left[\mathcal{L}_{ce} \left(y^s, p_{stu}^s \right) + \mathcal{L}_{dice} \left(y^s, p_{stu}^s \right) \right], \tag{3}$$

where $\mathcal{L}_{ce}$ and $\mathcal{L}_{dice}$ are cross-entropy loss and dice loss, respectively, and p_{stu}^s is the predictions of the student model on source labeled images x^s. Based on the above discussion, we integrate Eq. 1, Eq. 2 and Eq. 3, and the training objective for the student model is formulated as

$$\mathcal{L}_{stu} = \mathcal{L}_{stu}^{se.g.} + \lambda_{kd} \mathcal{L}_{tea}^{kd} + \lambda_{con} \mathcal{L}_{tea}^{con}, \tag{4}$$

where λ_{kd} and λ_{con} are the trade-off parameters with the associated losses. With the MT-UDA framework, we can distill the knowledge from *source-like domain* and *target-like domain* together for more accurate cross-modality image segmentation.

3   Experiments and Results

Dataset and Pre-processing. We evaluated our method on the Multi-Modality Whole Heart Segmentation (MM-WHS) 2017 dataset, consisting of unpaired 20 MR and 20 CT volumes with ground truth masks. We employed MR as source domain and CT as target domain. Following general UDA setting as in [5], each modality was first randomly split with 16 scans for training and 4 scans for testing. To validate the performance under the source-label scarcity scenario, we randomly selected 4 annotated MR scans for training in comparison experiments. For data pre-processing, following previous work [9], we cropped all the coronal slices into centering at the heart region after resampling with unit spacing. Four cardiac substructures, *i.e.*, ascending aorta (AA), left atrium blood cavity (LAC), left ventricle blood cavity (LVC), and myocardium of the left ventricle (MYO) were selected for segmentation.

Implementation Details. We followed [30] to optimize the proposed dual cycle alignment module for generating *source-like domain* images $\mathcal{D}_s^s$ and *target-like domain* images $\mathcal{D}_t^s$. Similar to [28], we verified our model on the transformed source-like images $x^{t \to s}$ instead of target domain images x^t, since our model was trained on source domain under source label scarcity. We implemented U-Net [18] as our network backbone for both student and teacher models in MT-UDA. We trained the framework for a total of 150 iterations and used Adam optimizer with the initial learning rate of 1×10^{-4}, momentum of 0.9, learning rate warm up over the first 20 iterations, and cosine decay of the learning rate with the SGD optimizer. Following [20], the EMA decay rate α was set to 0.999 for two teacher models, and hyperparameters λ_{con} and λ_{kd} were ramped up individually with

Table 1. Comparison results of different methods. suffix −4 or −16 after method names stand for the number of labelled source scans used for training.

Method		Dice ↑					ASD ↓				
		AA	LAC	LVC	MYO	Avg	AA	LAC	LVC	MYO	Avg
W/o Adaptation - 4		5.6	17.8	12.1	5.5	10.3	36.9	24.6	38.5	35.6	33.9
UDA-16	PnP-AdaNet [9]	74	68.9	61.9	50.8	63.9	12.8	6.3	17.4	14.7	12.8
	SIFA-v1 [5]	81.1	76.4	75.7	58.7	73	10.6	7.4	6.7	7.8	8.1
	SIFA-v2 [6]	81.3	79.5	73.8	61.6	74.1	7.9	6.2	5.5	8.5	7
UDA-4	DCAM	19.3	28.1	34.1	6.4	22	32.5	21.8	17.7	22.8	23.7
	SIFA-v2 [6]	50.5	59.6	31.9	28.9	42.7	8.8	7.3	15.8	13.2	11.3
SSL-4	MT [20]	3.6	26.8	14.5	4.6	12.4	34.5	22.7	**5.7**	17.6	20.1
	UA-MT [26]	20.1	40.5	2.5	11.3	18.6	40.1	23.3	43.2	20.9	31.9
UDA+SSL-4	DCAM+MT [20]	35.3	31.6	48.4	11.2	31.6	39.9	39.8	10.5	14.6	23.7
	DCAM+UA-MT [26]	61.3	59.7	46.5	19.2	46.7	5.6	8.3	8.2	10.6	8.2
	MT-UDA (Ours)	**72.7**	**71.4**	**60.7**	**41.7**	**61.6**	**5.3**	**5.7**	6.7	**6.1**	**5.9**

the sigmoid-shaped function $\lambda(t) = 0.01 \cdot e^{\left(-5(1-t/t_{\max})^2\right)}$, where t and t_{max} were the current and the last step, respectively. Data augmentation such as random rotation was applied in all the experiments for a fair comparison. We evaluate different methods on Dice score and average surface distance (ASD) with the largest 3D connected component of each substructure.

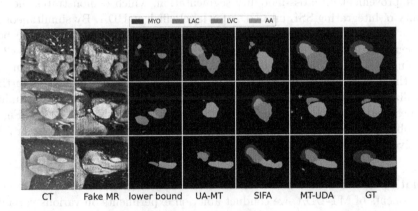

Fig. 3. Visualization of segmentation results generated by different methods.

Comparison with Other Methods. We compare our methods with the state-of-the-art UDA methods in cardiac segmentation, *i.e.,* Pnp-AdaNet [10] and SIFA [5,6], as well as two recent popular SSL approaches, including MT [20] and UA-MT [26]. In Table 1, we list the results of PnP-AdaNet and SIFA with 16 labeled source scans in cardiac segmentation. Since SIFA-v2 [6] obtains the best segmentation performance on each substructure, we further train SIFA-v2 on 4

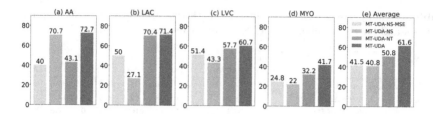

Fig. 4. Ablation results (Dice [%]) on different components.

labeled MR scans to simulate the source-label scarcity scenario. It is observed that SIFA-v2 obtains severely degraded performance on target domain when using 4 labeled source domain scans, which can be attributed to the source label scarcity. We also directly test the U-Net trained on 4 labeled MR scans from the source domain as our lower bound, referred as W/o Adaptation-4. By taking advantage of image-to-image translation *i.e.*, DCAM, a great improvement can be achieved when testing W/o Adaptation on fake MR images $x^{t \to s}$, but it is still not optimal with the average dice of merely 22% across the substructures. It is worth noting that MT and UA-MT can help improve the segmentation performance on target domain by leveraging unlabeled source domain images. Along with image appearance alignment, MT and UA-MT can achieve promising improvement on cross-modality segmentation, which demonstrates the feasibility of integrating SSL into UDA for label-efficient UDA. By simultaneously exploiting all available data sources, the proposed MT-UDA obtains the best segmentation results with the average dice of 61.6%, outperforming SIFA-v2 (4 training MR scans) by a large margin and achieving comparable performance with the state-of-the-art methods, but only requires 1/4 source labels. We further visualize the segmentation results on testing data of different methods including the best methods of UDA and SSL, *i.e.* SIFA-v2 and DCAM+UA-MT in Fig. 3. It is observed that our method can generate more reliable masks with fewer false positives than other methods.

Ablation Studies of Our Method. To evaluate the effectiveness of different components of MT-UDA, we conduct ablation experiments on various variants. Specifically, we remove one of the teacher models, separately, *i.e.*, W/o semantic knowledge transfer (MT-UDA-NS) and W/o structural knowledge transfer (MT-UDA-NT). We further implement the MSE loss in MT-UDA-NS to evaluate the efficacy of the structural loss, *i.e.*, MT-UDA-NS-MSE. Figure 4 demonstrates the ablation results of different substitutes. We can see that both types of knowledge transfer can benefit the model performance on unsupervised cross-domain segmentation. In comparison with the MSE loss in structural knowledge transfer, the proposed loss based on the weighted self-information can better improve the segmentation performance on some substructures such as AA, benefiting from the structural consistency across domains.

4   Conclusion

In this work, we present a novel label-efficient UDA framework, MT-UDA, which integrates SSL into UDA for cross-modality medical image segmentation under source label scarcity. By bridging both source and target domains to intermediate domains through knowledge transfer, the student model can leverage intra-domain semantic knowledge and exploit inter-domain structural knowledge concurrently, thereby mitigating both the domain discrepancy and source label scarcity. We evaluate the proposed MT-UDA on MM-WHS 2017 dataset, and demonstrate that our method outperforms the state-of-the-art UDA methods by a lot under the challenging source-label scarcity scenario.

Acknowledgement. This research is supported by Institute for Infocomm Research (I2R), Agency for Science, Technology and Research (A*STAR), Singapore.

References

1. Arjovsky, M., Chintala, S., Bottou, L.: Wasserstein generative adversarial networks. In: International Conference on Machine Learning, pp. 214–223. PMLR (2017)
2. Bai, W., et al.: Semi-supervised learning for network-based cardiac MR image segmentation. In: Descoteaux, M., Maier-Hein, L., Franz, A., Jannin, P., Collins, D.L., Duchesne, S. (eds.) MICCAI 2017. LNCS, vol. 10434, pp. 253–260. Springer, Cham (2017). https://doi.org/10.1007/978-3-319-66185-8_29
3. Chartsias, A., et al.: Disentangled representation learning in cardiac image analysis. Med. Image Anal. **58**, 101535 (2019)
4. Chen, C., Dou, Q., Chen, H., Heng, P.-A.: Semantic-aware generative adversarial nets for unsupervised domain adaptation in chest X-Ray segmentation. In: Shi, Y., Suk, H.-I., Liu, M. (eds.) MLMI 2018. LNCS, vol. 11046, pp. 143–151. Springer, Cham (2018). https://doi.org/10.1007/978-3-030-00919-9_17
5. Chen, C., Dou, Q., Chen, H., Qin, J., Heng, P.A.: Synergistic image and feature adaptation: towards cross-modality domain adaptation for medical image segmentation. In: Proceedings of the AAAI Conference on Artificial Intelligence, vol. 33, no. 01, pp. 865–872 (2019)
6. Chen, C., Dou, Q., Chen, H., Qin, J., Heng, P.A.: Unsupervised bidirectional cross-modality adaptation via deeply synergistic image and feature alignment for medical image segmentation. IEEE Trans. Med. Imaging **39**(7), 2494–2505 (2020)
7. Cheplygina, V., de Bruijne, M., Pluim, J.P.: Not-so-supervised: a survey of semi-supervised, multi-instance, and transfer learning in medical image analysis. Med. Image Anal. **54**, 280–296 (2019)
8. Cui, W., et al.: Semi-supervised brain lesion segmentation with an adapted mean teacher model. In: Chung, A.C.S., Gee, J.C., Yushkevich, P.A., Bao, S. (eds.) IPMI 2019. LNCS, vol. 11492, pp. 554–565. Springer, Cham (2019). https://doi.org/10.1007/978-3-030-20351-1_43
9. Dou, Q., et al.: PnP-AdaNet: plug-and-play adversarial domain adaptation network at unpaired cross-modality cardiac segmentation. IEEE Access **7**, 99065–99076 (2019)

10. Dou, Q., Ouyang, C., Chen, C., Chen, H., Heng, P.A.: Unsupervised cross-modality domain adaptation of convnets for biomedical image segmentations with adversarial loss. In: Proceedings of the 27th International Joint Conference on Artificial Intelligence, IJCAI 2018, pp. 691–697. AAAI Press (2018)
11. Ganin, Y., Lempitsky, V.: Unsupervised domain adaptation by backpropagation. In: International Conference on Machine Learning, pp. 1180–1189. PMLR (2015)
12. Goodfellow, I., et al.: Generative adversarial nets. In: Ghahramani, Z., Welling, M., Cortes, C., Lawrence, N., Weinberger, K.Q. (eds.) Advances in Neural Information Processing Systems, vol. 27. Curran Associates, Inc. (2014)
13. Hoffman, J., et al.: CyCADA: cycle-consistent adversarial domain adaptation. In: International Conference on Machine Learning, pp. 1989–1998. PMLR (2018)
14. Laine, S., Aila, T.: Temporal ensembling for semi-supervised learning. In: 5th International Conference on Learning Representations, ICLR 2017, Toulon, France, 24–26 April 2017, Conference Track Proceedings. OpenReview.net (2017)
15. Li, K., Wang, S., Yu, L., Heng, P.-A.: Dual-teacher: integrating intra-domain and inter-domain teachers for annotation-efficient cardiac segmentation. In: Martel, A.L., et al. (eds.) MICCAI 2020. LNCS, vol. 12261, pp. 418–427. Springer, Cham (2020). https://doi.org/10.1007/978-3-030-59710-8_41
16. Li, X., Yu, L., Chen, H., Fu, C.W., Xing, L., Heng, P.A.: Transformation-consistent self-ensembling model for semisupervised medical image segmentation. IEEE Trans. Neural Netw. Learn. Syst. **32**, 523–534 (2020)
17. Long, J., Shelhamer, E., Darrell, T.: Fully convolutional networks for semantic segmentation. In: Proceedings of the IEEE Conference on Computer Vision and Pattern Recognition, pp. 3431–3440 (2015)
18. Ronneberger, O., Fischer, P., Brox, T.: U-Net: convolutional networks for biomedical image segmentation. In: Navab, N., Hornegger, J., Wells, W.M., Frangi, A.F. (eds.) MICCAI 2015. LNCS, vol. 9351, pp. 234–241. Springer, Cham (2015). https://doi.org/10.1007/978-3-319-24574-4_28
19. Tajbakhsh, N., Jeyaseelan, L., Li, Q., Chiang, J.N., Wu, Z., Ding, X.: Embracing imperfect datasets: a review of deep learning solutions for medical image segmentation. Med. Image Anal. **63**, 101693 (2020)
20. Tarvainen, A., Valpola, H.: Mean teachers are better role models: weight-averaged consistency targets improve semi-supervised deep learning results. In: Proceedings of the 31st International Conference on Neural Information Processing Systems, NIPS 2017, pp. 1195–1204. Curran Associates Inc., Red Hook (2017)
21. Tarvainen, A., Valpola, H.: Mean teachers are better role models: weight-averaged consistency targets improve semi-supervised deep learning results. arXiv preprint arXiv:1703.01780 (2017)
22. Tzeng, E., Hoffman, J., Saenko, K., Darrell, T.: Adversarial discriminative domain adaptation. In: Proceedings of the IEEE Conference on Computer Vision and Pattern Recognition, pp. 7167–7176 (2017)
23. Vu, T.H., Jain, H., Bucher, M., Cord, M., Pérez, P.: ADVENT: adversarial entropy minimization for domain adaptation in semantic segmentation. In: Proceedings of the IEEE/CVF Conference on Computer Vision and Pattern Recognition, pp. 2517–2526 (2019)
24. Vu, T.H., Jain, H., Bucher, M., Cord, M., Pérez, P.: DADA: depth-aware domain adaptation in semantic segmentation. In: Proceedings of the IEEE/CVF International Conference on Computer Vision, pp. 7364–7373 (2019)

25. Yang, J., Dvornek, N.C., Zhang, F., Chapiro, J., Lin, M.D., Duncan, J.S.: Unsupervised domain adaptation via disentangled representations: application to cross-modality liver segmentation. In: Shen, D., et al. (eds.) MICCAI 2019. LNCS, vol. 11765, pp. 255–263. Springer, Cham (2019). https://doi.org/10.1007/978-3-030-32245-8_29

26. Yu, L., Wang, S., Li, X., Fu, C.-W., Heng, P.-A.: Uncertainty-aware self-ensembling model for semi-supervised 3D left atrium segmentation. In: Shen, D., et al. (eds.) MICCAI 2019. LNCS, vol. 11765, pp. 605–613. Springer, Cham (2019). https://doi.org/10.1007/978-3-030-32245-8_67

27. Zeng, Z., Xulei, Y., Qiyun, Y., Meng, Y., Le, Z.: SeSe-Net: Self-supervised deep learning for segmentation. Pattern Recogn. Lett. **128**, 23–29 (2019)

28. Zhang, Y., Miao, S., Mansi, T., Liao, R.: Task driven generative modeling for unsupervised domain adaptation: application to X-ray image segmentation. In: Frangi, A.F., Schnabel, J.A., Davatzikos, C., Alberola-López, C., Fichtinger, G. (eds.) MICCAI 2018. LNCS, vol. 11071, pp. 599–607. Springer, Cham (2018). https://doi.org/10.1007/978-3-030-00934-2_67

29. Zhao, Z., Zeng, Z., Xu, K., Chen, C., Guan, C.: DSAL: deeply supervised active learning from strong and weak labelers for biomedical image segmentation. IEEE J. Biomed. Health Inform. (2021)

30. Zhu, J.Y., Park, T., Isola, P., Efros, A.A.: Unpaired image-to-image translation using cycle-consistent adversarial networks. In: Proceedings of the IEEE International Conference on Computer Vision, pp. 2223–2232 (2017)

Context-Aware Virtual Adversarial Training for Anatomically-Plausible Segmentation

Ping Wang[1]($\boxtimes$), Jizong Peng[1], Marco Pedersoli[1], Yuanfeng Zhou[2],
Caiming Zhang[2], and Christian Desrosiers[1]

[1] Department of Software and IT Engineering, ETS, Montreal, Canada
`ping.wang.1@ens.etsmtl.ca`
[2] School of Software, Shandong University, Jinan, China

Abstract. Despite their outstanding accuracy, semi-supervised segmentation methods based on deep neural networks can still yield predictions that are considered anatomically impossible by clinicians, for instance, containing holes or disconnected regions. To solve this problem, we present a Context-aware Virtual Adversarial Training (CAVAT) method for generating anatomically plausible segmentation. Unlike approaches focusing solely on accuracy, our method also considers complex topological constraints like connectivity which cannot be easily modeled in a differentiable loss function. We use adversarial training to generate examples violating the constraints, so the network can learn to avoid making such incorrect predictions on new examples, and employ the REINFORCE algorithm to handle non-differentiable segmentation constraints. The proposed method offers a generic and efficient way to add any constraint on top of any segmentation network. Experiments on two clinically-relevant datasets show our method to produce segmentations that are both accurate and anatomically-plausible in terms of region connectivity.

1 Introduction

Due to the high complexity and cost of generating ground-truth annotations for medical image segmentation, a wide range of semi-supervised methods based on deep neural networks have been proposed for this problem. These methods, which leverage unlabeled data to improve performance, include distillation [1], attention learning [2], adversarial learning [3,4], entropy minimization [5], co-training [6,7], temporal ensembling [8,9], consistency-based regularization [10] and data augmentation [11,12]. When very few labeled images are available, however, it may be impossible for a segmentation network to learn the distribution of valid shapes, even when using a semi-supervised learning approach.

Electronic supplementary material The online version of this chapter (https://doi.org/10.1007/978-3-030-87193-2_29) contains supplementary material, which is available to authorized users.

M. de Bruijne et al. (Eds.): MICCAI 2021, LNCS 12901, pp. 304–314, 2021.
https://doi.org/10.1007/978-3-030-87193-2_29

As a result, the segmentation network can yield predictions that are considered anatomically impossible by clinicians [13]. Such predictions can severely impact downstream analyses which rely on anatomical measures, and often require a costly manual step to correct segmentation errors.

Various works have focused on incorporating constraints in semi-supervised or weakly-supervised segmentation methods [14–18]. The approach in [16] uses a simple L_2 penalty to impose size constraints on segmented regions in histopathology images. Kervadec et al. [14] proposed a similar differential loss to enforce inequality constraints on the size of segmented regions. Likewise, Zhou et al. [17] constrain the size of segmented regions with a loss function minimizing the KL divergence between the predicted class distribution and a target one. Despite showing the benefit of adding constraints in a segmentation model, these methods suffer from two important limitations. First, they are limited to simple constraints like region size or centroid position, which are insufficient to characterize the complex shapes found in medical imaging applications. Second, they require designing a problem-specific differentiable loss and, thus, have low generalizability.

Recent efforts have also been invested toward adding strong anatomical priors in segmentation networks. In [19], Oktay et al. present an anatomically constrained neural network (ACNN) using an autoencoder to reconstruct the segmentation mask of labeled images. The reconstruction loss of the autoencoder for a given image is then used as segmentation shape prior. As training the autoencoder requires a sufficient amount of labeled data, this approach is poorly suited to semi-supervised learning settings. The cardiac segmentation approach by Zotti et al. [20] improves accuracy by aligning a probabilistic shape atlas to the predicted segmentation during training. Likewise, Duan et al. [21] uses a multi-task approach to locate landmarks which guide an atlas-based label propagation during a refinement step. In spite of their added robustness, both theses approaches need large annotated datasets to learn the atlas and are sensitive to atlas registration errors. Recently, Painchaud et al. [13] proposed a segmentation method that uses a variational autoencoder to learn the manifold of valid segmentations. During inference, predicted segmentations are mapped to their nearest valid point in the manifold. While it offers strong anatomical guarantees, this post-processing method requires pre-computing an important number of valid points. Moreover, the projection of a predicted output on these points can lead to a segmentation considerably different from the ground-truth.

To address the above-mentioned limitations, we propose a Context-Aware Virtual Adversarial Training (CAVAT) method for semi-supervised segmentation, which considers complex constraints during training to learn an anatomically-plausible segmentation. Unlike existing approaches, which are limited to simple, differentiable constraints (e.g., region size, centroid position, etc.) and require designing a customized loss function, our method can be used out-of-the-box to add any constraint, differentiable or not, on top of a given segmentation model. Our detailed contributions are as follows:

- We propose a novel framework that helps obtain anatomically-plausible segmentations by considering complex anatomical priors in the learning process. Our framework is based on Virtual Adversarial Training (VAT) [22], which optimizes a minimax problem where adversarial examples are created from training samples so to maximize prediction divergence of the network. Unlike VAT, our method generates adversarial examples that maximize prediction divergence *as well as* constraint violation. The REINFORCE algorithm [23] is used to compute gradients for non-differentiable segmentation constraints.
- To our knowledge, our segmentation method is the first to consider complex anatomical priors in a general semi-supervised setting. In comparison, existing approaches require a large number of labeled images to learn a shape prior [13,19] or a complex and problem-specific step involving atlas registration [21,24]. Unlike these approaches, our method needs very few labeled examples and can be used with any segmentation network.

In the next section, we present our Context-aware Virtual Adversarial Training (CAVAT) method and show how it can be used to include connectivity constraints on the segmentation output. In our experiments, we demonstrate our semi-supervised segmentation method's ability to provide a higher accuracy and better constraint satisfaction when trained with very few labeled examples. Finally, we conclude with a summary of main contributions and results.

2 Proposed Method

We start by defining the semi-supervised segmentation problem considered in our work. Let $\mathcal{S} = \{(x_s, y_s)\}_{s=1}^{|\mathcal{S}|}$ be a small set of labeled examples, where each $x_s \in \mathbb{R}^{|\Omega|}$ is an image and $y_s \in \{0,1\}^{|\Omega| \times |\mathcal{C}|}$ is the corresponding ground-truth segmentation mask. Here, $\Omega \subset \mathbb{Z}^2$ denotes the set of image pixels and $\mathcal{C}$ the set of segmentation classes. Given labeled images $\mathcal{S}$ and a larger set of unlabeled images $\mathcal{U} = \{x_u\}_{u=1}^{|\mathcal{U}|}$, we want to learn a network f parameterized by weights θ which produces segmentations that are both accurate and anatomically-plausible.

An overview of the proposed method is shown in Fig. 1 (left). Our method is trained with both labeled and unlabeled data by optimizing the following objective:

$$\min_{\theta} \ \mathcal{L}_{\text{total}}(\theta; \mathcal{D}) \ = \ \mathcal{L}_{\text{sup}}(\theta; \mathcal{S}) + \lambda \mathcal{L}_{\text{CAVAT}}(\theta; \mathcal{U}) \tag{1}$$

The supervised loss $\mathcal{L}_{\text{sup}}(\cdot)$ encourages individual networks to predict segmentation outputs for labeled data that are close to the ground truth. In this work, we use the well-know cross-entropy loss:

$$\mathcal{L}_{\text{sup}}(\theta; \mathcal{S}) \ = \ -\frac{1}{|\mathcal{S}|} \sum_{(x,y) \in \mathcal{S}} \sum_{i \in \Omega} \sum_{j \in \mathcal{C}} y_{ij} \log\left(\mathbf{f}_{ij}(x, \theta)\right) \tag{2}$$

The context-aware VAT loss $\mathcal{L}_{\text{CAVAT}}(\cdot)$, which uses unlabeled images, increases the robustness of the model to adversarial noise and helps the model learn to produce valid segmentations with respect to the given constraints. This loss is detailed in the next section.

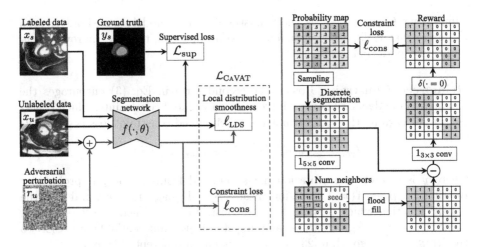

Fig. 1. Overview of the proposed method.

2.1 Context-Aware VAT Loss

A standard approach to incorporate constraints in a semi-supervised learning scenario is to add a loss term that penalizes the violation of these constraints [14, 16,17]. This simple approach poses three major problems. First, it may not be possible to model a given constraint with a function. For instance, testing region connectivity, which imposes each pair of points in a region to be connected by a path inside the region, requires running an algorithm. Second, even if such function exists, it may not be differentiable. This is often the case in segmentation due to the discrepancy between the continuous network output and the discrete segmentation on which the constraints are applied. Last, although both these conditions are satisfied, there is no guarantee that a given constraint will be violated during training, especially if it models a complex relationship. If the network never violates a constraint, it will not be able to learn how to satisfy it since the gradient from the constraint loss will be null.

To alleviate these problems, we define the following Context-aware VAT loss on unlabeled examples:

$$\mathcal{L}_{\text{CAVAT}}(\theta; \mathcal{U}) = \frac{1}{|\mathcal{U}|} \sum_{x_u \in \mathcal{U}} \max_{\|r\| \leq \epsilon} \left[\ell_{\text{LDS}}(x_u, r_u; \theta) + \gamma \, \ell_{\text{cons}}(x_u, r_u; \theta) \right] \quad (3)$$

This loss, which is composed of a local distribution smoothness (LDS) term ℓ_{LDS} and reinforced constraint term ℓ_{cons} is minimized with respect to network parameters θ and maximized with respect to the image perturbation r_u. γ is the weight balancing the two loss terms, which are described below.

Local Distributional Smoothness (LDS). The first term in Eq. (3) is the divergence-based LDS in the original VAT method [22] which is given by

$$\ell_{\text{LDS}}(x_u, r_u; \theta) = D_{\text{KL}}\big(\mathbf{f}(x_u; \theta) \,\|\, \mathbf{f}(x_u + r_u; \theta)\big) \quad (4)$$

Minimizing $\ell_{\text{LDS}}(\cdot)$ enhances the robustness of the model against adversarial examples that violates the virtual adversarial direction, thereby improving generalization performance.

Reinforced Constraint Loss. The second term in Eq. (3) encourages the production of adversarial examples leading to violated constraints, which is necessary for learning these constraints. The reinforced constraint loss is given by

$$\ell_{\text{cons}}(x_u, r_u; \theta) = -\mathbb{E}_{\widehat{y} \sim \mathbf{f}(x_u + r_u; \theta)} \Big[J(\widehat{y}) \Big] \tag{5}$$

where $\widehat{y}$ is discrete segmentation mask sampled from the output probability distribution $\mathbf{p}_u = \mathbf{f}(x_u + r_u; \theta)$ for an adversarial image $x_u + r_u$, and J is the reward function which outputs 1 if the constraint is satisfied else it returns 0. Since the discrete segmentation sampling step is non-differentiable, we resort to the REINFORCE algorithm [23] to convert it into a differentiable loss:

$$\nabla_\theta \ell_{\text{cons}} = -\sum_{\widehat{y}} J(\widehat{y}) \nabla_\theta p(\widehat{y}) \approx -\frac{1}{m} \sum_{\widehat{y}^{(s)} \sim \mathbf{p}_u, \, s=1}^{m} J(\widehat{y}^{(s)}) \nabla_\theta \log p(\widehat{y}^{(s)}) \tag{6}$$

where m is a given number of samples, empirically set to 10 in this paper. Assuming that outputs at different pixels are conditionally independent given the input image, i.e. $p(\widehat{y}^{(s)}) = \prod_i p(\widehat{y}_i^{(s)})$, the final loss can be expressed as

$$\ell_{\text{cons}}(x_u, r_u; \theta) = -\frac{1}{m} \sum_{\widehat{y}^{(s)} \sim \mathbf{p}_u, \, s=1}^{m} \sum_{i \in \Omega} J_i(\widehat{y}^{(s)}) \log p(\widehat{y}_i^{(s)}). \tag{7}$$

2.2   Local Connectivity Constraints

Although our method can be used with any differentiable or non-differentiable constraint, in this paper, we illustrate it on a well-known constraint with broad applicability: connectivity. Given a segmented region G, we say that G is connected if and only if there exists a path between each pair of pixels $p, q \in G$ such that all pixels in the path belong to G. Imposing connectivity in segmentation leads to a highly-complex problem which can only be solved for simplified cases, for example, by representing an image as a small set of superpixels [25]. However, considering connectivity over the whole image may not be practical since it is hard to achieve in the early training stages. For instance, having a single disconnected noisy pixel violates the constraint. To solve this problem, we relax the global constraint and instead consider connectivity at each local patch. Since satisfaction at local patches is a necessary condition for global satisfaction, enforcing it helps achieve our objective. Moreover, doing so provides a spatially-denser gradient since satisfaction can vary from one sub-region to another.

The reward computation process is illustrated in Fig. 1 (right) and detailed in Algorithm 1 of the Supplementary Materials. First, we generate discrete segmentations $\widehat{y}^{(s)}$, $s = 1, \ldots, m$, from the output probability map via multinomial

Table 1. Mean DSC, HD and non-connected pixels (N-conn) for segmenting the left ventricle (LV), right ventricle (RV) and myocardium (Myo) of ACDC, and segmenting prostate in Pʀᴏᴍɪsᴇ12. For each task, labeled data ratio and performance metric, we highlight the best method in **red** and the second best in **blue**.

Task/labeled%	Method	DSC (%) ↑	HD (mm) ↓	N-conn (%) ↓
ACDC LV/100%	Baseline	94.00 (0.09)	6.02 (1.13)	2.40 (0.72)
ACDC LV/3%	Baseline	88.14 (0.67)	20.58 (7.09)	6.90 (0.81)
	Entropy min	87.54 (0.83)	16.67 (0.75)	6.49 (0.67)
	VAT	88.65 (0.45)	20.66 (2.76)	6.32 (0.88)
	Co-training	88.81 (0.39)	12.36 (0.81)	6.35 (0.28)
	Mean teacher	90.91 (1.13)	**12.11 (1.97)**	**3.36 (0.99)**
	CᴀVAT ($r_u = 0$)	88.63 (0.69)	28.31 (5.78)	6.89 (0.86)
	CᴀVAT	89.18 (0.48)	22.62 (0.45)	4.75 (0.74)
	CoT + CᴀVAT	90.21 (0.47)	12.84 (3.56)	5.94 (0.71)
	MT + CᴀVAT	**91.04 (0.60)**	**9.52 (1.44)**	**3.20 (0.75)**
ACDC Myo/100%	Baseline	89.55 (0.09)	4.17 (0.19)	1.91 (0.58)
ACDC Myo/3%	Baseline	75.00 (2.55)	27.85 (3.51)	10.26 (2.23)
	Entropy min	74.01 (0.95)	22.06 (3.90)	11.68 (0.62)
	VAT	78.26 (0.62)	26.45 (6.69)	8.77 (0.77)
	Co-training	75.82 (0.39)	13.24 (1.02)	12.50 (0.71)
	Mean teacher	**82.56 (0.44)**	**11.62 (1.80)**	**4.26 (0.48)**
	CᴀVAT ($r_u = 0$)	78.44 (0.84)	27.16 (1.05)	6.48 (0.38)
	CᴀVAT	79.59 (0.30)	26.20 (0.59)	6.52 (1.13)
	CoT + CᴀVAT	79.25 (1.03)	12.34 (1.38)	8.92 (0.22)
	MT + CᴀVAT	**82.68 (0.43)**	**9.87 (0.74)**	**3.82 (1.56)**
ACDC RV/100%	Baseline	88.66 (0.31)	6.27 (0.38)	6.32 (0.97)
ACDC RV/5%	Baseline	63.17 (3.10)	17.90 (0.87)	27.50 (2.53)
	Entropy min	62.09 (1.22)	16.72 (1.73)	31.58 (2.70)
	VAT	69.52 (1.79)	20.46 (3.69)	25.81 (4.59)
	Co-training	63.97 (0.47)	17.30 (1.58)	29.07 (1.19)
	Mean teacher	**80.57 (0.65)**	**14.46 (1.81)**	**12.21 (1.05)**
	CᴀVAT ($r_u = 0$)	70.42 (1.87)	21.95 (2.75)	21.52 (1.12)
	CᴀVAT	72.88 (1.55)	21.06 (3.42)	20.43 (2.66)
	CoT + CᴀVAT	71.51 (1.89)	14.94 (2.04)	25.92 (1.57)
	MT + CᴀVAT	**80.70 (0.51)**	**11.90 (0.63)**	**11.45 (1.19)**
Pʀᴏᴍɪsᴇ12/100%	Baseline	87.99 (0.20)	5.04 (0.42)	6.87 (0.19)
Pʀᴏᴍɪsᴇ12/5%	Baseline	55.95 (1.80)	11.86 (5.11)	28.83 (2.18)
	Entropy min	56.39 (3.01)	10.95 (1.13)	26.70 (1.88)
	VAT	62.89 (4.20)	14.12 (2.06)	16.98 (4.31)
	Co-training	52.60 (0.67)	12.22 (2.91)	34.60 (2.33)
	Mean teacher	**71.09 (2.03)**	**6.76 (3.92)**	16.19 (3.58)
	CᴀVAT ($r_u = 0$)	63.68 (0.41)	15.57 (0.58)	15.12 (1.12)
	CᴀVAT	65.38 (2.24)	14.55 (4.42)	**11.55 (0.43)**
	CoT + CᴀVAT	66.65 (0.36)	15.29 (0.28)	11.57 (1.82)
	MT + CᴀVAT	**72.33 (2.57)**	**8.92 (0.06)**	12.33 (2.37)

sampling. For each sampled $\widehat{y}^{(s)}$, we then apply the flood-fill algorithm from a chosen seed pixel to produce the connected foreground region $C^{(s)}$. To select the seed pixel, we use a $1_{l \times l}$ convolution kernel on $\widehat{y}^{(s)}$ to compute the number of foreground pixels in a $l \times l$ window centered on each pixel of the image. Afterwards, we randomly choose a pixel with maximum value to favor selecting large connected components as reference region. For each patch of $k \times k$, we measure the number of foreground pixels that are not in $C^{(s)}$, using a simple convolution: $S^{(s)} = 1_{k \times k} \circledast (\widehat{y}^{(s)} - C^{(s)})$. Finally, we evaluate the constraint at pixel i as $J_i(\widehat{y}^{(s)}) = \delta(S_i^{(s)} = 0)$, where $\delta(\cdot)$ is the Kronecker delta.

3   Experimental Setup

We evaluate our CAVAT method on the Automated Cardiac Diagnosis Challenge (ACDC) dataset [26] and the Prostate MR Image Segmentation (PROMISE12) Challenge dataset [27]. Details on these datasets can be found in the Supplementary Materials. For ACDC, segmentation masks delineate three anatomic regions: left ventricle endocardium (LV), left ventricle myocardium (Myo) and right ventricle endocardium (RV). All these regions satisfy the connectivity constraint and have a single connected component. For PROMISE12, the goal is to segment the whole prostate which is also a connected region. We report three performance metrics: Dice similarity coefficient (DSC), Hausdorff distance (HD) and Non-Connectivity (N-conn). DSC emphasises on the overall overlap between a candidate segmentation and its ground truth; HD measures the maximum local disagreement between the two segmentation sets; the N-conn quantifies the percentage of foreground pixels which are not connected to a randomly-selected foreground seed. The hyper-parameters for computing the connectivity reward (see Sect. 2.2) were set empirically as follows: $l = 5$ and $k = 3$.

We tested labeled data ratios of 3% and 5% for each segmentation task, and compared our CAVAT method against using only the supervised loss $\mathcal{L}_{sup}$ (denoted as Baseline in our results) as well as four popular approaches for semi-supervised learning: Entropy minimization [5], Virtual Adversarial Training (VAT) [22], Co-training [6], and Mean Teacher [9]. Since our method can be used on top of any semi-supervised segmentation algorithm, we also evaluate its combination with Co-training (CoT + CAVAT) or Mean Teacher (MT + CAVAT). Last, we test our CAVAT model with the same loss as in Eq. (3) but no adversarial perturbation ($r_u = 0$).

For all tested approaches, we use ENet [28] as our segmentation backbone and train this network with a rectified Adam optimizer. The learning rate is initially set as to 1×10^{-5} and is updated by a warm-up and cosine decay strategy. We apply the same data augmentation as in [29]. The hyper-parameter balancing the two terms of Eq. (1) is set as follows: $\lambda = 1 \times 10^{-3}$ for LV, $\lambda = 5 \times 10^{-4}$ for Myo, $\lambda = 1 \times 10^{-4}$ for RV, and $\lambda = 1 \times 10^{-4}$ for PROMISE12. For all experiments, we report the mean performance (standard deviation) on 3 independent runs with different random seeds.

4 Experimental Results

Table 1 reports the DSC, HD and percentage of non-connected pixels (N-conn) on validation examples of the ACDC and PROMISE12 datasets. As can be seen, our CAVAT method boosts performance in all cases compared to the baseline using only labeled images (Baseline), with DSC improvements of 1.04% for ACDC LV, 4.59% for ACDC Myo, 9.71% for ACDC RV, and 9.43% for PROMISE12. Our method also significantly reduces the number of non-connected foreground pixels (N-conn) compared to the baseline, demonstrating its ability to learn the given constraint. Results also validate the benefit of generating constraint-specific adversarial examples, as seen from the better DSC, HD and N-conn scores of CAVAT compared to the setting with $r_u = 0$. Moreover, we also observe improvements when adding CAVAT to Co-training or Mean Teacher. In particular, our MT + CAVAT combination obtains the highest overall DSC and yields a lower N-conn than Mean Teacher, for all segmentation tasks. Additional results with 5% labeled data for ACDC LV and ACDC Myo, and with 8% labeled data for PROMISE12 can be found in Supplementary Materials.

In Fig. 2, we show examples of segmentations produced by the tested approaches for the three tasks, when using 5% of labeled data. As can be seen, adding CAVAT to the baseline or a semi-supervised learning method yields a more accurate segmentation and helps avoid disconnected regions. As last experiment, we performed a sensitivity analysis on hyper-parameter γ which controls the weight of the constraint loss in Eq. (3). The results and analysis for this experiment can be found in the Supplementary Materials.

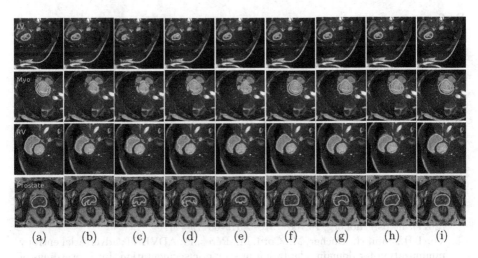

(a)   (b)   (c)   (d)   (e)   (f)   (g)   (h)   (i)

Fig. 2. Visual comparison of tested methods on the validation images. (a) Ground-truth; (b) Partial supervision baseline; (c) Entropy min; (d) VAT; (e) Co-training (CoT); (f) Mean Teacher (MT); (g) CAVAT; (h) CoT + CAVAT; (i) MT + CAVAT

5   Conclusion

We proposed CAVAT, a novel method for semi-supervised segmentation that can incorporate complex anatomical constraints on any segmentation model during training. Our method extends the virtual adversarial training (VAT) framework, making a network robust to adversarial perturbations, by generating examples which also cause the model to violate a given constraint. By improving its prediction for these adversarial examples, the network can thus learn to satisfy the constraint. To alleviate the need to define a specialized penalty function for the constraint, as well as to handle non-differentiable constraints, our method uses the REINFORCE algorithm. As a result, it can be used as a plug-in on any semi-supervised learning approach. Experiments on three segmentation tasks from the ACDC and PROMISE12 datasets reveal the effectiveness of our method in terms of both the accuracy and constraint satisfaction.

A potential limitation of the proposed method stems from its use of the REINFORCE algorithm, which requires sampling a sufficient number of discrete segmentations from the predicted probabilities otherwise optimization may be unstable. While we found that 10 samples gave good results, a larger number might be required for more complex constraints. Another drawback of our method is the computational cost of evaluating constraints during training, which might be prohibitive in some cases. As future work, we plan to extend our method to multi-class segmentation tasks. We will also investigate the combination of our method with other semi-supervised techniques and evaluate its usefulness for a broader range of constraints.

References

1. Radosavovic, I., Dollár, P., Girshick, R., Gkioxari, G., He, K.: Data distillation: towards omni-supervised learning. In: Proceedings of the IEEE Conference on Computer Vision and Pattern Recognition, pp. 4119–4128 (2018)
2. Min, S., Chen, X.: A robust deep attention network to noisy labels in semi-supervised biomedical segmentation. arXiv preprint arXiv:1807.11719 (2018)
3. Souly, N., Spampinato, C., Shah, M.: Semi supervised semantic segmentation using generative adversarial network. In: 2017 IEEE International Conference on Computer Vision (ICCV), pp. 5689–5697. IEEE (2017)
4. Zhang, Y., Yang, L., Chen, J., Fredericksen, M., Hughes, D.P., Chen, D.Z.: Deep adversarial networks for biomedical image segmentation utilizing unannotated images. In: Descoteaux, M., Maier-Hein, L., Franz, A., Jannin, P., Collins, D.L., Duchesne, S. (eds.) MICCAI 2017. LNCS, vol. 10435, pp. 408–416. Springer, Cham (2017). https://doi.org/10.1007/978-3-319-66179-7_47
5. Vu, T.H., Jain, H., Bucher, M., Cord, M., Pérez, P.: ADVENT: adversarial entropy minimization for domain adaptation in semantic segmentation. In: Proceedings of the IEEE Conference on Computer Vision and Pattern Recognition, pp. 2517–2526 (2019)
6. Peng, J., Estrada, G., Pedersoli, M., Desrosiers, C.: Deep co-training for semi-supervised image segmentation. Pattern Recogn. **107**, 107269 (2020)

7. Zhou, Y., et al.: Semi-supervised 3D abdominal multi-organ segmentation via deep multi-planar co-training. In: 2019 IEEE Winter Conference on Applications of Computer Vision (WACV), pp. 121–140. IEEE (2019)
8. Perone, C.S., Cohen-Adad, J.: Deep semi-supervised segmentation with weight-averaged consistency targets. In: Stoyanov, D., et al. (eds.) DLMIA/ML-CDS - 2018. LNCS, vol. 11045, pp. 12–19. Springer, Cham (2018). https://doi.org/10.1007/978-3-030-00889-5_2
9. Cui, W., et al.: Semi-supervised brain lesion segmentation with an adapted mean teacher model. In: Chung, A.C.S., Gee, J.C., Yushkevich, P.A., Bao, S. (eds.) IPMI 2019. LNCS, vol. 11492, pp. 554–565. Springer, Cham (2019). https://doi.org/10.1007/978-3-030-20351-1_43
10. Bortsova, G., Dubost, F., Hogeweg, L., Katramados, I., de Bruijne, M.: Semi-supervised medical image segmentation via learning consistency under transformations. In: Shen, D., et al. (eds.) MICCAI 2019. LNCS, vol. 11769, pp. 810–818. Springer, Cham (2019). https://doi.org/10.1007/978-3-030-32226-7_90
11. Chaitanya, K., Karani, N., Baumgartner, C.F., Becker, A., Donati, O., Konukoglu, E.: Semi-supervised and task-driven data augmentation. In: Chung, A.C.S., Gee, J.C., Yushkevich, P.A., Bao, S. (eds.) IPMI 2019. LNCS, vol. 11492, pp. 29–41. Springer, Cham (2019). https://doi.org/10.1007/978-3-030-20351-1_3
12. Zhao, A., Balakrishnan, G., Durand, F., Guttag, J.V., Dalca, A.V.: Data augmentation using learned transformations for one-shot medical image segmentation. In: Proceedings of the IEEE Conference on Computer Vision and Pattern Recognition, pp. 8543–8553 (2019)
13. Painchaud, N., Skandarani, Y., Judge, T., Bernard, O., Lalande, A., Jodoin, P.M.: Cardiac segmentation with strong anatomical guarantees. IEEE Trans. Med. Imaging 39(11), 3703–3713 (2020)
14. Kervadec, H., Dolz, J., Tang, M., Granger, E., Boykov, Y., Ben Ayed, I.: Constrained-CNN losses for weakly supervised segmentation. Med. Image Anal. 54, 88–99 (2019)
15. Pathak, D., Krahenbuhl, P., Darrell, T.: Constrained convolutional neural networks for weakly supervised segmentation. In: Proceedings of the IEEE International Conference on Computer Vision, pp. 1796–1804 (2015)
16. Jia, Z., Huang, X., Eric, I., Chang, C., Xu, Y.: Constrained deep weak supervision for histopathology image segmentation. IEEE Trans. Med. Imaging 36(11), 2376–2388 (2017)
17. Zhou, Y., et al.: Prior-aware neural network for partially-supervised multi-organ segmentation. arXiv preprint arXiv:1904.06346 (2019)
18. Masoud, S.N., Ghassan, H.: Incorporating prior knowledge in medical image segmentation: a survey. arxiv: abs/1607.01092 (2016)
19. Oktay, O., et al.: Anatomically constrained neural networks (ACNNs): application to cardiac image enhancement and segmentation. IEEE Trans. Med. Imaging 37(2), 384–395 (2017)
20. Zotti, C., Luo, Z., Lalande, A., Jodoin, P.M.: Convolutional neural network with shape prior applied to cardiac MRI segmentation. IEEE J. Biomed. Health Inform. 23(3), 1119–1128 (2018)
21. Duan, J., et al.: Automatic 3D bi-ventricular segmentation of cardiac images by a shape-refined multi-task deep learning approach. IEEE Trans. Med. Imaging 38(9), 2151–2164 (2019)
22. Takeru, M., Shin-ichi, M., Masanori, K., Shin, I.: Virtual adversarial training: a regularization method for supervised and semi-supervised learning. IEEE Trans. Pattern Anal. Mach. Intell. 41(8), 1979–1993 (2019)

23. Ronald, J.W.: Simple statistical gradient-following algorithms for connectionist reinforcement learning. Mach. Learn. **8**(3–4), 229–256 (1992)
24. Dong, S., et al.: Deep atlas network for efficient 3D left ventricle segmentation on echocardiography. Med. Image Anal. **61**, 101638 (2020)
25. Shen, R., Tang, B., Lodi, A., Tramontani, A., Ayed, I.B.: An ILP model for multi-label MRFs with connectivity constraints. IEEE Trans. Image Process. **29**, 6909–6917 (2020)
26. Bernard, O., Lalande, A., Zotti, C., Cervenansky, F., et al.: Deep learning techniques for automatic MRI cardiac multi-structures segmentation and diagnosis: is the problem solved? IEEE Trans. Med. Imaging **37**(11), 2514–2525 (2018)
27. Litjens, G., et al.: Evaluation of prostate segmentation algorithms for MRI: the PROMISE12 challenge. Med. Image Anal. **18**(2), 359–373 (2014)
28. Adam, P., Abhishek, C., Sangpil, K., Eugenio, C.: ENet: a deep neural network architecture for real-time semantic segmentation. arXiv: abs/1606.02147 (2016)
29. Peng, J., Kervadec, H., Dolz, J., Ayed, I.B., Pedersoli, M., Desrosiers, C.: Discretely-constrained deep network for weakly supervised segmentation. Neural Netw. **130**, 297–308 (2020)

Interactive Segmentation via Deep Learning and B-Spline Explicit Active Surfaces

Helena Williams[1,3,4(✉)], João Pedrosa[2], Laura Cattani[1], Susanne Housmans[1], Tom Vercauteren[3], Jan Deprest[1], and Jan D'hooge[4]

[1] Department of Obstetrics and Gynaecology, University Hospitals Leuven, Leuven, Belgium
helena.williams@kuleuven.be
[2] INESC TEC - Institute for Systems and Computer Engineering, Technology and Science, Porto, Portugal
[3] School of Biomedical Engineering and Imaging Sciences, King's College London, London, UK
[4] Department of Cardiovascular Sciences, KU Leuven, Leuven, Belgium

Abstract. Automatic medical image segmentation via convolutional neural networks (CNNs) has shown promising results. However, they may not always be robust enough for clinical use. Sub-optimal segmentation would require clinician's to manually delineate the target object, causing frustration. To address this problem, a novel interactive CNN-based segmentation framework is proposed in this work. The aim is to represent the CNN segmentation contour as B-splines by utilising B-spline explicit active surfaces (BEAS). The interactive element of the framework allows the user to precisely edit the contour in real-time, and by utilising BEAS it ensures the final contour is smooth and anatomically plausible. This framework was applied to the task of 2D segmentation of the levator hiatus from 2D ultrasound (US) images, and compared to the current clinical tools used in pelvic floor disorder clinic (4DView, GE Healthcare; Zipf, Austria). Experimental results show that: 1) the proposed framework is more robust than current state-of-the-art CNNs; 2) the perceived workload calculated via the NASA-TLX index was reduced more than half for the proposed approach in comparison to current clinical tools; and 3) the proposed tool requires at least 13 s less user time than the clinical tools, which was *significant* (p = 0.001).

1 Introduction

Medical image segmentation of anatomical structures can be used for disease diagnosis [6]. Manual segmentation requires expertise, time and is prone to error,

Electronic supplementary material The online version of this chapter (https://doi.org/10.1007/978-3-030-87193-2_30) contains supplementary material, which is available to authorized users.

The original version of this chapter was revised: the missing information in figure 2 was corrected. The correction to this chapter is available at
https://doi.org/10.1007/978-3-030-87193-2_70

M. de Bruijne et al. (Eds.): MICCAI 2021, LNCS 12901, pp. 315–325, 2021.
https://doi.org/10.1007/978-3-030-87193-2_30

therefore, automatic methods are desirable. Deep learning-based solutions with convolutional neural networks (CNNs) have been extensively explored [19,24]. However, their impressive average performance has not yet led to wide clinical adoption [4]. Medical images pose serious challenges to automatic methods, as they can be sensitive to small differences between training and testing data, due to such factors as image quality, imaging protocols (i.e. imaging acquisition discrepancies), pathology, and patient variation [7,13,23,27]. Therefore, it is important for clinical impact and acceptance, to be able to recover from a poor result and address the limitations of automatic segmentation. As the clinician remains liable for the measurements obtained for diagnosis, if the automatic method is incorrect, it is the responsibility of the clinician to identify the problem and correct the segmentation. Interactive segmentation with an intuitive mechanism, for smart correction of poor segmentation, may solve these problems, and give liability to the clinician without them having to manually re-segment, which is not time efficient and may cause frustration. This work is motivated to combine state-of-the-art CNN segmentation with an user interaction tool, which allows the clinician to view, correct (if needed) and save the desired segmentation.

An extensive range of CNN-based interactive methods have been proposed [21], exploiting bounding boxes [20], scribbles [16,26], extreme points [17] or clicks [12]. These achieved higher accuracy and robustness than their automatic counterparts, however, they can require a high cognitive load and understanding. In addition, the user still relies on the CNN to segment correctly and is not always able to edit the contour precisely, in an adequate and time efficient manner.

In this paper, an interactive segmentation tool for 2D semantic medical image segmentation is proposed. The tool is composed of three stages. In the first stage, a CNN automatically obtains an initial segmentation, this feeds as an initialisation to an active contour segmentation framework called B-spline explicit active surfaces (BEAS) [2], which smooths the contour to be more biologically plausible (acting as a post-processing step), and thirdly a novel algorithm which allows the user to interact with the contour in real-time was implemented.

The proposed approach is compared with manual tools used in clinic for 2D segmentation of the levator hiatus in pelvic floor disorder assessment: "Point" and "Trace" both available on the ultrasound (US) software 4DView (GE Healthcare; Zipf, Austria), and compared with a state-of-the-art scribble-based approach referred to as UGIR [26]. The segmentation methods are evaluated on 30 2D US images, the time taken to segment to a clinically acceptable level, and the perceived workload are measured and compared. The contributions of this work are four-fold: 1) A novel CNN-based interactive framework for 2D segmentation is proposed; 2) the interactive element works in real time and requires less user time and perceived workload than clinical methods and UGIR; 3) the method utilises the BEAS framework to ensure the final contour is more biologically plausible than the CNN segmentation, acting as a novel post-processing method; and 4) a new energy term is introduced that is dependent on the probability map of the CNN output, which has not been utilised in BEAS before.

2   Materials and Methods

2.1   Proposed Pipeline

The proposed pipeline is composed of three sequential parts: a 2D CNN which segments the target object (levator hiatus) from the US image; a BEAS-based post-processing method which smooths the CNN segmentation and represents the segmentation boundary as a B-spline explicit active surface; and a novel algorithm (implemented in a graphical user interface (GUI) referred to as Beyond), that allows the user to adapt the contour in real-time while benefiting from BEAS's active model properties. The framework is shown in Fig. 1. The first task of the pipeline, automatically defined the levator hiatus from the US image. This elaborates from previous work where 2D U-Net was used [6,15]. The segmentation is fed as input to the following task.

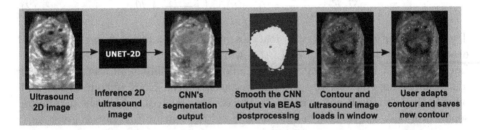

Fig. 1. The proposed pipeline that accepts a 2D US, segments the target object via CNN segmentation, smooths the segmentation and loads it in a window, to allow the user to adapt the contour via BEAS interaction.

BEAS-Based Smoothing. The second task utilises the BEAS framework [2]. A 2D version of BEAS is applied to the CNN segmentation after thresholding, to represent the CNN segmentation boundary as B-splines. The concept of BEAS, is to regard the boundary of a target object (i.e. the CNN segmentation) as an explicit function, where one of the coordinates of the boundary, is given explicitly as a function of the remaining coordinates. As the contour is a closed 2D object, the boundary can be represented in the polar domain, and the contour radius is represented as an explicit function of the polar angle, $i.e. \rho = \psi(\theta)$. Inspired by Bernard et al. [5], the explicit function ψ can be expressed as a linear combination of B-spline basis functions [2,5,25],

$$\rho = \sum_{\mathbf{k} \in \mathbb{Z}^{n-1}} c[\mathbf{k}] \beta^d \left(\frac{\theta}{h} - \mathbf{k} \right). \tag{1}$$

$\beta^d(.)$ is the uniform symmetric n-1-dimensional B-spline of degree d. $\psi(\theta)$ is separable and built as the product of n-1 1D B-splines. The knots of the B-splines are located on a regular grid defined on the polar coordinate system, with regular spacing given by h. The B-spline coefficients are gathered in $c[\mathbf{k}]$.

BEAS assumes that all coordinates of the boundary are visible from a fixed origin, which is a good approximation for this structure that tends to have a pearlike convex shape. Before refinement of the BEAS contour, the initial circular contour must be defined by parameters, such as the fixed origin and an initial radius. In this work, they are based on properties of the CNN segmentation output. The origin is defined as the center of mass of the CNN output, and the initial radius, r_c, is the average radius of the CNN segmentation output.

The initial contour can then refine and evolve towards the boundary of the CNN segmentation through the minimisation of a segmentation energy functional. To achieve this, a general localised region-based energy functional for level-set segmentation [14] was used. Barbosa et al. [2] adapted these localisation strategies for BEAS in terms of B-spline coefficients, and the expression of the energy gradient is given as,

$$\nabla_c E = \frac{\partial E}{\partial c[k]} = \int_\Gamma g(\theta) \beta^d \left(\frac{\theta - hk}{h} \right) d\theta. \tag{2}$$

The function $g(\theta)$ represents the features of the object to be segmented and is evaluated over the boundary Γ. In this work, the energy function used was the Localised Yezzi Energy, proposed by Lankton and Tannenbaum et al. [14,28]. This energy depends on the average intensity of the CNN output inside and outside the evolving B-spline contour. The contour evolves to have the maximum separation between them. For Localised Yezzi the feature function is given as,

$$g(\theta) = \frac{\left(I(\psi(\theta), \theta) - u_\theta \right)^2}{A_u} - \frac{\left(I(\psi(\theta), \theta) - v_\theta \right)^2}{A_v}. \tag{3}$$

A_u and A_v represent the areas inside and outside of the contour, respectively; and u_θ and v_θ are the mean intensities inside and outside the evolving contour at the polar angle, θ, respectively. $I(\psi(\theta), \theta)$ corresponds to the image value (i.e. CNN output) at position (ρ, θ). The Yezzi energy relies on the assumption that the interior and the exterior of the contour have the largest difference in average intensities. This is a good assumption for this work, as the goal is to represent the CNN segmentation output as a smooth B-spline contour. The final B-spline coefficients are saved and used in the following section.

Interaction Framework. Finally, the contour formed from the previous step and the corresponding US image are loaded in a window, where the user can interact with the contour. In the interactive framework, the energy function driving BEAS, is compounded of three energy terms: Localised Yezzi of the US image, E_U, Localised Yezzi of the CNN probability map, E_{CNN} and an interactive energy function, E_i. The total energy is given as:

$$E_{total} = \alpha E_U + \beta E_{CNN} + \gamma E_i, \tag{4}$$

where α, β and γ are hyper-parameters. Here, the initialised contour is determined by the evolved B-spline coefficients from the previous section, and it will evolve with each user-interaction to minimise the energy function defined.

Finally E_i, is based on a 2D version of reported work [3], where user-defined coordinates interact with the B-splines. The user can create markers where they want the contour to pass through, these act as anchors attracting the contour. A point introduced by the user can be expressed as $p_{user} = (\rho_{user}, \theta_{user})$. The energy function penalises the parametric distance, D, between the current boundary position and p_{user} at each B-spline knot. Therefore, D is defined as $D = (\rho - \rho_{user})^2$. The energy term driving the contour towards the user-defined points was proposed in 3D by Barbosa et al. [3], where its minimisation with respect to B-spline coefficients, $c[\mathbf{k}]$ was demonstrated. E_{user} is defined as:

$$E_{user} = \int_{\Gamma} \delta(\theta - \theta_{user})(\psi(\theta) - \rho_{user})^2 d\theta, \tag{5}$$

where $\delta(\theta - \theta_{user})$ corresponds to the Dirac function which is non-zero only at the position $\theta = \theta_{user}$. When multiple user-defined points are present, the sum of the parametric distance between the current contour, Γ and the user-defined points is used. This evolves the contour towards multiple user points, detailed information can be found in the paper by Barbosa et al. [3]. As the computational load is small, there is real-time feedback of the effect the modifications make.

2.2   Data Collection

Analysis of anonymised, archived US images was retrospective, so no ethics committee approval was required by the institute. The CNN was trained on a dataset of 444 2D US images from 213 patients, and corresponding ground truth labels of the levator hiatus. The training dataset comprised of two sets of archived clinical images with expert annotations, acquired by several operators. One dataset used for training, was a private dataset supplied by (GE Healthcare; Zipf, Austria) and the second dataset was a private dataset used in previous studies. 400 images were used for training and 44 were used for validation. The test data included a randomised selection of 30 anonymised 2D US images from 10 symptomatic women assessed at the pelvic floor clinic between March and May, 2019 at the pelvic floor disorder clinic at UZ Leuven, Leuven, Belgium. The US images were obtained from Transperineal US volumes acquired following the clinical protocol defined by Dietz et al. [8] on the Voluson E10 US system (GE Healthcare; Zipf, Austria). The 2D planes that were used to segment and assess the levator hiatus were manually determined by an expert clinician, at rest, during the Valsalva manoeuvre and contraction.

2.3   Experimental Details

Two clinical experts with over 4 years experience in pelvic floor US, participated in the experiment. They segmented the levator hiatus on 30 2D US images using 4DView Trace (GE Healthcare; Zipf, Austria) , 4DView Point (GE Healthcare; Zipf, Austria) and the proposed tool. Prior to the experiment the experts were given a tutorial how to use the new tool. 4DView Trace and Point can be found on the 'measure - generic area' function of 4DView (GE Healthcare; Zipf, Austria). In 4DView Trace (GE Healthcare; Zipf, Austria), the contour starts once the user clicks the US image, and it will follow the user's cursor around the levator hiatus until the user clicks on the US a second time. In 4DView Point (GE Healthcare; Zipf, Austria) the tool will trace the hiatus by the user defining multiple points around the levator hiatus with mouse clicks. The lines that connect the points are straight, therefore, the output segmentation is generalised and not anatomically accurate (i.e. sharp lines). 4DView Trace and 4DView Point (GE Healthcare; Zipf, Austria) may be referred to as Trace and Point respectively in this paper. Uncertainty-Guided Efficient Interactive Refinement (UGIR) utilises an interaction-based level set for fast refinement of segmentations [26], based on scribbles. The same CNN was used as the proposed model and scribbles were created in 3D Slicer [1,9].

The main aim was to compare the perceived subjective workload of the clinical tools and UGIR against the proposed tool. Therefore, half way through the experiment (after 15 segmentations) and at the end of the experiment, the perceived workload was subjectively evaluated by each expert and for each segmentation technique. To do this the National Aeronautics and Space Administration Task Load Index (NASA-TLX) was used [11]. Finally, the time taken for the expert to segment/edit the levator hiatus contour to a clinically accepted level was measured for each segmentation and compared.

2.4   Implementation Details

The proposed tool was implemented on a Windows desktop with a 24 GB NVIDIA Quadro P6000 (NVIDIA, California, United States). The CNN was implemented using NiftyNet [10], training and inference were ran on the GPU. The network architecture was an adaptation of 2D U-Net [22] with half the number of features $[32, 64, 128, 256, 512]$. An Adam optimiser, ReLU activation function, weighted decay factor of 10^{-5} and batch size of 64 were used. Whitening and histogram normalisation (i.e. when the image was set to have zero-mean and unit variance) were applied to reduce the effects of noise [18]. A Dice loss function was used, with a learning rate of 10^{-5}. The data augmentation used were: elastic deformation (deformation sigma $= 5$, number of control points $= 4$), random scaling (-20%, $+20\%$), vertical 'flipping' and an implementation of mixup [29]. Validation of the network training was performed every 250 epochs and the CNN trained for 12,000 epochs. The CNN model from epoch 10,000 was used at inference, as the validation loss function was lowest. The CNN hyperparameters were determined based on literature [6] and the performance of the training dataset.

BEAS optimisation was ran on the CPU. In task 2, the size of the neighbourhood used to estimate the local intensity of the image was set to 100 pixels (i.e. ≈ 30 mm), allowing the contour to recover from a bad initialisation. For both tasks the BEAS contour was discretised into 32 points (i.e. knots) along the polar angle direction, causing the scale parameter, h, to be implicitly fixed to 1. The B-spline coefficients, $c(\mathbf{k})$, are gathered in a 1D index array, spanning the polar domain with 32 B-spline coefficients. For interactive BEAS, in (4) $\alpha = 0.5$, $\beta = 0.3$ and $\gamma = 3$. The size of the neighbourhood used to estimate the local intensity was set to 10 pixels ($\approx 3\,\text{mm}$). This is low to avoid the contour evolving before user-interaction. Otherwise the contour may evolve towards bright regions of the US in order minimise the energy function. The hyper-parameters used for BEAS were determined by a grid search method where the range was guided by literature [2] and evaluated by assessing the performance of the training dataset.

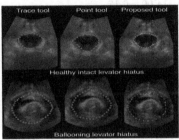

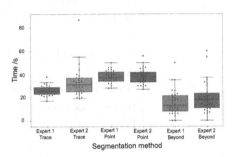

Fig. 2. Visual representation of the levator hiatus segmentation obtained with the clinical tools and proposed pipeline, in a healthy patient at contraction and a patient with ballooning hiatus at Valsalva.

Fig. 3. Time taken to delineate the levator hiatus to a clinically acceptable level with 4DView Trace, Point (GE Healthcare; Zipf, Austria) and the proposed tool.

3   Results

A demonstration video of the proposed tool being used to segment the levator hiatus (at rest, Valsalva and contraction) is available in the supplementary information. Figure 2 shows examples of the segmentation obtained via the clinical tools and the proposed pipeline. The experts agreed that the proposed tool accomplished a clinical acceptable standard for hiatal diagnosis for almost all 2D US images (29 images), thus the proposed tool achieved a 'clinical acceptability' of 97%. However, only 2 and 1 CNN + BEAS post-processing segmentation's required no editing from expert 1 and 2 respectively, equalling a 'clinical acceptability' of 5%. The 'clinical acceptability' of the CNN alone was 2%, and the 'clinical acceptability' of UGIR was 27%.

Figure 3 shows the time taken for expert 1 and 2 to delineate a segmentation of the levator hiatus to a clinically acceptable standard for diagnosis, using the clinical tools and proposed method. The recorded time of the proposed pipeline does not include CNN inference time, to be solely dependent on user interaction time and independent of the CNN's performance. The average CNN inference time was $4.90\pm1.58\,\text{s}$. The time taken using the UGIR method was $63.77\pm31.27\,\text{s}$. The time taken to edit the levator hiatus contour with the proposed tool was *'significantly lower'* (paired t-test, $p \leq 0.001$) than the clinical tools and state of the art method, UGIR. The time taken decreased over the experiment, as the

Table 1. The perceived weighted workload score from the NASA-TLX questionnaire and individual sub-scale scores that contribute to it are shown. A low weighted workload score corresponds to a tool that requires less effort, less frustration, less mental, temporal and physical demands, and a tool that has a higher perceived performance.

NASA-TLX	Average			
Weighted scores	Trace	Point	UGIR	Beyond
Effort	13.67	15.50	14.00	6.17
Frustration	12.00	11.00	21.34	2.17
Mental demand	14.33	10.84	8.67	6.17
Performance	11.00	18.34	28.34	5.34
Physical demand	6.67	4.67	4.00	3.67
Temporal demand	0.84	0.00	0.00	0.34
Total workload	58.47	60.34	76.33	23.84

combined mean of the first 15 segmentations was 19.33 ± 11.93 s and for the final 15 segmentations was 16.13 ± 11.40 s. Table 1 shows the mean NASA-TLX scores of both experts, and table S1 in the supplementary information shows the NASA-TLX scores half way through the experiment (attempt 1), and at the end of the experiment (attempt 2). Within the tables the individual weighted sub-scales are reported, as well as the total weighted work-load. The perceived workload was *'significantly lower'* (paired t-test, $p = 0.001$) in the proposed tool than the clinical tools perceived by both experts. It is worth noting a lower score for perceived performance in this table corresponds to a better perceived performance. Therefore, the experts found the proposed tool to perform better than the clinical tools and UGIR. Furthermore, the perceived mental workload of the proposed tool improved at the end of the experiment, showing improved performance after exposure to the tool.

4   Discussion

Figure 2 shows visually similar contours for all tools. The proposed method shows more anatomically plausible results than the Point tool. The proposed tool achieved visually 'clinically acceptable' results for almost all cases, however, only 2 and 1 CNN segmentations required no editing from expert 1 and 2 respectively. The sub-optimal segmentation was due to a poor US acquisition (that was noted as a clinically unacceptable US image), this reduced the visibility of the levator hiatal boundary, which severely impacted the CNN output. Retrospectively, the radius defining the initial BEAS contour in the second task was increased, and an optimal segmentation was obtained using the proposed tool. Thus, the tool is capable of a 'clinical acceptability' score of 100%, with further tuning.

Both experts achieved at least an average improvement of 13 s when using the proposed tool. The time measured did not include CNN inference time, to keep it independent to the experiment, and to allow for direct comparison with other segmentation tasks of different CNN architectures. It is assumed with optimisation the CNN inference time would reduce. The proposed tool nonetheless, is still quicker than the clinical tools and UGIR, and it may be assumed with further practice, the time taken would continue to decrease.

Following 30 levator hiatus segmentations, the NASA-TLX questionnaire demonstrated the tool improved perceived performance, reduced effort, frustration, mental and temporal demand. The proposed tool reduced the weighted perceived workload, by 36.50, 34.63 and 52.49 points on the NASA-TLX index scale, for Point, Trace and UGIR tools respectively. The performance improved by 13.00, 5.66, 23.00 points on the NASA-TLX weighted index scale compared to the Point, Trace and UGIR tools respectively. It may be assumed the performance of the Point tool was lower, due to the less anatomically accurate segmentation, shown in Fig. 2. Therefore, it may be assumed that this work could improve the clinical workflow. Table S1 showed that the perceived workload score was lower at the end of the experiment than at mid-experiment, highlighting that with increased exposure, the workload may continue to reduce.

The proposed tool is compounded of a post-processing filter and an interactive algorithm. It can be easily implemented on other 2D segmentation tasks, to improve the segmentation boundary and allow for easy editing of incorrect segmentation. There is scope to expand this work to 3D segmentation. Currently, the hyper-parameters used for BEAS (i.e. number of B-splines) requires manual optimisation. In future work, it would be beneficial to automate hyper-parameter selection dependent on the initial 2D segmentation, and compare performance for several segmentation tasks.

5   Conclusion

To conclude, in this work, a novel CNN-based interactive 2D segmentation tool was proposed. The interactive element works in real-time and requires less user time and perceived workload than current clinical methods, suggesting the proposed work may improve the current clinical workflow. The method utilised the BEAS framework, which ensured the final contour was more biologically plausible than CNN segmentation outputs. This framework can easily be implemented for other 2D segmentation tasks, to make the results more robust while improving the clinical acceptability and giving liability to clinicians.

Acknowledgments. We gratefully acknowledge General Electric Healthcare (Zif, Austria), for their continued research support.

References

1. 3D slicer image computing platform

2. Barbosa, D., Dietenbeck, T., Schaerer, J., D'hooge, J., Friboulet, D., Bernard, O.: B-Spline explicit active surfaces: an efficient framework for real-time 3-D region-based segmentation. IEEE Trans. Image Process. **21**, 241–51 (2012). a publication of the IEEE Signal Processing Society

3. Barbosa, D., et al.: Real-time 3D interactive segmentation of echocardiographic data through user-based deformation of B-Spline explicit active surfaces. Comput. Med. Imaging Graph. **38**, 57–67 (2014)

4. Benjamens, S., Dhunnoo, P., Meskó, B.: The state of artificial intelligence-based FDA-approved medical devices and algorithms: an online database. npj Digit. Med. **3** (2020). Article ID: 118. https://doi.org/10.1038/s41746-020-00324-0

5. Bernard, O., Friboulet, D., Thevenaz, P., Unser, M.: Variational B-Spline level-set: a linear filtering approach for fast deformable model evolution. IEEE Trans. Image Process. **18**(6), 1179–1191 (2009)

6. Bonmati, E., et al.: Automatic segmentation method of pelvic floor levator hiatus in ultrasound using a self-normalising neural network. J. Med. Imaging (Bellingham) **5**(2), 021206 (2018)

7. Dietz, H.P., Moegni, F., Shek, K.L.: Diagnosis of levator avulsion injury: a comparison of three methods. Ultrasound Obstet. Gynecol. **40**(6), 693–698 (2012)

8. Dietz, H.P., Shek, C., Clarke, B.: Biometry of the pubovisceral muscle and levator hiatus by three-dimensional pelvic floor ultrasound. Ultrasound Obstet. Gynecol. **25**(6), 580–585 (2005)

9. Fedorov, A., et al.: 3D slicer as an image computing platform for the quantitative imaging network. Magn. Reson. Imaging **30**(9), 1323–1341 (2012). Quantitative Imaging in Cancer

10. Gibson, E., et al.: NiftyNet: a deep-learning platform for medical imaging. CoRR, abs/1709.03485 (2017)

11. Hart, S.G.: Nasa-task load index (NASA-TLX); 20 years later. Proc. Hum. Factors Ergon. Soc. Ann. Meet. **50**(9), 904–908 (2006)

12. Jang, W.-D., Kim, C.-S.: Interactive image segmentation via backpropagating refinement scheme. In: Proceedings of The IEEE Conference on Computer Vision and Pattern Recognition (2019)

13. Kumar, S.N., Fred, A.L., Varghese, P.S.: An overview of segmentation algorithms for the analysis of anomalies on medical images. J. Intell. Syst. **29**(1), 612–625 (2020)

14. Lankton, S., Tannenbaum, A.: Localizing region-based active contours. IEEE Trans. Image Process. **17**, 12 (2008)

15. Li, X., Hong, Y., Kong, D., Zhang, X.: Automatic segmentation of levator hiatus from ultrasound images using U-net with dense connections. Phys. Med. Biol. **64**, 03 (2019)

16. Lin, D., Dai, J., Jia, J., He, K., Sun, J.: ScribbleSup: scribble-supervised convolutional networks for semantic segmentation. CoRR, abs/1604.05144 (2016)

17. Maninis, K., Caelles, S., Pont-Tuset, J., Gool, L.V.: Deep extreme cut: from extreme points to object segmentation. CoRR, abs/1711.09081 (2017)

18. Pal, K.K., Sudeep, K.S.: Preprocessing for image classification by convolutional neural networks. In: 2016 IEEE International Conference on Recent Trends in Electronics, Information Communication Technology (RTEICT), pp. 1778–1781 (2016)

19. Pham, D.L., Xu, C., Prince, J.L.: Current methods in medical image segmentation. Annu. Rev. Biomed. Eng. **2**(1), 315–337 (2000). PMID: 11701515

20. Rajchl, M., et al.: DeepCut: object segmentation from bounding box annotations using convolutional neural networks. CoRR, abs/1605.07866 (2016)

21. Ramadan, H., Lachqar, C., Tairi, H.: A survey of recent interactive image segmentation methods. Comput. Visual Media **6**(4), 355–384 (2020). https://doi.org/10.1007/s41095-020-0177-5
22. Ronneberger, O., Fischer, P., Brox, T.: U-Net: convolutional networks for biomedical image segmentation. CoRR, abs/1505.04597 (2015)
23. Sakinis, T., et al.: Interactive segmentation of medical images through fully convolutional neural networks. CoRR, abs/1903.08205 (2019)
24. Taghanaki, S.A., Abhishek, K., Cohen, J.P., Cohen-Adad, J., Hamarneh, G.: Deep semantic segmentation of natural and medical images: a review. CoRR, abs/1910.07655 (2019)
25. Unser, M.: Splines: a perfect fit for signal and image processing. IEEE Signal Process. Mag. **16**, 22–38 (1999)
26. Wang, G., Aertsen, M., Deprest, J., Ourselin, S., Vercauteren, T., Zhang, S.: Uncertainty-guided efficient interactive refinement of fetal brain segmentation from stacks of MRI slices. In: Martel, A.L., et al. (eds.) MICCAI 2020. LNCS, vol. 12264, pp. 279–288. Springer, Cham (2020). https://doi.org/10.1007/978-3-030-59719-1_28
27. Wang, G., et al.: Interactive medical image segmentation using deep learning with image-specific fine-tuning. CoRR, abs/1710.04043 (2017)
28. Yezzi, A., Tsai, A., Willsky, A.: A fully global approach to image segmentation via coupled curve evolution equations. J. Vis. Commun. Image Represent. **13**, 195–216 (2002)
29. Zhang, H., Cissé, M., Dauphin, Y.N., Lopez-Paz, D.: mixup: beyond empirical risk minimization. CoRR, abs/1710.09412 (2017)

Multi-compound Transformer for Accurate Biomedical Image Segmentation

Yuanfeng Ji[1], Ruimao Zhang[2], Huijie Wang[2], Zhen Li[2], Lingyun Wu[3],

Shaoting Zhang[3], and Ping Luo[1(✉)]

[1] The University of Hong Kong, Pok Fu Lam, Hong Kong
pluo@cs.hku.hk
[2] Shenzhen Research Institute of Big Data, The Chinese University of Hong Kong (Shenzhen),
Shenzhen, China
[3] SenseTime Research, Beijing, China

Abstract. The recent vision transformer (*i.e.* for image classification) learns non-local attentive interaction of different patch tokens. However, prior arts miss learning the cross-scale dependencies of different pixels, the semantic correspondence of different labels, and the consistency of the feature representations and semantic embeddings, which are critical for biomedical segmentation. In this paper, we tackle the above issues by proposing a unified transformer network, termed Multi-Compound Transformer (MCTrans), which incorporates rich feature learning and semantic structure mining into a unified framework. Specifically, MCTrans embeds the multi-scale convolutional features as a sequence of tokens, and performs intra- and inter-scale self-attention, rather than single-scale attention in previous works. In addition, a learnable proxy embedding is also introduced to model semantic relationship and feature enhancement by using self-attention and cross-attention, respectively. MCTrans can be easily plugged into a UNet-like network, and attains a significant improvement over the state-of-the-art methods in biomedical image segmentation in six standard benchmarks. For example, MCTrans outperforms UNet by 3.64%, 3.71%, 4.34%, 2.8%, 1.88%, 1.57% in Pannuke, CVC-Clinic, CVC-Colon, Etis, Kavirs, ISIC2018 dataset, respectively. Code is available at https://github.com/JiYuanFeng/MCTrans.

1 Introduction

Medical image segmentation, which aims to automatically delineate anatomical structures and other regions of interest from medical images, is essential for modern computer-assisted diagnosis (CAD) applications, such as lesion detection [1,2,6,11,17] and anatomical structure localization [8]. Recent advances in segmentation accuracy are primarily driven by the power of convolution neural networks (CNN) [10,18]. However, due to the local property of the convolutional kernels, the traditional CNN-based segmentation models (*e.g.* FCN [13]) lack the ability for modeling long-term dependencies. To address such an issue, various approaches have been exploited for powerful relation modeling. For example, the spatial pyramid based methods [5,9,23] adopt various sizes of convolutional kernels to aggregate contextual information from different

M. de Bruijne et al. (Eds.): MICCAI 2021, LNCS 12901, pp. 326–336, 2021.
https://doi.org/10.1007/978-3-030-87193-2_31

ranges in a single layer (Fig. 1(a)). The UNet [16] based encoder-decoder networks [12, 16, 24] merge the coarse-grained deep features and fine-grained shallow features with the same scales by applying skip-connection. Although these methods achieved great success in dense prediction, it is still limited by the inefficient non-local context modeling among arbitrary positions, making it bleak for further promoting the accuracy of complex views.

Recently, the Vision Transformer [19], which is built upon learning attentive interaction of different patch tokens, has achieved much attention in various vision tasks [3, 7, 21, 25]. For medical image segmentation, Chen *et al.* firstly propose Tran-sUNet [4], which adopts the self-attention mechanism to compute global context at the highest-level CNN features, ensuring various ranges dependencies in a specific scale (Fig. 1(c)). However, such a design is still sub-optimal for medical image segmentation for the following reasons. First, it only uses the self-attention mechanism for context modeling on a single scale but ignores the cross-scale dependency and consistency. The latter usually plays a critical role in the segmentation of lesions with dramatic size changes. Second, beyond the context modeling, how to learn the correlation between different semantic categories and how to ensure the feature consistency of the same category region are still not taken into account. But both of them have become critical for CNN-based segmentation scheme design [22].

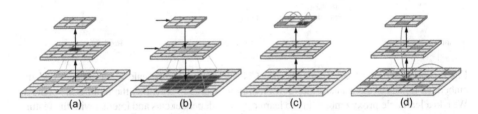

Fig. 1. Conceptual comparison of various mechanisms for context modeling for segmentation. In contrast to (a–c), MCTrans models pixel-wise relationships between multiple scales features, enabling more consistent and effective context encoding. The Prussian blue grids denote the target pixel while other color grids represent the support pixels. For simplicity, we only show a subset of the pathways between target pixels and support pixels. (Color figure online)

In this paper, attempting to overcome the limitations mentioned above, we propose the Multi-Compound Transformer (MCTrans), which incorporates rich context modeling and semantic relationship mining for accurate biomedical image segmentation. As illustrated in Fig. 2, MCTrans overcomes the limitations of conventional vision transformers by: (1) introducing the *Transformer-Self-Attention* (TSA) module to achieve cross-scale pixel-level contextual modeling via the self-attention mechanisms, leading to a more comprehensive feature enhancement for different scales. (2) developing the *Transformer-Cross-Attention* (TCA) to automatically learn the semantic correspondence of different semantic categories by introducing the proxy embedding. We further use such proxy embedding to interact with the feature representations via the cross-attention mechanism. By introducing auxiliary loss for the updated proxy embedding, we find that it could effectively improve feature correlations of the same category and the feature discriminability between different classes.

In summary, the main contributions of this paper are three folds. (1) We propose the MCTrans, which constructs cross-scale contextual dependencies and appropriates semantic relationships for accurate biomedical segmentation. (2) A novel learnable proxy embedding is introduced to build category dependencies and enhance feature representation through self-attention and cross-attention, respectively. (3) We plug the designed MCTrans into a UNet-like network and evaluate its performance on the six challenging segmentation datasets. The experiments show that MCTrans outperforms state-of-the-art methods by a significant margin with a slight computation increase in all tasks. These results demonstrate the effectiveness of all proposed network components.

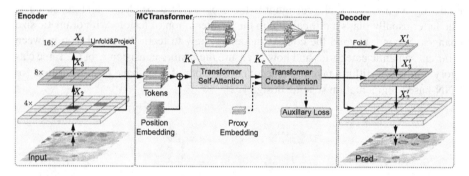

Fig. 2. The overview of MCTrans. We use CNN to extract multi-scale features, and feed the embedded tokens to the Transformer-Self-Attention module to construct the multi-scale context. We add a learnable proxy embedding to learn category dependencies and interact with the feature representations via the Transformer-Cross-Attention module. Finally, we fold the encoded tokens to several 2D feature maps and merge them progressively to generate segmentation results. For the details of the two modules, please refer to Fig. 3.

2   Related Work

Attention Mechanisms. Attention mechanisms have recently been used to construct pixel-level contextual representations. In specific, Oktay et al. [15] introduce an attention-based gate function to focus on the target and suppress irrelevant background. Lei et al. [14] further incorporate the feature-channel attention to model contextual dependencies in a more comprehensive manner. Moreover, Wang et al. [20] propose the non-local operations to connect each pair of pixels to accurately model their relationship. These methods establish context by modeling the semantic and spatial relationships between pixels in a single scale but neglect more rich information presented in other scales. In this paper, we utilize the transformer's power to construct pixel-level contextual dependencies between multiple-scale features, enabling flexible information exchange across different scales and producing more appropriate visual representations.

Transformer. The Transformer was proposed by Vaswani *et al.* [19] and first applied in the machine translation, which performs information exchange between all pairs of the inputs via the self-attention mechanism. Recently, Transformer has been proven its power in many computer vision tasks, including image classification [7], semantic segmentation [21], object detection and tracking [3,25], and so on. For medical image segmentation, our concurrent work TransUnet [4] employs Transformer-Encoder on the highest-level feature of UNet to collect long-range dependencies. Nonetheless, the methods mentioned above are not specifically designed for medical image segmentation. Our work focuses on carefully developing a better transformer-based approach, thoroughly leveraging the attention mechanism's advantages for medical image segmentation.

3   Multi-compound Transformer Network

As illustrated in Fig. 2, we introduce the MCTransformer between the classical UNet encoder and decoder architectures, which consists of the Transformer-Self-Attention (TSA) module and Transformer-Cross-Attention (TCA) module. The former is introduced to encode the contextual information between the multiple features, yielding rich and consistent pixel-level context. And the latter introduces learnable embedding for semantic relationship modeling and further enhances feature representations.

In practice, given an image $I \in \mathbb{R}^{H \times W}$, a deep CNN is adopted to extract multi-level features with different scales $\left\{ X_i \in \mathbb{R}^{\frac{H}{2^i} \times \frac{W}{2^i} \times C_i} \right\}$. For level i, features are unfolded with patch size of $P \times P$, where P is set to 1 in this paper, that is, each location of the i-th feature map will be considered as the "patch", yielding total $L_i = \frac{HW}{2^{2*i} \times P^2}$ patches. Next, different level of split patches are passed through to individual projection heads (i.e. 1×1 convolution layer) with the same output feature dimension C_e and attain the embedded tokens $T_i \in \mathbb{R}^{L_i \times C_e}$. In this paper, we concatenate the features of $i = 2, 3, 4$ level and form overall tokens $T \in \mathbb{R}^{L \times C}$, where $L = \sum_{i=2}^{4} L_i$. To compensate for missing position information, positional embedding $E_{pos} \in \mathbb{R}^{L \times C}$ is supplemented to the tokens to provide information about the relative or absolute position of the feature in the sequence, which can be formulated as $T = T + E_{pos}$. Next, we feed the tokens into the TSA module for multi-scale context modeling. The output enhanced tokens are further pass through the TCA module and interact with the proxy embedding $E_{pro} \in \mathbb{R}^{M \times C}$, where M is the number of categories of the dataset. Finally, we fold the encoded tokens back to pyramid features and merge them in a bottom-up style to obtain the final feature map for prediction.

3.1   Transformer-Self-attention

Given the 1D embedding tokens T as input, the TSA modules are employed to learn pixel-level contextual dependencies among multiple-scale features. As illustrated in Fig. 2, the TSA module consists of K_s layers, each of which consists of multi-head self-attention (MSA) and feed forward networks (FFN) (see Fig. 3(a)), layer normalization (LN) is applied before every block and residual connection after every block. The

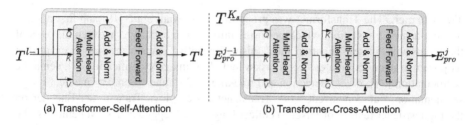

(a) Transformer-Self-Attention   (b) Transformer-Cross-Attention

Fig. 3. Illustration of the transformer-self-attention and transformer-cross-attention modules.

FFN contains two linear layers with a ReLU activation. For the l-th layers, the input to the self-attention is a triplet of (query, key, value) computed from the input T^{l-1} as:

$$\text{query} = T^{l-1}\mathbf{W}_Q^l, \text{key} = T^{l-1}\mathbf{W}_K^l, \text{value} = T^{l-1}\mathbf{W}_V^l \tag{1}$$

where $\mathbf{W}_Q^l \in \mathbb{R}^{C \times d_q}$, $\mathbf{W}_K^l \in \mathbb{R}^{C \times d_k}$, $\mathbf{W}_V^l \in \mathbb{R}^{C \times d_v}$ is the parameter matrices of different linear projections heads of l-th layer, and the d_q, d_k, d_v is the dimensions of three inputs. Self-Attention (SA) is then formulated as:

$$\text{SA}\left(T^{l-1}\right) = T^{l-1} + \text{Softmax}\left(\frac{T^{l-1}\mathbf{W}_Q^l\left(T^{l-1}\mathbf{W}_K^l\right)^\top}{\sqrt{d_k}}\right)\left(T^{l-1}\mathbf{W}_V^l\right) \tag{2}$$

MSA is an extension with h independent SA operations and project their concatenated outputs as:

$$\text{MSA}(T^{l-1}) = \text{Concat}\left(\text{SA}_1, \ldots, \text{SA}_h\right)W_O^l \tag{3}$$

where $\mathbf{W}_O \in \mathbb{R}^{hd_k \times C}$ is a parameter of output linear projection head. In this paper, we employ $h = 8$, $C = 128$ and d_q, d_k, d_v are equal to $C/h = 32$. As depicted in Fig. 3(a), the whole calculation can be formulated as:

$$T^l = \text{MSA}\left(T^{l-1}\right) + \text{FFN}\left(\text{MSA}\left(T^{l-1}\right)\right) \in \mathbb{R}^{L \times C} \tag{4}$$

We omitted the LN in the equation for simplicity. It should be noted the token T (flatten from multi-scale features) has an extremely long sequence length, and the quadratic computation complexity of MSA makes it not possible to handle. To this end, in this module, we use the Deformable Self Attention (DSA) mechanism proposed in [25] to replace the SA. As data-dependent sparse attention, which is not all-pairwise, DSA only attends to a sparse set of elements from the whole sequence regardless of its sequence length, which largely reduces computation complexity and allows the interactions of multi-level feature maps. For more details please refer to [25].

3.2   Transformer-Cross-attention

As figured in Fig. 2, beside the enhanced tokens T^{K_s}, a learnable proxy embedding E_{pro} is proposed to learn the global semantic relationship (i.e. intra-/inter- class) between categories. Like the TSA module, the TCA module consists of K_c layers but contains two multi-head self-attention blocks. In practice, for the j-th layer, the proxy embedding E_{pro}^{j-1} is transformed by various linear projection heads to yield inputs (query, key, value) of the first MSA block. Here, the MSA block's self-attention mechanism connects and interacts with each pair of categories, thus modeling the semantic correspondence of various labels. Next, the learned proxy embedding extracts and interacts with the features of the input tokens T^{K_s} via the cross attention in another MSA block, where the query input is the proxy embedding, key, and value inputs are the tokens T^{K_s}. Through the cross-attention, the features of tokens communicate with the learned global semantic relationship, comprehensively improving intra-class consistency and the inter-class discriminability of feature representation, yielding updated proxy embedding E_{pro}^{j}. Noted that the calculation of procedure two MSA block is equal to Eq. 2. Moreover, we introduce an auxiliary loss $Loss_{aux}$ to promote proxy embedding learning. In particular, the output $E_{pro}^{K_c}$ of the last layer of the TCA module is further passed to a linear projection head and yields a multi-class prediction $Pred_{aux} \in \mathbb{R}^M$. Base on the ground-truth segmentation mask, we find the unique elements to compute classification labels for supervision. In this way, the proxy embedding is driven to learn appropriate semantic relationship, and help to improve feature correlations of the same category and the feature discriminability between different categories. Finally, the encoded tokens T^{K_s} is fold back to 2D features and append the uninvolved features to form the pyramid features $\{X_0, X_1, X_2', X_3', X_4'\}$. We merge them progressively in regular bottom-up style with a $2\times$ upsampling layer and a 3×3 convolution to attain the final feature map for segmentation. For more details of the construction of multi-scale feature maps, please refer to Appendix.

Table 1. Ablation studies of core components of MCTrans. The performance is evaluated on Pannuke dataset. We estimate Flops and parameters by using $[1 \times 3 \times 256 \times 256]$ input. Note that, UNet+VIT-Enc network is equivalent to TransUNet.

Method	Params (M)	GFlops	Neo	Inflam	Conn	Dead	Epi	Ave
UNet [16]	7.853	14.037	82.86	66.16	62.45	38.10	75.02	64.92
UNet [16]+NonLocal [20]	8.379	14.172	82.67	67.48	62.63	40.44	76.41	65.93
UNet [16]+VIT-Enc [7]	27.008	18.936	83.34	68.33	63.18	38.11	77.25	66.04
MCTrans w/o TCA	7.115	18.061	83.87	68.54	64.68	44.25	78.30	67.93
MCTrans w/o TSA	6.167	11.589	83.39	67.82	63.94	44.35	76.31	67.16
MCTrans w/o Aux-Loss	7.642	18.065	83.92	67.92	64.22	45.16	78.14	67.87
MCTrans	7.642	18.065	83.99	68.24	64.95	46.39	78.42	68.40

4   Experiments

4.1   Datasets and Settings

The proposed MCTrans was evaluated on six segmentation datasets of three types. (1) Cell Segmentation [8]: Pannuke dataset (pathology, 7,904 cases, 6 classes), (2) PolyP

Segmentation [1,2,11,17]: CVC-Clinic dataset (colonoscopy, 612 cases, 2 classes), CVC-ColonDB dataset (380 cases, 2 classes), ETIS-Larib dataset (196 cases, 2 classes), Kvasir dataset (1,000 cases, 2 classes), (3) Skin Lesion Segmentation [6]: ISIC2018 dataset (dermoscopy, 2,594 cases, 2 classes). Each task has different data modalities, data sizes, and foreground classes, making them suitable for evaluating the effectiveness and generalization of the MCTrans. For cell segmentation, we report the results of the officially divided 3-fold cross-validation. For other tasks, since the annotation of test set is not publicly available, we report the 5-fold cross-validation results. Below, we mainly evaluate our approach on the Panunke dataset to show the effectiveness of different network components. Finally, we compare our MCTrans with the top methods on all of the datasets. We report all results in terms of the Dice Similarity Coefficient (DSC), and a better score indicates a better result.

We construct the MCTrans with the PyTorch toolkit. We adopt conventional CNN backbone networks, including VGG-Style [18] encoder and ResNet-34 [10], to extract multi-scale feature representations. For network optimization, we use the cross-entropy loss and dice loss to penalize the training error of segmentation and a cross-entropy loss with a weight of 0.1 for auxiliary supervision. We augment the training images with simple flipping. We use the Adam optimizer with an initial learning rate of 3e−4 to train the network. The learning rate is decayed linearly during the training. All models are trained on 1 V100 GPU. Please refer to the Appendix for more training details of specific datasets.

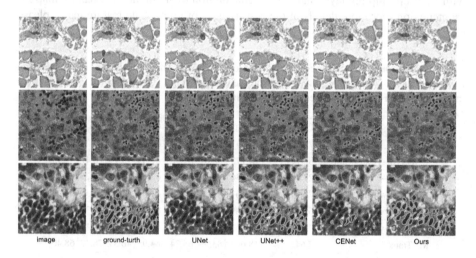

image   ground-turth   UNet   UNet++   CENet   Ours

Fig. 4. Segmentation results on the Pannuke dataset, which contains of five foreground classes: Neoplastic, Inflammatory, Connective, Dead, and Non-Neoplastic Epithelial. (Color figure online)

4.2   Ablation Studies

Analysis of the Network Components. We evaluate the importance of the core modules of MCTrans by the segmentation accuracy. We use the VGG-style network as the backbone. Compared to the UNet baseline which achieves a 64.92% dice score on the Pannuke dataset, MCTrans use TSA and TCA's power to achieve the accuracy of 68.40%. In Table 1, the performance is promoting to 67.93% by adding the TSA module to the Unet. To demonstrate the effectiveness of constructing multiple-scale pixel-level dependency, we employ the Non-local operation and Transformer-Encoder [7] on UNet's highest levels features to enable single-scale context propagation, yielding accuracies far behind our method. We further evaluate the influence of the TCA module. After adding the TCA, the learned semantic prior help to construct identified context dependencies and improve the score of Baseline and MCTrans to 67.16% and 68.40%, respectively. It indicates the effectiveness of learning semantic relationships to enhance the feature representations. We also investigate the case of removing auxiliary loss. Here, we only model semantic relationships among categories implicitly. This strategy degrades the performance to 67.87%.

Table 2. Sensitivity to the number of the TSA and TCA module.

N_s	2	4	6	8	N_c	2	4	6	8
DSC	67.25	67.67	**67.93**	67.50	–	68.15	**68.40**	68.31	68.11

Sensitivity to the Setting. We change the number of TSA and TCA modules and study the effect on the segmentation accuracy. We first increase the number N_s of the TSA module gradually to enlarge the modeling capacity. As shown in Table 2, we can see that when the size of TSA increases, the DSC score first increases and then decreases. After fixing N_s, we further plug the TCA and enlarge its size. We also discover that it reaches the top at $N_c = 4$ and then decreases. This indirectly shows that the capacity of transformer-based model is not as large as better when training on a small dataset.

4.3   Comparisons with State-of-the-Art Methods

In Table 3, we compare the MCTrans with the state-of-the-art methods on the Pannuke dataset. In the first group, we adopt a conventional VGG-Style network as feature extractor. Compared to other modeling mechanisms, our MCTrans achieves significant improvement by investing pixel-level dependencies across multiple-levels features. For a more comprehensive comparison, in the second group, we adopt a stronger features extractor (e.g., ResNet-34). Again, we achieve better accuracies than other methods. We provide the examples of the segmentation results in Fig. 4. In Table 4. We also report the results on five lesion segmentation, respectively. The results of our method still outperform other top methods by a significant margin. Such results demonstrate the versatility of the proposed MCTrans on various segmentation tasks.

We provide more details of the computational overheads (i.e. floating-point operations per second (Flops) and the number of parameters). As shown in Table 3, MCTrans achieves better results at the cost of reasonable computational overheads. Compared to the UNet baseline, MCTrans with almost identical parameters and a slight computation increase achieves a significant improvement of 3.64%. Note that the other top methods, such as UNet++, surpass MCTrans over much computation while yielding lower performance.

Table 3. Comparisons with other conventional methods on the Pannuke dataset.

Method	Params (M)	Flops (G)	Neo	Inflam	Conn	Dead	Epi	Ave
UNet [16]	7.853	14.037	82.86	66.16	62.45	38.10	75.02	64.92
UNet++ [24]	9.163	34.661	82.14	66.01	61.61	38.47	76.54	64.97
CENet [9]	17.682	18.779	83.05	66.92	62.41	38.021	76.44	65.37
AttentionUNet [15]	8.382	15.711	81.85	65.37	63.79	38.96	75.45	64.27
MCTrans	7.642	18.065	83.99	68.24	63.95	47.39	78.42	68.40
UNet [16]	24.563	38.257	82.85	65.48	62.29	40.11	75.57	65.26
UNet++ [24]	25.094	84.299	82.03	67.58	62.79	40.79	77.21	66.08
CENet [9]	34.368	41.389	82.73	**68.25**	63.15	41.12	77.27	66.50
AttentionUNet [15]	25.094	40.065	82.74	65.42	62.09	38.60	76.02	64.97
MCTrans	23.787	39.71	**84.22**	68.21	**65.04**	**48.30**	**78.70**	**68.90**

Table 4. Comparisons with other top methods on the five lesion segmentation datasets.

Method	CVC-Clinic	CVC-Colon	ETIS	Kavairs	ISIC2018
UNet [16]	88.59	82.24	80.89	84.32	88.78
UNet++ [24]	89.30	82.86	80.77	84.95	88.85
CENet [9]	91.53	83.11	75.03	84.92	89.53
AttentionUNet [15]	90.57	83.25	79.68	80.25	88.95
MCTrans	**92.30**	**86.58**	**83.69**	**86.20**	**90.35**

5   Conclusions

In this paper, we propose a powerful transformer-based network for medical image segmentation. Our method incorporates rich context modeling and semantic relationship mining via powerful attention mechanisms, effectively address the issues of cross-scale dependencies, the semantic correspondence of different categories, and so on. Our approach is effective and outperforms the state-of-the-art method such as TransUnet on several public datasets.

Acknowledgments. This work is partially supported by the General Research Fund of Hong Kong No. 27208720, the Open Research Fund from Shenzhen Research Institute of Big Data

No. 2019ORF01005, and the Research Donation from SenseTime Group Limited, the NSFC-Youth 61902335 and SRIBD Open Funding, the funding of Science and Technology Commission Shanghai Municipality No. 19511121400.

References

1. Bernal, J., Sánchez, F.J., Fernández-Esparrach, G., Gil, D., Rodríguez, C., Vilariño, F.: WM-DOVA maps for accurate polyp highlighting in colonoscopy: validation vs. saliency maps from physicians. Comput. Med. Imaging Graph. **43**, 99–111 (2015)
2. Bernal, J., Sánchez, J., Vilarino, F.: Towards automatic polyp detection with a polyp appearance model. Pattern Recogn. **45**(9), 3166–3182 (2012)
3. Carion, N., Massa, F., Synnaeve, G., Usunier, N., Kirillov, A., Zagoruyko, S.: End-to-end object detection with transformers. In: Vedaldi, A., Bischof, H., Brox, T., Frahm, J.-M. (eds.) ECCV 2020. LNCS, vol. 12346, pp. 213–229. Springer, Cham (2020). https://doi.org/10.1007/978-3-030-58452-8_13
4. Chen, J., et al.: TransUNet: transformers make strong encoders for medical image segmentation. arXiv preprint arXiv:2102.04306 (2021)
5. Chen, L.C., Papandreou, G., Kokkinos, I., Murphy, K., Yuille, A.L.: DeepLab: semantic image segmentation with deep convolutional nets, Atrous convolution, and fully connected CRFs. IEEE Trans. Pattern Anal. Mach. Intell. **40**(4), 834–848 (2017)
6. Codella, N., et al.: Skin lesion analysis toward melanoma detection 2018: a challenge hosted by the international skin imaging collaboration (ISIC). arXiv preprint arXiv:1902.03368 (2019)
7. Dosovitskiy, A., et al.: An image is worth 16x16 words: transformers for image recognition at scale. arXiv preprint arXiv:2010.11929 (2020)
8. Gamper, J., Alemi Koohbanani, N., Benet, K., Khuram, A., Rajpoot, N.: PanNuke: an open pan-cancer histology dataset for nuclei instance segmentation and classification. In: Reyes-Aldasoro, C.C., Janowczyk, A., Veta, M., Bankhead, P., Sirinukunwattana, K. (eds.) ECDP 2019. LNCS, vol. 11435, pp. 11–19. Springer, Cham (2019). https://doi.org/10.1007/978-3-030-23937-4_2
9. Gu, Z., et al.: CE-Net: context encoder network for 2D medical image segmentation. IEEE Trans. Med. Imaging **38**(10), 2281–2292 (2019)
10. He, K., Zhang, X., Ren, S., Sun, J.: Deep residual learning for image recognition. In: Proceedings of the IEEE Conference on Computer Vision and Pattern Recognition, pp. 770–778 (2016)
11. Jha, D., et al.: A comprehensive study on colorectal polyp segmentation with ResUNet++, conditional random field and test-time augmentation (2020)
12. Ji, Y., Zhang, R., Li, Z., Ren, J., Zhang, S., Luo, P.: UXNet: searching multi-level feature aggregation for 3D medical image segmentation. In: Martel, A.L., et al. (eds.) MICCAI 2020. LNCS, vol. 12261, pp. 346–356. Springer, Cham (2020). https://doi.org/10.1007/978-3-030-59710-8_34
13. Long, J., Shelhamer, E., Darrell, T.: Fully convolutional networks for semantic segmentation. In: Proceedings of the IEEE Conference on Computer Vision and Pattern Recognition, pp. 3431–3440 (2015)
14. Mou, L., et al.: CS-Net: channel and spatial attention network for curvilinear structure segmentation. In: Shen, D., et al. (eds.) MICCAI 2019. LNCS, vol. 11764, pp. 721–730. Springer, Cham (2019). https://doi.org/10.1007/978-3-030-32239-7_80
15. Oktay, O., et al.: Attention U-Net: learning where to look for the pancreas. arXiv preprint arXiv:1804.03999 (2018)

16. Ronneberger, O., Fischer, P., Brox, T.: U-Net: convolutional networks for biomedical image segmentation. In: Navab, N., Hornegger, J., Wells, W.M., Frangi, A.F. (eds.) MICCAI 2015. LNCS, vol. 9351, pp. 234–241. Springer, Cham (2015). https://doi.org/10.1007/978-3-319-24574-4_28

17. Silva, J., Histace, A., Romain, O., Dray, X., Granado, B.: Toward embedded detection of polyps in WCE images for early diagnosis of colorectal cancer. Int. J. Comput. Assist. Radiol. Surg. 9(2), 283–293 (2014)

18. Simonyan, K., Zisserman, A.: Very deep convolutional networks for large-scale image recognition. arXiv preprint arXiv:1409.1556 (2014)

19. Vaswani, A., et al.: Attention is all you need. arXiv preprint arXiv:1706.03762 (2017)

20. Wang, X., Girshick, R., Gupta, A., He, K.: Non-local neural networks. In: Proceedings of the IEEE Conference on Computer Vision and Pattern Recognition, pp. 7794–7803 (2018)

21. Xie, E., et al.: Segmenting transparent object in the wild with transformer. arXiv preprint arXiv:2101.08461 (2021)

22. Yu, C., Wang, J., Gao, C., Yu, G., Shen, C., Sang, N.: Context prior for scene segmentation. In: Proceedings of the IEEE/CVF Conference on Computer Vision and Pattern Recognition, pp. 12416–12425 (2020)

23. Zhao, H., Shi, J., Qi, X., Wang, X., Jia, J.: Pyramid scene parsing network. In: Proceedings of the IEEE Conference on Computer Vision and Pattern Recognition, pp. 2881–2890 (2017)

24. Zhou, Z., Rahman Siddiquee, M.M., Tajbakhsh, N., Liang, J.: UNet++: a nested U-Net architecture for medical image segmentation. In: Stoyanov, D., et al. (eds.) DLMIA/ML-CDS - 2018. LNCS, vol. 11045, pp. 3–11. Springer, Cham (2018). https://doi.org/10.1007/978-3-030-00889-5_1

25. Zhu, X., Su, W., Lu, L., Li, B., Wang, X., Dai, J.: Deformable DETR: deformable transformers for end-to-end object detection. arXiv preprint arXiv:2010.04159 (2020)

kCBAC-Net: Deeply Supervised Complete Bipartite Networks with Asymmetric Convolutions for Medical Image Segmentation

Pengfei Gu$^{(\boxtimes)}$, Hao Zheng, Yizhe Zhang, Chaoli Wang, and Danny Z. Chen

Department of Computer Science and Engineering, University of Notre Dame,
Notre Dame, IN 46556, USA
{pgu,hzheng3,yzhang29,chaoli.wang,dchen}@nd.edu

Abstract. Accurate and automatic medical image segmentation is challenging due to significant size and shape variations of objects (e.g., in multi-scales) and missing/blurring object borders. In this paper, we propose a new *deeply supervised k-complete-bipartite network with asymmetric convolutions* (kCBAC-Net) to exploit multi-scale features and improve the capability of standard convolutions for segmentation. (1) We leverage a generalized complete bipartite network to reuse multi-scale features, consolidate feature hierarchies at different scales, and preserve maximum information flow between encoder and decoder layers. (2) To further capture multi-scale information, we sequentially connect k complete bipartite network modules together to facilitate their processing in different image scales. (3) We replace the standard convolution by asymmetric convolution block to strengthen the central skeleton parts of standard convolution, enhancing the model's robustness on exploiting more discriminative features. (4) We employ auxiliary deep supervisions to boost information flow in the network and extract highly discriminative features. We evaluate our kCBAC-Net on three datasets (ultrasound lymph node segmentation (2D), 2017 ISIC Skin Lesion segmentation (2D), and MM-WHS CT (3D)), achieving state-of-the-art performance.

1 Introduction

Accurate image segmentation is critical to medical image analysis, disease diagnosis, and clinical applications (e.g., quantitative analysis of lymph node sizes and shapes [28], melanoma diagnosis [3], and cardiovascular surgical planning [18]). However, automatic medical image segmentation with high accuracy is a very non-trivial task due to multiple challenging factors, including (I) multiscale objects: the variations of object sizes and appearances can be very large, and in some extreme cases, the ratio between the largest and smallest objects could be hundreds; (II) missing/blurring object borders: due to low contrast or noisy background, the borders between objects or background may be missing or very ambiguous. Figure 1 gives some examples to illustrate such challenges.

© Springer Nature Switzerland AG 2021
M. de Bruijne et al. (Eds.): MICCAI 2021, LNCS 12901, pp. 337–347, 2021.
https://doi.org/10.1007/978-3-030-87193-2_32

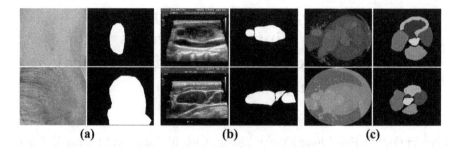

Fig. 1. Input image examples (left) and their ground truth (right) (best viewed in color). (a) 2017 ISIC Skin Lesion segmentation dataset [10]: Lesions vary a lot in sizes and shapes and their borders could be very ambiguous due to low contrast between lesion regions and surrounding skin. (b) Ultrasound lymph node images: Multi-scale objects and unclear borders increase the difficulty of distinguishing real lymph nodes and regions highly similar to lymph nodes. (c) 2017 MM-WHS CT dataset [31]: Different detailed structures are in various scales and boundaries are unclear. (Color figure online)

Many deep learning based image segmentation methods were proposed to address these challenges. There are three main types of methods for dealing with the issue of multi-scale objects. (1) Encoder-decoder structures: U-Net [20] and its variants [16, 30] utilized skip connections to fuse multi-scale features extracted from the encoder to the counterparts in the decoder. (2) Multi-path diagram: These frameworks [9, 21] fed multi-scale images to the same network (with shared weights) and captured features from various scale inputs. (3) Spatial pyramid design: DeepLabv3 [8] utilized an atrous spatial pyramid pooling module containing multiple parallel atrous convolution layers to extract multi-scale contextual information. An array of studies sought to tackle the missing/blurring border issue. In [5], two branches were used to segment the main regions-of-interest (ROIs) and the corresponding contours separately. In [27, 28], multiple sub-modules were built to encourage deep neural networks to learn richer and more comprehensive features. Despite yielding promising performance on various segmentation tasks, these known methods tended to emphasize on addressing one challenge while neglecting the other. It is highly desirable to tackle these two major issues simultaneously for accurate and robust segmentation.

In this paper, we propose a new *deeply supervised k-complete-bipartite network with asymmetric convolutions* (kCBAC-Net), aiming to extensively exploit multi-scale features and enhance the capability of standard convolutions on extracting discriminative features. Specifically, we develop kCBAC-Net with the following ideas. (1) A generalized complete bipartite network adopted from CB-Net [6] is employed to reuse and aggregate multi-scale features and boost information flow between encoder and decoder layers. (2) Following the structure of kU-Net [7], k complete bipartite network modules for processing different image scales are systematically connected to enhance and assimilate multi-scale information. (3) The standard convolution is incorporated with asymmetric con-

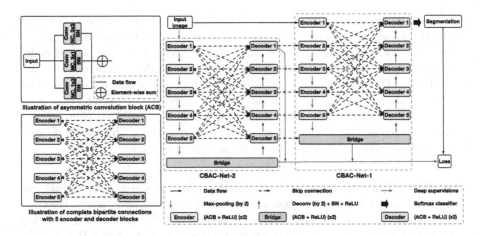

Fig. 2. An overview of kCBAC-Net with $k = 2$. The top-left box illustrates an asymmetric convolution block. The bottom-left box presents complete bipartite connections with 5 encoder and decoder blocks. The architecture is shown in 2D manner here; however, we also implement its 3D version. Best viewed in color. (Color figure online)

volutions [11] to strengthen the central skeleton portions of the standard convolution, reinforcing the network's robustness on exploiting discriminative features. (4) Auxiliary deep supervisions are applied to bolster the network on extracting discriminative features for tackling the missing/blurring border issue.

Our experiments on kCBAC-Net with two public datasets (2017 ISIC Skin Lesion segmentation (2D) [10] and MM-WHS CT (3D) [31]) and one in-house dataset (for lymph node segmentation in ultrasound images (2D)) show that kCBAC-Net outperforms state-of-the-art methods on these datasets.

2 Method

Figure 2 shows the architecture of our kCBAC-Net. It contains three main components: (1) a k-complete-bipartite network (kCB-Net) that extracts and reuses rich multi-scale features, using complete bipartite connections (CBC) and k sequentially connected network modules; (2) kCBAC-Net that exploits more discriminative features via asymmetric convolutions; (3) auxiliary deep supervisions employed at layers far from the layer for computing loss functions that benefit effective network training and highly discriminative feature capturing.

2.1 k-Complete-Bipartite Network (kCB-Net)

A common way to reuse multi-scale features is to leverage skip connections. Using skip connections, multi-scale features extracted by the encoder can be utilized by the counterparts of the decoder. But, such skip connections may not ensure full exploitation and reuse of multi-scale features captured by the encoder. To further exploit multi-scale features, we design kCB-Net with complete bipartite connections and the structure of sequentially connected k network modules.

In an encoder-decoder structure (e.g., U-Net) with l encoder and decoder blocks, let x_i and y_i denote the outputs of the ith encoder and decoder blocks, respectively. Then x_i (or y_i) can be computed by a transformation function from the output x_{i-1} (or y_{i-1}) of the previous encoder (or decoder) block as $x_i = E_i(x_{i-1})$ and $y_i = D_i(y_{i-1})$, where E_i (or $D_i(x)$) can be a composite of operations such as Max Pooling (Deconvolution (Deconv)), Convolution (Conv), Batch Normalization (BN), Rectified Linear Unit (ReLU), etc.

Complete Bipartite Connections (CBC). To better reuse multi-scale features and enhance information flow among encoder and decoder blocks, the complete bipartite network (CB-Net) [6] used the idea of CBC, introducing one skip connection between every pair of encoder and decoder blocks, while HR-Net [23] gradually expands in deeper stages of encoder. Specifically,

$$y_i = D_i(E_1(x_0) \odot \cdots \odot E_i(x_{i-1}) \odot \cdots \odot E_l(x_{l-1}) \odot y_{i-1}),$$

where $\odot$ denotes concatenation operation. An example of CBC is given in the bottom-left box of Fig. 2. With a skip connection between each pair of encoder/decoder blocks, multi-scale features can be consolidated in the network hierarchy, and can be effectively reused at the decoder blocks. In addition, information flow between encoder and decoder blocks can be enlarged by the introduced skip connections. These characteristics are essential for the network to exploit and reuse multi-scale information.

kCB-Net Organization. Another common deep learning based solution for handling multi-scale objects is a multi-path diagram. However, the design using the same network modules with shared weights may not be suitable for dealing with objects with significant size and appearance variations. In [7], kU-Net was proposed to handle multi-scale objects. Specifically, k U-Nets (without sharing parameters) working on different image scales are sequentially connected to extract multi-scale information. This structure has two compelling advantages. (1) The networks can view different-size regions of the same image with different scales, which facilitates processing multi-scale objects. (2) The multi-scale information extracted by one network can be propagated to the subsequent networks, which may assist further exploiting multi-scale information.

The organization of our kCB-Net is adopted from kU-Net [7]. First, an image of scale s_t ($t = 1, \ldots, k$) generated after $t - 1$ max pooling layers is processed by a network module CB-Net-t. Second, CB-Net-$(t - 1)$ takes every piece of information from CB-Net-t in the commensurate layers to well assimilate multi-scale information. We use this way to connect our CB-Net modules because such connections yield the best performance in our experiments. Note that the memory and time costs become larger when k increases. For example, on the lymph node dataset, for $k = 1, 2, 3, 4$, the F1 scores are 0.871, 0.897, 0.903, and 0.902 (see the F1 scores of CB-Net, kCB-Net, 3CB-Net, and 4CB-Net in Table 1), and the memory costs are 5.5 GB, 11.8 GB, 18 GB, and 24.4 GB, respectively. The performance saturates with larger k, and the costs increase largely. Considering the trade-off between performance and computation, we set $k = 2$ in our experiments.

Note that it is highly desirable to address the two major issues discussed in Sect. 1 simultaneously by the same model. Below we elaborate how to tackle the other issue, missing/blurring object borders, in our kCB-Net.

2.2 kCBAC-Net: Leveraging Asymmetric Convolutions

Recent studies attempting to tackle the missing/blurring border issue tended to focus on leveraging multiple modules to deal with the objects and boundaries separately, or encouraging the networks to learn richer and more distinguishable features. Such methods did not explore the possibility of identifying missing/blurring object borders by enhancing the network's capability via exploiting more discriminative features from the perspective of convolution operations.

An asymmetric convolution block (ACB) [11] adds additional $1 \times d$ and $d \times 1$ Conv layers on the basis of a standard $d \times d$ Conv layer, to strengthen the central skeleton parts of standard Conv, as:

$$ACB(x) = BN\left(Conv_{d \times d}(x)\right) + BN\left(Conv_{1 \times d}(x)\right) + BN\left(Conv_{d \times 1}(x)\right)$$

where d is an integer and $+$ denotes element-wise sum. It is more sophisticated than inception-v2 [22] that decomposes only standard Conv. The added asymmetric convolutions can enhance the network's robustness with respect to rotational distortions, thus improving its capability to learn more distinguishable and discriminative features. This extension is important for us to address missing/blurring borders, as it can reduce ambiguities of object borders more effectively than standard convolution. Hence, we replace the standard convolution layers in our kCB-Net by ACBs, and call the resulted network kCBAC-Net.

The architecture of kCBAC-Net is shown in Fig. 2. The structure of CBAC-Net is adopted from CB-Net [6], with five encoder and decoder blocks. Each encoder or decoder block contains two ACBs, with each ACB followed by a ReLU. Starting from the first encoder block, we double the number of feature channels, and half the number of feature channels in the decoder blocks. The structure of kCBAC-Net uses the organization of kCB-Net (see Sect. 2.1).

2.3 Deep Supervision (DS)

Effectively training deep learning networks (especially deeper networks with a large number of parameters) is non-trivial, as the gradient can be gradually vanishing when it is propagated back to early layers. Hence, the layers that are far away from the layer computing loss functions may not be trained well [2]. Deep supervision [13] is a technique to address this issue. However, simply adding deep supervisions to the network may collapse task-relevant information at shallow layers, and thus hurt the performance [24].

Table 1. Comparison of segmentation results on the ultrasound lymph node dataset. CB-Net w/o CBC is a deeper U-Net with one more scale of encoding and decoding block. The best results are marked in bold (the same for the other tables in this paper).

Method	IoU	Precision	Recall	F1 score
U-Net [20]	0.661	0.834	0.761	0.796
Zhang et al. [27]	0.810	0.901	0.889	0.895
kCB-Net	0.814	0.907	0.888	0.897
kCB-Net + ACB (kCBAC-Net)	0.819	0.900	0.900	0.900
kCBAC-Net + DS	**0.829**	**0.909**	**0.904**	**0.906**
Ablation study				
CB-Net w/o CBC (deeper U-Net)	0.757	0.869	0.854	0.861
CB-Net	0.771	0.893	0.849	0.871
3CB-Net	0.824	0.908	0.898	0.903
4CB-Net	0.822	0.911	0.894	0.902

Table 2. Comparison of segmentation results on the 2017 ISIC Skin Lesion dataset.

Method	Jaccard index	Dice	Sensitivity	Specificity
Yuan et al. [26]	0.765	0.849	0.825	0.975
Li et al. [15]	0.765	0.866	0.825	0.984
Lei et al. [14]	0.771	0.859	0.835	0.976
Mirikharaji et al. [17]	0.773	0.857	0.855	0.973
Xie et al. [25]	0.788	0.868	**0.884**	0.957
kCB-Net	0.784	0.881	0.827	0.984
kCBAC-Net	0.788	0.884	0.848	0.980
kCBAC-Net + DS	**0.794**	**0.887**	0.847	**0.984**

An important question is: *At which layers should we add deep supervisions?* It was shown experimentally [1] that (1) adding deep supervisions at layers far away from the layer computing loss functions can effectively address performance deteriorate, and (2) the last layer of the encoder in U-Net is its farthest layer to the layer computing loss functions. To boost information flow in our network and extract more discriminative features to further handle the missing/blurring border issue, we employ three auxiliary deep supervisions in our kCBAC-Net (see Fig. 2). Specifically, we first add two auxiliary deep supervisions at the last layer of the Bridge block, which is the farthest layer of each CBAC-Net; then we add an auxiliary deep supervision to the last layer of Decoder 1 of CBAC-Net-2 to further boost information flow. We denote the auxiliary losses at the last layer of the Bridge and Decoder 1 blocks of CBACNet-2 as $Loss_{aux1}$ and $Loss_{aux2}$, respectively, and the auxiliary loss at the last layer of the Bridge of CBACNet-1 as $Loss_{aux3}$. Adding these three auxiliary losses to the main loss ($Loss_{main}$), our total loss ($Loss_{total}$) becomes:

$$Loss_{total} = \sum_{i=1}^{3} \lambda_i Loss_{auxi} + Loss_{main}$$

where λ_i is a balancing weight, and $Loss_{auxi}$ and $Loss_{main}$ are computed using the auxiliary output and main output of kCBAC-Net. Our kCBAC-Net employs DS in multiple modules rather than only in a certain one (e.g., U-Net++ [30]).

3   Experiments and Results

Two 2D Datasets and One 3D Dataset. (1) **The ultrasound lymph node segmentation dataset:** This in-house ultrasound lymph node dataset contains 137 training and 100 test images. The task is to segment lymph nodes in 2D ultrasound images. (2) **The 2017 ISIC Skin Lesion segmentation dataset:** This public dataset [10] contains 2000 training, 150 validation, and 600 test images. The task is to segment lesion boundaries in 2D dermoscopic images. Following [25,26], we resize the images to 224×224, and apply a dual-threshold method to generate the final results. (3) **The 2017 MM-WHS CT dataset:** This public dataset [31] contains 20 unpaired 3D CT images. Similar to [29], the dataset is randomly split into 16 images and 4 images for training and test. The task is to segment seven cardiac structures: the left/right ventricle blood cavity (LV/RV), left/right atrium blood cavity (LA/RA), myocardium of the left ventricle (LV-myo), ascending aorta (AO), and pulmonary artery (PA).

Table 3. Comparison of segmentation results on the 2017 MM-WHS CT dataset. Note: "—" means that the corresponding results were not reported by that method.

Method	Metric	LV	RV	LA	RA	LV-myo	AO	PA	Mean
Payer et al. [19]	Dice	0.918	0.909	0.929	0.888	0.881	0.933	0.840	0.900
Dou et al. [12]	Dice	0.888	—	0.891	—	0.733	0.813	—	—
Chen et al. [4]	Dice	0.919	—	0.911	—	0.877	0.927	—	0.909
HFA-Net [29]	Dice	0.946	0.893	0.925	0.897	0.910	0.964	0.830	0.909
	Jaccard	0.898	0.810	0.861	0.816	0.836	0.930	0.722	0.839
	Hausdorff [voxel]	7.148	33.128	42.173	22.903	36.954	12.075	37.845	27.461
	ADB [voxel]	0.076	0.562	0.210	0.334	0.225	0.103	1.685	0.456
kCB-Net	Dice	0.950	0.894	0.929	**0.913**	0.916	0.968	0.837	0.915
	Jaccard	0.905	0.811	0.869	**0.841**	0.845	0.938	0.730	0.849
	Hausdorff [voxel]	8.887	**12.639**	14.630	24.229	10.642	20.713	33.663	17.915
	ADB [voxel]	0.074	0.338	0.205	**0.244**	0.122	0.074	1.467	0.361
kCBAC-Net	Dice	0.951	0.894	**0.939**	0.909	0.920	0.966	**0.843**	0.917
	Jaccard	0.907	0.812	**0.886**	0.834	0.851	0.934	**0.742**	0.852
	Hausdorff [voxel]	5.534	13.315	15.296	**14.670**	7.649	17.451	30.758	14.953
	ADB [voxel]	0.075	0.339	**0.139**	0.268	**0.115**	0.068	1.391	0.342
kCBAC-Net + DS	Dice	**0.951**	**0.902**	0.938	0.911	**0.922**	**0.974**	0.837	**0.919**
	Jaccard	**0.907**	**0.825**	0.883	0.838	**0.855**	**0.949**	0.734	**0.856**
	Hausdorff [voxel]	**5.500**	14.940	**12.403**	15.081	**7.337**	**6.848**	32.499	**13.516**
	ADB [voxel]	**0.074**	**0.285**	0.163	0.248	0.119	**0.059**	1.403	**0.336**

Implementation Details. Our new network is implemented with TensorFlow, trained on an Nvidia Tesla V100 Graphics Card with 32 GB GPU memory using the Adam optimizer ($\beta_1 = 0.9$, $\beta_2 = 0.999$, $\epsilon = 1e-10$). The "poly" learning rate policy, $L_r \times \left(1 - \frac{iter}{\#iter}\right)$, is applied, with the initial learning rate $= 5e-4$, and the maximum number of iterations is 140k for the ultrasound lymph node and 2017 ISIC Skin Lesion datasets, and 180k for the 2017 MM-WHS CT dataset (using about 26, 21, and 92 hours, respectively, for training). Standard data augmentation (e.g., image crop, flip, and rotation in 90, 180, and 270 degrees) is utilized to reduce overfitting. The balancing weights λ_1, λ_2, and λ_3 are set as 0.25, 0.50, and 0.75 for the ultrasound lymph node and 2017 MM-WHS CT datasets and 1.0, 1.0, and 1.0 for the 2017 ISIC Skin Lesion dataset.

Quantitative Results. Table 1 gives quantitative comparison of our method with U-Net [20] and the best-known method [27] on the ultrasound lymph node dataset. First, our kCB-Net already obtains a new state-of-the-art performance. We attribute this to the compelling advantages of CBC and the structure of k connected network modules (without sharing parameters) on exploiting multi-scale features. Second, by leveraging ACB and DS, our method further improves the state-of-the-art performance (1.9% IoU and 1.1% F1 score over Zhang et al. [27]), showing the capability of ACB and the employed DS for capturing more distinguishable and discriminative features to handle the missing/blurring border issue. Results on the 2017 ISIC Skin Lesion dataset are given in Table 2. We compare our results with five recent skin lesion segmentation methods, including the 2017 ISIC Challenge winner [26], dense deconvolution network [15], dense convolution U-Net [14], FCN with a star shape prior [17], and a state-of-the-art method [25]. First, kCB-Net attains comparable Jaccard index score (the only evaluation metric used by the challenge organizers to rank competitors) as the state-of-the-art method [25]. This indicates the effectiveness of CBC and the structure of k connected network modules on exploiting multi-scale information. Second, by utilizing ACB and DS, our method achieves the highest Jaccard index, Dice, and specificity, though its sensitivity is slightly lower than that of some other methods. We attribute the improvement to the effects of ACB and DS on handling missing/blurring borders. To further show the effectiveness of our method on 3D images, we experiment with the 2017 MM-WHS CT dataset, and compare with four representative methods. As shown in Table 3, our method yields better results than the other methods on all the evaluation measures and achieves new state-of-the-art performance. Qualitative results showing our strong capability of delineating missing/blurring borders and handling multi-scale objects are given in Fig. 3.

Ablation Study. We conduct ablation study using the ultrasound lymph node dataset to examine the importance of CBC and the structure of k connected network modules on exploiting multi-scale features. As shown in Table 1, (1) when removing CBC from CB-Net, the F1 score drops by 1%; (2) when experimenting with CB-Net (i.e., setting $k = 1$), the F1 score drops by 2.6%. These observations demonstrate that CBC and the structure of *k* connected network modules indeed play a meaningful role in exploiting multi-scale features.

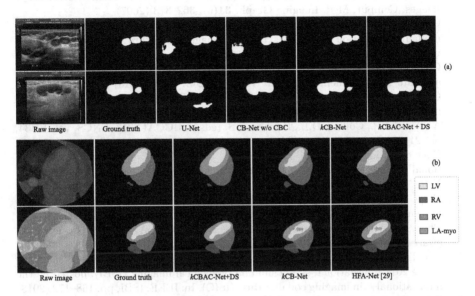

Fig. 3. Some visual qualitative results on the ultrasound lymph node dataset (a) and 2017 MM-WHS CT dataset (b), showing the capability of our new method (*k*CBAC-Net with deep supervisions) on handling the issues of multi-scale objects and missing/blurring borders. Magenta arrows mark some errors. (Color figure online)

4   Conclusions

In this paper, we presented a new *deeply supervised k-complete-bipartite network with asymmetric convolutions* (*k*CBAC-Net) for medical image segmentation. Our *k*CBAC-Net leverages a generalized complete bipartite network and the structure of *k* connected network modules to exploit and reuse multi-scale features for dealing with the multi-scale object issue. To further enhance the capability of our model on handling the missing/blurring border issue, ACB and auxiliary deep supervisions are employed. Experiments on two public datasets and one in-house dataset demonstrated the effectiveness of our *k*CBAC-Net.

Acknowledgement. This research was supported in part by NSF grants IIS-1455886, CCF-1617735, CNS-1629914, and IIS-1955395.

References

1. Ando, T., Hotta, K.: Cell image segmentation by feature random enhancement module. In: VISIGRAPP, pp. 520–527 (2021)

2. Bengio, Y., Simard, P., Frasconi, P.: Learning long-term dependencies with gradient descent is difficult. IEEE Trans. Neural Netw. **5**(2), 157–166 (1994)
3. Celebi, M.E., et al.: A methodological approach to the classification of dermoscopy images. Comput. Med. Imaging Graph. **31**(6), 362–373 (2007)
4. Chen, C., Dou, Q., Chen, H., Qin, J., Heng, P.A.: Unsupervised bidirectional cross-modality adaptation via deeply synergistic image and feature alignment for medical image segmentation. IEEE Trans. Med. Imaging **39**(7), 2494–2505 (2020)
5. Chen, H., Qi, X., Yu, L., Heng, P.A.: DCAN: deep contour-aware networks for accurate gland segmentation. In: CVPR, pp. 2487–2496 (2016)
6. Chen, J., Banerjee, S., Grama, A., Scheirer, W.J., Chen, D.Z.: Neuron segmentation using deep complete bipartite networks. In: Descoteaux, M., Maier-Hein, L., Franz, A., Jannin, P., Collins, D.L., Duchesne, S. (eds.) MICCAI 2017. LNCS, vol. 10434, pp. 21–29. Springer, Cham (2017). https://doi.org/10.1007/978-3-319-66185-8_3
7. Chen, J., Yang, L., Zhang, Y., Alber, M., Chen, D.Z.: Combining fully convolutional and recurrent neural networks for 3D biomedical image segmentation. In: NIPS, pp. 3036–3044 (2016)
8. Chen, L.C., Papandreou, G., Schroff, F., Adam, H.: Rethinking atrous convolution for semantic image segmentation. arXiv preprint arXiv:1706.05587 (2017)
9. Chen, L.C., Yang, Y., Wang, J., Xu, W., Yuille, A.L.: Attention to scale: scale-aware semantic image segmentation. In: CVPR, pp. 3640–3649 (2016)
10. Codella, N.C., et al.: Skin lesion analysis toward melanoma detection: a challenge at the 2017 international symposium on biomedical imaging (ISBI), hosted by the international skin imaging collaboration (ISIC). In: IEEE, ISBI, pp. 168–172 (2018)
11. Ding, X., Guo, Y., Ding, G., Han, J.: ACNet: strengthening the kernel skeletons for powerful CNN via asymmetric convolution blocks. In: ICCV, pp. 1911–1920 (2019)
12. Dou, Q., Ouyang, C., Chen, C., Chen, H., Heng, P.A.: Unsupervised cross-modality domain adaptation of ConvNets for biomedical image segmentations with adversarial loss. In: IJCAI, pp. 691–697 (2018)
13. Lee, C.Y., Xie, S., Gallagher, P., Zhang, Z., Tu, Z.: Deeply-supervised nets. In: AISTATS, pp. 562–570 (2015)
14. Lei, B., et al.: Skin lesion segmentation via generative adversarial networks with dual discriminators. Med. Image Anal. **64**, 101716 (2020)
15. Li, H., et al.: Dense deconvolutional network for skin lesion segmentation. IEEE J. Biomed. Health Inform. **23**(2), 527–537 (2019)
16. Liang, P., Chen, J., Zheng, H., Yang, L., Zhang, Y., Chen, D.Z.: Cascade decoder: a universal decoding method for biomedical image segmentation. In: IEEE, ISBI, pp. 339–342 (2019)
17. Mirikharaji, Z., Hamarneh, G.: Star shape prior in fully convolutional networks for skin lesion segmentation. In: Frangi, A.F., Schnabel, J.A., Davatzikos, C., Alberola-López, C., Fichtinger, G. (eds.) MICCAI 2018. LNCS, vol. 11073, pp. 737–745. Springer, Cham (2018). https://doi.org/10.1007/978-3-030-00937-3_84
18. Pace, D.F., Dalca, A.V., Geva, T., Powell, A.J., Moghari, M.H., Golland, P.: Interactive whole-heart segmentation in congenital heart disease. In: Navab, N., Hornegger, J., Wells, W.M., Frangi, A.F. (eds.) MICCAI 2015. LNCS, vol. 9351, pp. 80–88. Springer, Cham (2015). https://doi.org/10.1007/978-3-319-24574-4_10
19. Payer, C., Štern, D., Bischof, H., Urschler, M.: Multi-label whole heart segmentation using CNNs and anatomical label configurations. In: Pop, M., et al. (eds.) STACOM 2017. LNCS, vol. 10663, pp. 190–198. Springer, Cham (2018). https://doi.org/10.1007/978-3-319-75541-0_20

20. Ronneberger, O., Fischer, P., Brox, T.: U-Net: convolutional networks for biomedical image segmentation. In: Navab, N., Hornegger, J., Wells, W.M., Frangi, A.F. (eds.) MICCAI 2015. LNCS, vol. 9351, pp. 234–241. Springer, Cham (2015). https://doi.org/10.1007/978-3-319-24574-4_28

21. Sun, C., Paluri, M., Collobert, R., Nevatia, R., Bourdev, L.: ProNet: learning to propose object-specific boxes for cascaded neural networks. In: CVPR, pp. 3485–3493 (2016)

22. Szegedy, C., Vanhoucke, V., Ioffe, S., Shlens, J., Wojna, Z.: Rethinking the inception architecture for computer vision. In: CVPR, pp. 2818–2826 (2016)

23. Wang, J., et al.: Deep high-resolution representation learning for visual recognition. IEEE Trans. Pattern Anal. Mach. Intell. **43**(10), 3394–3364 (2020)

24. Wang, Y., Ni, Z., Song, S., Yang, L., Huang, G.: Revisiting locally supervised learning: an alternative to end-to-end training. In: ICLR (2021)

25. Xie, Y., Zhang, J., Lu, H., Shen, C., Xia, Y.: SESV: accurate medical image segmentation by predicting and correcting errors. IEEE Trans. Med. Imaging **40**(1), 286–296 (2021)

26. Yuan, Y., Lo, Y.C.: Improving dermoscopic image segmentation with enhanced convolutional-deconvolutional networks. IEEE J. Biomed. Health Inform. **23**(2), 519–526 (2019)

27. Zhang, Y., Ying, M.T.C., Chen, D.Z.: Decompose-and-integrate learning for multi-class segmentation in medical images. In: Shen, D., et al. (eds.) MICCAI 2019. LNCS, vol. 11765, pp. 641–650. Springer, Cham (2019). https://doi.org/10.1007/978-3-030-32245-8_71

28. Zhang, Y., Ying, M.T., Yang, L., Ahuja, A.T., Chen, D.Z.: Coarse-to-fine stacked fully convolutional nets for lymph node segmentation in ultrasound images. In: IEEE, BIBM, pp. 443–448 (2016)

29. Zheng, H., et al.: HFA-Net: 3D cardiovascular image segmentation with asymmetrical pooling and content-aware fusion. In: Shen, D., et al. (eds.) MICCAI 2019. LNCS, vol. 11765, pp. 759–767. Springer, Cham (2019). https://doi.org/10.1007/978-3-030-32245-8_84

30. Zhou, Z., Rahman Siddiquee, M.M., Tajbakhsh, N., Liang, J.: UNet++: a nested U-Net architecture for medical image segmentation. In: Stoyanov, D., et al. (eds.) DLMIA/ML-CDS -2018. LNCS, vol. 11045, pp. 3–11. Springer, Cham (2018). https://doi.org/10.1007/978-3-030-00889-5_1

31. Zhuang, X., Shen, J.: Multi-scale patch and multi-modality atlases for whole heart segmentation of MRI. Med. Image Anal. **31**, 77–87 (2016)

Multi-frame Attention Network for Left Ventricle Segmentation in 3D Echocardiography

Shawn S. Ahn[1]([✉]), Kevinminh Ta[1], Stephanie Thorn[2], Jonathan Langdon[4], Albert J. Sinusas[2,3,4], and James S. Duncan[3,4]

[1] Department of Biomedical Engineering, Yale University, New Haven, CT, USA
shawn.ahn@yale.edu
[2] Section of Cardiovascular Medicine, Department of Internal Medicine, Yale University, New Haven, CT, USA
[3] Department of Electrical Engineering, Yale University, New Haven, CT, USA
[4] Department of Radiology and Biomedical Imaging, Yale University, New Haven, CT, USA

Abstract. Echocardiography is one of the main imaging modalities used to assess the cardiovascular health of patients. Among the many analyses performed on echocardiography, segmentation of left ventricle is crucial to quantify the clinical measurements like ejection fraction. However, segmentation of left ventricle in 3D echocardiography remains a challenging and tedious task. In this paper, we propose a multi-frame attention network to improve the performance of segmentation of left ventricle in 3D echocardiography. The multi-frame attention mechanism allows highly correlated spatiotemporal features in a sequence of images that come after a target image to be used to augment the performance of segmentation. Experimental results shown on 51 in vivo porcine 3D+time echocardiography images show that utilizing correlated spatiotemporal features significantly improves the performance of left ventricle segmentation when compared to other standard deep learning-based medical image segmentation models.

Keywords: 3D echocardiography · Multi-frame attention · Segmentation

1 Introduction

Cardiovascular diseases (CVDs) are the number one cause of death globally with more than 17 million deaths each year [21]. CVDs include a wide range of disorders of the heart and vessels including coronary artery disease, valvular diseases, and cerebrovascular diseases. Echocardiography is one of the main imaging modalities used routinely by cardiologists to assess the functional status of the heart. 2D echocardiograhy is typically used to observe the motion of the heart during the cardiac cycle as well as to quantify the ejection fraction by measuring

© Springer Nature Switzerland AG 2021
M. de Bruijne et al. (Eds.): MICCAI 2021, LNCS 12901, pp. 348–357, 2021.
https://doi.org/10.1007/978-3-030-87193-2_33

the size of the left ventricle cavity at end-diastole and end-systole. 3D echocardiography has been gaining more utility in clinical practice, but measuring and segmenting the left ventricle cavity in 3D volume images are still challenging tasks that take considerable time even for experts.

Left ventricle segmentation in 3D echocardiography remains a difficult task due to intrinsic limitations such as low image resolution and high dimensionality. Also, there is a lack of ground truth labels for all time frames within a given 3D echocardiography sequence. Sonographers and cardiologists typically assess only the end-diastole (ED) and end-systole (ES) frames in clinical settings to analyze ejection fraction, but segmentation of left ventricle in all time frames is often critical to do further analysis in motion tracking and strain analysis [1,3,13,14,17,19].

Many recent deep learning algorithms have shown remarkable results with high expert-level accuracy. While these networks have shown promising results in Computed Tomography (CT) and Magnetic Resonance imaging (MRI), there is still a relative lack of progress in 3D cardiac segmentation in echocardiography [2]. Also, most of the previous neural network frameworks for cardiac segmentation tasks are based on single image inputs [4,5]. However, unlike CT and MR, echocardiography is inherently a spatiotemporal dataset. These make it difficult to transfer established frameworks for their use in images with low signal-to-noise ratio (SNR) like echocardiography. Although many works in left ventricle segmentation in 2D echocardiography have been done [6,8,12,18,26], full 3D volume characterization is still needed to accurately assess cardiac function using measurements like ejection fraction. Others have looked at non-deep learning approaches using spatiotemporal consistencies for left ventricle echocardiography segmentation [7]. However, they require manual tracings to initialize the sparse representations and appearance dictionaries, which makes it computationally expensive and slow. Previous works in computer vision and natural language processing communities have also looked at the use of temporal consistencies using attention mechanisms to focus in on regions of interest during the training of neural network models. Some examples of these attention mechanisms include global attention [25], local attention [11,25], and co-attention [10,22,24], which have all shown improvements in their respective domains. Attention mechanisms have also been shown to have improvement from standard neural network approaches in medical imaging segmentation [9,16].

Given the inherent spatiotemporal nature of an echocardiography image sequence, a notable characteristic of echocardiography is that there is inter-frame consistency among the images due to high temporal resolution. In addition, the myocardium is usually placed in the center of the image and is the only object that deforms throughout the time sequence. Therefore, we hypothesize that utilizing the spatiotemporal features with attention will improve the overall segmentation task performance in ultrasound. Thus, in this paper, we introduce a multi-frame attention architecture for segmentation of left ventricle in 3D echocardiography. The proposed attention module can be easily scalable to include multiple times frames to utilize spatiotemporal features in a time series medical image dataset. To our knowledge, this work is the first to success-fully implement a spatiotemporal attention module using multiple time frames to segment the left ventricle in a volumetric echocardiography sequence.

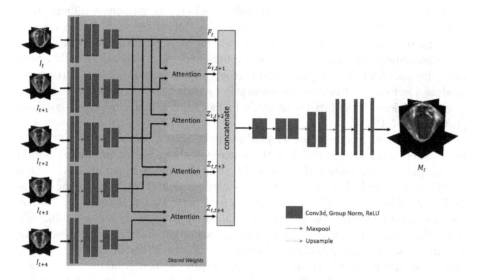

Fig. 1. Proposed Multi-frame Attention Network for segmentation of 3D left ventricle in echocardiography.

2   Methods

2.1   Multi-Frame Attention Network for Segmentation

We propose to extract and utilize spatiotemporal features in multiple time frames to segment out the left ventricle myocardium in 3D+time echocardiography. Using multiple time frames allows us to take advantage of the high temporal res-olution of echocardiography by looking at spatiotemporal consistencies among many time frames. The proposed multi-frame attention segmentation architec-ture is illustrated in Fig. 1. The idea of attention mechanism has been proposed by others previously [9,10,22] for use in segmenting out objects in a 2D video sequence. In this work, we adopt the idea of attention in a 3D+time dataset and expand its use by incorporating multiple time frames.

Given a target volume frame I_t, the volume frames that come immediately after within the same sequence will be selected, $I_{(t+i)}$, where i = 1,..., n-1, and n is the total number of frames in a given sequence of volumes. The net-work can adapt to incorporate as many time frames as hardware limitations allow. Figure 1 shows an instance where the inputs to the model are a target frame and 4 reference frames. This allows the target volume frame to utilize a series of 4 volume frames that follow to learn the spatiotemporal attention features. Each volume from a selected series will go through a standard encoder architecture with shared weights. For a 5-frame input, at the end of the encoder phase, there will be 5 feature maps, F_t, F_{t+1}, F_{t+2}, F_{t+3}, F_{t+4}. Each feature map corresponding to the input volume has the dimension of $c \times w \times h \times d$ where c = channel, w = width, h = height, and d = depth. The reference feature maps, F_{t+1}, F_{t+2}, F_{t+3}, F_{t+4}, will each be used with the target feature map, F_t, to calculate the attention maps.

The attention map leverages the correlation between the features extracted from a target feature map and a reference feature map. First, we can represent the correlation map as $C = F_t^T \otimes F_{t+i}$, where $\otimes$ represents the matrix multiplication. To utilize the matrix multiplication, F_t and F_{t+i}, where i = 1, ..., 4, are reshaped to the dimension of $c \times whd$. Thus, the correlation map has the dimension of $whd \times whd$.

$$C(F_t, F_{t+i}) = F_t^T \otimes F_{t+i} \in \mathbb{R}^{hwd \times hwd} \tag{1}$$

The correlation map is then normalized with a softmax function to calculate the attention map. Specifically, each entry (j, k) of the attention weight matrix represents the spatiotemporal attention feature between the j location of target feature map, F_t, and the k location of reference feature map, F_{t+i}.

$$A_{t,t+i} = softmax(C) \in \mathbb{R}^{hwd \times hwd} \tag{2}$$

Once the attention map, $A_{t,t+i}$, is calculated, it is again matrix multiplied by the target feature map, F_t, to ultimately focus in on the regions of the target feature map that show high spatiotemporal correlation with the reference feature map. Ultimately, the post-attention target feature map, $Z_{t,t+i}$ represents refined target feature map given a reference frame in the volume sequence. Figure 2 shows a visual illustration of the attention module.

$$Z_{t,t+i} = F_t \otimes A_{t,t+i} \in \mathbb{R}^{c \times hwd} \tag{3}$$

Once all the post-attention feature maps are calculated, they are concatenated with the original target feature map, F_t, and go forward in a standard decoder phase. As a result, the multi-frame attention maps are trained end-to-end and learn how the reference volume frames can inform the target volume frame to locate the object that is consistently moving through time.

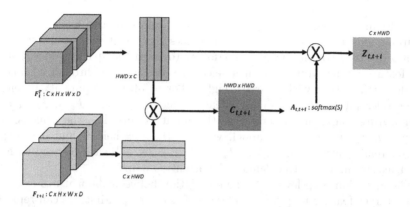

Fig. 2. Visual illustration of Multi-frame Attention Module.

3   Experiments and Results

3.1   Dataset

We evaluate our multi-frame attention approach using in vivo porcine 3D echocardiography dataset. The following volumes were acquired during myocardial infarction studies in pigs. All studies were performed in compliance with the Institutional Animal Care and Use Committee policies. Specifically, the dataset consists of 51 3D+time volumetric echocardiography sequences acquired from 14 pigs. Each pig was imaged at different time points corresponding to different physiological states (baseline rest, 3-day post-balloon occlusion at mid left anterior descending artery (LAD), 7-day post-balloon occlusion). Volumetric sequences were acquired using Philips iE33 ultrasound machine with the X7-2 probe. Each 3D image was taken in the apical view, placing the left ventricle at the center of the image. The endocardium and epicardium boundaries were labeled at the end-systole (ES) and end-diastole (ED) frames by a trained expert. Thus, in total, there are 102 full 3D myocardium labeled in the 51 volumetric sequences. The original image volume size was $224 \times 208 \times 208$. All images were resampled to $64 \times 64 \times 64$ as input to our models.

3.2   Implementation Details

In training our models, we divide our 102 fully labeled in vivo 3D echocardiography frames into 60 training, 20 validation, and 22 testing set. Each sample is either the end-systole (ES) frame or the end-diastole (ED) frame with additional 4 frames that come after the ES or ED frame. The model was trained using a SGD optimizer, batch size of 1, learning rate of 0.0001. A total of 100 epochs were trained. The network is trained by minimizing the binary cross entropy loss function. The model was implemented using PyTorch and the experiments were conducted on a GTX 1080 Ti GPU. Once the model is trained, the inference time is approximately 20 s.

3.3 Experimental Studies

We evaluate our multi-frame attention network by comparing with a standard U-Net [15]. To fit the U-Net for our specific 3D echocardiography dataset, we modify the batch normalization into group normalization since our batch size is 1. Group normalization was found to have comparable performance with batch normalization [23], so we believe that the performance of the models would not change significantly when using a higher batch size and batch normalization. We also compare our model with the co-attention network proposed by [10] since it utilizes a similar spatiotemporal attention framework. However, key differences between our proposed method and [10] are (1) our proposed method utilizes multiple time frames and (2) the output of our proposed method is a single frame segmentation while [10] predicts segmentation masks from both inputs. We also change the number of reference frames to evaluate whether more reference frames would lead to better performance. We conduct our experiments with the same proposed method but with 2 reference frames (3-frame input), 3 reference frames (4-frame input), and 4 reference frames (5-frame input).

To evaluate the accuracy of our proposed method, we adopt the generalized Dice Similarity Coefficient (DSC) and the Volumetric Similarity (VS) [20]. DSC measures the overlap between the predicted mask and the ground truth, while VS considers the how much the volumes of the segments differ. Thus, we selected these metrics that represent different aspects of segmentation accuracy. We also evaluate the correlation coefficient between the predicted ejection fraction and the ground truth ejection fraction using the endocardial boundaries to look at the clinical relevance of our work.

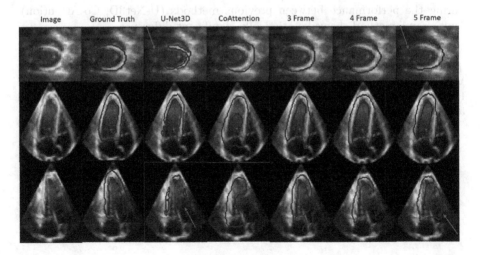

Fig. 3. Cross sectional display of echocardiography images and left ventricle masks (blue) at 3 different viewing planes. The results shown in this figure are from the same volumetric image. Red arrow shows that incorporating multiple time frames as reference frames allows the model to identify regions that are difficult to pick up from intensity information alone. (Color figure online)

3.4   Results and Discussion

Figure 3 shows a visual illustration of our results. Looking at the U-Net3D prediction and our proposed 5-frame attention model, there is noticeable improvement
in the walls where the signal intensity is relatively low compared to other regions
(red arrow). This is most likely due to U-Net3D learning the boundaries based on
intensity gradient of a single image volume while our proposed method takes into
account the movement of the myocardial boundaries to guide the location of the
walls even when the signal intensity is low. Table 1 reports the complete findings
of our results. Our findings show that utilizing multiple time frames in a given
sequence to segment the left ventricle at our target frame improves the overall
performance. It is likely that the high temporal resolution of echocardiography
gives the model better context as to where the walls are at any one time point by
utilizing other frames. Moreover, increasing the number of reference frames to
capture additional spatiotemporal attention features augments the performance
even further. This is also seen in Fig. 3, where in the short axis view images (first
row) among 3 Frame, 4 Frame, 5 Frame, the 5 Frame model completes the dough-
nut shape that matches well with the ground truth. Table 2 reports the analysis
of statistical significance between the evaluated methods based on a paired t-test.
It shows that there is a statistically significant improvement ($p < 0.05$) in the
Dice Similarity Coefficient of multi-frame (5 frames) method when compared
to multi-frame (4 frames) method. Since multi-frame (4 frames) outperforms
U-Net3D and Co-Attention methods, we can infer that multi-frame (5 frames)
performs the best based on the p-value. The Volumetric Similarity (VS) values
did not show statistically significant improvement when comparing the different
number of reference frames in the multi-frame method. However, when comparing the performance between previous methods (U-Net3D, Co-Attention),
multi-frame method showed statistically significant difference.

Table 1. Summary of segmentation evaluation metrics. DSC = Dice Score Cofficient,
VS = Volumetric Similarity, EF Corr Coeff = Ejection Fraction Correlation Coefficient.
Higher number is better for all metrics. Myocardium refers to the combined endocardial
and epicardial boundaries of the left ventricle.

Methods	DSC (Myocardium)	VS (Myocardium)	EF correlation coeff.
U-Net3D	0.451 (±0.08)	0.61 (±0.15)	0.231
Co-Attention ([10])	0.545 (±0.06)	0.43 (±0.11)	0.257
Multi-frame (3 frames)	0.582 (±0.14)	0.88 (±0.08)	0.305
Multi-frame (4 frames)	0.572 (±0.16)	0.86 (±0.14)	0.486
Multi-frame (5 frames)	**0.641 (±0.14)**	**0.90 (±0.08)**	**0.673**

To analyze the clinical relevance and use of our proposed method, we also look
at the correlation coefficient between the ground truth ejection fraction (EF)
and the predicted ejection fraction. Our results show that the EF correlation
coefficient increases with the use of multi-frame attention features.

3.5 Limitations and Future Works

Although the proposed multi-frame attention network shows promise in improving segmentation tasks in a spatiotemporal dataset such as 3D echocardiography, it has much room for improvement. First, the current study evaluated up to 5 time frames as maximum into the model. We will explore incorporating more time frames, potentially including all time frames in a given echocardiography sequence. Second, our dataset is currently limited to the porcine in vivo models that were acquired during a myocardial infarction studies. These images have a pigtail catheter inserted in the left ventricle which may introduce artifacts in the image. Also, some images were acquired during open chest thoracotomy procedure, while other images were acquired during a closed chest state. Therefore, in future studies, we will incorporate 3D echocardiography sequences from patients to extend our framework in a fully standardized clinical setting.

Table 2. Summary of paired t-test for statistical significance between methods.

Comparison	DSC	VS
U-Net3D vs. CoAttention [10]	p = 0.02	p = 0.001
Co-Attention ([10]) vs. Multi-frame (3 frames)	p = 0.08	p = 5.27e–10
Multi-frame (3 frames) vs. Multi-frame (4 frames)	p = 0.89	p = 0.37
Multi-frame (4 frames) vs. Multi-frame (5 frames)	p = **0.01**	p = 0.25

4 Conclusion

In this paper, we present a multi-frame attention architecture for segmentation of left ventricle in 3D echocardiography. The multi-frame attention network successfully utilized the spatiotemporal features of echocardiography image volumes in a given sequence, and it extracted highly correlated features between target volume and reference volume to guide the location of the myocardial boundaries even when the signal intensity was low. Our proposed framework is also easily scalable to incorporate as many frames as reference volumes. The results indicate that taking advantage of the spatiotemporal nature of images can augment the performance of 3D left ventricle segmentation. Future work will expand our validations to include clinical 3D echocardiography datasets to train our models more robustly. Also, we will evaluate the effectiveness of attention using public 2D echocardiography datasets to see the feasibility of multi-frame attention network for different data dimensions.

Acknowledgments. We would like to thank the technical assistance provided by the staff of the Yale Translational Research Imaging Center. This work was supported in part by the following grants: R01HL121226, R01HL137365, F30HL158154, and Medical Scientist Training Program Grant T32GM007205.

References

1. Ahn, S.S., Ta, K., Lu, A., Stendahl, J.C., Sinusas, A.J., Duncan, J.S.: Unsupervised motion tracking of left ventricle in echocardiography. In: Medical Imaging 2020: Ultrasonic Imaging and Tomography, vol. 11319, p. 113190Z. International Society for Optics and Photonics (2020)
2. Chen, C., et al.: Deep learning for cardiac image segmentation: a review. Front. Cardiovasc. Med. **7**, 25 (2020)
3. Compas, C.B., et al.: Radial basis functions for combining shape and speckle tracking in 4d echocardiography. IEEE Trans. Med. Imaging **33**(6), 1275–1289 (2014)
4. Dong, S., et al.: Deep atlas network for efficient 3d left ventricle segmentation on echocardiography. Med. Image Anal. **61**, 101638 (2020)
5. Dong, S., Luo, G., Wang, K., Cao, S., Li, Q., Zhang, H.: A combined fully convolutional networks and deformable model for automatic left ventricle segmentation based on 3d echocardiography. BioMed Res. Int. **2018** (2018)
6. Ghorbani, A., et al.: Deep learning interpretation of echocardiograms. NPJ Dig. Med. **3**(1), 1–10 (2020)
7. Huang, X., et al.: Contour tracking in echocardiographic sequences via sparse representation and dictionary learning. Med. Image Anal. **18**(2), 253–271 (2014)
8. Leclerc, S., et al.: Deep learning for segmentation using an open large-scale dataset in 2d echocardiography. IEEE Trans. Med. Imaging **38**(9), 2198–2210 (2019)
9. Liu, F., Wang, K., Liu, D., Yang, X., Tian, J.: Deep pyramid local attention neural network for cardiac structure segmentation in two-dimensional echocardiography. Med. Image Anal. **67**, 10187 (2021)
10. Lu, X., et al.: See more, know more: unsupervised video object segmentation with co-attention siamese networks. In: Proceedings of the IEEE Conference on Computer Vision and Pattern Recognition, pp. 3623–3632 (2019)
11. Luong, M.T., et al.: Effective approaches to attention-based neural machine translation. arXiv preprint arXiv:1508.04025 (2015)
12. Ouyang, D., et al.: Video-based ai for beat-to-beat assessment of cardiac function. Nature **580**(7802), 252–256 (2020)
13. Papademetris, X., Sinusas, A.J., Dione, D.P., Constable, R.T., Duncan, J.S.: Estimation of 3-d left ventricular deformation from medical images using biomechanical models. IEEE Trans. Med. Imaging **21**(7), 786–800 (2002)
14. Parajuli, N., et al.: Flow network tracking for spatiotemporal and periodic point matching: Applied to cardiac motion analysis. Med. Image Anal. **55**, 116–135 (2019)
15. Ronneberger, O., Fischer, P., Brox, T.: U-Net: convolutional networks for biomedical image segmentation. In: Navab, N., Hornegger, J., Wells, W.M., Frangi, A.F. (eds.) MICCAI 2015. LNCS, vol. 9351, pp. 234–241. Springer, Cham (2015). https://doi.org/10.1007/978-3-319-24574-4_28
16. Schlemper, J., et al.: Attention gated networks: learning to leverage salient regions in medical images. Med. Image Anal. **53**, 197–207 (2019)
17. Shi, P., Sinusas, A.J., Constable, R.T., Ritman, E., Duncan, J.S.: Point-tracked quantitative analysis of left ventricular surface motion from 3-d image sequences. IEEE Trans. Med. Imaging **19**(1), 36–50 (2000)
18. Stendahl, J.C., et al.: Regional myocardial strain analysis via 2d speckle tracking echocardiography: validation with sonomicrometry and correlation with regional blood flow in the presence of graded coronary stenoses and dobutamine stress. Cardiovasc. Ultrasound **18**(1), 1–16 (2020)

19. Ta, K., Ahn, S.S., Lu, A., Stendahl, J.C., Sinusas, A.J., Duncan, J.S.: A semi-supervised joint learning approach to left ventricular segmentation and motion tracking in echocardiography. In: 2020 IEEE 17th International Symposium on Biomedical Imaging (ISBI), pp. 1734–1737. IEEE (2020)
20. Taha, A.A., Hanbury, A.: Metrics for evaluating 3d medical image segmentation: analysis, selection, and tool. BMC Med. Imaging **15**(1), 1–28 (2015)
21. Virani, S.S., et al.: Heart disease and stroke statistics-2020 update: a report from the American heart association. Circulation **141**, E139–E596 (2020)
22. Wu, L., et al.: Deep coattention-based comparator for relative representation learning in person re-identification. IEEE Trans. Neural Netw. Learn. Syst. **32**, 722–735 (2020)
23. Wu, Y., He, K.: Group normalization. In: Proceedings of the European conference on computer vision (ECCV), pp. 3–19 (2018)
24. Xu, H., Saenko, K.: Ask, attend and answer: exploring question-guided spatial attention for visual question answering. In: Leibe, B., Matas, J., Sebe, N., Welling, M. (eds.) ECCV 2016. LNCS, vol. 9911, pp. 451–466. Springer, Cham (2016). https://doi.org/10.1007/978-3-319-46478-7_28
25. Xu, K., et al.: Show, attend and tell: neural image caption generation with visual attention. In: International Conference on Machine Learning, pp. 2048–2057 (2015)
26. Zhang, J., et al.: Fully automated echocardiogram interpretation in clinical practice: feasibility and diagnostic accuracy. Circulation **138**(16), 1623–1635 (2018)

Coarse-To-Fine Segmentation of Organs at Risk in Nasopharyngeal Carcinoma Radiotherapy

Qiankun Ma[1], Chen Zu[2], Xi Wu[3], Jiliu Zhou[1,3], and Yan Wang[1(✉)]

[1] School of Computer Science, Sichuan University, Chengdu, China
[2] Department of Risk Controlling Research, JD.com, Beijing, China
[3] School of Computer Science, Chengdu University of Information Technology, Chengdu, China

Abstract. Accurate segmentation of organs at risk (OARs) from medical images plays a crucial role in nasopharyngeal carcinoma (NPC) radiotherapy. For automatic OARs segmentation, several approaches based on deep learning have been proposed, however, most of them face the problem of unbalanced foreground and background in NPC medical images, leading to unsatisfactory segmentation performance, especially for the OARs with small size. In this paper, we propose a novel end-to-end two-stage segmentation network, including the first stage for coarse segmentation by an encoder-decoder architecture embedded with a target detection module (TDM) and the second stage for refinement by two elaborate strategies for large- and small-size OARs, respectively. Specifically, guided by TDM, the coarse segmentation network can generate preliminary results which are further divided into large- and small-size OARs groups according to a preset threshold with respect to the size of targets. For the large-size OARs, considering the boundary ambiguity problem of the targets, we design an edge-aware module (EAM) to preserve the boundary details and thus improve the segmentation performance. On the other hand, a point cloud module (PCM) is devised to refine the segmentation results for small-size OARs, since the point cloud data is sensitive to sparse structures and fits the characteristic of small-size OARs. We evaluate our method on the public Head&Neck dataset, and the experimental results demonstrate the superiority of our method compared with the state-of-the-art methods. Code is available at https://github.com/DeepMedLab/Coarse-to-fine-segmentation.

Keywords: Image segmentation · Organs at risk · Nasopharyngeal carcinoma · Coarse-to-fine

1 Introduction

Nasopharyngeal carcinoma (NPC) is a common but fatal malignant tumor arising from nasopharynx or upper throat [1]. For NPC patients, radiation therapy is one of the main treatments. During radiotherapy planning, delineating organs at risk (OARs) is a crucial step to avoid potential radiation risks to normal tissues. Currently, the OARs are always delineated by radiation oncologists manually based on computed tomography (CT) scans, which is extremely time-consuming and subjective [2, 3]. Thus, it is highly

© Springer Nature Switzerland AG 2021
M. de Bruijne et al. (Eds.): MICCAI 2021, LNCS 12901, pp. 358–368, 2021.
https://doi.org/10.1007/978-3-030-87193-2_34

desirable to develop an automatic OARs segmentation approach for NPC patients to deliver efficient and accurate radiotherapy planning.

With the rise of deep learning, a number of approaches have been proposed to delineate OARs from NPC CT images [4–12]. The current OARs segmentation methods can be divided into two categories: 1) organ-specific segmentation based on target region and 2) multi-organ segmentation based on entire CT image. Specifically, the organ-specific segmentation aims to design an exclusive segmentation model to achieve the best segmentation performance for each OAR [4, 5]. Nevertheless, such methods inevitably bring tremendous computational overhead due to the multiple specific models, thus limiting their applicability. In contrast, the multi-organ segmentation based on entire CT image targets at training a deep model which can segment multiple OARs simultaneously. For instance, Tong et al. [12] proposed a fully convolutional neural network (FCNN) with a shape representation model to learn the shape of segmentation targets and delineate the OARs of NPC. Zhu et al. [8] developed a 3D Squeeze-and-Excitation U-Net for OARs segmentation. Gao et al. [9] presented a model named FocusNet, which locates the center points of multiple OARs respectively to extract the corresponding 3D image patches and further segments them. Tang et al. [10] explored the 3D UNet to get the region of interests (ROIs) of each organ by embedding an OAR detection module. Liang et al. [11] proposed a multi-view spatial aggregation framework using 2D ROIs detection module to assist segmentation.

Although current OARs segmentation methods have achieved promising progress, the performance is still somewhat unsatisfactory due to the following challenges. First, compared with other types of cancer which only have a small number of OARs, the number of OARs for NPC patients is up to more than ten, as shown in Fig. 1. Second, it is obvious in Fig. 1 that the size of OARs is highly variable, making the segmentation models prone to segment the large-size OARs (e.g., brain stem, mandible) but neglect the small-size ones (e.g., optical nerves, optical chiasm). Third, most of the OARs only occupy small volumes in CT images, causing the problem of target sparsity and regional imbalance between foreground and background. The ratio between the background and the smallest organ can even reach nearly 10^5:1 in some extreme cases [9]. Fourth, the boundary ambiguity of OARs in NPC CT images is also a sore point.

In this paper, to overcome the above-mentioned challenges, we propose a novel end-to-end coarse-to-fine segmentation model to automatically segment multiple OARs in CT images. Specifically, the entire framework consists of a coarse stage and a fine stage. The coarse stage tries to generate rough segmentation results using an encoder-decoder architecture embedded with a target detection module (TDM). According to the size of targets, the preliminary results from the coarse stage are further categorized into large- and small-size OARs groups using a preset threshold. In the fine stage, we design two exclusive refinement networks for the large- and small-size OARs, respectively. Particularly, to tackle the boundary ambiguity problem of the large-size targets, an edge-aware module (EMA) is devised for capturing the boundary details for performance refinement. Moreover, considering that the point cloud data is sensitive to sparse structures and fits the characteristic of small-size OARs, we explore a point cloud module (PCM) to refine the segmentation results of small-size OARs. We evaluate our method on the public Head&Neck dataset [14]. The experimental results demonstrate that our proposed method achieves better performance than other state-of-the-art methods in both qualitative and quantitative measures.

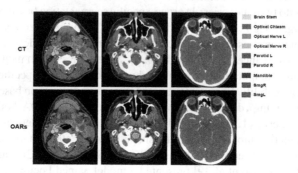

Fig. 1. Illustration of CT images and typical nine OARs to be delineated in NPC.

2 Methodology

The architecture of the proposed method is illustrated in Fig. 2, which consists of two stages. The first stage is the TDM guided coarse segmentation network which employs the encoder-decoder architecture as backbone while the second stage is responsible for the refinement by two elaborate modules for large-size and small-size OARs respectively. The details of our model and the objective function will be introduced in the following sub-sections.

2.1 Architecture

TDM Guided Coarse Segmentation Network: As shown in Fig. 2(a), the coarse segmentation network in the first stage is a U-Net-like network embedded with TDM. With the guidance of TDM, the network can locate the target areas of OARs and neglect irrelevant background regions. Specifically, taking a CT volume as input, the encoder is equipped with four down-sampling blocks to extract latent representative features, each with two residual sub-blocks based on 3D convolution and a max-pooling layer applied to halve the resolution. After four down-sampling operations, the final extracted feature maps are fed into the TDM for ROI extraction and cropping. Particularly, the TDM contains two separate heads, one for bounding box regression to indicate the location and size of ROI for each OAR, and the other for binary classification to judge whether the corresponding OAR class is detected correctly by the TDM. Then, we adopt the ROI Align layer [15] in order to get feature maps with fixed dimensions. At the end of TDM, two fully connected layers are subsequently applied to predict the class of each OAR proposal and further regress coordinates and size offsets of its bounding box, respectively. Then, these generated ROI proposals are down-sampled to the same size as the features from each encoder layer to crop them, aiming to reduce the interference of irrelevant background regions. In decoder, each up-sampling block initially adopts the trilinear up-sampling to double the size of inputs, and then applies a $3 \times 3 \times 3$ convolution followed by local contrast normalization. After each up-sampling operation, the cropped feature maps from the encoder are concatenated with the feature maps of the decoder. At the last decoder layer, we apply a $1 \times 1 \times 1$ convolution to the final feature maps to generate the coarse segmentation results. Finally, we calculate the size of the segmented multiple OARs and according to a threshold which is preset to 1000 [9] in this paper, the coarse segmentation results are divided into large- and small-size OAR groups and respectively sent to EAM and PCM of the second stage for refinement.

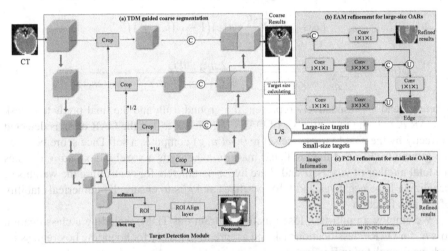

Fig. 2. Overview of our network architecture, including (a) TDM guided coarse segmentation (b) EAM refinement for large-size OARs and (c) PCM refinement for small-size OARs. 'Conv' denotes the convolutional layer, 'C' and 'U' denote concatenation and up-sampling.

Edge-Aware Module (EAM): Aiming to solve the boundary ambiguity problem of segmentation, we design an EAM to refine the coarse segmentation results of large-size OARs. Considering that the rich edge information is mainly contained in the low-level feature maps, we only employ the cropped features of the first two encoder blocks in the first stage as the input of the EAM, as shown in Fig. 2(b). Then, these two low-level feature maps undergo the processing of convolution, concatenation and up-sampling, to obtain the corresponding edge maps which represent the boundaries of OARs. Finally, the edge maps and the coarse segmentation results are fused by a $1 \times 1 \times 1$ convolution to obtain refined segmentation results.

Point Cloud Module (PCM): In view of the advantages of point cloud networks in handling the sparse data, we create a PCM for small-size OARs refinement, as shown in Fig. 2(c). Our PCM is mainly based on the encoder-decoder architecture proposed by Balsiger et al. [13]. Specifically, for each small-size OAR, the volumetric coarse segmentation result is converted into a point cloud $P = [p_1, p_2, \ldots, p_K]$ with K (set to 2048) points $p_i \in R^3$. Moreover, we additionally extract the image information by following [13] and concatenate the extracted image information with the point cloud P as the input of our PCM. Finally, the outputs of PCM are utilized to replace the values of the corresponding points in the coarse segmentation result to obtain the refined segmentation result.

2.2 Objective Function

The objective function consists of four parts: segmentation loss, TDM loss, EAM loss and PCM loss. Specifically, the segmentation loss function can be expressed as:

$$Loss_{seg} = \sum_{c=1}^{C} I(c)\left(1 - \varphi(m^c, g^c)\right)$$

$$\varphi(m, g) = \frac{\sum_{i=1}^{N} m_i g_i}{\sum_{i=1}^{N} m_i g_i + \alpha \sum_{i=1}^{N} m_i(1 - g_i) + \beta \sum_{i=1}^{N}(1 - m_i)g_i + \varepsilon}, \quad (1)$$

where g^c and m^c respectively represent the ground truth and the final predicted mask of OAR c. C is the total number of OARs. $I(c)$ is set to 1 if the OAR class is detected correctly by the TDM, otherwise set to 0. $\varphi(m, g)$ computes a soft Dice score between the ground truth g and the predicted mask m, where i is a voxel index and N denotes the total number of voxels. α and β are hyper-parameters for controlling the weights of penalizing false negatives and false positives and ε is to ensure the numerical stability of the loss function.

We employ a multi-task loss function to train our TDM, including a classification loss L_c for OARs classification task and a regression loss L_r for bounding box regression task, as formulated in Eq. 2:

$$Loss_{TDM} = \frac{1}{M} \sum_i L_c\left(P_i, P_i^*\right) + \lambda \frac{1}{M} \sum_i L_r\left(t_i, t_i^*\right), \quad (2)$$

where L_c adopts the cross entropy (CE) loss and L_r uses the smooth L1 loss. λ is a hyper-parameter to balance these two terms. M is the total number of anchors participating in the calculation, while P_i^*, t_i^* denote the predicted class label and box parameter, respectively, and P_i, t_i are their ground truths.

Aiming to capture sufficient edge information, the EAM loss is introduced to constrain the edge map and can be expressed as:

$$Loss_{EAM} = \frac{1}{C} \sum_{c \in C} \overline{\Delta Jc}(m(c)),$$

$$m_i(c) = \begin{cases} 1 - p_i(c) \text{ if } c = y_i(c) \\ p_i(c) \text{ otherwise} \end{cases}, \quad (3)$$

where C denotes the total number of OARs. For the pixel i of OAR c, $y_i(c)$ denotes the ground truth binary mask while $p_i(c)$ denotes the predicted probability between 0 and 1. $\overline{\Delta Jc}$ represents the Lovasz extension of the Jaccard index [16].

For PCM, we employ the binary cross entropy (BCE) loss to constrain the output of the point cloud network. The loss of PCM is defined as:

$$Loss_{PCM} = L_{BCE}(S, S^*), \qquad (4)$$

where the S and S^* are the output and ground truth of PCM.

The total loss is defined as:

$$Loss_{total} = Loss_{seg} + \mu_1 Loss_{TDM} + \mu_2 Loss_{EAM} + \mu_3 Loss_{PCM}, \qquad (5)$$

where μ_1, μ_2 and μ_3 are balance terms.

2.3 Training Details

Our network is trained on PyTorch framework and an NVIDIA GeForce GTX 1080Ti with 11 GB memory. Specifically, we take Adaptive moment estimation (Adam) optimizer with a momentum of 0.9 to optimize the network. The proposed network is trained for 200 epochs and the batch size is set to 1. The initial learning rate is set to 0.001, and decays to 0.0001 and 0.00001 when the epoch respectively reaches 100 and 150. Based on our trial studies, α and β in Eq. 1 both equal to 0.5, λ in Eq. (2) is set to 1. μ_1, μ_2 and μ_3 in Eq. (5) are set to 1, 2 and 2, respectively.

3 Experiment and Analysis

3.1 Dataset and Evaluation

Our proposed method is evaluated on the MICCAI 2015 Head&Neck Auto Segmentation Challenge dataset, which contains a set of CT volumes for 48 NPC patients with the image size varying from $512 \times 512 \times 39$ to $512 \times 512 \times 181$. Each patient involves nine OARs, including brain stem, mandible, optic chiasm, optic nerve (both left and right), parotid (both left and right), submandibular gland (both left and right). The dataset was split by the Challenge, where 33 subjects are used as training set and the remaining 15 subjects are used as the test set. All samples are preprocessed to fit the maximum input size of our model, i.e., $240 \times 240 \times 112$. To evaluate and analyze the experimental results, we adopt three common metrics, including dice similarity coefficient (DSC), 95th percentile Hausdorff distance (95% HD), and average surface distance (ASD).

Table 1. Quantitative comparisons terms of DSC with state-of-the-art methods. † denotes p < 0.05 through paired t-test.

OAR	AnatomyNet[8]	FocusNet[9]	U_aNet[10]	Multi-view[11]	Proposed
Brain Stem	86.7 ± 2†	87.5 ± 2.6	87.5 ± 2.5†	**92.3 ± 1.0**	87.9 ± 2.4
Mandible	92.5 ± 2†	93.5 ± 1.9	**95.0 ± 0.8**	94.1 ± 0.7	94.5 ± 0.7
Optical Chiasm	53.2 ± 15†	59.6 ± 18.1	61.5 ± 10.2†	**71.3 ± 8.3**	65.9 ± 9.6

(*continued*)

Table 1. (*continued*)

OAR	AnatomyNet[8]	FocusNet[9]	U_aNet[10]	Multi-view[11]	Proposed
Optical Nerve L	72.1 ± 6†	73.5 ± 9.6	74.8 ± 7.1†	73.8 ± 4.6	**75.3 ± 6.8**
Optical Nerve R	70.6 ± 10†	74.4 ± 7.2	72.3 ± 5.9†	73.4 ± 5.1	**74.7 ± 5.3**
Parotid L	88.1 ± 2†	86.3 ± 3.6	88.7 ± 1.9†	88.2 ± 1.3	**89.2 ± 1.5**
Parotid R	87.4 ± 4†	87.9 ± 3.1	87.5 ± 5.0†	87.0 ± 1.5	**88.4 ± 4.9**
SMG L	81.4 ± 4†	79.8 ± 8.1	82.3 ± 5.2†	81.5 ± 2.9	**82.9 ± 4.8**
SMG R	81.3 ± 4†	80.1 ± 6.1	81.5 ± 4.5	80.0 ± 3.4	**81.5 ± 3.7**

3.2 Comparison with State-Of-The-Art Methods

To validate the advancement of our method, we compare it with four state-of-the-art (SOTA) OARs segmentation methods, including AnatomyNet [8], FocusNet [9], UaNet [10], Multi-view [11]. Table 1 gives the DSC results of the nine OARs segmented by different methods. As observed, our method achieves the best performance in six OARs. To study the statistical significance of our proposed method, we also perform paired t-tests (only for the methods with public available code) to compare the SOTAs against our method, through which we can find that almost all p-values are less than 0.05, demonstrating the statistical significance of the achieved improvement. Moreover, the quantitative results regarding 95% HD and ASD are shown in Fig. 3, from where we can see that our method obtains the competitive performance with the SOTAs.

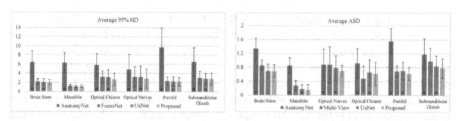

Fig. 3. Quantitative comparisons with the state-of-the-art methods in terms of 95% HD and ASD.

For qualitative comparison, we display the visual comparison results in Fig. 4. Note that, the visual results of the FocusNet and Multi-view are not given here since their code has not yet been released. Here, we also display the result of 3D Unet for comparison, due to its widely application in medical image segmentation tasks. As observed, the 3D Unet presents the worst segmentation results as the gap between the prediction and the ground truth is the largest. Compared with AnatomyNet and UaNet, our method gives more precise segmentation results with less false positive predictions.

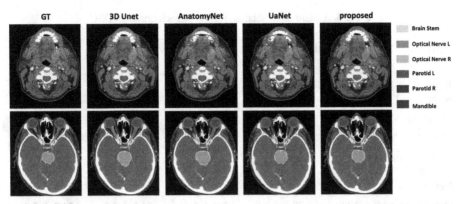

Fig. 4. Qualitative comparisons with the state-of-the-art methods. The areas with obvious improvement achieved by our method are circled by the red ellipsoids. (Color figure online)

3.3 Ablation Study

To evaluate the contributions of the components in the proposed method, we perform the ablation study in a progressive way using the 3D U-net (i.e., the first-stage without TDM) as the baseline. To be specific, our experimental settings include: (1) 3D U-net, (2) 3D U-net + TDM, (3) 3D U-net + TDM + EAM, (4) 3D U-net + TDM + PCM, (5) 3D U-net + TDM + EAM + PCM (proposed).

Table 2. Quantitative ablation study results. ✓ stands for the addition of corresponding module.

3D Unet	TDM	EAM	PCM	DSC [%]	95% HD	ASD
✓				75.42 ± 9.6	6.25 ± 3.7	1.64 ± 0.8
✓	✓			80.51 ± 5.3	2.76 ± 1.7	0.80 ± 0.4
✓	✓	✓		81.21 ± 4.9	2.47 ± 1.2	0.65 ± 0.3
✓	✓		✓	81.42 ± 4.8	2.54 ± 1.3	0.71 ± 0.3
✓	✓	✓	✓	**82.25 ± 4.4**	**2.38 ± 1.2**	**0.57 ± 0.2**

The quantitative ablation results are shown in Table 2. By comparing the first and second row, we find that DSC significantly improves from 75.42% to 80.51%, the 95% HD drops from 6.25 to 2.76 and the ASD drops from 2.46 to 0.89, respectively, demonstrating the necessity of TDM of the proposed method. Similarly, by further incorporating the EAM or PCM, the performance becomes better in terms of all three metrics. Undoubtedly, the completed model with TDM, EAM and PCM achieves the best performance, with the highest DSC and lowest 95%HD and ASD. Table 3 shows the DSC results of nine OARs based on different experimental settings. Specifically, by incorporating the TDM, the segmentation results were improved for all organs. With the introduction of the EAM and PCM, the 3D Unet + TDM + EAM achieves the best dice results on three large organs, (i.e., brainstem, parotid, and SMG R) and the 3D Unet + TDM + PCM

Table 3. Quantitative ablation results of each organ on DSC

OAR	3D Unet	3D Unet + TDM	3D Unet + TDM + EAM	3D Unet + TDM + PCM	Proposed
Brain Stem	84.5 ± 5.6	87 ± 2.6	$\mathbf{88.6 \pm 2.0}$	86.7 ± 2.9	87.9 ± 2.4
Mandible	89.3 ± 3.8	93.9 ± 1.1	94.4 ± 0.7	93.6 ± 1.8	$\mathbf{94.5 \pm 0.7}$
Optical Chiasm	48.4 ± 21.4	60.9 ± 11.2	59.8 ± 12.6	$\mathbf{67.3 \pm 8.0}$	65.9 ± 9.6
Optical Nerve L	68.1 ± 9.7	73.8 ± 7.3	73.5 ± 7.9	$\mathbf{75.8 \pm 5.4}$	75.3 ± 6.8
Optical Nerve R	64.9 ± 10.8	72.6 ± 6.8	72.8 ± 5.4	74.5 ± 5.2	$\mathbf{74.7 \pm 5.3}$
Parotid L	84.4 ± 8.5	87.9 ± 2.9	$\mathbf{89.5 \pm 1.4}$	87.2 ± 3.1	89.2 ± 1.5
Parotid R	82.8 ± 9.1	86.6 ± 5.6	88.2 ± 4.8	85.8 ± 6.7	$\mathbf{88.4 \pm 4.9}$
SMG L	77.5 ± 8.9	81.5 ± 5.5	82.6 ± 5.3	81.8 ± 5.2	$\mathbf{82.9 \pm 4.8}$
SMG R	78.9 ± 8.6	80.4 ± 4.8	$\mathbf{81.5 \pm 3.6}$	80.1 ± 4.8	81.5 ± 3.7
Average	75.42 ± 9.6	80.51 ± 5.3	81.21 ± 4.9	81.42 ± 4.8	$\mathbf{82.25 \pm 4.4}$

achieves the best dice results on two small organs, (i.e., Optical Chiasm and Optical Nerve L). These experimental results show that the introduction of EAM and PCM is helpful for the improvement of large- and small-size OARs.

The qualitative ablation results shown in Fig. 5 also demonstrate the effectiveness of each component we proposed. Particularly, from the results of (2) and (3), we can see that the EAM indeed could produce the segmentation results with more accurate boundaries, as indicated by the red arrows. By comparing (2) with (4), we can conclude that the PCM could refine the segmentation performance for the small-size OARs, as shown in the red squares. The visual effect in (5) yielded by our complete model has the smallest difference from the ground truth for both large- and small-size targets, supporting the findings in the statistical data in Table 2.

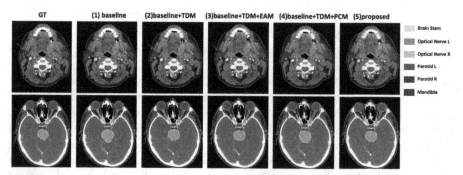

Fig. 5. Qualitative ablation study results. 'Optical Nerve L' and 'Optical Nerve R' are small-size OAR, the other four OARs appearing in the images are all large-size OAR.

4   Conclusion

In this paper, we propose a novel end-to-end two-stage segmentation network to automatically segment multiple OARs in NPC, including the first stage for coarse segmentation and the second stage for refinement by two elaborate modules. Concretely, in the first stage, we construct a well-performed target detection module (TDM) to locate and crop the general area for each OAR, thus eliminating the interference of large background area and making the network pay more attention to the OARs. In the second stage, an edge-aware module (EAM) is established to focus on the segmentation boundary of large-size targets and alleviate the boundary ambiguity problem. For small-size targets, since the point cloud data is sensitive to sparse structures, a point cloud module (PCM) is employed to further refine the segmentation performance. Experiments on the public Head&Neck dataset show that our method achieves competitive results compared with the state-of-the-art methods.

Acknowledgments. This work is supported by National Natural Science Foundation of China (NSFC 62071314) and Sichuan Science and Technology Program (2021YFG0326, 2020YFG0079).

References

1. Zhong, T., Huang, X., Tang, F., et al.: Boosting-based cascaded convolutional neural networks for the segmentation of CT organs-at-risk in nasopharyngeal carcinoma. Med. Phys. **46**(12), 5602–5611 (2019)
2. Nelms, B.E., Tomé, W.A., Robinson, G., et al.: Variations in the contouring of organs at risk: test case from a patient with oropharyngeal cancer. Int. J. Radiat. Oncol. Biol. Phys. **82**(1), 368–378 (2012)
3. Brouwer, C.L., Steenbakkers, R.J.H.M., van den Heuvel, E., et al.: 3D variation in delineation of head and neck organs at risk. Radiat. Oncol. **7**(1), 1–10 (2012)
4. Ren, X., Xiang, L., Nie, D., et al.: Interleaved 3D-CNN s for joint segmentation of small-volume structures in head and neck CT images. Med. Phys. **45**(5), 2063–2075 (2018)
5. Ibragimov, B., Xing, L.: Segmentation of organs-at-risks in head and neck CT images using convolutional neural networks. Med. Phys. **44**(2), 547–557 (2017)
6. Raudaschl, P.F., et al.: Evaluation of segmentation methods on head and neck CT: auto-segmentation challenge 2015. Med. Phys. **44**(5), 2020–2036 (2017)
7. Wang, Z., Wei, L., Wang, L., et al.: Hierarchical vertex regression-based segmentation of head and neck CT images for radiotherapy planning. IEEE Trans. Image Process. **27**(2), 923–937 (2017)
8. Zhu, W., Huang, Y., Tang, H., et al.: Anatomynet: deep 3d squeeze-and-excitation u-nets for fast and fully automated whole-volume anatomical segmentation. Med. Phys. **46**(2), 576–589 (2019)
9. Gao, Y., et al.: Focusnet: imbalanced large and small organ segmentation with an end-to-end deep neural network for head and neck ct images. In: Shen, D., Liu, T., Peters, T.M., Staib, L.H., Essert, C., Zhou, S., Yap, P.-T., Khan, A. (eds.) MICCAI 2019. LNCS, vol. 11766, pp. 829–838. Springer, Cham (2019). https://doi.org/10.1007/978-3-030-32248-9_92
10. Tang, H., Chen, X., Liu, Y., et al.: Clinically applicable deep learning framework for organs at risk delineation in CT images. Nat. Mach. Intell. **1**(10), 480–491 (2019)

11. Liang, S., Thung, K.H., Nie, D., et al.: Multi-view spatial aggregation framework for joint localization and segmentation of organs at risk in head and neck CT images. IEEE Trans. Med. Imaging **39**(9), 2794–2805 (2020)
12. Tong, N., Gou, S., Yang, S., et al.: Fully automatic multi-organ segmentation for head and neck cancer radiotherapy using shape representation model constrained fully convolutional neural networks. Med. Phys. **45**(10), 4558–4567 (2018)
13. Balsiger, F., Soom, Y., Scheidegger, O., Reyes, M.: Learning shape representation on sparse point clouds for volumetric image segmentation. In: Shen, D., Liu, T., Peters, T.M., Staib, L.H., Essert, C., Zhou, S., Yap, P.-T., Khan, A. (eds.) MICCAI 2019. LNCS, vol. 11765, pp. 273–281. Springer, Cham (2019). https://doi.org/10.1007/978-3-030-32245-8_31
14. Raudaschl, P.F., Zaffino, P., Sharp, G.C., et al.: Evaluation of segmentation methods on head and neck CT: auto-segmentation challenge 2015. Med. Phys. **44**(5), 2020–2036 (2017)
15. He, K., Gkioxari, G., Dollár, P., et al.: Mask R-CNN. In: IEEE International Conference on Computer Vision (ICCV), pp. 2980–2988 (2017)
16. Berman, M., Triki, A.R., Blaschko, M.B.: The lovász-softmax loss: a tractable surrogate for the optimization of the intersection-over-union measure in neural networks. In: Proceedings of the IEEE Conference on Computer Vision and Pattern Recognition, pp. 4413–4421 (2018)

Joint Segmentation and Quantification of Main Coronary Vessels Using Dual-Branch Multi-scale Attention Network

Hongwei Zhang, Dong Zhang, Zhifan Gao$^{(\boxtimes)}$ ⓘ, and Heye Zhang

School of Biomedical Engineering, Sun Yat-sen University, Shenzhen, China
gaozhifan@mail.sysu.edu.cn

Abstract. Joint segmentation and quantification of main coronary vessels are important to the diagnosis and intraoperative treatment of coronary artery disease. They can help clinicians decide whether to carry out coronary revascularization and choose the interventional stent. However, joint segmentation and quantification in a framework is still challenging because of intrinsic distinction of optimization objects for these two tasks. In this paper, we propose a dual-branch multi-scale attention network (DMAN) to achieve synergistic optimization process in a framework. Our DMAN consists of a nested residual module and a attentive regression module. The nested residual module is used to extract and aggregate multi-level and multi-scale features. The attentive regression module introduces a two-phase attention block to express interactive correlation of separated regions and capture the informativeness of the important region in the image. Our DMAN is evaluated over 1893 X-ray coronary angiography images collected from 529 subjects. DMAN achieves the dice coefficient of 0.916 for segmentation and the MAE of 1.30 ± 0.62 mm for quantification.

1 Introduction

Joint segmentation and quantification of main coronary vessels (i.e., estimation of the morphological indices of the stenosis) on X-ray angiography images are of great significance in the diagnosis and interventional therapy of coronary artery disease. The main coronary vessels are the main blood perfusion suppliers of the heart myocardium, meanwhile, they are the frequent sites for the formulation of stenosis lesions in contrast to their branches [1,2]. The blood flow is limited in the main coronary vessels with stenosis lesion, especially under conditions of increased demand [3]. The serious absence of blood flow in the main coronary vessels may jeopardize a large mass of myocardium, even leads to myocardial infarction [4]. The accurate segmentation of coronary vessels can provide visual inspection and pixel-level semantic interpretation. However, only the segmentation results can not provide the comprehensive assessment for the treatment of coronary artery disease. The quantification of stenosis indices also

© Springer Nature Switzerland AG 2021
M. de Bruijne et al. (Eds.): MICCAI 2021, LNCS 12901, pp. 369–378, 2021.
https://doi.org/10.1007/978-3-030-87193-2_35

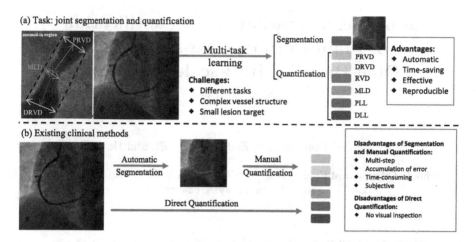

Fig. 1. (a) The proposed method simultaneously segments the main coronary and quantifies the multiple stenosis lesion indices by multi-task learning. (b) Traditional methods achieve the quantification indices based on segmentation results. This leads to the accumulation of error. And direct quantification methods can attain accurate quantification but they cannot provide the visual inspection.

plays an important role in the treatment decision-making, regarding coronary revascularization and stent selection [5]. The morphological indices (namely, lesion length, minimum lumen diameter, and reference vessel diameter, etc., as shown in Fig. 1(a)) can provide valuable assessment information for the stenosis lesion [6]. For instance, the diameter narrow has a great influence on blood perfusion. Lesion length deicides the choice of stent size [7].

The ability to joint segment and quantify the main coronary vessels is especially valuable in medicine but the challenges remain as follows. The beneficial correlation of segmentation and quantification is difficult to model in a framework. Segmentation is a pixel-level classification task, while quantification is an image-level regression task. Their dimension, distribution, and optimization types are different. For segmentation, the complex anatomical structure of coronary vessels hinders the learning of discriminative feature. For example, the same main coronary vessels from different subjects vary wildly in a fixed observation viewpoint and the recognition of main vessels from a subject is also subject to the similarity between main vessels and other branches. For quantification, it is difficult to achieve the complex nonlinear mapping from the image to the feature vector. Because stenosis lesion is relatively small in the whole X-ray coronary angiography image and its feature of the lesion target is not obvious.

There are no reports in the literature that joint segmentation and quantification of main coronary vessels are solved in a framework. As shown in Fig. 1(b), most methods focus on the segmentation of coronary vessels [8–11]. All the vessels are treated equally without distinguishing more valuable main vessels. Most quantitative methods use two-stage models [12–14]. They need to reconstruct or segment vessels first and obtain the quantitative indices through additional calculation. The accumulation of error patterns of segmentation leads to larger quantification errors. Direct quantification is the mapping between the medical image and the quantitative measurement without segmentation or reconstruction [15,16]. Although accurate estimation of the quantitative indices can be obtained, it cannot provide the visual inspection and hence limits the clinical application perspective.

In this paper, we propose a dual-branch multi-scale attention network (DMAN) to effectively integrate the joint segmentation and quantification for the coronary main vessels. DMAN consists of a nested residual module and an attention regression module. Specifically, the nested residual module is a two-level feature extractor [17]. On the bottom level, different sizes of receptive fields are mixed in order to capture more intra-stage information from different scales without degrading the resolution of the feature map. On the top level, the connection of multiple nested units is beneficial for the further extraction of the coronary feature. The attention regression module integrates self-attention and context attention. The self-attention simulates the relationship between the widely separated spatial areas in the feature map [18]. The context attention can extract the information about the important areas in the image [19]. Beneficial from the multi-task cooperation network, the dual-task branches of the shared representation provide complementary information to each other to achieve accurate segmentation and multi-indices estimation.

The main contributions of our method are as follows: (1) We develop a clinical tool for joint segmentation and quantification of main coronary vessels to provide visual inspection and necessary lesion information to guide diagnosis and treatment process. (2) Segmentation and quantification are formulated in a framework to utilize their similarities and differences by multi-task representation learning. Segmentation can hint the location of stenosis lesions to obtain better quantitative accuracy, while quantification can be a global constraint to improve segmentation performance. (3) The nested residual module is introduced to promote the extraction of the multi-level and multi-scale features. Two attention mechanisms are embedded in the indices estimation process to extract the small lesion feature by weighting the importance of different regions.

2   Dual-Branch Multi-scale Attention Network for Joint Segmentation and Quantification

Our DMAN is a dual-branch neural network model to harmoniously integrate the learning of segmentation and quantification. It consists of a nested residual module and an attentive regression module. The workflow of our DMAN is summarized as Fig. 2.

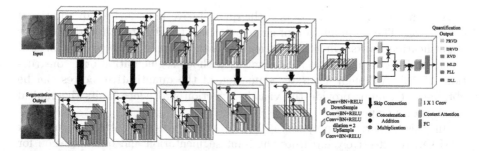

Fig. 2. The architecture of DMAN for joint segmentation and quantification of main coronary vessels.

2.1   Problem Formulation

Joint segmentation and quantification of main coronary vessels can be solved simultaneously because of their relevance through a common low dimensional representation and specialized predictors for different tasks. The acquisition of common representation and predictors can be formulated as an inductive bias learning problem [20,21]. Its aim is to find an optimal hypothesis $\mathbf{h}^* = \{h_1^*, \ldots, h_N^*\}$ for the N tasks from hypothesis space H. The optimal hypothesis h_i^* to task T_i can be expressed as $\hat{y} = g_i(f(x))$, where f is representation mapped from the input image x and used across different tasks, g_i is a predictor specialized to the task T_i at hand. Then, our optimal hypothesis can get solved by minimizing the empirical error on the training dataset $z = (x, y)$

$$\mathbf{h}^* = \arg\min_{\mathbf{h} \in H} \hat{e}_z(\mathbf{h}) = \arg\min_{\mathbf{h} \in H} \frac{1}{N} \sum_{i=1}^{N} \frac{1}{M} \sum_{j=1}^{M} L\left(h_i\left(x_j\right), y_{ij}\right) \tag{1}$$

where y is the label, N is the number of tasks, M is the number of training samples for each task and L is the loss function.

In our setting, $\mathbf{h}^* = \{h_1^*, h_2^*\}$. h_1^* is the segmentation task and h_2^* is the quantification task.

2.2   Multi-scale Segmentation

Local and global contextual information is very important for segmenting the main vessels. So we introduce a base unit namely residual Ublock to extract multi-scale information. It is a variant of the residual block and the main difference between them is that RSUB replaces the plain, single-stream convolution with a U structure, and replaces the original feature with the local feature transformed by a weight layer. The structure change brings about more pooling operations and larger range of receptive fields. This makes our RSUB more capable of perceiving richer local and global features directly. Our segmentation network is two-level nested stacking of multiple RSUBs to form a big U-structure, including a six stages encoder and a five stages decoder. The depth of RSUB in each stage is configured properly. Considering the shallow stages with the feature maps of large height and width, the deeper RSUBs are more suitable to capture more large-scale information. The depth of RSUBs in the stages with low resolution is set shallower to avoid loss of useful context caused by further downsampling, hence we replace pooling and upsampling with dilated convolutions to keep the same resolution of intermediate feature maps as the input feature maps. Six multi-scale output feature maps are generated from different stages. These maps are upsampled to the input image size and fused with a concatenation operation. Then a 1×1 convolution layer and a sigmoid function are applied to generate the final feature map. The binary cross-entropy is used to optimize the segmentation target function. It can be formulated as follows.

$$h_1^* = \arg\min_{h \in H} -\frac{1}{M} \sum_{j=1}^{M} [y_{1j} \log(h_1^*(x_j)) + (1 - y_{1j}) \log(1 - h_1^*(x_j))] \quad (2)$$

2.3 Attentive Quantification

The attentive regression module shares encoder with the segmentation network and merges skip connections. This forces the learning of the shared representation of segmentation and quantification to produce beneficial interactions for the prediction of the stenosis indices. Then we connect the encoder to two attentive modules to attain the important features, especially the subtle feature of stenosis.

The learning of detail features is vital but not easy to our quantification task because the stenosis lesion is relatively small in the whole XRA image. In order to capture detail feature of stenosis we introduce a self-attention module. It can learn long-range relationship between stenosis and other regions based on the perception of the different regions' impact on stenosis. In particular, the module first maps the image features $x \in R^{C \times M}$ into three feature spaces f, g and h by implementing 1×1 convolutions $f(x) = W_f x, g(x) = W_g x, h(x) = W_h x$, where C is the channel numbers, M is the number of spatial locations, and $W_f, W_g \in R^{\frac{C}{8} \times C}, W_h \in R^{C \times C}$ are the weight. Besides, $\alpha_{j,i} = \exp(S_{ij})/\sum_{i=1}^{M} \exp(S_{ij})$ is used to indicate the i_{th} location' contribution when synthesizing the j_{th} region. The attentive output is denoted as $o = (o_1, o_2, \ldots, o_i, \ldots, o_M) \in R^{C \times M}$, and $o_i = x_i + \gamma \sum_{i=1}^{M} \alpha_{j,i} h(x_i)$. γ is a learnable scalar initialized as 0, it works by learning to assign more weight to the non-local region.

However, the contributions of different regions are obviously not equal to the representation of the image meaning. So it is necessary to extract the major contributing regions for the input image and aggregate the informative representation of these regions to generate a valuable feature vector. In particular, we first attain the hidden feature by one layer perception $u_r = \tanh(W_r x_r + b_r)$. Then in order to measure the informativeness of the region, we apply the region context vector u_s to the hidden feature u_r through softmax function to get importance

| Raw | GT | Our | Jun[8] | Yang[22] | Yang[23] | Zhu[24] | Xian[25] |

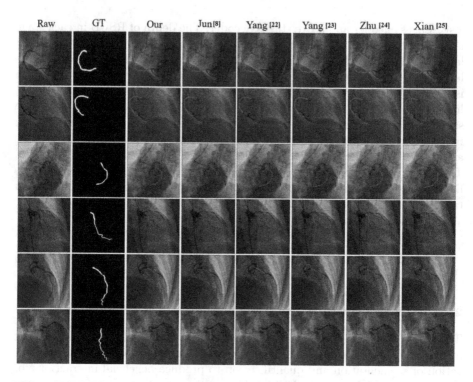

Fig. 3. Comparison with the state-of-the-art methods. Representative results show our DMAN has better segmentation performance. The second column "GT" is the ground truth.

weight $\beta_r = \exp\left(u_r^\top u_s\right)/\sum_r \exp\left(u_r^T u_s\right)$. Finally, we compute weighted sum of the regions to attain representation based on $x = \sum_r \beta_r x_r$. The vector x summarizes the important information in the current image.

The output feature of two attentive modules denoted as K is fed into two fully connected layers with 512 and 6 units, each employing a LeakyReLU activation. The output of regression module is $f(x_i) = W_o K + b_o$, where W_o and b_o are the weight matrix and bias respectively. Here we use mean absolute error (MAE) as our optimization loss function.

$$h_2^* = \arg\min_{h \in H} \frac{1}{d \times M} \sum_{j=1}^{M} |h_2^*(x_j) - y_{2j}| \qquad (3)$$

3   Experiments and Results

3.1   Experiment Setup

Dataset. A total of 529 patients were retrospectively enrolled in our study. They have completed coronary angiography using a clinical angiographic X-ray system

Table 1. Performance of our model DMAN and other methods for the segmentation main coronary vessels. Dice Similarity Coefficient (DSC), Precision (Pr) and Sensitivity (Se) are used for the segmentation evaluation criterion.

Method	ALL			LAD			LCX			RCA		
	DSC	Pr	Se	DSC	Pr	Se	DSC	Pr	Se	DSC	Pr	Se
Jun et al. [8]	0.810	0.861	0.786	0.826	0.868	0.805	0.786	0.874	0.746	0.854	0.910	0.823
Yang et al. [22]	0.866	0.917	0.833	0.858	0.915	0.813	0.832	0.906	0.792	0.913	0.932	0.833
Yang et al. [23]	0.896	0.906	0.893	0.908	0.925	0.897	0.888	0.908	0.873	0.937	0.943	0.935
Zhu et al. [24]	0.884	0.901	0.873	0.869	0.892	0.851	0.866	0.891	0.848	0.922	0.924	0.924
Xian et al. [25]	0.870	0.879	0.902	0.906	0.888	0.929	0.891	0.888	0.901	0.917	0.919	0.920
Our DMAN	**0.916**	**0.921**	**0.913**	0.907	0.912	0.904	**0.898**	0.907	0.892	**0.940**	**0.943**	**0.941**

(Philips Allura XPER). Catheterization was performed through the femoral or radial routes using standard catheters and coronary angiograms were digitally recorded. The selection of angiography images was according to the standard clinical procedure [14]. First, the XRA image sequences of these patients were selected from the appropriate viewpoints. Then, for the selected XRA image sequence, the image keyframe in the end-diastole phase with good image quality and full contrast agent penetration was selected and the image unable to recognize the coronary structures like chronic total occlusion was excluded. A dataset of 1893 images is built and 563 are LAD, 619 are LCX, and 711 are RCA. The main vessels in the keyframes were manually segmented and quantified by an experienced radiologist. The segmentation area of each main coronary vessel is set from ostium to the far distal. For RCA, the distal end of the segmented area is the bifurcation point between posterior descending artery (PDA) and postero-lateral artery (PL). As a result, the main vessel angiography images and labels are obtained.

Training Setup. Our DMAN was trained and tested on a NVIDIA Titan Xp 12 GB GPU. For training, we used the RMSProp optimizer with a batch of 8 images per step and an initial learning rate of 0.0002. The decay rate was set to 0.95. In our experiments, patient-wise 10-fold cross-validation is employed on our dataset.

Comparison with the State-of-the-Art Methods. Our DMAN was compared with nine single-task methods where five methods [8, 22–25] concern the segmentation of the main coronary vessels, and the other four methods [14, 26–28] concern the quantification of the stenosis lesions. The evaluation metrics used to assess the predictability of our models are Dice Similarity Coefficient, Precision and Sensitivity for the segmentation task and MAE and Pearson Correlation Coeffcient for the quantification task.

Table 2. Performance of our DMAN and four quantification methods for multi-index coronary stenosis estimation. Average Mean Absolute Error (MAE) and Pearson Correlation Coefficient are used for the quantification evaluation criterion.

Method	MAE	Pearson(%)	PRVD	DRVD	RVD	MLD	PLL	DLL
Wan et al. [14]	1.34 ± 0.65	90.80 ± 11.85	0.69	0.71	0.65	0.57	2.84	2.62
Krizhevsky et al. [26]	1.72 ± 0.73	87.91 ± 13.10	0.98	1.06	1.03	0.87	3.18	3.18
Zhen et al. [27]	1.41 ± 0.63	90.55 ± 12.10	0.72	0.74	0.67	0.58	2.86	2.90
Xue et al. [28]	1.43 ± 0.65	88.66 ± 12.64	0.76	0.80	0.78	0.70	2.80	2.73
Our DMAN	**1.30 ± 0.62**	**91.30 ± 11.10**	0.75	**0.68**	0.67	**0.57**	**2.59**	**2.58**

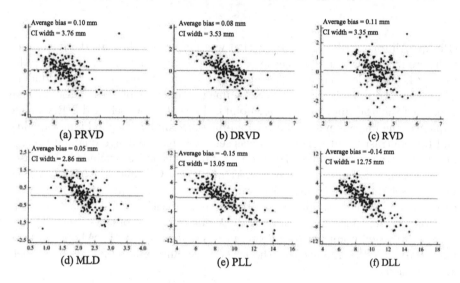

Fig. 4. High agreement between our DMAN and the ground truth (GT) with respect to six clinical indices. The blue dashed lines indicate the 95% confidence intervals of the bias. The green dashed lines indicate the bias. (Color figure online)

3.2 Results and Analysis

Accurate Main Coronary Segmentation. Figure 3 visually shows the accuracy of segmentation on the representative images. The main coronary vessel regions detected by our approach have high consistency with the ground truth. Table 1. also shows that DMAN has a high segmentation performance with a dice coefficient of 91.6%, a precision of 92.1% and a sensitivity of 91.3%. Compared with the state-of-the-art methods, Dice Similarity Coefficient, Precision and Sensitivity increase 0.02–0.106, 0.001-0.06 and 0.011–0.02 in total.

Precise Stenosis Index Quantification. Table 2 presents the high quantitative accuracy of our approach. The average MAE is 1.30 ± 0.62 mm and the Pearson correlation coefficient is 91.30% ± 11.10%. For each quantitative indice, the MAE of the PRVD, DRVD, RVD, MLD, PLL and DLL are 0.75 mm, 0.68 mm, 0.67 mm. 0.57 mm. 2.59 mm and 2.58 mm. The Bland-Altman plots of all six indices in Fig. 4 also show that the predictive stenosis indices are very close to the actual values. Comparing Row 2 to 5 of Table 2, it shows our DMAN obtains more accurate multi-index stenosis lesion quantification. These results demonstrate the superiority of our approach.

4 Conclusion

In this study, we have proposed a simultaneous segmentation and quantification model (namely DMAN) for main coronary vessels by multi-scale multi-task learning. The proposed method was conducted on 529 subjects and yielded better segmentation and quantification results, by comparing with five segmentation methods and four quantification methods. Experiment results demonstrate that our model can aid in the clinical diagnosis of coronary disease.

Acknowledgement. This work was supported by the National Natural Science Foundation of China under Grant U1908211 and Key Program for International Cooperation Projects of Guangdong Province under Grant 2018A050506031.

References

1. Sianos, G., et al.: The SYNTAX score: an angiographic tool grading the complexity of coronary artery disease. EuroIntervention **1**(2), 219–227 (2005)
2. Halon, D.A., Sapoznikov, D., Lewis, B.S., Gotsman, M.S.: Localization of lesions in the coronary circulation. Am. J. Cardiol. **52**(8), 921–926 (1983)
3. Vogel, R.A.: Assessing stenosis significance by coronary arteriography: are the best variables good enough? J. Am. Coll. Cardiol. **12**(3), 692–693 (1988)
4. Fioranelli, M., Gonnella, C., Tonioni, S., D'Errico, F., Carbone, M.: Clinical anatomy of the coronary circulation. In: Imaging Coronary Arteries, pp. 1–11 (2013)
5. Rittger, H., Schertel, B., Schmidt, M., Justiz, J., Brachmann, J., Sinha, A.M.: Three-dimensional reconstruction allows accurate quantification and length measurements of coronary artery stenoses. EuroIntervention **5**(1), 127–132 (2009)
6. Tomasello, S. D., Costanzo, L., Galassi, A. R.: Quantitative coronary angiography in the interventional cardiology. In: Advances in the Diagnosis of Coronary Atherosclerosis (2011)
7. Garrone, P., et al.: Quantitative coronary angiography in the current era: principles and applications. J. Intervent. Cardiol **22**(6), 527–536 (2009)
8. Jun, T.J., Kweon, J., Kim, Y.H., Kim, D.: T-net: nested encoder-decoder architecture for the main vessel segmentation in coronary angiography. Neural Netw. **128**, 216–233 (2020)
9. Qin, B., et al.: Accurate vessel extraction via tensor completion of background layer in X-ray coronary angiograms. Pattern Recogn. **87**, 38–54 (2019)
10. Cervantes-Sanchez, F., Cruz-Aceves, I., Hernandez-Aguirre, A., Hernandez-Gonzalez, M.A., Solorio-Meza, S.E.: Automatic segmentation of coronary arteries in X-ray angiograms using multiscale analysis and artificial neural networks. Appl. Sci. **9**(24), 5507 (2019)

11. Hao, D., et al.: Sequential vessel segmentation via deep channel attention network. Neural Netw. **128**, 172–187 (2020)
12. Cong, W., Yang, J., Ai, D., Chen, Y., Liu, Y., Wang, Y.: Quantitative analysis of deformable model-based 3-D reconstruction of coronary artery from multiple angiograms. IEEE Trans. Biomed. Eng. **62**(8), 2079–2090 (2015)
13. Janssen, J.P., Rares, A., Tuinenburg, J.C., Koning, G., Lansky, A.J., Reiber, J.H.: New approaches for the assessment of vessel sizes in quantitative vascular X-ray analysis. Int. J. Cardiovasc. Imaging **26**(3), 259–271 (2010)
14. Wan, T., Feng, H., Tong, C., Li, D., Qin, Z.: Automated identification and grading of coronary artery stenoses with X-ray angiography. Comput. Methods Prog. Biomed. **167**, 13–22 (2018)
15. Zhang, D., Yang, G., Zhao, S., Zhang, Y., Zhang, H., Li, S.: Direct quantification for coronary artery stenosis using multiview learning. In: International Conference on Medical Image Computing and Computer-Assisted Intervention, pp. 449–457 (2019)
16. Zhang, D., et al.: Direct quantification for coronary artery stenosis using multiview learning. IEEE Trans. Med. Imaging **39**(12), 4322–4334 (2020)
17. Qin, X., Zhang, Z., Huang, C., Dehghan, M., Zaiane, O.R., Jagersand, M.: U2-Net: going deeper with nested U-structure for salient object detection. Pattern Recogn. **106**, 107404 (2020)
18. Zhang, H., Goodfellow, I., Metaxas, D., Odena, A.: Self-attention generative adversarial networks. In: International Conference on Machine Learning, pp. 7354–7363 (2019)
19. Yang, Z., Yang, D., Dyer, C., He, X., Smola, A., Hovy, E.: Hierarchical attention networks for document classification. In: The 2016 Conference of the North American Chapter of the Association for Computational Linguistics: Human Language Technologies, pp. 1480–1489 (2016)
20. Baxter, J.: A model of inductive bias learning. J. Artif. Intell. Res. **12**, 149–198 (2000)
21. Maurer, A., Pontil, M., Romera-Paredes, B.: The benefit of multitask representation learning. J. Mach. Learn. Res. **17**(81), 1–32 (2016)
22. Yang, S., Kweon, J., Kim, Y. H.: Major vessel segmentation on x-ray coronary angiography using deep networks with a novel penalty loss function. In: International Conference on Medical Imaging with Deep Learning (2019)
23. Yang, S., et al.: Deep learning segmentation of major vessels in X-ray coronary angiography. Sci. Rep. **9**(1), 1–11 (2019)
24. Zhu, X., Cheng, Z., Wang, S., Chen, X., Lu, G.: Coronary angiography image segmentation based on PSPNet. Comput. Methods Prog. Biomed. **200**, 105897 (2020)
25. Xian, Z., Wang, X., Yan, S., Yang, D., Chen, J., Peng, C.: Main coronary vessel segmentation using deep learning in smart medical. Math. Prob. Eng. **2020**, 1–9 (2020)
26. Krizhevsky, A., Sutskever, I., Hinton, G.E.: Imagenet classification with deep convolutional neural networks. Adv. Neural Inf. Process. Syst. **25**, 1097–1105 (2012)
27. Zhen, X., Wang, Z., Islam, A., Bhaduri, M., Chan, I., Li, S.: Direct estimation of cardiac bi-ventricular volumes with regression forests. In: International Conference on Medical Image Computing and Computer-Assisted Intervention, pp. 586–593 (2014)
28. Xue, W., Islam, A., Bhaduri, M., Li, S.: Direct multitype cardiac indices estimation via joint representation and regression learning. IEEE Trans. Med. Imaging **36**(10), 2057–2067 (2017)

A Spatial Guided Self-supervised Clustering Network for Medical Image Segmentation

Euijoon Ahn[1]([⊠]), Dagan Feng[1,2], and Jinman Kim[1]

[1] School of Computer Science, The University of Sydney, Darlington, NSW, Australia
euijoon.ahn@sydney.edu.au
[2] Med-X Research Institute, Shanghai Jiao Tong University, Shanghai, China

Abstract. The segmentation of medical images is a fundamental step in automated clinical decision support systems. Existing medical image segmentation methods based on supervised deep learning, however, remain problematic because of their reliance on large amounts of labelled training data. Although medical imaging data repositories continue to expand, there has not been a commensurate increase in the amount of annotated data. Hence, we propose a new spatial guided self-supervised clustering network (SGSCN) for medical image segmentation, where we introduce multiple loss functions designed to aid in grouping image pixels that are spatially connected and have similar feature representations. It iteratively learns feature representations and clustering assignment of each pixel in an end-to-end fashion from a single image. We also propose a context-based consistency loss that better delineates the shape and boundaries of image regions. It enforces all the pixels belonging to a cluster to be spatially close to the cluster centre. We evaluated our method on 2 public medical image datasets and compared it to existing conventional and self-supervised clustering methods. Experimental results show that our method was most accurate for medical image segmentation.

Keywords: Self-supervised learning · Clustering · Convolutional neural network · Medical image segmentation

1 Introduction

Supervised deep learning methods allow the derivation of image features for a variety of image analysis problems using underlying algorithms and large-scale labelled data [1]. In the medical imaging domain, however, there is a paucity of labelled data due to the cost and time entailed in manual delineation by imaging experts, inter- and intra-observer variability amongst these experts and then the complexity of the images themselves where there may be many different appearances based on, for instance, bone and soft tissues windows on computed tomography (CT) and different sequences on Magnetic Resonance (MR) and noise on Ultrasound (US) images.

Researchers have employed many different approaches to help solve these challenges including deep learning with transferable knowledge across different domains and fine-tuning those knowledges with a relatively smaller amount of labelled image data (i.e.,

© Springer Nature Switzerland AG 2021
M. de Bruijne et al. (Eds.): MICCAI 2021, LNCS 12901, pp. 379–388, 2021.
https://doi.org/10.1007/978-3-030-87193-2_36

domain adaptation). Other approaches use unsupervised feature learning [2, 3] where the aim is to learn invariant local image features using algorithms such as sparse coding and auto-encoder. Recently, self-supervised learning, a form of unsupervised learning, where the data themselves generate supervisory signals for the feature learning, has shown great success in many computer vision [4] and medical image segmentation tasks [5–7]. For example, Zhuang et al. [8] trained a convolutional neural network (CNN) in a self-supervised manner by predicting the spatial transformation of 3D CT scan images for brain tumour segmentation. Similarly, Tajbakhsh et al. [9] constructed supervisory signals by predicting colour, rotation and noise for lung lobe segmentation in CT scans. Another approach to construct a supervisory signal is to use image clustering [10–12]. The key concept of self-supervised clustering is to use clustering assignments (i.e., cluster labels) as surrogate labels to learn the parameters of CNNs. Caron et al. [10] proposed a DeepCluster that iteratively learns and improvs the feature representation and clustering assignment of image features during CNN training. Similarly, Ji et al. [11] improved the clustering by maximising the mutual information between spatially transformed image patches (Invariant Information Clustering (IIC)). In medical imaging, Moriya et al. [13] used k-means clustering to group similar pixels on micro-CT images and learned the feature representation of pixels using the cluster labels. While these self-supervised clustering methods have shown to be effective in various medical image segmentation tasks, they are limited by the manual selection of cluster size (e.g., k from k-means) and may fail when there are complicated regions with fuzzy boundaries, various shapes, artifacts and noise.

In this paper, we propose a new spatial guided self-supervised clustering network (SGSCN) for medical image segmentation, where we iteratively learn optimal cluster size and improve the feature representation of each pixel. We also design a context-based consistency loss as a differentiable loss function that aids in segmenting image regions with fuzzy boundaries and noise. The context-based consistency loss enforces all the pixels belonging to a cluster to be spatially close to the cluster centre. We validated our approach on 2 public datasets and compared it to other unsupervised clustering and self-supervised clustering methods.

2   Materials and Methods

2.1   Materials

We used 2 public datasets for our experimental analysis. Each dataset was used to access the performance of our method on 2 different problems.

Skin lesion segmentation – we used PH2 [14] public dataset. It provides 200 dermoscopic image studies, including 80 common nevi, 80 atypical nevi and 40 melanomas. Manually annotated lesions from expert dermatologists were available from the PH2 and used as the ground truth data. The PH2 dataset provides various images with complex skin conditions.

Liver tumour segmentation – we used Sun Yat-sen University US (SYSU-US) public dataset [15]. It provides 20 sets of 2D US image sequences containing 10–30 images of abdomen with liver tumour. The ground truth images were annotated by experts and provided by the SYSU. A subset of 100 US images were used for the evaluation of our method on single slice image segmentation and this was done by randomly selecting 5 US images from each image sequence, consistent with other research [15].

2.2 Overview of the SGSCN

A schematic of our method is shown in Fig. 1. Given an image x_n, we train a convolutional segmentation network F parameterized by θ_f that generates segmentation map $S_n = F(x_n; \theta_f)$. We then normalise the segmentation map S_n such that $\hat{S}_n$ has zero mean and unit variance. The final segmentation map (cluster label C_n of each pixel) is obtained by selecting the channel-wise dimension that has maximum response value in $\hat{S}_n$ (i.e., *argmax* function). The parameters θ_f were trained by minimizing the cross-entropy loss between $\hat{S}_n$ and C_n. Sparse spatial loss and context-based consistency loss are also used to enhance the clustering assignments by understanding the spatial relationships of image pixels and regions. The SGSCN iteratively learns and improves feature representation and clustering assignment of each pixel in an end-to-end fashion.

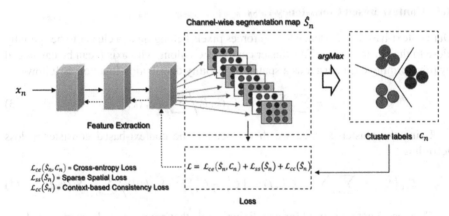

Fig. 1. A schematic of our SGSCN.

2.3 Cross-Entropy Loss for Self-supervised Clustering

As in a standard CNN, the parameters of the convolutional segmentation network F are learned by calculating the cross-entropy loss between $\hat{S}_n$ and C_n, which is defined as follows:

$$\mathcal{L}_{ce}\left(\hat{S}_n, C_n\right) = \sum_{n=1}^{N} \sum_{p=1}^{i} -\delta(p - C_n) \ln \hat{S}_{n-1} \qquad (1)$$

where p is the cluster index ($p = 1, \cdots, i$) and $\delta(\cdot)$ is the indicator function that assigns the value 1 when ($p - C_n$) is equal to 0 and the value 0 for otherwise. The cross-entropy loss is minimized using stochastic gradient descent (SGD).

2.4 Sparse Spatial Loss

Classic cross-entropy loss ignores the spatial relationships of image pixels or regions, which generate sub-optimal parameters during CNN training. Unfortunately, this makes it difficult to group spatially connected pixels (e.g., edges) or regions.

L_1-norm based regularization techniques are used widely in image restoration and denoising and has proven effectiveness when dealing with sparse data [16, 17]. It has also shown to less penalize subtle edges or spatially connected regions in images. Thus, we use the L_1-norm to measure the vertical and horizontal differences of the segmentation map $\hat{S}_n$ to better understand the spatial relationships of image pixels and regions. This is defined as follows:

$$\mathcal{L}_{ss}\left(\hat{S}_n\right) = \sum_{k=1}^{W-1} \sum_{l=1}^{H-1} \left\| \hat{S}_{k+1,l} - \hat{S}_{k,l} \right\|_1 + \left\| \hat{S}_{k,l+1} - \hat{S}_{k,l} \right\|_1 \tag{2}$$

where W and H are the width and the height of the input image and $\hat{S}_{k,l}$ denotes the pixel value at (k, l) coordinate in the segmentation map $\hat{S}_n$.

2.5 Context-Based Consistency Loss

Our context-based consistency loss enforces pixels belonging to a cluster to be spatially close to the cluster centre. The cluster centre of C_n along with axis k can be calculated by transforming each cluster as a spatial probability distribution function as follows:

$$C_n^k = \sum_n \sum_{k,l} k \cdot \hat{S}_n(k, l) \Big/ \sum_{k,l} \hat{S}_n(k, l). \tag{3}$$

Using the cluster centres, we then calculate the context-based consistency loss according to:

$$\mathcal{L}_{cc}\left(\hat{S}_n\right) = \sum_n \sum_{k,l} \left\| (k, l) - \left(C_n^k - C_n^l \right) \right\|^2 \cdot \hat{S}_n(k, l) \Big/ \sum_{k,l} \hat{S}_n(k, l). \tag{4}$$

This loss function is used for penalizing pixels that are spatially located away from the cluster centre.

2.6 Training via Backpropagation

The overall loss function for the convolutional segmentation network F is the sum of the cross-entropy loss, sparse spatial loss and context-based consistency loss. We set an arbitrary maximum number of possible clusters i (as in Eq. 1) and this is iteratively minimised using the overall loss function. Similar image pixels would be assigned to same clusters during the course of training, making the unique number of cluster smaller than initially defined maximum cluster size. This process is repeated until the clustering and the loss become stable.

3   Experimental Setup

3.1   Evaluation

We evaluated our method by comparing it to other unsupervised and self-supervised clustering methods. As the baseline, we compared our method to well-established unsupervised k-means clustering algorithm. We also compared it to the state-of-the-art self-supervised clustering methods – DeepCluster [10] and IIC [11]. We used the three standard metrics including Dice similarity coefficient (DSC), Hammoude distance (HM) and XOR which are routinely used among researchers to assess segmentation performance. A higher DSC score corresponds to a better result. In contrast, lower scores for HM and XOR correspond to better results. Since the background regions, such as dark corners, in dermoscopic images and in US images are also considered as one of largest segments within an image, we only considered the predicted segment (i.e., cluster) that had the largest overlap with the GT segment in our evaluation.

3.2   Implementation Details

For our convolutional segmentation network F, we adopted a shallow CNN architecture, comprises 3 convolutional layers, each of which has ReLU activation and batch normalisation function. Each convolutional layer has the following network parameters: kernel size of 3×3, stride of 1, pad size of 1 and, filter size of 100. Here the filter size of 100 is equivalent to the maximum number of possible clusters i (as defined in Eq. 1). We set the uniform learning rate of 0.1 with a momentum of 0.9 for skin lesion segmentation task and a smaller learning rate of 0.05 with a momentum of 0.9 for liver tumour segmentation. We trained our network on a GeForce GTX 1080 Ti GPU (11 GB memory). We used an empirical process to discover appropriate number of clusters (3 to 8) for k-means, DeepCluster and IIC. For k-means, $k = 3$ had the highest accuracy for skin lesion segmentation and $k = 6$ for liver tumour segmentation. We set $k = 3$ for DeepCluster and IIC in our all experiments.

4   Results

The results of skin lesion segmentation from dermoscopic images are shown in Table 1. Our SGSCN generated higher scores and had the best overall DSC average (83.4%) and the best XOR (28.2%) accuracy. The results of liver tumour segmentation from US images are outlined in Table 2 and they show that our method had the highest DSC value (63.2%), and the best HM (46.2%) and XOR (52.3%) scores. We show the results for four selected images in Fig. 2 with two dermoscopic images in the 1st and 2nd rows and two US images in the 3rd and 4th rows. Figure 3 shows the relative improvement in segmentation accuracy due to the use of sparse spatial and context-based consistency losses, in addition to the cross-entropy loss. We also show the sample segmentation results of using sparse spatial and context-based consistency loss in Fig. 4.

Table 1. The segmentation results on PH2 (skin lesion segmentation) compared to other methods

Mean%	k-means ($k = 3$)	k-means ($k = 4$)	DeepCluster	IIC	Our method
DSC	71.3	67.7	79.6	81.2	**83.4**
HM	130.8	165.2	35.8	35.3	**32.3**
XOR	41.3	44.9	31.3	29.8	**28.2**

Table 2. The segmentation results on SYSU-US (liver tumour segmentation) compared to other methods

Mean%	k-means ($k = 5$)	k-means ($k = 6$)	DeepCluster	IIC	Our method
DSC	37.1	39.3	60.9	58.3	**63.2**
HM	93.2	95.2	47.9	56.2	**46.2**
XOR	75.1	73.3	55.2	55.8	**52.3**

5 Discussion

Our findings indicate that our SGSCN a) outperformed other unsupervised and self-supervised clustering methods; b) was able to locate the lesion and its proximity to image boundaries (see Fig. 2); c) progressively improved the feature representation and the clustering assignments of each pixel and d) enhanced the clustering assignment of each pixel using our sparse spatial and context-based consistency loss (see Fig. 3 and Fig. 4).

Our SGSCN had higher accuracy than other recent self-supervised clustering methods. This is attributed to our sparse spatial loss that helps in segmenting spatially connected regions in images (see Fig. 4, row1). The context-based consistency loss further enhances the segmentation by focusing on segments that are spatially close to the cluster centre (see Fig. 4, row2).

The quality of clustering used in the conventional unsupervised k-means was not as robust as those from recent self-supervised clustering methods such as DeepCluster and IIC. DeepCluster and IIC also iteratively learned feature representation and clustering assignments and hence had a higher accuracy in skin lesion and liver tumour segmentation when compared to k-means. Their performance, however, varied widely depending on the initial selection of fixed cluster size and do not incorporate the spatial relationships of image pixels and regions during CNN training.

In the segmentation of skin lesion, our SGSCN outperformed all other unsupervised and self-supervised clustering methods. The DeepCluster and IIC were the next closest to ours in overall DSC and XOR scores. DeepCluster, however, did not consider the spatial relationships of image regions and therefore was not able to accurately segment lesions with fuzzy boundaries (see Fig. 2, row1).

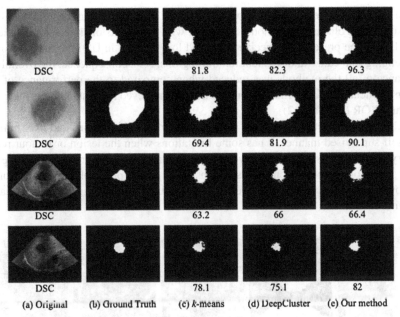

Fig. 2. Segmentation results from 4 study examples (top two rows: dermoscopic images and bottom two rows: US images), where (a)–(e) represent the original image in column 1, ground truth in column 2, and the segmentation results from column 3 to column 7 for k-means, DeepCluster and our method.

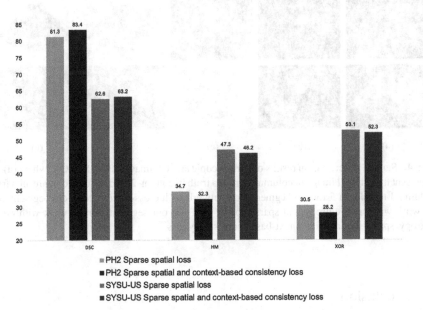

Fig. 3. DSC, HM and XOR scores of our SGSCN with sparse spatial loss and our SGSCN with both sparse spatial and context-based consistency loss.

In liver tumour segmentation from US images, our SGSCN outperformed all other self-supervised clustering methods, consistent with the results of skin lesion segmentation. Due to the presence of speckle noise and low contrast in US images, it was more challenging to localise and segment tumour regions. As a consequence, the overall performance of all the methods was reduced when compared to the performance of skin lesion segmentation. Nevertheless, our method still achieved the best DSC and the lowest HM and XOR (see Table 2).

Although our method improved the segmentation of skin lesions and liver tumours in a self-supervised manner, it has some limitations when the lesion or tumour region is very small, not visually distinctive and has incomplete boundary. It is possible that other spatial or geometric constraints such as affine or thin plate spline grid may further improve the medical image feature representation, and we will investigate this as part of our future work.

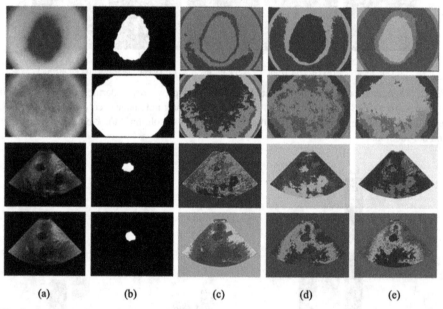

Fig. 4. Sample segmentation results of dermoscopic and US images using SGSCN, where (a)-(e) represent the original image in column 1, ground truth in column 2, and the clustering results from column 3 to column 5 for our segmentation network with cross-entropy loss, our segmentation network with cross-entropy and sparse spatial loss and our segmentation network with cross-entropy, sparse spatial and context-based consistency loss.

6   Conclusion

In this work, we developed a self-supervised clustering network that iteratively learns image feature representation and clustering of image pixels by characterising the spatial relationships of image regions. We compared our method to other unsupervised and self-supervised clustering methods on 2 public datasets and showed that our method outperformed other methods. Our findings indicate that our method can be applied to various medical image data.

References

1. Tan, M., Le, Q.: Efficientnet: rethinking model scaling for convolutional neural networks. In: International Conference on Machine Learning, pp. 6105–6114. PMLR (2019)
2. Ahn, E., Kumar, A., Fulham, M., Feng, D., Kim, J.: Convolutional sparse kernel network for unsupervised medical image analysis. Med. Image Anal. **56**, 140–151 (2019)
3. Ahn, E., Kumar, A., Fulham, M., Feng, D., Kim, J.: Unsupervised domain adaptation to classify medical images using zero-bias convolutional auto-encoders and context-based feature augmentation. IEEE Trans. Med. Imaging **39**, 2385–2394 (2020)
4. Misra, I., Maaten, L.v.d.: Self-supervised learning of pretext-invariant representations. In: Proceedings of the IEEE/CVF Conference on Computer Vision and Pattern Recognition, pp. 6707–6717 (2020)
5. Bai, W., et al.: Self-supervised learning for cardiac MR image segmentation by anatomical position prediction. In: Shen, D., Liu, T., Peters, T.M., Staib, L.H., Essert, C., Zhou, S., Yap, P.-T., Khan, A. (eds.) MICCAI 2019. LNCS, vol. 11765, pp. 541–549. Springer, Cham (2019). https://doi.org/10.1007/978-3-030-32245-8_60
6. Blendowski, M., Nickisch, H., Heinrich, M.P.: How to learn from unlabeled volume data: self-supervised 3D context feature learning. In: Shen, D., Liu, T., Peters, T.M., Staib, L.H., Essert, C., Zhou, S., Yap, P.-T., Khan, A. (eds.) MICCAI 2019. LNCS, vol. 11769, pp. 649–657. Springer, Cham (2019). https://doi.org/10.1007/978-3-030-32226-7_72
7. Xia, X., Kulis, B.: W-net: a deep model for fully unsupervised image segmentation. arXiv preprint arXiv:1711.08506 (2017)
8. Zhuang, X., Li, Y., Yifan, H., Ma, K., Yang, Y., Zheng, Y.: Self-supervised feature learning for 3D medical images by playing a rubik's cube. In: Shen, D., Liu, T., Peters, T.M., Staib, L.H., Essert, C., Zhou, S., Yap, P.-T., Khan, A. (eds.) MICCAI 2019. LNCS, vol. 11767, pp. 420–428. Springer, Cham (2019). https://doi.org/10.1007/978-3-030-32251-9_46
9. Tajbakhsh, N., et al.: Surrogate supervision for medical image analysis: effective deep learning from limited quantities of labeled data. In: 2019 IEEE 16th International Symposium on Biomedical Imaging (ISBI 2019), pp. 1251–1255. IEEE (2019)
10. Caron, M., Bojanowski, P., Joulin, A., Douze, M.: Deep clustering for unsupervised learning of visual features. In: Ferrari, V., Hebert, M., Sminchisescu, C., Weiss, Y. (eds.) Computer Vision – ECCV 2018. LNCS, vol. 11218, pp. 139–156. Springer, Cham (2018). https://doi.org/10.1007/978-3-030-01264-9_9
11. Ji, X., Henriques, J.F., Vedaldi, A.: Invariant information clustering for unsupervised image classification and segmentation. In: Proceedings of the IEEE/CVF International Conference on Computer Vision, pp. 9865–9874 (2019)
12. Ahn, E., Kumar, A., Feng, D., Fulham, M., Kim, J.: Unsupervised feature learning with K-means and an ensemble of deep convolutional neural networks for medical image classification. arXiv preprint arXiv:1906.03359 (2019)
13. Moriya, T., Roth, H.R., Nakamura, S., Oda, H., Nagara, K., Oda, M., Mori, K.: Unsupervised segmentation of 3D medical images based on clustering and deep representation learning. In: Medical Imaging 2018: Biomedical Applications in Molecular, Structural, and Functional Imaging, p. 1057820. International Society for Optics and Photonics (2018)

14. Mendonça, T., Ferreira, P.M., Marques, J.S., Marcal, A.R., Rozeira, J.: PH 2-A dermoscopic image database for research and benchmarking. In: 2013 35th Annual International Conference of the IEEE Engineering in Medicine and Biology Society (EMBC), pp. 5437–5440. IEEE (2013)
15. Lin, L., Yang, W., Li, C., Tang, J., Cao, X.: Inference with collaborative model for interactive tumor segmentation in medical image sequences. IEEE Trans. Cybern. **46**, 2796–2809 (2015)
16. Fu, H., Ng, M.K., Nikolova, M., Barlow, J.L.: Efficient minimization methods of mixed l2–l1 and l1–l1 norms for image restoration. SIAM J. Sci. Comput. **27**, 1881–1902 (2006)
17. Zhang, M., Desrosiers, C.: High-quality image restoration using low-rank patch regularization and global structure sparsity. IEEE Trans. Image Process. **28**, 868–879 (2018)

Comprehensive Importance-Based Selective Regularization for Continual Segmentation Across Multiple Sites

Jingyang Zhang[1,2,4], Ran Gu[3,4], Guotai Wang[3], and Lixu Gu[1,2,4(✉)]

[1] School of Biomedical Engineering, Shanghai Jiao Tong University, Shanghai,, China
`gulixu@sjtu.edu.cn`
[2] Institute of Medical Robotics, Shanghai Jiao Tong University, Shanghai, China
[3] School of Mechanical and Electrical Engineering, University of Electronic Science and Technology of China, Chengdu, China
[4] SenseTime Research, Shanghai, China

Abstract. In clinical practice, a desirable medical image segmentation model should be able to learn from sequential training data from multiple sites, as collecting these data together could be difficult due to the storage cost and privacy restriction. However, existing methods often suffer from catastrophic forgetting problem for previous sites when learning from images from a new site. In this paper, we propose a novel comprehensive importance-based selective regularization method for continual segmentation, aiming to mitigate model forgetting by maintaining both shape and reliable semantic knowledge for previous sites. Specifically, we define a comprehensive importance weight for each model parameter, which consists of shape-aware importance and uncertainty-guided semantics-aware importance, by measuring how a segmentation's shape and reliable semantic information is sensitive to the parameter. When training model on a new site, we adopt a selective regularization scheme that penalizes changes of parameters with high comprehensive importance, avoiding the shape knowledge and reliable semantics related to previous sites being forgotten. We evaluate our method on prostate MRI data sequentially acquired from six institutes. Results show that our method outperforms many continual learning methods for relieving model forgetting issue. Code is available at https://github.com/jingyzhang/CISR.

Keywords: Continual learning · Multi-site segmentation · Comprehensive importance · Selective regularization

1 Introduction

Convolutional neural networks have achieved remarkable performance in medical image segmentation [9]. These architectures require a large number of training images, which are commonly acquired from multiple sites (or hospitals), to

---

J. Zhang and R. Gu—The authors contributed equally to this work.

© Springer Nature Switzerland AG 2021
M. de Bruijne et al. (Eds.): MICCAI 2021, LNCS 12901, pp. 389–399, 2021.
https://doi.org/10.1007/978-3-030-87193-2_37

improve model generalization capability. However, it is impractical to aggregate together such large multi-site datasets due to the expensive storage cost and the privacy restriction across institutes. An alternative way is to train a model with a sequential stream of multi-site data rather than a consolidated set, where data of different sites arrives in sequence without storing and access to old data of previous sites. In this setting, a naive continuous model fine-tuning scheme concerning only the new incoming site would cause considerable performance degradation on previously learned sites, called *catastrophic forgetting* [15], due to the data distribution discrepancy across multiple sites with different acquisition protocols. It is desired yet challenging to enable a model to continually segment on a new site without sacrificing the performance on previous sites.

Much effort has been directed at continual learning for mitigating model forgetting[18]. For example, experience replay methods are proposed to strengthen memory of old knowledge by explicitly storing old raw data [13,19] or implicitly training generative models [21], which yet requires additional replay storage and selection criterion. Besides, dynamically expandable networks [23] augment architecture with new modules (e.g., gating autoencoders [2] and batch normalization layers [5]) to accommodate new knowledge, contributing to zero forgetting yet causing quadratic parameter increase and requiring task label for each sample at test time. In a task-agnostic manner with fixed network architecture, selective regularization methods [1,7,25] explore model parameters that are important for preserving old knowledge, and then minimize their alterations when learning new knowledge. Despite of their success in image-level classification, a naive translation of these methods to continual segmentation would yield sub-optimal performance [4] for two-fold reasons. First, the selected important parameters are not aware of shape information that is abundant in structural dense segmentation predictions, aggravating model forgetting especially for shape knowledge. Second, the segmentation reliability is ignored, which misguides the parameter selection to fit and even remember semantic noise in the segmentation results.

In this paper, we propose a novel comprehensive importance-based selective regularization (CISR) method for continual multi-site segmentation, which mitigates model forgetting by simultaneously preserving shape information and reliable semantics for previously learned sites. To prioritize parameter usage in the model related to shape and reliable semantic information, we propose a comprehensive importance (CI) weight for each parameter that consists of shape-aware importance (SpAI) and uncertainty-guided semantics-aware importance (USmAI). Concretely, SpAI is measured by the parameter sensitivity to shape-relevant predictions (i.e., a level set representation and a segmentation embedding), accounting for the complementary shape information with local boundary and global topology. USmAI is estimated for each parameter based on the sensitivity to only confident segmentation predictions with reliable semantic information instead of uncertain ones with potential noise, by exploiting uncertainty estimation with Monte Carlo Dropout. Finally, when training model on a new site, we utilize a selective regularization scheme that penalizes changes of

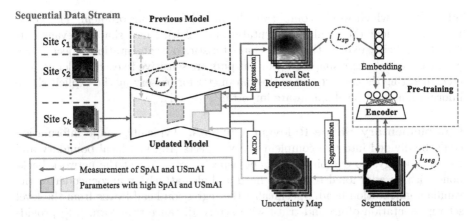

Fig. 1. Overview of our framework. For each model parameter, we define a comprehensive importance (CI) consisting of: (1) shape-aware importance (SpAI) considering complementary shape information by a level set representation and a segmentation embedding (Sect. 2.1); and (2) uncertainty-guided semantics-aware importance (USmAI) considering reliable semantic information without uncertain segmentation results (Sect. 2.2). When fine-tuning the model on a new incoming site in the sequential data stream, a selective regularization loss L_{sr} penalizes changes of important parameters with high accumulated CI, mitigating model forgetting for previously learned sites (Sect. 2.3).

important parameters with high CI, preventing both shape and semantic knowledge for previous sites being overwritten and forgotten. We have evaluated our method with the application of prostate MRI segmentation, using a sequential stream of public datasets acquired from six institutes with different acquisition protocols. The results validate that our method effectively alleviates model forgetting issue and outperforms many state-of-the-art continual learning methods.

2 Methods

In our problem setting, we are given a sequential stream of images from K sites, which are sequentially used to train a segmentation model. In round $k \in [1, K]$ of this continual learning procedure, we can only obtain images and ground truths $\{(x_n, y_n)\}_{n=1}^{N}$ from a new incoming site ζ_k without access to old data from previous sites $\{\zeta_i\}_{i=1}^{k-1}$. Figure 1 illustrates our proposed comprehensive importance-based selective regularization (CISR) method, with an objective to consecutively learn on a new site without sacrificing the performance on previous sites.

2.1 Shape-Aware Importance (SpAI)

Different from the classification task considering sample-wise isolated accuracy, the segmentation task requires structural dense predictions with abundant shape information, which would complicate the model forgetting problem. Therefore, we first augment the vanilla segmentation model with new shape-relevant outputs to explicitly exploit the shape characteristics of segmentation object. Then we measure how these shape-relevant outputs are sensitive to each model parameter, reflecting the importance of this parameter for preserving the explicit shape information, i.e., the shape-aware importance (SpAI).

Complementary Shape-Relevant Outputs. Typically, shape information could be divided into two complementary aspects [16], i.e., local boundary and global topology. To exploit both shape aspects, we augment the network backbone with an additional regression head and also attach an autoencoder to the vanilla segmentation head. Specifically, the regression head learns from the level set representation of ground truth with signed distance transform [14], providing rich boundary delineation. The autoencoder on the top of segmentation head has encoder-decoder components, learning an intermediate representation from which the input segmentation can be reconstructed. Internally, encoder component is designed to be undercomplete [17] with a last fully-connected layer. It, pretrained by ground truth[1], can compress segmentation result into a compact embedding with highly reduced dimension, thus encoding global topology in it.

Formally, based on the mean squared error L_{mse}, we define a joint shape loss L_{sp} for shape-relevant outputs with trade-off parameter α_r and α_e:

$$L_{sp} = \alpha_r L_r + \alpha_e L_e, \quad \text{with} \quad L_r = L_{mse}\left(r_n, \mathcal{T}_r(y_n)\right), \; L_e = L_{mse}\left(\mathcal{T}_e(s_n), \mathcal{T}_e(y_n)\right), \tag{1}$$

where $\mathcal{T}_r(y_n)$ denotes the level set representation of ground truth y_n, as defined by a signed distance transform [14]. Loss L_r encourages the regression output r_n to be a predicted level set representation of the segmentation target in image x_n, which delineates the detailed object boundary. Besides, $\mathcal{T}_e(s_n)$ and $\mathcal{T}_e(y_n)$ are the segmentation embedding and ground truth embedding, respectively, by passing the segmentation result s_n and ground truth y_n through the encoder component of autoencoder. Loss L_e enables $\mathcal{T}_e(s_n)$ to exploit the global topology of segmentation target in x_n. Therefore, after convergence, r_n and $\mathcal{T}_e(s_n)$ are regarded as shape-relevant outputs, characterizing complementary shape information.

Measurement of SpAI. To mitigate model forgetting for shape information, shape-relevant outputs are the targets that need to be preserved when learning on a new site. Therefore, motivated by [1], we measure the sensitivity of shape-relevant outputs r_n and $\mathcal{T}_e(s_n)$ with respect to a change of each model parameter:

$$\Omega_{ij}^{sp} = \frac{1}{N} \sum_{n=1}^{N} \frac{\beta_r \partial \|r_n\|_2^2 + \beta_e \partial \|\mathcal{T}_e(s_n)\|_2^2}{\partial \theta_{ij}}, \tag{2}$$

---

[1] Before each round of continual learning, the encoder component is pretrained and consecutively fine-tuned with the coupled decoder component, by minimizing a reconstruction loss with ground truth mask inputs. It should be frozen [24] in the later to avoid being corrupted by incomplete shape predictions due to model forgetting.

where $\frac{\partial \|r_n\|_2^2}{\partial \theta_{ij}}$ and $\frac{\partial \|\mathcal{T}_e(s_n)\|_2^2}{\partial \theta_{ij}}$ denote the gradient of squared L_2 norm of r_n and $\mathcal{T}_e(s_n)$ with respect to parameter θ_{ij}. β_r and β_e are trade-off parameters. Sensitivity Ω_{ij}^{sp} is obtained by averaging gradients over all N images. Intuitively, it reflects that how much a small perturbation to θ_{ij} would change the shape-relevant outputs r_n and $\mathcal{T}_e(s_n)$. Therefore, sensitivity Ω_{ij}^{sp} can also be regarded as SpAI, measuring parameter importance for preserving shape knowledge. Parameters with high SpAI should be unchanged to avoid forgetting shape knowledge when training on subsequent sites, while parameters with small SpAI can be updated without constraints since they slightly affect the shape-relevant outputs.

2.2   Uncertainty-Guided Semantics-Aware Importance (USmAI)

Besides shape knowledge, segmentation semantics is also crucial since it accounts for the pixel-wise predictions with inherent image property. However, considering the low contrast and inhomogeneous appearance of medical images [22], segmentation results may be noisy and unreliable, misguiding the selective regularization with important parameters for semantic noise that is commonly around segmentation boundary. To solve this problem, we estimate uncertainty for each segmentation prediction, and then propose a uncertainty-guided scheme to measure the parameter importance regarding only confident segmentation results with reliable semantic information, i.e., the uncertainty-guided semantics-aware importance (USmAI).

Uncertainty Estimation. Given image x_n and ground truth y_n, the network learns to predict semantic segmentation via segmentation head with loss:

$$L_{seg} = L_{ce}(s_n, y_n) + L_{dice}(s_n, y_n), \tag{3}$$

where s_n is the segmentation result of x_n. L_{ce} and L_{dice} denote the cross-entropy loss and dice loss, respectively. At the same time, we estimate uncertainty for s_n by Monte Carlo Dropout (MCDO) [6]. Specifically, given the same input x_n, we perform D times forward passes through the network backbone and segmentation head with the activation of dropout operations, leading to D-fold segmentation predictions $\{\tilde{s}_n^d\}_{d=1}^D$. The average of them is denoted by μ_n, and then used to calculate the entropy as the estimated uncertainty u_n:

$$\mu_n = \frac{1}{D}\sum_{d=1}^{D} \tilde{s}_n^d, \quad \text{and} \quad u_n = -\mu_n \log \mu_n. \tag{4}$$

Measurement of USmAI. Under the guidance of uncertainty u_n, we select only confident predictions from s_n and then measure their sensitivity concerning a change of each model parameter, which is defined as USmAI for this parameter:

$$\Omega_{ij}^{sm} = \frac{1}{N}\sum_{n=1}^{N} \frac{\partial \|\mathbb{I}(u_n < T)\, s_n\|_2^2}{\partial \theta_{ij}}, \tag{5}$$

where $\mathbb{I}(\cdot)$ denotes the indicator function. T is a threshold for u_n to select confident targets (low uncertainty) in s_n with reliable semantics. Ω_{ij}^{sm} is measured by averaging the gradients of these confident targets concerning parameter θ_{ij}. A higher Ω_{ij}^{sm} indicates that even a small perturbation for θ_{ij} would largely change reliable segmentation results, implying a higher importance of θ_{ij} for preserving meaningful semantic information. Therefore, the changes of parameters with high USmAI should be penalized to overcome forgetting for reliable semantics.

2.3   Comprehensive Importance-Based Selective Regularization

SpAI and USmAI are combined into a comprehensive importance (CI), prioritizing the parameter usage for keeping shape knowledge and reliable semantics:

$$\Omega_{ij}^c = \Omega_{ij}^{sp} + \Omega_{ij}^{sm}. \tag{6}$$

When training model on a new site, in addition to the inter-site supervised loss $L_{sp} + L_{seg}$ for this site by Eq. (1) and Eq. (3), we design a selective regularization loss L_{sr} that penalizes changes of parameters with high accumulated CI to avoid forgetting shape knowledge and reliable semantics for previous sites:

$$L = L_{sp} + L_{seg} + \lambda L_{sr}, \quad \text{with} \quad L_{sr} = \sum_{i,j} \Omega_{ij}^{c^*} (\theta_{ij} - \theta_{ij}^*)^2, \tag{7}$$

where λ is a trade-off parameter. Notably, Ω_{ij}^c is computed in each learning round for a specific incoming site, and accumulated over all previously learned sites by moving average, as denoted by $\Omega_{ij}^{c^*}$. It is used as a weight for the change between current parameter θ_{ij} and old parameter θ_{ij}^* (as determined by optimizing Eq. (7) for the previous site in the sequence), formulating the selective regularization loss L_{sr} to avoid changing parameters with high $\Omega_{ij}^{c^*}$ that are important for previous sites. In this way, both shape and reliable semantic knowledge can be effectively preserved, and thus catastrophic forgetting problem would be mitigated.

3   Experiments

Dataset. We employed a well-established multi-site prostate T2-weighted MRI dataset [12], including 30 cases with in/through plane resolution 0.6–0.625/3.6–4 mm from RUNMC [3] (Site A), 30 cases with resolution 0.4/3mm from BMC [3] (Site B), 19 cases with resolution 0.67–0.79/1.25 mm from HCRUDB [8] (Site C), 13 cases with resolution 0.325–0.625/3-3.6 mm from UCL [10] (Site D), 12 cases with resolution 0.25/2.2–3 mm from BIDMC [10] (Site E), and 12 cases with resolution 0.625/3.6 mm from HK [10] (Site F). We organized this multi-site dataset in a sequential stream ordered by Site A→B→C→D→E→F. For pre-processing, we resized all images to size 384×384 in the axial plane and normalized them to zero mean and unit variance. For each site, we used images from 60%, 15% and 25% of cases for training, validation and testing.

Implementation. We adopted 2D-UNet [20] as network backbone due to the large variance on through-plane resolution among different sites [11]. Weight λ was empirically set as a large value 10^5 [18]. Parameter $\alpha_r, \alpha_e, \beta_r$ and β_e were set as 0.001, 0.1, 0.1, 0.001 for suitable trade-off. The autoencoder was designed with mirrored encoder-decoder components, where the encoder component was used as an embedding network containing three cascaded blocks with Conv 3×3 (16, 32, 64 kernels) using stride 2 and 1 in each, followed by a flatten operation and a fully-connected layer with 64 hidden units. In each learning round with data from a new site, we optimized the objective function Eq. (7) by AdamOptimizer with learning rate 5×10^{-4}, batch size 5 and epoch number 200.

Evaluation Metrics. We evaluate the segmentation performance by dice similarity coefficient (DSC) and average symmetric surface distance (ASD). After the model finishes continual learning on the last site ζ_K, we compute DSC and ASD on all sites $\{\zeta_i\}_{i=1}^K$ (including the current and previous sites), leading to two sets of results $\{D_{K,i}\}_{i=1}^K$ and $\{A_{K,i}\}_{i=1}^K$, respectively. Entry $D_{K,i}$ and $A_{K,i}$ denote the test DSC and ASD on site ζ_i after learning on the last site ζ_K. Based on them, we define several specialized metrics for continual learning, i.e., the *average* of DSC and ASD ($\text{DSC}^{ave} = \sum_{i=1}^K D_{K,i}/K$, $\text{ASD}^{ave} = \sum_{i=1}^K A_{K,i}/K$) for the generic evaluation; the *backward transfer* [13] of DSC and ASD ($\text{DSC}^{bwt} = \sum_{i=1}^{K-1}(D_{K,i} - D_{i,i})/(K-1)$, $\text{ASD}^{bwt} = \sum_{i=1}^{K-1}(A_{K,i} - A_{i,i})/(K-1)$) that particularly reflect model forgetting for previous sites. Notably, an advanced continual learning method should have a high DSC^{ave} and DSC^{bwt} with a low ASD^{ave} and ASD^{bwt}.

Comparison with State-of-the-art Continual Learning Methods. We compare our CISR with several state-of-the-art continual learning methods, including elastic weight consolidation (EWC) method [7] that preserves old knowledge based on the Fisher information, synaptic intelligence (SI) method [25] that updates network memory in an online manner, and memory-aware synapses (MAS) method [1] considering plain semantics without the uncertainty-guided scheme. We also implement a naive continuous fine-tuning (FT) scheme as baseline method with only the inter-site supervised loss considered in Eq. (7).

Table 1 lists the quantitative evaluation after continual learning finished on the last site F. FT suffers from the severe model forgetting problem (the worst *bwt* measures of DSC^{bwt} and ASD^{bwt}) and the poor segmentation results (the worst *ave* measures of DSC^{ave} and ASD^{ave}). Classical methods of EWC, SI and MAS improves the continual learning performance over FT, while their advantages are still relatively limited. Importantly, our proposed CISR outperforms all these methods, e.g., an advantage over MAS by 3.13% DSC^{ave}, 5.23% DSC^{bwt}, 0.22 mm ASD^{ave}, 0.37 mm ASD^{bwt}, indicating its superiority for continual learning without sacrificing performance on previous sites.

We also report in Fig. 2 how the test performance on site A changes during the entire continual learning. Once finishing learning on new sites, FT immediately forgets previously learned knowledge for site A, leading to evident performance decrease on it. EWC, SI and MAS mitigate this problem to some extent, yet still

Table 1. Evaluation after the model finishes continual learning on the last site F.

Method	$DSC^{ave}(\%)\uparrow$	$DSC^{bwt}(\%)\uparrow$	$ASD^{ave}(mm)\downarrow$	$ASD^{bwt}(mm)\downarrow$
FT (baseline)	40.78 ± 28.26	-50.54 ± 21.58	13.66 ± 8.67	13.75 ± 6.52
EWC [7]	68.83 ± 17.51	-23.90 ± 14.09	4.43 ± 3.34	3.94 ± 3.14
SI [25]	75.97 ± 14.11	-15.27 ± 10.05	3.87 ± 2.40	2.97 ± 2.96
MAS [1]	76.81 ± 11.18	-13.23 ± 8.24	3.64 ± 3.10	2.28 ± 3.32
USmAI	78.11 ± 10.60	-11.42 ± 7.92	3.49 ± 2.80	2.06 ± 3.02
$SpAI_r$	77.85 ± 10.96	-12.10 ± 8.07	3.51 ± 2.75	2.16 ± 2.76
$SpAI_e$	77.53 ± 11.04	-12.74 ± 8.34	3.57 ± 2.85	2.36 ± 2.69
SpAI	79.31 ± 8.86	-9.89 ± 6.19	3.44 ± 2.89	2.11 ± 2.89
CISR (**Ours**)	$\mathbf{79.94 \pm 7.71}$	$\mathbf{-8.00 \pm 6.15}$	$\mathbf{3.42 \pm 2.83}$	$\mathbf{1.91 \pm 2.86}$

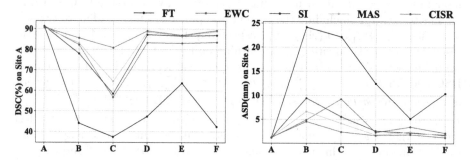

Fig. 2. Changes of test performance on Site A by different methods during continual learning, where the model is sequentially trained on sites A→B→C→D→E→F.

challenged by model forgetting especially after training on site C, since most images from this site contain prostate cancer [8] with visible appearance difference for previous site A. Our CISR maintains consistently the least performance decrease on site A, showing its best capability for reducing model forgetting.

Furthermore, we visualize in Fig. 3 the segmentation results on site A, D and F before and after the model finishes learning on site F. By training on site F that is newly incoming in sequence, all methods achieve highly improved performance on this site (the third row), indicating their sufficient adaptation capability for new site. However, FT leads to severe performance degradation on previously learned site A and D. Among all methods, our CISR achieves the most accurate results on site A and D after training on site F, and maintains the highest overlap ratio with previously obtained results (the first two rows). Besides, high uncertainty focuses on object ambiguous boundaries with semantic noise, explaining the feasibility of our uncertainty guidance for filtering out them.

Ablation Study. We validate the role of SpAI (including $SpAI_r$ using only r_n for boundary delineation and $SpAI_e$ using only $\mathcal{T}_e(s_n)$ for topology awareness)

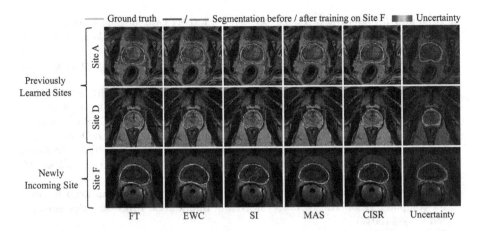

Fig. 3. Visualization of segmentation examples on site A, D and F before (i.e., after training on site A, D and E, respectively) and after training on the newly incoming site F. Also, the uncertainty maps of our CISR is visualized in the last column.

and USmAI for reliable semantics in our method. Table 1 shows that USmAI outperforms MAS owing to our uncertainty-guided scheme for filtering out semantic noise. Preserving shape knowledge by $SpAI_r$ and $SpAI_e$ has an advantage over MAS, and a combination of them in SpAI even improves performance over each single component. Beside, incorporating SpAI and USmAI in our CISR further facilitates more advantages owing to the jointly preserved shape and semantics.

4   Conclusion

This paper proposes a comprehensive importance-based selective regularization method for continual multi-site segmentation. We propose to reduce model forgetting by strengthening network memory for both shape knowledge (with complementary boundary and topology cues) and reliable semantics (with only confident predictions). Experiments show the effectiveness on prostate segmentation with sequential multi-site data. In the further, it would be of interest to study the application for other continual segmentation tasks with longer site sequences.

Acknowledgments. This research is partially supported by the National Key research and development program (No. 2016YFC0106200), Beijing Natural Science Foundation-Haidian Original Innovation Collaborative Fund (No. L192006), and the funding from Institute of Medical Robotics of Shanghai Jiao Tong University as well as the 863 national research fund (No. 2015AA043203).

References

1. Aljundi, R., Babiloni, F., Elhoseiny, M., Rohrbach, M., Tuytelaars, T.: Memory aware synapses: learning what (not) to forget. In: Proceedings of the European Conference on Computer Vision, pp. 139–154 (2018)

2. Aljundi, R., Chakravarty, P., Tuytelaars, T.: Expert gate: lifelong learning with a network of experts. In: Proceedings of the IEEE Conference on Computer Vision and Pattern Recognition, pp. 3366–3375 (2017)
3. Bloch, N., et al.: NCI-ISBI 2013 challenge: automated segmentation of prostate structures. Cancer Imaging Arch. (2015)
4. Douillard, A., Chen, Y., Dapogny, A., Cord, M.: PLOP: learning without forgetting for continual semantic segmentation. arXiv preprint arXiv:2011.11390 (2020)
5. Karani, N., Chaitanya, K., Baumgartner, C., Konukoglu, E.: A lifelong learning approach to brain MR segmentation across scanners and protocols. In: Frangi, A.F., Schnabel, J.A., Davatzikos, C., Alberola-López, C., Fichtinger, G. (eds.) MICCAI 2018. LNCS, vol. 11070, pp. 476–484. Springer, Cham (2018). https://doi.org/10.1007/978-3-030-00928-1_54
6. Kendall, A., Gal, Y.: What uncertainties do we need in bayesian deep learning for computer vision? In: Advances in Neural Information Processing Systems, pp. 5574–5584 (2017)
7. Kirkpatrick, J., et al.: Overcoming catastrophic forgetting in neural networks. Proc. Natl. Acad. Sci. 114(13), 3521–3526 (2017)
8. Lemaître, G., Martí, R., Freixenet, J., Vilanova, J.C., Walker, P.M., Meriaudeau, F.: Computer-aided detection and diagnosis for prostate cancer based on mono and multi-parametric MRI: A review. Comput. Biol. Med. 60, 8–31 (2015)
9. Litjens, G., Litjens, G., et al.: A survey on deep learning in medical image analysis. Med. Image Anal. 42, 60–88 (2017)
10. Litjens, G., et al.: Evaluation of prostate segmentation algorithms for MRI: the PROMISE12 challenge. Med. Image Anal. 18(2), 359–373 (2014)
11. Liu, Q., Dou, Q., Yu, L., Heng, P.A.: MS-Net: multi-site network for improving prostate segmentation with heterogeneous MRI data. IEEE Trans. Med. Imaging 39(9), 2713–2724 (2020)
12. Liu, Q., Dou, Q., Heng, P.-A.: Shape-aware meta-learning for generalizing prostate MRI segmentation to unseen domains. In: Martel, A.L., et al. (eds.) MICCAI 2020. LNCS, vol. 12262, pp. 475–485. Springer, Cham (2020). https://doi.org/10.1007/978-3-030-59713-9_46
13. Lopez-Paz, D., Ranzato, M.: Gradient episodic memory for continual learning. In: Advances in Neural Information Processing Systems, pp. 6467–6476 (2017)
14. Ma, J., He, J., Yang, X.: Learning geodesic active contours for embedding object global information in segmentation CNNs. IEEE Trans. Med. Imaging 40(1), 93–104 (2021)
15. McCloskey, M., Cohen, N.J.: Catastrophic interference in connectionist networks: the sequential learning problem. In: Psychology of Learning and Motivation, vol. 24, pp. 109–165. Elsevier (1989)
16. Navarro, F., et al.: Shape-aware complementary-task learning for multi-organ segmentation. In: Suk, H.-I., Liu, M., Yan, P., Lian, C. (eds.) MLMI 2019. LNCS, vol. 11861, pp. 620–627. Springer, Cham (2019). https://doi.org/10.1007/978-3-030-32692-0_71
17. Oktay, O., et al.: Anatomically constrained neural networks (ACNNs): application to cardiac image enhancement and segmentation. IEEE Trans. Med. Imaging 37(2), 384–395 (2017)
18. Parisi, G.I., Kemker, R., Part, J.L., Kanan, C., Wermter, S.: Continual lifelong learning with neural networks: a review. Neural Netw. 113, 54–71 (2019)
19. Rebuffi, S.A., Kolesnikov, A., Sperl, G., Lampert, C.H.: ICARL: incremental classifier and representation learning. In: Proceedings of the IEEE Conference on Computer Vision and Pattern Recognition, pp. 2001–2010 (2017)

20. Ronneberger, O., Fischer, P., Brox, T.: U-Net: convolutional networks for biomedical image segmentation. In: Navab, N., Hornegger, J., Wells, W.M., Frangi, A.F. (eds.) MICCAI 2015. LNCS, vol. 9351, pp. 234–241. Springer, Cham (2015). https://doi.org/10.1007/978-3-319-24574-4_28

21. Shin, H., Lee, J.K., Kim, J., Kim, J.: Continual learning with deep generative replay. In: Advances in Neural Information Processing Systems, pp. 2990–2999 (2017)

22. Wang, G., et al.: Interactive medical image segmentation using deep learning with image-specific fine tuning. IEEE Trans. Med. Imaging **37**(7), 1562–1573 (2018)

23. Yoon, J., Yang, E., Lee, J., Hwang, S.J.: Lifelong learning with dynamically expandable networks. arXiv preprint arXiv:1708.01547 (2017)

24. Yue, Q., Luo, X., Ye, Q., Xu, L., Zhuang, X.: Cardiac segmentation from LGE MRI using deep neural network incorporating shape and spatial priors. In: Medical Image Computing and Computer-Assisted Intervention, pp. 559–567 (2019)

25. Zenke, F., Poole, B., Ganguli, S.: Continual learning through synaptic intelligence. Proc. Mach. Learn. Res. **70**, 3987 (2017)

ReSGAN: Intracranial Hemorrhage Segmentation with Residuals of Synthetic Brain CT Scans

Miika Toikkanen, Doyoung Kwon, and Minho Lee[✉]

Department of Artificial Intelligence, Kyungpook National University,
Daegu, South Korea
mholee@knu.ac.kr

Abstract. Intracranial hemorrhage (ICH) is a dangerous condition of bleeding within the skull that calls for rapid and precise diagnosis due to potentially fatal consequences. In this paper, we propose Residual Segmentation with Generative Adversarial Networks (ReSGAN) to accurately localize the hemorrhage from computerized tomography (CT) scans with a GAN-based model. Although convolutional neural networks have shown success in the ICH segmentation task, precise localization remains challenging due to in-balance and scarcity of labeled training data. Synthetic samples from generative models, and aligned templates as reference from brain atlas have been demonstrated to alleviate the issues. We consider synthetic templates as another candidate and solve the problem by directly applying a generative model to segmentation. Our ReSGAN learns a distribution of pseudo-normal brain CT scans, that through residuals, reliably delineates the hemorrhaging areas. We perform experiments on two datasets and compare our model against a well established baseline, that consistently shows significant improvements, therefore demonstrating the validity of our novel method.

Keywords: Semantic image synthesis · Segmentation · Brain · Intracranial hemorrhage · Non-contrast CT

1 Introduction

Intracranial hemorrhage (ICH) is a potentially fatal form of internal bleeding that occurs within the skull as a consequence of ruptured blood vessels. This can be a result of physical trauma to the head or structurally weakened veins due to disease. ICH requires urgent diagnosis and treatment because the buildup of blood may deform and damage the brain tissue, and restrict the normal blood flow inside the brain, leading to further complications [19]. The condition is usually diagnosed with the help of Computerized Tomography (CT), which offers

Electronic supplementary material The online version of this chapter (https://doi.org/10.1007/978-3-030-87193-2_38) contains supplementary material, which is available to authorized users.

© Springer Nature Switzerland AG 2021
M. de Bruijne et al. (Eds.): MICCAI 2021, LNCS 12901, pp. 400–409, 2021.
https://doi.org/10.1007/978-3-030-87193-2_38

a low image acquisition time, suitable to the time-critical situation [2]. Consequently, development of automated methods for localizing the affected tissue can reduce the mortality through rapid and accurate diagnosis.

Convolutional neural networks have offered good results in the localization of ICH from CT scans [8,10]. However, the task remains challenging due to imbalance and scarcity of labeled training data. Generative adversarial networks (GAN) [5] have been investigated as a solution to create synthetic training samples [1], and have been successfully applied in medical image modalities as well [3,15,16]. Another promising solution is a template based model, where the contrast between the sample and a healthy exemplar is considered. The Siamese U-Net [11] demonstrated performance gains by comparing learned representations of a patient CT scan and an aligned healthy template from a normal brain atlas. Motivated by these results, we are interested in learning to synthesize the distribution of normal brain templates that can be used to segment hemorrhage directly, without aligned templates or data-augmentation.

The GauGAN [12] introduced spatially-adaptive de-normalization (SPADE), which modulates normalized layer activations of GAN using semantic labels. This enabled the conditional synthesis of photo-realistic images, which previously suffered from loss of semantic information in the normalization layers. The class-adaptive normalization (CLADE) [17,18] provided a computationally lighter model with similar performance based on the observation that the benefits are mostly caused by semantic awareness instead of spatial awareness. The method has also been used to synthesize brain-tumor CT scans for data augmentation [16].

In this paper, we introduce the Residual Segmentation with Generative Adversarial Networks (ReSGAN), a novel approach for ICH localization. ReSGAN learns a distribution of pseudo-normal CT scans that captures the differences of patient images and ICH labels. By viewing normal and abnormal brain tissue as semantic classes, we develop an algorithm for manipulating CT scans and obtaining the ICH delineation from residual images. Our contributions are, the first trial for semantic manipulation of CT-scans to produce normal samples, its application to ICH segmentation and a demonstration of its feasibility by improved segmentation performance against baselines.

2   Method

Our goal is to obtain accurate segmentation labels by considering the difference in the appearance of normal and abnormal classes. The hemorrhage can be seen in CT scans as a brighter tone of pixel intensities and deformation of the brain tissue due to blood buildup. Our training data do not contain aligned normal-abnormal data pairs or examinations of healthy individuals, therefore we ignore the structural deformation caused by ICH and instead focus on synthesising the appearance of the semantic classes correctly. Hence, we refer to our synthetic templates as pseudo-normal. Following sections describe our method in detail.

2.1 Learning to Synthesize Templates from CT Scans

Given a brain CT image $I^{src} \in R^{1 \times H \times W}$ and a desired semantic layout $M^{tgt} \in R^{C \times H \times W}$ of background, bone, normal and abnormal tissues, we synthesize a plausible CT image $I^{fake} \in R^{1 \times H \times W}$, where the content of the image agrees with M^{tgt} and maintains the unique structure of the subjects brain. During training, we also utilize a ground truth semantic label $M^{src} \in R^{C \times H \times W}$, that describes the class-identities of I^{src}. Here C corresponds to the number of classes, while H and W refer to the height and width of the image, respectively. Our ReSGAN generator architecture and training procedure are illustrated in Fig. 1, for more details, see supplementary material. We use the GauGAN [12] with CLADE [17] as our starting point and make several required improvements.

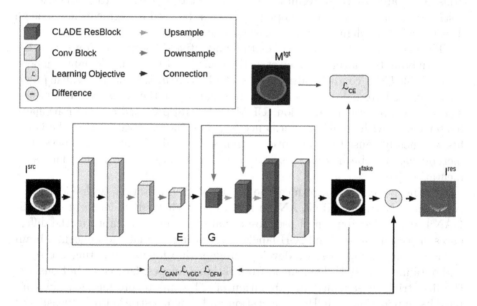

Fig. 1. A simplified diagram of the ReSGAN generator architecture. E encodes the image with conv blocks consisting of convolution, normalization and down-sampling operations. G combines CLADE residual blocks and up-sampling to form an encoder-decoder network. From I^{src} we threshold the normal target M^{tgt} and predict I^{fake}, which yields the residual I^{res}. Perceptual losses $\mathcal{L}_{VGG}$, $\mathcal{L}_{DFM}$ as well as adversarial loss $\mathcal{L}_{GAN}$ are computed from real and synthetic images, while semantic class loss $\mathcal{L}_{CE}$ is computed from the synthetic image and target label. For simplicity, only two downsampling steps are shown and the critic networks D and S are omitted.

Class-Agnostic Encoder. We train an encoder network E to achieve image-to-image translation. Our generator can be thought of as an auto-encoder with class-conditional decoder. This is different from GauGAN [12], which conditions on the input images in a stochastic manner using a variational auto-encoder. In order to synthesize a normal brain from an abnormal one, we need to implicitly detect the hemorrhage. This requires that E learns class-agnostic encoding, and the decoder G learns class conditional synthesis. Since we want to generate an image without ICH, the semantic label M^{tgt} does not contain abnormal tissue. The background and skull correspond to the minimum and maximum pixel intensities, and consequently, the label is easily obtained by a thresholding. The synthetic image can then be generated using Eq. (1).

$$I^{fake} = G(E(I^{src}), M^{tgt}) \qquad (1)$$

Semantic Class Critic. The encoder-decoder network will minimize its learning objective by reconstructing the original input. To gain control via semantic labels, we introduce the semantic class critic S that incurs a penalty for synthesizing incorrect classes. We employ the U-Net [13] to predict semantic labels from images and minimize the class weighted cross-entropy loss $\mathcal{L}_{CE}(M^{src}, S(I^{src}))$ described by Eq. (2), where w_c is the inverse of the class probabilities.

$$\mathcal{L}_{CE}(y, x) = -\sum_{c=1}^{C} w_c \sum_{i=1}^{H} \sum_{j=1}^{W} y_{c,i,j} log(x_{c,i,j}) \qquad (2)$$

Similarly, for the generator we optimize $\mathcal{L}_{CE}(M^{tgt}, S(I^{fake}))$. During training, M^{tgt} is obtained from the hemorrhage ground truth labels, however we randomly swap the abnormal label to normal. This encourages G to obey the class label, however, since the real class identities are not available to E, it also encourages E to ignore class specific information.

Masked Perceptual Losses. Following GauGAN, we use the adversarial hinge-loss $\mathcal{L}_{GAN}$, perceptual losses $\mathcal{L}_{DFM}$ from the discriminator activations [14] and $\mathcal{L}_{VGG}$ from pre-trained classifier network activations [9]. However, we apply masking to the images before computing the perceptual losses. This is intended to prevent them from conflicting $\mathcal{L}_{CE}$. Using the absolute difference of the source and target semantic labels, we compute the mask $M^{diff} \in R^{1 \times H \times W}$, that equals to 0 if $M_{i,j}^{src} \neq M_{i,j}^{tgt}$, for each pixel coordinate i, j. While the discriminator network is only trained on $\mathcal{L}_{GAN}$, the full learning objective for the generator is $\mathcal{L}$ with weights λ_{VGG}, λ_{DFM} and λ_{CE} to balance the terms.

$$\mathcal{L} = \mathcal{L}_{GAN} + \lambda_{CE}\mathcal{L}_{CE} + \lambda_{DFM}\mathcal{L}_{DFM} + \lambda_{VGG}\mathcal{L}_{VGG} \qquad (3)$$

2.2   Segmentation Through Residuals

Using a slice from the patient CT scan, and the synthetic template that represents the corresponding normal brain, the residual $I^{res} = I^{fake} - I^{src}$ is computed to highlight the hemorrhaging areas. This is analogous to anomaly detection with an auto-encoder. Our model reconstructs the image according

to the label, in which we assume to be no abnormal tissue. However, if the
input image has abnormal areas, the residual is much greater than zero at those
locations. Figure 2 shows examples obtained using ReSGAN.

Thresholding. Due to the information bottleneck of the encoder-decoder, perfect reconstruction is not possible. The non-hemorrhaging areas in the residual
contain noise. To deal with the issue, we include additional filtering and morphological operations in our pipeline. After suppressing high-frequency noise with a
median filter, we apply a binary threshold to segment the image and clean any
remaining spurious outputs with morphological opening and closing.

Learned Transformation. We find it more effective to learn the best post-
processing instead of searching for a fixed threshold. Although additional computation is required, this is an attractive solution since the threshold can be
decided locally using a larger surrounding context. For this purpose we use the
cross-entropy loss to learn an additional U-Net [13] that refines the residuals
into more accurate predictions. Figure 2 clearly demonstrates the benefit gained
from this method. The thresholding approach misses some areas and erroneously
predicts tissue outside the skull as ICH, while the learned transformation suffers
less from those issues.

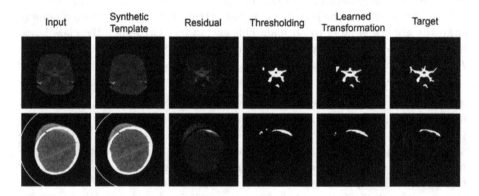

Fig. 2. Examples of segmentation results from two datasets using ReSGAN. The residual highlights the hemorrhage, but thresholding or using a learned transformation
results in an accurate segmentation mask.

2.3   Experiments

Datasets. We make use of two different non-contrast head CT scan datasets.
First, a set of 275 examinations collected from Kyungpook National University
Hospital, which we refer to as KNUH. In this data, the skulls are removed
using simple intensity thresholding. In addition, we experiment on the CT-ICH

[4,7,8], a dataset of 75 subjects, which includes the skulls and some noise for anonymization of patients. We randomly split both datasets into equal 5 folds for cross-validation and additional one for testing. Labels for various types of ICH are annotated, but in this work we only consider a single class of it either being absent or present.

Training Details. On the KNUH dataset, we train the ReSGAN for 200 epochs and use 3 down-sampling steps, while on the CT-ICH dataset, we use 4 down-sampling steps and train for 300 epochs. Over the last half of the training, the learning rate is linearly decayed to zero. We use batch size of 8 and automatic mixed precision training with Adam optimizer for ReSGAN and stochastic gradient descent (SGD) optimizer for the learned transformation. Images are loaded at 540×540 pixels and randomly cropped to 512×512 for additional variation. Random vertical and horizontal flipping are also used. The loss weights are set as $\lambda_{VGG} = \lambda_{DFM} = \lambda_{CE} = 10$ based on previous works and a coarse search of values, while the class weighing w_c is computed for each cross validation fold separately. All of the code is written in PyTorch, and the experiments are performed using a single NVIDIA Quadro GV100 GPU.

3 Results

In this section, we demonstrate the effectiveness of our method by comparisons against previously established baselines. We select the best ReSGAN model based on the perceptual similarity between real and synthetic images, as well as the correspondence of semantic classes to the desired target, measured by the Fréchet Inception Distance (FID) [6] and Dice score, respectively. For the threshold-based model we conduct a coarse-to-fine parameter search, while for the baselines as well as our learned transformation model, we pick the model with highest validation dice score. The reported numbers are obtained by computing the mean and standard deviation between the best model from each cross-validation fold on the test set. We report the Dice score (DCS) and the 95th percentile Hausdorff distance (HSD) as the final segmentation result. Cases where HSD is not defined are ignored in the computation.

KNUH Dataset. We compare our approach against the U-Net and the Siamese U-Net baselines from previous work [11]. Under the same conditions, we train an additional baseline, Cascaded U-Net, where two U-Nets are joined back to back. Following previous work, the experiment is performed on mid-axial slices. Based on the results summarized in Table 1, ReSGAN creates accurate predictions by thresholding the residuals and furthermore, the learned transformation improves the performance considerably over using a fixed threshold. The improved segmentation quality is qualitatively confirmed by the example outputs in Fig. 3. Clearly the ReSGAN with learned transformation best matches the target mask. Moreover, in failure cases, such as creating two blobs instead of one in the bottom row, the result still corresponds well to the appearance of the input image.

Table 1. Quantitative comparison between the baselines and the variations of our model from KNUH dataset experiment. The arrows (↓) and (↑) indicate whether low or high value is preferred.

KNUH	DCS (↑)	HSD (↓)
U-Net	0.609 ± 0.012	86.795 ± 3.254
Siamese U-Net	0.661 ± 0.015	32.473 ± 3.286
Cascaded U-Net	0.645 ± 0.008	53.473 ± 2.229
ReSGAN w/Tresholding	0.696 ± 0.005	34.201 ± 1.605
ReSGAN w/Learned Transformation	**0.726 ± 0.011**	**25.490** ± 2.863

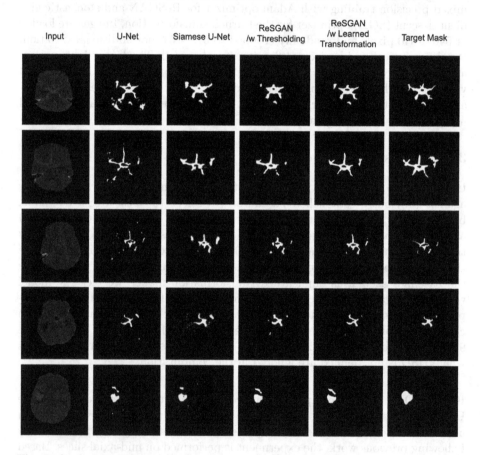

Fig. 3. Qualitative comparison between the baselines and our model variants from KNUH dataset experiment.

CT-ICH Dataset. In this experiment we compare our model against an U-Net baseline. Only the slices that contain at least one pixel labeled as abnormal are used. This yields data that is more diverse and imbalanced, but similar in size to the KNUH dataset. The results summarized in Table 2 are in agreement with the previous experiment. In this dataset the skulls and some noise for anonymization are present and have to be reconstructed correctly, slightly reducing the performance gain. Regardless, our model outperforms the baseline in this setting and further improves the result with learned transformation. For a qualitative comparison of outputs, see supplementary material.

Table 2. Quantitative comparison between U-Net baseline and our model variants from CT-ICH dataset experiment with abnormal samples. The arrows (↓) and (↑) indicate whether low or high value is preferred.

CT-ICH	DCS (↑)	HSD (↓)
U-Net	0.612 ± 0.029	78.788 ± 15.490
ReSGAN w/Tresholding	0.653 ± 0.012	68.070 ± 15.136
ReSGAN w/Learned Transformation	**0.661** ± 0.027	**67.254** ± 11.457

4   Clinical Considerations

In this paper, we solved the problem of ICH segmentation by considering a plausible appearance of the individual's brain without hemorrhage. This is done without access to any samples of healthy individuals. Therefore our pseudo-normal templates ignore the structural difference between normal and abnormal brain caused by blood buildup. However, as we empirically demonstrate, it is not crucial for accurate delineation. ReSGAN successfully learns the distribution of pseudo-normal templates that captures the correct segmentation label through residuals and therefore serves as an useful segmentation method.

5   Conclusion

We introduced a novel generative approach to ICH segmentation from CT scan slices. Our proposed model, ReSGAN, correctly learns the distribution of pseudo-normal brain CT scans that delineate the hemorrhaging areas through residuals. We show that ReSGAN can reach competitive performance with simple thresholding of the residuals and further outperform our baselines by learning a transformation from the residuals to segmentation labels. In future work, we would like to investigate the applicability of our algorithm to other modalities such as magnetic resonance imaging and full three dimensional CT-scans. The current architecture could be expanded to 3D by predicting the residuals slice by slice, and then applying a 3D network for post-processing. An exploration of our model as an data-augmentation tool may also yield interesting results.

Acknowledgment. This work was supported by the National Research Foundation of Korea(NRF) grant funded by the Korea government(MSIT) (NRF-No. 2021R1A2C3011169)[30%], Electronics and Telecommunications Research Institute(ETRI) grant funded by the Korean government[21ZS1100, Core Technology Research for Self-Improving Integrated Artificial Intelligence System][30%], and the Industrial Strategic Technology Development Program(20011875, Development of AI based diagnostic technology for medical imaging devices) funded By the Ministry of Trade, Industry & Energy(MOTIE, Korea)[40%]

References

1. Antoniou, A., Storkey, A., Edwards, H.: Data augmentation generative adversarial networks. arXiv preprint arXiv:1711.04340 (2017)
2. Bhadauria, H., Singh, A., Dewal, M.: An integrated method for hemorrhage segmentation from brain CT imaging. Comput. Electric. Eng. **39**(5), 1527–1536 (2013)
3. Frid-Adar, M., Diamant, I., Klang, E., Amitai, M., Goldberger, J., Greenspan, H.: Gan-based synthetic medical image augmentation for increased cnn performance in liver lesion classification. Neurocomputing **321**, 321–331 (2018)
4. Goldberger, A.L., et al.: PhysioBank, PhysioToolkit, and PhysioNet: components of a new research resource for complex physiologic signals. Circulation **101**(23), e215–e220 (2000). circulation Electronic Pages: http://circ.ahajournals.org/content/101/23/e215.full PMID:1085218, https://doi.org/10.1161/01.CIR.101.23.e215
5. Goodfellow, I., et al.: Generative adversarial nets. In: Advances in Neural Information Processing Systems, pp. 2672–2680 (2014)
6. Heusel, M., Ramsauer, H., Unterthiner, T., Nessler, B., Hochreiter, S.: Gans trained by a two time-scale update rule converge to a local nash equilibrium. In: Advances in Neural Information Processing Systems, pp. 6626–6637 (2017)
7. Hssayeni, M.: Computed tomography images for intracranial hemorrhage detection and segmentation (2020). https://doi.org/10.13026/4nae-zg36
8. Hssayeni, M.D., Croock, M.S., Salman, A.D., Al-khafaji, H.F., Yahya, Z.A., Ghoraani, B.: Intracranial hemorrhage segmentation using a deep convolutional model. Data **5**(1), 14 (2020)
9. Johnson, J., Alahi, A., Fei-Fei, L.: Perceptual losses for real-time style transfer and super-resolution. In: Leibe, B., Matas, J., Sebe, N., Welling, M. (eds.) ECCV 2016. LNCS, vol. 9906, pp. 694–711. Springer, Cham (2016). https://doi.org/10.1007/978-3-319-46475-6_43
10. Kuo, W., Häne, C., Yuh, E., Mukherjee, P., Malik, J.: Patchfcn for intracranial hemorrhage detection. arXiv preprint arXiv:1806.03265 (2018)
11. Kwon, D., et al.: Siamese U-Net with healthy template for accurate segmentation of intracranial hemorrhage. In: Shen, D., et al. (eds.) MICCAI 2019. LNCS, vol. 11766, pp. 848–855. Springer, Cham (2019). https://doi.org/10.1007/978-3-030-32248-9_94
12. Park, T., Liu, M.Y., Wang, T.C., Zhu, J.Y.: Semantic image synthesis with spatially-adaptive normalization. In: Proceedings of the IEEE/CVF Conference on Computer Vision and Pattern Recognition, pp. 2337–2346 (2019)
13. Ronneberger, O., Fischer, P., Brox, T.: U-Net: convolutional networks for biomedical image segmentation. In: Navab, N., Hornegger, J., Wells, W.M., Frangi, A.F. (eds.) MICCAI 2015. LNCS, vol. 9351, pp. 234–241. Springer, Cham (2015). https://doi.org/10.1007/978-3-319-24574-4_28

14. Salimans, T., Goodfellow, I., Zaremba, W., Cheung, V., Radford, A., Chen, X.: Improved techniques for training gans. In: Advances in Neural Information Processing Systems, pp. 2234–2242 (2016)
15. Sandfort, V., Yan, K., Pickhardt, P.J., Summers, R.M.: Data augmentation using generative adversarial networks (cyclegan) to improve generalizability in ct segmentation tasks. Sci. Rep. **9**(1), 1–9 (2019)
16. Shin, H.C., et al.: Medical image synthesis for data augmentation and anonymization using generative adversarial networks. In: Gooya, A., Goksel, O., Oguz, I., Burgos, N. (eds.) SASHIMI 2018. LNCS, vol. 11037, pp. 1–11. Springer, Cham (2018). https://doi.org/10.1007/978-3-030-00536-8_1
17. Tan, Z., et al.: Efficient semantic image synthesis via class-adaptive normalization. IEEE Trans. Pattern Anal. Mach. Intell. (2021)
18. Tan, Z., et al.: Rethinking spatially-adaptive normalization. arXiv preprint arXiv:2004.02867 (2020)
19. Xi, G., Keep, R.F., Hoff, J.T.: Mechanisms of brain injury after intracerebral haemorrhage. Lancet Neurol. **5**(1), 53–63 (2006)

Refined Local-imbalance-based Weight for Airway Segmentation in CT

Hao Zheng[1,2,3], Yulei Qin[1,3], Yun Gu[1,3,5(✉)], Fangfang Xie[4], Jiayuan Sun[4], Jie Yang[1,3(✉)], and Guang-Zhong Yang[2,3]

[1] Institute of Image Processing and Pattern Recognition, Shanghai Jiao Tong University,
Shanghai, China
{geron762,jieyang}@sjtu.edu.cn
[2] School of Biomedical Engineering, Shanghai Jiao Tong University, Shanghai, China
[3] Institute of Medical Robotics, Shanghai Jiao Tong University, Shanghai, China
[4] Shanghai Chest Hospital, Shanghai, China
[5] Shanghai Center for Brain Science and Brain-Inspired Technology, Shanghai, China

Abstract. As 3D navigated bronchoscopy is increasingly used for the biopsy and treatment of peripherally located lung cancer lesions, accurate segmentation of distal small airways plays an important role in both pre- and intra-operative navigation. When adopting CNN-based methods in this task, the gradients to these peripheral branches may disappear before arriving at the bottom layers. Firstly, this is closely related to the ratio of the foreground gradient to the background gradient. Generally, small ratios can lead to the erosion of the surface while the consequence is more serious for the distal small airways. To accurately segment the branches of different sizes, we propose a local-imbalance-based weight that adjusts the gradient ratios according to the quantification of local class imbalance. In addition, if the features of some under-represented areas are not learned in the first few epochs, the gradients to these regions may be filtered out by the last activation layer in the following training. To resolve this problem, we propose in this paper a BP-based weight enhancement strategy that restarts the training with refined weight maps. The largest connected domain in our results achieves a tree length detected rate of 95% with a precision of 92% in the Binary Airway Segmentation Dataset. The code is publicly available at https://github.com/haozheng-sjtu/Local-imbalance-based-Weight.

This research was partly supported by National Key R&D Program of China (No. 2019YFB1311503), Committee of Science and Technology, Shanghai, China (No. 19510711200), Shanghai Sailing Program (No. 20YF1420800), National Nature Science Foundation of China (No. 62003208), Shanghai Municipal of Science and Technology Project (No. 20JC1419500), and Science and Technology Commission of Shanghai Municipality (No. 20DZ2220400).

Electronic supplementary material The online version of this chapter (https://doi.org/10.1007/978-3-030-87193-2_39) contains supplementary material, which is available to authorized users.

M. de Bruijne et al. (Eds.): MICCAI 2021, LNCS 12901, pp. 410–419, 2021.
https://doi.org/10.1007/978-3-030-87193-2_39

Keywords: Airway segmentation · Local-imbalance-based weight ·
BP-based weight enhancement

1   Introduction

As thinner bronchoscopes are increasingly used clinically, 3D navigated bronchoscopy is now common for the diagnosis and treatment of peripherally located
lung cancer lesions. To reach to these peripheral targets, the navigation system needs a virtual lung model with a detailed bronchial tree structure, ideally
extending to alveoli. Due to the large number of airway branches, manual delineation is laborious and time-consuming. Therefore, it is necessary to develop an
accurate segmentation model especially for the distal small airways.

Over the years, CNN-based methods [1,8,10,15–17] have been widely used
in this task. Juarez *et al.* [6] train a 3D UNet [2] to delineate the airway tree,
and they also combine the CNN model with several GNN layers in the following
work [5]. Qin *et al.* [10] develop their AirwayNet and further propose to detect
more peripheral bronchi by feature recalibration and attention distillation [12].
When adopting CNN-based models to segment such complicated tree structures,
the class imbalance between foreground and background (inter-class imbalance)
as well as that between large and small airways (intra-class imbalance) can be a
major challenge for detecting more distal branches. Gradient erosion and dilation are proposed by Zheng *et al.* [18] to elucidate the influence of inter-class
imbalance. With large weights in the loss function, the peripheral bronchi can
be detected while the prediction is thicker than the ground truth. As the weight
reduces, the prediction becomes thinner and breakages occur among the small
airways. Zheng *et al.* claim that this phenomenon is caused by the noisy information cumulated during back-propagation (BP) and has a more serious impact
on the voxels with severe local class imbalance.

To identify what happens during BP, we build a simple 3D UNet and check
the attention of the gradient flow, which is defined as

$$G^m = \sum_{c=1}^{N} |g_c^m|, \tag{1}$$

where G^m denotes the gradient attention of the mth convolutional layer, g_c^m
means the cth channel in the gradient to this layer, N is the total channel number.
G^m is further normalized by G^m/G_{max}^m, where G_{max}^m is the maximum in this
map. As shown in Fig. 1, for an input patch mainly consisting of small airways,
after trained five epochs with Dice loss [9], the network predicts the branch
highlighted by the blue circle as background. After passing the Sigmoid layer
$S(x)$, since $S'(x) = S(x) \times (1 - S(x))$, the gradient to this branch is filtered out.
Besides, the gradient attention to other small branches is diluted during the BP.
In contrast, if we use Tversky loss [13] with a hyper-parameter $\alpha = 0.2$, the
network can find most of this branch and the gradient attention to this part is
kept until the first convolutional layer. But there is still an undetected branch (in

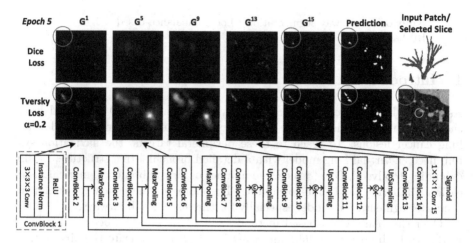

Fig. 1. The intermediate gradient attention maps in an UNet trained by Dice loss or Tversky loss ($\alpha = 0.2$). The detailed structure of the UNet is shown in the bottom part. Each ConvBlock consists of a $3 \times 3 \times 3$ convolutional layer, an instance normalization layer, and a ReLU layer. The input 3D patch and the selected slice are demonstrated on the right. The attention maps of the input gradients to five layers as well as the predictions after five epochs during training are shown in the upper part.

orange circle) that gets no gradient attention. From these observations, we find two important factors when adopting CNNs to segment the peripheral bronchi. First, it is necessary to adjust the gradients to different branches according to their sizes. Compared with Dice loss, Tversky loss ensures a higher ratio of the foreground gradient $\frac{\partial T}{\partial p_f}$ to background gradient $\frac{\partial T}{\partial p_b}$,

$$\left| \frac{\partial T}{\partial p_f} \middle/ \frac{\partial T}{\partial p_b} \right| = \frac{1}{\alpha} \frac{1}{1-T} - 1 \geq \frac{1-\alpha}{\alpha}. \tag{2}$$

During BP, the foreground gradients interact with their background neighborhood through the 3D convolutional kernels. When the gradient ratio is small, the foreground gradients may be diluted through this process. For large airways, this leads to an eroded surface, while for peripheral bronchi, the erosion evolves into breakages and undetected branches. Thus, the airway points with severe local class imbalance need larger gradient ratios to prevent the gradient erosion problem. To this end, in this paper, we propose to alleviate the gradient erosion and dilation problem based on the quantification of local class imbalance. Second, the first few epochs are of great importance. If a pattern is not learned at the beginning of training, the gradients to the corresponding areas can be filtered out by the last activation layer. Then it is hard for the network to learn the related features from other branches. In chest CT images, the thickness, shape and intensity of the airway wall are easily affected by pulmonary diseases, which becomes one of the major reasons for the breakages of segmentation.

To overcome this problem, we further develop a BP-based weight enhancement strategy to reduce the number of unlearnable false negatives during training.

The main contributions of this work are as follows:

1. To mitigate the impact of class imbalance on airway segmentation, we propose to adjust the gradient ratio of each airway point based on the quantification of local class imbalance.
2. To learn the patterns of the hard-to-segment regions, we design a BP-based weight enhancement strategy which further fine-tunes the gradient ratios.
3. Extensive experiments are implemented on an airway segmentation dataset including 90 cases. We achieve the state-of-the-art length detected and branch detected rates while keeping a precision of more than 92%.

2 Method

In this paper, we propose to strengthen the learning of detailed structures by keeping suitable gradients to the peripheral branches during BP with refined weight maps. The following sections provide a detailed explanation of the local-imbalance-based weight and BP-based weight enhancement strategy.

2.1 Local-imbalance-based Weight

To accurately segment more peripheral bronchi, the gradient erosion and dilation problem cannot be neglected. For large airways, the related features are easy to learn while the slight erosion or dilation around the surface does not affect the clinical use. However, for the segmental bronchi, a small gradient ratio directly leads to the breakages and undetected parts. If we assign large weights to all these branches, although an increased sensitivity can be obtained, the airway walls of the relatively thick branches would be misclassified as the lumen, while wrong connections between the adjacent branches may appear after the bifurcation or trifurcation area. Therefore, it is vital to adjust the gradient ratios according to the size of different branches. In this paper, we achieve this based on the local class imbalance between foreground and background, which is measured by the local foreground rate

$$FR_p = \frac{1}{N} \sum_{x_i \in \mathbf{B}} y_i, \qquad (3)$$

where FR_p denotes the foreground rate of point p, $\mathbf{B}$ is a pre-defined neighborhood with N voxels centered at p, x_i denotes a voxel within $\mathbf{B}$ and y_i is the corresponding label. We adopt a cubic neighborhood with a size of $L \times L \times L$.

To mitigate the gradient erosion and dilation based on this information, we need a bridge between the gradient ratio and the quantification results, which can be achieved by the weight map in the loss function. We adopt the General Union loss [18] as our segmentation loss, which is defined as

$$L = 1 - \frac{\sum_{i=1}^{N} w_i p_i^{r_e} g_i}{\sum_{i=1}^{N} w_i (\alpha p_i + \beta g_i)}, \qquad (4)$$

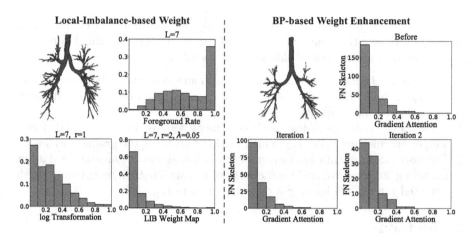

Fig. 2. The left part demonstrates the distributions of the foreground rate, weight map after logarithm transformation as well as the final local-imbalance-based weight map for a given example. The right part illustrates the histograms of the gradient attention values of the skeleton points belonging to false negatives before and after the BP-based weight enhancement for another example. The attention values are calculated based on the gradients to the first convolutional layer of WingsNet after 5 epochs.

where $\alpha + \beta = 1$, r_e controls the extent of the element-wise focal effect, N is the total number of voxels, p_i is the prediction of each voxel, g_i is the corresponding ground truth and w_i is the weight. If we assign $w_{b,j} = 1$ to all background voxels and $w_{f,j}$ to a foreground point $p_{f,j}$, the gradient ratio is

$$\left| \frac{\partial L}{\partial p_{f,j}} \middle/ \frac{\partial L}{\partial p_b} \right| = w_{f,j} \left(\frac{r_e p_{f,j}^{r_e - 1}}{\alpha} \frac{1}{1 - L} - 1 \right). \tag{5}$$

It is seen that the gradient ratio can be directly controlled by $w_{f,j}$. To convert the FR into weight maps, we first adopt a logarithm transformation

$$f_L(x_p) = -log_{10} FR_p, \tag{6}$$

where L is the hyper-parameter of the size of the neighborhood. But only with this parameter, the flexibility is limited. Thus, we further enrich the transformation by power functions. Before that, a clamp function f_c is applied to prevent the exponential growth of the values greater than 1,

$$f_c(x) = \begin{cases} x, & x \leq 1 \\ 1, & x > 1. \end{cases} \tag{7}$$

Then the power function and a lower bound λ are added,

$$f_{L,r,\lambda}(x_p) = (1 - \lambda)(f_c(-log_{10} FR_p))^r + \lambda, \tag{8}$$

where r means the root. In this paper, we experimentally chose $L = 7$, $\lambda = 0.05$, and r was randomly selected from $\{r \mid 2 + \frac{i}{N}, i = 0, 1, ..., N\}$ during training to generate the weight map W_{LIB} for airway segmentation. The corresponding experiment results are shown in the supplementary materials. Figure 2 illustrates an example of the FR and weight maps. It is seen that after this transformation, only a small number of the airway voxels are assigned large weights.

2.2 BP-based Weight Enhancement

In addition to the class imbalance problem, the segmentation of peripheral branches is also affected by the disease lesions as well as the complicated background texture. If these less common patterns are not learned at the beginning of training, the gradients to these areas may be filtered out by the last activation layer in the following epochs. To deal with this problem, we propose a strategy in which the training is iteratively restarted with refined weight maps.

We adopt WingsNet [18] as the segmentation network and check the gradient attention of the first convolutional layer after five epochs. As shown in the left top histogram in Fig. 2, at this time, nearly 200 undetected skeleton points [7] of a given example receive gradient attention less than 0.1, which account for 5.4% of the tree skeleton. To strengthen the learning of these hard-to-segment regions, we further increase the weights by

$$w_i^{BP} = (1 - GA_i^{cls})^2 (1 - \frac{d_i}{d_{max}})^2, \qquad (9)$$

where GA_i^{cls} is the gradient attention value of the closest skeleton voxel for an airway point p_i, d_i is the Euclidean distance between these two points and d_{max} is the maximal distance within this airway tree. We set $GA = 1$ for the detected skeleton points, thus only the voxels close to the undetected centerlines receive this compensation. In Eq. (9), the first item assigns the weight based on the gradients to the bottom layer during BP, while the second item is used to alleviate the influence of branch size. Then the training is restarted with the new weight map $W_{new} = W_{LIB} + \gamma W_{BP}$, where γ controls the extent of enhancement. Moreover, we can repeat this process to iteratively update the W_{BP}. As shown in Fig. 2, the number of false-negative skeleton points reduces significantly after two iterations. In this paper, we chose to refine the W_{BP} after five epochs, while this parameter is closely related to the network architecture and training strategy.

In summary, to find a suitable gradient ratio for each airway point, we first design a local-imbalance-based weight to resolve the class imbalance problem. This map is further refined based on the learning process, reducing the unlearnable false negatives. These weights can be directly inserted into the loss function.

3   Experiments and Results

The Binary Airway Segmentation Dataset [11] contained 90 CT scans. The pixel spacing ranged from 0.5 to 0.82 mm, and the slice thickness varied from 0.5 to

Table 1. Comparison with other methods in the Binary Airway Segmentation Dataset. The last column lists the inference time. 'WE' denotes the BP-based weight enhancement. 'Iter2 Epoch3' means the enhancement is repeated twice based on the model trained by 3 epochs. 'FN_WE' means $GA_i = 0$ for all FN skeleton points in Eq. (9).

Method	Branch(%)	Length(%)	Precision(%)	Time
Jin et al. [4] 2017	83.07 ± 11.49	85.38 ± 10.39	93.93 ± 1.92	10 min
Juarez et al. [6] 2018	82.11 ± 12.35	84.13 ± 8.59	91.41 ± 2.53	32 s
Qin et al. [10] 2019	81.38 ± 13.84	83.59 ± 10.37	95.76 ± 1.84	42 s
Wang et al. [15] 2019	83.52 ± 11.23	86.25 ± 8.54	93.36 ± 2.12	43 s
Juarez et al. [5] 2019	60.49 ± 23.88	68.03 ± 21.06	**96.41 ± 1.81**	34 s
Qin et al. [12] 2020	87.64 ± 9.18	91.82 ± 5.31	91.54 ± 2.89	46 s
Zheng et al. [18] 2021	88.68 ± 7.87	92.53 ± 4.45	91.41 ± 3.29	45 s
clDice [14] 2021	86.67 ± 8.75	90.45 ± 7.04	91.57 ± 3.14	45 s
WE	90.97 ± 6.90	93.79 ± 4.42	91.59 ± 2.98	45 s
LIB Weight	92.35 ± 6.02	94.40 ± 3.59	92.17 ± 2.72	45 s
LIB Weight + WE	93.00 ± 5.16	94.92 ± 3.08	92.14 ± 2.81	45 s
LIB Weight + FN_WE	**93.96 ± 4.64**	**95.64 ± 2.55**	91.03 ± 3.08	45 s
LIB Weight + WE Iter2	92.92 ± 5.24	94.82 ± 3.36	91.85 ± 2.88	45 s
LIB Weight + WE Iter3	93.68 ± 4.80	95.31 ± 2.87	91.42 ± 2.95	45 s
LIB Weight + WE Epoch3	93.75 ± 5.19	95.31 ± 3.24	90.31 ± 3.41	45 s
LIB Weight + WE Epoch10	92.89 ± 5.59	94.83 ± 3.27	92.20 ± 2.73	45 s

1.0 mm. The 90 cases were split into the training set, validation set, and test set with 50, 20, and 20 images respectively. During preprocessing, the voxel values were clipped to $[-1000, 600]$ HU and then rescaled to $[0, 255]$. Besides, the intensities of the voxels outside the chest were unified to 255.

During training, we sampled 16 patches with a size of $[128, 128, 128]$ from each scan in each epoch. We adopted the same two-stage training strategy [18]. SGD optimizer was used with a momentum of 0.9 and a weight decay of 0.0001. Random crop and random rotation were performed for data augmentation. In the first stage, the network was trained by 100 epochs. The initial learning rate was 0.01, and it was divided by 10 in the 60th and 90th epochs. In the second stage, the network was trained by 50 epochs. The initial lr was 0.01 and decreased in the 40th and 45th epochs. The hyper-parameters were tuned in the validation set. In the loss function, we chose $\alpha = 0.05$ and 0.1 in the two stages. Besides, we used $N = 5$ and $\gamma = 0.5$ when generating the weight map. The framework was implemented in Pytorch 1.4 and running on 2 NVIDIA Tesla V100 32G GPUs.

The quantitative results are illustrated in Table 1. We adopt three evaluation metrics including tree length detected rate [3], branch detected rate[3], and precision. AirwayNet is proposed by Qin et al. [10] to enhance the connectivity between airway points, and in their recent work [12], the branch detected rate is further improved from 81.4% to 87.6% by feature recalibration and attention distillation. Juarez et al. integrate a graph neural network module into the 3D UNet architecture [5]. However, in our experiments, it is found that the GNN module has a negative effect in several cases, leading to larger deviations (21.06% and

23.88%) in both tree length and branch detected rates. Radial distance loss is developed by Wang *et al.* [15] to train their spatial fully connected network. This loss weakens the impact of branch size by reducing the weights for each voxel based on their distances to the centerlines, achieving a length detected rate of 86.3%. We also compare with clDice loss [14], but it is found that the iterative min- and max-pooling cannot obtain an accurate airway centerline, which limits its performance. Zheng *et al.* [18] build their WingsNet with group supervision to enhance the training of shallow layers, while General Union loss is designed to resolve the intra-class imbalance. We adopted this method as our baseline and first replaced the distance-based weight in General Union loss with the proposed LIB weight. This can improve the branch detected rate from 88.7% to 92.4%. After performing the BP-based weight enhancement strategy, this value is further increased to 93.0%. The corresponding results are demonstrated in Fig. 3. It is seen that the false negatives mainly distribute around the surface of large airways, which means this region needs larger weight to avoid the gradient erosion problem. Besides, the FNs are also caused by the breakages as we only take the largest connected domain into account. As for the false positives, some unannotated peripheral bronchi can be detected by the network, leading to a trade-off between recall and precision. The dilation parts of some distal branches account for the rest of the FPs.

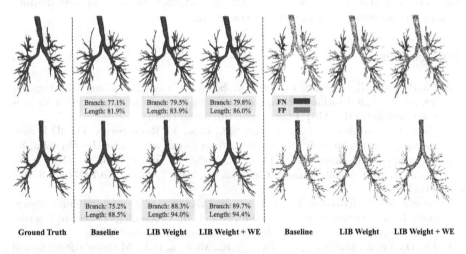

Fig. 3. Illustration of the airway segmentation results. We demonstrate the network predictions of three methods in the left part. Besides, to provide a better analysis of the performance, the false positives (FP) and false negatives (FN) are colored in red and green respectively in the right section. (Color figure online)

More ablation study is also shown in Table 1. The BP-based weight enhancement was performed on distance-based weights [18], which can boost the branch detected rate by 2.3%. We also set $GA_i = 0$ for all FN skeleton points in Eq. (9), which means the enhancement is equally given to all FN skeleton voxels. This leads to a trade-off between sensitivity and specificity. Besides, we conducted the weight enhancement for more iterations. It is seen that there is no significant change in the branch and length detected rates after two iterations, while the precision drops a little. The sensitivity further increases after three iterations but the precision gradually decreases to 91.4%. A suitable weight compensation contributes to the recall of segmentation without hurting the precision, but excessive enhancement can cause the gradient dilation problem. Finally, we generated the W_{BP} based on the model trained by three and ten epochs. The predictions after three epochs include more false negatives, thus the stronger enhancement leads to an increase in recall but a decrease in precision. In contrast, the difference between the results of W_{BP}^{epoch5} and $W_{BP}^{epoch10}$ is statistically insignificant. Therefore, we chose this hyper-parameter as five to save the implementation time.

4   Conclusion

In this paper, to accurately segment both the large and small airways, we alleviate the gradient erosion and dilation problem in the CNN-based method by inserting the local-imbalance-based weight into the loss function. We further develop a BP-based weight enhancement strategy to strengthen the learning of the hard-to-segment regions. The overall performance of the proposed demonstrates the potential clinical value of the method.

References

1. Charbonnier, J., Rikxoort, E., Setio, A., Schaefer-Prokop, C., Ciompi, F.: Improving airway segmentation in computed tomography using leak detection with convolutional networks. Med. Image Anal. **36**, 52–60 (2016)
2. Çiçek, Ö., Abdulkadir, A., Lienkamp, S.S., Brox, T., Ronneberger, O.: 3D U-Net: learning dense volumetric segmentation from sparse annotation. In: Ourselin, S., Joskowicz, L., Sabuncu, M.R., Unal, G., Wells, W. (eds.) MICCAI 2016. LNCS, vol. 9901, pp. 424–432. Springer, Cham (2016). https://doi.org/10.1007/978-3-319-46723-8_49
3. Feuerstein, M., Kitasaka, T., Mori, K.: Adaptive branch tracing and image sharpening for airway tree extraction in 3-d chest ct. In: Proceedings of Second International Workshop on Pulmonary Image Analysis, pp. 273–284 (2009)
4. Jin, D., Xu, Z., Harrison, A.P., George, K., Mollura, D.J.: 3d convolutional neural networks with graph refinement for airway segmentation using incomplete data labels. In: Machine Learning in Medical Imaging, pp. 141–149 (2017)
5. Garcia-Uceda Juarez, A., Selvan, R., Saghir, Z., de Bruijne, M.: A joint 3D UNet-graph neural network-based method for airway segmentation from chest CTs. In: Suk, H.-I., Liu, M., Yan, P., Lian, C. (eds.) MLMI 2019. LNCS, vol. 11861, pp. 583–591. Springer, Cham (2019). https://doi.org/10.1007/978-3-030-32692-0_67
6. Garcia-Uceda Juarez, A., Tiddens, H.A.W.M., de Bruijne, M.: Automatic airway segmentation in chest CT using convolutional neural networks. In: Stoyanov, D., et al. (eds.) RAMBO/BIA/TIA -2018. LNCS, vol. 11040, pp. 238–250. Springer, Cham (2018). https://doi.org/10.1007/978-3-030-00946-5_24

7. Lee, T., Kashyap, R., Chu, C.: Building skeleton models via 3-d medial surface axis thinning algorithms. CVGIP Graph. Models Image Process. **56**(6), 462–478 (1994). https://doi.org/10.1006/cgip.1994.1042

8. Meng, Q., Roth, H., Kitasaka, T., Oda, M., Mori, K.: Tracking and segmentation of the airways in chest ct using a fully convolutional network. In: International Conference on Medical Image Computing and Computer-Assisted Intervention (2017)

9. Milletari, F., Navab, N., Ahmadi, S.: V-net: fully convolutional neural networks for volumetric medical image segmentation. In: 2016 Fourth International Conference on 3D Vision (3DV), pp. 565–571 (Oct 2016)

10. Qin, Y., et al.: Airwaynet: a voxel-connectivity aware approach for accurate airway segmentation using convolutional neural networks. In: International Conference on Medical Image Computing and Computer-Assisted Intervention (2019)

11. Qin, Y., Gu, Y., Zheng, H., Chen, M., Yang, J., Zhu, Y.: Airwaynet-se: a simple-yet-effective approach to improve airway segmentation using context scale fusion. In: 2020 IEEE 17th International Symposium on Biomedical Imaging (ISBI), pp. 809–813 (2020)

12. Qin, Y., et al.: Learning bronchiole-sensitive airway segmentation CNNs by feature recalibration and attention distillation. In: Martel, A.L., et al. (eds.) MICCAI 2020. LNCS, vol. 12261, pp. 221–231. Springer, Cham (2020). https://doi.org/10.1007/978-3-030-59710-8_22

13. Salehi, S., Erdogmus, D., Gholipour, A.: Tversky loss function for image segmentation using 3d fully convolutional deep networks. CoRR abs/1706.05721 (2017)

14. Shit, S., et al.: cldice - a novel topology-preserving loss function for tubular structure segmentation. In: Proceedings of the IEEE/CVF Conference on Computer Vision and Pattern Recognition (CVPR), pp. 16560–16569 (2021)

15. Wang, C., et al.: Tubular structure segmentation using spatial fully connected network with radial distance loss for 3D medical images. In: Shen, D., et al. (eds.) MICCAI 2019. LNCS, vol. 11769, pp. 348–356. Springer, Cham (2019). https://doi.org/10.1007/978-3-030-32226-7_39

16. Yun, J., et al.: Improvement of fully automated airway segmentation on volumetric computed tomographic images using a 2.5 dimensional convolutional neural net. Med. Image Anal. **51**, 13–20 (2019)

17. Zhao, T., Yin, Z., Wang, J., Gao, D., Chen, Y., Mao, Y.: Bronchus segmentation and classification by neural networks and linear programming. In: Shen, D., et al. (eds.) MICCAI 2019. LNCS, vol. 11769, pp. 230–239. Springer, Cham (2019). https://doi.org/10.1007/978-3-030-32226-7_26

18. Zheng, H., et al.: Alleviating class-wise gradient imbalance for pulmonary airway segmentation. IEEE Trans. Med. Imaging (2021). https://doi.org/10.1109/TMI.2021.3078828

Selective Learning from External Data for CT Image Segmentation

Youyi Song[1](✉), Lequan Yu[2], Baiying Lei[3], Kup-Sze Choi[1], and Jing Qin[1]

[1] Center for Smart Health, School of Nursing,
The Hong Kong Polytechnic University, Hong Kong, China
`youyisong.song@connect.polyu.hk`
[2] Department of Statistics and Actuarial Science, The University of Hong Kong,
Hong Kong, China
[3] School of Biomedical Engineering, Shenzhen University, Shenzhen, China

Abstract. Learning from external data is an effective and efficient way of training deep networks, which can substantially alleviate the burden on collecting training data and annotations. It is of great significance in improving the performance of CT image segmentation tasks, where collecting a large amount of voxel-wise annotations is expensive or even impractical. In this paper, we propose a generic selective learning method to maximize the performance gains of harnessing external data in CT image segmentation. The key idea is to learn a weight for each external data such that 'good' data can have large weights and thus contribute more to the training loss, thereby implicitly encouraging the network to mine more valuable knowledge from informative external data while suppressing to memorize irrelevant patterns from 'useless' or even 'harmful' data. Particularly, we formulate our idea as a constrained non-linear programming problem, solved by an iterative solution that alternatively conducts weights estimating and network updating. Extensive experiments on abdominal multi-organ CT segmentation datasets show the efficacy and performance gains of our method against existing methods. The code is publicly available (Released at https://github.com/YouyiSong/Codes-for-Selective-Learning).

Keywords: Selective learning · External data · Constrained non-linear programming · CT image segmentation

1 Introduction

Deep neural networks have achieved remarkable success on a wide range of CT image segmentation tasks [1–6]. However, their success heavily relies on a large amount of training data which is not always available. Collecting CT data, in practice, is time-consuming and expensive; besides the image itself, collecting voxel-wise annotations requires tedious efforts from domain experts, with multiple rounds to correct annotation errors for reaching a consensus among experts.

© Springer Nature Switzerland AG 2021
M. de Bruijne et al. (Eds.): MICCAI 2021, LNCS 12901, pp. 420–430, 2021.
https://doi.org/10.1007/978-3-030-87193-2_40

To circumvent this difficulty, a possible way is to use external data [7–9], shared in websites or by other institutions, by jointly training the network on the internal and external data. However, external data may have a different distribution from the internal data. They may vary hugely in image quality and relevance, and even not be a reliable reflection of the task to be learned [10]. Therefore, by the naive joint training, external data often cannot be effectively exploited.

More advanced techniques then attempt to select informative external data [11–18] or assign an importance weight to each external data [19–28]. The key idea of data selection is to drop some 'useless' or 'harmful' external data, and is often conducted by discarding external data that have a large loss value during training. This type of methods hence favors to learn easy patterns and ignores numerous informative data with hard patterns which have been known to make deep models more accurate and robust [29,30]. Furthermore, how to judge the loss value to be large enough for selecting external data remains elusive and is often done in a heuristic manner.

Instead of completely ignoring 'bad' external data, data weighting-based methods attempt to assign a small weight to them, just making their contribution less to the training loss. Earlier works include mainly importance sampling [19,20], boosting [21,22], and hard example mining [23]. These works, however, cannot distinguish informative external data with hard patterns from outliers. Recently, Ren et al. [27] proposed a meta-learning framework to learn external data' weights by using the gradient descent direction of the mini-batch data. This method, however, requires to compute second order gradients of the network, being computationally expensive.

In this paper, we investigate data weighting problem and propose a constrained non-linear optimization technique to learn external data' weights. We jointly learn the weights and network parameters for maximally leveraging their complementary benefits. We also put a hard constraint that enforces the network to learn better than without using external data. We solve this constrained joint learning problem by an iterative solution that alternatively conducts weights estimating and network updating. We assess the proposed method on the abdominal multi-organ segmentation task, and obtained positive results that show the efficacy of our method against existing methods.

2 Methodology

2.1 Problem Setup

Let $\mathcal{X}$ and $\mathcal{Y}$ be the input space and output space for the segmentation task, respectively. Given a segmentation network with the function space of $\mathcal{F}$, training this network is to find a function $f \in \mathcal{F}$ that best approximates the unknown target segmentation mapping by solving the optimization problem below

$$\underset{f \in \mathcal{F}}{\operatorname{argmin}} \, \mathbb{E}_{p(x,y)}[\ell(f(x), y)], \quad (x, y) \in \mathcal{X} \times \mathcal{Y}, \tag{1}$$

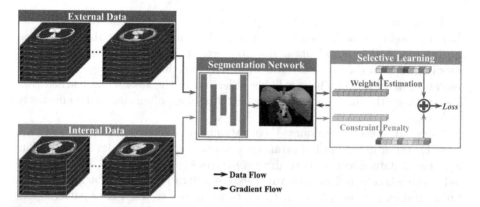

Fig. 1. The illustrative pipeline of how we selectively learn from external data: for each external data, we learn a weight to adjust its importance in the training loss, while for internal data, we put a hard constraint to enforce the network to learn better than without using external data.

where $\mathbb{E}_{p(x,y)}$ denotes the expectation over the underlying joint density $p(x,y)$ and $\ell : \mathcal{Y} \times \mathcal{Y} \to \mathbb{R}_+$ denotes the loss function. We here fold in any regularization terms into ℓ for simplicity.

When learning from external data, the above canonical learning paradigm (Eq. 1), however, may become biased, as external data may have a different distribution from the internal data. We hence aim at learning a weight w_k for each external data (x_k, y_k) to correct the distribution discrepancy, for helping the network to mine more valuable knowledge from informative external data while suppressing to memorize irrelevant patterns from 'useless' or even 'harmful' data; an illustrative pipeline of our method is shown in Fig. 1.

2.2   Selective Learning

We formulate our weights learning as a constrained non-linear programming problem. Let $\{x_n, y_n\}_{n=1}^N$ and $\{x_m, y_m\}_{m=1}^M$ be the external and internal datasets, respectively. We jointly learn external data' weights and the network by solving the following problem

$$\underset{\substack{f \in \mathcal{F}, \\ \mathbf{w} \in [0,1]^N}}{\mathrm{argmin}} \left(\frac{1}{N} \sum_{n=1}^N w_n \ell_n(f) + \frac{1}{M} \sum_{m=1}^M \ell_m(f) \right), \qquad (2a)$$

$$\mathrm{s.t.} \sum_{m=1}^M \ell_m(f) \leq \sum_{m=1}^M \ell_m(\hat{f}_M^*), \qquad (2b)$$

$$\mathbf{1}^T \mathbf{w} > 0, \qquad (2c)$$

where $\mathbf{w} := (w_1, \cdots, w_N)^T$, $\ell_i(f) := \ell(f(x_i), y_i)$, and $\hat{f}_M^*$ is the optimum trained on internal data only; $\hat{f}_M^* \in \mathrm{argmin}_{f \in \mathcal{F}} \frac{1}{M} \sum_{m \in [M]} \ell_m(f)$. The constraint (2b) is to first guarantee the learning performance on the internal data and second to prevent the network to over-fit the external data. The constraint (2c) is to prevent degenerate solutions of $\mathbf{w}$, *e.g.*, $\mathbf{w} = \mathbf{0}$ where not all external data are correctly utilized.

Feasibility Analysis. Our problem certainly has feasible solution. To see this, assuming that we have known the optima of weights, $\mathbf{w}^*$, our problem in that case degenerates into a standard learning problem with the weighted loss for external data; the constraints (2b) and (2c) here are satisfied almost surely. The main concern is whether our problem is well formulated such that its optima corresponds to the best function that we can learn. We check this by using reductio ad absurdum. Assume that $\{\hat{f}^*, \hat{\mathbf{w}}^*\}$ is the optima and f^* is the best learnable function. If $\hat{f}^*$ behaves worse than f^*, then by altering $\hat{\mathbf{w}}^*$, our objective function (2a) can have a smaller value by replacing $\hat{f}^*$ with f^*, which means $\{\hat{f}^*, \hat{\mathbf{w}}^*\}$ is not the optima, contradicted. This result means that our problem has been well formulated; solving it results in getting the best learnable function.

2.3   Optimization

Figure 1 shows an illustrative pipeline of how we solve our problem. We use block coordinate decent that splits the problem into two subproblems: (1) weights estimating and (2) network updating. We alternatively conduct weights estimating where the network is fixed and network updating where the weights are fixed.

Weights Estimating. It is done by solving the subproblem below

$$\mathbf{w}^{(t+1)} \in \underset{\mathbf{w} > 0}{\operatorname{argmin}} \left(\sum_{n=1}^{N} w_n d(\mathcal{D}_n, \mathcal{D}|f^{(t)}) + \lambda \mathbf{w}^T \mathbf{w} \right), \quad \text{s.t.} \sum_{n=1}^{N} w_n = N, \quad (3)$$

where t and $d(\mathcal{D}_n, \mathcal{D}|f^{(t)})$ denote the updating step and the discrepancy between the distribution $\mathcal{D}_n$ and $\mathcal{D}$ measured by the network at step t. Here $\mathcal{D}_n$ and $\mathcal{D}$ stand for distributions for the external data (x_n, y_n) and the internal dataset $\{x_m, y_m\}_{m=1}^{M}$. In addition, $\lambda > 0$ is a hyperparameter to balance two terms.

The first term in (3) is small when external data with a small discrepancy value having been assigned a large weight. This is expected, as external data similar to the internal data should contribute more to the network updating. The second term has a small value when all external data have been assigned similar weights, and hence it encourages to learn from as many external data as possible. In addition, note that when the constraint here is satisfied, the constraint (2c) is also satisfied.

Finally, we define $d(\mathcal{D}_n, \mathcal{D}|f) = |\ell_n(f) - \bar{\ell}_{[M]}(f)|$, where $\bar{\ell}_{[M]}(f)$ denotes the average of $\ell_m(f)$ on the internal dataset $\{x_m, y_m\}_{m=1}^{M}$. As expected, it has a large value when the network performs differently on the external data (x_n, y_n) and the internal dataset $\{x_m, y_m\}_{m=1}^{M}$, while a small value when performing similarly.

Network Updating. Given the weights $\mathbf{w}^{(t+1)}$, we update the network by

$$f^{(t+1)} = f^{(t)} - \gamma^{(t)} \nabla \left(\sum_{i=1}^{b} w_i^{(t+1)} \ell_i(f^{(t)}) + \xi^{(t)} \mathcal{C}(f^{(t)}) \right), \qquad (4)$$

where $\gamma^{(t)}$ and b stand for the learning rate and batch size, respectively. The only difference from the canonical network updating with the weighted loss is that we convert the constraint (2b) to the term $\mathcal{C}(f^{(t)})$ with the penalty coefficient $\xi^{(t)}$.

We define $\mathcal{C}(f^{(t)}) = \sum_{i \in [M]} \max \left(\ell_i(f^{(t)}) - \ell_i(\hat{f}_M^*), \ 0 \right)$. We can see that if (2b) is satisfied, then $\mathcal{C}(f^{(t)}) = 0$, and thus it has no effect on the network updating. Otherwise, *i.e.* $\mathcal{C}(f^{(t)}) > 0$, then the updating step will be enforced to primarily eliminate the violation by using a large $\xi^{(t)}$.

We initialize ξ to a very small value at the early stage, as it is normal that the network behaves worse than $\hat{f}_M^*$ at that stage. We then increase it when meeting violations, using the update rule: $\xi^{(t+1)} = \xi^{(t)} + \gamma_\xi^{(t)} \mathcal{C}(f^{(t)})$, where $\gamma_\xi^{(t)}$ stands for the increasing rate; we get this update rule by using sub-gradients [31], as $\mathcal{C}(f^{(t)})$ is non-differentiable.

3   Experiments

3.1   Experimental Setup

Dataset. The proposed method was assessed on the abdominal multi-organ segmentation task from two CT datasets [32–34][1]. We term them as BTCV and TCIA for simplicity. BTCV has 47 CT volumes with resolutions of 0.6–0.9 mm (in-plane) and 0.5–5.0 mm (inter-slice), while TCIA has 43 CT volumes with resolutions of 0.6–0.9 mm (in-plane) and 1.5–2.5 mm (inter-slice). Each CT volume has 122.85 slices on average in the BTCV dataset while 238.02 in the TCIA dataset. They provide voxel-wise annotations for 8 organs: (1) Duodenum, (2) Esophagus, (3) Gallbladder, (4) Liver, (5) Left Kidney, (6) Pancreas, (7) Spleen, and (8) Stomach.

Network Architecture. We used the original 2D U-Net [35], except for 5 modifications: (1) we replaced batch normalization with instance normalization [36], (2) we replaced ReLU with leaky ReLU [37] with the slope of 0.01, (3) we implemented down-sampling and up-sampling by strided and transposed convolutions, (4) we added dropout [38] with the probability of 0.1 after each convolutional layer, and (5) we used softmax function to normalize network's output.

---

[1] Available on https://zenodo.org/record/1169361#.XSFOm-gzYuU.

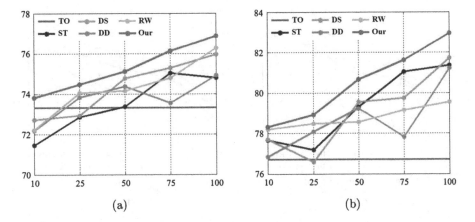

Fig. 2. The test mean DSC of methods with 5 amounts (%) of external data on (a) the BTCV dataset where we sampled external data from the TCIA datset and (b) the TCIA dataset where we sampled external data from the BTCV dataset.

Implementation Details. We clipped the HU values to $[-200, 250]$ and then normalized to $[0, 255]$. We resized all the slices with a size of 256×256. We used Dice loss to train the network by Adam [39] with the learning rate of 0.0003. We set the batch size to 16 and λ to 5. We initialized ξ as 0.05 and fixed γ_ξ as 0.0001. We ran 40 epochs; in the first 10 epochs, we adopted the standard training for the warm-up purpose, and put the hard constraint after 20 epochs. We generated segmentation results by assigning the categorical label with the highest prediction value to voxels.

Competitors. We compared our method against three methods, denoted by DS [15], RW [27], and DD [28], respectively. DS belongs to data selection-based methods, while RW and DD belong to data weighting-based methods. DS selects k data with the smallest loss value from the batch at each iteration to update the network; we set $k = 10$ in our experiments. RW assigns data weights by minimizing the weighted loss on the validation set; we used 20% of the given internal training data for the validation. DD assigns data weights based on the distribution discrepancy between external data and internal data; we here assumed that all external data are sampled from the same distribution. For a fair comparison, we used the same experimental setup when implementing these methods, and their hyperparameters were set to the recommended values by the authors.

3.2   Experimental Results

Performance Improvement. We first compare the performance improvement of different methods by using external data. We use Dice Similarity Coefficient (DSC) as the performance metric. The test mean DSC of 8 organs is presented in Fig. 2; (a) on the BTCV dataset and (b) on the TCIA dataset. In each dataset,

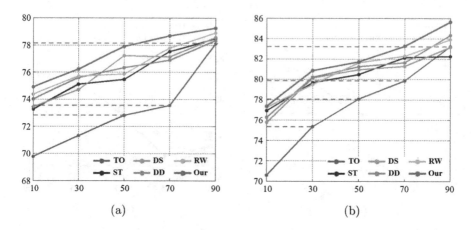

Fig. 3. The test mean *DSC* of methods with 5 amounts (%) of internal training data when using all external data on (a) the BTCV dataset and (b) the TCIA dataset.

we randomly selected 50% data (CT volumes not 2D slices) for training and the remaining 50% for testing, and sampled external data from another dataset with five different amounts: 10%, 25%, 50%, 75%, and 100%.

From Fig. 2(a), we can see that our method consistently works better than all competitors in all scenarios, showing the efficacy of our method; TO and ST in the figure stand for target only, without using external data, and standard training, the naive joint training, respectively. We also can see that not all methods can obtain performance gains by using external data; when using 10% of TCIA data, only our method beats TO. This evidence suggests that our method is less dependent on the amount of external data, more flexible to selectively learn from external data. Furthermore, it is clear that not all methods can obtain a bigger performance gain by using more external data; there are performance drops of ST and DD. This may indicate that when learning from external data not all data helps and thus learning should be done selectively.

Similar trends are also found in the TCIA dataset, as shown in Fig. 2(b), though the extent is slightly different. The main new finding is that DS, RW, and DD behave differently compared to ST, while our method is consistent; they generally work better than ST in the BTCV dataset while not here. This finding may suggest that these methods are sensitive to external data, not robust as our method.

Internal Data Alleviation. We here compare methods' ability to alleviate internal training data by using external data. To simulate the segmentation performance varying with the increasing of internal training data, we randomly selected 10% internal data for testing and 5 amounts for training (10%, 30%, 50%, 70%, and 90%). As for external data, here we used all. The test mean *DSC* is presented in Fig. 3.

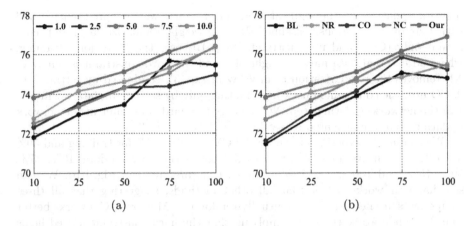

Fig. 4. The test mean *DSC* of our method on the BTCV dataset with (a) different values of the hyperparameter λ and (b) different variations.

We can see that our method yields the best performance in both datasets in all scenarios, which means that our method alleviates the internal training data most, showing the efficacy of our method again. We also can see that all methods outperform TO, except for ST in the TCIA dataset when using 90% internal data for training, and our method has the biggest performance gain over TO, though all methods have a narrowing gain gap when more training data are given. This observation implies that our method is more effective to leverage informative external data and suppress 'useless' or 'harmful' data. Furthermore, among all methods, only RW and Our consistently work better than ST, while other methods are with mixing results; RW in fact slightly works worse than ST in the TCIA dataset when using 30% internal data for training. This suggests that RW and our method are more robust while our method is more accurate.

3.3   Analysis of Our Method

Hyperparameter Selection. Our method has a hyperparameter, λ, in Eq. (3), a larger value of it resulting external data to be assigned weights that are more similar. To decide the best value of it, we experimented with 5 values: 1.0, 2.5, 5.0, 7.5 and 10.0, on the BTCV dataset with 50% internal data for training and the remaining 50% for testing (randomly selected). Figure 4(a) shows the test mean *DSC* with 5 different amounts of external data are used (10%, 25%, 50%, 75%, and 100% of the TCIA dataset). We can see that when $\lambda = 5$ our method yields the highest *DSC*; we hence set λ to 5 in all other experiments. We also can see that a large value (7.5 or 10.0) generally works better than a small value (1.0 or 2.5).

Ablation Study. To investigate the effect of three main components of our method: (1) data weighting, (2) weight regularization, and (3) constrained network updating, on the segmentation performance, we compared our method with four of it variations: baseline (BL), no weight regularization (NR), constraint only (CO), and no constraint (NC). BL uses the naive joint training, where all external data have a weight of 1 and the hard constraint is removed. NR removes the regularization term $\mathbf{w}^T\mathbf{w}$ in Eq. (3), conducted by setting λ to 0. CO uses the constrained network updating only by setting all data weights to 1 in the network updating step. NC removes the hard constraint in the network updating step, implemented by setting $\xi^{(t)}$ in Eq. (4) to 0.

We conducted experiments on the BTCV dataset, 50% for training and 50% for testing, with five amounts of TCIA dataset as the external data: 10%, 25%, 50%, 75%, and 100%. The test mean DSC is presented in Fig. 4(b). We can see that Our works better than all other methods, suggesting that all three components are necessary and mutually reinforcing. Moreover, CO works better than BL while worse than NC, implying that the hard constraint indeed helps while the key of our method is data weighting.

4   Conclusion

There is tremendous motivation for training deep networks with as few effort spent on training data collection as possible. In this paper, we have investigated a simple, generic, and effective method to allow deep networks to selectively learn from external data, in order to maximize the performance gains of harnessing external data in CT image segmentation tasks. It is conceptually simple, learning a weight for each external data for learning selectively. It is generic, applicable to different network backbones and loss functions. It is also effective, yielding positive results on various experimental scenarios with consistent performance gains over existing methods.

Acknowledgement. The work described in this paper is supported by two grants from the Hong Kong Research Grants Council under General Research Fund scheme (Project No. PolyU 152035/17E and 15205919), and a grant from HKU Startup Fund and HKU Seed Fund for Basic Research (Project No. 202009185079).

References

1. Litjens, G., Kooi, T., Bejnordi, B., et al.: A survey on deep learning in medical image analysis. Med. Image Anal. **42**, 60–88 (2017)
2. Shen, D., Wu, G., Suk, H.: Deep learning in medical image analysis. Ann. Rev. Biomed. Eng. **19**, 221–248 (2017)
3. Kakeya, H., Okada, T., Oshiro, Y.: 3D U-JAPA-Net: mixture of convolutional networks for abdominal multi-organ CT segmentation. In: International Conference on Medical Image Computing and Computer-Assisted Intervention, pp. 426–433 (2018)
4. Song, Y., Yu, Z., Zhou, T., et al.: Learning 3D features with 2D CNNs via surface projection for CT volume segmentation. In: International Conference on Medical Image Computing and Computer-Assisted Intervention, pp. 176–186 (2020)

5. Huang, R., Zheng, Y., Hu, Z., et al.: Multi-organ segmentation via co-training weight-averaged models from few-organ datasets. In: International Conference on Medical Image Computing and Computer-Assisted Intervention, pp. 146–155 (2020)
6. Isensee, F., Jaeger, P., Kohl, S., et al.: nnU-Net: a self-configuring method for deep learning-based biomedical image segmentation. Nat. Methods **18**, 1–9 (2020)
7. Mansour, Y., Mohri, M., Suresh, A., et al.: A theory of multiple-source adaptation with limited target labeled data. arXiv preprint arXiv:2007.09762 (2020)
8. Cortes, C., Mohri, M., Suresh, A., et al.: Multiple-source adaptation with domain classifiers. arXiv preprint arXiv:2008.11036 (2020)
9. Konstantinov, N., Frantar, E., Alistarh, D., et al.: On the sample complexity of adversarial multi-source PAC learning. In: International Conference on Machine Learning, pp. 5416–5425 (2020)
10. Charikar, M., Steinhardt, J., Valiant, G.: Learning from untrusted data. In: Proceedings of 49th Annual ACM SIGACT Symposium on Theory of Computing, pp. 47–60 (2017)
11. Qiao, M., Valiant, G.: Learning discrete distributions from untrusted batches. arXiv preprint arXiv:1711.08113 (2017)
12. Awasthi, P., Blum, A., Haghtalab, N., et al.: Efficient PAC learning from the crowd. In: International Conference on Learning Theory, pp. 127–150 (2017)
13. Hendrycks, D., Mazeika, M., Wilson, D., et al.: Using trusted data to train deep networks on labels corrupted by severe noise. arXiv:1802.05300 (2018)
14. Han, B., Yao, Q., Yu, X., et al.: Co-teaching: robust training of deep neural networks with extremely noisy labels. In: Advances in Neural Information Processing Systems, pp. 8536–8546 (2018)
15. Ghadikolaei, H., Ghauch, H., Fischione, C., et al.: Learning and data selection in big datasets. In: International Conference on Machine Learning, pp. 2191–2200 (2019)
16. Jain, A., Orlitsky, A.: A general method for robust learning from batches. arXiv preprint arXiv:2002.11099 (2020)
17. Zhang, C., Yao, Y., Liu, H., et al.: Web-supervised network with softly update-drop training for fine-grained visual classification. In: AAAI Conference on Artificial Intelligence, pp. 12781–12788 (2020)
18. Zhang, C., Yao, Y., Shu, X., et al.: Data-driven meta-set based fine-grained visual classification. arXiv preprint arXiv:2008.02438 (2020)
19. Bugallo, M., Elvira, V., Martino, L., et al.: Adaptive importance sampling: the past, the present, and the future. IEEE Signal Process. Mag. **34**(4), 60–79 (2017)
20. Katharopoulos, A., Fleuret, F.: Not all samples are created equal: deep learning with importance sampling. In: International Conference on Machine Learning, pp. 2525–2534 (2018)
21. Sun, Y., Kamel, M., Wong, A., et al.: Cost-sensitive boosting for classification of imbalanced data. Pattern Recogn. **40**(12), 3358–3378 (2007)
22. Liu, Q., Ihler, A., Fisher, J.: Boosting crowdsourcing with expert labels: local vs. global effect. In: International Conference on Information Fusion, pp. 9–14 (2015)
23. Malisiewicz, T., Gupta, A., Efros, A.: Ensemble of exemplar-svms for object detection and beyond. In: International Conference on Computer Vision, pp. 89–96 (2011)
24. Dumitrache, A., Aroyo, L., Welty, C.: Crowdsourcing ground truth for medical relation extractio. arXiv preprint arXiv:1701.02185 (2017)

25. Yan, K., Cai, J., Zheng, Y., et al.: Learning from multiple datasets with heterogeneous and partial labels for universal lesion detection in CT. IEEE Trans. Med. Imaging (2020)
26. Luo, L., Yu, L., Chen, H., et al.: Deep mining external imperfect data for chest X-ray disease screening. IEEE Trans. Med. Imaging **39**(11), 3583–3594 (2020)
27. Ren, M., Zeng, W., Yang, B., et al.: Learning to reweight examples for robust deep learning. In: International Conference on Machine Learning, pp. 4334–4343 (2018)
28. Konstantinov, N., Lampert, C.: Robust learning from untrusted sources. In: International Conference on Machine Learning, pp. 3488–3498 (2019)
29. Lin, T., Goyal, P., Girshick, R., et al.: Focal loss for dense object detection. In: IEEE International Conference on Computer Vision, pp. 2980–2988 (2017)
30. Song, H., Kim, M., Lee, J.: Selfie: refurbishing unclean samples for robust deep learning. In: International Conference on Machine Learning, pp. 5907–5915 (2019)
31. Nandwani, Y., Pathak, A., Singla, P.: A primal dual formulation for deep learning with constraints. In: Advances in Neural Information Processing Systems, pp. 1–9 (2019)
32. Gibson, E., Giganti, F., Hu, Y., et al.: Automatic multi-organ segmentation on abdominal CT with dense v-networks. IEEE Trans. Med. Imaging **37**(8), 1822–1834 (2018)
33. Landman, B., Xu, Z., Eugenio, I., et al.: MICCAI multi-atlas labeling beyond the cranial vault-workshop and challenge (2015)
34. Roth, H., Farag, A., Turkbey, E., et al.: Data from pancreas-CT. Cancer Imaging Arch. (2015)
35. Ronneberger, O., Fischer, P., Brox, T.: U-Net: convolutional networks for biomedical image segmentation. In: Navab, N., Hornegger, J., Wells, W.M., Frangi, A.F. (eds.) MICCAI 2015. LNCS, vol. 9351, pp. 234–241. Springer, Cham (2015). https://doi.org/10.1007/978-3-319-24574-4_28
36. Ulyanov, D., Vedaldi, A., Lempitsky, V.: Instance normalization: The missing ingredient for fast stylization. arXiv preprint arXiv:1607.08022 (2016)
37. Xu, B., Wang, N., Chen, T., et al.: Empirical evaluation of rectified activations in convolutional network. arXiv preprint arXiv:1505.00853 (2015)
38. Srivastava, N., Hinton, G., Krizhevsky, A., et al.: Dropout: a simple way to prevent neural networks from overfitting. J. Mach. Learn. Res. **15**(1), 1929–1958 (2014)
39. Kingma, D., Ba, J.: Adam: A method for stochastic optimization. arXiv preprint arXiv:1412.6980 (2014)

Projective Skip-Connections for Segmentation Along a Subset of Dimensions in Retinal OCT

Dmitrii Lachinov[1,2](✉), Philipp Seeböck[1], Julia Mai[1,2], Felix Goldbach[1], Ursula Schmidt-Erfurth[1], and Hrvoje Bogunović[1,2]

[1] Department of Ophthalmology and Optometry, Medical University of Vienna, Vienna, Austria
[2] Christian Doppler Laboratory for Artificial Intelligence in Retina, Department of Ophthalmology and Optometry, Medical University of Vienna, Vienna, Austria
dmitrii.lachinov@meduniwien.ac.at

Abstract. In medical imaging, there are clinically relevant segmentation tasks where the output mask is a projection to a subset of input image dimensions. In this work, we propose a novel convolutional neural network architecture that can effectively learn to produce a lower-dimensional segmentation mask than the input image. The network restores encoded representation only in a subset of input spatial dimensions and keeps the representation unchanged in the others. The newly proposed projective skip-connections allow linking the encoder and decoder in a UNet-like structure. We evaluated the proposed method on two clinically relevant tasks in retinal Optical Coherence Tomography (OCT): geographic atrophy and retinal blood vessel segmentation. The proposed method outperformed the current state-of-the-art approaches on all the OCT datasets used, consisting of 3D volumes and corresponding 2D en-face masks. The proposed architecture fills the methodological gap between image classification and ND image segmentation.

1 Introduction

The field of medical image segmentation is dominated by neural network based solutions. The convolution neural networks (CNNs), notably U-Net [24] and its variants, demonstrate state-of-the-art performance on a variety of medical benchmarks like BraTS [1], LiTS [3], REFUGE [23], CHAOS [13] and ISIC [6]. Most of these benchmarks focus on the problem of segmentation, either 2D or 3D, where the input image and the output segmentation mask are of the same dimension. However, a few medical protocols like OCT for retina [16,17], OCT and Ultrasound for skin [21,25], Intravascular ultrasound (IVUS) pullback images for vasculature [26], CT for diaphragm analysis [29] and online tumor

Electronic supplementary material The online version of this chapter (https://doi.org/10.1007/978-3-030-87193-2_41) contains supplementary material, which is available to authorized users.

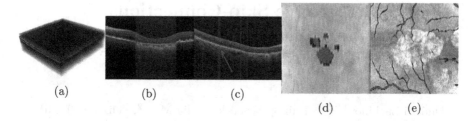

Fig. 1. (a) 3D retinal OCT volume. Ground truth annotations are shown in 2D cross-sectional slices for (b) GA and (c) blood vessels in yellow. Predictions of the proposed method are visualized in en-face projection images for (d) GA and (e) blood vessels. Green region - true positives, orange - false positives, dark red - false negatives. (Color figure online)

tracking [22] require the segmentation to be performed on the image projection, resulting in the output segmentation of lower spatial dimension than the input image. It introduces the problem of dimensionality reduction into the segmentation, for example, segmenting a flat 2D en-face structures on the projection of a 3D volumetric image.

This scenario has so far not been sufficiently explored in the literature and it is currently not clear what CNN architectures are the most suitable for this task. Here, we propose a new approach for $ND \to MD$ segmentation, where $M < N$, and evaluate it on two clinically relevant tasks of 2D Geographic Atrophy (GA) and Blood Vessels segmentation in 3D optical coherence tomography (OCT) retinal images.

1.1   Clinical Background

Geographic atrophy (GA) is an advanced form of age-related macular degeneration (AMD). It corresponds to localized irreversible and progressive loss of retinal photoreceptors and leads to a devastating visual impairment. 3D optical coherence tomography (OCT) is a gold standard modality for retinal examination in ophthalmology. In OCT, GA is characterized by hypertransmission of OCT signal below the retina. As GA essentially denotes a loss of tissue, it does not have "thickness", and it is hence delineated as a 2D en-face area (Fig. 1).

Retinal blood vessels provide oxygen and nutrition to the retinal tissues. Retinal vessels are typically examined using 2D fundus photographs where they can be seen as dark lines. In OCT volumes, retinal blood vessels can be detected in individual slices (B-scans) as interconnected morphological regions, dropping shadows on the underlying retinal layer structures (Fig. 1).

1.2   Related Work

Recently, several methods dealing with dimensionality reduction in segmentation tasks have been proposed [11,16,17]. Liefers et al. [17] introduced a fully convolutional neural network to address the problem of $ND \rightarrow MD$ segmentation. The authors evaluated their method on $2D \rightarrow 1D$ tasks of Geographic Atrophy (GA) and Retinal Layer segmentation. In contrast to our method, their approach is limited by the fixed size of the input image and shortcut networks that are prone to overfitting. This is caused by the full compression of the image in 2nd dimension, leading to a large receptive field but completely removing the local context.

Recently, Li et al. [16] proposed an image projection network IPN designed for 3D-to-2D image segmentation, without any explicit shortcuts or skip-connections. Pooling is only performed for a subset of dimensions, resulting in a highly anisotropic receptive field. This limits the amount of context the network can use for segmentation.

Ji et al. [11] proposed a deep voting model for GA segmentation. First, multiple fully-connected classifiers are trained on axial depth scans (A-scans), producing a single probability value for each A-scan (1D-to-0D reduction). During the inference phase, the predictions of the trained classifiers are concatenated to form the final output. This approach doesn't account for neighboring AScans, thus lacks spatial context, and requires additional postprocessing.

Existing approaches suffer from two main limitations: On the one hand, they are designed to handle images of a fixed size, making patch based training difficult. This is of particular relevance as it has been shown that patch based training improves overall generalization as it serves as additional data augmentation [15]. On the other hand, existing $ND \rightarrow MD$ methods have a large receptive field due to a high number of pooling operations or large pooling kernel sizes, that fail to capture local context. To illustrate this issue, we conducted a compared receptive fields of state-of-the-art segmentation [2,7,12,19,28,30] and classification methods [5,18,20] for medical imaging across different benchmarks [1,3,6] with popular architectures designed for natural images classification [8,9] (Fig. 1–2 in the supplement). We observed that current state-of-the art methods for segmentation in a subset of input dimensions [16,17] *fall out of the cluster with methods designed for medical tasks*, and have a receptive field comparable to the networks designed for natural images classification. However, the amount of training data available for medical tasks differs by the orders of magnitude from the natural images data. Alternatively, the $ND \rightarrow MD$ segmentation task can be solved using $ND \rightarrow ND$ methods. It can be attempted by either first projecting the input image to the output MD space, which looses context, or by running ND segmentation first, which is memory demanding, requires additional postprocessing and is not effective.

Our approach is explicitly designed to overcome these limitations. We propose a segmentation network that can handle arbitrary sized input and is on par with state-of-the-art $ND \rightarrow ND$ medical segmentation methods in terms of receptive field size.

1.3   Contribution

In this paper we introduce a novel CNN architecture for $ND \to MD$ segmentation. The proposed approach has a decoder that operates in the same dimensionality space as the encoder, and restores the compressed representation of the bottleneck layer only in a subset of the input dimensions. The contribution of this work is threefold: (1) We propose projective skip-connections addressing the general problem of segmentation in a subset of dimensions; (2) We provide clear definitions and instructions on how to reuse the method for arbitrary input and output dimensions; (3) We perform an extensive evaluation with three different datasets on the task of Geographic Atrophy and Retinal Blood Vessel $3D \to 2D$ segmentation in retinal OCTs. The results demonstrate that the proposed method clearly outperforms the state-of-the-art.

2   Method

Let the image $I \in \mathbb{R}^{\prod_{d=1}^{N} n_d}$, where N is the number of dimensions, and n_d is the size of the image in the corresponding dimension d. We want to find such function f that $f : I \longrightarrow O$, where $O \in \mathbb{R}^{\prod_{d=1}^{M} n_d}$ is output segmentation and M is its dimensionality. We focus on the case where $M \leq N$ with dimensions $d \leq M$ being target dimensions, where we perform segmentation; and $M < d \leq N$ being reducible dimensions.

We parameterize f by a convolutional neural network. The CNN architecture we propose to use for the segmentation along dimensions M follows a U-Net [24,32] architecture. The encoder consists of multiple blocks that sequentially process and downsample the input volume. Unlike the other methods, we don't perform the global pooling in the network bottleneck in the dimensions outside of M. Instead, we freeze the size of these dimensions and propagate the features through the decoder into the final classification layer. The decoder restores the input resolution only across those target dimensions M where we perform the segmentation. The decoder keeps feature maps in the remaining reducible dimensions $M < d \leq N$ compressed. Since the sizes of encoder and decoder feature maps do not match, we propose *projective skip-connections* to link them.

At the last stage of the network, each location of the output MD image has a corresponding feature tensor of compressed representation in $(N - M)D$ (refer to Fig. 2). We perform Global Average Pooling (GAP) and a regular convolution to obtain MD logits.

2.1   Projective skip-connections

The purpose of projective skip-connections is to compress the reducible dimensions $M < d \leq N$ to the size of bottleneck and leave the target dimensions $d \leq M$ unchanged. In contrast to GAP, we don't completely reduce the target dimensions to size of 1 (Sect. 3.2). Instead, we use average pooling with varying kernel size (Fig. 2). Keeping in mind that the feature map size along dimension d

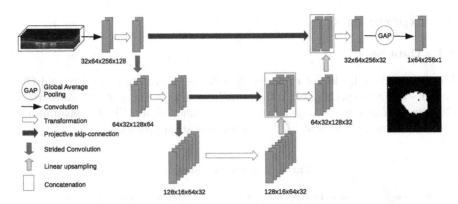

Fig. 2. The proposed CNN architecture solving $ND \rightarrow MD$ segmentation problem, where $N = 3$ and $M = 2$. *Transformation* can be any sort of operation, residual blocks in our case. The input is a 3D crop from the SD-OCT image. The final feature map is averaged (GAP) in dimensions that are not in M, only last dimension in our case. Then the feature map is processed with final convolution, that reduces the number of channels from 32 to 1. The output is a 2D en-face segmentation mask. The network in this figure has a configuration of $l = 3$, $C = \{2, 4, 8\}$, and $B = \{1, 1, 1\}$.

generated by our proposed network f of depth l with input image I at the level j of the encoder will be $\frac{n_d}{2^{j-1}}$, the feature map size at the level j of the **decoder** will be n_d^j (1).

$$n_d^j = \begin{cases} \frac{n_d}{2^{j-1}}, & \text{if } d \le M \\ \frac{n_d}{2^{l-1}}, & \text{otherwise.} \end{cases} \quad (1) \qquad k_d^j = \begin{cases} 1, & \text{if } d \le M \\ 2^{l-j}, & \text{otherwise,} \end{cases} \quad (2)$$

We propose to perform the average pooling of the encoder feature maps with kernel size and stride k_d^j (2), where d is the corresponding dimension, and j is the corresponding layer of the network. After the average pooling is performed, we concatenate the encoder and decoder feature maps.

The proposed CNN can be explicitly described with the depth or the number of total network levels l, the number of channels per convolution for each level $C = \{c_0 \times 2^{i-1} : 1 \le i \le l\}$, and the number of residual blocks at each level $B = \{b_i : i \le l\}$. The schematic representation of the proposed approach is shown in Fig. 2. The network has a configuration of $l = 3$, $C = \{2, 4, 8\}$ and $B = \{1, 1, 1\}$. The input is a 3D volume patch $I \in \mathbb{R}^{64 \times 128 \times 256}$, so $N = 3$. We are interested in the en-face segmentation, so $M = 2$. In the extreme case where $M = N$, the proposed CNN is equivalent to a U-Net architecture with residual blocks. If $M = 0$, the proposed CNN is equivalent to N-D ResNet with skip-connections in the last residual blocks.

3   Experiments

3.1   Data Sets

We use a rich collection of three medical datasets with annotated retinal OCT volumes for our experiments (Table 1). All our datasets consist of volumetric 3D OCT images and corresponding en-face 2D annotations (Fig. 1), meaning that the used datasets represent $3D \rightarrow 2D$ segmentation problems.

Table 1. Datasets description

Dataset	Device	Volumes	Eyes	Patients	Depth, mm	En-face area, mm^2	Depth spacing, μm	En-face spacing, μm^2	Depth resolution, px	En-face resolution, px^2
GA 1	Spectralis	192	70	37	1.92	6.68 × 6.86	3.87	156.46 × 7.91	496	43 × 867
GA 2	Spectralis	260	147	147	1.92	6.02 × 6.03	3.87	122.95 × 11.16	496	49 × 540
Vessel 1	TopconDR	1	1	1	2.3	6 × 6	2.6	46.87 × 11.72	885	128 × 512
	Cirrus	40	40	40	2	6.05 × 6.01	1.96	47.24 × 11.74	1024	128 × 512
	Spectralis	43	43	43	1.92	5.88 × 5.82	3.87	60.57 × 5.68	496	97 × 1024
	Topcon2000	9	9	9	2.3	6 × 6	2.6	46.88 × 11.72	885	128 × 512
	Topcon-Triton	32	32	32	2.57	7 × 7	2.59	27.34 × 13.67	992	256 × 512

The following are the three datasets. **Dataset 'GA1':** The scans were taken at multiple time points from an observational longitudinal study of natural GA progression. Annotations of GA were performed by retinal experts on 2D Fundus Auto Fluorescence (FAF) images and transferred by image registration to the OCT, resulting in 2D en-face OCT annotations. **Dataset 'GA2':** With the OCT scans originating from a clinical trial that has en-face annotations of GA that were annotated directly on the OCT BScans by a retinal expert. **Dataset 'Vessel 1':** A diverse set of OCT scans across device manufacturers of patients with AMD where the retinal blood vessels were directly annotated on OCT en-face projections by retinal experts.

3.2   Baseline Methods

We exhaustively compared our approach against multiple $ND \rightarrow ND$ baselines: *UNet 3D* [32] operating on OCT volume; *UNet 2D* [24] operating on OCT volume projections; *UNet++* [31] operating on OCT volume projections. And $ND \rightarrow MD$ methods: *SD* or Selected Dimensions [17]; *FCN* or Fully Convolutional Network, used as a baseline in [17]; *IPN* or Image Projection Network [16];

Ablation Experiment: 3D2D is a variation of the proposed method with 3D encoder and 2D decoder networks. The bottleneck and skip-connections perform global pooling in the dimensions $d \geq M$. This ablation experiment was included to study the effect of propagating deep features to the last stages of the network.

3.3 Training Details

As a part of preprocessing, we flattened the volume along the Bruch's Membrane, and cropped a 3D retina region with the size of 128 pixels along AScan (vertical direction) and keep the full size in the rest of dimensions. We resampled the images to have $119.105 \times 5.671\ \mu m^2$ en-face spacing. Before processing the data, we performed z-score normalization in cross-sectional plane. For generating OCT projections, we employed the algorithm introduced by Chen et al. [4].

For the task of Geographic Atrophy segmentation, the *proposed* method has the following configuration: $l = 4$, $B = \{1, 1, 1, 1\}$, $C = \{32, 64, 128, 256\}$. We use residual blocks [8] with 3D convolutions with kernel size of 3. Instead of Batch Normalization [10], we employ Instance Normalization [27], due to small batch size. The *3D2D* network has the same configuration as the proposed method. *SD* and *FCN* are reproduced as in the paper [17], but with the AScan size equals to 128 and 32 channels in the first convolution. *IPN* network is implemented as in the paper [16] for $3D \rightarrow 2D$ case with 32 channels in the first convolution. *UNet 2D* is reproduced as in the paper [24], but with residual blocks. *UNet 3D* [32] and *UNet++* [31] were reproduced by following the corresponding papers. *UNet 3D* output masks were converted to 2D by pooling the reducible dimensions and thresholding. *For Retinal Vessel segmentation*, we used the networks of the same configuration, but we set the number of channels in the first layer equal to 4.

The models were trained with Adam [14] optimizer for $3 \cdot 10^4$ and 10^4 iterations with weight decay of 10^{-5}, learning rate of 10^{-3} and decaying with a factor of 10 at iteration $2 \cdot 10^4$ and $6 \cdot 10^3$ for GA and blood vessels segmentation respectively. For GA segmentation, the batch size was set to 8 and patch size to $64 \times 256 \times 64$. For the *SD* and *FCN* models we used patch size of $64 \times 512 \times 512$ due to architecture features described in the paper [17]. For blood vessels segmentation, the batch size was set to 8 and patch size was $32 \times 128 \times 256$. The optimizer was chosen empirically for each model. We used Dice score as a loss function for both tasks.

3.4 Evaluation Details

For all three datasets (*GA1*, *GA2*, *Vessel 1*) we conducted a 5-fold cross-validation for evaluation. The splits were made on patient level with stratification by baseline GA size for *GA* datasets and by device manufacturer for *Vessel 1* dataset. The results marked as validation are average of all samples from all the validation splits. The *Dice score* and 95th percentile of Hausdorff distance averaged across scans (Dice and HD95) were used. To test for significant differences between the proposed method and baselines, we conducted two-sided Wilcoxon signed-rank tests, using $\alpha = 0.05$. In addition to cross-validation results, we report performances on hold out (*external*) datasets: models trained on *GA1* were evaluated on *GA2* serving as an external test set and vice versa.

4 Results and Discussion

Cross-validation results are reported in the Table 2. The qualitative examples of segmentations can be found in Supplementary materials. For all the tasks, the proposed methods achieved a significant improvement in both the mean *Dice* and *Hausdorff* distance. This improvement is also reflected in the box-plots, showing higher mean *Dice* scores and lower variance consistently across all datasets (Fig. 3a). Of note, *SD* approach outperformed *FCN*, successfully reproducing their results reported in [17]. In Table 2 we can see a significant margin in the performance between ND methods like *2D UNet, 3D UNet, UNet++* and $3D \rightarrow 2D$ methods. This indicates that the problem of $3D \rightarrow 2D$ segmentation cannot be solved efficiently with existing 2D or 3D methods.

External test set results are provided in Table 2. The proposed method outperforms all other approaches, with significant improvements in *GA2*. The improvement over *SD* [17] on GA 1 dataset is slightly below the significance threshold, p-value = 0.063, but the boxplots (Fig. 3b) highlight the lower variance and higher median *Dice* of the proposed architecture.

Table 2. Experiments results. Left: cross-validation, right: external test set. (*: p-value≤ 0.05; **: p-value$\leq 10^{-5}$; ***: p-value$\leq 10^{-10}$).

Dataset	GA 1		GA 2		Vessel 1	
Model	Dice	HD95	Dice	HD95	Dice	HD95
UNet 3D [32]	0.71***	0.71***	0.85***	0.58***	–	–
UNet 2D [24]	0.73***	1.43***	0.82***	1.18***	0.51***	0.63***
UNet++ [31]	0.75***	1.22***	0.83***	1***	0.50***	0.65***
FCN [17]	0.769***	0.533***	0.890***	0.114***	0.358***	0.934***
IPN [16]	0.793***	0.523**	0.910***	0.116***	0.630***	0.482*
SD [17]	0.788***	0.591***	0.919***	0.092***	0.654***	0.447***
3D2D	0.791***	0.563**	0.927***	0.070**	0.650***	0.451*
Proposed	**0.820**	**0.336**	**0.935**	**0.052**	**0.684**	**0.367**

Dataset	GA 1		GA 2	
Model	Dice	HD95	Dice	HD95
FCN [17]	0.761***	0.426***	0.878***	0.141***
IPN [16]	0.796***	0.549**	0.903***	0.094**
SD [17]	0.817	0.404	0.905***	0.113**
3D2D	0.819*	0.382	0.906***	0.113**
Proposed	**0.824**	**0.310**	**0.915**	**0.079**

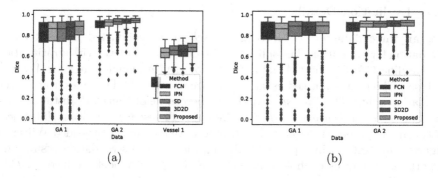

(a)   (b)

Fig. 3. Segmentation performance evaluated with (a) cross-validation and (b) external test set.

5   Conclusion

The problem of image segmentation in a subset of dimensions is a characteristic of a few relevant clinical applications. It can not be efficiently solved with well-studied 2D-to-2D or 3D-to-3D segmentation methods. In this paper, we first analyzed and discussed the existing approaches designed for $ND \rightarrow MD$ segmentation. For instance, existing methods share the same features, like large receptive field and encoder performing dimensionality reduction. The skip-connections, however, are implemented as subnetworks or not used at all.

Based on this analysis we proposed a novel convolutional neural network architecture for image segmentation in a subset of input dimensions. It consists of encoder that doesn't completely reduce any of the dimensions; the decoder that restores only the dimensions where the segmentation is needed; and projective skip-connections that help to link the encoder and the decoder of the network. The proposed method was tested on three medical datasets and it clearly outperformed the state of the art in two $3D \rightarrow 2D$ retinal OCT segmentation tasks.

Acknowledgements. The financial support by the Austrian Federal Ministry for Digital and Economic Affairs, the National Foundation for Research, Technology and Development and the Christian Doppler Research Association is gratefully acknowledged.

References

1. Bakas, S., et al.: Identifying the Best machine learning algorithms for brain tumor segmentation, progression assessment, and overall survival prediction in the BRATS challenge. arXiv arXiv:abs/1811.02629 (2018)
2. Bi, L., Kim, J., Kumar, A., Feng, D.: Automatic liver lesion detection using cascaded deep residual networks. arXiv arXiv:abs/1704.02703 (2017)
3. Bilic, P., et al.: The liver tumor segmentation benchmark (LiTS). arXiv arXiv:abs/1901.04056 (2019)
4. Chen, Q., Niu, S., Shen, H., Leng, T., de Sisternes, L., Rubin, D.L.: Restricted summed-area projection for geographic atrophy visualization in SD-OCT images. Transla. Vis. Sci. Technol. **4**(5), 2 (2015). https://doi.org/10.1167/tvst.4.5.2
5. Díaz, I.G.: Incorporating the knowledge of dermatologists to convolutional neural networks for the diagnosis of skin lesions. Computer Vision and Pattern Recognition. arXiv arXiv:1703.01976 (2017)
6. Gutman, D., et al.: Skin lesion analysis toward melanoma detection: a challenge at the 2017 International symposium on biomedical imaging (ISBI), hosted by the international skin imaging collaboration (ISIC). In: 2018 IEEE 15th International Symposium on Biomedical Imaging, ISBI 2018, pp. 168–172 (2018)
7. Han, X.: Automatic liver lesion segmentation using a deep convolutional neural network method. arXiv arXiv:abs/1704.07239 (2017)
8. He, K., Zhang, X., Ren, S., Sun, J.: Deep residual learning for image recognition. In: 2016 IEEE Conference on Computer Vision and Pattern Recognition (CVPR), pp. 770–778 (June 2016)

9. Huang, G., Liu, Z., Weinberger, K.Q.: Densely connected convolutional networks. 2017 IEEE Conference on Computer Vision and Pattern Recognition (CVPR), pp. 2261–2269 (2017)

10. Ioffe, S., Szegedy, C.: Batch normalization: accelerating deep network training by reducing internal covariate shift. arXiv arXiv:abs/1502.03167 (2015)

11. Ji, Z., Chen, Q., Niu, S., Leng, T., Rubin, D.L.: Beyond retinal layers: a deep voting model for automated geographic atrophy segmentation in SD-OCT images. Transl. Vis. Sci. Technol. **7**(1), 1 (2018)

12. Jiang, Z., Ding, C., Liu, M., Tao, D.: Two-stage cascaded U-Net: 1st place solution to BraTS challenge 2019 segmentation task. In: BrainLes@MICCAI (2019)

13. Kavur, A.E., et al.: CHAOS challenge - combined (CT-MR) healthy abdominal organ segmentation. arXiv arXiv:abs/2001.06535 (2020)

14. Kingma, D.P., Ba, J.: Adam: a method for stochastic optimization. CoRR abs/1412.6980 (2015)

15. Krizhevsky, A., Sutskever, I., Hinton, G.E.: ImageNet classification with deep convolutional neural networks. In: CACM (2017)

16. Li, M., et al.: Image projection network: 3D to 2D image segmentation in octa images. IEEE Trans. Med. Imaging **39**(11), 3343–3354 (2020)

17. Liefers, B., González-Gonzalo, C., Klaver, C., van Ginneken, B., Sánchez, C.: Dense segmentation in selected dimensions: application to retinal optical coherence tomography. In: Proceedings of Machine Learning Research, London, United Kingdom, 08–10 July 2019, vol. 102, pp. 337–346. PMLR (2019)

18. Matsunaga, K., Hamada, A., Minagawa, A., Koga, H.: Image classification of melanoma, nevus and seborrheic keratosis by deep neural network ensemble. arXiv arXiv:abs/1703.03108 (2017)

19. McKinley, R., Rebsamen, M., Meier, R., Wiest, R.: Triplanar ensemble of 3D-to-2D CNNs with label-uncertainty for brain tumor segmentation. In: BrainLes@MICCAI (2019)

20. Menegola, A., Tavares, J., Fornaciali, M., Li, L., Avila, S., Valle, E.: RECOD Titans at ISIC challenge 2017. arXiv arXiv:abs/1703.04819 (2017)

21. Mlosek, R.K., Malinowska, S.: Ultrasound image of the skin, apparatus and imaging basics. J. Ultrasonography **13**(53), 212–221 (2013). https://doi.org/10.15557/JoU.2013.0021

22. Murphy, M.J.: Tracking moving organs in real time. Semin. Radiat. Oncol. **14**(1), 91–100 (2004)

23. Orlando, J.I., et al.: REFUGE challenge: a unified framework for evaluating automated methods for glaucoma assessment from fundus photographs. Med. Image Anal. **59**, 101570 (2020)

24. Ronneberger, O., Fischer, P., Brox, T.: U-Net: convolutional networks for biomedical image segmentation. In: Navab, N., Hornegger, J., Wells, W.M., Frangi, A.F. (eds.) MICCAI 2015. LNCS, vol. 9351, pp. 234–241. Springer, Cham (2015). https://doi.org/10.1007/978-3-319-24574-4_28

25. Srivastava, R., Yow, A.P., Cheng, J., Wong, D.W.K., Tey, H.L.: Three-dimensional graph-based skin layer segmentation in optical coherence tomography images for roughness estimation. Biomed. Opt. Exp. **9**(8), 3590–3606 (2018)

26. Sun, S., Sonka, M., Beichel, R.R.: Graph-based IVUS segmentation with efficient computer-aided refinement. IEEE Trans. Med. Imaging **32**(8), 1536–1549 (2013). https://doi.org/10.1109/TMI.2013.2260763

27. Ulyanov, D., Vedaldi, A., Lempitsky, V.: Instance normalization: the missing ingredient for fast stylization. arXiv arXiv:abs/1607.08022 (2016)

28. Vorontsov, E., Chartrand, G., Tang, A., Pal, C., Kadoury, S.: Liver lesion segmentation informed by joint liver segmentation. 2018 IEEE 15th International Symposium on Biomedical Imaging, ISBI 2018, pp. 1332–1335 (2018)
29. Yalamanchili, R., et al.: Automatic segmentation of the diaphragm in non-contrast CT images. In: 2010 IEEE International Symposium on Biomedical Imaging: From Nano to Macro, pp. 900–903 (2010)
30. Zhao, Y.-X., Zhang, Y.-M., Liu, C.-L.: Bag of tricks for 3D MRI brain tumor segmentation. In: Crimi, A., Bakas, S. (eds.) BrainLes 2019. LNCS, vol. 11992, pp. 210–220. Springer, Cham (2020). https://doi.org/10.1007/978-3-030-46640-4_20
31. Zhou, Z., Siddiquee, M.M.R., Tajbakhsh, N., Liang, J.: UNet++: a nested U-Net architecture for medical image segmentation. In: Deep Learning in Medical Image Analysis and Multimodal Learning for Clinical Decision Support : 4th International Workshop, DLMIA 2018, and 8th International Workshop, ML-CDS 2018, held in conjunction with MICCAI 2018, Granada, Spain, vol. 11045, pp. 3–11 (2018)
32. Çiçek, Ö., Abdulkadir, A., Lienkamp, S.S., Brox, T., Ronneberger, O.: 3D U-Net: learning dense volumetric segmentation from sparse annotation. In: MICCAI (2016)

MouseGAN: GAN-Based Multiple MRI Modalities Synthesis and Segmentation for Mouse Brain Structures

Ziqi Yu[1,2], Yuting Zhai[1,2], Xiaoyang Han[1,2], Tingying Peng[3], and Xiao-Yong Zhang[1,2(✉)]

[1] Institute of Science and Technology for Brain-Inspired Intelligence, Fudan University, Shanghai 200433, People's Republic of China
xiaoyong_zhang@fudan.edu.cn
[2] Key Laboratory of Computational Neuroscience and Brain-Inspired Intelligence, Ministry of Education, Fudan University, Shanghai 200433, China
[3] Helmholtz AI, Helmholtz Zentrum, Munich, Germany

Abstract. Automatic segmentation of mouse brain structures in magnetic resonance (MR) images plays a crucial role in understanding brain organization and function in both basic and translational research. Due to fundamental differences in contrast, image size, and anatomical structure between the human and mouse brains, existing neuroimaging analysis tools designed for the human brain are not readily applicable to the mouse brain. To address this problem, we propose a generative adversarial network (GAN)-based network, named MouseGAN, to synthesize multiple MRI modalities and to segment mouse brain structures using a single MRI modality. MouseGAN contains a modality translation module to project multi-modality image features into a shared latent content space that encodes modality-invariant brain structures and a modality-specific attribute. In addition, the content encoder learned from the modality translation module is reused for the segmentation module to improve the structural segmentation. Our results demonstrate that MouseGAN can segment up to 50 mouse brain structures with an averaged dice coefficient of 83%, which is a 7–10% increase compared to baseline U-Net segmentation. To the best of our knowledge, it is the first Atlas-free tool for segmenting mouse brain structures from MRI data. Another benefit is that with the help of the shared encoder, MouseGAN can handle missing MRI modalities without significant sacrifice of the performance. We will release our code and trained model to promote its free usage for neuroimaging applications.

Keywords: Segmentation · Mouse brain · MRI · Generative adversarial network · Disentangled representations

1 Introduction

Accurate extraction of brain structures from magnetic resonance (MR) images is a cornerstone for delineating pathological regions, analyzing brain functions and understanding the relationships between morphology and physiology. As one of the most important

M. de Bruijne et al. (Eds.): MICCAI 2021, LNCS 12901, pp. 442–450, 2021.
https://doi.org/10.1007/978-3-030-87193-2_42

model organisms, the mouse serves as an irreplaceable bridge for basic research, drug discovery, preclinical translation and neuroscience.

Currently, the existing conventional MR image processing toolkits [1, 2] are well developed for human brains. However, when they apply to mouse brain analysis, the performance remarkably suffers from the difference of image contrast, image size, and anatomical structures. For example, a popular method to segment human brain structures is multi-atlas segmentation [3], in which multiple high-resolution atlases with well-labeled brain structures are required and non-rigid registration is performed to register a brain to be segmented to these atlases. This is impractical for mouse brain segmentation, since it is already extremely expensive and time-consuming to establish even one single MRI-based mouse brain atlas. Several deep-learning (DL) based atlas-free methods were reported to apply to mouse brain MRI, but they only focused on skull stripping. So far, no deep learning-based methods have been reported to segment fine mouse brain structures using MRI data. One major reason is that accurate segmentation of fine brain structures requires multi-modality MRI that provides much richer feature information than single modality [4]. However, unlike human brain MRI, multi-modality mouse brain MRI data of high-resolution are lacking since collecting multiple MR modalities for each mouse is impractical in most preclinical MR centers. Therefore, based on existing single modality mouse brain MRI data, an automatic generation of missing MRI modality is highly desirable for efficient segmentation.

Nowadays, with the advancement of deep learning, synthetic modalities can be obtained by an image-to-image translation procedure, e.g., based on a generative adversarial network (GAN). Particularly, a cycle-GAN architecture also enables the training of the translation using unpaired data using cycle-consistency loss function in combination with adversarial loss [5]. However, typical cycle-GAN deals with one-to-one domain translation, which limits its application in analyzing MRI with more than two modalities. More recently, image-to-image translation with disentangled representation for content and style, e.g., [6] has attracted increased attention in medical image analysis as it can filter out modality-specific information and leaves only a modality-invariant representation that better reflects morphological or pathological tissue structure. A few successful examples can be found for segmenting brain tumours [7] and abdominal organs [8, 9].

In this paper, we take this promising approach to use a disentangled generative neural work, termed as MouseGAN, to tackle the challenging mouse brain structural segmentation task using a single MRI modality. MouseGAN is a unified model that combines modality synthesis and structural segmentation: it contains a modality translation module to project multi-modality image features into a shared latent content space that encodes modality-invariant brain structures and a modality-specific attribute. Then, the latent content can be combined with other modality-specific attributes to impute images of other modalities. The content encoder learned from the modality translation module is reused for the segmentation module, which greatly reduces the parameter size of the segmentation module and allows for effective training with a small number of segmentation annotations. The shared encoder between translation and segmentation modules is a novel component of MouseGAN, differing from the previous methods [7–9] that mostly use decoupled networks for image translation and segmentation, separately. We demonstrate in the later sections that the shared encoder disentangles modality-specific features and is an indispensable component to enhance segmentation performance.

There are a few key contributions in MouseGAN. First, to the best of our knowledge, MouseGAN is the first atlas-free tool to be used for the segmentation of up to 50 brain structures using MRI data. Second, MouseGAN can generate segmentation using only one modality from the five common modalities, T1, T2, T2-star, QSM, magnitude MR images as input and is therefore very flexible. Last but not least, with the help of a translation module and shared content space, MouseGAN can impute missing modalities from existing ones and improve segmentation performance.

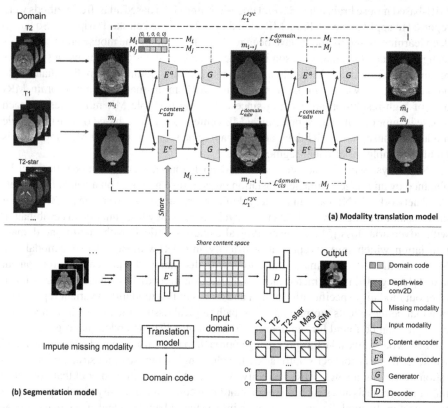

Fig. 1. Schematic of MouseGAN: given an unpaired dataset including images and one-hot codes of all modality domains (T1, T2, T2-star, ...), we first train a modality translation module to generate all domain images from the various missing input domains (a). Then, we employ this modality translation module as an auxiliary network in the subsequent segmentation pipeline by reusing the content encoder parameters and imputing missing modalities. (b). We anticipate that the learning of features shared between different modalities in a self-supervised manner can better characterize brain structure, thereby leading to better segmentation.

2 Methods

We illustrate MouseGAN as a schematic plot of Fig. 1 and give a detailed explanation of each component below.

2.1 Modality Translation Model

Synthesis of different modality images is carried out using the modality-to-modality translation generative adversarial network derived from DRIT++ [5], as shown in Fig. 1(a). Let $\{M_i\}_{i=1\sim k}$ be the descriptor for each domain, $(m_i, m_j) \in M$ be the two randomly-sampled images from two exemplary domains, respectively, and their domain codes in one-hot format $\left(z_{m_i}^d, z_{m_j}^d\right)$, where $z^d \subset \mathbb{R}^k$. The model consists of domain-specific content encoders $\left\{E_{m_i}^c, E_{m_j}^c\right\}$, attribute encoders $\left\{E_{m_i}^a, E_{m_j}^a\right\}$, generators $\left\{G_{m_i}^c, G_{m_j}^c\right\}$, and domain discriminators $\left\{D_{m_i}^c, D_{m_j}^c\right\}$ as well as a content discriminator D_{adv}^c. In the training stage, the input domain images are encoded onto a shared content space C, which is served as a common latent space to characterize style-independent for modeling the characterization of data variations, i.e., inherent variations of brain structures among different mouses in our application. The attribute encoders embed images with domain codes into a domain-specific spaces $\{A_i\}_{i=1\sim k}$. Then the forward image style translation $M_i \rightarrow M_j$ is achieved by generator via swapping the attribute domain codes, as $m_{i \rightarrow j} = G_{M_j}\left(z_{m_i}^c, z_{m_j}^a, z_{m_j}^d\right)$, where z_m^c and z_m^a are content and attribute representations yielding via corresponding encoders, respectively. It should be noted that the structure of the source image m_i remains unchanged as we keep the same content $Z_{m_i}^c$ in the newly generated image $m_{i \rightarrow j}$. In a similar way, we synthesize $m_{j \rightarrow i}$ and both the two synthetic images $\{m_{i \rightarrow j}, m_{j \rightarrow i}\}$ are then fed to discriminators that must discern between real or fake images, and must also recognize the domains of the generated images. To deal with un-paired training images, we reconstruct synthetic images back via a second domain transfer step and enforce $\{\hat{m}_i, \hat{m}_j\}$ to match the real input images $\{m_i, m_j\}$ with cross-cycle consistency loss.

To compel the task of multi-modality translation during training, we use a self-reconstruction loss $\mathcal{L}_1^{self-recon}$ to constrain encoders and decoders for generation high quality translations, latent regression loss $\mathcal{L}_1^{latent}$ to explicitly encourage the connection between image and the latent code to be invertible. Domain classification loss $\mathcal{L}_{cls}^{domain}$, cross-cycle consistency loss $\mathcal{L}_1^{cc}$, domain adversarial loss $\mathcal{L}_{adv}^{domain}$ and KL loss are also implemented to the training process and more details can be found in [5].

2.2 Segmentation Model

For segmentation, the pretrained modality translation model is served as an auxiliary network by imputing missing modalities in the input domain since it is too expensive and unrealistic to collect all modality data for every mouse. After modality translation, input images and synthesized images are fed to a depth-wise convolution before an encoder to preprocess the context modality-by-modality (Fig. 1(b)). We reuse the architecture and the parameters of the content encoder obtained in the modality translation training stage as the encoder of the segmentation module and only train the decoder. This is motivated by the fact that the content encoder in the modality translation model learns domain-invariant anatomical features in a self-supervised manner and could extract and distil these representative features in a shared latent content space, leading to a better segmentation of different anatomical structures. Thus, in contrast to the typical model training from scratch, our segmentation model can utilize all unpaired domain data for both training and testing, gaining advantage on complementary information of multi-modality synthetic images.

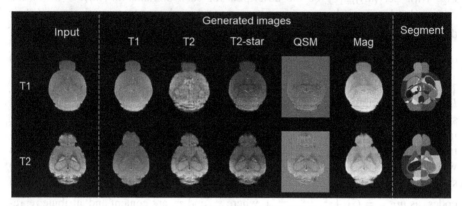

Fig. 2. Multi-modality translation and segmentation results. We show example results with only T1 or T2 images as input domain to synthesize other modalities and to generate segmentation masks.

3 Results and Discussion

3.1 Dataset Description

The mouse brain structural images were acquired in multiple modalities with the 11.7T MRI system, including T1-weighted imaging (T1), T2-weighted imaging (T2), T2-star, quantitative susceptibility mapping (QSM) and magnitude MR images (Mag), covering 75 mouse brains with 56,970 slices. All MR images were preprocessed by skull stripping and bias field correction. Those MR scans have in-plane (intra-slice) resolution about 0.07 mm × 0.07 mm and the slice thickness (inter-slice) is 0.07 mm. The ground truth of 50 brain structures was generated via an atlas-based method and then manually corrected by a brain anatomy expert.

3.2 Experiment Details

In the preprocessing process, each slice was resized to 256 × 256 matrix size and the intensity distribution is normalized into zero mean and unit variance. We use one-hot domain codes to indicate each modality attribute. For training translation module with the unpaired data, we use 80% scans as training set and 20% scans as testing set in each modality. For training segmentation model, we first freeze the parameters of the encoder obtained from the previously trained content encoder in the translation module and then train only the decoder and depth-wise convolution filter. We use categorical cross entropy as a loss function for the segmentation task and use the dice coefficient and average surface distance (ASD) as our evaluation metrics. For all training procedures, we use Adam optimizer with a learning rate of 0.0001, and set the size of the batch to 4. Finally, we fine tune the segmentation model with a learning rate of 0.00001 to update and refine all parameters including the content encoder.

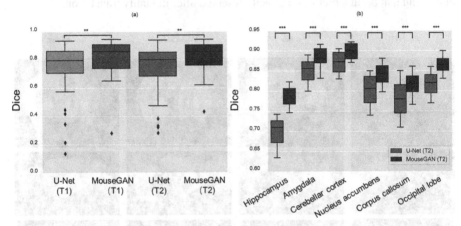

Fig. 3. MouseGAN outperforms baseline U-Net in segmentation. (a) Averaged dice coefficients of all 50 structures with different input domains (shown in brackets). (b) Performance of several critical structures using T2 as input domain (For mouse brain structural imaging, T2 is the most common modality). (** < 0.01, *** < 0.001)

Table 1. Quantitative results of different ablation settings.

Methods	Overview		Hippocampus	
	Dice	ASD	Dice	ASD
U-Net 2D (T2)	0.7476	0.1047	0.6963	0.1085
U-Net 2D (T2) + share content encoder	0.7834	0.0895	0.7631	0.0810
U-Net 2D (T2) + share content encoder + translation module = MouseGAN (T2)	**0.8273**	**0.0721**	**0.7925**	**0.0681**

3.3 Evaluation on Image Synthesis

After training both modules, our objective is to inspect whether the structural properties or biomarkers provided by one input modality are sufficient to generate other modalities. We find that this is indeed the case as demonstrated in Fig. 2 where the input domain only contains single modality (T1 or T2). We observe that our translation module can successfully capture the key divergences between T1 and T2 images: for example, corpus callosum and lateral ventricle have low contrast in T1 images, however, the synthetic T2 images have relatively high contrast views, similar to the real T2 images. On the other hand, the brain structures in synthetic images show equivalent contours and sizes as the original images, despite immense differences between their intensity distributions, suggesting that brain structures are well preserved after modality translation.

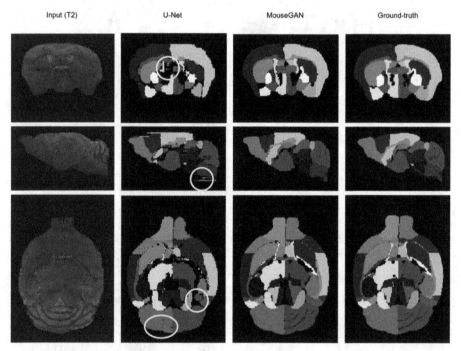

Fig. 4. Visualization of exemplary segmentation results. Top row: coronal view, middle row: sagittal view, bottom row: axial view. The circles in white color indicate some representative regions which mismatch the ground truth. By contrast, MouseGAN shows much better performance in these regions.

3.4 Evaluation on Structural Segmentation

In order to define the best combination of modality translation module and trained structural segmentation module, we tried various settings by different input domains and combinations of modality synthesis, as follows: (a) we used only T2 modality as the input MouseGAN, impute all other four modalities for the segmentation module. (b) alternatively, input only T1 modality. Additionally, we also trained a baseline U-Net (the same backbone as our segmentation module) from scratch without an auxiliary module from modality transfer, using T1 or T2 modality as input (same as MouseGAN). As shown in Fig. 3, MouseGAN outperforms baseline U-Net by a large margin, with an averaged dice of 0.82/0.83 for T1/T2, as compared to 0.75/0.75 for U-Net segmentation. Particularly, the segmentation performance of hippocampus, one of the most important brain structures that is related to Alzheimer disease, increases from 0.70 to 0.79. In fact, a boost of 5–10% was found in most key brain structures (Fig. 3(b)). The performance boost of MouseGAN as compared to baseline U-Net indicates that our translation module and content encoder have successfully learned and disentangled the latent properties that are shared between different modalities. We can efficiently use these properties for downstream tasks. Figure 4 shows a few exemplary segmentation masks of MouseGAN and U-Net, along with the ground-truth labels. As highlighted in the dashed white circles, U-Net turns to miss-segment a few fine brain structures, whereas MouseGAN, utilizing complementary information in multi-modality MRI, can generate more consistent structures.

3.5 Ablation Studies

To qualitatively analyze the contribution of each individual module of MouseGAN, we first set up a baseline model without shared content encoder weight and translation module. Then we add those two modules one by one into the baseline model. To be consistent, we always use T2 modality as the input. Table 1 summaries our obtained results. As expected, both the shared content encoder and translation module bring significant performance advancement. It is important to note that the average surface distance (ASD) is distinctly reduced, which indicates that even with limited information of a single input modality, MouseGAN can still reconstruct the essential representations for missing modalities.

4 Conclusion

In summary, we propose a novel generative adversarial network (GAN)-based network, MouseGAN, for simultaneous image synthesis and segmentation for mouse brain MRI. Based on a disentangled representation of content and style attribute, MouseGAN is able to synthesize multiple MR modalities from single ones in a structure-preserving manner and can hence to handle cases with missing modalities. Furthermore, it uses the learned modality-invariant information to improve structural segmentation. Our results demonstrate that MouseGAN can segment up to 50 mouse brain structures with a 7–10% increase in Dice coefficient, as compared to baseline U-Net segmentation. Since MouseGAN is the first atlas-free tool for the segmentation of mouse brain structures using MRI data, we will release our code and trained model[1] to promote its free usage and hence a large research community that studies mouse brain can be benefited.

---

[1] https://github.com/yu56500/MouseGAN.

Acknowledgements. This study was supported in part by Shanghai Municipal Science and Technology Major Project (No. 2018SHZDZX01), ZJLab, and Shanghai Center for Brain Science and Brain-Inspired Technology, the National Natural Science Foundation of China (81873893), Science and Technology Commission of Shanghai Municipality (20ZR1407800), the 111 Project (B18015), and the Major Research plan of the National Natural Science Foundation of China (KRF201923).

References

1. Jenkinson, M., Beckmann, C.F., Behrens, T.E.J., Woolrich, M.W., Smith, S.M.: FSL. Neuroimage **62**, 782–790 (2012). https://doi.org/10.1016/j.neuroimage.2011.09.015
2. Friston, K.J., Holmes, A.P., Worsley, K.J., Poline, J.-P., Frith, C.D., Frackowiak, R.S.J.: Statistical parametric maps in functional imaging: a general linear approach. Hum. Brain Mapp. **2**, 189–210 (1994). https://doi.org/10.1002/hbm.460020402
3. Iglesias, J.E., Sabuncu, M.R.: Multi-atlas segmentation of biomedical images: a survey. Med. Image Anal. **24**, 205–219 (2015). https://doi.org/10.1016/j.media.2015.06.012
4. Zhou, T., Ruan, S., Canu, S.: A review: deep learning for medical image segmentation using multi-modality fusion. Array **3–4**, 100004 (2019). https://doi.org/10.1016/j.array.2019.100004
5. Lee, H.-Y., Tseng, H.-Y., Huang, J.-B., Singh, M., Yang, M.-H.: Diverse image-to-image translation via disentangled representations. In: Ferrari, V., Hebert, M., Sminchisescu, C., Weiss, Y. (eds.) ECCV 2018. LNCS, vol. 11205, pp. 36–52. Springer, Cham (2018). https://doi.org/10.1007/978-3-030-01246-5_3
6. Zhu, J.-Y., Park, T., Isola, P., Efros, A.A.: Unpaired image-to-image translation using cycle-consistent adversarial networks. arXiv:1703.10593 [cs] (2020)
7. Chen, C., Dou, Q., Jin, Y., Chen, H., Qin, J., Heng, P.-A.: Robust multimodal brain tumor segmentation via feature disentanglement and gated fusion. In: Shen, D., et al. (eds.) MICCAI 2019. LNCS, vol. 11766, pp. 447–456. Springer, Cham (2019). https://doi.org/10.1007/978-3-030-32248-9_50
8. Yang, J., Dvornek, N.C., Zhang, F., Chapiro, J., Lin, M., Duncan, J.S.: Unsupervised domain adaptation via disentangled representations: application to cross-modality liver segmentation. In: Shen, D., et al. (eds.) MICCAI 2019. LNCS, vol. 11765, pp. 255–263. Springer, Cham (2019). https://doi.org/10.1007/978-3-030-32245-8_29
9. Jiang, J., Veeraraghavan, H.: Unified cross-modality feature disentangler for unsupervised multi-domain mri abdomen organs segmentation. In: Martel, A.L., et al. (eds.) MICCAI 2020. LNCS, vol. 12262, pp. 347–358. Springer, Cham (2020). https://doi.org/10.1007/978-3-030-59713-9_34

Style Curriculum Learning for Robust Medical Image Segmentation

Zhendong Liu[1,2,3], Van Manh[1,2,3], Xin Yang[1,2,3], Xiaoqiong Huang[1,2,3],
Karim Lekadir[4], Víctor Campello[4], Nishant Ravikumar[5,6],
Alejandro F. Frangi[1,5,6,7], and Dong Ni[1,2,3]([✉])

[1] National-Regional Key Technology Engineering Laboratory
for Medical Ultrasound, School of Biomedical Engineering, Health Science Center,
Shenzhen University, Shenzhen, China
nidong@szu.edu.cn
[2] Medical Ultrasound Image Computing (MUSIC) Lab, Shenzhen University,
Shenzhen, China
[3] Marshall Laboratory of Biomedical Engineering,
Shenzhen University, Shenzhen, China
[4] Artificial Intelligence in Medicine Lab (BCN-AIM), Department de Matemàtiques i
Informàtica, Universitat de Barcelona, Barcelona, Spain
[5] Centre for Computational Imaging and Simulation Technologies in Biomedicine
(CISTIB), School of Computing and School of Medicine,
University of Leeds, Leeds, UK
[6] Leeds Institute of Cardiovascular and Metabolic Medicine,
University of Leeds, Leeds, UK
[7] Medical Imaging Research Center (MIRC), KU Leuven, Leuven, Belgium

Abstract. The performance of deep segmentation models often degrades due to distribution shifts in image intensities between the training and test data sets. This is particularly pronounced in multi-centre studies involving data acquired using multi-vendor scanners, with variations in acquisition protocols. It is challenging to address this degradation because the shift is often not known *a priori* and hence difficult to model. We propose a novel framework to ensure robust segmentation in the presence of such distribution shifts. Our contribution is three-fold. First, inspired by the spirit of curriculum learning, we design a novel style curriculum to train the segmentation models using an easy-to-hard mode. A style transfer model with style fusion is employed to generate the curriculum samples. Gradually focusing on complex and adversarial style samples can significantly boost the robustness of the models. Second, instead of subjectively defining the curriculum complexity, we adopt an automated gradient manipulation method to control the hard and adversarial sample generation process. Third, we propose the Local Gradient Sign strategy to aggregate the gradient locally and stabilise training during gradient manipulation. The proposed framework can generalise to unknown distribution without using any target data. Extensive experiments on the public M&Ms Challenge dataset demonstrate that our

---

Z. Liu and V. Manh—Contribute equally to this work.

© Springer Nature Switzerland AG 2021
M. de Bruijne et al. (Eds.): MICCAI 2021, LNCS 12901, pp. 451–460, 2021.
https://doi.org/10.1007/978-3-030-87193-2_43

proposed framework can generalise deep models well to unknown distributions and achieve significant improvements in segmentation accuracy.

Keywords: Image segmentation · Style transfer · Curriculum learning

1 Introduction

Recent studies have witnessed the great success of deep models in medical image segmentation [11]. However, these deep models often suffer from a drop in performance when applied to new data distributions, different from the training data (see Fig. 1). It is expensive and practically impossible to collect a large amount of manually annotated data from each new distribution to retrain the model in clinical practice. Therefore, a general and retraining-free framework that ensures model robustness to distribution shifts is highly desired in the clinic. To date, several approaches have been proposed to address this issue.

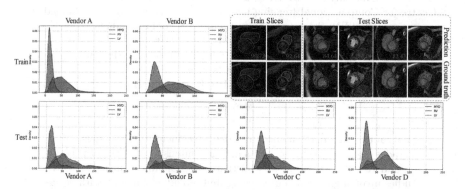

Fig. 1. Remarkable distribution shift between the training set and test set across different vendors. The results of the performance drop are shown in the upper right corner. The yellow digits denote the average DSC of the MYO (green), LV (red) and RV (blue). (Color figure online)

Data Augmentation. Data augmentation (DA) using spatial/intensity transformations is commonly employed to expand the distribution of training data and improve model generalisation. Zhang et al. [19] proposed *Mixup*, where, two inputs and their corresponding labels are proportionally interpolated to augment the training data. Similarly, Yun et al. [18] cut and paste local patches from one image into another. Although DA methods can mitigate model overfitting to some degree, they cannot guarantee the ability of deep models to generalise to multi-centre, multi-vendor data, typically encountered in real clinical scenarios.

Domain Adaptation. GAN-based domain adaptation methods often focus on learning domain-invariant representations [4,8] or aligning the feature space of different distributions [17,21] and have shown promise for dealing with variations in data distributions. Curriculum-based domain adaptation methods typically define curriculum complexity subjectively, such as using the average distance of the domain feature space [13] and developing a 'simpler' task than semantic segmentation [20]. As these methods rely on the availability of unlabelled target data, they are restricted to the task of domain-mapping/adaptation. Hence, they may not generalise well to new 'unseen' data distributions.

Distribution Generalisation. Compared to the domain adaptation setting, distribution generalisation methods tested on unseen data distributions, are better suited to address the challenges encountered in real clinical scenarios. Some studies employ style transfer methods to remove the distribution shift in test data [12,14]. Although these approaches are novel, the selection of style data is subjective. Other studies have utilised adversarial training to augment the training set [2,16]. Forcing networks to learn from the most difficult samples can enhance model robustness to a certain degree. Still, it also makes the training more difficult to optimise, even causing catastrophic overfitting [9].

This work proposes a retraining-free and general framework to improve the ability of segmentation models to generalise to unknown distributions. Our contribution is three-fold. *First*, inspired by the advantages of curriculum learning in improving model generalisation, we design a novel style curriculum, structured in an easy-to-hard adversarial learning strategy, to train the model. Gradually forcing the model to learn from progressively harder adversarial samples allows the model to generalise well to new distributions. Specifically, a style transfer model with style fusion is adopted to generate samples for the curriculum. *Second*, instead of defining the complexity order subjectively, we employ a novel gradient manipulation method to automatically control the sample generation process following the increasingly harder direction. *Third*, since the gradient manipulation-based training is difficult to optimise, we propose Local Gradient Sign (LGS), which locally aggregates the gradient to increase the complexity of the generated samples gradually, thus making the training more stable. Extensive experiments on public data from the Multi-Centre, Multi-Vendor and Multi-Disease Cardiac Segmentation (M&Ms) Challenge [3] show that our method outperforms all 2D methods, and is on par with the top performing approaches of the challenge.

2   Methodology

Our framework is shown in Fig. 2. First, we adopt a style transfer model with a style fusion operation to generate the curriculum samples z_i. Then, we employ the gradient manipulation method to update the learning weight Γ, so the curriculum samples are arranged from easy to hard. To stabilise the training, we further propose Local Gradient Sign to operate the gradient locally.

2.1 Curriculum Learning for Robustness

Deep neural networks are typically trained using a sequence of unordered samples. Curriculum learning methods [1,6,20] can guide the model to learn better by organising the training samples in a meaningful order. A common curriculum needs to address three challenges: (1) Decide the curriculum samples. (2) Arrange the samples in an easy-to-hard order. (3) Ensure model stability during training. Using curriculum learning, deep models can leverage information learned from easy examples, to ease learning of new and harder samples. Gradually concentrating on learning the harder tasks can make the deep model more robust. Creating an effective curriculum is critical to design a reasonable samples generator, a meaningful learning strategy, and a stable learning method.

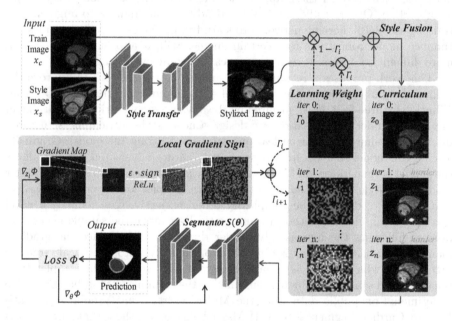

Fig. 2. Schematic view of our proposed framework.

2.2 Style Transfer Based Sample Generation

Many works [7,10] argued that the images' style could be described as the distribution of colours, edges, smoothness, etc. Since style transfer (ST) methods are powerful tools to render one image's style onto another, we employ this approach to produce large sets of stylised images. This expands the training set and increases the richness of the information available for training the segmentation model. ST is only an optional module of our system to generate initial samples.

A reasonable samples generator of the curriculum requires the ST method to balance transfer quality and efficiency. Therefore, we follow WaveCT-AIN [12] and re-implement their high-quality ST. To better control the stylised degree, we propose a style fusion operation to modulate the content of training images and the style of stylised images. The style fusion weight is updated during back-propagation to control the degree of stylisation.

2.3   Gradient Manipulation Based Learning Strategy

ST is a powerful technique that can expand style curriculum samples, but the random transformation lacks control for generating samples in a progressive, easy-to-hard fashion. Following the work from Goodfellow et al. [5], we adopt a novel gradient manipulation adversarial strategy to phase curriculum learning in increasing order of complexity. They proposed the Fast Gradient Sign Method to generate adversarial examples through the following formula.

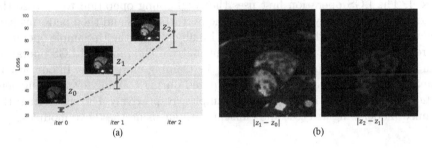

Fig. 3. (a) The loss statistics of curriculum samples at different iterations in training. (b) The absolute difference of samples z_i between iter 0 and 1, iter 1 and 2.

$$x' = x + \epsilon sign(\nabla_x J(f_\theta(x), y)) \tag{1}$$

Where θ are the parameters of the model, x is the input of the model, y is the label of x, ϵ is a small perturbation vector, x' is the generated adversarial samples, and $J(f_\theta(x), y)$ is the loss function used to train the model. The method adjusts the input data by a small step (ϵ) in the gradient direction of the cost function that maximises the loss. With this adversarial approach, the method can quickly generate more difficult samples for the deep models. Similarly, we iteratively accumulate the gradient of the cost function to produce increasingly harder adversarial samples (see step 7 in **Algorithm** 1). Moreover, in contrast with the typical gradient manipulation adversarial methods [5,9,16], which add the perturbation into the samples directly, we use it as the learning weight to modulate the training image and stylised image. This ensures control over the adversarial samples generated using our approach (see step 5 in **Algorithm** 1). To verify the effectiveness of our method, we visualise the loss of three curriculum samples iteratively generated by each training sample (Fig. 3 (a)) and the image differences of the adjacent curriculum samples (we perform pseudo-color processing in Fig. 3 (b) for better visualisation). The increase of the training loss shows that the method generates harder samples as the iterations progress.

2.4   Local Gradient Smoothing for Stability

Although the proposed gradient manipulation method guarantees the curriculum's learning order, it is not easy to optimise the model by introducing the whole adversarial perturbations. Therefore, we further propose Local-Gradient-Sign (LGS), to reduce the influence of adversarial perturbations on the model. The LGS formula is defined as follows.

$$LGS(\Phi, \epsilon, size) = US(ReLu(\epsilon * sign(AP(\nabla_{z_i}\Phi, size)))), \quad i = 0, 1, \cdots, n \quad (2)$$

Here, $US(\cdot)$ is an UpSample operation, $AP(\cdot)$ is an AvgPooling operation, z_i is curriculum sample and ϵ is a decimal from 0 to 1. In our curriculum setting, we set the max learning iter n to 3, learning step ϵ to 0.25 and the pooling size to 4×4. The LGS operation first uses the AvgPooling operation to average the gradient matrix locally for smoothing the perturbations and then performs partial truncation using the ReLu function. Finally, it uses the UpSample operation to restore the original size of the gradient matrix. The proposed LGS operation can alleviate the perturbations to make the training more stable. In our design, the ST and LGS operations are only required during training for data augmentation and attack mitigation of hard samples, respectively. During testing, the trained model is lightweight and efficient without using these two modules.

Algorithm 1. Local Style Curriculum Learning (LSCL)

Notation: Segmentation model $f(\cdot)$; Style transfer model $S(\cdot)$; max iter n; learning rate α; segmentation model parameters θ; full zeros matrix O and full ones matrix I
Input: Content data $X_c \in \mathbb{R}^{b \times h \times w}$ and label $Y_c \in \mathbb{R}^{b \times h \times w}$; style data $X_s \in \mathbb{R}^{b \times h \times w}$
Output: Optimal θ^*

1: **for** $(x_c, y_c) \in (X_c, Y_c)$, $x_s \in X_s$ **do**
2:     Initialization: $z = S(x_c, x_s)$, $\Gamma_0 = O \in \mathbb{R}^{h \times w}$
3:     **for** $i = 0, \ldots, n$ **do**
4:         Update stylised learning sample: $z_i = \Gamma_i * z + (I - \Gamma_i) * x_c$
5:         Compute loss function: $\Phi = J(f(z_i, \theta), y_c)$
6:         Accumulate learning weight: $\Gamma_{i+1} = \Gamma_i + LGS(\nabla_{z_i}\Phi)$
7:         Update model parameters: $\theta \leftarrow \theta - \alpha\nabla_\theta\Phi$
8: **return** θ as θ^*

---

3   Experimental Results

Our experimental data came from the M&Ms Challenge. This challenge cohort had 375 patients scanned in clinical centres using four different magnetic resonance scanner vendors (A, B, C and D). The training set contained 150 images from two different vendors (75 each of vendor A and B), in which only the end-diastolic (ED) and end-systolic (ES) phases are annotated. The images have been segmented by experienced clinicians from the respective institutions, including contours for the left (LV), right ventricle (RV) blood pools, and the left ventricular myocardium (MYO). All experiments are evaluated on 50 new studies from vendor A, B, C and 50 additional studies from the unseen vendor D.

We totally obtained 3,284 training slices from the annotated ED & ES phases and 10,607 style slices cross the short-axis view of vendor A and B. We first used the training slices to train the segmentation model [15], employing Adam optimiser (learning rate of 10^{-3}–10^{-5}). Based on the well-trained segmentation model and pre-trained style transfer model, we randomly sampled the slices of vendors A, B as content-style input. Then we adopted the LSCL algorithm to finetune the segmentation model for 10 epochs using SGD-momentum optimiser (learning rate of 10^{-5}). To further improve the model robustness, we finally adopted the Test-time Augmentation (TTA), including three rotation operations ($90°$, $180°$, $270°$) to aggregate the multiple predictions. We used a segmentation loss comprising cross-entropy loss (weight of 0.6) and dice loss (weight of 0.4).

3.1 Result Details

We performed the comparative experiment and ablation experiment using indicators of Dice similarity coefficient (DSC), Jaccard index (JAC), Hausdorff distance (HD, [mm]) and Average symmetric surface distance (ASSD, [mm]). In all experiments, we adopted the same evaluation criteria and ranking method as the M&Ms Challenge performs [3], including the *Min-max Score*. Due to the space limitations, we computed the averaged metrics over the LV, RV and MYO.

Table 1. The mean (std) results of DSC and HD on different vendors of all patients.

Methods	Vendor A		Vendor B		Vendor C		Vendor D		DSC score	HD score	Min-max score
	DSC ↑	HD ↓	DSC ↑	HD ↓	DSC ↑	HD ↓	DSC ↑	HD ↓			
U-net	0.858	17.113	0.835	13.392	0.822	17.616	0.835	17.153	0.8345	16.674	0.000
	(0.034)	(13.218)	(0.056)	(7.541)	(0.067)	(13.544)	(0.045)	(14.328)			
Mixup	0.872	10.441	0.859	11.483	0.837	12.596	0.837	11.904	0.8465	11.821	0.528
	(0.034)	(3.315)	(0.047)	(3.605)	(0.065)	(6.010)	(0.040)	(4.506)			
P3	0.878	12.587	0.887	9.872	0.865	10.461	0.864	13.969	0.8705	11.887	0.779
	(0.042)	(12.279)	(0.053)	(3.607)	(0.055)	(6.499)	(0.050)	(16.845)			
LSCL	0.877	9.888	0.865	10.746	0.857	11.224	0.866	11.479	0.8647	11.007	0.789
	(0.033)	(3.042)	(0.051)	(3.315)	(0.063)	(5.445)	(0.036)	(5.027)			
P2	0.884	12.461	0.892	9.796	0.870	9.548	0.867	13.432	0.8750	11.370	0.869
	(0.039)	(12.237)	(0.051)	(3.369)	(0.046)	(3.294)	(0.049)	(14.845)			
LSCL-TTA	0.884	9.802	0.876	10.238	0.865	10.592	0.868	11.195	0.8710	10.602	0.890
	(0.032)	(3.529)	(0.047)	(2.979)	(0.059)	(5.192)	(0.034)	(4.890)			
P1	0.889	12.072	0.893	9.482	0.876	9.465	0.877	13.091	0.8813	11.111	0.958
	(0.042)	(12.641)	(0.046)	(3.343)	(0.042)	(3.562)	(0.042)	(14.838)			

Method Comparison. The quantitative comparisons among our methods and others on M&Ms data are shown in Table 1. U-net is the baseline and Mixup

Table 2. Training and inference time of different methods.

Method	Training time	Inference time	GPU device
LSCL-TTA	5 h	0.2 s	GTX 1080 Ti
P1	60 h	≈1 s	Titan XP
P2	48 h	4.8 s	Tesla V100
P3	4–5 days	N/A	TITAN V100

Table 3. Ablation experiment on ED and ES volumes.

Methods	ED				ES			
	DSC ↑	JAC ↑	HD ↓	ASSD ↓	DSC ↑	JAC ↑	HD ↓	ASSD ↓
WaveCT-AIN	0.847	0.747	11.469	1.146	0.827	0.713	12.386	1.413
SCL	0.861	0.766	10.743	1.038	0.843	0.736	11.042	1.159
LSCL	0.876	0.787	10.543	0.920	0.853	0.751	11.459	1.146
LSCL-TTA	0.882	0.797	10.246	0.870	0.861	0.762	10.898	1.077

is a common data augmentation method. P3-P1 are the top three methods on the leaderboard of the M&Ms Challenge. Table 1 shows our method helps the U-net significantly improve the performance by 3.65% and 6.072 mm in terms of the DSC and HD metrics. Smaller standard deviations show that our method is more robust and stable against distribution shifts. Table 1 and Table 2 show that

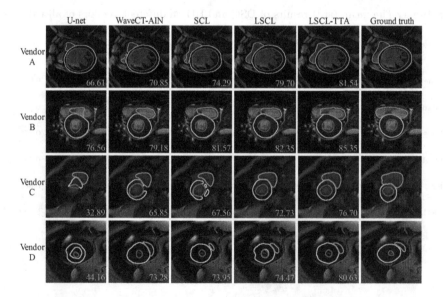

Fig. 4. Visualisation of ablation methods at different vendors.

our proposed method (LSCL-TTA) achieves competitive performance among the top ranking methods, while presents the lowest computational costs.

Ablation Experiment. To thoroughly evaluate our proposed framework, we conducted two ablation experiments. The first experiment uses only the style transfer method *WaveCT-AIN* [12] to generate random style transformations and then feeds the stylised outputs to train the segmentation model. The second experiment does not introduce the LGS method, which we name SCL. As shown in Table 3, the WaveCT-AIN shows the worst performance since the model learns the stylised samples arbitrarily, causing the model to be trapped in local minima. SCL, instead, can make the model learn increasingly harder tasks. We further proposed LSCL to stabilise the adversarial training. Figure 4 visualises the segmentation results of test samples corresponding to different ablation methods. The proposed LSCL-TTA shows the best segmentation results on each vendor.

4 Conclusion

This paper proposes a novel style curriculum learning framework to ensure segmentation models are robust against the distribution shift. Extensive experiments show that our proposed framework can significantly improve the generalisation. The proposed framework is universal and retraining-free, which makes it compelling for use in clinical practice.

Acknowledgement. This work was supported by the National Key R&D Program of China (No. 2019YFC0118300), Shenzhen Peacock Plan (No. KQTD2016053112051497, KQJSCX20180328095606003), Royal Academy of Engineering under the RAEng Chair in Emerging Technologies (CiET1919/19) scheme, EPSRC TUSCA (EP/V04799X/1), the Royal Society CROSSLINK Exchange Programme (IES/NSFC/201380), European Union's Horizon 2020 research and innovation program under grant agreement number 825903 (euCanSHare project), Spanish Ministry of Science, Innovation and Universities under grant agreement RTI2018-099898-B-I00.

References

1. Bengio, Y., Louradour, J., Collobert, R., Weston, J.: Curriculum learning. In: Proceedings of the 26th Annual International Conference on Machine Learning, pp. 41–48 (2009)
2. Cai, Q.Z., Liu, C., Song, D.: Curriculum adversarial training. In: Proceedings of the 27th International Joint Conference on Artificial Intelligence, pp. 3740–3747 (2018)
3. Campello, V.M., et al.: Multi-centre, multi-vendor and multi-disease cardiac segmentation: The m&ms challenge. IEEE Trans. Med. Imaging (2021)
4. Chen, C., et al.: Unsupervised multi-modal style transfer for cardiac MR segmentation. In: Pop, M., et al. (eds.) STACOM 2019. LNCS, vol. 12009, pp. 209–219. Springer, Cham (2020). https://doi.org/10.1007/978-3-030-39074-7_22
5. Goodfellow, I.J., Shlens, J., Szegedy, C.: Explaining and harnessing adversarial examples. arXiv preprint arXiv:1412.6572 (2014)

6. Hacohen, G., Weinshall, D.: On the power of curriculum learning in training deep networks. In: International Conference on Machine Learning, pp. 2535–2544. PMLR (2019)

7. Huang, E., Gupta, S.: Style is a distribution of features. arXiv preprint arXiv:2007.13010 (2020)

8. Huang, X., Liu, M.-Y., Belongie, S., Kautz, J.: Multimodal unsupervised image-to-image translation. In: Ferrari, V., Hebert, M., Sminchisescu, C., Weiss, Y. (eds.) ECCV 2018. LNCS, vol. 11207, pp. 179–196. Springer, Cham (2018). https://doi.org/10.1007/978-3-030-01219-9_11

9. Li, B., Wang, S., Jana, S., Carin, L.: Towards understanding fast adversarial training. arXiv preprint arXiv:2006.03089 (2020)

10. Li, Y., Wang, N., Liu, J., Hou, X.: Demystifying neural style transfer. arXiv preprint arXiv:1701.01036 (2017)

11. Liu, S., et al.: Deep learning in medical ultrasound analysis: a review. Engineering 5(2), 261–275 (2019)

12. Liu, Z., et al.: Remove appearance shift for ultrasound image segmentation via fast and universal style transfer. In: 2020 IEEE 17th International Symposium on Biomedical Imaging (ISBI), pp. 1824–1828. IEEE (2020)

13. Liu, Z., et al.: Open compound domain adaptation. In: Proceedings of the IEEE/CVF Conference on Computer Vision and Pattern Recognition, pp. 12406–12415 (2020)

14. Ma, C., Ji, Z., Gao, M.: Neural style transfer improves 3D cardiovascular MR image segmentation on inconsistent data. In: Shen, D., et al. (eds.) MICCAI 2019. LNCS, vol. 11765, pp. 128–136. Springer, Cham (2019). https://doi.org/10.1007/978-3-030-32245-8_15

15. Ronneberger, O., Fischer, P., Brox, T.: U-Net: convolutional networks for biomedical image segmentation. In: Navab, N., Hornegger, J., Wells, W.M., Frangi, A.F. (eds.) MICCAI 2015. LNCS, vol. 9351, pp. 234–241. Springer, Cham (2015). https://doi.org/10.1007/978-3-319-24574-4_28

16. Volpi, R., Namkoong, H., Sener, O., Duchi, J., Murino, V., Savarese, S.: Generalizing to unseen domains via adversarial data augmentation. arXiv preprint arXiv:1805.12018 (2018)

17. Yan, W., et al.: The domain shift problem of medical image segmentation and vendor-adaptation by Unet-GAN. In: Shen, D., et al. (eds.) MICCAI 2019. LNCS, vol. 11765, pp. 623–631. Springer, Cham (2019). https://doi.org/10.1007/978-3-030-32245-8_69

18. Yun, S., Han, D., Oh, S.J., Chun, S., Choe, J., Yoo, Y.: CutMix: regularization strategy to train strong classifiers with localizable features. In: Proceedings of the IEEE/CVF International Conference on Computer Vision, pp. 6023–6032 (2019)

19. Zhang, H., Cisse, M., Dauphin, Y.N., Lopez-Paz, D.: mixup: beyond empirical risk minimization. arXiv preprint arXiv:1710.09412 (2017)

20. Zhang, Y., David, P., Gong, B.: Curriculum domain adaptation for semantic segmentation of urban scenes. In: Proceedings of the IEEE International Conference on Computer Vision, pp. 2020–2030 (2017)

21. Zhang, Z., Yang, L., Zheng, Y.: Translating and segmenting multimodal medical volumes with cycle-and shape-consistency generative adversarial network. In: Proceedings of the IEEE Conference On Computer Vision and Pattern Recognition, pp. 9242–9251 (2018)

Towards Efficient Human-Machine Collaboration: Real-Time Correction Effort Prediction for Ultrasound Data Acquisition

Yukun Ding[1], Dewen Zeng[1], Mingqi Li[2], Hongwen Fei[2], Haiyun Yuan[2],
Meiping Huang[2], Jian Zhuang[2], and Yiyu Shi[1(✉)]

[1] University of Notre Dame, Notre Dame, USA
yshi4@nd.edu
[2] Guangdong Provincial People's Hospital, Guangzhou, China

Abstract. One of the foremost challenges of using Deep Neural Network-based methods for a fully automated segmentation in clinics is the lack of performance guarantee. In the foreseeable future, a feasible and promising way is that radiologists sign off the machine's segmentation results and make corrections if needed. As a result, the human effort for image segmentation that we try to minimize will be dominated by segmentation correction. While such effort can be reduced by the advance of segmentation models, for ultrasound a novel direction can be explored: optimizing the data acquisition. We observe a substantial variation of segmentation quality among repetitive scans of the same subject even if they all have high visual quality. Based on this observation, we propose a framework to help sonographers obtain ultrasound videos that not only meet the existing quality standard but also result in better segmentation results. The promising result demonstrates the feasibility of optimizing the data acquisition for efficient human-machine collaboration.

1 Introduction

Deep Neural Network (DNNs) are largely black-box models and have worse generalizability than human experts, which makes it challenging to deploy them for fully-automated medical image segmentation in clinics. Therefore, human-machine collaboration becomes a promising way to harness the advantages of both, while addressing their respective limitations [6]. For segmentation, a quality assurance process is applied where radiologists inspect the machine's segmentation results and make corrections if needed. Because most segmentations do not need correction, we can enjoy the machine segmentation while having satisfactory accuracy and trustworthiness from human verification. Because the

Electronic supplementary material The online version of this chapter (https://doi.org/10.1007/978-3-030-87193-2_44) contains supplementary material, which is available to authorized users.

M. de Bruijne et al. (Eds.): MICCAI 2021, LNCS 12901, pp. 461–470, 2021.
https://doi.org/10.1007/978-3-030-87193-2_44

segmentation after verification and correction has satisfactory quality, the ultimate goal is to reduce the human effort. Although the verification effort is irreducible (every image needs to be verified), the correction effort can be reduced by improving the segmentation quality.

While developing better segmentation models is the most straightforward way to reduce correction effort, we approach the goal from a novel perspective: selecting better input data for a given segmentation model. Compared with other imaging modalities, ultrasound has a less automated image acquisition routine and the data quality highly depends on the skills and objectivity of operators [1]. The variation among patients and scanners makes it even more challenging to obtain high-quality images [2]. Many approaches have been proposed to provide real-time feedback about image quality [1,15,18]. If the system determines that the current image does not meet a quality standard, the operator is prompted to repeat predefined protocols [13]. Despite various methods, all of them only consider the visual quality for humans to get desired information. However, there is a difference between human vision and the machine vision supported by DNNs in terms of "image quality" [11]. Images that are considered high-quality by radiologists do not necessarily result in high-quality segmentation because of the limited performance of DNNs.

We collect a Myocardial Contrast Echocardiography (MCE) dataset with repetitive scans for the same subject that all have satisfactory visual quality for diagnosis and segment the myocardial by a state-of-the-art model (more details in Sect. 3). The percentage of frames that require correction of the segmentation is denoted as *failure rate*, and reported in Fig. 1. For most subjects, there is a gap between the maximum and minimum failure rates among the multiple scan of the same subject which indicates a large variation. Meanwhile, a retake of ultrasound scan is relatively easy but correction of failed segmentations is much more time-consuming. This leads to our key motivation: *the data acquisition process can be exploited to provide preferred input for DNN-based segmentation models that results in better segmentation and lower manual correction effort.*

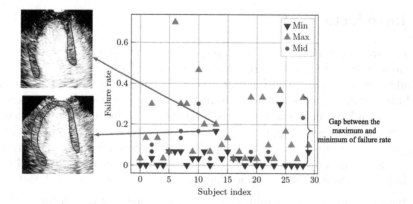

Fig. 1. Segmentation failure rate. The images on the left are an example showing a frame with segmentation failed in one scan but not in another scan which causes the difference in failure rate of two scans of the same subject.

Traditionally, sonographers acquire data that are high-quality for human vision. We propose a framework that enables sonographers to acquire data that not only meets all existing quality standards but also produces better segmentation results and consequently lower correction effort. The key component is a real-time correction effort prediction model. For a given segmentation model, it informs the sonographer of the predicted correction effort for the current data. Once a high correction effort is predicted which means a segmentation failure, sonographers can do a re-scan to replace the unsatisfactory data. In this way, the failed segmentation can be effectively avoided as early as the data acquisition stage. It is tempting to focus on conventional quality metrics such as Dice Similarity Coefficient (DSC) as did in segmentation quality prediction [4,9,10,14,19] because higher segmentation quality would lead to lower correction effort. However, we show that these quality metrics correlate poorly with the correction effort estimated by radiologists. In addition, objectives of existing works are around improving reliability by detecting potential failures for further inspection while we try to actively avoid bad segmentation from the beginning. We further provide a general cost model that can estimate benefits of the proposed approach without extensive experiments. The substantial effort reduction observed in the simulated experiments demonstrates that optimizing the data acquisition for not only human vision but also the machine vision, and directly predicting human effort instead of conventional quality metrics, have great potential for more efficient human-machine collaboration in medical image computing. The dataset we collected will be released to public to help future research in this area.

2   Method

2.1   Framework

The overall framework is shown in Fig. 2. The scan consists of the normal clinical routine of data acquisition including quality assurance to ensure high quality for human diagnosis. After segmentation, the real-time correction effort prediction model predicts the effort needed for manual correction by radiologists. A real-time effort prediction model is needed because a frame-by-frame check is time-consuming and sonographers may not have enough expertise. If the predicted effort is lower than the predefined threshold, the exam is finished. Otherwise, the sonographer will be prompted to do a re-scan (a max number of re-scans can be set to avoid too many trials). Because the prediction model helps to replace the data that leads to segmentation failure, the effort of manual correction can be reduced. The overhead will be a slightly increased number of scans, which we will show later is minimal compared with the cost of segmentation correction.

2.2   Prediction Target

In this work, we consider human effort as the time needed to correct the segmentation results, but the framework proposed allows more customized modeling of

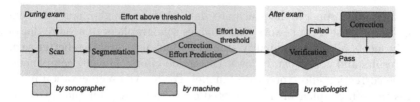

Fig. 2. Framework overview.

effort. Although we are trying to minimize the time, directly predicting the exact time is problematic because it has significant variation among radiologists with different levels of experience or using different annotation softwares/tools which makes it hard to predict and generalize.

It is tempting to use conventional quality metrics such as DSC as the indicator of correction effort since they seem to be more stable than the correction time. These metrics are based on the deviation from a ground truth annotation. However, due to the intra- and interobserver variability, deviation from a given expert's annotation that leads to a lower quality score does not necessarily mean a correction is required [3,5,12]. Moreover, some deviations are naturally more easy to be corrected than others. Examples are shown in Fig. 3(a). In the first row, the obvious error that needs to be corrected does not lead to a low DSC. In the second row, the lower DSC is caused by the inherent ambiguity in boundary while the segmentation is clinically acceptable. The first row also gives a comparison of different types of deviations.

For a quantitative evaluation, we have an experienced cardiologist to inspect the segmentation results of 3000 frames in the MCE dataset which will be detailed in Sect. 3. The cardiologist is asked to rate the effort needed to correct the results in three classes: "no effort", "low effort", and "high effort". The probability density histograms of DSC are shown in Fig. 3. It is clear that DSC

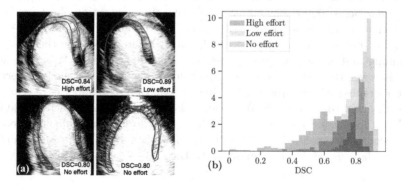

Fig. 3. (a) Disagreement between DSC and correction effort. Labels and segmentations are in red and blue respectively. (b) Probability density of DSC.

does not reliably indicate the cardiologist's rating. Therefore, we propose to use this categorical effort label as the prediction target to be accurate and robust.

2.3   Prediction Network

We use ResNet-18 [8] as the backbone of the effort prediction network. We concatenate the input image and the predicted segmentation mask as a 4-channel image. Motivated by the use of uncertainty map in the literature [4, 7], we also add the pixel-wise softmax uncertainty map to the input as the fifth channel. Then we add an attention module inspired by Residual Attention Network [17] to our effort prediction model. The motivation is that, for the effort prediction task, the model can attend to selected areas where segmentation errors might occur. The network structure is shown in Fig. 4. We add the attention module at the first convolution layer and the second residual unit. The attention module uses an encoder-decoder structure to increase the reception field. See [17] for the details of this structure. The model takes 10 ms and 4 ms per-frame on a GTX 1080Ti GPU using PyTorch with and without attention module respectively. This makes it feasible to provide real-time feedback.

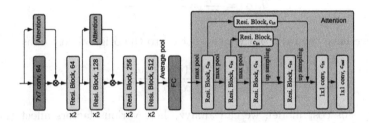

Fig. 4. Left: overall network structure. Right: the attention module. Resi. Block is the standard 2-layer block used in ResNet. c_{in} and c_{out} denote the numbers of channels of the input and the output respectively.

2.4   Cost Model

We analyze the benefits and overhead of the proposed framework via a cost model, so that people can quickly estimate the cost reduction in their applications. Because the cost for the initial scan and verification are not affected by the proposed method, we focus on the correction cost and the re-scan cost below.

We consider the segmentation on the video obtained in a scan fails if any correction is needed, and denote α as the failure probability. α can be estimated by looking at the segmentation results of interest. However, a fixed α may be inaccurate in practice because each subject may have a unique condition that makes it easier or more difficult to obtain images that can be segmented well. Therefore, we consider each subject has a unique α and repetitive scans for the same subject are independent. Denote the probability density function for the distribution of α as $f(\alpha)$.

Denote the average cost of manually correcting a failed segmentation as c_c and the average cost of a re-scan as c_s. Denoting C_α as the expected cost with a given α, we have $C_\alpha = \alpha c_c$. The original overall cost is $C = c_c \int \alpha f(\alpha) d\alpha$.

We use a prediction model to identify failed segmentations. Denote its precision and recall as p and r respectively. Then C'_α, the expected cost at given α using the proposed method can be written as below. The first term is for predicted passed segmentation and the second term is for predicted failed segmentation.

$$C'_\alpha = \alpha(1-r)c_c + \frac{\alpha r}{p}(c_s + C'_\alpha). \tag{1}$$

Solving Eq. (1) we get

$$C'_\alpha = \frac{p\alpha c_c - p\alpha r c_c + \alpha r c_s}{p - \alpha r} \tag{2}$$

Comparing C_α and C'_α, we have $C'_\alpha < C_\alpha$ if and only if $p > \alpha + \frac{c_s}{c_c}$. This indicates a lower bound of the precision to be able to reduce the cost. We further define $h(\alpha) = \frac{C'_\alpha}{C_\alpha}$ and then $1 - h(\alpha)$ is the cost reduction rate:

$$h(\alpha) = \frac{C'_\alpha}{C_\alpha} = \frac{p - pr + r\frac{c_s}{c_c}}{p - \alpha r} \tag{3}$$

Finally we have the ratio of the new cost C' to the original cost C as

$$\frac{C'}{C} = \frac{\int \alpha f(\alpha) h(\alpha) d\alpha}{\int \alpha f(\alpha) d\alpha} \tag{4}$$

Using this cost model, we can analyze how various factors affect the cost-saving of the proposed framework. $f(\alpha)$ accounts for the property of the data which should be estimated for specific applications. Higher α leads to higher $\frac{C'}{C}$ because it takes more scans to obtain good segmentation. $h(\alpha)$ is determined by $\frac{c_s}{c_c}$, p, and r. $\frac{c_s}{c_c}$ is the ratio of the re-scan cost to the correction cost which can be easily estimated by sonographers and radiologists. Smaller $\frac{c_s}{c_c}$ means smaller $\frac{C'}{C}$ and thus more cost reduction. An accurate prediction model is crucial for the cost reduction as reflected by p and r. Before the experimental validation in Sect. 3, the cost model shows a substantial possible cost reduction. Numerical examples of the cost reduction is given in the supplementary.

3   Experiments and Results

As a proof of concept, we use a Myocardial Contrast Echocardiography (MCE) dataset we collect in house. To facilitate research in this area, we release the dataset in https://github.com/dewenzeng/effort_prediction_mce. Each scan is a sequence of 30 frames and the myocardial is segmented for perfusion analysis. The first part of the dataset includes one scan for each of the 100 subjects. We train a state-of-the-art video segmentation model RVOS [16] to segment the myocardial. We obtained segmentation results and effort labels from radiologists for 60 subjects and use them to train the effort prediction model. Repetitive scans are not needed here as we only train the model to predict the frame-wise correction effort. For both the segmentation model and our effort prediction model, the first 60 subjects are used for training, while the later 40 subjects are used for validation. Then we use the trained models to test the whole proposed framework on the test data. The test data includes additional 30 subjects. Each subject has 2–6 repetitive scans of satisfactory visual quality as determined by sonographers. Check the supplementary for more details of the experiment setting. In all figures below, the shaded area indicates the 90% confidence interval calculated from 5 repetitive experiments.

3.1   Per-Frame Prediction

We first validate our effort prediction network for per-frame correction effort prediction. For the simplicity of comparison, we merge the "low effort" and "high effort" into one class and train the model to do a binary classification of whether a correction is needed. The results are shown in Fig. 5. When comparing different methods, the method that aggregates pixel-level uncertainty adopted from [10] has much worse performance than DNN-based prediction models. We can see that training with effort label achieves significantly higher accuracy than training with DSC. In both cases, the attention module improves performance. Because using attention increases the number of parameters from 11.2M to 11.7M, we compare it with two baselines that expand the backbone network to approximately the same number of parameters. Baseline A widens all layers with a fixed widen ratio. Baseline B widens the layers aligned with the attention module. The results show the effectiveness of attention modules.

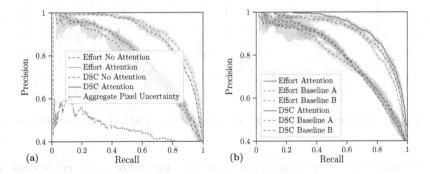

Fig. 5. (a) Comparison of methods. (b) Effect of attention modules.

3.2   Simulated Experiments

We validate the proposed framework via a simulated experiment because the data was acquired in advance. Following the workflow in Fig. 2, for each subject, we take the next pre-acquired scan available in our dataset when a scan is conducted. If no more data available in the dataset, the scan with the lowest predicted effort is selected. To simulate the unknown order of multiple scans for a subject, for each subject we average the results from all possible permutations of available scans in the dataset. More details are included in the supplementary.

The results of the simulated experiment are shown in Fig. 6(a). The x-axis is the average number of scans for all 30 subjects. The y-axis is the percentage of frames that require correction in the data acquired. The curve is generated by enumerating different thresholds on the predicted effort which triggers a re-scan. When no re-scan is conducted, which is the baseline, the average number of scans is 1. As the threshold decreases, the average number of scans increases. As indicated by the blue curve, we are able to reduce the failure rate from 10.3% to 5.5% while the number of scans is only increased to 1.4. Note that the lowest possible failure rate on the test data is 3.8% because some subjects have failed segmentations in all available scans.

One concern might be the segmentation quality on the selected image sequences. We validate this by looking at the percentage of frames that need "high effort" correction in frames that need correction. As shown in Fig. 6(b), such a percentage decreases which means a decreased average correction effort for failed segmentatio. This means the actual effort reduction will be more than the failure rate reduction. Similar to the observations in the per-frame prediction, using DSC as the prediction target leads to larger variation and worse performance.

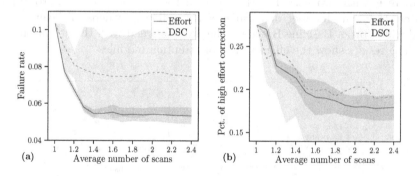

Fig. 6. (a) Segmentation failure rate. (b) Percentage of high effort correction.

We additionally try using 3-class effort label instead of 2-class effort label for effort prediction and get even higher performance. Specifically, when the average number of scans is 1.6, the failure rate is reduced to 5.0%. This means a 51% reduction in the number of images that need correction, and the percentage of "high effort" correction decreases from 27% to 20%. More details are given in the supplementary. While using 3-class labels improves upon using 2-class labels, a more fine-grained label means more ambiguity and variability in the labeling by radiologists which may hurt the generalization of the prediction model.

Meanwhile, the marginal benefits of effort reduction decrease. Therefore, 2-class or 3-class effort labels could be sufficient in practice.

Although the time measurement is less generalizable than the failure rate due to the variance in sonographers, patients, machines, radiologists etc., we measure time in our case for completeness. For the MCE and segmentation for perfusion analysis, each re-scan takes 14 s on average by an experienced sonographer including probe adjustment and recording. The average per-frame correction time by an experienced cardiologist is measured to be 19 s and 45 s for low effort correction and high effort correction respectively. When using 3-class labels and the average number of scans is 1.6, the average correction time per-subject is reduced from 78 s to 36 s. The average scan time is increased by 8.4 s which is only a small portion of the reduction on correction time.

4 Conclusion

The proposed framework provides real-time correction effort prediction as assistance to sonographers to avoid segmentation failure during data acquisition. The method is orthogonal to the development of segmentation models and could be extended to other applications. The promising results demonstrate the feasibility and benefits of optimizing the data acquisition for not only human vision but also machine vision, and directly predicting human effort instead of conventional quality metrics.

References

1. Abdi, A.H., et al.: Automatic quality assessment of echocardiograms using convolutional neural networks: feasibility on the apical four-chamber view. IEEE Trans. Med. Imaging **36**(6), 1221–1230 (2017)
2. Akkus, Z., et al.: A survey of deep-learning applications in ultrasound: artificial intelligence-powered ultrasound for improving clinical workflow. J. Am. Coll. Radiol. **16**(9), 1318–1328 (2019)
3. Bridge, P., Fielding, A., Rowntree, P., Pullar, A.: Intraobserver variability: should we worry? J. Med. imaging Radiat. Sci. **47**(3), 217–220 (2016)
4. DeVries, T., Taylor, G.W.: Leveraging uncertainty estimates for predicting segmentation quality. arXiv preprint arXiv:1807.00502 (2018)
5. Zeng, D., et al.: Segmentation with multiple acceptable annotations: a case study of myocardial segmentation in contrast echocardiography. In: Feragen, A., Sommer, S., Schnabel, J., Nielsen, M. (eds.) IPMI 2021. LNCS, vol. 12729, pp. 478–491. Springer, Cham (2021). https://doi.org/10.1007/978-3-030-78191-0_37
6. Ding, Y., et al.: Hardware design and the competency awareness of a neural network. Nat. Electron. **3**(9), 514–523 (2020)
7. Ding, Y., et al.: Uncertainty-aware training of neural networks for selective medical image segmentation. In: Medical Imaging with Deep Learning, pp. 156–173. PMLR (2020)
8. He, K., Zhang, X., Ren, S., Sun, J.: Deep residual learning for image recognition. In: CVPR, pp. 770–778 (2016)

9. Hoebel, K., et al.: An exploration of uncertainty information for segmentation quality assessment. In: Medical Imaging 2020: Image Processing, vol. 11313, p. 113131K. International Society for Optics and Photonics (2020)

10. Jungo, A., Meier, R., Ermis, E., Herrmann, E., Reyes, M.: Uncertainty-driven sanity check: application to postoperative brain tumor cavity segmentation. arXiv preprint arXiv:1806.03106 (2018)

11. Liu, Z., et al.: Machine vision guided 3d medical image compression for efficient transmission and accurate segmentation in the clouds. In: Proceedings of the IEEE/CVF Conference on Computer Vision and Pattern Recognition, pp. 12687–12696 (2019)

12. McErlean, A., et al.: Intra-and interobserver variability in CT measurements in oncology. Radiology **269**(2), 451–459 (2013)

13. Østvik, A., Smistad, E., Aase, S.A., Haugen, B.O., Lovstakken, L.: Real-time standard view classification in transthoracic echocardiography using convolutional neural networks. Ultrasound Med. Biol. **45**(2), 374–384 (2019)

14. Robinson, R., et al.: Real-time prediction of segmentation quality. In: Frangi, A.F., Schnabel, J.A., Davatzikos, C., Alberola-López, C., Fichtinger, G. (eds.) MICCAI 2018. LNCS, vol. 11073, pp. 578–585. Springer, Cham (2018). https://doi.org/10.1007/978-3-030-00937-3_66

15. Snare, S.R., Torp, H., Orderud, F., Haugen, B.O.: Real-time scan assistant for echocardiography. IEEE Trans. Ultrason. Ferroelectr. Freq. Control **59**(3), 583–589 (2012)

16. Ventura, C., Bellver, M., Girbau, A., Salvador, A., Marques, F., Giro-i Nieto, X.: RVOS: end-to-end recurrent network for video object segmentation. In: Proceedings of the IEEE/CVF Conference on Computer Vision and Pattern Recognition, pp. 5277–5286 (2019)

17. Wang, F., et al.: Residual attention network for image classification. In: Proceedings of the IEEE Conference on Computer Vision and Pattern Recognition, pp. 3156–3164 (2017)

18. Wu, L., Cheng, J.Z., Li, S., Lei, B., Wang, T., Ni, D.: FUIQA: fetal ultrasound image quality assessment with deep convolutional networks. IEEE Trans. Cybern. **47**(5), 1336–1349 (2017)

19. Zhang, R., Chung, A.C.S.: A fine-grain error map prediction and segmentation quality assessment framework for whole-heart segmentation. In: Shen, D., et al. (eds.) MICCAI 2019. LNCS, vol. 11765, pp. 550–558. Springer, Cham (2019). https://doi.org/10.1007/978-3-030-32245-8_61

Residual Feedback Network for Breast Lesion Segmentation in Ultrasound Image

Ke Wang[1,2], Shujun Liang[1,2], and Yu Zhang[1,2(✉)]

[1] School of Biomedical Engineering,
Southern Medical University,
Guangzhou 510515, Guangdong, China
yuzhang@smu.edu.cn
[2] Guangdong Provincial Key Laboratory of Medical Image Processing, Southern
Medical University, Guangzhou 510515, Guangdong, China

Abstract. Accurate lesion segmentation in breast ultrasound (BUS) images is of great significance for the clinical diagnosis and treatment of breast cancer. However, precise segmentation on missing/ambiguous boundaries or confusing regions remains challenging. In this paper, we proposed a novel residual feedback network, which enhances the confidence of the inconclusive pixels to boost breast lesion segmentation performance. In the proposed network, a residual representation module is introduced to learn the residual representation of missing/ambiguous boundaries and confusing regions, which promotes the network to make more efforts on those hardly-predicted pixels. Moreover, a residual feedback transmission strategy is designed to update the input of the encoder blocks by combining the residual representation with original features. This strategy could enhance the regions including hardly-predicted pixels, which makes the network can further correct the errors in initial segmentation results. Experimental results on three datasets (3813 images in total) demonstrate that our proposed network outperforms the state-of-the-art segmentation methods. Our code is available at https://github.com/mniwk/RF-Net.

Keywords: Ultrasound image · Breast lesion segmentation · Residual feedback network

1 Introduction

Breast cancer is the leading cancer in women, which disrupts the lives of millions of women in the world [1]. Early screening of breast cancer can reduce mortality by up to 20% [2]. Ultrasound imaging is one of the popularly screening methods

Electronic supplementary material The online version of this chapter (https://doi.org/10.1007/978-3-030-87193-2_45) contains supplementary material, which is available to authorized users.

to detect breast lesions due to its advantages of non-radiation, real-time visualization, cost-effectiveness and non-invasive diagnosis [3,4]. Segmenting breast lesion is a crucial step in computer-aided diagnosis system, which can provide interpretive information for different functional tissues discrimination and breast cancer diagnosis [4,5]. However, automatic precise lesion segmentation in BUS images is a challenging task due to (1) missing/ambiguous boundaries caused by speckle noise and the similar visual appearance between lesions and non-lesion background; (2) large variety of breast lesions, which come in various shapes, sizes and locations; (3) significant individual differ-ences in breast structures between patients [5–7].

Automated segmentation of breast lesions in BUS images has been studied for many years. The existing methods can be roughly divided into two categories: traditional methods and deep learning methods [8,9]. Manual cropping tumor-centered regions of interest (ROIs) and complex preprocessing and/or postprocess steps are commonly required in traditional methods [10,11]. Luo Y et al. [10] performed the interaction to firstly determine the ROI and combined region- and edge-based information to overcome the problem of over-segmentation and under-segmentation in BUS images. Huang Q et al. [11] also cropped the ROIs, utilized simple linear iterative clustering and k-nearest neighbor algorithms to obtain the final segmentation. Deep-learning-based methods have triumphed in segmentation tasks [12]. For BUS images, Han L et al. [13] proposed a semi-supervised segmentation method using a dual-attention-fusion block to extract features of the lesion regions and non-lesion regions separately, which enhanced the discriminative ability to discriminate lesions. Vakanski A et al. [14] introduced attention blocks into a U-Net architecture and learned feature representations that prioritize spatial regions with high saliency levels. Zhu L et al. [15] proposed a novel second-order sub-region pooling network (S^2P-Net) for boosting the breast lesion segmentation in ultrasound images, which aggregated the multi-context information from the whole image and multiple sub-regions. Despite promising segmentation results of existing methods, precise segmentation in the confusing regions and unclear boundaries of the breast BUS images segmentation is still challenging.

In this paper, we propose a novel residual feedback network based on encoder-decoder architecture to boost breast lesion segmentation by learning residual representation of hardly-predicted pixels, and feeding it into encoder blocks, to enhance the confidence of the hardly-predicted pixels. The contributions of our work can be summarized as follows: (1) A supervised residual representation module is designed to learn the residual representations of missing/ambiguous boundaries and confusing regions. (2) A residual feedback transmission strategy is designed to update the input of encoder blocks by feeding residual representations back and combining it with original features. This strategy could enhance the weight of regions including hardly-predicted pixels, which makes the network can further correct the errors in initial segmentation results. (3) Experimental results on three public datasets (3813 images in total) demonstrate that our proposed network outperforms the state-of-the-art segmentation methods.

2   Method

2.1   Overview

Figure 1 illustrates the architecture of the proposed residual feedback network for lesion segmentation in BUS images. In the proposed network, an encoder-decoder architecture is used as the baseline. The proposed network includes two steps: (1) in the first step, we perform an encoder-decoder architecture to generate initial segmentation results under the input of BUS images. And then, a residual representation module is designed to take the features of decoder blocks as inputs and learn the residual information about low-confident and error-predicted pixels, which is supervised by the residual masks (the difference between the ground truths and the initial segmentation results). (2) In the second step, a residual feedback transmission strategy provides an opportunity to correct the errors by utilizing the residual representation generated during the first step, and then the preferable residual-guided segmentation results can be obtained by reusing the encoder-decoder baseline. It worth to note that the proposed network is trained in an end-to-end manner without introducing any extra-parameter.

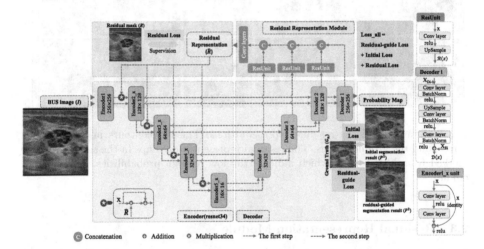

Fig. 1. The overview of our proposed network for lesion segmentation in BUS images. ResNet-34 is adopted as the Encoder path with pre-trained parameters. Our network includes two steps. The first step (gray dotted arrows) is used to generate the initial segmentation results and learn the residual representation of missing/ambiguous boundaries and confusing regions. The second step is used to (red dotted arrows) feed the residual representation into encoder path and generate more precise segmentation results. (Color figure online)

2.2 Encoder-Decoder Baseline

In this study, we modify the ResNet-34 as the encoder path by retaining the first five blocks, to extract detailed context features of BUS images. We also apply the skip connection in U-Net architecture to transport the detailed information from encoder to decoder, aiming to remedy the information loss owing to consecutive pooling and striding convolutional operations [18]. As shown in Fig. 1, each encoder block contains x^i Encoderi_x units (where $x^i \in \{3, 4, 6, 3\}$ and $i \in \{2, 3, 4, 5\}$). Each Encoderi_x unit consists of two 3×3 convolutional layers, except Encoder1 unit which contains one 7×7 convolutional layer. The decoder path comprises four convolutional blocks. As illustrated in Fig. 1, each decoder block contains an element-wise addition operation, a 1×1 convolution layer, an upsampling layer, a 3×3 convolution layer, and a 1×1 convolution layer, consecutively. And the upsampling layer is used to ensure the same scale with corresponding encoder blocks by linear interpolation.

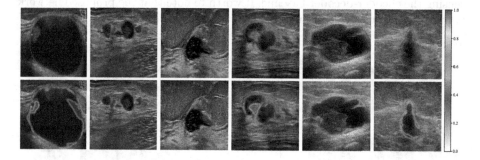

Fig. 2. Examples of the residual representations. The red contours in the first row correspond the ground truths. The residual representation maps in the second row are marked from blue to red, which indicate the low to high probabilities. (Color figure online)

2.3 Residual Representation Module

In general, the predicted probabilities of missing/ambiguous boundaries or confusing regions are low in segmented map, but salient in the residual image (the difference between the ground truths and the segmented results, see Fig. 2). The goal of the proposed residual representation module is to learn the residual representations of missing/ambiguous boundaries and confusing regions, and encourage the network to explore more information about hardly-predicted pixels.

In the first step, we obtain the initial segmentation results P^1 with the input of BUS images I. Then, we introduce the residual representation module, which integrates the features of multiple decoder blocks to learn the residual representations ($\hat{R}$) under the supervision of the residual masks (R, the difference between the ground truths G and the segmented results P^1), and generate confident representation maps for the inconclusive pixels.

The architecture of the proposed residual representation module is shown in Fig. 1, and the residual units are used to extract the context information from decoder blocks. Each residual unit (ResUnit) contains a 3×3 convolutional layer and an upsampling layer that makes the feature maps with the same size of 256×256. Then the residual representations can be generated by two consecutive convolutional layers with kernel sizes of 3×3 and 1×1. It can be observed in Fig. 2 that the missing/ambiguity boundaries and the confusing regions can be easily localized under the guidance of the residual representations.

2.4 Residual Feedback Transmission and Loss Function

Existing segmentation methods principally design a module to predict the boundaries or a loss function for boundaries restriction [19–22], which is a forward processing and only focuses on the boundary pixels. In this study, the missing/ambiguity boundaries and the confusing regions can be easily located with the help of the residual representation module. Under the inspiration of [20], we adopt a residual feedback transmission strategy to make the proposed network further correct the errors in initial segmentation results.

After the residual representations $\hat{R}$ and the initial segmentation results P^1 being obtained in the first step, the residual representations are fed into each encoder block to weight the low-confidence and wrongly predicted pixels in P^1 by combining with the original image features with element-wise addition and channel-wise multiplication (see the second step (red dotted arrows) in Fig. 1). Then the second step is still performed on the same network architecture as the first step, and outputs the presents residual-guided segmentation results P^2.

For training our network in an end-to-end manner, the final loss L consists of three sub-losses for the initial segmentation results ($L_{wbl}\{G, P^1\}$), the residual-guided segmentation results ($L_{wbl}\{G, P^2\}$), and the residual representation $L_{wbce}\{R, \hat{R}\}$ respectively:

$$L = L_{wbl}\{G, P^1\} + L_{wbl}\{G, P^2\} + L_{wbce}\{R, \hat{R}\} \tag{1}$$

where L_{wbl} and L_{wbce} denote the weighted-balanced loss function [27] and the weighted binary cross-entropy loss function, respectively. And L_{wbl} and L_{wbce} are expressed as follows:

$$L_{wbl} = 1 - w \frac{\sum_{n=1}^{N_1} p_n y_n}{\sum_{n=1}^{N_1} (p_n + y_n)} - (1 - w) \frac{\sum_{n=1}^{N_0} (1 - p_n)(1 - y_n)}{\sum_{n=1}^{N_0} (2 - p_n - y_n)} \tag{2}$$

$$L_{wbce} = -w \sum_{n=1}^{N_1} y_n log(p_n) - (1 - w) \sum_{n=1}^{N_0} (1 - y_n) log(1 - p_n) \tag{3}$$

where $w = N_0/(N_1 + N_0)$ is the weighted-balanced parameter proposed in our previous work [27]. Specifically, y_n and p_n denote the ground truth and predicted probability of pixels, respectively. N_0 and N_1 denote the number of the pixels where $y_n = 0$ and $y_n = 1$. For each BUS image in the test phase, the segmentation results can be generated recurrently for more precise performance owning to the proposed feedback transmission strategy.

Implementation Details. We implement our network on Pytorch and train it on one GeForce Titan GPU with the mini-batch size of 16. We adopt an adaptive gradient algorithm (Adagrad) optimizer with momentum of 0.9 to optimize and update the network parameters with 70 epochs. The learning rate is $2e^{-4} \times 0.95^{epoch_i}$, where $epoch_i$ is the i-th epoch. The network parameters are initialized by "he_normal" [23].

3 Experiments

Datasets. Three BUS image datasets are used in this study to evaluate the effectiveness of the proposed network. The first one (datasetA) contains two public datasets: (a) the dataset is available at https://www.ultrasoundcases. info/ derived from SonoSkills and Hitachi Medical Systems Europe, which has a total of 2499 tumor images from 554 patients, including benign lesions (1061 BUS images) and malignant lesions (1438 BUS images) (b) BUSI [24], from the Baheya Hospital for Early Detection and Treatment of Women's Cancer (Cairo, Egypt), which contains 210 images with benign lesions and 437 images with malignant lesions and 133 normal images without lesions. The second one, UDIAT (datasetB), containing 110 images with benign lesions and 53 images with malignant lesions, is a public dataset provided by [5], which is collected from the UDIAT Diagnostic Center of the Parc Taulfi Corporation, Sabadell (Spain). The third one, datasetC, is available at https://radiopaedia.org/ and contains 504 tumor images, including benign lesion (242 BUS images) and malignant lesions (262 BUS images). We perform five-fold cross-validation on datasetA and report the validated results, then use datasetB and datasetC as two test datasets to illustrate the effectiveness of the proposed model based on five models trained on datasetA.

Evaluation Metrics. For quantitatively comparing different methods, we employ widely-used segmentation metrics, including dice similarity coefficient (DSC), jaccard index (JI), recall, precision, Hausdorff distance (HD, in pixel) and average surface distance (ASD, in pixel) [7,9,25]. For DSC, JI, Recall, and Precision, larger values indicate better performance, while HD and ASD vice versa. We evaluate the overlap areas between the predicted lesions and the delineated lesions by radiologists with DSC, JI, recall, and precision, and measure the Euclidean distance of boundaries with HD and ASD [7].

3.1 Segmentation Performance

We compare our proposed method with four state-of-the-art methods, including U-Net [17], DeepLabV3 [26], CE-Net [18], and S^{2}P-Net [15]. For a fair comparison, we adopt the same encoder (ResNet-34) for DeepLabV3 and CE-Net, and validate their performance with/without pretrain parameters [28] (DeepLabV3$^{w/wo}$ and CE-Net$^{w/wo}$ in Table 1, 2, 3). Our proposed method and the comparison methods are trained and validated on datasetA (reporting mean results of the five-fold cross-validation), and tested on datasetB and datasetC. All the networks are fine-tuned for best training parameters.

Table 1. Comparison with different methods on datasetA (mean ± variance).

Methods	DSC	JI	Recall	Precision	HD	ASD
U-Net	84.33 ± 0.99	76.42 ± 1.01	86.01 ± 0.81	87.10 ± 1.54	30.12 ± 1.55	11.1 ± 1.50
DeepLabV3^{wo}	84.16 ± 0.48	75.69 ± 0.68	86.29 ± 0.49	85.70 ± 0.58	28.08 ± 0.87	9.03 ± 0.39
DeepLabV3^w	85.97 ± 0.71	77.85 ± 0.95	87.49 ± 0.64	87.40 ± 1.61	25.58 ± 0.99	8.36 ± 0.30
CE-Net^{wo}	84.59 ± 0.97	76.38 ± 1.14	86.46 ± 1.30	86.27 ± 2.11	28.34 ± 1.58	8.62 ± 0.24
CE-Net^w	86.32 ± 0.75	78.39 ± 0.86	88.02 ± 0.72	87.23 ± 1.27	26.30 ± 1.48	7.58 ± 0.55
S^{2}P-Net	84.65 ± 0.72	76.63 ± 0.64	85.94 ± 0.70	87.41 ± 0.99	29.23 ± 1.69	10.7 ± 1.33
Our method	**86.91 ± 0.55**	**79.24 ± 0.61**	**88.22 ± 0.77**	**88.40 ± 1.55**	**24.44 ± 0.94**	**7.68 ± 0.59**

Table 2. Comparison with different methods on datasetB (mean ± variance).

Methods	DSC	JI	Recall	Precision	HD	ASD
U-Net	74.20 ± 1.02	65.99 ± 1.13	83.54 ± 1.33	78.01 ± 2.81	46.83 ± 4.07	33.62 ± 5.62
DeepLabV3^{wo}	74.87 ± 2.28	65.58 ± 2.48	85.09 ± 1.51	71.97 ± 2.65	25.26 ± 3.17	9.24 ± 1.27
DeepLabV3^w	78.94 ± 1.49	69.70 ± 1.66	87.02 ± 0.65	77.13 ± 2.11	23.39 ± 1.62	10.89 ± 1.92
CE-Net^{wo}	74.13 ± 1.94	65.65 ± 1.95	86.24 ± 1.95	71.25 ± 2.29	28.84 ± 3.25	11.23 ± 1.57
CE-Net^w	79.61 ± 1.76	71.00 ± 1.87	88.62 ± 1.34	77.07 ± 2.50	24.74 ± 3.53	9.65 ± 1.98
S^{2}P-Net	77.42 ± 1.13	69.04 ± 1.10	85.42 ± 2.15	77.30 ± 1.95	29.57 ± 2.18	16.36 ± 2.41
Our method	**81.79 ± 0.76**	**73.09 ± 0.64**	**90.07 ± 1.00**	**78.61 ± 0.97**	**18.06 ± 0.87**	**6.54 ± 0.29**

From the results in Table 1, 2, 3, our proposed method achieves highest performance in terms of DSC, JI, Recall and Precision, and outperforms all the competitors. In addition, our method also achieves the competitive values of HD and ASD with a high level of robustness, which demonstrates that our model can achieve more accurate boundary localization. The improvement of these metrics on all three datasets indicates that our model also has better generalization ability than all the competitors.

Figure 3 shows the segmentation results produced by different methods. Apparently, our method can detect missing/ambiguity boundaries accurately and better segment the whole breast lesion regions. The superior performance of our method, with significantly improved results (p < 0.05 for paired t-test in Tables 1, 2 and 3), indicates that our residual representation module and the residual feedback transmission strategy can help distinguish some pseudo-lesions from positive lesions, and generate more precise boundaries. Moreover, it is noteworthy that our model achieves superior performance on malignant lesion with posterior acoustic shadowing in the second row of Fig. 3.

Table 3. Comparison with different methods on datasetC (mean ± variance).

Methods	DSC	JI	Recall	Precision	HD	ASD
U-Net	83.42 ± 0.65	75.53 ± 0.66	85.82 ± 1.01	87.08 ± 0.76	34.11 ± 2.13	14.44 ± 1.98
DeepLabV3^{wo}	83.80 ± 0.91	75.53 ± 1.16	86.75 ± 1.06	84.70 ± 0.93	27.96 ± 1.16	8.50 ± 0.89
DeepLabV3^{w}	85.20 ± 0.53	77.33 ± 0.74	87.79 ± 0.51	86.02 ± 1.12	25.80 ± 1.09	8.13 ± 0.96
CE-Net^{wo}	84.17 ± 0.55	75.98 ± 0.63	87.16 ± 1.25	85.01 ± 1.20	28.34 ± 0.91	8.01 ± 0.40
CE-Net^{wo}	86.13 ± 0.43	78.46 ± 0.47	88.94 ± 0.68	86.41 ± 1.16	26.30 ± 1.48	7.58 ± 0.55
S^{2}P-Net	84.01 ± 0.35	76.10 ± 0.41	85.93 ± 0.45	87.36 ± 0.44	32.92 ± 0.87	13.63 ± 0.95
Our method	**87.00 ± 0.23**	**79.54 ± 0.23**	**88.99 ± 0.79**	**87.87 ± 0.79**	**23.66 ± 0.51**	**6.77 ± 0.51**

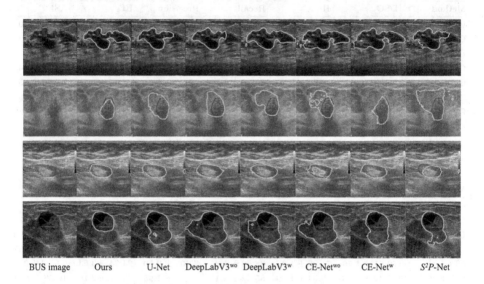

BUS image   Ours   U-Net   DeepLabV3^{wo} DeepLabV3^{w} CE-Net^{wo} CE-Net^{w} S^2P-Net

Fig. 3. The segmentation results (yellow) produced by different methods and the ground truths (red) of random test subjects. (Color figure online)

3.2   Ablation Study

Three ablation studies experiments are conducted to verify the setups in our network design. Using the encoder-decoder architecture (in Fig. 1) as baseline, and we add three setups which boost the segmentation performance in BUS images to the baseline in a step-by-step manner: 1) the encoder based on ResNet34 with pretrain parameters [28] is denoted as 'Pretrain'; 2) the residual representation module (denoted as 'RSM'); 3) the residual feedback transmission strategy (denoted as 'RFT') which is the guidance for more accurate segmentation results.

From the segmentation results in Table 4, it can be seen that all three setups help improve the segmentation performance on three datasets (p < 0.05, paired t-test). Compared with the baseline, the model with pretrained parameters shows better performance, especially on the testing datasets (improvement approximately 5.77% and 1.98% for DSC on datasetB and datasetC). This reflects

Table 4. Segmentation results on three datasets to validate the effectiveness of three setups in our method (mean±variance).

Pretrain		×	√	√	√
RSM		×	×	√	√
RFT		×	×	×	√
DatasetA	DSC	84.60 ± 0.58	85.76 ± 0.49	86.46 ± 0.48	**86.91 ± 0.55**
	Recall	86.63 ± 0.96	86.95 ± 0.43	87.65 ± 0.37	**88.22 ± 0.77**
	Precision	86.12 ± 0.91	87.89 ± 1.24	88.25 ± 0.84	**88.34 ± 1.55**
	HD	28.09 ± 1.13	26.55 ± 0.68	25.54 ± 0.56	**24.44 ± 0.94**
DatasetB	DSC	74.06 ± 0.83	79.83 ± 0.89	80.75 ± 0.44	**81.79 ± 0.76**
	Recall	84.41 ± 1.54	87.13 ± 0.35	89.12 ± 0.87	**90.07 ± 1.00**
	Precision	72.20 ± 1.54	78.54 ± 1.93	78.36 ± 0.92	**78.61 ± 0.97**
	HD	30.63 ± 3.25	22.64 ± 3.17	20.49 ± 2.14	**18.06 ± 0.87**
DatasetC	DSC	84.04 ± 0.41	86.02 ± 0.28	86.44 ± 0.22	**87.00 ± 0.23**
	Recall	86.33 ± 0.54	87.71 ± 0.56	88.26 ± 0.64	**88.99 ± 0.79**
	Precision	85.73 ± 0.75	87.71 ± 0.78	87.74 ± 0.60	**87.87 ± 0.79**
	HD	28.38 ± 0.50	26.15 ± 0.98	25.12 ± 1.85	**23.66 ± 0.51**

that the models pretrained on ImageNet dataset have sensitive and transferable parameters, which are more robust on different testing datasets. In addition, we observe that employing the RSM yields significant improvements in term of recall (0.70%, 1.99%, 0.55% on datasetA, datasetB, datasetC). This achievement can be explained by learning the representations of missing/ambiguous boundaries and confusing regions in residual representation module, and encouraging the network to explore more information about hardly-predicted pixels. Further, observed from Table 4, our model adopting RFT achieve more accurate segmentation results in terms of all metrics, especially on the boundary-related metrics HD. It proves that the RFT, which feedbacks the learned representation into the encoder, can enhance the weight of the hardly-predicted pixels, and make the network iteratively correct the errors in segmentation process.

4   Discussion and Conclusion

This paper proposes a novel residual feedback network for breast lesion segmentation on ultrasound image and aims to focus on hardly-predicted pixels. The key idea is to design a supervised residual representation module, which extracts multi-scale features from the decoder blocks and generates residual representations representing the semantic context of inconclusive pixels. And a residual feedback transmission strategy is designed to embed the residual representations into the encoder blocks to increase the saliency of hardly-predicted pixels, which makes our proposed network can further correct the errors in initial segmentation results. Experimental results have evaluated on three datasets, and significantly outperforms the state-of-the-art methods. Specifically, the DSC score is 86.91% for datasetA, 81.79% for datasetB, and 87.00% for datasetC, which proves our model has better generalization ability on testing datasets (datasetB and datasetC). Detailed discussion of typical failed cases of our model will be included in Supplementary Material.

Acknowledgements. This work was supported in part by the National Natural Science Foundation of China under Grant 61971213 and Grant 61671230, in part by the Basic and Applied Basic Research Foundation of Guangdong Province under Grant 2019A1515010417, and in part by the Guangdong Provincial Key Laboratory of Medical Image Processing under Grant No.2020B1212060039. The authors have no relevant conflicts of interest to disclose.

References

1. Ahmad, A. (ed.): Breast Cancer Metastasis and Drug Resistance. AEMB, vol. 1152. Springer, Cham (2019). https://doi.org/10.1007/978-3-030-20301-6
2. Berg, W.A., et al.: Combined screening with ultrasound and mammography vs mammography alone in women at elevated risk of breast cancer. JAMA **299**(18), 2151–2163 (2008)
3. Sahiner, B., et al.: Malignant and benign breast masses on 3D US volumetric images: effect of computer-aided diagnosis on radiologist accuracy. Radiology **242**(3), 716–724 (2007)
4. Xu, Y., et al.: Medical breast ultrasound image segmentation by machine learning. Ultrasonics **91**, 1–9 (2019)
5. Yap, M.H., et al.: Automated breast ultrasound lesions detection using convolutional neural networks. IEEE J. Biomed. Health Inform. **22**(4), 1218–1226 (2018)
6. Lei, B., et al.: Segmentation of breast anatomy for automated whole breast ultrasound images with boundary regularized convolutional encoder-decoder network. Neurocomputing **321**, 178–186 (2018)
7. Xing, J., et al. Lesion segmentation in ultrasound using semi-pixel-wise cycle generative adversarial nets. IEEE ACM Trans. Comput. Biol. Bioinform. (2020). https://ieeexplore.ieee.org/abstract/document/9025227
8. Saeed, J.N.: A Survey of Ultrasonography Breast Cancer Image Segmentation Techniques. Infinite Study (2020)
9. Xian, M., et al.: Automatic breast ultrasound image segmentation: a survey. Pattern Recognit. **79**, 340–355 (2018)
10. Luo, Y., et al.: A novel segmentation approach combining region- and edge-based information for ultrasound images. BioMed Res. Int. **2017** (2017). https://www.hindawi.com/journals/bmri/2017/9157341/
11. Huang, Q., et al.: Segmentation of breast ultrasound image with semantic classification of superpixels. Med. Image Anal. **61**, 101657 (2020)
12. Havaei, M., Davy, A., Warde-Farley, D., et al.: Brain tumor segmentation with deep neural networks. Med. Image Anal. **35**, 18–31 (2017)
13. Han, L., et al.: Semi-supervised segmentation of lesion from breast ultra-sound images with attentional generative adversarial network. Comput. Methods Programs Biomed. **189**, 105275 (2020)
14. Vakanski, A., et al.: Attention-enriched deep learning model for breast tumor segmentation in ultrasound images. Ultrasound Med. Biol. **46**(10), 2819–2833 (2020)

15. Zhu, L., et al.: A second-order subregion pooling network for breast lesion segmentation in ultrasound. In: Martel, A.L., et al. (eds.) MICCAI 2020. LNCS, vol. 12266, pp. 160–170. Springer, Cham (2020). https://doi.org/10.1007/978-3-030-59725-2_16

16. He, K., et al.: Deep residual learning for image recognition. In: Proceedings of the IEEE Conference on Computer Vision and Pattern Recognition, pp. 770–778 (2016)

17. Ronneberger, O., Fischer, P., Brox, T.: U-Net: convolutional networks for biomedical image segmentation. In: Navab, N., Hornegger, J., Wells, W.M., Frangi, A.F. (eds.) MICCAI 2015. LNCS, vol. 9351, pp. 234–241. Springer, Cham (2015). https://doi.org/10.1007/978-3-319-24574-4_28

18. Gu, Z., et al.: Ce-net: context encoder network for 2d medical image segmentation. IEEE Trans. Med. Imaging **38**(10), 2281–2292 (2019)

19. Wang, S., et al.: Boundary coding representation for organ segmentation in prostate cancer radiotherapy. IEEE Trans. Med. Imaging **40**(1), 310–320 (2020)

20. Feng, M., et al.: Attentive feedback network for boundary-aware salient object detection. In: Proceedings of the IEEE Conference on Computer Vision and Pattern Recognition, pp. 1623–1632 (2019)

21. Wei, J., et al.: F^{3}Net: Fusion, Feedback and Focus for Salient Object Detection. In: Proceedings of the AAAI Conference on Artificial Intelligence, vol. 34, no. 07, pp. 12321–12328 (2020)

22. Zhang, R., Li, G., Li, Z., Cui, S., Qian, D., Yu, Y.: Adaptive context selection for polyp segmentation. In: Martel, A.L., et al. (eds.) MICCAI 2020. LNCS, vol. 12266, pp. 253–262. Springer, Cham (2020). https://doi.org/10.1007/978-3-030-59725-2_25

23. He, K.M., et al.: Delving deep into rectifiers: surpassing human-level performance on imagenet classification. In: International Conference on Computer Vision, pp. 1026–1034 (2015)

24. Al-Dhabyani, W., et al.: Dataset of breast ultrasound images. Data Brief **28**, 104863 (2020)

25. Guo, Y., et al.: A novel breast ultrasound image segmentation algorithm based on neutrosophic similarity score and level set. Comput. Methods Programs Biomed. **123**, 43–53 (2016)

26. Chen, L.C., et al.: Rethinking atrous convolution for semantic image segmentation. preprint arXiv:1706.05587 (2017)

27. Wang, K., Liang, S.J., et al.: Breast ultrasound image segmentation: a coarse-to-fine fusion convolutional neural network. Med. Phys. (2021). https://doi.org/10.1002/mp.15006

28. Xie, S.N., et al.: Aggregated residual transformations for deep neural networks. In: Proceedings of the IEEE Conference on Computer Vision and Pattern Recognition, pp. 1492–1500 (2017)

Learning to Address Intra-segment Misclassification in Retinal Imaging

Yukun Zhou[1,2,3](✉), Moucheng Xu[1,2], Yipeng Hu[1,2,5], Hongxiang Lin[1,3,6], Joseph Jacob[1,7], Pearse A. Keane[3,8], and Daniel C. Alexander[1,4]

[1] Centre for Medical Image Computing, University College London, London, UK
yukun.zhou.19@ucl.ac.uk
[2] Department of Medical Physics and Biomedical Engineering, UCL, London, UK
[3] NIHR Biomedical Research Centre at Moorfields Eye Hospital, London, UK
[4] Department of Computer Science, University College London, London, UK
[5] Wellcome/EPSRC Centre for Interventional and Surgical Sciences, London, UK
[6] Research Center for Healthcare Data Science, Zhejiang Lab, Hangzhou, China
[7] UCL Respiratory, University College London, London, UK
[8] UCL Institute of Ophthalmology, University College London, London, UK

Abstract. Accurate multi-class segmentation is a long-standing challenge in medical imaging, especially in scenarios where classes share strong similarity. Segmenting retinal blood vessels in retinal photographs is one such scenario, in which arteries and veins need to be identified and differentiated from each other and from the background. Intra-segment misclassification, i.e. veins classified as arteries or vice versa, frequently occurs when arteries and veins intersect, whereas in binary retinal vessel segmentation, error rates are much lower. We thus propose a new approach that decomposes multi-class segmentation into multiple binary, followed by a binary-to-multi-class fusion network. The network merges representations of artery, vein, and multi-class feature maps, each of which are supervised by expert vessel annotation in adversarial training. A skip-connection based merging process explicitly maintains class-specific gradients to avoid gradient vanishing in deep layers, to favor the discriminative features. The results show that, our model respectively improves F1-score by 4.4%, 5.1%, and 4.2% compared with three state-of-the-art deep learning based methods on DRIVE-AV, LES-AV, and HRF-AV data sets. Code: https://github.com/rmaphoh/Learning-AVSegmentation

Keywords: Multi-class segmentation · Intra-segment misclassification · Retinal vessel · Binary-to-multi-class fusion network

Electronic supplementary material The online version of this chapter (https://doi.org/10.1007/978-3-030-87193-2_46) contains supplementary material, which is available to authorized users.

1   Introduction

Measurements of retinal vessel morphology are vital for computer-aided diagnosis of ophthalmological and cardiovascular diseases, such as diabetes, hypertension, and arteriosclerosis [6,28]. Retinal vessel segmentation on fundus images plays a key role in retinal vessel characterisation, by providing rich morphological features that are sensitive for diagnosis. However, manual annotation of retinal vessels is highly time-consuming, thus automated vessel segmentation is in high demand to improve the efficiency of the diagnosis, with or without computer assistance, of a wide range of diseases.

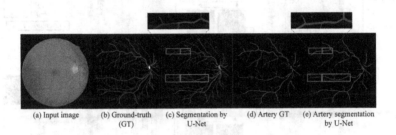

(a) Input image   (b) Ground-truth   (c) Segmentation by   (d) Artery GT   (e) Artery segmentation
                         (GT)   U-Net   by U-Net

Fig. 1. Illustration of retinal vessel segmentation maps highlighting common errors arising at vessel intersections. In (c), the green boxes highlight two partial arteries that are misclassified as veins in a multi-class segmentation algorithm, while correctly classified as arteries in the binary segmentation task (e).

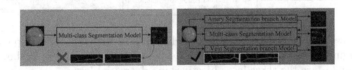

Fig. 2. Strategy of the binary-to-multi-class fusion.

In automatic multi-class retinal vessel segmentation, i.e. segmenting veins and arteries from retinal fundus images, there are two main categories, namely, graph-based methods and feature-based methods. Graph-based methods utilise the vessel topological and structural knowledge to search for the vessel tree structures and classify them into artery and vein [3,4,24,27,30,32]. Feature-based methods design various feature extraction tools to discriminate the vessel category [11,17]. The methods that have achieved state-of-the-art performance have come from end-to-end deep-learning-based models [7–9,16,20,26,29,31]. However, the intra-segment misclassification that occurs around inter-class vessel intersections, as illustrated in Fig. 1, is a well-recognised common issue across all the deep-learning-based models.

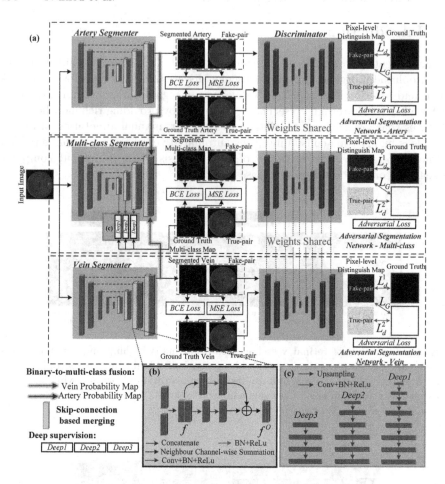

Fig. 3. Flowchart of the binary-to-multi-class fusion network. The skip-connection based merging is highlighted in (b). The deep supervision structure is shown in (c). L_d^1, L_d^2, and L_G are the three loss objectives in adversarial loss.

In this paper, we focus on intra-segment misclassification in multi-class retinal vessel segmentation [13]. More specifically, we observed and identified that a type of intra-segment misclassification frequently happens when the inter-class vessels intersect, as illustrated in Fig. 1(c), and has been reported in [8,9,15,16,18,26,29,31]. However, the binary segmentation, e.g., extracting only arteries from images, illustrated in Fig. 1(e), was able to correctly identify the single class arteries, probably due to the simplified task that has less uncertainty pixel labels, such as those at inter-class vessel intersections. This has motivated a novel multi-class segmentation algorithm that decomposes the problem into a combination of multiple binary segmentation tasks followed by a binary-to-multi-class fusion step, as shown in Fig. 2. This avoids direct learning on ambiguous inter-class vessel intersections. We thus hypothesise that the network specifically designed will substantially improve performance in local areas where inter-class vessel intersections confuse multi-class classification.

We summarise our contributions as follows. First, we identified that the intra-segment misclassification happens at inter-class intersections in multi-class segmentation, while not in the binary. Second, we propose a deep learning model to address intra-segment misclassification, following the binary-to-multi fusion strategy. The flowchart is shown in Fig. 3. Lastly, we report experiments including ablation studies on three real-world clinical data sets, to demonstrate the effectiveness of the proposed model.

2   Methods

We devise a deep learning based model, consisting of two components, an Adversarial Segmentation Network, as the backbone of the fusion network, and a Binary-to-multi-class Fusion Network. The Adversarial Segmentation Network detects and corrects higher-order inconsistencies between expert annotation maps and the ones produced by the generator, termed a segmenter here [14,23]. The binary-to-multi-class fusion network merges the representations learnt from individual vessel types into a multi-class feature, to supplement the discriminative information of each single class surrounding the inter-class intersections.

2.1   Adversarial Segmentation Network

The Adversarial Segmentation Network compromises a segmenter and a pixel-level discriminator. The segmenter is a variant of U-Net while the discriminator is vanilla U-Net [20]. The input of segmenter is retinal fundus image and output is multi-class segmentation map. For discriminator, we concatenate the segmentation map and fundus image as input and obtain a pixel-level distinguish map as output, as shown in Fig. 3.

Skip-Connection Based Merging Process in Segmenter. This structure is employed to alleviate the gradient vanishing, as well as maintaining high resolution information. The structure is shown in Fig. 3(b). We start with a concatenation of the up- and down-sampling feature maps at the same level into a combined feature $f \in \mathbb{R}^{W \times H \times N}$, where W, H, and N respectively represent width, height, and channel of the spatial feature, which is divisible by output channel C. The merging process consist of a main path and a skip path. An operation in the main path $\phi : \mathbb{R}^{W \times H \times N} \to \mathbb{R}^{W \times H \times C}$ denotes two blocks, one of which comprises a convolution operation, batch normalisation, and activation function like ReLU in this case. Another operation in the skip path ψ consists of batch normalisation and ReLU. Then the kth channel of the output feature f^O, denoted as f_k^O, is

$$f_k^O = \phi(f)_k + \psi \left(\Sigma_{i=1}^{N/C} f_{(k-1)N/C+i} \right), \tag{1}$$

where $\sum_{i=1}^{N/C} f_{(k-1)N/C+i}$ indicates the Neighbour channel-wise summation, separately operating on the up- and down-sampling features.

Adversarial Training. The adversarial training is employed to prompt the segmentation capability [2,22]. Specifically, there are two reasons to adapt a U-Net as discriminator. First, the adversarial loss is pixel-wise, thus every pixel value in a segmented map mimics that in the experts annotation, thereby obtaining clear and sharp segmentation with high resolution information [22]. Second, the U-Net is more robust than the combination of down-sampling convolution blocks in the discriminator [21], thus better converges towards the Nash equilibrium [5]. The input of discriminator is the concatenation of the fundus image and segmentation map. In order to achieve convergence, the segmenter should produce realistic segmentation maps to minimise the $\mathbb{E}_{x \sim p_x(x)}[log(1 - D(x, G(x)))]$, corresponding to minimise L_G in Fig. 3(a). The pixel-level discriminator is trained to maximise the $\mathbb{E}_{y \sim p_{data}(y)}[log D(x, y)] + \mathbb{E}_{x \sim p_x(x)}[log(1 - D(x, G(x)))]$, i.e., minimising the L_d^1 and L_d^2, where x indicates fundus image and y represents the experts annotation. The loss function over the segmentation network is

$$Loss_{seg} = \alpha \cdot L_{GAN} + \beta \cdot L_{BCE} + \gamma \cdot L_{MSE} \tag{2}$$

which combines the adversarial loss, binary cross entropy (BCE), and Mean square error (MSE) with weight hyperparameters α, β, and γ. The BCE and MSE measure the distance between $G(x)$ and y, as shown in Fig. 3(a).

2.2 Binary-to-multi-class Fusion Network

Considering the preferable accuracy in binary segmentation, e.g., artery segmentation shown in Fig. 1(e), we devise a binary-to-multi-class fusion network to merge representations of the artery, vein, and multi-class feature to improve the multi-class segmentation performance at the intersections. The whole structure is shown in Fig. 3(a). The main segmenter outputs the multi-class vessel map. All three segmenters share the same discriminator. The artery segmentation map f_a, multi-class feature map f_m, and vein segmentation map f_v are concatenated to generate fused feature maps for the final convolution operation, to generate the multi-class segmentation map.

Additionally, we utilise the deep supervisions in up-sampling stages to avoid gradient vanishing in deep layers [12], thus enhancing the discriminative information. The structure is shown in Fig. 3(c). The feature maps in each up-sampling stages $\{f^i, i = 1, 2, 3\}$ are through the structure to respectively obtain f_{s1}, f_{s2}, f_{s3}, to get the deep supervision loss $\{L_{BCE_i}, L_{MSE_i} \mid i \in (s1, s2, s3)\}$. Combining Eq. 2, the fusion loss function combination is

$$Loss_{fusion} = Loss_{seg} + \sum_{i=1}^{3} \frac{1}{2^i}(\beta L_{BCE_{s_i}} + \gamma L_{MSE_{s_i}}), \tag{3}$$

where the $\frac{1}{2^i}$ works as the weights for the deep supervision of the three side outputs. The larger weight is allocated to the deeper layer as the risk of gradient vanishing increases.

3 Experiments

3.1 Experiment Setting

We include three clinical data sets in experiments, the DRIVE-AV [10, 25], LES-AV [19], and HRF-AV [1, 9]. DRIVE-AV includes 40 colour retinal fundus images

Table 1. Multi-class vessel segmentation results on DRIVE-AV, LES-AV, and HRF-AV data sets compared with state-of-the-art methods. Since the TR-GAN is the most competitive compared method, the P-value of Mann–Whitney U test between TR-GAN and the proposed method is calculated to show the statistic significance.

DRIVE-AV					
Methods	Sen	F1	ROC	PR	MSE
U-net [20]	53.6±0.42	65.57±0.58	79.84±0.21	65.92±0.17	3.71±0.01
CS-Net [18]	52.03±0.16	63.68±0.20	81.96±0.12	63.04±0.11	3.91±0.02
MNNSA [15]	59.88±0.82	61.8±0.33	78.89±0.34	65.43±0.19	3.62±0.02
TR-GAN [2]	67.84±1.24	65.63±0.71	80.93±0.57	67.44±0.63	3.63±0.01
Ensemble	66.92±0.67	67.13±0.6	81.07±0.29	68.34±0.17	3.49±0.01
Proposed	**69.87±0.11**	**70.03±0.03**	**84.13±0.05**	**71.17±0.03**	**3.09±0.01**
P-value	2.46e-4	1.83e-4	1.83e-4	1.83e-4	1.83e-4
LES-AV					
Methods	Sen	F1	ROC	PR	MSE
U-net [20]	58.02±0.49	61.6±0.41	78.51±0.23	65.71±0.3	2.46±0.03
CS-Net [18]	49.88±0.88	58.58±0.83	74.12±0.16	64.99±0.14	2.51±0.04
MNNSA [15]	51.31±0.58	57.16±0.22	76.75±0.33	58.86±0.22	2.82±0.05
TR-GAN [2]	59.33±0.72	61.14±0.57	77.57±0.19	65.4±0.26	2.54±0.07
Ensemble	58.81±0.58	62.51±0.34	78.58±0.28	66.31±0.13	2.59±0.07
Proposed	**62.94±0.93**	**66.69±0.47**	**81.03±0.04**	**68.71±0.47**	**2.17±0.05**
P-value	1.83e-4	1.83e-4	1.83e-4	1.83e-4	1.83e-4
HRF-AV					
Methods	Sen	F1	ROC	PR	MSE
U-net [20]	63.43±1.12	67.12±0.85	79.98±0.77	72.2±0.42	2.49±0.01
CS-Net [18]	62.5±1.47	66.07±0.66	79.12±0.86	69.18±0.86	2.64±0.01
MNNSA [15]	62.81±0.92	64.3±0.63	78.83±0.79	69.06±0.46	2.78±0.02
TR-GAN [2]	63.14±0.97	67.5±0.39	80.67±0.74	72.45±0.38	2.37±0.02
Ensemble	64.29±1.39	68.58±0.87	80.33±0.41	71.09±0.59	2.24±0.01
Proposed	**67.68±1.57**	**71.7±0.44**	**83.44±0.75**	**73.96±0.31**	**1.9±0.01**
P-value	1.83e-4	1.83e-4	1.83e-4	1.83e-4	1.83e-4

with the size of $(565, 584)$, containing labels for artery, vein, crossing, and uncertainty. The LES-AV data set contains 22 images with the size of $(1620, 1444)$. The HRF-AV data set includes 45 images with the size of $(3504, 2336)$. In DRIVE-AV, 20 images are for training and 20 images for testing. In LES-AV, we sample 11 images for training, and leave 11 images for testing. In HRF-AV, 24 images are used for training and 21 images for testing. 10% of training images are used for validation.

Table 2. The ablation study of binary-to-multi-class fusion network. Adversarial segmentation network (*Seg*), deep supervision (*Deep*), and binary-to-multi-class fusion (*BF*). The metrics AUC-ROC (*ROC*) and F1-score (*F1*). Growth is shown in bracket.

Components			DRIVE-AV		LEV-AV		HRF-AV	
Seg	*Deep*	*BF*	*ROC*	*F1*	*ROC*	*F1*	*ROC*	*F1*
			79.84	65.57	78.51	61.6	79.98	67.12
✓			$81.77_{(1.93)}$	$67.45_{(1.88)}$	$79.44_{(0.93)}$	$63.78_{(2.18)}$	$81.68_{(1.7)}$	$69.41_{(2.29)}$
✓	✓		$82.3_{(0.53)}$	$68.52_{(1.07)}$	$79.95_{(0.51)}$	$64.93_{(1.15)}$	$82.31_{(0.63)}$	$70.16_{(0.75)}$
✓	✓	✓	$84.13_{(1.83)}$	$70.03_{(1.51)}$	$81.03_{(1.08)}$	$66.69_{(1.76)}$	$83.44_{(1.13)}$	$71.7_{(1.54)}$

For implementation and training details, the images in DRIVE-AV are zero-padded to the size of $(592, 592)$. The images in LES-AV and HRF-AV are respectively resized to $(800, 720)$ and $(880, 592)$ in the training, for computational consideration [8,13]. In validation, the images are resized back to the original size for metrics calculation. We set learning rate 0.0008, and batch size 2 across experiments. α, β, and γ in Eq. 2 are set at 0.08, 1.1, and 0.5 respectively. The hyper-parameter search is in Supplementary Materials Sect. 1. Total training time of a model is approximately 12 h for 1500 epochs on one Tesla T4.

For evaluation metrics, the retinal vessel multi-task segmentation is an unbalanced task. Some metrics like accuracy and specificity are always high, thus losing practical value. We employ AUC (Area Under The Curve)-ROC (Receiver Operating Characteristics), sensitivity, F1-score, AUC-PR (Precision-Recall), and MSE to comprehensively evaluate the methods. For example, the F1-score computation is $F = (n_a \cdot F_a + n_v \cdot F_v + n_u \cdot F_u)/(n_a + n_v + n_u)$, where n_a, n_v, and n_u respectively represent the pixels number belonging to artery, vein, and uncertainty in label. F_a, F_v, and F_u are the F1-score scores in each binary measurement, e.g., artery pixels versus all other pixels. The Mann–Whitney U test is employed to test the statistic significance of the difference across methods. For fair comparison, we implemented all baselines with the same pre-processing, input size, and evaluation methods.

3.2   Experiment Results

Comparing with Most Recent Methods. The compared methods CS-Net [18], MNNSA [15], and TR-GAN [2] are recent related works. The U-Net is a baseline. We list the overall multi-class vessel segmentation performance in Table 1. The P-value of Mann–Whitney U test is smaller than 0.05, showing statistic significance. The ROC and PR curves on the DRIVE-AV test are shown in Fig. 4. Based on the results, the proposed method enhances performance, e.g., F1-scores respectively increase by 4.4%, 5.09%, 4.2% in DRIVE-AV, LES-AV, and HRF-AV. For all segmentation maps, the reader is referred to Supplementary Materials Sect. 2.

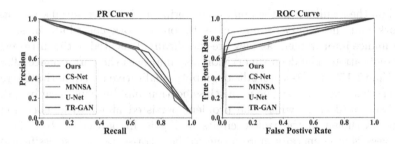

Fig. 4. ROC and PR curves of various multi-class segmentation methods.

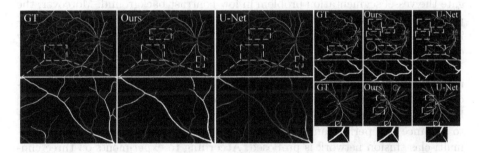

Fig. 5. The visualisation results with the proposed method and U-Net. The White dotted boxes show solved intra-segment misclassification, while yellow circle shows some residual issues. Better view with zoom in.

Ablation Study. We evaluate the contribution of each component from the binary-to-multi-class fusion network, including with and without the adversarial segmentation network, deep supervision, and binary-to-multi-class fusion. As shown in Table 2, the adversarial segmentation network performs better due to skip-connection based merging and the pixel-level adversarial learning, when compared with vanilla U-Net [20] in first line. The binary-to-multi-class fusion network improves the overall metrics, with the largest improvement 1.83% in AUC-ROC and 1.76% in F1-score. Observing the benefits brought by BF, the benefit of introducing artery and vein binary branches is highlighted. With the "Ensemble" method in Table 1, we separately train two binary adversarial segmentation networks with artery and vein labels, and then concatenate the artery and vein maps to get the multi-class segmentation map, aiming to verify the effectiveness of our learning based merging strategy.

4   Discussion

Based on the results from the ablation experiment, the adversarial segmentation network and the binary-to-multi-class fusion network contribute most of the performance improvement as they are specifically dedicated to the intra-segment misclassification. The deep supervision also improves the performance albeit no more than 1.15% in F1-score. Although both qualitative and quantitative results suggest that the intra-segment misclassification has been noticeably reduced, there were still cases where the problem persisted albeit to a much reduced extent, as shown in the yellow circles in Fig. 5. In particular, branches with small vessels are generally more prone to the intra-segment misclassification.

We would explore two aspects for future direction. Firstly, topological regularisation may be embedded in the model via various approaches, such as postprocessing, customised convolution filter, and the feature merging between topological and representative structures. Secondly, it may be interesting to investigate the vessel segmentation problem in low-contrast background. Moreover, the proposed methods may be further evaluated by embedding the downstream clinical applications, such as the hypertension, diabetes, and atherosclerosis early diagnosis and grading.

In this work, we first identify a key point for improving the multi-class segmentation task in a real clinical application, that is the intra-segment misclassification caused by inter-class vessel intersections. Based on the hypothesis that the binary branch segmenter offers representation to the multi-class main segmenter to enhance the performance around the inter-class intersections, a binary-to-multi-class fusion network is proposed. According to experiments on three clinical data sets, our hypothesis is verified as the proposed network achieves new state-of-the-art performance.

Acknowledgements. This work is supported by EPSRC grants EP/M020533/1 EP/R014019/1 and EP/V034537/1 as well as the NIHR UCLH Biomedical Research Centre. Dr Keane is supported by a Moorfields Eye Charity Career Development Award (R190028A) and a UK Research & Innovation Future Leaders Fellowship (MR/T019050/1). Dr Jacob is funded by a Wellcome Trust Clinical Research Career Development Fellowship 209553/Z/17/Z.

References

1. Budai, A., Bock, R., Maier, A., Hornegger, J., Michelson, G.: Robust vessel segmentation in fundus images. Int. J. Biomed. Imaging **2013** (2013)

2. Chen, W., et al.: TR-GAN: topology ranking GAN with triplet loss for retinal artery/vein classification. In: Martel, A.L., et al. (eds.) MICCAI 2020. LNCS, vol. 12265, pp. 616–625. Springer, Cham (2020). https://doi.org/10.1007/978-3-030-59722-1_59

3. Dashtbozorg, B., Mendonça, A.M., Campilho, A.: An automatic graph-based approach for artery/Vein classification in retinal images. IEEE Trans. Image Process. **23**(3), 1073–1083 (2013)

4. Estrada, R., Allingham, M.J., Mettu, P.S., Cousins, S.W., Tomasi, C., Farsiu, S.: Retinal artery-vein classification via topology estimation. IEEE Trans. Med. Imaging **34**(12), 2518–2534 (2015)

5. Fedus, W., Rosca, M., Lakshminarayanan, B., Dai, A.M., Mohamed, S., Goodfellow, I.: Many paths to equilibrium: Gans do not need to decrease a divergence at every step. In: International Conference on Learning Representations (2018)

6. Fraz, M.M., et al.: Blood vessel segmentation methodologies in retinal images-a survey. Comput. Methods Programs Biomed. **108**(1), 407–433 (2012)

7. Fu, H., Xu, Y., Lin, S., Kee Wong, D.W., Liu, J.: DeepVessel: retinal vessel segmentation via deep learning and conditional random field. In: Ourselin, S., Joskowicz, L., Sabuncu, M.R., Unal, G., Wells, W. (eds.) MICCAI 2016. LNCS, vol. 9901, pp. 132–139. Springer, Cham (2016). https://doi.org/10.1007/978-3-319-46723-8_16

8. Galdran, A., Meyer, M., Costa, P., Campilho, A., et al.: Uncertainty-aware artery/vein classification on retinal images. In: 2019 IEEE 16th International Symposium on Biomedical Imaging (ISBI 2019), pp. 556–560. IEEE (2019)

9. Hemelings, R., Elen, B., Stalmans, I., Van Keer, K., De Boever, P., Blaschko, M.B.: Artery-vein segmentation in fundus images using a fully convolutional network. Comput. Med. Imaging Graph. **76**, 101636 (2019)

10. Hu, Q., Abràmoff, M.D., Garvin, M.K.: Automated separation of binary overlapping trees in low-contrast color retinal images. In: Mori, K., Sakuma, I., Sato, Y., Barillot, C., Navab, N. (eds.) MICCAI 2013. LNCS, vol. 8150, pp. 436–443. Springer, Heidelberg (2013). https://doi.org/10.1007/978-3-642-40763-5_54

11. Huang, F., Dashtbozorg, B., ter Haar Romeny, B.M.: Artery/vein classification using reflection features in retina fundus images. Mach. Vis. Appl. **29**(1), 23–34 (2018)

12. Lee, C.Y., Xie, S., Gallagher, P., Zhang, Z., Tu, Z.: Deeply-supervised nets. In: Artificial intelligence and statistics, pp. 562–570. PMLR (2015)

13. Li, L., Verma, M., Nakashima, Y., Kawasaki, R., Nagahara, H.: Joint learning of vessel segmentation and artery/vein classification with post-processing. In: Medical Imaging with Deep Learning (2020)

14. Luc, P., Couprie, C., Chintala, S., Verbeek, J.: Semantic segmentation using adversarial networks. arXiv preprint arXiv:1611.08408 (2016)

15. Ma, W., Yu, S., Ma, K., Wang, J., Ding, X., Zheng, Y.: Multi-task Neural Networks with Spatial Activation for Retinal Vessel Segmentation and Artery/Vein Classification. In: Shen, D., et al. (eds.) MICCAI 2019. LNCS, vol. 11764, pp. 769–778. Springer, Cham (2019). https://doi.org/10.1007/978-3-030-32239-7_85

16. Meyer, M.I., Galdran, A., Costa, P., Mendonça, A.M., Campilho, A.: Deep convolutional artery/Vein classification of retinal vessels. In: Campilho, A., Karray, F., ter Haar Romeny, B. (eds.) ICIAR 2018. LNCS, vol. 10882, pp. 622–630. Springer, Cham (2018). https://doi.org/10.1007/978-3-319-93000-8_71

17. Mirsharif, Q., Tajeripour, F., Pourreza, H.: Automated characterization of blood vessels as arteries and veins in retinal images. Comput. Med. Imaging Graph. **37**(7–8), 607–617 (2013)

18. Mou, L., et al.: CS-Net: channel and spatial attention network for curvilinear structure segmentation. In: Shen, D., et al. (eds.) MICCAI 2019. LNCS, vol. 11764, pp. 721–730. Springer, Cham (2019). https://doi.org/10.1007/978-3-030-32239-7_80

19. Orlando, J.I., Barbosa Breda, J., van Keer, K., Blaschko, M.B., Blanco, P.J., Bulant, C.A.: Towards a glaucoma risk index based on simulated hemodynamics from fundus images. In: Frangi, A.F., Schnabel, J.A., Davatzikos, C., Alberola-López, C., Fichtinger, G. (eds.) MICCAI 2018. LNCS, vol. 11071, pp. 65–73. Springer, Cham (2018). https://doi.org/10.1007/978-3-030-00934-2_8

20. Ronneberger, O., Fischer, P., Brox, T.: U-Net: convolutional networks for biomedical image segmentation. In: Navab, N., Hornegger, J., Wells, W.M., Frangi, A.F. (eds.) MICCAI 2015. LNCS, vol. 9351, pp. 234–241. Springer, Cham (2015). https://doi.org/10.1007/978-3-319-24574-4_28

21. Schonfeld, E., Schiele, B., Khoreva, A.: A u-net based discriminator for generative adversarial networks. In: Proceedings of the IEEE/CVF Conference on Computer Vision and Pattern Recognition, pp. 8207–8216 (2020)

22. Son, J., Park, S.J., Jung, K.H.: Retinal vessel segmentation in fundoscopic images with generative adversarial networks. arXiv preprint arXiv:1706.09318 (2017)

23. Son, J., Park, S.J., Jung, K.H.: Towards accurate segmentation of retinal vessels and the optic disc in fundoscopic images with generative adversarial networks. J. Digit. Imaging 32(3), 499–512 (2019)

24. Srinidhi, C.L., Aparna, P., Rajan, J.: Automated method for retinal artery/vein separation via graph search metaheuristic approach. IEEE Trans. Image Process. 28(6), 2705–2718 (2019)

25. Staal, J., Abràmoff, M.D., Niemeijer, M., Viergever, M.A., Van Ginneken, B.: Ridge-based vessel segmentation in color images of the retina. IEEE Trans. Med. Imaging 23(4), 501–509 (2004)

26. Wang, Z., Lin, J., Wang, R., Zheng, W.: Retinal artery/vein classification via rotation augmentation and deeply supervised u-net segmentation. In: Proceedings of the 2019 4th International Conference on Biomedical Signal and Image Processing (ICBIP 2019), pp. 71–76 (2019)

27. Wang, Z., Jiang, X., Liu, J., Cheng, K.T., Yang, X.: Multi-task siamese network for retinal artery/vein separation via deep convolution along vessel. IEEE Trans. Med. Imaging 39(9), 2904–2919 (2020)

28. Wong, T.Y., Klein, R., Klein, B.E., Tielsch, J.M., Hubbard, L., Nieto, F.J.: Retinal microvascular abnormalities and their relationship with hypertension, cardiovascular disease, and mortality. Surv. Ophthalmol. 46(1), 59–80 (2001)

29. Wu, Y., Xia, Y., Song, Y., Zhang, Y., Cai, W.: Nfn+: a novel network followed network for retinal vessel segmentation. Neural Netw. 126, 153–162 (2020)

30. Xie, J.: Classification of retinal vessels into artery-vein in OCT angiography guided by fundus images. In: Martel, A.L., et al. (eds.) MICCAI 2020. LNCS, vol. 12266, pp. 117–127. Springer, Cham (2020). https://doi.org/10.1007/978-3-030-59725-2_12

31. Xu, X., et al.: Simultaneous arteriole and venule segmentation with domain-specific loss function on a new public database. Biomed. Opt. Express 9(7), 3153–3166 (2018)

32. Zhao, Y., et al.: Retinal vascular network topology reconstruction and artery/vein classification via dominant set clustering. IEEE Trans. Med. Imaging 39(2), 341–356 (2019)

Flip Learning: Erase to Segment

Yuhao Huang[1,2,3], Xin Yang[1,2,3], Yuxin Zou[1,2,3], Chaoyu Chen[1,2,3],
Jian Wang[1,2,3], Haoran Dou[4], Nishant Ravikumar[4,5],
Alejandro F. Frangi[1,4,5,6], Jianqiao Zhou[7], and Dong Ni[1,2,3(✉)]

[1] National-Regional Key Technology Engineering Laboratory for Medical
Ultrasound, School of Biomedical Engineering, Health Science Center,
Shenzhen University, Shenzhen, China
nidong@szu.edu.cn
[2] Medical Ultrasound Image Computing (MUSIC) Lab, Shenzhen University,
Shenzhen, China
[3] Marshall Laboratory of Biomedical Engineering, Shenzhen University,
Shenzhen, China
[4] Centre for Computational Imaging and Simulation Technologies in Biomedicine
(CISTIB), University of Leeds, Leeds, UK
[5] Leeds Institute of Cardiovascular and Metabolic Medicine, University of Leeds,
Leeds, UK
[6] Medical Imaging Research Center (MIRC), KU Louven, Leuven, Belgium
[7] Department of Ultrasound Medicine, Ruijin Hospital, School of Medicine,
Shanghai Jiaotong University, Shanghai, China

Abstract. Nodule segmentation from breast ultrasound images is challenging yet essential for the diagnosis. Weakly-supervised segmentation (WSS) can help reduce time-consuming and cumbersome manual annotation. Unlike existing weakly-supervised approaches, in this study, we propose a novel and general WSS framework called Flip Learning, which only needs the box annotation. Specifically, the target in the label box will be erased gradually to flip the classification tag, and the erased region will be considered as the segmentation result finally. Our contribution is three-fold. First, our proposed approach erases superpixel level using a Multi-agent Reinforcement Learning framework to exploit the prior boundary knowledge and accelerate the learning process. Second, we design two rewards: classification score and intensity distribution reward, to avoid under- and over-segmentation, respectively. Third, we adopt a coarse-to-fine learning strategy to reduce the residual errors and improve the segmentation performance. Extensively validated on a large dataset, our proposed approach achieves competitive performance and shows great potential to narrow the gap between fully-supervised and weakly-supervised learning.

Keywords: Ultrasound · Weakly-supervised segmentation · Reinforcement learning

---

Y. Huang and X. Yang—Contribute equally to this work.

© Springer Nature Switzerland AG 2021
M. de Bruijne et al. (Eds.): MICCAI 2021, LNCS 12901, pp. 493–502, 2021.
https://doi.org/10.1007/978-3-030-87193-2_47

1 Introduction

Nodule segmentation in breast ultrasound (US) is important for quantitative diagnostic procedures and treatment planning. Image segmentation's performance has been significantly advanced by the recent availability of fully-supervised segmentation methods [7,8]. However, training such methods usually relies on the availability of pixel-level masks laboriously and manually annotated by sonographers. Hence, designing an automatic weakly-supervised segmentation (WSS) based system that only requires coarse labels, e.g. bounding box (BBox), is desirable to ease the pipeline of manual annotation and save time for sonographers.

Breast cancer occurs in the highest frequency in women among all cancers and is also one of the leading causes of cancer death worldwide [15]. Thus, extracting the nodule boundary is essential for detecting and diagnosing breast cancer at its early stage. As shown in Fig. 1, segmenting the nodule's boundary from the US image with weak annotation, i.e., BBox, is still very challenging. First, nodules of the same histological type may present completely different US image characteristics because of variances in their disease differentiation and stage. Second, nodules of different types have extremely high inter-class differences and often display varied appearance patterns, making designing machine learning algorithms difficult. The third challenge is that different tissues of US images have different echo characteristics. Therefore, the intensity distribution of foreground and background in different US images also has great diversity.

Fig. 1. Breast nodule images in 2D US with annotated box and boundary.

In the WSS literature, Class Activation Mapping (CAM) based methods [2,14,22] were proposed to visualise the most discriminative features and regions of interest obtained by the classifier. Wei et al. [19] proposed to erase the CAM area predicted by the classifier constantly to optimise its performance on the WSS task. However, classifiers are only responsive to small and sparse discriminative regions from the object of interest, which deviates from the segmentation task requirement that needs to localise dense, interior and integral regions for pixel-wise inference. Therefore, the above methods ignored the pairing relationship between pixels in the image and strongly relied on the CAMs with well positioning and coverage performance. Except for the image-level WSS approaches based on inaccurate CAMs, some methods employed box annotations to obtain high-quality prediction masks at a small annotation cost. However, most of them, such as BoxSup [3], SDI [5] and Box2Seg [6], highly rely on pseudo-mask generation algorithms (e.g., MCG [10] or GrabCut [12] based on object shape priors). Thus, they may not suit US image segmentation tasks.

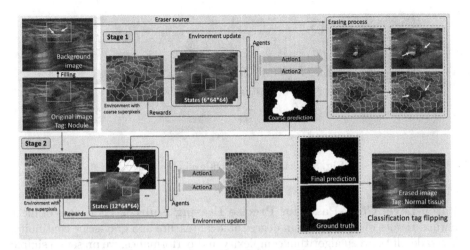

Fig. 2. The workflow of our proposed framework.

In this study, we propose a novel and general Flip Learning framework for BBox based WSS. We believe our proposed framework is totally different from the current existing WSS methods. Our contribution is three-fold. First, the erasing process via Multi-agent Reinforcement Learning (MARL) is based on superpixels, which can capture the prior boundary information and improve learning efficiency. Second, we carefully design two rewards for guiding the agents accurately. Specifically, the classification score reward (CSR) is used for pushing the agents' erasing for label flipping, while the intensity distribution reward (IDR) is employed to limit the over-segmentation. Third, we employ a coarse-to-fine (C2F) strategy to simplify agents' learning for residuals decreasing and segmentation performance improvement. Validation experiments demonstrated that the proposed Flip Learning framework could achieve high accuracy in the nodule segmentation task.

2   Method

Figure 2 shows the workflow of our proposed framework for nodule segmentation in 2D US images. The proposed Flip Learning framework is based on MARL, in which the agents erase the nodule from its BBox. The classification score of the nodule will decrease progressively, and the tag will be flipped. The erased region will be taken as the final segmentation prediction. In our erase-to-segment system, we first generate a background image to provide the eraser source to fill the erased region suitably to be indistinguishable from normal tissue. We further generate superpixels in the BBox for prior boundary extraction and thus improve learning efficiency. Then, a C2F strategy is employed to obtain an accurate segmentation result effectively.

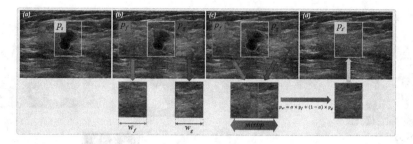

Fig. 3. Example of patch-level copy-paste.

Patch-Level Copy-Paste for Precise Erasing Source. US images of most normal tissues display a certain degree of local texture and gray-scale continuity. To trade-off between algorithm complexity and performance, we present a simple and feasible method to fill the BBox area. As shown in Fig. 3: (a)-(b) According to the context information, we first obtain the patches (p_f and p_g with width w_f and w_g, respectively) in the original image as region proposals. These patches are similar to the background around the annotated box. (c) Next, we adopt a *mixup* [21] to copy and paste these region proposals to generate the pseudo-background image (see Eq. 1). (d) Finally, to achieve a higher quality fusion between the pseudo-background and the real background, we adopt a local-edge mean filter to optimise the copy-paste edge area $p_{s'}$.

$$p_{s'}(x,y) = \alpha \times p_f(x,y) + (1 - \alpha) \times p_g(x,y), \quad \alpha = \frac{x}{(w_f + w_g)/2} \tag{1}$$

MARL Framework for Efficient and Accurate Erasing. Recently, RL-based methods have shown great potential in various medical imaging tasks [20, 23]. To accelerate the learning process, we propose to use a MARL framework with two agents erasing the object simultaneously. We further define the MARL erasing framework using these elements:

-Environment: The *Environment* is defined as the BBox area of the original US image. It contains the object to be segmented and limits the agents' moving. However, it is noted that erasing based on pixel-level is difficult and inefficient due to the weak supervised signal. Thus, we use superpixel algorithms for enhancing the supervised signal, which group pixels with similar properties into perceptually meaningful atomic regions while considering spatial constraints. Additionally, it preserves the edge information of the original image during enhancing the local consistency. As explained above, in this study, we generate superpixels for the BBox area to obtain a superpixel-level environment before erasing. Specifically, we adopt Superpixels Extracted via EnergyDriven Sampling (SEEDS) to obtain superpixel-blocks (details refer to [1]).

-Agents: The *Agent* is to learn a policy for segmentation by interacting with the superpixel-level environment. In our study, two agents $Agent_{k, k=1,2}$

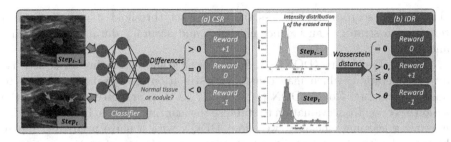

Fig. 4. Details of (a) CSR and (b) IDR. Note that we first train a classifier to classify 'nodule' and 'normal tissue', and obtain CSR by calculating the differences of nodule scores between $step_{t-1}$ and $step_t$.

share the parameters in the convolution layers for knowledge sharing, and have independent fully connected layers for decision-making. They traverse the whole BBox with each of them handling approximately half of the total superpixels, and the centre superpixel of BBox is set as start point for each agent.

-States: The *State* of one agent can be defined as a 64×64 area, with its centre indicates the agent's position. We further define *States* as the last six areas observed by two agents, with the size of $6 \times 64 \times 64$. States concatenating can provide rich information, which can promote agents' learning.

-Actions: The *Action* space of our framework contains only two operations, including *erasing* and *passing*. Note that the action is taken on the superpixel level. The action *erasing* indicates that this superpixel will be erased, and the region will be filled according to the generated background image. The action *passing* represents the agents choose not to erase this superpixel region.

-Rewards: The *Reward* guides the agents' erasing process. As shown in Fig. 4, we design two rewards, including 1) CSR and 2) IDR. Specifically, CSR is a basic reward used to guide the agents to erase the object from the BBox for classification tag flipping. However, using such reward separately may cause over-segmentation because the agents may tend to fill the whole BBox to make a high score. Thus, we introduce additional IDR to overcome this problem. Specifically, if one superpixel is erased in $step_t$ and makes the intensity distribution (I_t) of the erased area highly different from that in $step_{t-1}$, the agent will be punished. The difference between the two distributions is calculated using the Wasserstein distance [18]. The total reward R_k for each $Agent_k$ can be defined as:

$$R_k = sgn(Sc_{t-1} - Sc_t) + thr(W(I_{t-1}, I_t), \theta), \tag{2}$$

where sgn and thr are the sign and threshold function, respectively. Sc_t represents the nodule scores in $step_t$. $W(\cdot)$ calculates the Wasserstein distance between the erased region's intensity distribution in $step_{t-1}$ and $step_t$, and $\theta = 25$ is the pre-set threshold. More details can be seen in Fig. 4.

-Terminal Signals: The terminal signal represents when to terminate the agent-environment interaction. We adopt two types of terminated strategy in our study: 1) attaining the maximum number of traversals $\mathcal{N} = 2$ and 2) classification score of nodule less than a pre-set threshold $\beta = 0.05$. Such a termination strategy can balance efficiency and accuracy during both training and testing.

In deep Q-network (DQN) [9], both selecting and evaluating an action use a *max* operation, which may cause over-estimation for Q-values and lead to unstable training. Double DQN (DDQN) [17] is then proposed to decouple the action selection from evaluation to stabilise the learning process. In this study, for improving learning efficiency, we adopt DDQN with a naive prioritised replay buffer [13] with the size of $\mathcal{M}$, which contains sequences of state s, action a, reward r and next state s'. The q_{target} can be calculated by:

$$q_{target} = r + \gamma Q(s', \underset{a'}{argmax}\, Q(s', a'; \omega); \omega'), \qquad (3)$$

$Q(\omega)$ and $Q(\omega')$ represent current and target Q-networks, respectively. γ is the discount factor that balances the importance of current and future rewards. Then, with the uniform sampling $U(\cdot)$, and the loss function can be written by:

$$L = \mathbb{E}_{s,r,a,s' \sim U(\mathcal{M})}(q_{target} - Q(s, a; \omega))^2 \qquad (4)$$

C2F Learning Strategy. Finer superpixel implies more accurate boundary prior knowledge, leading to a higher upper bound of segmentation performance. However, learning on the BBox with fine-superpixel directly may be too difficult for the agents. The agents may get into local optimality easily due to the weak supervised signal. Thus, the segmentation result may contain residual errors. Inspired by [16], we adopt a two-stage strategy for reducing the residual errors and improving the segmentation performance. Specifically, in the first stage of learning, the agents learn on a coarse superpixel BBox and output a coarse prediction (see *Stage* 1 in Fig. 2). Such coarse predictions are leveraged to make a channel-wise concatenation with original images as the second stage's input, which can provide additional information and help the agent learn on the BBox with fine-superpixel efficiently (see *Stage* 2 in Fig. 2).

3   Experimental Result

Materials and Implementation Details. We validated our proposed Flip Learning method on segmenting the breast nodule in 2D US images. Approved by the local IRB, a total of 1,723 images (with size 448×320) were collected from 1,129 patients. Each image's nodule was annotated with the mask and BBox by sonographers manually (see Fig. 1). The dataset was split into 1278, 100 and 345 images for training, validation and independent testing at the patient level with no overlap. The classifier and agents used the same training, validation and testing set. In this study, we implemented our method in *Pytorch*, using an NVIDIA TITAN 2080 GPU. We first trained the classifier with architecture of ResNet18 using AdamW optimiser with learning rate = 1e-3 and batch-size = 128. Then, the proposed MARL system (ResNet18 backbone) was trained with Adam optimiser in 100 epochs, costing about 1.5 days (learning rate = 5e-5

Table 1. Method comparison (mean±std). The best WSS results are shown in bold.

	DICE↑	JAC↑	CON↑	HD↓	ASD↓
U-net	93.44 ± 3.76	87.91 ± 5.02	84.76 ± 9.22	15.22 ± 8.99	2.68 ± 1.79
GrabCut	1.66 ± 8.91	1.07 ± 5.96	$< 10^3$	38.11 ± 124.32	8.11 ± 27.32
Saliency	43.02 ± 9.74	27.89 ± 7.93	−199.90 ± 189.29	85.08 ± 20.04	25.86 ± 4.52
Grad-CAM	52.66 ± 14.34	38.77 ± 12.81	−153.95 ± 565.80	61.71 ± 27.44	24.91 ± 7.08
Grad-CAM++	61.99 ± 11.14	45.94 ± 11.13	−40.17 ± 129.12	57.11 ± 24.23	22.91 ± 6.81
SP-RG	62.66 ± 14.98	47.31 ± 15.78	−64.76± 519.77	61.99 ± 20.60	14.17 ± 9.22
Ours	**91.12 ± 2.79**	**83.81 ± 4.62**	**80.31 ± 6.90**	**16.95 ± 7.29**	**2.80 ± 1.33**

Table 2. Ablation study (mean±std). The best results are shown in bold.

Strategy			DICE↑	JAC↑	CON↑	HD↓	ASD↓
MARL	IDR	C2F					
✗	✗	✗, F	47.95 ± 19.17	33.26 ± 16.59	−314.65 ± 1150.81	27.78 ± 9.28	5.21 ± 4.78
✗	✗	✗, C	77.33 ± 10.22	65.09 ± 12.77	38.11 ± 10.88	21.81 ± 9.47	3.43 ± 1.96
✓	✗	✗, F	46.27 ± 15.32	31.99 ± 16.72	−177.89 ± 820.77	28.32 ± 9.87	5.09 ± 3.71
✓	✗	✗, C	77.64 ± 12.33	65.23 ± 12.08	37.33 ± 12.05	22.08 ± 9.31	3.62 ± 1.45
✓	✓	✗, F	79.08 ± 8.77	69.87 ± 11.03	41.67 ± 14.99	21.73 ± 7.14	3.68 ± 1.77
✓	✓	✗, C	83.41 ± 9.52	72.66 ± 10.72	53.66 ± 12.83	20.01 ± 5.32	3.21 ± 1.41
✓	✗	✓	83.98 ± 11.27	74.85 ± 11.22	56.72 ± 13.20	19.55 ± 6.05	3.00 ± 1.59
✓	✓	✓	**91.12 ± 2.79**	**83.81 ± 4.62**	**80.31 ± 6.90**	**16.95 ± 7.29**	**2.80 ± 1.33**

and batch-size = 64). The superpixels are generated by *OpenCV* function and indexed from 1 to S. Both agents start from the centre superpixel with index S/2. One agent traverses the superpixels from S/2 to S, the other one traverses reversely from index S/2 to 1. The replay buffer has a size of 8000, and the target Q-network copied the parameters of the current Q-network every 1200 iterations.

Quantitative and Qualitative Analysis. We evaluated the segmentation performance by five metrics, including indicators of Dice similarity coefficient (DICE-%), Jaccard index (JAC-%), Conformity(CON-%), Hausdorff distance (HD-pixel) and Average surface distance (ASD-pixel). These metrics can provide an objective evaluation between the ground truth (GT) and the prediction.

As shown in Table 1, we reported the quantitative results of Flip Learning (Ours) and six other methods (with the same BBox annotations for labels as ours) including, U-net [11], Grad-CAM [14], Grad-CAM++ [2], GrabCut [12], Saliency [4], and superpixel-level region growing based on intensity distribution (SP-RG). It can be seen that our methods outperform all the traditional methods and the most common WSS approaches (i.e., Grad-CAM and Grad-CAM++). It is also noted that the results of our proposed method are very closed to that of U-net, which is a *fully-supervised* method.

In the ablation studies, we conducted different experiments to test the superiority of our proposed MARL, IDR, and C2F. CSR is set as the basic reward for all the methods. As shown in Table 2, MARL may not boost the performance in

Table 3. Impact of annotation box shift (mean ± std).

	DICE↑	JAC↑	CON↑	HD↓	ASD↓
0–10 pixels	89.22 ± 4.12	81.22 ± 3.22	72.28 ± 12.18	17.02 ± 7.87	2.88 ± 1.31
10–20 pixels	88.16 ± 6.23	79.56 ± 5.32	69.73 ± 14.11	16.53 ± 7.19	2.92 ± 2.15
20–30 pixels	86.72 ± 8.33	63.44 ± 6.33	53.09 ± 11.28	23.84 ± 9.88	4.12 ± 2.13

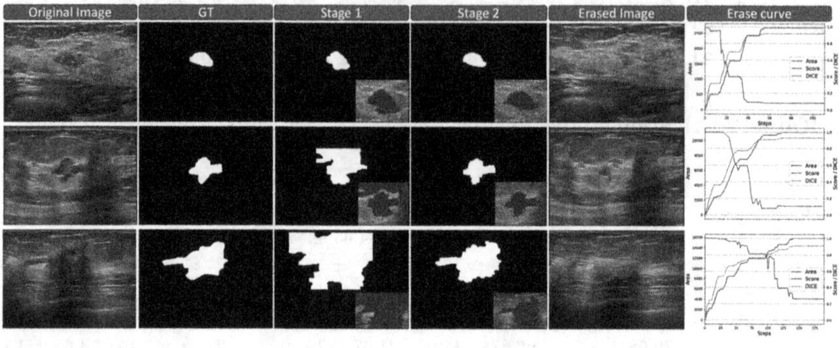

Fig. 5. Typical cases of flip learning. Mask predictions have been post-processed. Note that the classification tags of original images are 'nodule', and after two-stage erasing, their tags will flip to 'normal tissue'. The erase curves show the variation of classification scores during erasing. (Color figure online)

both fine and coarse superpixel-based environment (row 1–4). However, it can save almost 50% running times compared with the single-agent situation. Interesting to see that interacting with the fine-superpixel-based environment directly (i.e., one-stage) makes agents learn difficultly. Without enough information for guiding agents' action in the huge search space, they may fail to segment accurately. Thus, there will remain many segmentation residuals, which can cause obvious performance degradation (row 1–6). The contribution of IDR and C2F can be observed in the last 4 rows: equipping each of them separately can boost the accuracy, while combining them will obtain a great improvement in all the evaluation metrics. To test the sensitivity of our approach to box annotation, we validated it on different box shifting levels, including 1) 0–10 pixels, 2) 10–20 pixels and 3) 20–30 pixels. The results reported in Table 3 indicate that our methods can perform well though the box's centre is shifting.

Figure 5 shows three typical cases of our proposed method. Compared with the result of stage one, it can be seen that the final result of stage two obtains a more accurate boundary and overall mask, which is very close to the GT. The erase curves shown in the last column indicate the relationship among erased area size (green), DICE (yellow), and classification score (red). It can be observed that through erasing, the DICE and erased area are gradually increasing, and their variation is nearly synchronous. Moreover, the classification score curve will decrease continuously, and the classification tag will be flipped from 'nodule' to 'normal tissue', which proves the effectiveness of our Flip Learning approach.

4   Conclusion

We propose a novel Flip Learning framework for nodule segmentation in 2D US images. We use MARL to erase the nodule from the superpixel-based BBox to flip its classification tag. We develop two rewards, including CSR and IDR, for overcoming the under- and over-segmentation, respectively. Moreover, we propose to adopt a C2F learning strategy in two stages, which can achieve more accurate results than a one-stage method. Experiments on our large in-house dataset validate the efficacy of our method. Our patch-level copy-paste filling strategy is limited in some cases. Thus, in the future, we will explore a more general background filling approach (e.g. GAN), to generate a more accurate background for different types of images.

Acknowledgment. This work was supported by the National Key R&D Program of China (No. 2019YFC0118300), Shenzhen Peacock Plan (No. KQTD20160-53112051497, KQJSCX20180328095606003), Royal Academy of Engineering under the RAEng Chair in Emerging Technologies (CiET1919/19) scheme, EPSRC TUSCA (EP/V04799X/1) and the Royal Society CROSSLINK Exchange Programme (IES/NSFC/201380).

References

1. Van den Bergh, M., Boix, X., Roig, G., de Capitani, B., Van Gool, L.: SEEDS: Superpixels extracted via energy-driven sampling. In: Fitzgibbon, A., Lazebnik, S., Perona, P., Sato, Y., Schmid, C. (eds.) ECCV 2012. LNCS, vol. 7578, pp. 13–26. Springer, Heidelberg (2012). https://doi.org/10.1007/978-3-642-33786-4_2
2. Chattopadhay, A., Sarkar, A., Howlader, P., Balasubramanian, V.N.: Gradcam++: generalized gradient-based visual explanations for deep convolutional networks. In: 2018 IEEE Winter Conference on Applications of Computer Vision (WACV), pp. 839–847. IEEE (2018)
3. Dai, J., He, K., Sun, J.: Boxsup: exploiting bounding boxes to supervise convolutional networks for semantic segmentation. In: Proceedings of the IEEE International Conference on Computer Vision, pp. 1635–1643. IEEE (2015)
4. Hou, X., Zhang, L.: Saliency detection: a spectral residual approach. In: 2007 IEEE Conference on Computer Vision and Pattern Recognition, pp. 1–8. IEEE (2007)
5. Khoreva, A., Benenson, R., Hosang, J., Hein, M., Schiele, B.: Simple does it: weakly supervised instance and semantic segmentation. In: Proceedings of the IEEE Conference on Computer Vision and Pattern Recognition, pp. 876–885. IEEE (2017)
6. Kulharia, V., Chandra, S., Agrawal, A., Torr, P., Tyagi, A.: Box2Seg: attention weighted loss and discriminative feature learning for weakly supervised segmentation. In: Vedaldi, A., Bischof, H., Brox, T., Frahm, J.-M. (eds.) ECCV 2020. LNCS, vol. 12372, pp. 290–308. Springer, Cham (2020). https://doi.org/10.1007/978-3-030-58583-9_18
7. Liu, S., et al.: Deep learning in medical ultrasound analysis: a review. Engineering 5(2), 261–275 (2019)
8. Minaee, S., Boykov, Y., Porikli, F., Plaza, A., Kehtarnavaz, N., Terzopoulos, D.: Image segmentation using deep learning: A survey. arXiv preprint arXiv:2001.05566 (2020)

9. Mnih, V., et al.: Human-level control through deep reinforcement learning. Nature **518**(7540), 529–533 (2015)

10. Pont-Tuset, J., Arbelaez, P., Barron, J.T., Marques, F., Malik, J.: Multiscale combinatorial grouping for image segmentation and object proposal generation. IEEE Trans. Pattern Anal. Mach. Intell. **39**(1), 128–140 (2016)

11. Ronneberger, O., Fischer, P., Brox, T.: U-net: convolutional networks for biomedical image segmentation. In: Navab, N., Hornegger, J., Wells, W.M., Frangi, A.F. (eds.) MICCAI 2015. LNCS, vol. 9351, pp. 234–241. Springer, Cham (2015). https://doi.org/10.1007/978-3-319-24574-4_28

12. Rother, C., Kolmogorov, V., Blake, A.: "Grabcut" interactive foreground extraction using iterated graph cuts. ACM Trans. Graph. (TOG) **23**(3), 309–314 (2004)

13. Schaul, T., Quan, J., Antonoglou, I., Silver, D.: Prioritized experience replay. In: ICLR (Poster) (2016)

14. Selvaraju, R.R., Cogswell, M., Das, A., Vedantam, R., Parikh, D., Batra, D.: Gradcam: visual explanations from deep networks via gradient-based localization. In: Proceedings of the IEEE International Conference on Computer Vision, pp. 618–626. IEEE (2017)

15. Siegel, R.L., Miller, K.D., Fuchs, H.E., Jemal, A.: Cancer statistics, 2021. CA: Cancer J. Clin. **71**(1), 7–33 (2021)

16. Tu, Z.: Auto-context and its application to high-level vision tasks. In: 2008 IEEE Conference on Computer Vision and Pattern Recognition, pp. 1–8. IEEE (2008)

17. Van Hasselt, H., Guez, A., Silver, D.: Deep reinforcement learning with double q-learning. In: Thirtieth AAAI Conference on Artificial Intelligence (2016)

18. Villani, C.: Optimal transport: old and new, vol. 338. Springer, Cham (2008)

19. Wei, Y., Feng, J., Liang, X., Cheng, M.M., Zhao, Y., Yan, S.: Object region mining with adversarial erasing: A simple classification to semantic segmentation approach. In: Proceedings of the IEEE Conference on Computer Vision and Pattern Recognition, pp. 1568–1576. IEEE (2017)

20. Yang, X., et al.: Searching collaborative agents for multi-plane localization in 3d ultrasound. Med. Image Anal. **72**, 102119 (2021)

21. Zhang, H., Cissé, M., Dauphin, Y.N., Lopez-Paz, D.: mixup: Beyond empirical risk minimization. arXiv preprint arXiv:1710.09412 (2017)

22. Zhou, B., Khosla, A., Lapedriza, A., Oliva, A., Torralba, A.: Learning deep features for discriminative localization. In: Proceedings of the IEEE Conference on Computer Vision and Pattern Recognition, pp. 2921–2929. IEEE (2016)

23. Zhou, S.K., Le, H.N., Luu, K., Nguyen, H.V., Ayache, N.: Deep reinforcement learning in medical imaging: A literature review. arXiv preprint arXiv:2103.05115 (2021)

DC-Net: Dual Context Network for 2D Medical Image Segmentation

Rongtao Xu[1,3], Changwei Wang[1,3], Shibiao Xu[2(✉)], Weiliang Meng[1,3], and Xiaopeng Zhang[1,3]

[1] National Laboratory of Pattern Recognition, Institute of Automation, Chinese Academy of Sciences, Beijing, China
[2] School of Artificial Intelligence, Beijing University of Posts and Telecommunications, Beijing, China
shibiaoxu@bupt.edu.cn
[3] School of Artificial Intelligence, University of Chinese Academy of Sciences, Beijing, China

Abstract. Medical image segmentation is essential for disease diagnosis analysis. There are many variants of U-Net that are based on attention mechanism and dense connections have made progress. However, CNN-based U-Net lacks the ability to capture the global context, and the context information of different scales is not effectively integrated. These limitations lead to the loss of potential context information. In this work, we propose a Dual Context Network (DC-Net) to aggregate global context and fuse multi-scale context for 2D medical image segmentation. In order to aggregate the global context, we present the Global Context Transformer Encoder (GCTE), which reshapes the original image and the multi-scale feature maps into a sequence of image patches, and combines the advantages of Transformer Encoder on global context aggregation to improve the performance of encoder. For the fusion of multi-scale context, we propose the Adaptive Context Fusion Module (ACFM) to adaptively fuse context information by learning Adaptive Spatial Weights and Adaptive Channel Weights to improve the performance of decoder. We apply our DC-Net with GCTE and ACFM to skin lesion segmentation and cell contour segmentation tasks, experimental results show that our method can outperform other advanced methods and get state-of-the-art performance.

Keywords: Medical image segmentation · Visual transformer · Multi-scale context fusion.

1 Introduction

In recent years, deep learning has been successfully applied to medical image segmentation tasks such as skin lesion segmentation, vascular wind, lung segmentation, and cell segmentation. In particular, U-Net [16] and its variants [5,8,

---

R. Xu and C. Wang—Contributed equally

© Springer Nature Switzerland AG 2021
M. de Bruijne et al. (Eds.): MICCAI 2021, LNCS 12901, pp. 503–513, 2021.
https://doi.org/10.1007/978-3-030-87193-2_48

14,15,23] that are based on fully convolutional neural network (FCN) [13] have achieved great success in many medical image segmentation tasks. Basically, U-Net consists of an encoder, a decoder and a skip connection between them. The skip connection can add details to the decoder that the encoder loses. However, the convolutional layer of the above methods is only a local operation and lacks the integration of context information, while medical image segmentation usually benefits from a wide range of context information. Additionally, U-Net can obtain limited context aggregation ability by continuously stacking downsampling encoders, which is far from the optimal solution.

To improve the CNN's awareness of context, non-local Neural Networks [19] uses self-attention mechanisms to model long-distance dependencies between pixels to obtain global context information. On the other hand, DeeplabV3 [3] and CE-Net [8] integrate multi-scale context information, these context information from different receptive fields can further enhance the expressive ability of features. Inspired by the above works, we try to combine these two ways of context augmented to make U-Net gain stronger context awareness.

In this work, we present DC-Net, powered by Global Context Transformer Encoder (GCTE) and Adaptive Context Fusion Module (ACFM) to mitigate the above issues, the contributions are given as below: First, we designed the GCTE based on the VIT [21] inspiration, and aggregated global context information by modeling the long-range dependence between pixels. Second, we designed the ACFM to adaptively fuse context information of different scales to further take advantage of U-Net's inherent feature hierarchy. Our proposed method has enhanced U-Net's ability to perceive context information from the two aspects of aggregating global context and fusing multi-scale context, thereby greatly improving the accuracy of segmentation. Finally, we verified the effectiveness of our method on the ISIC 2018 and ISBI 2012 datasets. Experiments show that our method surpasses the existing state-of-the-art methods on both datasets.

2   Related Works

Medical Image Segmentation Based on Deep Learning. Fully Convolutional Network (FCN) [13] frameworks based U-Net [16] and DeepLab [3,4] are widely used for image segmentation. There are also some variants of U-Net with good performance for medical image segmentation, such as Attention U-Net [15], U-Net++ [23], Inf-Net [6]. Although these methods usually use the attention mechanism and dense connections to improve the feature representation ability of U-Net, they do not integrate the global context or multi-scale context, which are crucial for medical image segmentation.

Global Context Augmented. The self-attention mechanism [19] is proposed based on capturing the dependencies between long-range features, which has been used for image classification [19], image generation [22] and image segmentation [20], etc. No-local U-Net proposes a global aggregation block based on the self-attention mechanism to gather global information to obtain more accurate segmentation results. Transformer [17], which has been successfully applied in NLP tasks recently, was introduced into vision tasks by Visual Transformer [21]. Visual Transformer provides a purely self-attention-based pipeline by converting images into patch sequences to further enhance the network's ability to model the global context. In this work, we propose a Global Context Transformer Encoder based on Visual Transformer to augment U-Net's perception of the global context.

Cross-Scale Context Aggregation. Integrating multi-scale context information is a common method to improve the expressive ability of features in many visual tasks [3,8,12,18]. DeepLabV3 [3] designed an Atrous Spatial Pyramid Pooling module to integrate the context information of multiple receptive fields. CE-Net [8] proposed dense atrous convolution block and residual multi-kernel pooling to capture multi-scale context information. In this work, we try to fuse the multi-scale context information output by the U-Net decoding layer through an adaptive weighting method to obtain a more powerful feature representation.

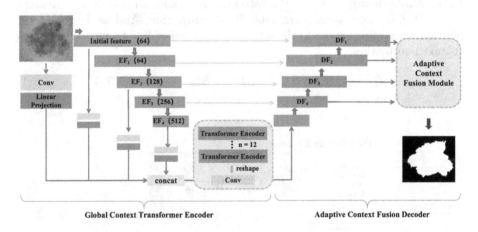

Fig. 1. The structure of the proposed DC-Net with Global Context Transformer Encoder (GCTE) and Adaptive Context Fusion Module (ACFM).

3   Our Proposed Method

We propose a Dual-context network (DC-Net) for medical image segmentation tasks. As shown in Fig. 1, the proposed DC-Net consists of two main parts: Global Context Transformer Encoder and Adaptive Context Fusion Module. In Sect. 3.1, our transformer is combined to encode the feature representation of the image patches decomposed by the multi-scale feature maps. In Sect. 3.2, different proportions of context information are adaptively fused for decoding.

3.1   Global Context Transformer Encoder

Multi-scale Feature Serialization. Different from Vision Transformer [21], our Global Context Transformer Encoder first performs multi-scale feature Serialization. As shown in Fig. 1, the initial feature maps is obtained once the original image undergoes the first pooling. Then the Initial feature maps are processed by four feature extraction blocks of ResNet-34 [9] in turn to obtain the first Encode Feature maps (EF_1), the second Encode Feature maps (EF_2), the third Encode Feature maps (EF_3), and the fourth Encode Feature maps (EF_4). To aggregate global context information, we perform linear projection and patch embeddings to the original image, initial feature maps, EF_2 and EF_4, respectively. Specifically, in order to process 2D images with a resolution of (H, W) and feature maps of various scales, we reshape the input $x \in \mathbb{R}^{H \times W \times C}$ into a series of flattened 2D patches $x_p^i \in \mathbb{R}^{P^2 \times C}$ where P is the size of each patch, the value of i is an integer ranging from 1 to N. The length of the input sequence is obtained by $N = \frac{H \times W}{P^2}$.

Patch Embedding. We map the flattened image and multi-scale feature maps to the D dimension using a trainable linear projection. And we learn specific position embeddings and add them to the patch embedding to encode the patch context information:

$$Z_0^i = \left[x_p^1 E^i; x_p^2 E^i; \cdots x_p^N E^i\right] + E_{pos}^i \qquad i = 1, 2, 3, 4 \qquad (1)$$

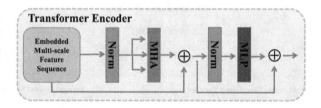

Fig. 2. The structure of the Transformer Encoder, including Multi-Head Attention (MHA) and Multi-Layer Perceptron (MLP) blocks.

Global Context Transformer as Encoder. We use the encoded image patches and the multi-scale feature maps patches as the input of Transformer Encoder [21]. The structure of the Transformer Encoder is shown in Fig. 2. The Transformer based on the self-attention mechanism can solve the limitation that CNN cannot model long-range dependence. For specific details, we suggest that readers review [21]. As an advantage, CNN has inherent hierarchical feature maps, and feature maps at different levels contain different information. The high-level feature maps contain more semantic information and the low-level feature maps contain more detailed information. In order to combine the advantages of CNN and transformer, we fuse multi-level information to obtain better image representation, which is fed into the transformer. Better feature representation feeds can further stimulate the global context modeling ability of transformer. Compared with simply applying transformer directly, our Global Context Transformer Encoder can achieve a higher segmentation score. We will discuss this in the ablation experiment.

3.2 Decoder with Adaptive Context Fusion Module

Effectively recovering the high-level semantic feature maps extracted from the encoder is very important to the decoder, which can affect the quality of the segmentation results significantly. Due to different receptive fields, feature maps of different scales contain different levels of context information. However, the inconsistency between different feature scales limits the fusion of context information of each scale. The lack of cross-scale context leads to inflexible in processing lesions with complex boundaries because it is difficult to adapt to changes in the scale and pathological environment of the lesion or cell. In order to solve the above problems, we propose the Adaptive Context Feature Module (ACFM), which can make full use of the context and semantic information of high-level feature maps and the detailed and local information of low-level feature maps. The learnable weight parameters allow ACFM to adaptively fuse cross-scale context information with low computing cost.

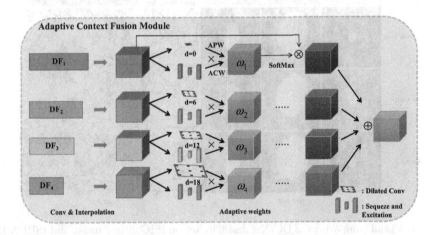

Fig. 3. The details of the Adaptive Context Fusion Module.

The detailed structure of ACFM is shown in Fig. 3. We cascade four feature decoding blocks, each feature decoding block contains a 1×1 convolution, a 3×3 deconvolution, and a 1×1 convolution. Then we obtain the Decode Feature maps DF_1, DF_2, DF_3, and DF_4 after the skip connection and the feature decoding blocks. For adaptive learning context fusion, we decouple adaptive weights (ω_i)

into Adaptive Spatial Weights (APW) and Adaptive Channel Weights (ACW). We learn the APW and ACW of different scales through carefully designed modules, and multiply them with the feature maps of the corresponding resolutions and then concatenate the results. Specifically, in order to obtain context information of different receptive fields, we use a 1×1 convolution, and three dilated convolutions with dilation rate of 6, 12, and 18 respectively to process the DF_1, DF_2, DF_3, and DF_4. As shown in Fig. 3, then the obtained initial spatial weights are softmax normalized to obtain the APW. Similarly, we learn the corresponding ACW by applying the Sequeze and Excitation[10] to DF_1, DF_2, DF_3, and DF_4. ω_i is obtained by multiplying APW and ACW. This softmax process is shown in the following formula:

$$\omega_i = \frac{e^{\omega_i}}{\sum_i^4 e^{\omega_i}} \tag{2}$$

The four feature maps obtained by multiplication are added to obtain adaptive context aggregation feature maps, and the final segmentation results are obtained after two convolution layers processing.

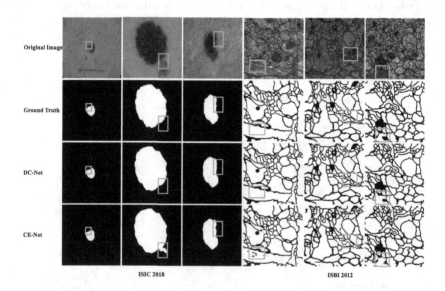

Fig. 4. Visual comparison of DC-Net and CE-Net on ISIC 2018 dataset and ISBI 2012 dataset. From top to bottom are the original image, the ground truth, the segmentation result of our DC-Net, and the segmentation result of CE-Net.

4 Experiments and Discussion

4.1 Experimental Settings

Dataset and Evaluation. The ISIC 2018 skin lesion segmentation dataset [2] is annotated by experienced dermatologists, and the goal is to automatically segment melanoma from dermoscopic images. This data set contains 2594 skin lesion images and their corresponding ground truth. We randomly divide the data set into training set, validation set and test set by 70%, 10%, and 20%. For all experiments, we apply simple data augmented methods such as flipping and random rotation.

The ISBI 2012 [1] is a cell segmentation dataset. This dataset with ground truth contains 30 images (512×512 pixels). We augmented all 30 images of the ISBI training set to obtain 300 images. We use 240 of them as the training set and 60 as the testing set. For all experiments, we use the same data set settings. DICE and IoU are used to evaluate our proposed method.

Loss Function and Implementation Details. We use the Dice loss function, which is widely used in the field of medical image segmentation.

$$L_{dice} = 1 - \frac{2|X \cap Y|}{|X| + |Y|} \tag{3}$$

Where $|X|$ and $|Y|$ represent the number of ground truth and predicted image elements respectively. We use ViT [21]'s Transformer Encoder with 12 layers and hidden layer $D = 768$. The model is trained with adam optimizer with learning rate of 0.0002. For the original image, initial feature maps, EF_2 and EF_4, patch sizes P are set to 32, 8, 4, 1 respectively to make the length of the input sequence consistent. All experiments are run on a single NVIDIA TITAN V.

4.2 Evaluation on ISIC 2018 and ISBI 2012

We first apply our DC-Net to skin lesion segmentation. Due to the low contrast of skin lesions and the huge variation of melanoma, the task of skin lesion segmentation is extremely challenging. In order to verify the superiority of our proposed DC-Net, we compared the state-of-the-arts on Table 1 under the same experimental environment and data set settings. Compared with CE-Net, the Dice

Table 1. Evaluation on ISIC 2018 and ISBI 2012.

ISIC 2018						
Method	U-Net [16]	Deeplabv3+ [3]	Attention UNet [15]	CE-Net [8]	CA-Net [7]	DC-Net
Dice	0.885	0.893	0.898	0.922	0.921	**0.943**
IoU	0.778	0.790	0.801	0.873	0.871	**0.896**

ISBI 2012							
Method	U-Net [16]	Deeplabv3+ [3]	[15]	U-Net++ [23]	CE-Net [8]	Inf-Net [6]	DC-Net
Dice	0.935	0.873	0.933	0.939	0.939	0.945	**0.963**
IoU	0.878	0.775	0.874	0.884	0.886	0.896	**0.930**

score of our DC-Net increased from 0.921 to 0.943, and the IoU score increased from 0.871 to 0.896 on Table 1. This proves the effectiveness of our proposed method for the skin lesion segmentation task (Fig. 4).

In addition to the segmentation at the skin level, the segmentation at the cell level is also very important for evaluating the performance of the model. We apply our Dual Context Network to the task of cell contour segmentation and compare DC-Net with a series of state-of-the-arts. From Table 1, our DC-Net got the highest Dice and IoU, reaching 0.963 and 0.930 respectively. Compared with CE-Net, the segmentation performance of DC-Net has been greatly improved, and IoU has increased by 4.96%. In addition, the average hausdorff distance [11] of our DC-Net is more competitive than that of advanced Inf-Net and CE-Net on ISIC 2018 (27.85mm vs 32.34mm vs 37.82mm). This further proves the advantages of our DC-Net with GCTE and ACFM over other advanced methods. Some visualization examples are shown in Fig. 5. Due to the global context enhancement of the GCTE and the effective integration of cross-scale information by ACFM, our DC-Net suppresses conflicting information, and the obtained results have more accurate boundaries and less noise.

4.3   Ablation Study

In this section, we prove through experiments on ISIC 2018 and ISBI 2012 that our proposed GCTE and ACFM can improve the performance of encoding and decoding, respectively, and our DC-Net can greatly improve the performance of medical image segmentation.

Table 2. Ablation study on ISIC 2018 and ISBI 2012.

Method	ISIC 2018			ISBI 2012		
	Dice	IoU	ACC	Dice	IoU	ACC
Backbone (Baseline)	0.9146	0.8515	0.8992	0.9382	0.8814	0.8398
Backbone + DAC & RMP (CE-Net cite[8])	0.9224	0.8735	0.9226	0.9391	0.8860	0.8475
Backbone + VIT	0.9253	0.8770	0.9253	0.9531	0.9103	0.8619
Backbone + GCTE	0.9298	0.8831	0.9302	0.9618	0.9269	0.8740
Backbone + ACFM without APW	0.9291	0.8826	0.9291	0.9622	0.9277	0.8745
Backbone + ACFM without ACW	0.9392	0.8918	0.9354	0.9624	0.9282	0.8752
Backbone + ACFM	0.9418	0.8945	0.9403	0.9628	0.9288	0.8743
Our DC-Net	**0.9429**	**0.8960**	**0.9437**	**0.9634**	**0.9302**	**0.8761**

The first four feature extraction blocks of ResNet-34 [9] are used to replace the U-Net encoder to derive our Backbone. As observed from Table 2, whether it is using GCTE alone or ACFM alone, the performance of the two datasets has

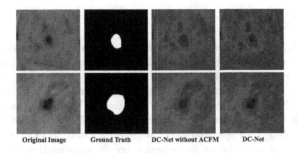

Original Image   Ground Truth   DC-Net without ACFM   DC-Net

Fig. 5. Comparison of visual attention weight maps of DF_2 between DC-Net without ACFM and DC-Net.

been greatly improved. Backbone + VIT [21] refers to the last feature layer of the encode of backbone as the input of VIT. The result proves that the performance of our proposed GCTE is better than Backbone + VIT. Backbone + ACFM without APW means to add ACFM without Adaptive Spatial Weights on the basis of the backbone. Similarly, Backbone + ACFM without ACW means that the added ACFM without Adaptive Channel Weights. Both ACW and APW can improve segmentation performance significantly, and the APW contributes a lot to the improvement of network performance. The ablation experiment verified the importance of effective global context aggregation and adaptive feature fusion, indicating that our DC-Net with GCTE and ACFM significantly improves the performance of the baseline model.

5   Conclusion

In this work, we propose the DC-Net with Global Context Transformer Encoder and Adaptive Context Fusion Module for medical image segmentation. As pointed out, CNN-based U-Net lacks a summary of the global context, and the integration of multi-scale contexts is also worthy of further improvement. We carefully designed Global Context Transformer Encoder to capture the global context with transformer pipeline and fusion inherent multi-level feature maps of encoder. We propose Adaptive Context Fusion Module to adaptively fuse multi-scale context information to obtain better feature representation. Experiments on multiple medical image segmentation tasks show that our DC-Net significantly exceeds the previous methods.

Acknowledgement. This work is supported by National Key R&D Program of China (No. 2020YFC2008500, No. 2020YFC2008503), and the Open Research Fund of Key Laboratory of Space Utilization, Chinese Academy of Sciences (No. LSU-KFJJ-2020-04), National Natural Science Foundation of China (61971418, 91646207).

References

1. Cardona, A., et al.: An integrated micro-and macroarchitectural analysis of the drosophila brain by computer-assisted serial section electron microscopy. PLoS Biol. **8**(10), e1000502 (2010)
2. Challenge, I.: Isic challenge. https://challenge2018.isic-archive.com/
3. Chen, L.C., Papandreou, G., Schroff, F., Adam, H.: Rethinking atrous convolution for semantic image segmentation. arXiv preprint arXiv:1706.05587 (2017)
4. Chen, L.C., Zhu, Y., Papandreou, G., Schroff, F., Adam, H.: Encoder-decoder with atrous separable convolution for semantic image segmentation. In: The European Conference on Computer Vision (ECCV) (September 2018)
5. Çiçek, Ö., Abdulkadir, A., Lienkamp, S.S., Brox, T., Ronneberger, O.: 3D U-net: learning dense volumetric segmentation from sparse annotation. In: Ourselin, S., Joskowicz, L., Sabuncu, M.R., Unal, G., Wells, W. (eds.) MICCAI 2016. LNCS, vol. 9901, pp. 424–432. Springer, Cham (2016). https://doi.org/10.1007/978-3-319-46723-8_49
6. Fan, D.P., et al.: Inf-net: automatic covid-19 lung infection segmentation from CT images. IEEE Trans. Med. Imaging **39**, 2626–2637 (2020)
7. Gu, R., et al.: Ca-net: Comprehensive attention convolutional neural networks for explainable medical image segmentation. IEEE Trans. Med. Imaging **40**(2), 699–711 (2020)
8. Gu, Z., et al.: Ce-net: context encoder network for 2d medical image segmentation. IEEE Trans. Med. Imaging **38**(10), 2281–2292 (2019)
9. He, K., Zhang, X., Ren, S., Sun, J.: Deep residual learning for image recognition. In: Proceedings of the IEEE Conference on Computer Vision and Pattern Recognition, pp. 770–778 (2016)
10. Hu, J., Shen, L., Sun, G.: Squeeze-and-excitation networks. In: Proceedings of the IEEE Conference on Computer Vision and Pattern Recognition, pp. 7132–7141 (2018)
11. Huttenlocher, D.P., Klanderman, G.A., Rucklidge, W.J.: Comparing images using the hausdorff distance. IEEE Trans. Pattern Anal. Mach. Intell. **15**(9), 850–863 (1993)
12. Lin, T.Y., Dollár, P., Girshick, R., He, K., Hariharan, B., Belongie, S.: Feature pyramid networks for object detection. In: Proceedings of the IEEE Conference on Computer Vision and Pattern Recognition, pp. 2117–2125 (2017)
13. Long, J., Shelhamer, E., Darrell, T.: Fully convolutional networks for semantic segmentation. In: The IEEE Conference on Computer Vision and Pattern Recognition (CVPR) (June 2015)
14. Milletari, F., Navab, N., Ahmadi, S.: V-net: fully convolutional neural networks for volumetric medical image segmentation. CoRR abs/1606.04797 (2016)
15. Oktay, O., et al.: Attention u-net: Learning where to look for the pancreas. arXiv preprint arXiv:1804.03999 (2018)
16. Ronneberger, O., Fischer, P., Brox, T.: U-net: convolutional networks for biomedical image segmentation. In: Navab, N., Hornegger, J., Wells, W.M., Frangi, A.F. (eds.) MICCAI 2015. LNCS, vol. 9351, pp. 234–241. Springer, Cham (2015). https://doi.org/10.1007/978-3-319-24574-4_28
17. Vaswani, A., et al.: Attention is all you need. arXiv preprint arXiv:1706.03762 (2017)
18. Wang, C.Y., Bochkovskiy, A., Liao, H.Y.M.: Scaled-yolov4: Scaling cross stage partial network. arXiv preprint arXiv:2011.08036 (2020)

19. Wang, X., Girshick, R., Gupta, A., He, K.: Non-local neural networks. In: Proceedings of the IEEE Conference on Computer Vision and Pattern Recognition, pp. 7794–7803 (2018)
20. Wang, Z., Zou, N., Shen, D., Ji, S.: Non-local u-nets for biomedical image segmentation. In: Proceedings of the AAAI Conference on Artificial Intelligence, vol. 34, pp. 6315–6322 (2020)
21. Wu, B., et al.: Visual transformers: Token-based image representation and processing for computer vision. arXiv preprint arXiv:2006.03677 (2020)
22. Zhang, H., Goodfellow, I., Metaxas, D., Odena, A.: Self-attention generative adversarial networks. In: International Conference on Machine Learning, pp. 7354–7363. PMLR (2019)
23. Zhou, Z., Siddiquee, M.M.R., Tajbakhsh, N., Liang, J.: Unet++: A nested u-net architecture for medical image segmentation. CoRR abs/1807.10165 (2018)

LIFE: A Generalizable Autodidactic Pipeline for 3D OCT-A Vessel Segmentation

Dewei Hu[1], Can Cui[1], Hao Li[1], Kathleen E. Larson[2], Yuankai K. Tao[2], and Ipek Oguz[1(✉)]

[1] Department of Electrical Engineering and Computer Science,
Vanderbilt University, Nashville, USA
`ipek.oguz@vanderbilt.edu`

[2] Department of Biomedical Engineering, Vanderbilt University, Nashville, TN, USA

Abstract. Optical coherence tomography (OCT) is a non-invasive imaging technique widely used for ophthalmology. It can be extended to OCT angiography (OCT-A), which reveals the retinal vasculature with improved contrast. Recent deep learning algorithms produced promising vascular segmentation results; however, 3D retinal vessel segmentation remains difficult due to the lack of manually annotated training data. We propose a learning-based method that is only supervised by a self-synthesized modality named local intensity fusion (LIF). LIF is a capillary-enhanced volume computed directly from the input OCT-A. We then construct the local intensity fusion encoder (LIFE) to map a given OCT-A volume and its LIF counterpart to a shared latent space. The latent space of LIFE has the same dimensions as the input data and it contains features common to both modalities. By binarizing this latent space, we obtain a volumetric vessel segmentation. Our method is evaluated in a human fovea OCT-A and three zebrafish OCT-A volumes with manual labels. It yields a Dice score of 0.7736 on human data and 0.8594 ± 0.0275 on zebrafish data, a dramatic improvement over existing unsupervised algorithms.

Keywords: OCT angiography · Self-supervised · Vessel segmentation

1 Introduction

Optical coherence tomography (OCT) is a non-invasive imaging technique that provides high-resolution volumetric visualization of the retina [19]. However, it offers poor contrast between vessels and nerve tissue layers [9]. This can be overcome by decoupling the dynamic blood flow within vessels from stationary nerve tissue by decorrelating multiple cross-sectional images (B-scans) taken at the same spatial location. By computing the variance of these repeated B-scans, we obtain an OCT angiography (OCT-A) volume that has better visualization of retinal vasculature than traditional OCT [15]. In contrast to other techniques

© Springer Nature Switzerland AG 2021
M. de Bruijne et al. (Eds.): MICCAI 2021, LNCS 12901, pp. 514–524, 2021.
https://doi.org/10.1007/978-3-030-87193-2_49

such as fluorescein angiography (FA), OCT-A is advantageous because it both provides depth-resolved information in 3D and is free of risks related to dye leakage or potential allergic reaction [9]. OCT-A is popular for studying various retinal pathologies [4,14]. Recent usage of the vascular plexus density as a disease severity indicator [12] highlights the need for vessel segmentation in OCT-A.

Unlike magnetic resonance angiography (MRA) and computed tomography angiography (CTA), OCT-A suffers from severe speckle noise, which induces poor contrast and vessel discontinuity. Consequently, unsupervised vessel segmentation approaches [2,3,22,28,33] developed for other modalities do not translate well to OCT-A. Denoising OCT/OCT-A images has thus been an active topic of research [5,13,24]. The noise is compounded in OCT-A due to the unpredictable patterns of blood flow as well as artifacts caused by residual registration errors, which lead to insufficient suppression of stationary tissue. This severe noise level, coupled with the intricate detail of the retinal capillaries, leads to a fundamental roadblock to 3D segmentation of the retinal blood vessels: the task is too challenging for unsupervised methods, and yet, obtaining manual segmentations to train supervised models is prohibitively expensive. For instance, a single patch capturing only about 5% of the whole fovea (Fig. 4f) took approximately 30 h to manually segment. The large inter-subject variability and the vast inter-rater variability which is inevitable in such a detailed task make the creation of a suitably large manual training dataset intractable.

As a workaround, retinal vessel segmentation attempts have been largely limited to 2D images with better SNR, such as the depth-projection of the OCT-A [10]. This only produces a single 2D segmentation out of a whole 3D volume, evidently sacrificing the 3D depth information. Similar approaches to segment inherently 2D data such as fundus images have also been reported [17]. Recently, Liu et al. [20] proposed an unsupervised 2D vessel segmentation method using two registered modalities from two different imaging devices. Unfortunately, multiple scans of a single subject are not typically available in practice. Further, the extension to 3D can be problematic due to inaccurate volumetric registration between modalities. Zhang et al. proposed the optimal oriented flux [18] (OOF) for 3D OCT-A segmentation [32], but, as their focus is shape analysis, neither a detailed discussion nor any numerical evaluation on segmentation are provided.

We propose the local intensity fusion encoder (LIFE), a self-supervised method to segment 3D retinal vasculature from OCT-A. LIFE requires neither manual delineation nor multiple acquisition devices. To our best knowledge, it is the first label-free learning method with quantitative validation of 3D OCT-A vessel segmentation. Fig. 1 summarizes the pipeline. Our **novel contributions** are:

- An *en-face* denoising method for OCT-A images via local intensity fusion (LIF) as a new modality (Sect. 2.1)
- A variational auto-encoder network, the local intensity fusion encoder (LIFE), that considers the original OCT-A images and the LIF modality to estimate the latent space which contains the retinal vasculature (Sect. 2.2)
- Quantitative and qualitative evaluation on human and zebrafish data (Sect. 3)

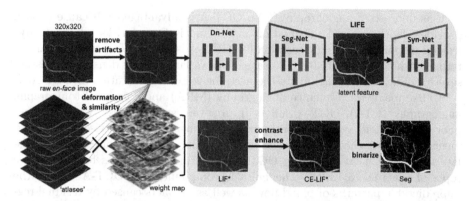

Fig. 1. Overall pipeline. * indicates LIF and CE-LIF provide supervision for Dn-Net and LIFE in training process respectively.

2 Methods

2.1 Local Intensity Fusion: LIF

Small capillaries have low intensity in OCT-A since they have slower blood flow, and are therefore hard to distinguish from the ubiquitous speckle noise. We exploit the similarity of vasculature between consecutive *en-face* OCT-A slices to improve the image quality. This local intensity fusion (LIF) technique derives from the Joint Label Fusion [29] and related synthesis methods [7,24,27].

Joint label fusion (JLF) [29] is a well-known multi-atlas label fusion method for segmentation. In JLF, a library of K atlases with known segmentations $(\boldsymbol{X}_k, \boldsymbol{S}_k)$ is deformably registered to the target image $\boldsymbol{Y}$ to obtain $(\hat{\boldsymbol{X}}_k, \hat{\boldsymbol{S}}_k)$. Locally varying weight maps are computed for each atlas based on the local residual registration error between $\boldsymbol{Y}$ and $\hat{\boldsymbol{X}}_k$. The weighted sum of the $\hat{\boldsymbol{S}}_k$ provides the consensus segmentation $\boldsymbol{S}$ on the target image.

JLF has been extended to joint intensity fusion (JIF), an image synthesis method that does not require atlas segmentations. JIF has been used for lesion in-painting [7] and cross-modality synthesis [27]. Here, we propose a JIF variant, LIF, performing fusion between the 2D *en-face* slices of a 3D OCT-A volume.

Instead of an external group of atlases, for each 2D *en-face* slice $\boldsymbol{X}$ of a 3D OCT-A volume, adjacent slices within an R-neighborhood $\{\boldsymbol{X}_{-R}, \ldots, \boldsymbol{X}_{+R}\}$ are regarded as our group of 'atlases' for $\boldsymbol{X}$. Note that the atlas X_0 is the target X itself, represented as the image with a red rim in Fig. 1. We perform registration using the *greedy* software [31]. While closely related, we note that the self-fusion method reported in [24] for tissue layer enhancement is not suitable for vessel enhancement as it tends to substantially blur and distort blood vessels [13].

Similar to a 1-D Gaussian filter along the depth axis, LIF has a blurring effect that improves the homogeneity of vessels without dilating their thickness in the *en-face* image. Further, it can also smooth the speckle noise in the background while raising the overall intensity level, as shown in Fig. 2a/2b. In order to make

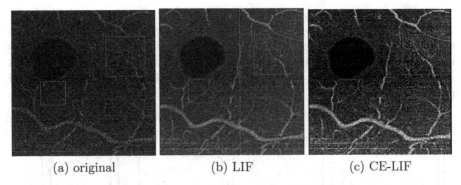

(a) original   (b) LIF   (c) CE-LIF

Fig. 2. Modalities of *en-face* OCT-A. **Large red box** highlights improvement in capillary visibility. **Small red box** points out a phantom vessel.

vessels stand out better, we introduce the contrast enhanced local intensity fusion (CE-LIF)[1] in Fig. 2c. However, intensity fusion of *en-face* images sacrifices the accuracy of vessel diameter in the depth direction. Specifically, some vessels existing exclusively in neighboring images are inadvertently projected on the target slice. For example, the small red box in Fig. 2 highlights a phantom vessel caused by incorrect fusion. As a result, LIF and CE-LIF are not appropriate for direct use in application, in spite of the desirable improvement they offer in visibility of capillaries (e.g., large red box). In the following section, we propose a novel method that allows us to leverage LIF as an auxiliary modality for feature extraction during which these excessive projections will be filtered out.

2.2   Cross-Modality Feature Extraction: LIFE

Liu et al. [20] introduced an important concept for unsupervised feature extraction. Two depth-projected 2D OCT-A images, M_1 and M_2, are acquired using different devices on the same retina. If they are well aligned, then aside from noise and difference in style, the majority of the anatomical structure would be the same. A variational autoencoder (VAE) is set as a pix2pix translator from M_1 to M_2 in which the latent space L_{12} keeps full resolution.

$$L_{12} = f_e(M_1) \quad \text{and} \quad M_2' = f_d(L_{12}) \tag{1}$$

If M_2 is well reconstructed ($M_2' \approx M_2$), then the latent feature map L_{12} can be regarded as the common features between M_1 and M_2, namely, vasculature. The encoder f_e is considered a segmentation network (Seg-Net) and the decoder f_d a synthesis network (Syn-Net).

Unfortunately, this method has several drawbacks in practice. Imaging the same retina with different devices is rarely possible even in research settings and unrealistic in clinical practice. Furthermore, the 3D extension does not appear straightforward due to the differences in image spacing between OCT devices and

---

[1] https://pillow.readthedocs.io/en/stable/reference/ImageEnhance.html.

the difficulty of volumetric registration in these very noisy images. In contrast, we propose to use a single OCT-A volume X and its LIF, X_{LIF}, as the two modalities. This removes the need for multiple devices or registration, and allows us to produce a 3D segmentation by operating on individual *en-face* OCT-A slices rather than a single depth-projection image. We call the new translator network local intensity fusion encoder (LIFE).

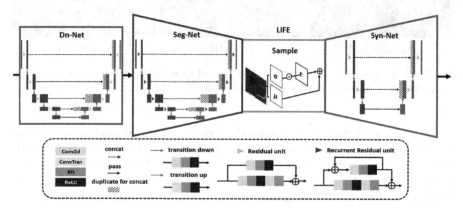

Fig. 3. Network architecture

Figure 3 shows the network architecture. To reduce the influence of speckle noise, we train a residual U-Net as a denoising network (Dn-Net), supervised by LIF. For the encoder, we implement a more complex model (R2U-Net) [1] than Liu et al. [20], supervised by CE-LIF. As the decoder we use a shallow, residual U-Net to balance computational power and segmentation performance. The reparameterization trick enables gradient back propagation when sampling is involved in a deep network [16]. This sampling is achieved by $S = \mu + \sigma \cdot \epsilon$, where $\epsilon \in \mathcal{N}(0, 1)$, and μ and σ are mean and standard deviation of the latent space. The intensity ranges of all images are normalized to $[0, 255]$. To introduce some blurring effect, both L_1 and L_2 norm are added to the VAE loss function:

$$Loss = a \sum_{i,j} |\boldsymbol{Y}(i,j) - \boldsymbol{Y}'(i,j)| + \frac{b}{N} \sum_{i,j} (\boldsymbol{Y}(i,j) - \boldsymbol{Y}'(i,j))^2 \qquad (2)$$

where (i,j) are pixel coordinates, N is the number of pixels, $\boldsymbol{Y}$ is CE-LIF and $\boldsymbol{Y}'$ is the output of Syn-Net. $a = 1$ and $b = 0.05$ are hyperparameters. Equation 2 is also used as the loss for the Dn-Net, with the LIF image as $\boldsymbol{Y}$, $a = 1$ and $b = 0.01$.

As discussed above, LIF enhances the appearance of blood vessels but also introduces phantom vessels because of fusion. The set of input vessel features $\mathcal{V}$ in X will thus be a subset of $\mathcal{V}_{LIF}$ in X_{LIF}. Because LIFE works to extract $\mathcal{V} \cap \mathcal{V}_{LIF}$, the phantom features that exist only in $\mathcal{V}_{LIF}$ will be cancelled out as long as the model is properly trained without suffering from overfitting.

2.3   Experimental Details

Preprocessing for Motion Artifact Removal. Decorrelation allows OCT-A to emphasize vessels while other tissue types get suppressed (Fig. 4a). However, this requires the repeated OCT B-scans to be precisely aligned. Any registration errors cause motion artifacts, such that stationary tissue is not properly suppressed (Fig. 4b). These appear as horizontal artifacts in *en-face* images (Fig. 4c). We remove these artifacts by matching the histogram of the artifact B-scan to its closest well-decorrelated neighbor (Fig. 4d).

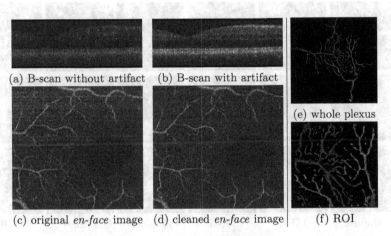

Fig. 4. **(a-d)** Horizontal motion artifacts and their removal. **(e,f)** Three manually labelled vessel plexuses. Only the branches contained within the ROI were fully segmented; any branches outside the ROI and all other trees were omitted for brevity.

Binarization. To binarize the latent space L_{12} estimated by LIFE, we apply the 2nd Perona-Malik diffusion equation [26] followed by the global Otsu threshold [25]. Any islands smaller than 30 voxels are removed.

Dataset. The OCT volumes were acquired with $2560 \times 500 \times 400 \times 4$ pix. (spectral $\times$ lines $\times$ frames $\times$ repeated frames) [6,23]. OCT-A is performed on motion-corrected [11] OCT volumes using singular value decomposition. We manually crop the volume to only retain the depth slices that contain most of the vessels near the fovea, between the ganglion cell layer (GCL) and inner plexiform layer (IPL). Three fovea volumes are used for training and one for testing. As the number of slices between GCL and IPL is limited, we aggressively augment the dataset by randomly cropping and flipping 10 windows of size [320, 320] for each *en-face* image. To evaluate on vessels differing in size, we labeled 3 interacting plexus near the fovea, displayed in Fig. 4e. A smaller ROI ($120 \times 120 \times 17$) cropped in the center (Fig. 4f) is used for numerical evaluation.

To further evaluate the method, we train and test our model on OCT-A of zebrafish eyes, which have a simple vessel structure ideal for easy manual labeling. This also allows us to test the generalizability of our method to images

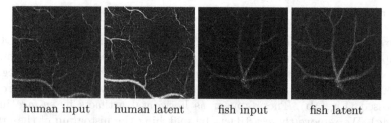

human input   human latent   fish input   fish latent

Fig. 5. The latent image L_{12} from LIFE considerably improves vessel appearance.

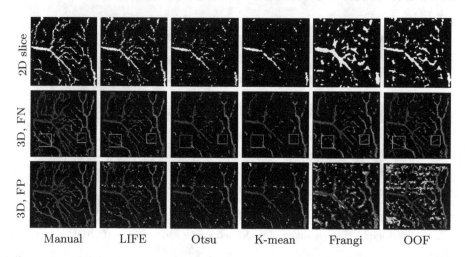

Manual   LIFE   Otsu   K-mean   Frangi   OOF

Fig. 6. 2D slice and 3D marching cubes rendering of segmentation results on human retina, with Gaussian smoothing, $\sigma = 0.70$. **Red, green, blue** show three different branches; **yellow** highlights false positives. LIFE is the only method that can recover the 3D structure and connectivity of the capillaries outside the largest vessels without causing excessive FP (yellow). **White boxes** highlight LIFE's improved sensitivity. (Color figure online)

from different species. Furthermore, the fish dataset contains stronger speckle noise than the human data, which allows us to test the robustness of the method to high noise. 3 volumes ($480 \times 480 \times 25 \times 5$ each) are labeled for testing and 5 volumes are used for training. All manual labelling is done on ITKSnap [30].

Baseline Methods. Due to the lack of labeled data, no supervised learning method is applicable. Similar to our approach that follows the enhance + binarize pattern, we apply Frangi's multi-scale vesselness filter [8] and optimally oriented flux (OOF) [18,32] respectively to enhance the artifact-removed original image, then use the same binarization steps described above. We also present results using Otsu thresholding and k-means clustering.

Implementation details. All networks are trained on an NVIDIA RTX 2080TI 11GB GPU for 50 epochs with batch size set to 2. For the first 3 epochs, the entire network uses the same Adam optimizer with learning rate of 0.001. After that, LIFE and decoder are separately optimized with starting learning rates of 0.002 and 0.0001 respectively in order to distribute more workload on the LIFE. Both networks decay every 3 epochs with at a rate of 0.5.

3   Results

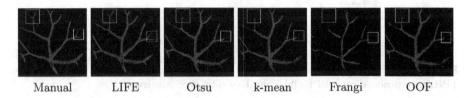

| Manual | LIFE | Otsu | k-mean | Frangi | OOF |

Fig. 7. Segmentation result of zebrafish retina with the same rendering setting in Fig. 6.

Figure 5 displays examples of extracted latent images. It is visually evident that LIFE successfully highlights the vasculature. Compared with the raw input, even delicate capillaries show improved homogeneity and separability from the background. Figure 6 illustrates 2D segmentation results within the manually segmented ROI, where LIFE can be seen to have better sensitivity and connectivity than the baseline methods. Figure 6 also shows a 3D rendering (via marching cubes [21]) of each method. In the middle row, we filtered out the false positives (FP) to highlight the false negatives (FN). These omitted FP areas are highlighted in yellow in the bottom row. It is easy to see that these FPs are often distributed along horizontal lines, caused by unresolved motion artifacts. Hessian-based methods appear especially sensitive to motion artifacts and noise; hence Frangi's method and OOF introduce excessive FP. Clearly, LIFE achieves the best preservation in small capillaries, such as the areas highlighted in white boxes, without introducing too many FPs.

Figure 7 shows that LIFE has superior performance on the zebrafish data. The white boxes highlight that only LIFE can capture smaller branches.

Figure 8 shows quantitative evaluation across B-scans, and Table 1 across the whole volume. Consistent with our qualitative assessments, LIFE significantly ($p \ll 0.05$) and dramatically (over 0.20 Dice gain) outperforms the baseline methods on both human and fish data.

Finally, we directly binarize LIF and CE-LIF as additional baselines. The Dice scores on the human data are 0.5293 and 0.4892, well below LIFE (0.7736).

4   Discussion and Conclusion

We proposed a method for 3D segmentation of fovea vessels and capillaries from OCT-A volumes that requires neither manual annotation nor multiple image

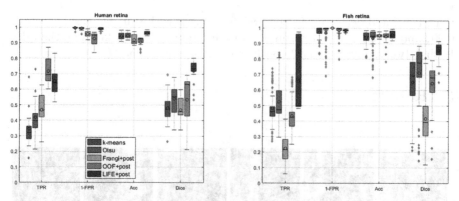

Fig. 8. Quantitative result evaluation for **(left)** human and **(right)** zebrafish data. TPR: true positive rate, FPR: false positive rate, Acc: accuracy.

Algorithm	TPR		FPR		Accuracy		Dice	
	Human	Fish	Human	Fish	Human	Fish	Human	Fish
k-means	0.3633	0.4167	**0.0042**	0.0303	0.9440	0.9228	0.5152	0.6249
Otsu	0.4403	0.4356	0.0076	0.0346	0.9472	0.9191	0.5772	0.6399
Frangi+bin	0.4900	0.2117	0.0419	0.0002	0.9198	**0.9489**	0.5002	0.4212
OOF+bin	**0.6826**	0.3775	0.0748	0.0165	0.9053	0.9350	0.5414	0.6247
LIFE+bin	0.6613	**0.4999**	0.0104	**0.0153**	**0.9627**	0.9386	**0.7736**	**0.8594**

Table 1. Quantitative evaluation of human and zebrafish segmentation. TPR: true positive rate, FPR: false positive rate. Bold indicates the best score per column.

acquisitions to train. The introduction of the LIF modality brings many benefits for the method. Since LIF is directly computed from the input data, no inter-volume registration is needed between the two modalities input to LIFE. Further, rather than purely depending on image intensity, LIF exploits local structural information to enhance small features like capillaries. Still, there are some disadvantages to overcome in future research. For instance, LIFE cannot directly provide a binarized output and hence the crude thresholding method used for binarization influences the segmentation performance.

Acknowledgements. This work is supported by NIH R01EY031769, NIH R01EY030490 and Vanderbilt University Discovery Grant Program.

References

1. Alom, M.Z., Hasan, M., Yakopcic, C., Taha, T.M., Asari, V.K.: Recurrent residual CNN based on u-net (r2u-net) for medical image segmentation. arXiv preprint arXiv:1802.06955 (2018)

2. Aylward, S.R., Bullitt, E.: Initialization, noise, singularities, and scale in height ridge traversal for tubular object centerline extraction. IEEE Trans. Med. Imaging **21**(2), 61–75 (2002). https://doi.org/10.1109/42.993126

3. Bozkurt, F., Köse, C., Sarı, A.: A texture-based 3d region growing approach for segmentation of ica through the skull base in cta. Multimedia Tools Appl. **79**(43), 33253–33278 (2020)

4. Burke, T.R., et al.: Application of oct-angiography to characterise the evolution of chorioretinal lesions in acute posterior multifocal placoid pigment epitheliopathy. Eye **31**(10), 1399–1408 (2017)

5. Devalla, S.K., et al.: A deep learning approach to denoise OCT images of the optic nerve head. Sci. Rep. **9**(1), 1–13 (2019)

6. El-Haddad, M.T., Bozic, I., Tao, Y.K.: Spectrally encoded coherence tomography and reflectometry: Simultaneous en face and cross-sectional imaging at 2 gigapixels per second. J. Biophotonics **11**(4), e201700268 (2018)

7. Fleishman, G.M., et al.: Joint intensity fusion image synthesis applied to MS lesion segmentation. In: MICCAI BrainLes Workshop, pp. 43–54 (2017)

8. Frangi, A.F., Niessen, W.J., Vincken, K.L., Viergever, M.A.: Multiscale vessel enhancement filtering. In: Wells, W.M., Colchester, A., Delp, S. (eds.) MICCAI 1998. LNCS, vol. 1496, pp. 130–137. Springer, Heidelberg (1998). https://doi.org/10.1007/BFb0056195

9. Gao, S., et al.: Optical coherence tomography angiography. IOVS **57**(9), OCT27-OCT36 (2016)

10. Giarratano, Y., et al.: Automated and network structure preserving segmentation of optical coherence tomography angiograms. arXiv preprint arXiv:1912.09978 (2019)

11. Guizar-Sicairos, M., Thurman, S.T., Fienup, J.R.: Efficient subpixel image registration algorithms. Opt. Lett. **33**(2), 156–158 (2008)

12. Holló, G.: Comparison of peripapillary oct angiography vessel density and retinal nerve fiber layer thickness measurements for their ability to detect progression in glaucoma. J. glaucoma **27**(3), 302–305 (2018)

13. Hu, D., Malone, J., Atay, Y., Tao, Y., Oguz, I.: Retinal OCT denoising with pseudo-multimodal fusion network. In: MICCAI OMIA, pp. 125–135 (2020)

14. Ishibazawa, A., et al.: OCT angiography in diabetic retinopathy: a prospective pilot study. Am. J. Ophthalmol. **160**(1), 35–44 (2015)

15. Jia, Y., et al.: Split-spectrum amplitude-decorrelation angiography with optical coherence tomography. Opt. Express **20**(4), 4710–4725 (2012)

16. Kingma, D.P., Welling, M.: Auto-encoding variational bayes. arXiv preprint arXiv:1312.6114 (2013)

17. Lahiri, A., Roy, A.G., Sheet, D., Biswas, P.K.: Deep neural ensemble for retinal vessel segmentation in fundus images towards achieving label-free angiography. In: IEEE EMBC, pp. 1340–1343. IEEE (2016)

18. Law, M.W.K., Chung, A.C.S.: Three dimensional curvilinear structure detection using optimally oriented flux. In: Forsyth, D., Torr, P., Zisserman, A. (eds.) ECCV 2008. LNCS, vol. 5305, pp. 368–382. Springer, Heidelberg (2008). https://doi.org/10.1007/978-3-540-88693-8_27

19. Li, M., Idoughi, R., Choudhury, B., Heidrich, W.: Statistical model for oct image denoising. Biomed. Opt. Express **8**(9), 3903–3917 (2017)

20. Liu, Y., et al.: Variational intensity cross channel encoder for unsupervised vessel segmentation on oct angiography. In: SPIE Medical Imaging 2020: Image Processing, vol. 11313, p. 113130Y (2020)

21. Lorensen, W.E., Cline, H.E.: Marching cubes: a high resolution 3d surface construction algorithm. SIGGRAPH Comput. Graph. **21**(4), 163–169 (1987)
22. Lorigo, L.M., et al.: CURVES: curve evolution for vessel segmentation. Med. Image Anal. **5**(3), 195–206 (2001)
23. Malone, J.D., El-Haddad, M.T., Yerramreddy, S.S., Oguz, I., Tao, Y.K.: Handheld spectrally encoded coherence tomography and reflectometry for motion-corrected ophthalmic OCT and OCT-A. Neurophotonics **6**(4), 041102 (2019)
24. Oguz, I., Malone, J.D., Atay, Y., Tao, Y.K.: Self-fusion for OCT noise reduction. In: SPIE Medical Imaging 2020: Image Processing, vol. 11313, p. 113130C (2020)
25. Otsu, N.: A threshold selection method from gray-level histograms. IEEE Trans. Syst. Man Cybern. **9**(1), 62–66 (1979)
26. Perona, P., Malik, J.: Scale-space and edge detection using anisotropic diffusion. IEEE Trans. Pattern Anal. Mach.Intell. **12**(7), 629–639 (1990)
27. Ufford, K., Vandekar, S., Oguz, I.: Joint intensity fusion with normalized cross-correlation metric for cross-modality MRI synthesis. In: SPIE Medical Imaging 2020: Image Processing, vol. 11313 (2020)
28. Vasilevskiy, A., Siddiqi, K.: Flux maximizing geometric flows. IEEE Trans. Pattern Anal. Mach. Intell. **24**, 1565–1578 (2001)
29. Wang, H., Suh, J.W., Das, S.R., Pluta, J.B., Craige, C., Yushkevich, P.A.: Multi-atlas segmentation with joint label fusion. IEEE PAMI **35**(3), 611–623 (2012)
30. Yushkevich, P.A., et al.: User-guided 3D active contour segmentation of anatomical structures: significantly improved efficiency and reliability. Neuroimage **31**(3), 1116–1128 (2006)
31. Yushkevich, P.A., Pluta, J., Wang, H., Wisse, L.E., Das, S., Wolk, D.: Fast automatic segmentation of hippocampal subfields and medial temporal lobe subregions in 3T and 7T T2-weighted MRI. Alzheimer's Dement. **7**(12), P126–P127 (2016)
32. Zhang, J., et al.: 3d shape modeling and analysis of retinal microvasculature in oct-angiography images. IEEE TMI **39**(5), 1335–1346 (2020)
33. Zhao, S., Tian, Y., Wang, X., Xu, P., Deng, Q., Zhou, M.: Vascular extraction using mra statistics and gradient information. Mathematical Problems in Engineering 2018 (2018)

Superpixel-Guided Iterative Learning from Noisy Labels for Medical Image Segmentation

Shuailin Li[1]([✉]), Zhitong Gao[1], and Xuming He[1,2]

[1] ShanghaiTech University, Shanghai, China
{lishl,gaozht,hexm}@shanghaitech.edu.cn
[2] Shanghai Engineering Research Center of Intelligent Vision and Imaging,
Shanghai, China

Abstract. Learning segmentation from noisy labels is an important task for medical image analysis due to the difficulty in acquiring high-quality annotations. Most existing methods neglect the pixel correlation and structural prior in segmentation, often producing noisy predictions around object boundaries. To address this, we adopt a superpixel representation and develop a robust iterative learning strategy that combines noise-aware training of segmentation network and noisy label refinement, both guided by the superpixels. This design enables us to exploit the structural constraints in segmentation labels and effectively mitigate the impact of label noise in learning. Experiments on two benchmarks show that our method outperforms recent state-of-the-art approaches, and achieves superior robustness in a wide range of label noises. Code is available at https://github.com/gaozhitong/SP_guided_Noisy_Label_Seg.

Keywords: Learning with noisy labels · Semantic segmentation

1 Introduction

Semantic segmentation of medical images, a fundamental task in computer-aided clinical diagnoses, has recently achieved remarkable progress thanks to the effective feature learning based on deep neural networks (DNN) [3,13,19]. Training such DNNs for segmentation typically requires a large dataset with pixelwise annotations that accurately delineate object boundaries. In the medical domain, however, acquiring such high-quality annotations is often difficult due to lack

S. Li and Z. Gao—Equal contribution.
This work was supported by Shanghai Science and Technology Program 21010502700 and by the ShanghaiTech-UII Joint Lab.

Electronic supplementary material The online version of this chapter (https://doi.org/10.1007/978-3-030-87193-2_50) contains supplementary material, which is available to authorized users.

© Springer Nature Switzerland AG 2021
M. de Bruijne et al. (Eds.): MICCAI 2021, LNCS 12901, pp. 525–535, 2021.
https://doi.org/10.1007/978-3-030-87193-2_50

of experienced annotators and/or visual ambiguity in object boundaries [6,11]. Consequently, the annotated datasets often include a varying amount of label noise in practice, ranging from small boundary offsets to large region errors. Learning from those noisy annotations has been particularly challenging for deep segmentation networks due to the memorization effect [2].

There have been several attempts to tackle the problem of training segmentation networks from noisy labels, which can be largely grouped into two categories. The first type of methods view the annotation of each image as either clean or corrupted, and iteratively select or reweight image samples during training [23,26]. In particular, Zhu et al. [26] implicitly reweight image losses by simultaneously training a label evaluation network and a segmentation network, while Xue et al. [23] explicitly select a subset of images by extending the Co-teaching [8] scheme into a tri-network framework. Such image-level weighting strategies, however, are less robust under severe noise settings as they are unable to fully utilize the pixels with clean annotations in each image. To address this limitation, the second group of training methods consider the segmentation as a pixel-wise classification task [24,25], and perform pixel-wise sample selection or label refinement based on the state-of-the-art robust classifier learning strategies, such as the confidence learning technique [16] and the Co-teaching method with tri-networks [25]. Despite their better use of annotations, the pixel-level approaches ignore the pixel correlation and spatial prior in image segmentation, and hence tend to produce noisy prediction around object boundaries.

In this work, we propose a novel robust learning strategy for semantic image segmentation, aiming to exploit the structural prior of images and correlation in pixel labels. To this end, we adopt a superpixel representation and develop an iterative learning scheme that combines noise-aware training of segmentation network and noisy label refinement, both guided by the superpixels. Such integration allows us to better utilize the structural constraint in segmentation labels for model learning, which can effectively mitigate the impact of label noise. We note that while superpixel has been employed in recent work [12], they only use it to correct noisy labels and ignore the impact of noise during training.

Specifically, in each iteration, we first jointly train two deep networks using selected subsets of superpixels with small loss values, following the multi-view learning framework [8,22]. As in the Co-teaching method, such a multi-view learning strategy regularizes the network training via the predictions of the peer networks. Here we treat each superpixel as a data sample in selection, which enables us to enforce spatial smoothness and provide better object boundary cues in network training. To avoid overfitting to label noise, we design an automatic stopping criterion for the joint learning based on the loss statistics of superpixels. After the network training, we then use the network predictions to estimate the reliability of superpixel labels and relabel a subset of most unreliable ones. Such label refinement allows us to improve the label quality for the subsequent model training. The network and label updates are repeated until no further improvement can be achieved for the label refinement.

We evaluate our method on two public benchmarks, ISIC skin lesion dataset [7] and JSRT chest x-ray dataset [5,20], under extensive noise settings. Empirical results show that our method consistently outperforms the previous state of the art and demonstrates training robustness in a wide range of label noises.

2   Method

We now introduce our robust learning strategy for semantic segmentation, which aims to exploit the structural constraints in the label masks and to fully utilize reliable pixel-level labels for effective learning. To achieve this, we adopt a superpixel-based data representation, and develop an iterative learning method that jointly optimizes the network parameters and refines noisy labels.

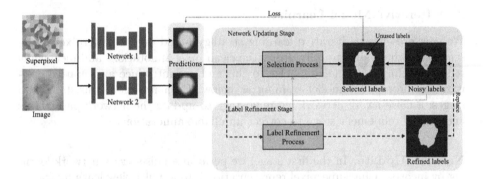

Fig. 1. Overview of our robust training process. We use superpixels as our guidance in an iterative learning process which jointly updates network parameters and refines noisy labels. Each iteration selects superpixels with small losses to update two networks and relabels a set of superpixels based on network outputs.

Specifically, given a target network and noisy training data, we first compute the superpixels of the input images. Based on such pixel groupings, our iterative learning procedure alternates between a noise-aware network training stage and a label refinement stage until no improvement can be achieved. For the network training stage, we adopt the multi-view learning framework which jointly trains two instances of the segmentation network. For the label refinement stage, we use the output of two trained networks to estimate the reliability of superpixel labels and to update the unreliable labels. An overview of our training pipeline is shown in Fig. 1. Below we first present our superpixelization procedure in Sect. 2.1, followed by the two stages in the iterative learning in Sect. 2.2.

2.1   Superpixel Representation

To exploit the image structural prior and spatial correlation in pixel labels, we first compute a superpixel representation for the training images. Such superpixel representation has been shown effective for different medical image modalities in literature, e.g. [17] for CT, [21] for MR and [4] for US images. Specifically, we use the off-the-shelf superpixelization method, SLIC [1], to partition each image into a set of homogeneous regions. For color images, we adopt the CIE-lab color space to represent pixel features, while for other modalities, such as X-ray images, we use both the pixel intensity and deep features from a U-net trained with a noise-aware method, e.g. [22].[1]

We assume the pixels share similar groundtruth labels in each superpixel, which enable us to enforce the structural constraints on the label masks and better preserve object boundaries. More importantly, we treat each superpixel as a data sample in the subsequent robust network learning as well as the label refinement. This allows us to estimate the noise level of pixel annotations in a more reliable manner by pooling the pixel information from each superpixel.

2.2   Iterative Model Learning

We now present our iterative learning strategy based on the superpixel representation, which aims to fully utilize the clean pixel annotations and meanwhile reduce the impact of noisy labels. To this end, we introduce an iterative optimization process for model training as below. Each iteration consists of two stages: a noise-aware network learning stage to update the network parameters and a label refinement stage to correct unreliable annotations.

Network Update. In the first stage, we perform a noise-aware network learning by incorporating superpixel representation into a multi-view learning framework. Specifically, following the Co-teaching strategy [8,10,18], we jointly train two instances of the target segmentation network using partial data with small losses [2]. To better select data samples with clean labels, we design a superpixel-wise loss that combines the loss values of two networks with an agreement-based regularization [22] on superixels. Our loss provides a reliable guidance to sample selection thanks to the structural prior encoded in the superpixels.

Formally, given an image $\mathbf{X}$, we denote its annotation as $\mathbf{Y} = \{Y_i\}_{i=1}^M, Y_i \in \{1, \cdots, C\}$ where C is the number of semantic classes and M is the number of pixels. The superpixel map is represented by $\mathbf{S} = \{S_i\}_{i=1}^M$ where $S_i \in \{1, 2, \cdots, K\}$ and K is the number of superpixels. Here $S_j = k$ means that pixel j belongs to superpixel k.

We aim to train two deep neural networks denoted by $f(\cdot, \theta_1)$ and $f(\cdot, \theta_2)$. To define a loss for each image, we first generate the predicted probability maps from two networks, denoted by $\mathbf{P}^1, \mathbf{P}^2 \in \mathbb{R}^{C \times M}$, where $\mathbf{P}^i = f(\mathbf{X}, \theta_i), i = 1, 2$. We then compute the superpixel-wise probabilities $\mathbf{P}_s^i \in \mathbb{R}^{C \times K}, i = 1, 2$ and the corresponding soft labels $\mathbf{Y}_s \in [0, 1]^{C \times K}$ by averaging over each superpixel:

---

[1] On the chest x-ray dataset [5,20], our superpixels achieve undersegmentation errors lower than 0.32 for 800 superpixels per image, which is comparable to the natural image setting[1].

$$\mathbf{P}_s^i(c,k) = \frac{1}{N(k)} \sum_{j:S_j=k} \mathbf{P}^i(c,j), \qquad \mathbf{Y}_s(c,k) = \frac{1}{N(k)} \sum_{j:S_j=k} \mathbb{1}(Y_j = c) \qquad (1)$$

where $N(k) = |\{j : S_j = k\}|$ is the size of the superpixel. Inspired by [22], we define our superpixel-wise loss function ℓ^{sp} by considering both classification losses and prediction agreement on each superpixel:

$$\ell^{sp} = (1 - \lambda) * (\ell_{ce}(\mathbf{P}_s^1, \mathbf{Y}_s) + \ell_{ce}(\mathbf{P}_s^2, \mathbf{Y}_s)) + \lambda * \ell_{kl}(\mathbf{P}_s^1, \mathbf{P}_s^2) \qquad (2)$$

where ℓ_{ce} is the cross-entropy loss with soft labels, ℓ_{kl} is the symmetric Kullback-Leibler(KL) Divergence, and λ is a balance factor. By considering both two terms in the small loss criterion, we aim to select and update on training data with low label noise while maximizing networks' agreement. Denote R as the ratio of pixels being selected, we perform the small-loss selection by choosing the superpixel set $\hat{\mathcal{D}}_s$ as follows:

$$\hat{\mathcal{D}}_s = \arg \min_{\mathcal{D}_s:N(\mathcal{D}_s) \geq R \cdot M} \sum_{k \in \mathcal{D}_s} \ell_k^{sp} \qquad (3)$$

where $N(\mathcal{D}_s) = \sum_{k \in \mathcal{D}_s} N(k)$ is the total number of pixels in the superpixel set. Given the small-loss selection, we train the network based on the average loss:

$$\mathcal{L} = \frac{1}{N(\hat{\mathcal{D}}_s)} \sum_{S_i \in \hat{\mathcal{D}}_s} \ell_i \qquad (4)$$

where ℓ has the same form as Eq. 2 except it is defined on the pixel level. Here we skip the superpixel-level pooling as in Eq. 1 for more efficient back propagation.

Stopping Criterion. While the selection strategy enables the network training with mostly clean-labeled data, some noisy labels are inevitably selected and gradually affect model performance. To tackle the problem, we propose a criterion to stop network training before such overfitting. Our criterion is defined based on the loss gap G_l between the selected data and the rest of the training set as follows:

$$G_l = \frac{1}{K - |\hat{\mathcal{D}}_s|} \sum_{k \notin \hat{\mathcal{D}}_s} \ell_k^{sp} - \frac{1}{|\hat{\mathcal{D}}_s|} \sum_{k \in \hat{\mathcal{D}}_s} \ell_k^{sp} \qquad (5)$$

Intuitively, the model tends to first learn the relatively simple patterns in clean data, then starts to overfit to label noise [2]. Consequently, G_l first gradually increases, then starts to decrease. Based on this observation, we stop the model training when G_l reached the maximum before decreasing.

After training, we observe that the outputs of two peer networks are very similar to each other. Consequently, we arbitrarily choose one network to make predictions during test/deployment.

Label Refinement. In this stage, we use the trained networks to estimate the reliability of the superpixel annotations and relabel a subset of unreliable ones. Specifically, we choose superpixels with large losses, which indicates strong inconsistency between model predictions and their labels. We then relabel them according to predicted class labels. Formally, we define the unreliable superpixel set $\hat{\mathcal{D}}_u$ based on the superpixel losses and compute the predicted superpixel labels $\hat{\mathbf{Y}} = \{\hat{Y}_i\}_{i=1}^{K}, \hat{Y}_i \in \{1, \cdots, C\}$ as below,

$$\hat{\mathcal{D}}_u = \arg\max_{\mathcal{D}_u : N(\mathcal{D}_u) \leq (1-R)\cdot M} \sum_{k \in \mathcal{D}_u} \ell_k^{sp} \tag{6}$$

$$\hat{Y}_k = \arg\max_c \frac{1}{2}(\mathbf{P}_s^1(c,k) + \mathbf{P}_s^2(c,k)) \tag{7}$$

where ℓ_k^{sp} is defined in Eq. 2, R is the selection ratio mentioned before. Finally, we update the pixel-wise label map $\mathbf{Y}' = \{Y_i'\}_{i=1}^{M}, Y_i' \in \{1, \cdots, C\}$ as

$$Y_i' = \mathbb{1}(S_i \in \hat{\mathcal{D}}_u)\hat{Y}_{S_i} + \mathbb{1}(S_i \notin \hat{\mathcal{D}}_u)Y_i \tag{8}$$

After the label refinement, we replace $\mathbf{Y}$ by $\mathbf{Y}'$, increase R by a fixed ratio γ and start the next iteration.

3   Experiment

We validate our method on two public datasets, ISIC [7] and JSRT [5,20], which consists of images from two different modalities. We follow the literature and use simulated label noises as no public benchmark with real label noises is available.

3.1   Dataset

ISIC Dataset. ISIC 2017 dataset [7] is a public large-scale dataset of dermoscopy images, acquired from a variety of devices used at multiple sites. This dataset contains 2000 training and 600 test images with corresponding segmentation masks. We resize all images to 128×128 in resolution.

JSRT Dataset. JSRT dataset [5,20] is a public chest x-ray dataset containing three classes of annotations: lung, heart and clavicle. There are 247 chest radiographs in total, with unified resolution 2048×2048. We split them into a training set of 197 images and a test set of 50 images, and resize them into 256×256 [2].

Noise Patterns. To simulate manual noisy annotations, we randomly select a ratio α of samples from the training data to apply morphological [23–26] or affine transformation with noise level controlled by β. For affine transformation, we use a combination of rotation and translation to imitate other real-world noise patterns. Unlike prior works, we use the relative size w.r.t the target object region

---

[2] Clavicles are particularly small in chest x-ray images. To facilitate fine-grained segmentation and reduce consuming time, we crop their region of interest by statistics on the training set.

when controlling the noise level β, as people usually annotate target object in a favorable field of view by zooming in or out images. We investigate our algorithm in several noisy settings with α being $\{0.3, 0.5, 0.7, 1.0\}$ and β being $\{0.5, 0.7\}$. Some noisy examples are shown in the supplementary.

3.2 Experiment Setup

Comparisons. We compare our method with several state-of-the-art approaches, including Co-teaching [8], Tri-network [25] and JoCoR [22], which employ the robust learning at the pixel level. We do not include methods such as [15,24] as they rely on a clean validation set. For fair comparison, we re-implement these methods with the same network backbone and training policy.

Table 1. Quantitative comparisons of noisy-labeled segmentation methods on ISIC dataset, where the metric is Dice[%] over the last 10 epochs. α and β control the noise ratio and noise level, respectively.

	Baseline	Co-teaching[8]	Tri-network[25]	JoCoR [22]	Ours
Original data	82.49	82.72	82.96	83.64	**84.26**
$\alpha = 0.3, \beta = 0.5$	80.75	81.44	81.50	82.65	**84.00**
$\alpha = 0.3, \beta = 0.7$	79.46	81.47	80.73	81.58	**83.34**
$\alpha = 0.5, \beta = 0.5$	78.95	81.22	80.94	82.41	**83.90**
$\alpha = 0.5, \beta = 0.7$	75.44	80.06	80.24	81.06	**83.19**
$\alpha = 0.7, \beta = 0.5$	76.61	79.61	79.55	80.55	**83.83**
$\alpha = 0.7, \beta = 0.7$	71.51	78.50	76.61	79.05	**83.12**
$\alpha = 1.0, \beta = 0.5$	71.13	76.69	75.61	78.43	**82.23**
$\alpha = 1.0, \beta = 0.7$	63.71	73.68	70.01	74.30	**81.39**

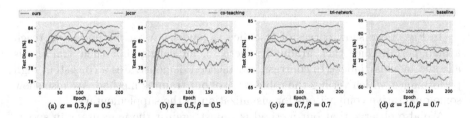

(a) $\alpha = 0.3, \beta = 0.5$ (b) $\alpha = 0.5, \beta = 0.5$ (c) $\alpha = 0.7, \beta = 0.7$ (d) $\alpha = 1.0, \beta = 0.7$

Fig. 2. Curves of test dice vs. epoch on four different noise settings.

Implementation Details. We adopt nnU-Net [9] as the segmentation network. Following [14], we take two networks sharing the same architecture yet with different initializations. Following [8], the noise rate is assumed to be known, and we set initial selection ratio $\mathcal{R}$ as $(1 - \text{noise rate})$ and γ as 1.1. The balance factor λ is 0.65. We train our model by a SGD optimizer, with a constant learning rate 0.005. The batch sizes are 32 for ISIC dataset and 8 for JSRT dataset. We implement the code framework with PyTorch on TITAN Xp GPU.

Evaluation Metric. During testing, we use the standard metric Dice coefficient (Dice) to evaluate the quality of predicted masks. We stop iterative learning when label refinement cannot bring any benefit, i.e., G_l no longer shows the rising trend for training. To make fair comparisons, we train all methods for maximum 200 epochs and report average Dice over the last 10 epochs.

3.3   Experiments on ISIC Dataset

Table 1 reports a summary of quantitative results of ISIC dataset. At the mild noise setting ($\alpha = 0.3, \beta = 0.5$), we achieve 84.00% Dice and outperform recent methods more than 1.35% Dice. As the noise increases, the performance of baseline decreases sharply, indicating the significant impact of label noise. Other methods mitigate this impact to some extent, but their performance still drop notably. By contrast, our method consistently outperforms them and maintains high performance, validating its robustness to different noise settings. Remarkably, in the extreme noise setting ($\alpha = 1.0, \beta = 0.7$), our method achieves 81.39% Dice and outperforms JoCoR (7.09% Dice), Co-teaching (7.71% Dice) and Tri-Network (11.38% Dice)[3].

Table 2. Ablation study on our model components.

Method	Superpixel	Selection	Label Refinement	Dice [%]
Ours	✓	✓	✓	**83.12**
	✗	✓	✓	81.15
	✓	✗	✓	79.32
	✓	✓	✗	80.56

In Fig. 2, we show curves of test dice vs. epochs. Most methods first reach a high performance then gradually decrease, indicating that their training is affected by noisy labels. In contrast, our method demonstrates a consistent high performance, which verifies the robustness of our training method. We also show some qualitative comparisons for visualization in the supplementary.

We also observe that our method is robust against the inaccuracy in superpixelization. For the ISIC dataset, we use 100 superpixels per image with relative high undersegmentation error (1.0), and our superpixel selection can potentially discard inaccurate superpixels in the noise-aware learning.

---

[3] We also observe that our method outperforms the baseline with 84.26% Dice on the original dataset, likely due to the noise in manual annotations [7].

3.4   Ablation Study

We first verify the effect of superpixel representation based on a set of experiments on ISIC dataset under the noise setting ($\alpha = 0.7, \beta = 0.7$), whose results are shown in Table 2. Row #1 is our method which achieves 83.12% Dice. Changing superpixel to pixel representation brings a performance drop of 1.97% Dice in row #2. This demonstrates the advantage of superpixel representation in learning with noisy labels.

In addition, to analyze the effect of selection module and label refinement module in iterative learning, we take a drop-one-out manner at the same setting. Ablating selection module from the model leads to a decrease of 3.80% Dice in row #3, meanwhile, removing label refinement module makes the performance drop 2.56% Dice in row #4. It is evident that both modules are essential for our robust iterative learning strategy. We also validate the effectiveness of adaptive stopping criterion and report the quality of refined labels in the supplementary.

3.5   Experiments on JSRT Dataset

To explore the generalization capability of our method, we also conduct experiments on JSRT dataset. Figure 3 presents the average results of three classes, and the table in supplementary reports detailed values for each class. Our method outperforms other methods consistently on all three classes.

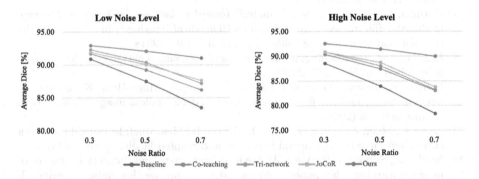

Fig. 3. Average results on JSRT dataset with different noise settings: low noise level $\beta = 0.5$ and high noise level $\beta = 0.7$, respectively.

4   Conclusion

In this paper, we propose a robust learning strategy for medical image segmentation. Unlike previous methods, we exploit structural prior and pixel correlation for segmentation model learning, which significantly mitigate the impact of label noise. We develop an iterative learning scheme based on superpixel representation. In each iteration, we jointly train two deep networks using selected subsets of superpixels, and also relabel a subset of unreliable superpixels. Evaluation on two benchmarks with simulated noises demonstrates that our learning strategy achieves the state-of-the-art performance and robustness in extensive noise settings. We note that learning with realistic label errors is an important future research topic and building a benchmark with such label noises is a crucial step.

References

1. Achanta, R., Shaji, A., Smith, K., Lucchi, A., Fua, P., Süsstrunk, S.: Slic superpixels compared to state-of-the-art superpixel methods. IEEE Trans. Pattern Anal. Mach. Intell. (2012)
2. Arpit, D., et al.: A closer look at memorization in deep networks. In: ICML (2017)
3. Ching, T., Himmelstein, D.S., et al.: Opportunities and obstacles for deep learning in biology and medicine. J. Royal Soc. Interface **15**, 20170387 (2018)
4. Daoud, M.I., Atallah, A.A., Awwad, F., Al-Najjar, M., Alazrai, R.: Automatic superpixel-based segmentation method for breast ultrasound images. Expert Systems with Applications (2019)
5. Ginneken, B., Stegmann, M.B., Loog, M.: Segmentation of anatomical structures in chest radiographs using supervised methods: a comparative study on a public database. Medical image analysis (2006)
6. Gurari, D., Theriault, D., et al.: How to collect segmentations for biomedical images? a benchmark evaluating the performance of experts, crowdsourced nonexperts, and algorithms. In: 2015 IEEE Winter Conference on Applications of Computer Vision (2015)
7. Gutman, D., et al.: Skin lesion analysis toward melanoma detection: a challenge at the 2017 International Symposium on Biomedical Imaging (isbi), Hosted by the International Skin Imaging Collaboration (isic). ISBI (2018)
8. Han, B., et al.: Co-teaching: robust training of deep neural networks with extremely noisy labels. In: NeurIPS (2018)
9. Isensee, F., Jaeger, P.F., Kohl, S.A.A., Petersen, J., Maier-Hein, K.: nnu-net: a self-configuring method for deep learning-based biomedical image segmentation. Nature methods (2020)
10. Jiang, L., Zhou, Z., Leung, T., Li, L.J., Fei-Fei, L.: Mentornet: learning data-driven curriculum for very deep neural networks on corrupted labels. ArXiv (2018)
11. Kohli, M., Summers, R., Geis, J.: Medical image data and datasets in the era of machine learning–whitepaper from the 2016 c-mimi meeting dataset session. J. Digital Imaging **30**, 392–399 (2017)
12. Li, Y., Jia, L., Wang, Z., Qian, Y., Qiao, H.: Un-supervised and semi-supervised hand segmentation in egocentric images with noisy label learning. Neurocomputing (2019)
13. Litjens, G., et al.: A survey on deep learning in medical image analysis. Medical Image Analysis (2017)
14. Malach, E., Shalev-Shwartz, S.: Decoupling "when to update" from "how to update". In: NIPS (2017)
15. Mirikharaji, Z., Yan, Y., Hamarneh, G.: Learning to segment skin lesions from noisy annotations. ArXiv (2019)
16. Northcutt, C.G., Jiang, L., Chuang, I.: Confident learning: Estimating uncertainty in dataset labels. ArXiv (2019)

17. Qin, W., et al.: Superpixel-based and boundary-sensitive convolutional neural network for automated liver segmentation. Physics in Medicine & Biology (2018)
18. Ren, M., Zeng, W., Yang, B., Urtasun, R.: Learning to reweight examples for robust deep learning. In: ICML (2018)
19. Shen, D., Wu, G., Suk, H.: Deep learning in medical image analysis. Annual review of biomedical engineering (2017)
20. Shiraishi, J., et al.: Development of a digital image database for chest radiographs with and without a lung nodule: receiver operating characteristic analysis of radiologists' detection of pulmonary nodules. AJR, American journal of roentgenology (2000)
21. Tian, Z., Liu, L., Zhang, Z., Fei, B.: Superpixel-based segmentation for 3d prostate mr images. IEEE Trans. Med. Imaging **35**, 791–801 (2015)
22. Wei, H., Feng, L., Chen, X., An, B.: Combating noisy labels by agreement: a joint training method with co-regularization. CVPR (2020)
23. Xue, C., Deng, Q., Li, X., Dou, Q., Heng, P.: Cascaded robust learning at imperfect labels for chest x-ray segmentation. In: MICCAI (2020)
24. Zhang, M., et al.: Characterizing label errors: confident learning for noisy-labeled image segmentation. In: MICCAI (2020)
25. Zhang, T., Yu, L., Hu, N., Lv, S., Gu, S.: Robust medical image segmentation from non-expert annotations with tri-network. In: MICCAI (2020)
26. Zhu, H., Shi, J., Wu, J.: Pick-and-learn: automatic quality evaluation for noisy-labeled image segmentation. In: MICCAI (2019)

A Hybrid Attention Ensemble Framework
for Zonal Prostate Segmentation

Mingyan Qiu[1,2], Chenxi Zhang[1,2(✉)], and Zhijian Song[1,2(✉)]

[1] Digital Medical Research Center, School of Basic Medical Sciences, Fudan University,
Shanghai 200032, China
{chenxizhang,zjsong}@fudan.edu.cn
[2] Shanghai Key Laboratory of Medical Imaging Computing and Computer-Assisted
Intervention, Shanghai 200032, China

Abstract. Accurate and automatic segmentation of the prostate sub-regions is of great importance for the diagnosis of prostate cancer and quantitative analysis of prostate. By analyzing the characteristics of prostate images, we propose a hybrid attention ensemble framework (HAEF) to automatically segment the central gland (CG) and peripheral zone (PZ) of the prostate from a 3D MR image. The proposed attention bridge module (ABM) in the HAEF helps the Unet to be more robust for cases with large differences in foreground size. In order to deal with low segmentation accuracy of the PZ caused by small proportion of PZ to CG, we gradually increase the proportion of voxels in the region of interest (ROI) in the image through a multi-stage cropping and then introduce self-attention mechanisms in the channel and spatial domain to enhance the multi-level semantic features of the target. Finally, post-processing methods such as ensemble and classification are used to refine the segmentation results. Extensive experiments on the dataset from NCI-ISBI 2013 Challenge demonstrate that the proposed framework can automatically and accurately segment the prostate sub-regions, with a mean DSC of 0.881 for CG and 0.821 for PZ, the 95% HDE of 3.57 mm for CG and 3.72 mm for PZ, and the ASSD of 1.08 mm for CG and 0.96 mm for PZ, and outperforms the state-of-the-art methods in terms of DSC for PZ and average DSC of CG and PZ.

Keywords: 3D Prostate MRI · Zonal segmentation · Attention mechanism

1 Introduction and Related Work

Prostate cancer is one of the most common malignant tumours in the world, and it usually ranks first in the incidence of male malignancies in Europe and the United States [1]. The prostate can be divided into two regions on the MR image: the central gland (CG) and the peripheral zone (PZ). About seventy percent of prostate cancers arise in the PZ, which is a target for prostate biopsy. The ratio of CG volume to the whole prostate (WG) is a valuable indicator for monitoring prostate hyperplasia. Therefore, accurate segmentation of the prostate CG and PZ region is of great value for diagnosis of prostate cancer and assessment of benign prostate hyperplasia.

M. de Bruijne et al. (Eds.): MICCAI 2021, LNCS 12901, pp. 536–547, 2021.
https://doi.org/10.1007/978-3-030-87193-2_51

Martin et al. [2] proposed an automatic segmentation of prostate sub-regions based on probabilistic atlas. Litjens et al. [3] proposed a pattern recognition approach incorporating features of anatomy, intensity and texture to classify multi-parametric MRI voxels in prostate region. Makni et al.

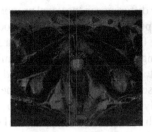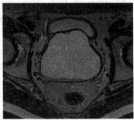

Fig. 1. Visualization of two prostate examples.

[4] presented an evidential C-Means (ECM) classifier-based method to extract prostate zonal anatomy, where a relaxation step was introduced to integrate voxels' spatial neighbourhood information. Toth et al. [5] used active appearance models with multiple coupled level sets to automatically segment prostate sub-regions. Their method requires high quality data and took more than 3 min to segment a given image. Qiu et al. [6] reported a dual optimization-based method to extract the WG, CG and PZ simultaneously, which yielded high accuracy results for both 3D body- and endo-coil MR images. This method still requires some user-chosen boundary points for initialization. In recent years, methods based on convolutional neural network have achieved remarkable performance in prostate zonal segmentation [6]. Clark et al. [8] proposed an architecture with four continuous 2D convolutional neural networks (CNNs) to automatically extract prostate gland and transition zone (TZ) in diffusion-weighted imaging (DWI). Meyer et al. [9] reported an anisotropic network model based on 3D Unet [10], which can simultaneously segment TZ, PZ, distal prostatic urethra (DPU) and anterior fibromuscular stroma (AFS) from axial T2w MRI. Liu et al. [11] used a 2D fully CNN with a feature pyramid attention mechanism for segmentation of TZ and PZ on T2w MR slices which yields good accuracy on both internal testing and external testing datasets.

It has always been a great challenge to accurately segment the CG and PZ at the same time. Due to different imaging protocols, the proportion of fore-ground voxels among different samples varies greatly. As shown in Fig. 1, the proportion of foreground voxels to the whole MR image ranges from 0.58% to 10.9%, which greatly increases the difficulty of accurate prostate zonal segmentation. The volume of the PZ is much smaller than that of the CG, leading to the segmentation accuracy of the PZ is far inferior to that of the CG. Additionally, owing to the low degree of recognition of the apex and base parts of the prostate, model is sometimes confused about which Z slice to start or end in prostate region and yields low accuracy in the apex and base of the prostate.

In order to solve the above problems, we propose a hybrid attention ensemble framework (HAEF), to automatically and accurately segment CG and PZ from T2-weighted MR images. To address the difference in the voxel proportion of the whole prostate among different samples, we propose an attention bridge module (ABM) to fill the semantic gap between the shallow, fine-grained encoder features and the deep, coarse-grained decoder features. By fusing the multi-scale feature information from the encoder, the ABM assists the decoder to better locate the target and recover the details of the target. In order to deal with low segmentation accuracy of the PZ caused by small proportion of PZ to CG, we gradually increase the proportion of voxels in the region of interest (ROI) in the image through a multi-stage cropping and then introduce the self-attention mechanism in the channel and spatial domain to enhance the multi-level semantic features of the target. Finally, multi-stage segmentation results are integrated to enhance the segmentation results of the PZ. In order to solve the issue of the mis-segmentation at apex and base of the prostate, we propose to classify the ROIs on each slice, and use the classification results to optimize the integrated segmentation results.

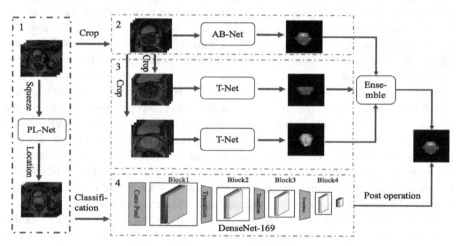

Fig. 2. Diagram of HAEF architecture.

Encoded features

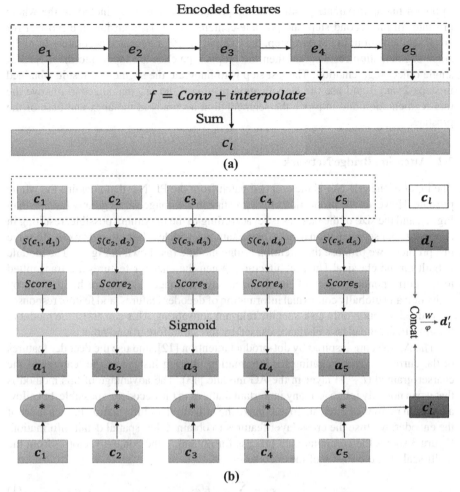

(a)

(b)

Fig. 3. Attention bridge module. (a) The five-layer encoder features are convoluted and interpolated to get the same size and number of channels as the l-layer feature map, and summed to obtain the context vector cl. (b) Diagram of the five-layer attention bridge modules.

2 Methods

2.1 Hybrid Attention Ensemble Framework

As shown in Fig. 2, the proposed hybrid attention ensemble framework (HAEF) is mainly composed of four parts: prostate location network (PL-Net), attention bridge network (AB-Net), targeted segmentation network (T-Net), ensemble and classification. All segmentation models used in the proposed framework are based on the classical Unet architecture with five-layer symmetric encoder-decoder, and adopt Leaky Relu with a negative slope of 0.01 as the activation function. The first step is to locate the prostate. The whole MR image of the prostate is input into the Unet for rough segmentation to obtain the approximate position of the prostate and the patch including the whole prostate. In the second step, the prostate-centered patch is input into the AB-Net for zonal segmentation. In the third step, the CG and the PZ are located accurately from the zonal segmentation results, and then the cropped patches are input into the T-Net for single object segmentation. The last step is to integrate the segmented results obtained in steps 2 and 3 and use the classification results of the original slices to remove the mis-segmentation and improve the segmentation accuracy at the apex and base of the prostate.

2.2 Attention Bridge Network

The input of the AB-Net is the cropped patch from the PL-Net that contains the whole prostate. However, the proportion of the prostate in the image varies greatly (as shown in Fig. 2), and the size of the patches is set according to the maximum possible, resulting in relatively poor segmentation results for the samples with less foreground voxels. To solve this problem, we propose the attention bridge module (as shown in Fig. 3). The module is built on the classical Unet architecture. When the encoder features are transmitted to the corresponding layer of the decoder, salient image regions are highlighted by activation and globally contextual information of decoder features, and feature responses are pruned to retain only task-specific relevant activations. This can assist the model to better mitigate the class imbalance caused by too small foreground objects.

The AB-Net are inspired by dot-product attention [12], and use the decoder features of the current l-layer as gating for information filtering instead of the features of the coarser-grained $(l-1)$ layer in the AG module [13]. The advantage of this method is that it can not only retain as many important features in the decoder as possible, but allow a focus on regions of interest. To maximize the utilization of information in each layer of the encoder, we fuse the cross-layer features to obtain richer spatial detail information. Figure 3 shows the computational process for obtaining the context vector c_l from the multi-scale feature fusion of the encoder.

$$c_l = \sum_{\hat{l}=1}^{L} f_{\hat{l}}^l e_{\hat{l}},\tag{1}$$

where L denotes the depth of Unet, $f_{\hat{l}}^l$ denotes the interpolation operation and the $1 \times 1 \times 1$ convolution operation with a step size of 1, $e_{\hat{l}}$ denotes the $\hat{l}$-layer encoder feature map. The output size is equal to the size of the current l-layer feature map, and the number of output channels is equal to the number of channels in the current l-layer.

The attention bridge of each layer is shown in Fig. 3(b). The similarity score S is calculated by comparing the context vector c_l of the current layer with the eigenvector d_l of the corresponding decoding layer, then the attention vector a_l is derived:

$$a_l = Sigmoid(S(c_l, d_l)) = \frac{1}{1 + e^{-S(c_l, d_l)}}.\tag{2}$$

We take into account four different alternatives for calculating similarity scores:

$$S(c_l, d_l) = \begin{cases} c_l^T d_l, & dot, \\ c_l^T W_a d_l, & general, \\ \frac{c_l d_l}{c_l d_l}, & cosine, \\ v_a^T relu(W_a[c_l; d_l]) & concat. \end{cases} \quad (3)$$

The active eigenvector c_l' is obtained by multiplying the attention coefficient a_l as a weight with the contextual vector c_l.

$$c_l' = a_l c_l. \quad (4)$$

Like traditional skip connections, we employ a simple concatenation layer to combine the information from the two vectors to produce the following feature map:

$$d'_l = \varphi\left(W_l\left[c_l'; d_l\right]\right), \quad (5)$$

where φ is defined as group normalization and nonlinear function relu.

Finally, in the last layer of Unet, the prediction probability distribution of each voxel can be acquired through softmax processing on the output feature map, then the final mask of prostate can be obtained.

2.3 Targeted Segmentation Network

The CG and PZ are extracted from the original images using AB-Net and then the CG- and PZ-centered patches are input into T-Net separately for fine segmentation of single object. T-Net is a classical Unet based network that integrates channels and spatial attention mechanisms [14], which focus on both "what" and "where" information.

We take the intermediate feature image obtained after convolution and other operations as input, from which a 1D channel attention map and a 3D spatial attention map are inferred sequentially. Then attention maps are multiplied by the input feature map element-wisely and broadcasted along the channel and the spatial dimension respectively. The channel ($Ac(F)$) and spatial attention ($As(F)$) are calculated as follows:

$$Ac(F) = \sigma(MLP(AvgPool(F)) + MLP(MaxPool(F))), \quad (6)$$

$$As(F) = \sigma\left(conv\left([AvgPool(F); MaxPool(F)]\right)\right), \quad (7)$$

where σ denotes the sigmoid function, MLP denotes the multi-layer perceptron, F represents the input feature map, and $conv$ represents the convolution operation.

2.4 Post-processing

Ensemble. The prostate multi-region mask output by AB-Net and the CG and PZ mask output by T-Net are integrated, and the final result is obtained by voting. Voxels can be classified as 0 (back ground)/1(PZ)/2 (CG) in ABNet, 0/1 in TNet-PZ and 0/2 in TNet-CG. The category of each voxel is determined by the number of dominant votes in three outputs for that voxel. E.g., if results of a pixel from ABNet, TNet-PZ and TNet-CG are $(1, 1, 2)$, the voting result of the pixel is 1, i.e. PZ. If there is a tie, according to the results of Table 3, the tie vote rule is that the result of TNet-PZ is given priority, followed by ABNet (e.g. for $(0, 1, 2)$, the final result is 1). And for the case of $(0, 1, 0)$, the TNet-CG result does not determine the PZ category and is ignored. The other two results $(0, 1)$ are compared. Then it becomes a tie case, and the final result is 1. For case of $(0, 0, 2)$, the result of TNet-PZ is ignored and the voting result is 0.

Classification. The segmentation model is sometimes confused about which Z slice to start or end in the prostate region and yields low accuracy at the apex and base of the prostate. Therefore, we propose to use a classification network to classify all 2D MR slices, and optimize the integrated segmentation results with the help of the slice classification results to remove the mis-segmented sub-regions at both ends. All MR slices are divided into four types: only background (0), only PZ (1), only CG (2), and PZ + CG (3). The comparative experimental results of several classical classification network models show that the DenseNet-169 network achieves the highest classification accuracy, which is about 95.9%. Therefore, the DenseNet-169 is used to post-process the segmentation results. We also found that if a slice contains the PZ, it generally contains the CG, and the CG is more likely to appear at both ends. The main cases of mis-segmentation are (1) the start or end slice without prostate is predicted to contain CG or CG + PZ; (2) the slice only contain the CG is predicted to contain CG + PZ. Therefore, when the number of categories of the segmentation result in the slice is more than that of the classification result, the redundant category regions compared with the classification result are removed.

3   Experiments and Results

3.1   Data and Implementation

We performed extensive experiments to evaluate our method on a dataset from NCI-ISBI 2013 Challenge which consists of 1.5T endo-rectal coil T2w MR images (40 cases) and 3T body-coil T2w MR images (40 cases) [15]. Two cases were removed due to incomplete labels. The final training dataset includes 68 cases, and the test dataset includes 10 cases. The T2w MR sequences were acquired with 4 mm or 3 mm thickness, while the voxel size varies from 0.4 mm to 0.625 mm. The image size is from 320×320 to 400×400.

We implemented our framework using PyTorch (Paszke et al., 2017), and ran all experiments on an NVIDIA GTX 1080Ti GPU. All images were processed with intensity normalization and contrast limited adaptive histogram equalization (CLAHE) and then resampled to a fixed voxel size of $0.625 \times 0.625 \times 1.5$ mm^3. Due to the limitation of the memory capacity, the MR image was first resized to a size of $160 \times 160 \times 48$ for coarse segmentation of the prostate. Then the prostate-centered area of $128 \times 128 \times 48$ voxels was cropped from the original image for zonal segmentation. Random transformations mainly including random rotation, shear, zoom and flip operations were performed in each training iteration for data enhancement. The loss function adopted the soft dice loss and cross-entropy loss, $Loss = \lambda L_{Dice} + \lambda L_{CE}$, λ equals 1. The segmentation network initial learning rate was set to $2e^{-4}$, and iterative optimization was conducted using the optimization algorithm of Adabound. The total training time of all steps was about 24 h, and the average test time was about 1 s per case. The evaluation metrics of the Dice similarity coefficient (DSC), 95% Hausdorff distance (95% HDE) and the Average Symmetric Surface Distance (ASSD) were used to measure the effectiveness of our loss function.

3.2 Results

Determination of Similarity Function. We experimented with different similarity functions (cosine, dot, general and concat) mentioned in Sect. 2.2 for prostate zonal segmentation to determine the optimal similarity function for attention bridge model. Since the dot and general functions are calculated by dot product which requires high computational cost, only the attention bridge model is used in the first three layers of the decoder of Unet. The results are shown in Table 1. It is observed that the performance of the general function is much better than that of other three similarity functions. There-fore, we choose the general function as the similarity score function in AB-Net to obtain the activation weight of the feature maps in the encoder.

Table 1. Comparison of similarity functions in AB-NET.

Similarity Functions	DSC (%)		95% HDE (mm)		ASSD (mm)	
	PZ	CG	PZ	CG	PZ	CG
cosine	61.60 ± 14.37	80.65 ± 8.05	6.43 ± 1.72	5.53 ± 1.90	2.00 ± 0.65	1.90 ± 0.73
dot	76.85 ± 4.26	85.34 ± 7.42	4.93 ± 1.54	5.61 ± 3.32	1.28 ± 0.30	1.57 ± 0.92
general	**78.24 ± 5.07**	**87.52 ± 4.27**	**4.32 ± 1.68**	**3.72 ± 1.38**	**1.18 ± 0.38**	**1.27 ± 0.54**
concat	76.81 ± 5.07	87.08 ± 5.83	4.81 ± 1.53	4.12 ± 2.03	1.20 ± 0.29	1.32 ± 0.75

Comparison with Competing Segmentation Models. Table 2 shows the quantitative results of our method and other classical segmentation models in the prostate zonal segmentation task. Our method yields a mean DSC of 0.881 for CG and 0.821 for PZ, the HDE95 of 3.57 mm for CG and 3.72 mm for PZ, and the MDE of 1.08 mm for CG and 0.96 mm for PZ and outperforms all the compared methods in terms of both DSC and distance-based metrics. In particular, compared with other methods, our method greatly improves the segmentation accuracy of PZ. Figure 4 shows the segmentation results of different segmentation models on slices in apex, mid-gland and base subregions. It is observed that our method produces more realistic results which are smooth at the boundary of PZ (as indicated by blue arrows).

Table 2. Quantitative comparison to different models on segmentation of PZ and CG.

Model	DSC (%)		95% HDE (mm)		ASSD (mm)	
	PZ	CG	PZ	CG	PZ	CG
Vnet [16]	72.88 ± 7.49	84.57 ± 7.47	5.69 ± 2.64	4.70 ± 2.79	1.63 ± 0.81	1.53 ± 0.81
DenseVoxelNet [17]	71.31 ± 6.18	84.42 ± 5.65	5.61 ± 2.19	5.27 ± 2.38	1.57 ± 0.42	1.47 ± 0.54
Resnet50 [18]	70.25 ± 7.68	83.53 ± 6.27	4.95 ± 1.44	4.52 ± 1.37	1.55 ± 0.40	1.59 ± 0.56
Unet [10]	74.94 ± 6.67	84.86 ± 7.99	5.58 ± 2.06	6.34 ± 4.88	1.44 ± 0.45	1.80 ± 1.22
Attention Unet [13]	75.73 ± 7.74	86.01 ± 6.52	5.15 ± 1.36	4.73 ± 2.52	1.29 ± 0.43	1.47 ± 0.78
HAEF (ours)	**82.13 ± 4.63**	**88.08 ± 5.25**	**3.72 ± 1.66**	**3.57 ± 1.56**	**0.96 ± 0.46**	**1.08 ± 0.67**

Ablation Study. In order to evaluate the effectiveness of components presented in our HAEF framework, we conducted a set of ablation experiments for the HAEF framework. Quantitative comparison results are reported in Table 3. Fine-tuning Unet is used as a baseline. cl. represents slice classification module. Experimental results indicate that introducing ABM to Unet can improve both the segmentation results of CG and PZ. T-Net yields better results as well, especially the results of PZ. The ensemble models of AB-Net and T-Net with classification module for post-processing achieve the best performance in all metrics.

Table 3. Ablation analysis of proposed HAEF framework.

Model	DSC (%)		95% HDE (mm)		ASSD (mm)	
	PZ	CG	PZ	CG	PZ	CG
Unet	74.94 ± 6.67	84.86 ± 7.99	5.58 ± 2.06	6.34 ± 4.88	1.44 ± 0.45	1.08 ± 1.22
AB-Net	78.24 ± 5.07	87.52 ± 4.27	4.32 ± 1.68	3.72 ± 1.38	1.18 ± 0.38	1.27 ± 0.54
T-Net	81.30 ± 4.98	87.03 ± 4.96	4.00 ± 1.73	4.26 ± 1.96	1.01 ± 0.45	1.25 ± 0.48
AB-Net + T-Net	81.97 ± 4.42	87.79 ± 5.06	3.82 ± 1.58	3.70 ± 1.42	0.97 ± 0.44	1.14 ± 0.64
AB-Net + cl	78.92 ± 5.03	87.91 ± 4.49	3.93 ± 1.64	3.72 ± 1.38	1.12 ± 0.41	1.19 ± 0.59
AB-Net + T-Net + cl	82.13 ± 4.63	88.08 ± 5.25	3.72 ⊥ 1.66	3.57 ± 1.56	0.96 ± 0.45	1.08 ± 0.67

Comparison with Previous Works. A comparison of our method with previous works on segmentation of zonal prostate is given in Table 4. For the CG, the method reported by Isensee et al. [7] produces a DSC of 0.896 which is slightly higher than our method's DSC of 0.881, and ranks the first; while for the PZ, our method generates a DSC of 0.821 and performs best. And our method produces the best average DSC of CG and PZ (0.851).

Table 4. Quantitative comparison to previous works on segmentation of PZ and CG.

Work	Input	DSC (%)		
		Mean	PZ	CG
Semi-Autom				
Lijtens [3]	mpMRI	82.0	75.0 ± 7.0	89.0 ± 3.0
Makni [4]	mpMRI	81.5	76.0 ± 6.0	87.0 ± 4.0
Qiu [6]	T2w	75.7	69.1 ± 6.9	82.2 ± 3.0
Automatic				
Toth [5]	T2w	73.5	68.0 ± ?	79.0 ± ?
Clark [8]	mpMRI	–	–	84.7 ± ?
Liu [11]	T2w	80.0	74.0 ± 8.0	86.0 ± 7.0
Isensee [7]	T2w	82.7	75.8 ± ?	**89.6 ± ?**
Meyer [9]	T2w	83.7	79.8 ± 5.1	87.6 ± 6.6
HAEF (ours)	T2w	**85.1**	**82.1 ± 4.6**	88.1 ± 5.3

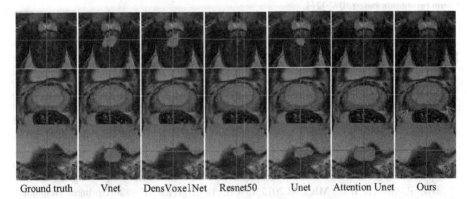

| Ground truth | Vnet | DensVoxelNet | Resnet50 | Unet | Attention Unet | Ours |

Fig. 4. Segmentation results of six segmentation models on slices in apex (top row), mid-gland (middle row) and base (bottom row) sub-regions. Green represents CG, and red is PZ. (Color figure online)

4 Discussion and Conclusions

We present a hybrid attention ensemble framework (HAEF) for automated prostate zonal segmentation from MR images. The HAEF consists of four components: prostate location network (PL-Net), attention bridge network (AB-Net), targeted segmentation network (T-Net) and module of ensemble and classification. The AB-Net maximizes the utilization of fine-grained encoder features by fusing the multi-scale feature information from the encoder, which is more robust for cases with large differences in foreground size. Unlike Ref. [19], which leverages the encoder multi-layer features to refine the features at each encoder individual layer, our method filters the encoder multi-layer features by gating signal from decoder features d_l, with the ultimate goal of obtaining a finer segmentation result. With the attention modules in channel and spatial dimension, T-Net can significantly improve the segmentation performance of single target, especially the PZ. The experimental results indicate that AB-Net and T-Net can complement each other, and integration of both modules is able to deal with the details of the segmentation target better. The classification for the slice is proposed for the first time as a post-processing method, which can effectively reduce the mis-segmentation of the model at apex and base of the prostate by comparing the classification result with the segmentation result category. Experimental results demonstrate that the proposed method is able to provide an accurate and automatic segmentation for prostate CG and PZ regions, with a mean DSC of 0.881 for CG and 0.821 for PZ, the 95% HDE of 3.57 mm for CG and 3.72 mm for PZ, and the ASSD of 1.08 mm for CG and 0.96 mm for PZ. Its good performance and no need for human interaction make it a promising method for image-guided prostate interventions and computer-aided diagnosis of prostate cancer. Future work is expected to be evaluation of our framework on more 3D volumetric data and study on improving segmentation accuracy at apex and base of the prostate with attention mechanism.

Acknowledgments. This work was supported by the Science and Technology Innovation Action Plan of Shanghai [grant number: 19511121302], and National Natural Science Foundation of China [grant number: 82072021].

References

1. Siegel, R.L., Miller, K.D., Jemal, A.: Cancer Statistics, 2017. CA Cancer J. Clin. **60**(5), 277–300 (2010)
2. Martin, S., Troccaz, J., Daanen, V.: Automated segmentation of the prostate in 3D MR images using a probabilistic atlas and a spatially constrained deformable model. Med. Phys. **37**(4), 1579–1590 (2010)
3. Litjens, G., Debats, O., van de Ven, W., Karssemeijer, N., Huisman, H.: A pattern recognition approach to zonal segmentation of the prostate on MRI. In: Ayache, N., Delingette, H., Golland, P., Mori, K. (eds.) MICCAI 2012. LNCS, vol. 7511, pp. 413–420. Springer, Heidelberg (2012). https://doi.org/10.1007/978-3-642-33418-4_51
4. Makni, N., Iancu, A., Colot, O., Puech, P., Mordon, S., et al.: Zonal segmentation of prostate using multispectral magnetic resonance images. Med. Phys. **38**(11), 6093–6105 (2011)
5. Toth, R., Ribault, J., Gentile, J., Sperling, D., Madabhushi, A.: Simultaneous segmentation of prostatic zones using active appearance models with multiple coupled levelsets. Comput. Vis. Image Underst. **117**(9), 1051–1060 (2013)
6. Qiu, W., Yuan, J., Ukwatta, E., Sun, Y., Rajchl, M., et al.: Dual optimization based prostate zonal segmentation in 3D MR images. Med. Image Anal. **18**(4), 660–673 (2014)
7. Isensee, F., Jaeger, P.F., Kohl, S.A.A., Petersen, J., Maier-Hein, K.H.: nnU-Net: a self-configuring method for deep learning-based biomedical image segmentation. Nat. Methods **18**, 203–211 (2021)
8. Clark, T., Zhang, J., Baig, S., Wong, A., Haider, M.A., Khalvati, F.: Fully automated segmentation of prostate whole gland and transition zone in diffusion-weighted MRI using convolutional neural networks. J. Med. Imag. **4**(4), 041307 (2017)
9. Meyer, A., Rak, M., Schindele, D., Blaschke, S., Schostak, M., Fedorov, A., Hansen, C.: Towards patient-individual PI-RADS v2 sector map: CNN for automatic segmentation of prostatic zones from T2-weighted MRI. In: ISBI, pp. 696–700. IEEE (2019)

10. Ronneberger, O., Fischer, P., Brox, T.: U-Net: convolutional networks for biomedical image segmentation. In: Navab, N., Hornegger, J., Wells, W.M., Frangi, A.F. (eds.) MICCAI 2015. LNCS, vol. 9351, pp. 234–241. Springer, Cham (2015). https://doi.org/10.1007/978-3-319-24574-4_28

11. Liu, Y., Yang, G., Mirak, S.A., Hosseiny, M., Azadikhah, A., Zhong, X.: Automatic prostate zonal segmentation using fully convolutional network with feature pyramid attention. IEEE Access 7, 163626–163632 (2019)

12. Luong, M. T., Pham, H., Manning, C.D.: Effective approaches to attention-based neural machine translation. Comput. Sci. (2015)

13. Oktay, O., et al.: Attention U-Net: learning where to look for the pancreas. In: MIDL (2018)

14. Woo, S., Park, J., Lee, J.Y., Kweon, I.S.: CBAM: convolutional block attention module. In: Ferrari, V., Hebert, M., Sminchisescu, C., Weiss, Y. (eds.) ECCV 2018, LNCS, vol. 11211, pp. 3–19. Springer, Cham (2018)

15. Bloch, N., et al.: NCI-ISBI 2013 challenge: automated segmentation of prostate structures. Cancer Imag. Arch (2015). https://doi.org/10.7937/K9/TCIA.2015.zF0vlOPv

16. Milletari, F., Navab, N., Ahmadi, S.A.: V-Net: fully convolutional neural networks for volumetric medical image segmentation. In: 2016 Fourth International Conference on 3D Vision (3DV), pp. 565–571. IEEE (2016)

17. Yu, L., et al.: Automatic 3D cardiovascular MR segmentation with densely-connected volumetric ConvNets. In: Descoteaux, M., Maier-Hein, L., Franz, A., Jannin, P., Collins, D.L., Duchesne, S. (eds.) MICCAI 2017. LNCS, vol. 10434, pp. 287–295. Springer, Cham (2017). https://doi.org/10.1007/978-3-319-66185-8_33

18. Hara, K., Kataoka, H., Satoh, Y.: Can Spatiotemporal 3D CNNs retrace the history of 2D CNNs and ImageNet?. In: CVPR, pp. 6546–6555. IEEE (2018)

19. Wang, Y., et al.: Deep attentional features for prostate segmentation in ultrasound. In: Frangi, A.F., Schnabel, J.A., Davatzikos, C., Alberola-López, C., Fichtinger, G. (eds.) MICCAI 2018. LNCS, vol. 11073, pp. 523–530. Springer, Cham (2018). https://doi.org/10.1007/978-3-030-00937-3_60

3D-UCaps: 3D Capsules Unet for Volumetric Image Segmentation

Tan Nguyen[1(⊠)], Binh-Son Hua[1], and Ngan Le[2]

[1] VinAI Research, Hanoi, Vietnam
{v.tannh10,v.sonhb}@vinai.io
[2] Department of Computer Science and Computer Engineering,
University of Arkansas, Fayetteville 72701, USA
thile@uark.edu

Abstract. Medical image segmentation has been so far achieving promising results with Convolutional Neural Networks (CNNs). However, it is arguable that in traditional CNNs, its pooling layer tends to discard important information such as positions. Moreover, CNNs are sensitive to rotation and affine transformation. Capsule network is a data-efficient network design proposed to overcome such limitations by replacing pooling layers with dynamic routing and convolutional strides, which aims to preserve the part-whole relationships. Capsule network has shown a great performance in image recognition and natural language processing, but applications for medical image segmentation, particularly volumetric image segmentation, has been limited. In this work, we propose 3D-UCaps, a 3D voxel-based Capsule network for medical volumetric image segmentation. We build the concept of capsules into a CNN by designing a network with two pathways: the first pathway is encoded by 3D Capsule blocks, whereas the second pathway is decoded by 3D CNNs blocks. 3D-UCaps, therefore inherits the merits from both Capsule network to preserve the spatial relationship and CNNs to learn visual representation. We conducted experiments on various datasets to demonstrate the robustness of 3D-UCaps including iSeg-2017, LUNA16, Hippocampus, and Cardiac, where our method outperforms previous Capsule networks and 3D-Unets. Our code is available at https://github.com/VinAIResearch/3D-UCaps.

Keywords: Capsule network · CapsNet · Medical image segmentation

1 Introduction

Medical image segmentation (MIS) is a visual task that aims to identify the pixels of organs or lesions from background medical images. It plays a key role in medical analysis, computer-aided diagnosis, and smart medicine due to the great improvement in diagnostic efficiency and accuracy. Thanks to recent advances of deep learning, convolutional neural networks (CNNs) can be used to extract hierarchical feature representation for segmentation, which is robust to image

© Springer Nature Switzerland AG 2021
M. de Bruijne et al. (Eds.): MICCAI 2021, LNCS 12901, pp. 548–558, 2021.
https://doi.org/10.1007/978-3-030-87193-2_52

degradation such as noise, blur, contrast, etc. Among many CNNs-based segmentation approaches, FCN [19], Unet [5], and Auto-encoder-like architecture have become the desired models for MIS. Particularly, such methods achieved impressive performance in brain tumor [15,16], liver tumor [2,18], optic disc [23,28], retina [17], lung [12,26], and cell [8,20]. However, CNNs are limited in their mechanism of aggregating data at pooling layers. Notably, pooling summarizes features in a local window and discards important information such as pose and object location. Therefore, CNNs with consecutive pooling layers are unable to perverse the spatial dependencies between objects parts and wholes. Moreover, the activation layer plays an important role in CNNs; however, it is not interpretable and has often been used as a black box. MIS with CNNs is thus prone to performance degradation when data undergoes some transformations such as rotations. A practical example is during an MRI scan, subject motion causes transformations to appear in a subset of slices, which is a hard case for CNNs [29].

To overcome such limitations by CNNs, Sabour et al. [24] developed a novel network architecture called Capsule Network (CapsNet). The basic idea of CapsNet is to encode the part-whole relationships (e.g., scale, locations, orientations, brightnesses) between various entities, i.e., objects, parts of objects, to achieve viewpoint equivariance. Unlike CNNs which learn all part features of the objects, CapsNet learns the relationship between these features through dynamically calculated weights in each forward pass. This optimization mechanism, i.e., dynamic routing, allows weighting the contributions of parts to a whole object differently at both training and inference. CapsNet has been mainly applied to image recognition; its performance is still limited compared to the state-of-the-art by CNNs-based approaches. Adapting CapsNet for semantic segmentation, e.g., SegCaps [13,14], receives even less attention. In this work, we propose an effective 3D Capsules network for volumetric image segmentation, named 3D-UCaps. Our 3D-UCaps is built on both 3D Capsule blocks, which take temporal relations between volumetric slices into consideration, and 3D CNNs blocks, which extract contextual visual representation. Our 3D-UCaps contains two pathways, i.e., encoder and decoder. Whereas encoder is built upon 3D Capsule blocks, the decoder is built upon 3D CNNs blocks. We argue and show empirically that using deconvolutional Capsules in the decoder pathway not only reduces segmentation accuracy but also increases model complexity.

In summary, our contributions are: (1) An effective 3D Capsules network for volumetric image segmentation. Our 3D-UCaps inherits the merits from both 3D Capsule block to preserve spatial relationship and 3D CNNs block to learn better visual representation. (2) Extensive experiments on various datasets and ablation studies that showcase the effectiveness and robustness of 3D-UCaps for MIS.

2   Background

In CNNs, each filter of convolutional layers works like a feature detector in a small region of the input features and as we go deeper in a network, the detected low-level features are aggregated and become high-level features that can be used to distinguish between different objects. However, by doing so, each feature map only contains information about the presence of the feature, and the network relies on fixed learned weight matrix to link features between layers. It leads to the problem that the model cannot generalize well to unseen changes in the input image and usually perform poorly in that case.

CapsNet [24] is a new concept that strengthens feature learning by retaining more information at aggregation layer for pose reasoning and learning the part-whole relationship, which makes it a potential solution for semantic segmentation and object detection tasks. Each layer in CapsNet aims to learn a set of entities (i.e., parts or objects) with their various properties and represent them in a high-dimensional form, particularly vector in [24]. The length of this vector indicates the presence of the entity in the input while its orientation encodes different properties of that entity. An important assumption in CapsNet is the entity in previous layer are simple objects and based on an agreement in their votes, complex objects in next layer will be activated or not. This setting helps CapsNet reflect the changes in input through the activation of properties in the entity and still recognize the object successfully based on a dynamic voting between layers. Let $\{c_1^l, c_2^l, \ldots, c_n^l\}$ be the set of capsules in layer l, $\{c_1^{l+1}, c_2^{l+1}, \ldots, c_m^{l+1}\}$ be the set of capsule in layer $l + 1$, the overall procedure will be:

$$c_j^{l+1} = \text{squash}\left(\sum_i r_{ij} v_{j|i}\right), \qquad v_{j|i} = W_{ij} c_i^l \tag{1}$$

where W_{ij} is the learned weight matrix to linear mapping features of capsule c_i^l in layer l to feature space of capsule c_j^{l+1} in layer $l + 1$. The r_{ij} are coupling coefficients between capsule i and j that are dynamically assigned by a routing algorithm in each forward pass such that $\sum_j r_{ij} = 1$.

SegCaps [13,14], a state-of-the-art Capsule-based image segmentation, has made a great improvement to expand the use of CapsNet to the task of object segmentation. This method functions by treating an MRI image as a collection of slices, each of which is then encoded and decoded by capsules to output the segmentation. However, SegCaps is mainly designed for 2D still images, and it performs poorly when being applied to volumetric data because of missing temporal information. Our work differs in that we build the CapsNet to consume 3D data directly so that both spatial and temporal information can be fully used for learning. Furthermore, our 3D-UCaps is able to take both advantages of CapsNet and 3D CNNs into consideration.

3   Our Proposed 3D-UCaps Network

In this work, we propose a hybrid 3D-UCaps network, which inherits the merits from both CapsNet and 3D CNNs. Our proposed 3D-UCaps follows Unet-like architecture [5] and contains three main components as follows.

Visual Feature Extractor: We use a set of dilated convolutional layers to convert the input to high-dimensional features that can be further processed

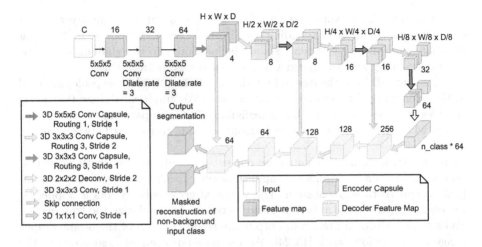

Fig. 1. Our proposed 3D-UCaps architecture with three components: visual feature extraction; capsule encoder, and convolution decoder. Number on the blocks indicates number of channels in convolution layer and dimension of capsules in capsule layers.

by capsules. It contains three convolution layers with the number of channels increased from 16 to 32 then 64, kernel size $5 \times 5 \times 5$ and dilate rate set to 1, 3, and 3, respectively. The output of this part is a feature map of size $H \times W \times D \times 64$.

Capsule Encoder: The visual feature from the previous component can be cast (reshaped) into a grid of $H \times W \times D$ capsules, each represented as a single 64-dimensional vector. Our capsule layer is a 3D convolutional capsule, which consumes volumetric data directly instead of treating it as separate slices as in SegCaps [13]. The advantage of our 3D capsule layer is that contextual information along the temporal axis can be included in the feature extraction. In additional to increasing the dimensionality of capsules as we ascend the hierarchy [24], we suggest to use more capsule types in low-level layers and less capsule types in high-level layers. This is due to the fact that low-level layers represent simple object while high-level layers represent complex object and the clustering nature of routing algorithm [9]. The number of capsule types in the encoder path of our network are set to $(16, 16, 16, 8, 8, 8)$, respectively. This is in contrast to the design in SegCaps, where the numbers of capsules are increasing $(1, 2, 4, 4, 8, 8)$ along the encoder path. We make sure that the number of capsule types in the last convolutional capsule layer is equal to the number of categories in the segmentation, which can be further supervised by a margin loss [24]. The output from a convolution capsule layer has the shape $H \times W \times D \times C \times A$, where C is the number of capsule types and A is the dimension of each capsule.

Convolutional Decoder: We use the decoder of 3D Unet [5] which includes deconvolution, skip connection, convolution and BatchNorm layers [10] to generate the segmentation from features learned by capsule layers. Particularly, we reshape the features to $H \times W \times D \times (C \star A)$ before passing them to the next convolution layer or concatenating with skip connections. The overall architecture can be seen in Fig. 1. Note that in our design, we only use capsule layers in the contracting path but not expanding path in the network. Sabour et al. [24] point out that "routing-by-agreement" should be far more effective than max-pooling, and max-pooling layers only exist in the contracting path of Unet.

This contradicts to the design by LaLonde et al. [13], where capsules are used in the expanding path in the network as well. We empirically show that using capsules in the expanding path has negligible effects compared to the traditional design while incurring high computational cost due to routing between capsule layers.

Training Procedure. We supervise our network with ground truth segmentation as follows. The margin loss is applied at the capsule encoder with downsampled ground truth segmentation. The weighted cross entropy loss is applied at the decoder head to optimize the entire network. To regularize the training, we also use an additional branch to output the reconstruction of the original input image as in previous work [13,24]. We use masked mean-squared error for the reconstruction. The total loss is the weighted sum of the three losses.

4   Experimental Results

Evaluation Setup. We perform experiments on various MIS datasets to validate our method. Specifically, we experiment with iSeg-2017 [29], LUNA16 [1], Hippocampus, and Cardiac [25]. iSeg is a MRI dataset of infant brains that requires to be segmented into white matter (WM), gray matter (GM), and cerebrospinal fluid (CSF). A recent analysis [29] shows that previous methods tend to perform poorly on subjects with movement and unusual poses. We follow the experiment setup by 3D-SkipDenseSeg [3] to conduct the report on this dataset where 9 subjects are used for training and 1 subject (subject #9) is for testing.

Additionally, we experiment on LUNA16, Hippocampus, and Cardiac [25] to compare with other capsule-based networks [13,27]. We follow a similar experiment setup in SegCaps [13] to conduct the results on LUNA16. We also use 4-fold cross validation on training set to conduct the experiments on Hippocampus and Cardiac.

Implementation Details. We implemented both 3D-SegCaps and 3D-UCaps in Pytorch. The input volumes are normalized to [0, 1]. We used patch size set as $64 \times 64 \times 64$ for iSeg and Hippocampus whereas patch size set as $128 \times 128 \times 128$ on LUNA16 and Cardiac. Both 3D-SegCaps and 3D-UCaps networks were trained without any data augmentation methods. We used Adam optimization with an initial learning rate of 0.0001. The learning rate is decayed by a factor of 0.05 if Dice score on validation set not increased for 50,000 iterations and early stopping is performed with a patience of 250,000 iterations as in [13]. Our models were trained on NVIDIA Tesla V100 with 32GB RAM, and it takes from 2-4 days depends on the size of the dataset.

Performance and Comparison. In this section, we compare our 3D-UCaps with both SOTA 3D CNNs-based segmentation approaches and other existing SegCaps methods. Furthermore, we have implemented 3D-SegCaps which is an

Table 1. Comparison on iSeg-2017 dataset. The first group is 3D CNN-based networks. The second group is Capsule-based networks. The best performance is in **bold**.

Method	Depth	Dice Score			
		WM	GM	CSF	Average
Qamar et al. [22]	82	90.50	**92.05**	**95.80**	**92.77**
3D-SkipDenseSeg [3]	47	**91.02**	91.64	94.88	92.51
VoxResNet [4]	25	89.87	90.64	94.28	91.60
3D-Unet [5]	18	89.83	90.55	94.39	91.59
CC-3D-FCN [21]	34	89.19	90.74	92.40	90.79
DenseVoxNet [11]	32	85.46	88.51	91.26	89.24
SegCaps (2D) [13]	16	82.80	84.19	90.19	85.73
Our 3D-SegCaps	16	86.49	88.53	93.62	89.55
Our 3D-UCaps	17	**90.95**	**91.34**	**94.21**	**92.17**

extension version of 2D-SegCaps [13] on volumetric data to prove the effectiveness of incorporating deconvolution layers into 3D-UCaps. Our 3D-SegCaps share similar network architecture with 2D-SegCaps [13] and implemented with 3D convolution layers. This section is structured as follows: We first provide a detailed analysis on iSeg with different criteria such as segmentation accuracy, network configurations, motion artifact, and rotation invariance capability. We then report segmentation accuracy on various datasets, including LUNA16, Hippocampus, and Cardiac.

Accuracy: The comparison between our proposed 3D-SegCaps, 3D-UCaps with SOTA segmentation approaches on iSeg dataset [29] is given in Table 1. Thanks to taking both spatial and temporal into account, both 3D-SegCaps, 3D-UCaps outperforms 2D-SegCaps with large margin on iSeg dataset. Moreover, our 3D-UCaps consisting of Capsule encoder and Deconvolution decoder obtains better results than 3D-SegCaps, which contains both Capsule encoder and Capsule decoder. Compare to SOTA 3D CNNs networks our 3D-UCaps achieves compatible performance while our network is much shallower i.e. our

Table 2. Performance of 3D-UCaps on iSeg with different network configurations

Method	Dice Score			
	WM	GM	CSF	Average
Change number of capsule (set to 4)	89.02	89.78	89.95	89.58
Without feature extractor	89.15	89.66	90.82	89.88
Without margin loss	87.62	88.85	92.06	89.51
Without reconstruction loss	88.50	88.96	90.18	89.22
3D-UCaps	**90.95**	**91.34**	**94.21**	**92.17**

3D-UCaps contains only 17 layers compare to 82 layers in [22]. Compare to SOTA 3D CNNs networks which has similar depth, i.e. our 3D-UCaps with 18 layers, our 3D-UCaps obtains higher Dice score at individual class and on average.

Network Configuration: To prove the effectiveness of the entire network architecture, we trained 3D-UCaps under various settings. The results are given in Table 2. We provide a baseline where we change the number of capsules at the first layer from 16 capsules (our setting in Sect. 3) to 4 capsules (similar to Seg-Caps). We also examine the contribution of each component by removing feature extraction layer, margin loss, reconstruction loss, respectively. The result shows that each change results in accuracy drop, which validates the competence of our network model.

Table 3. Performance on iSeg with motion artifact on different axis. The experiment was conducted 5 times and report average number to minimize the effect of randomization

Method	x-axis			y-axis			z-axis		
	CSF	GM	WM	CSF	GM	WM	CSF	GM	WM
3D-SkipDenseSeg [3]	83.93	88.76	88.52	78.98	87.80	87.89	82.88	88.38	88.27
SegCaps (2D) [13]	88.11	83.01	82.01	86.43	81.80	80.91	89.36	83.99	82.76
Our 3D-SegsCaps	90.70	86.15	84.24	87.75	84.21	82.76	89.77	85.54	83.92
Our 3D-UCaps	**91.04**	**88.87**	**88.62**	**90.31**	**88.21**	**88.12**	**90.86**	**88.65**	**88.55**

Moving Artifact: Motion artifact caused by patient moving when scanning was reported as a hard case in [29]. We examine the influence of motion artifact to our 3D-UCaps in Table 3. In this table, motion artifact at each axis was simulated by randomly rotating 20% number of slices along the axis with an angle between -5 and 5 degree. As can be seen, 3D-based capsules (3D-SegCaps and 3D-UCaps) both outperforms SegCaps in all classes in all rotations.

Fig. 2. Comparison on iSeg with test object rotated about z-axis from zero to 90 degree. Best view in zoom.

Rotation Invariance: To further study rotation equivariance and invariance properties in our 3D-UCaps, we trained our network without any rotation data augmentation. During testing, we choose an axis to rotate the volume, and apply the rotation with angle values fixed to 5, 10, 15, .., 90 degrees. Here we conduct the experiment on iSeg and choose z-axis as the rotation axis. We choose 3D SkipDense [3] as 3D CNNs-based segmentation method and compare robustness to rotations between our 3D-UCaps, our 3D-SegCaps, 2D-SegCaps, and 3D Skip-Dense [3]. The segmentation accuracy of rotation transformation on each target class is reported in Fig. 2. We found that the performance tends to drop slightly when the rotation angles increases. Except 2D-SegCaps, there is no significant difference in performance between 3D CNNs-based network and Capsule-based networks even though traditional 3D CNN-based network is not equipped with learning rotation invariance property. This could be explained by that the networks perform segmentation on a local patch of the volume at a time, making them resistant to local changes. Further analysis of the robustness of capsule network on the segmentation task would be necessary, following some recent analysis on the classification task [6,7].

Results on Other Datasets: Besides iSeg, we continue benchmarking our 3D-UCaps on other datasets. The performance of 3D-UCaps on LUNA16, Hippocampus, and Cardiac is reported in Table 4, 5, 6. Different from other datasets, LUNA16 was annotated by an automated algorithm instead of a radiologist. When conducting the report on LUNA16, SegCaps [13] removed 10 scans with exceedingly poor annotations. In Table 4, we compare our performance in two cases: full dataset and remove 10 exceedingly poor annotations. The results show that our 3D-UCaps outperforms previous methods and our 3D-SegCaps baseline, respectively.

Table 4. Comparison on LUNA16 in two cases where * indicates full dataset. The best score is in **bold**.

Method	Split-0	Split-1	Split-2	Split-3	Average
SegCaps (2D) [13]	**98.50**	98.52	98.46	98.47	98.48
Our 3D-UCaps	98.49	**98.61**	**98.72**	**98.76**	**98.65**
SegCaps* (2D) [27]	98.47	98.19	98.07	98.24	98.24
Our 3D-UCaps*	**98.48**	**98.60**	**98.70**	**98.76**	**98.64**

Table 5. Comparison on Hippocampus dataset with 4-fold cross validation.

Method	Anterior			Posterior		
	Recall	Precision	Dice	Recall	Precision	Dice
Multi-SegCaps (2D) [13]	80.76	65.65	72.42	84.46	60.49	70.49
EM-SegCaps (2D) [27]	17.51	20.01	18.67	19.00	34.55	24.52
Our 3D-SegCaps	94.70	75.41	83.64	93.09	73.20	81.67
Our 3D-UCaps	**94.88**	**77.48**	**85.07**	**93.59**	**74.03**	**82.49**

Table 6. Comparison on Cardiac dataset with 4-fold cross validation.

Method	Recall	Precision	Dice
SegCaps (2D) [13]	**96.35**	43.96	60.38
Multi-SegCaps (2D) [27]	86.89	54.47	66.96
Our 3D-SegCaps	88.35	56.40	67.20
Our 3D-UCaps	92.69	**89.45**	**90.82**

5   Conclusion

In this work, we proposed a novel network architecture that can both utilize 3D capsules for learning features for volumetric segmentation while retaining the advantage of traditional convolutions in decoding the segmentation results. Even though we use capsules with dynamic routing [13,24] only in the encoder of a simple Unet like architecture, we can achieve competitive result with the state-of-the-art models on iSeg-2017 challenge while outperforming SegCaps [13] on different complex datasets. Exploring hybrid architecture between Capsule-based and traditional neural network is therefore a promising approach to medical image analysis while keeping model complexity and computation cost plausible.

References

1. Armato, S.G., III., et al.: The lung image database consortium (lidc) and image database resource initiative (idri): a completed reference database of lung nodules on CT scans. Med. Phys. **38**(2), 915–931 (2011)
2. Bilic, P., et al.: The liver tumor segmentation benchmark (lits). arXiv preprint arXiv:1901.04056 (2019)
3. Bui, T.D., Shin, J., Moon, T.: Skip-connected 3d densenet for volumetric infant brain MRI segmentation. Biomed. Signal Process. Control **54**, 101613 (2019)
4. Chen, H., Dou, Q., Yu, L., Qin, J., Heng, P.A.: Voxresnet: deep voxelwise residual networks for brain segmentation from 3d MR images. NeuroImage **170**, 446–455 (2018)
5. Çiçek, Özgün., Abdulkadir, Ahmed, Lienkamp, Soeren S.., Brox, Thomas, Ronneberger, Olaf: 3D U-Net: Learning Dense Volumetric Segmentation from Sparse Annotation. In: Ourselin, Sebastien, Joskowicz, Leo, Sabuncu, Mert R.., Unal, Gozde, Wells, William (eds.) MICCAI 2016. LNCS, vol. 9901, pp. 424–432. Springer, Cham (2016). https://doi.org/10.1007/978-3-319-46723-8_49
6. Gu, J., Tresp, V.: Improving the robustness of capsule networks to image affine transformations. In: Proceedings of the IEEE/CVF Conference on Computer Vision and Pattern Recognition, pp. 7285–7293 (2020)
7. Gu, J., Tresp, V., Hu, H.: Capsule network is not more robust than convolutional network. In: Proceedings of the IEEE/CVF Conference on Computer Vision and Pattern Recognition, pp. 14309–14317 (2021)
8. Hatipoglu, N., Bilgin, G.: Cell segmentation in histopathological images with deep learning algorithms by utilizing spatial relationships. Med. Biological Eng. Comput. **55**(10), 1829–1848 (2017)

9. Hinton, G.E., Sabour, S., Frosst, N.: Matrix capsules with em routing. In: International conference on learning representations (2018)
10. Ioffe, S., Szegedy, C.: Batch normalization: accelerating deep network training by reducing internal covariate shift. In: International Conference on Machine Learning, pp. 448–456. PMLR (2015)
11. Jégou, S., Drozdzal, M., Vazquez, D., Romero, A., Bengio, Y.: The one hundred layers tiramisu: fully convolutional densenets for semantic segmentation. In: Proceedings of the IEEE Conference on Computer Vision and Pattern Recognition Workshops, pp. 11–19 (2017)
12. Jin, D., Xu, Z., Tang, Y., Harrison, A.P., Mollura, D.J.: Ct-realistic lung nodule simulation from 3d conditional generative adversarial networks for robust lung segmentation. In: International Conference on Medical Image Computing and Computer-Assisted Intervention, pp. 732–740. Springer (2018)
13. LaLonde, R., Bagci, U.: Capsules for object segmentation. arXiv preprint arXiv:1804.04241 (2018)
14. LaLonde, R., Xu, Z., Irmakci, I., Jain, S., Bagci, U.: Capsules for biomedical image segmentation. Med. Image Anal. **68**, 101889 (2021)
15. Le, N., Gummadi, R., Savvides, M.: Deep recurrent level set for segmenting brain tumors. In: International Conference on Medical Image Computing and Computer-Assisted Intervention (MICCAI), pp. 646–653 (2018)
16. Le, N., Le, T., Yamazaki, K., Bui, T.D., Luu, K., Savides, M.: Offset curves loss for imbalanced problem in medical segmentation. In: 25th International Conference on Pattern Recognition (ICPR), pp. 6189–6195 (2021)
17. Le, N., Yamazaki, K., Gia, Q.K., Truong, T., Savvides, M.: A multi-task contextual atrous residual network for brain tumor detection & segmentation. In: 25th International Conference on Pattern Recognition (ICPR), pp. 5943–5950 (2021)
18. Li, X., Chen, H., Qi, X., Dou, Q., Fu, C.W., Heng, P.A.: H-denseunet: hybrid densely connected unet for liver and tumor segmentation from CT volumes. IEEE Trans. Med. Imaging **37**(12), 2663–2674 (2018)
19. Long, J., Shelhamer, E., Darrell, T.: Fully convolutional networks for semantic segmentation. In: Proceedings of the IEEE Conference on Computer Vision and Pattern Recognition, pp. 3431–3440 (2015)
20. Moshkov, N., Mathe, B., Kertesz-Farkas, A., Hollandi, R., Horvath, P.: Test-time augmentation for deep learning-based cell segmentation on microscopy images. Sci. Rep. **10**(1), 1–7 (2020)
21. Nie, D., Wang, L., Adeli, E., Lao, C., Lin, W., Shen, D.: 3-d fully convolutional networks for multimodal isointense infant brain image segmentation. IEEE Trans. Cybernetics **49**(3), 1123–1136 (2018)
22. Qamar, S., Jin, H., Zheng, R., Ahmad, P., Usama, M.: A variant form of 3d-unet for infant brain segmentation. Future Generation Comput. Syst. **108**, 613–623 (2020)
23. Ramani, R.G., Shanthamalar, J.J.: Improved image processing techniques for optic disc segmentation in retinal fundus images. Biomed. Signal Process. Control **58**, 101832 (2020)
24. Sabour, S., Frosst, N., Hinton, G.E.: Dynamic routing between capsules. arXiv preprint arXiv:1710.09829 (2017)
25. Simpson, A.L., et al.: A large annotated medical image dataset for the development and evaluation of segmentation algorithms. arXiv preprint arXiv:1902.09063 (2019)
26. Souza, J.C., Diniz, J.O.B., Ferreira, J.L., da Silva, G.L.F., Silva, A.C., de Paiva, A.C.: An automatic method for lung segmentation and reconstruction in chest x-ray using deep neural networks. Comput. Methods Programs Biomed. **177**, 285–296 (2019)

27. Survarachakan, S., Johansen, J.S., Aarseth, M., Pedersen, M.A., Lindseth, F.: Capsule nets for complex medical image segmentation tasks. In: Colour and Visual Computing Symposium (2020)

28. Veena, H., Muruganandham, A., Kumaran, T.S.: A review on the optic disc and optic cup segmentation and classification approaches over retinal fundus images for detection of glaucoma. SN Appl. Sci. **2**(9), 1–15 (2020)

29. Wang, L., et al.: Benchmark on automatic six-month-old infant brain segmentation algorithms: the iseg-2017 challenge. IEEE Trans. Med. Imaging **38**(9), 2219–2230 (2019)

HRENet: A Hard Region Enhancement Network for Polyp Segmentation

Yutian Shen[1], Xiao Jia[2], and Max Q.-H. Meng[1,3(✉)]

[1] Department of Electronic Engineering, The Chinese University of Hong Kong,
Sha Tin, Hong Kong
max.meng@sustech.edu.cn
[2] Department of Radiation Oncology, Stanford University, Stanford, CA, USA
[3] Department of Electronic and Electrical Engineering,
Southern University of Science and Technology, Shenzhen, China

Abstract. Automatic polyp segmentation in the screening system is of great practical significance for the diagnosis and treatment of colorectal cancer. However, accurate segmentation in the colonoscopy images still remains a challenge. In this paper, we propose a hard region enhancement network (HRENet) based on an encoder-decoder framework. Specifically, we design an informative context enhancement (ICE) module to explore and intensify the features from the lower-level encoder with explicit attention on hard regions. We also develop an adaptive feature aggregation (AFA) module to select and aggregate the features from multiple semantic levels. In addition, we train the model with a proposed edge and structure consistency aware loss (ESCLoss) to further boost the performance. Extensive experiments on three public datasets show that our proposed algorithm outperforms the state-of-the-art approaches in terms of both learning ability and generalization capability. In particular, our HRENet achieves a *mIoU* of 92.11% and a *Dice* of 92.56% on Kvasir-SEG dataset. And the model trained with Kvasir-SEG and CVC-Clinic DB retains a high inference performance on the unseen dataset CVC-Colon DB with a *mIoU* of 88.42% and a *Dice* of 85.26%. The code is available at: https://github.com/CathySH/HRENet.

Keywords: Colonoscopy · Polyp segmentation · Deep learning · Feature enhancement · Hard region

1 Introduction

Colorectal cancer has become the second leading cause of cancer-related deaths around the globe in 2020 [18]. Most of the cases develop from tiny growth on the inner lining of the colon or rectum known as polyps, thereby making segmentation of precancerous polyps in screening process very vital. However, polyp segmentation has always been a challenging task for two main reasons: (i) the polyps vary in shape, size, color and texture, (ii) the boundary between a polyp and surrounding mucosa is usually blurred due to the low-intensity contrast.

M. de Bruijne et al. (Eds.): MICCAI 2021, LNCS 12901, pp. 559–568, 2021.
https://doi.org/10.1007/978-3-030-87193-2_53

In recent years, success in deep learning field greatly stimulates automatic polyp segmentation in colonoscopy images. The Fully Convolutional Network (FCN) [12] is a commonly used architecture in semantic segmentation. Brandao et al. [3] adopted the FCN with a pre-trained VGG model for handling polyp segmentation. Later, Akbari et al. [1] utilized the modified FCN8s to further improve the accuracy. UNet [17] is another prevalent architecture and has become a baseline network for most medical image segmentation tasks. Based on UNet, UNet++ [24] added a series of nested, dense skip pathways to form a densly-connected encoder-decoder network with deep supervision. ACSNet [23] extracted global context from the bottom encoder and delivered local context with spatial attention to cope with the diversity of polyp size and shape. PraNet [5] roughly located the polyps with high-level features and mined discriminate regions through an erasing foreground object manner to achieve better learning and generalization ability. HarDNet-MSEG [9] adopted a pre-trained HarDNet68 as encoder and cascaded partial inspired decoder with receptive field blocks to segment the polyp with more precise edge. However, existing methods tend to predominantly focus on areas that are easy to segment while overlooking those difficult pixels, such as tiny polyps or uncertain boundaries. Therefore, our main motivation is incorporating features with more attention on the hard region so as to improve the segmentation accuracy.

In this paper, we propose a novel hard region enhancement network (HRENet) for polyp segmentation by exploiting features on the hard regions. To achieve this, we first design an informative context enhancement (ICE) module to intensify the lower-level encoder features under the guidance of hard-region attention. We then develop an adaptive feature aggregation (AFA) module to dynamically aggregate the multi-level features and deliver them to the decoder block. Meanwhile, we design an edge and structure consistency aware deep loss (ESCLoss) to further improve the segmentation performance. Extensive experiments over three public datasets have demonstrated the strong learning and generalization ability of HRENet.

2   Methodology

The architecture of the proposed HRENet is illustrated in Fig. 1. The ResNet-34 [7] is adopted as the encoder sub-network, which comprises five encoding blocks. The decoder branch also has five blocks that each consists of two Conv-BN-ReLu combinations and an upsample operation.

The ICE and AFA modules are placed between two consecutive decoding blocks. Specifically, the ICE module mines hard region attention and accordingly enhances the features from the shallower encoder layer. Then the AFA module dynamically aggregates the features from multiple semantic layers. The whole network is trained in an end-to-end manner with the ESCLoss.

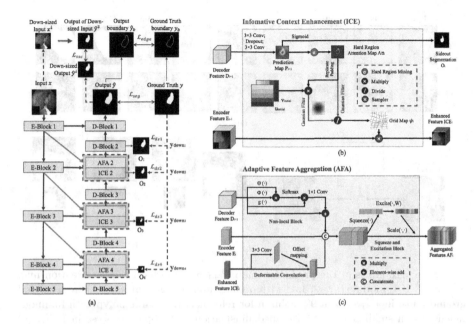

Fig. 1. (a) Overview of our proposed HRENet. (b) Illustration of informative context enhancement (ICE) module. (c) Illustration of adaptive feature aggregation (AFA) module. (Color figure online)

2.1   Informative Context Enhancement (ICE)

Learning informative context on hard regions holds the potential to complement the features for better segmentation. As shown in Fig. 1(b), the ICE module utilizes the hidden features from the previous decoding block to mine hard regions. Then guided by the hard region attention, we sample the fine-grained features from the lower-level encoder to form the enhanced features.

To be specific, the attention map of the ith ICE module Att_i is determined by the prediction map P_{i+1} via:

$$Att_i^j = \alpha \times (1 - cos(2\pi \times \sigma(P_{i+1}^j))) + (1 - \alpha) \tag{1}$$

where Att_i^j and P_{i+1}^j denote the value of jth pixel of the ith hard region attention map and prediction map from $i + 1^{th}$ decoder block. σ represents the sigmoid activation, then hard-to-classify pixels will have activation values $\sigma(P)$ close to 0.5, resulting in high attention value Att, and vice versa. α is set as 0.9 to give attention score for even very confident area. Examples of prediction map P_{i+1} and the resulting Att_i in the ith ICE module ($i = 2, 3, 4$) are visualized in Fig. 2(a).

The mapping between the enhanced features and the shallow encoder features $\Psi_i : E_{i-1} \rightarrow ICE_i$ can be formulated as two function $u(x, y)$ and $v(x, y)$ such that: $ICE_i(x, y) = E_{i-1}(u(x, y), v(x, y))$. Then for enhancement purpose, (u, v)

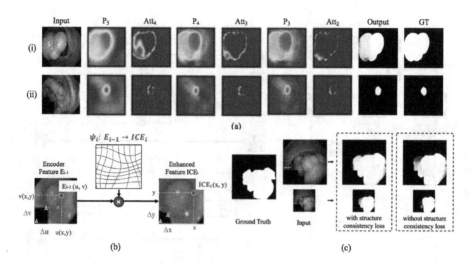

Fig. 2. (a) Examples of the prediction map P_{i+1} and the hard region attention map Att_i in i^{th} ICE module ($i = 2, 3, 4$) with orginal input image, output segmentation and ground truth in $1^{st}, 8^{th}$ and 9th column for reference: (i) a large polyp with luminous mucus, (ii) a small polyp. (b) Detailed illustration of mapping process in ICE$_i$. (c) Comparison of predicted segmentation maps for an input with different scales. (Color figure online)

should be more densely distributed in hard regions and sparsely in confident area. The problem can be further approximated to find u and v such that:

$$\int_0^{v(x,y)} \int_0^{u(x,y)} Att_i(u', v') du' dv' = xy \tag{2}$$

where $u(x, y)$, $v(x, y)$, x and y are normalized into the range from 0 to 1. As illustrated in Fig. 2(b), for the pixels with low attention value (e.g. the points marked with green triangles in black region), Δx and Δy correspond to a much larger increment Δu and Δv, which leads to sparse sampling.

To solve the above problem of finding u and v, we borrow the idea from [16] that each pixel (u', v') is pulling other pixels with a force $Att_i(u', v')$. By introducing a distance kernel $k((x, y), (u', v'))$ as a regularizer, u and v can be formulated as:

$$u(x, y) = \frac{\sum_{u',v'} Att_i(u', v') k((x, y), (u', v')) u'}{\sum_{u',v'} Att_i(u', v') k((x, y), (u', v'))} \tag{3}$$

$$v(x, y) = \frac{\sum_{u',v'} Att_i(u', v') k((x, y), (u', v')) v'}{\sum_{u',v'} Att_i(u', v') k((x, y), (u', v'))} \tag{4}$$

As in [16,21,22], we use a Gaussian kernel as the distance kernel and pad the afore-calculated attention map with its border values to avoid the undesirable bias effect that will sample towards the image center.

2.2 Adaptive Feature Aggregation (AFA)

AFA is developed to aggregate features from different semantic levels, including the enhanced features of ICE module and those passed from encoder and the previous decoder block.

As illustrated in Fig. 1(c), we utilize a non-local block [20] to intensify features from the previous decoder layer D_{i+1} with long-range dependency by computing the response at each position as a weighted sum of features at all positions. Since the enhanced features ICE_i is deformed to some extent during mapping in ICE module, we use a deformable convolution block [4] consisting of two DeformConv-BN-ReLU combinations to extract local context information. Here, a basic convolution layer is utilized to learn the offset and then the deformable convolution layer shifts the sampling points accordingly. Finally, after concatenation with encoder feature E_i, a squeeze and excitation block [8] is incorporated to adaptively recalibrate the channel-wise feature responses by modelling interdependencies between channels. The concatenated features are squeezed to a single vector that is successively learnt by fully connected layers and excited by activation layers. The learnt channel-wise weights are used to enhance informative context feature channels and suppress the uninstructive ones.

2.3 Edge and Structure Consistency Aware Loss (ESCLoss)

As shown in Fig. 1, we propose the ESCLoss that comprises four terms: segmentation loss $\mathcal{L}_{seg}$, deep supervision loss $\mathcal{L}_{ds}$, edge penalty loss $\mathcal{L}_{edge}$, and structure consistency loss $\mathcal{L}_{sc}$. It is defined as: $\mathcal{L}_{ESC} = \mathcal{L}_{seg} + \mathcal{L}_{ds} + \mathcal{L}_{edge} + \mathcal{L}_{ssc}$.

$\mathcal{L}_{seg}$ is the combination of a binary cross entropy (BCE) loss and a dice loss: $\mathcal{L}_{seg}(y, \hat{y}) = \mathcal{L}_{BCE+Dice}(y, \hat{y})$, where y and $\hat{y}$ denote the ground truth segmentation and the final prediction respectively. $\mathcal{L}_{BCE}$ encourages the model to predict the right class label of each pixel while $\mathcal{L}_{Dice}$ can predict and correct higher-order inconsistencies. $\mathcal{L}_{ds}$ is incorporated to optimize the hidden layer features from decoders for more precise hard region attention maps: $\mathcal{L}_{ds} = \sum_{i=1}^{4} \mathcal{L}_{ds_i}(y_{down_i}, O_i) = \sum_{i=1}^{4} \mathcal{L}_{BCE+Dice}(y_{down_i}, O_i)$, where y_{down_i} and O_i represent the ith down-sized ground truth map and the ith side-out prediction map.

Apart from $\mathcal{L}_{Seg}$ and $\mathcal{L}_{ds}$, which mainly consider the area, $\mathcal{L}_{edge}$ is introduced to attract attention on boundary accuracy [14]. The $\mathcal{L}_{edge}$ is defined as a focal loss [11] between the pixels on the true boundary and those on predicted boundary: $\mathcal{L}_{edge}(y_b, \hat{y}_b) = -\frac{1}{N} \sum_{i=1}^{N} (\alpha(1-\hat{y}_{b_i})^{\gamma} y_{b_i} log(\hat{y}_{b_i}) + (1-\alpha)\hat{y}_{b_i}^{\gamma}(1-y_{b_i})log(1-\hat{y}_{b_i}))$, where y_b and $\hat{y}_b$ denote the boundary of ground truth and the prediction map respectively. The boundaries are inferred as in [13]. α is set as 0.25 to mitigate the imbalance between edge pixels and background pixels, γ is set as 2 to focus more on misclassified pixels.

In order to learn more information on polyp structure and to boost generalization ability for different input scales, we further incorporate a $\mathcal{L}_{ssc}$ defined as: $\mathcal{L}_{ssc}(\hat{y}^{\Downarrow}, \hat{y}^{\downarrow}) = \frac{1}{N} \sum_{i=1}^{N} (\alpha \frac{1-SSIM(\hat{y}_i^{\Downarrow}, \hat{y}_i^{\downarrow})}{2} + (1-\alpha)|\hat{y}_i^{\Downarrow} - \hat{y}_i^{\downarrow}|)$, where $\hat{y}_i^{\downarrow}$ denotes i^{th} pixel of the down-sized predicted segmentation map from a normal input image and $\hat{y}_i^{\Downarrow}$ denotes i^{th} pixel of the predicted segmentation from the same image with down-scaled size. SSIM represents the single scale structure similarity [6]. As shown in Fig. 2(c), with the proposed structure consistency loss, the HRENet can adapt to different scales.

3   Experiments

3.1   Experimental Settings

Datasets. Experiments are conducted on three benchmark polyp segmentation datasets: Kvasir-SEG [10], CVC-Clinic DB [2] and CVC-Colon DB [19]. The first one is the largest-scale challenging dataset released in 2020, which contains 1000 images with various polyp regions. The second and last dataset separately consist of 612 image frames extracted from 29 different colonoscopy sequences and 300 annotated frames from 15 short colonoscopy sequences.

Implementation Details & Evaluation Metrics. We use Pytorch to implement our algorithm. The model is optimized by SGD with batch size of 4, momentum of 0.9 and weight decay of 10^{-5}. Additionally, we use a learning rate initialized to be 0.001 and decreased by a factor $\frac{epoch}{num_{epoch}^{0.9}}$ to train the network with 150 epochs. During training, each image is resized with random horizontal and vertical flips, rotation and random cropping. In the inference stage, input images are resized and fed into the model without any post-processing. All experiments are run on an NVIDIA GeForce RTX 2080 Ti GPU.

We adopt eight commonly used metrics [10] for medical image segmentation tasks, including *Recall, Specificity, Precision, Dice Coefficient, Intersection-over-Union for Polyp (IoUp), Intersection-over-Union for Background (IoUb), Mean IoU (mIoU) and Accuracy.*

3.2   Comparison with State-of-the-arts

We conduct a series of experiments to compare the proposed HRENet with classical medical segmentation baselines (*i.e.,* UNet [17] and UNet++ [24]) as well as current state-of-the-art algorithms (*i.e.,* Akbari *et al.* [1], ACSNet [23], PraNet [5] and HarDNet-MSEG [9,15]) specially developed for polyp segmentation. For fair comparison, we follow the same experiment settings as in [23], [5] and [9]. All models are trained and tested on the same split training/validation/test sets with same data augmentation mentioned in Sec. 3.1. Meanwhile, the comparisons are implemented based on the released code and settings in original papers.

Experiments on the Kvasir-SEG Dataset. In this section, we refer to settings in [23] that 60% of the Kvasir-SEG dataset is used as training set, 20% as validation set, and the remaining 20% as test set. All the images are set to a fixed size of 320×320. As shown in Table 1, our method achieves the best performance with a *Dice* of 92.56% and a *mIoU* of 92.11%.

Table 1. Quantitative results and ablation study on the Kvasir-SEG dataset.

Method	Rec	Spec	Prec	Dice	IoUp	IoUb	mIoU	Acc
UNet (MICCAI'15) [17]	87.91	97.30	87.27	84.70	76.77	94.26	85.52	95.33
Akbari et al. (EMBC'18) [1]	90.17	98.18	89.74	88.15	80.94	95.62	88.28	96.56
UNet++ (TMI'19) [24]	84.33	98.31	89.27	83.49	76.21	94.48	85.34	95.48
ACSNet (MICCAI'20) [23]	92.72	98.03	92.39	91.30	85.92	96.31	91.12	97.08
PraNet (MICCAI'20) [5]	91.41	98.64	92.22	90.49	84.70	96.53	90.62	97.22
HarDNet-MSEG (arXiv'21) [9]	87.67	98.37	89.38	86.26	78.43	95.03	86.73	96.11
HRENet (Ours)	92.83	**98.76**	**93.78**	**92.56**	**87.33**	**96.89**	**92.11**	**97.64**
HRENet_w/o_ICE	92.26	98.14	92.10	90.53	85.25	96.22	90.74	96.92
*HRENet_w/o_ICE**	92.68	98.07	91.15	90.47	84.85	96.07	90.46	96.86
HRENet_w/o_AFA	**93.08**	98.44	92.75	91.91	86.54	96.77	91.66	97.46
HRENet_w/o_ESCLoss	92.00	98.14	91.24	89.92	84.20	96.00	90.10	96.78

*Settings of the additional experiment are illustrated in Sec.3.3.

Figure 2(a)(i) shows an example of hard region attention maps for a large polyp with luminous mucus, where our model first draws high attention value on the boundary and the illuminated area of the polyp and then progressively modifies the attention while learning. In the last ICE module, only a little amount of pixels near the boundary remains as hard region. As for the small polyp shown in Fig. 2(a)(ii), the area is well identified with confidence, so the hard region attention is mainly distributed on the boundary pixels. In such way, our HRENet is able to segment polyps more accurately by mining hard regions and accordingly leveraging informative contexts. Some examples of visual comparison between the state-of-the-arts and our method is shown in Fig. 3. UNet [17], UNet++ [24], PraNet [5] and HarD-MSEG [9] exhibit errors in the presence of illuminated spot and are weak in distinguishing polyps from background. Akbari et al. [1] and ACSNet [23] are really sensitive to textures and tend to focus on easy-to-classify regions thus overlooking the less-prominent polyp region. Among these methods, our proposed model consistently segment polyps with more precise area and boundary, showing robustness and superiority.

Experiments on the Cross Dataset. In this section, we follow the same dataset settings as in [5] and [9] to test the model's generalization capability. The Kvasir-SEG and the CVC-Clinic DB serve as the seen dataset with 90% as training set and the remaining 10% as validation set, while CVC-Colon DB serves as test set. All the images are set to 352×352. As shown in Table 2, our proposed method again outperforms all state-of-the-art methods with an improvement of 3.64% in *Dice* and 2.35% in *mIoU*, showing good model generalization ability.

3.3 Ablation Study

The ablation experiments are conducted on the Kvasir-SEG dataset to verify the effectiveness and necessity of each module. As shown in Table 1, our algorithm has witnessed performance degradation without either ICE modules, AFA modules or the ESCLoss, decreasing the *Dice* by 2.03%, 0.65% and 2.64% respectively.

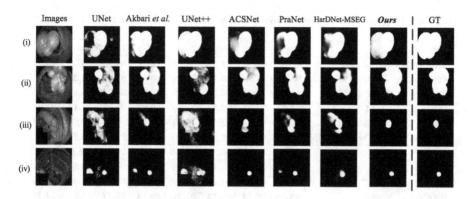

Fig. 3. Examples of polyp segmentation results on Kvasir-SEG test images: (i) polyp with illuminated spot, (ii) polyp with folds and illuminated spot, (iii) small polyp with obscure boundary, (iv) tiny polyp. (Color figure online)

Table 2. Quantitative results on the cross dataset.

Method	Rec	Spec	Prec	Dice	IoUp	IoUb	mIoU	Acc
UNet (MICCAI'15) [17]	80.62	98.12	73.74	72.87	62.20	96.98	79.59	97.12
Akbari *et al.* [1]	82.42	98.91	82.36	78.94	69.28	97.86	83.57	97.96
UNet++ (TMI'19)[24]	77.05	95.65	63.75	63.85	54.01	94.60	74.31	94.81
ACSNet (MICCAI'20) [23]	86.92	98.37	84.35	81.62	74.36	97.77	86.06	97.86
PraNet (MICCAI'20) [5]	82.87	**99.43**	85.58	80.98	71.83	98.49	85.16	98.56
HarDNet-MSEG (arXiv'21) [9]	80.96	99.21	82.75	76.55	67.04	98.03	82.54	98.12
HRENet (Ours)	**89.07**	99.04	**85.80**	**85.26**	**78.32**	**98.50**	**88.41**	**98.57**

As the ICE module's optimization over parallel skip connection adopted in baseline UNet is two-folds: 1) Down-concatenations are added that features from shallow encoder layers are delivered to deeper decoders to enrich the feature representations. 2) Hard region attention guided feature enhancement is proposed to efficiently utilize the shallow features. We further conduct an experiment *HRENet_w/o_ICE** where down-concatenations are adopted, and the results show that the performance improvement is boosted by our proposed method of hard region feature enhancement rather than the effect of simply adding multi-scale features.

4   Conclusion

In this work, we propose a novel hard region enhancement network (HRENet) for polyp segmentation. Our model is able to learn multi-level features with careful attention on hard regions by proposing the ICE modules and AFA modules. We also design a ESCLoss incorporating segmentation accuracy evaluation, boundary and structure-consistency constraints to further boost the performance. Extensive experiments and ablation studies have demonstrated the strong learning and generalization ability of the proposed algorithm.

Acknowledgements. This work was supported by National Key R&D program of China with Grant No.2019YFB1312400, Hong Kong RGC CRF grant C4063-18G and Hong Kong RGC GRF grant # 14211420.

References

1. Akbari, M., Mohrekesh, M., Nasr-Esfahani, E., Soroushmehr, S.R., Karimi, N., Samavi, S., Najarian, K.: Polyp segmentation in colonoscopy images using fully convolutional network. In: 2018 40th Annual International Conference of the IEEE Engineering in Medicine and Biology Society (EMBC), pp. 69–72. IEEE (2018)
2. Bernal, J., Sánchez, F.J., Fernández-Esparrach, G., Gil, D., Rodríguez, C., Vilariño, F.: Wm-dova maps for accurate polyp highlighting in colonoscopy: Validation vs. saliency maps from physicians. Computerized Med. Imaging Graph. **43**, 99–111 (2015)
3. Brandao, P., et al.: Fully convolutional neural networks for polyp segmentation in colonoscopy. In: Medical Imaging 2017: Computer-Aided Diagnosis, vol. 10134, p. 101340F. International Society for Optics and Photonics (2017)
4. Dai, J., et al.: Deformable convolutional networks. In: Proceedings of the IEEE International Conference on Computer Vision, pp. 764–773 (2017)
5. Fan, D.-P., Ji, G.-P., Zhou, T., Chen, G., Fu, H., Shen, J., Shao, L.: PraNet: parallel reverse attention network for polyp segmentation. In: Martel, A.L., Abolmaesumi, P., Stoyanov, D., Mateus, D., Zuluaga, M.A., Zhou, S.K., Racoceanu, D., Joskowicz, L. (eds.) MICCAI 2020. LNCS, vol. 12266, pp. 263–273. Springer, Cham (2020). https://doi.org/10.1007/978-3-030-59725-2_26
6. Godard, C., Mac Aodha, O., Brostow, G.J.: Unsupervised monocular depth estimation with left-right consistency. In: Proceedings of the IEEE Conference on Computer Vision and Pattern Recognition, pp. 270–279 (2017)
7. He, K., Zhang, X., Ren, S., Sun, J.: Deep residual learning for image recognition. In: Proceedings of the IEEE Conference on Computer Vision and Pattern Recognition, pp. 770–778 (2016)
8. Hu, J., Shen, L., Sun, G.: Squeeze-and-excitation networks. In: IEEE Conference on Computer Vision and Pattern Recognition (2018)
9. Huang, C.H., Wu, H.Y., Lin, Y.L.: Hardnet-mseg: a simple encoder-decoder polyp segmentation neural network that achieves over 0.9 mean dice and 86 fps. arXiv preprint arXiv:2101.07172 (2021)
10. Jha, D., et al.: Kvasir-seg: A segmented polyp dataset. In: International Conference on Multimedia Modeling, pp. 451–462. Springer (2020)
11. Lin, T.Y., Goyal, P., Girshick, R., He, K., Dollár, P.: Focal loss for dense object detection. In: Proceedings of the IEEE International Conference on Computer Vision, pp. 2980–2988 (2017)
12. Long, J., Shelhamer, E., Darrell, T.: Fully convolutional networks for semantic segmentation. In: Proceedings of the IEEE Conference on Computer Vision and Pattern Recognition, pp. 3431–3440 (2015)
13. Luo, Z., Mishra, A., Achkar, A., Eichel, J., Li, S., Jodoin, P.M.: Non-local deep features for salient object detection. In: Proceedings of the IEEE Conference on Computer Vision and Pattern Recognition, pp. 6609–6617 (2017)
14. Murugesan, B., Sarveswaran, K., Shankaranarayana, S.M., Ram, K., Sivaprakasam, M.: Joint shape learning and segmentation for medical images using a minimalistic deep network. arXiv preprint arXiv:1901.08824 (2019)

15. paperswithcode.com: Medical image segmentation on kvasir-seg. [EB/OL]. https://paperswithcode.com/sota/medical-image-segmentation-on-kvasir-seg Accessed 3 Mar 2021
16. Recasens, A., Kellnhofer, P., Stent, S., Matusik, W., Torralba, A.: Learning to zoom: a saliency-based sampling layer for neural networks. In: Proceedings of the European Conference on Computer Vision (ECCV), pp. 51–66 (2018)
17. Ronneberger, O., Fischer, P., Brox, T.: U-net: convolutional networks for biomedical image segmentation. In: International Conference on Medical Image Computing and Computer-Assisted Intervention, pp. 234–241. Springer (2015)
18. Siegel, R.L., et al.: Colorectal cancer statistics, 2020. CA: a cancer journal for clinicians **70**(3), 145–164 (2020)
19. Tajbakhsh, N., Gurudu, S.R., Liang, J.: Automated polyp detection in colonoscopy videos using shape and context information. IEEE Trans. Med. Imaging **35**(2), 630–644 (2015)
20. Wang, X., Girshick, R., Gupta, A., He, K.: Non-local neural networks. In: Proceedings of the IEEE Conference on Computer Vision and Pattern Recognition, pp. 7794–7803 (2018)
21. Xing, X., Yuan, Y., Meng, M.Q.H.: Diagnose like a clinician: Third-order attention guided lesion amplification network for wce image classification. In: 2020 IEEE/RSJ International Conference on Intelligent Robots and Systems (IROS), pp. 10145–10151. IEEE (2020)
22. Xing, X., Yuan, Y., Meng, M.Q.H.: Zoom in lesions for better diagnosis: attention guided deformation network for wce image classification. IEEE Trans. Med. Imaging **39**(12), 4047–4059 (2020)
23. Zhang, R., Li, G., Li, Z., Cui, S., Qian, D., Yu, Y.: Adaptive context selection for polyp segmentation. In: Martel, A.L., Abolmaesumi, P., Stoyanov, D., Mateus, D., Zuluaga, M.A., Zhou, S.K., Racoceanu, D., Joskowicz, L. (eds.) MICCAI 2020. LNCS, vol. 12266, pp. 253–262. Springer, Cham (2020). https://doi.org/10.1007/978-3-030-59725-2_25
24. Zhou, Z., Siddiquee, M.M.R., Tajbakhsh, N., Liang, J.: Unet++: Redesigning skip connections to exploit multiscale features in image segmentation. IEEE transactions on medical imaging **39**(6), 1856–1867 (2019)

A Novel Hybrid Convolutional Neural Network for Accurate Organ Segmentation in 3D Head and Neck CT Images

Zijie Chen[1,2,3,4], Cheng Li[5], Junjun He[1,2,6,7], Jin Ye[1,2], Diping Song[1,2], Shanshan Wang[5,8,9], Lixu Gu[6,7], and Yu Qiao[1,2(✉)]

[1] Shenzhen Key Lab of Computer Vision and Pattern Recognition, SIAT-SenseTime Joint Lab, Shenzhen Institute of Advanced Technology, Chinese Academy of Sciences, Shenzhen, Guangdong, China
yu.qiao@siat.ac.cn
[2] Shanghai AI Lab, Shanghai, China
[3] Shenzhen Yino Intelligence Techonology Co., Ltd., Shenzhen, Guangdong, China
[4] Shenying Medical Technology (Shenzhen) Co., Ltd., Shenzhen, Guangdong, China
[5] Paul C. Lauterbur Research Center for Biomedical Imaging, Shenzhen Institute of Advanced Technology, Chinese Academy of Sciences, Shenzhen, Guangdong, China
[6] School of Biomedical Engineering, Shanghai Jiao Tong University, Shanghai, China
[7] Institute of Medical Robotics, Shanghai Jiao Tong University, Shanghai, China
[8] Peng Cheng Laboratory, Shenzhen, Guangdong, China
[9] Pazhou Lab, Guangzhou, Guangdong, China

Abstract. Radiation therapy (RT) is widely employed in the clinic for the treatment of head and neck (HaN) cancers. An essential step of RT planning is the accurate segmentation of various organs-at-risks (OARs) in HaN CT images. Nevertheless, segmenting OARs manually is time-consuming, tedious, and error-prone considering that typical HaN CT images contain tens to hundreds of slices. Automated segmentation algorithms are urgently required. Recently, convolutional neural networks (CNNs) have been extensively investigated on this task. Particularly, 3D CNNs are frequently adopted to process 3D HaN CT images. There are two issues with naïve 3D CNNs. First, the depth resolution of 3D CT images is usually several times lower than the in-plane resolution. Direct employment of 3D CNNs without distinguishing this difference can lead to the extraction of distorted image features and influence the final segmentation performance. Second, a severe class imbalance problem exists, and large organs can be orders of times larger than small organs. It is difficult to simultaneously achieve accurate segmentation for all the organs. To address these issues, we propose a novel hybrid

Z. Chen, C. Li and J. He—Contributed equally to this work.

Electronic supplementary material The online version of this chapter (https://doi.org/10.1007/978-3-030-87193-2_54) contains supplementary material, which is available to authorized users.

© Springer Nature Switzerland AG 2021
M. de Bruijne et al. (Eds.): MICCAI 2021, LNCS 12901, pp. 569–578, 2021.
https://doi.org/10.1007/978-3-030-87193-2_54

CNN that fuses 2D and 3D convolutions to combat the different spatial resolutions and extract effective edge and semantic features from 3D HaN CT images. To accommodate large and small organs, our final model, named OrganNet2.5D, consists of only two instead of the classic four downsampling operations, and hybrid dilated convolutions are introduced to maintain the respective field. Experiments on the MICCAI 2015 challenge dataset demonstrate that OrganNet2.5D achieves promising performance compared to state-of-the-art methods.

Keywords: Segmentation of organs-at-risks · Hybrid 2D and 3D convolutions · 3D HaN CT images

1 Introduction

Head and neck (HaN) cancers, such as oral cavity and nasopharynx, are one of the most prevalent cancer types worldwide [19]. Treatment of HaN cancers relies primarily on radiation therapy. To prevent possible post-treatment complications, accurate segmentation of organs-at-risks (OARs) is vital during the treatment planning [6]. In the clinic, computed tomography (CT)-based treatment planning is routinely conducted because of its high efficiency, high spatial resolution, and the ability to provide relative electron density information. Manual delineation of OARs in CT images is still the primary choice regardless of the time-consuming and tedious process. Several hours are required to process the images of only one patient [7]. Besides, it subjects to high inter- and intra-observer variations, which can significantly influence the prognosis of the treatment [1]. Automatic segmentation methods are in urgent need to speed up the process and achieve robust outcomes.

The low contrast of soft tissues in HaN CT images and the large volume size variations of different organs make it challenging to achieve automatic and accurate segmentation of all OARs in an end-to-end fashion. Conventional learning approaches often rely on one or multiple atlases or require the extraction of hand-crafted image features [2,21], which is difficult to be enough comprehensive and distinctive for the segmentation task. Deep neural networks, especially convolutional neural networks (CNNs), have proved to be highly effective for medical image segmentation in different applications [9,13]. Many efforts have been devoted to CNN-based segmentation of OARs in HaN CT images. To deal with the class imbalance issue caused by the differently sized organs, image patches based on certain prior knowledge were extracted before conducting CNN-based segmentation [8,15]. Two-step CNNs consisting of a region detector and a segmentation unit were also employed [12,17]. To make full use of the image information, a joint localization and segmentation network with a multi-view spatial aggregation framework was proposed [10]. The inputs to these models were either 3D image patches lacking the global features or 2D images without the depthwise information. AnatomyNet was designed to specifically process whole-volume 3D HaN CT images [22]. The major contributions of AnatomyNet include a novel network architecture for effective feature extraction and a combined loss function to combat the class imbalance problem. Following AnatomyNet, FocusNet was proposed to better handle the segmentation of both large and small organs with a delicate network structure design [4].

Despite the inspiring results achieved, several issues exist in the developed approaches. First, some studies dealt with only 2D inputs and thus, did not fully exploit the 3D image information [8,10,12]. Others conducted 3D convolutions but without paying attention to the different in-plane and depth resolutions [4,15,22]. The in-plane resolution of 3D HaN CT images is normally several times higher than the depth resolution. The direct employment of 3D convolutions can probably lead to the extraction of distorted image features, which might not be optimal for the segmentation task. Anisotropic convolutions have been proposed to solve this issue but without distinguishing the low-level and high-level features [11]. Second, for networks processing whole volume 3D CT images (AnatomyNet and FocusNet), only one downsampling layer was used to preserve the information of small anatomies. Consequently, the receptive fields of these networks are limited. To increase the receptive field, DenseASPP with four dilation rates (3, 6, 12, and 18) was introduced to FocusNet [4]. However, when the dilation rates of cascaded dilated convolutions have a common factor relationship, the gridding issue may appear that influence the segmentation accuracy [20]. Besides, pure 3D networks also suffer from increased parameters and computational burden issues, which also limit the network depth and performance.

To address these issues, a hybrid convolutional neural network, Organ-Net2.5D, is proposed in this work to improve the segmentation performance of OARs in HaN CT images. OrganNet2.5D integrates 2D convolutions with 3D convolutions to simultaneously extract clear low-level edge features and rich high-level semantic features. The hybrid dilated convolution (HDC) module is introduced to OrganNet2.5D as a replacement for the DenseASPP in FocusNet. HDC module is able to increase the network receptive field without decreasing the image resolutions and at the same time, avoid the gridding issue. Organ-Net2.5D has three blocks: the 2D convolution block for the extraction of clear edge image features, the coarse 3D convolution block for the extraction of coarse high-level semantic features with a limited receptive field, and the fine 3D convolution block for the extraction of refined high-level semantic features with an enlarged receptive field through the utilization of HDC. Similar to AnatomyNet and FocusNet [4,22], a combined loss of Dice loss and focal loss is employed to handle the class imbalance problem. The effectiveness of the proposed Organ-Net2.5D is evaluated on two datasets. On the publicly available MICCAI Head and Neck Auto Segmentation Challenge 2015 dataset (MICCAI 2015 challenge dataset), promising performance is achieved by OrganNet2.5D compared to state-of-the-art approaches.

2   Method

2.1   Dataset

We evaluate the performance of our proposed model on two datasets. The first dataset is collected from two resources of 3D HaN CT images (the Head-Neck

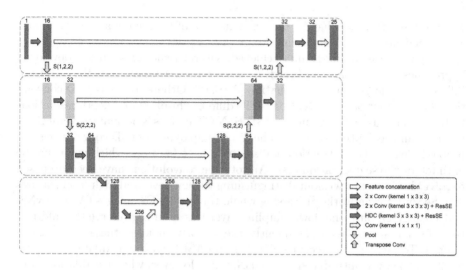

Fig. 1. An illustration of the proposed OrganNet2.5D network architecture. The blue, yellow, and green boxes indicate the 2D convolution block, the coarse 3D convolution block, and the fine 3D convolution block, respectively. (Color figure online)

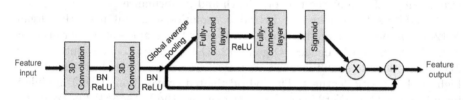

Fig. 2. The $2 \times$ Conv + ResSE unit. "$\times$" refers to element-wise multiplication and "$+$" is element-wise summation.

Cetuximab collection (46 samples) [3] and the Martin Vallières of the Medical Physics Unit, McGill University, Montreal, Canada (261 samples)[1]. This first dataset is utilized to validate the effectiveness of the different blocks of our model. Segmentation annotations of 24 OARs are provided by experienced radiologists with quality control management. We randomly grouped the 307 samples into a training set of 240 samples, a validation set of 20 samples, and a test set of 47 samples. To compare the performance of our proposed method to the existing approaches, we utilize the MICCAI 2015 challenge dataset [14]. There are 48 samples, among which 33 samples are provided as the training set, 10 as the offset test set, and the remaining 5 as the onsite test set. Manual segmentation of 9 OARs is available for the 33 training samples and 10 offset test samples. Similar to previous studies, we optimize our model with the training samples and report the model performance on the 10 offset test samples.

---

[1] https://wiki.cancerimagingarchive.net/display/Public/Head-Neck-PET-CT.

2.2   Network Architecture

The overall network architecture of our proposed OrganNet2.5D is shown in Fig. 1. OrganNet2.5D follows the classic encoder-decoder segmentation network structure [16]. The inputs to our network are the whole volume 3D HaN CT images and the outputs are the segmentation results of the 25 categories for the first dataset (24 OARs and background) or 10 categories (9 OARs and background) for the MICCAI 2015 challenge dataset. OrganNet2.5D contains three major blocks, the 2D convolution block, the coarse 3D convolution block, and the fine 3D convolution block.

2D Convolution Block. The 2D convolution block is designed for the extraction of clear edge image features. It is widely accepted that during image encoding, the low-level features extract the geometric information and the high-level features extract the semantic information. Therefore, in our model, only the first two convolutions near the inputs and the corresponding last two convolutions near the outputs are replaced with 2D convolutions. Without the direct application of 3D convolutions, distorted image edge feature extraction can be avoided. Meanwhile, considering the different in-plane and depth image resolutions, in-plane downsampling is conducted with the 2D convolution block to calibrate the image features for the following 3D convolution operations.

Coarse 3D Convolution Block. The 2D convolution block is followed by the coarse 3D convolution block. To prevent information loss, especially for the small anatomies, only one downsampling is preserved. The coarse 3D convolution block is designed to extract rich semantic features that are important for the pixel-wise distinction task. Following the successful practice of existing methods, the basic unit of our coarse 3D convolution block is composed of two standard 3D convolution layers and one squeeze-and-excitation residual module (ResSE module, Fig. 2). The ResSE module is responsible for feature filtering to highlight the important features and suppress the irrelevant ones. With the filtered image features, the final segmentation step can concentrate more on the important features and better results can be expected.

Fine 3D Convolution Block. With the 2D convolution block and coarse 3D convolution block, clear edge and rich semantic image features are extracted. However, since only two downsampling layers are used (one 2D downsampling and one 3D downsampling), the receptive field of the network is limited. Without the global image information, the segmentation accuracy may be compromised. As such, a series of hybrid dilated convolution (HDC) modules is employed to integrate the global image information with the semantic features and at the same time, to prevent the gridding issue [20]. Moreover, by using different dilation rates, multi-scale image features are extracted, which can better process the OARs of different sizes.

2.3   Loss Function

A combination of focal loss and Dice loss is employed to prevent the model from biasing the large objects.

Focal loss forces the network to focus on the hard samples, which refers to the samples predicted by the network with high uncertainty. It is improved from the cross-entropy loss with both fixed and dynamic loss weighting strategies. The focal loss is calculated as:

$$L_{focal} = -\frac{1}{N} \sum_{n=1}^{N} \sum_{c=1}^{C} \alpha_c (1 - p_n^c)^\gamma y_n^c \log p_n^c \tag{1}$$

where N refers to the sample size, C refers to the different categories (25 for the first dataset and 10 for the second), α_c is the fixed loss weight of the cth OAR, $p \in [0,1]$ is the network prediction, $(1 - p_n^c)^\gamma$ is the dynamic loss weight, and $y \in \{0,1\}$ is the manual label.

Dice loss deals with the class imbalance problem by minimizing the distribution distance between the network prediction and the manual segmentation. For multi-class segmentation, one Dice loss should be calculated for each class and the final Dice loss is the average over all the classes. In this work, the average Dice loss is calculated as:

$$L_{avgdice} = 1 - \frac{1}{C} \sum_{c=1}^{C} \sum_{n=1}^{N} \frac{2 \times p_n^c \times y_n^c}{p_n^c + y_n^c} \tag{2}$$

The final loss function for our network training is a weighted summation of the two losses:

$$L = L_{focal} + \lambda L_{avgdice} \tag{3}$$

For our experiments, we empirically set $\gamma = 2$ and $\lambda = 1.0$. The fixed weights α_c in the focal loss for the first dataset are 0.5, 1.0, 1.0, 1.0, 4.0, 4.0, 4.0, 4.0, 4.0, 1.0, 1.0, 4.0, 1.0, 1.0, 3.0, 3.0, 1.0, 1.0, 1.0, 1.0, 1.0, 1.0, 3.0, 1.0, and 1.0 for the 25 categories (background, brain stem, eye left, eye right, lens left, lens right, optic nerve left, optic nerve right, optic chiasma, temporal lobes left, temporal lobes right, pituitary, parotid gland left, parotid gland right, inner ear left, inner ear right, middle ear left, middle ear right, tongue, temporomandibular joint left, temporomandibular joint right, spinal cord, mandible left, and mandible right), and for the second dataset are 0.5, 1.0, 4.0, 1.0, 4.0, 4.0, 1.0, 1.0, 3.0, and 3.0 for the 10 categories (background, brain stem, optic chiasma, mandible, optic nerve left, optic nerve right, parotid gland left, parotid gland right, submandibular left, submandibular right).

2.4   Implementation Details

All our models are implemented with PyTorch on an NVIDIA GeForce GTX 1080Ti GPU (11G) with a batch size of 2. The inputs to the networks are resized to $256 \times 256 \times 48$. Adam optimizer is utilized to train the models. The

Table 1. Segmentation performance on the first dataset averaged over the 24 OARs with different network configurations

Models	3DUNet-SE	3DUNet-SE-2D	3DUNet-SE-2D-C2	3DUNet-SE-2D-DC2	Proposed
DSC (%)	83.9 ± 2.0	84.4 ± 2.0	84.2 ± 2.1	84.3 ± 2.1	**84.6 ± 1.9**
95HD	3.38	3.39	3.24	3.41	**3.07**

Table 2. DSC of 10 small organs on the first dataset with different network configurations (%)

Models	3DUNet-SE	3DUNet-SE-2D	3DUNet-SE-2D-C	3DUNet-SE-2D-DC	Proposed
Lens L	82.7 ± 7.8	**84.5 ± 6.8**	83.5 ± 7.5	83.2 ± 6.5	84.4 ± 6.6
Lens R	82.6 ± 5.6	**84.1 ± 5.8**	83.3 ± 6.9	83.9 ± 5.4	83.9 ± 5.3
Opt. Ner. L	70.9 ± 10.0	71.0 ± 10.1	70.7 ± 10.2	70.5 ± 10.8	**71.1 ± 10.3**
Opt. Ner. R	70.1 ± 8.7	71.7 ± 8.9	**71.8 ± 9.3**	70.4 ± 9.1	71.2 ± 9.2
Opt. Chiasm	57.2 ± 14.3	59.6 ± 15.1	57.8 ± 15.4	58.0 ± 14.9	**59.8 ± 14.8**
Pituitary	74.4 ± 11.6	75.1 ± 13.0	74.3 ± 12.4	75.1 ± 11.4	**75.6 ± 12.3**
Mid. Ear L	86.4 ± 5.1	86.6 ± 4.8	86.5 ± 5.1	86.7 ± 4.9	**86.9 ± 5.1**
Mid. Ear R	85.4 ± 4.1	85.5 ± 4.3	85.8 ± 4.6	85.6 ± 4.7	**85.9 ± 4.3**
T.M.J. L	83.5 ± 7.2	83.8 ± 7.5	83.8 ± 7.1	82.7 ± 7.8	**83.8 ± 7.2**
T.M.J. R	82.1 ± 8.3	81.6 ± 8.8	82.8 ± 7.9	82.7 ± 8.2	**82.9 ± 7.9**

step decay learning rate strategy is used with an initial learning rate of 0.001 that is reduced by a factor of 10 every 50 epochs until it reaches 0.00001. Two evaluation metrics are calculated to characterize the network performance, the Dice score coefficient (DSC) and the 95% Hausdorff distance (95HD).

3   Experimental Results

3.1   Results on the Collected Public Dataset

Ablation studies regarding our network design are conducted. Average DSC and 95HD on the test set are listed in Table 1. DSC values of the 10 small organs are presented in Table 2. See supplementary material for results on all 24 organs. Four network configurations are involved. 3DUNet-SE refers to the baseline where 3D UNet is combined with the ResSE module. 3DUNet-SE includes only the coarse 3D convolution block in Fig. 1. Introducing the 2D convolution block to 3DUNet-SE, we obtain the 3DUNet-SE-2D model. 3DUNet-SE-2D-C replaces the HDC module in the proposed OrganNet2.5D (Fig. 1) with standard 3D convolutions, and 3DUNet-SE-2D-DC replaces the HDC module with dilated convolutions of the same dilation rate of 2.

Overall, our proposed model achieves the highest mean DSC and lowest mean 95HD. Statistical analysis confirms that our model performs significantly better than the other network configurations ($p < 0.05$ with paired t-tests of the DSC values). These results reflect that both the 2D convolution block and the fine 3D convolution block can enhance the segmentation results. Furthermore, our proposed OrganNet2.5D gives excellent performance on small organ segmentation by generating the best results for 7 of the 10 small organs (Table 2).

Table 3. Segmentation results on the MICCAI 2015 challenge dataset

Models	MICCAI 2015	AnatomyNet [22]	FocusNet [4]	SOARS [5]	SCAA [18]	Proposed
Brain Stem	88.0	86.7 ± 2	87.5 ± 2.6	87.6 ± 2.8	**89.2 ± 2.6**	87.2 ± 3.0
Opt. Chiasm	55.7	53.2 ± 15	59.6 ± 18.1	64.9 ± 8.8	62.0 ± 16.9	**66.3 ± 7.4**
Mandible	93.0	92.5 ± 2	93.5 ± 1.9	95.1 ± 1.1	**95.2 ± 1.3**	92.2 ± 2.1
Opt. Ner. L	64.4	72.1 ± 6	73.5 ± 9.6	75.3 ± 7.1	**78.4 ± 6.1**	75.0 ± 7.8
Opt. Ner. R	63.9	70.6 ± 10	74.4 ± 7.2	74.6 ± 5.2	**76.0 ± 7.5**	74.1 ± 5.1
Parotid L	82.7	88.1 ± 2	86.3 ± 3.6	88.2 ± 3.2	**89.3 ± 1.5**	86.7 ± 2.6
Parotid R	81.4	87.4 ± 4	87.9 ± 3.1	88.2 ± 5.2	**89.2 ± 2.3**	85.8 ± 4.9
Subman. L	72.3	81.4 ± 4	79.8 ± 8.1	**84.2 ± 7.3**	83.2 ± 4.9	82.1 ± 5.8
Subman. R	72.3	81.3 ± 4	80.1 ± 6.1	**83.8 ± 6.9**	80.7 ± 5.2	82.1 ± 4.1
Mean DSC	74.9	79.2	80.3	82.4	**82.6**	81.3

3.2   Results on MICCAI 2015 Challenge Dataset

We compare the performance of our proposed model to the state-of-the-art methods on the MICCAI 2015 challenge dataset (Table 3). It should be noted that in the table, the MICCAI 2015 results were the best results obtained for each OAR possibly by different methods. AnatomyNet was trained with additional samples except for the MICCAI 2015 challenge dataset. All the results of existing methods are adopted from the respective papers without method re-implementation to avoid implementation biases.

Segmentation results show that our proposed model achieves better performance than the three most prevalent methods in the field (MICCAI 2015, AnotomyNet, and FocusNet) indicated by the mean DSC, which confirms the effectiveness of the proposed network. Compared to the two recently published methods, SOARS and SCAA, our method is slightly worse. However, it should be noted that SOARS utilized neural network search to find the optimal network architecture [5], which is more computationally intensive. SCAA combined the 2D and 3D convolutions with a very complicated network design [18]. Nevertheless, with the simple and easy-to-implement architecture, our OrganNet2.5D still performs the best when segmenting the smallest organ, optic chiasma. This observation reflects the suitability of our network modifications and training strategy for our task. Visual results lead to similar conclusions as to the quantitative results (See supplementary material for details).

4   Conclusion

In this study, we present a novel network, OrganNet2.5D, for the segmentation of OARs in 3D HaN CT images, which is a necessity for the treatment planning of radiation therapy for HaN cancers. To fully utilize the 3D image information, deal with the different in-plane and depth image resolutions, and solve the difficulty of simultaneous segmentation of large and small organs, Organ-Net2.5D consists of a 2D convolution block to extract clear edge image features, a coarse 3D convolution block to obtain rich semantic features, and a fine 3D convolution block to generate global and multi-scale image features. The effectiveness of OrganNet2.5D was evaluated on two datasets. Promising performance was achieved by our proposed OrganNet2.5D compared to the state-of-the-art approaches, especially on the segmentation of small organs.

Acknowledgements. This research is partially supported by the National Key Research and Development Program of China (No. 2020YFC2004804 and 2016YFC0106200), the Scientific and Technical Innovation 2030- "New Generation Artificial Intelligence" Project (No. 2020AAA0104100 and 2020AAA0104105), the Shanghai Committee of Science and Technology, China (No. 20DZ1100800 and 21DZ1100100), Beijing Natural Science Foundation-Haidian Original Innovation Collaborative Fund (No. L192006), the funding from Institute of Medical Robotics of Shanghai Jiao Tong University, the 863 national research fund (No. 2015AA043203), Shenzhen Yino Intelligence Techonology Co., Ltd., Shenying Medical Technology (Shenzhen) Co., Ltd., the National Natural Science Foundation of China (No. 61871371 and 81830056), the Key-Area Research and Development Program of GuangDong Province (No. 2018B010109009), the Basic Research Program of Shenzhen (No. JCYJ20180507182400762), and the Youth Innovation Promotion Association Program of Chinese Academy of Sciences (No. 2019351).

References

1. Brouwer, C.L., Steenbakkers, R.J.H.M., Heuvel, E.V.d., et al.: 3D variation in delineation of head and neck organs at risk. Radiat. Oncol. **7**(1), 32 (2012)
2. Chen, A., Niermann, K.J., Deeley, M.A., Dawant, B.M.: Evaluation of multiple-atlas-based strategies for segmentation of the thyroid gland in head and neck CT images for IMRT. Phys. Med. Biol. **57**(1), 93–111 (2012)
3. Clark, K., Vendt, B., Smith, K., et al.: The cancer imaging archive (TCIA): maintaining and operating a public information repository. J. Digit. Imaging **26**(6), 1045–1057 (2013)
4. Gao, Y., et al.: FocusNet: imbalanced large and small organ segmentation with an end-to-end deep neural network for head and neck CT images. In: Shen, D., et al. (eds.) MICCAI 2019. LNCS, vol. 11766, pp. 829–838. Springer, Cham (2019). https://doi.org/10.1007/978-3-030-32248-9_92
5. Guo, D., et al.: Organ at risk segmentation for head and neck cancer using stratified learning and neural architecture search. In: 2020 IEEE Conference on Computer Vision and Pattern Recognition (CVPR), pp. 4223–4232. Virtual Conference (2020)
6. Han, X., et al.: Atlas-based auto-segmentation of head and neck CT images. In: Metaxas, D., Axel, L., Fichtinger, G., Székely, G. (eds.) MICCAI 2008. LNCS, vol. 5242, pp. 434–441. Springer, Heidelberg (2008). https://doi.org/10.1007/978-3-540-85990-1_52
7. Harari, P.M., Song, S., Tome, W.A.: Emphasizing conformal avoidance versus target definition for IMRT planning in head-and-neck cancer. Int. J. Radiat. Oncol. Biol. Phys. **77**(3), 950–958 (2010)

8. Ibragimov, B., Xing, L.: Segmentation of organs-at-risks in head and neck CT images using convolutional neural networks. Med. Phys. **44**(2), 547–557 (2017)

9. Li, C., Sun, H., Liu, Z., Wang, M., Zheng, H., Wang, S.: Learning cross-modal deep representations for multi-modal MR image segmentation. In: Shen, D., et al. (eds.) MICCAI 2019. LNCS, vol. 11765, pp. 57–65. Springer, Cham (2019). https://doi.org/10.1007/978-3-030-32245-8_7

10. Liang, S., Thung, K.-H., Nie, D., Zhang, Y., Shen, D.: Multi-view spatial aggregation framework for joint localization and segmentation of organs at risk in head and neck CT images. IEEE Trans. Med. Imaging **39**(9), 2794–2805 (2020)

11. Liu, S., et al.: 3D anisotropic hybrid network: transferring convolutional features from 2D images to 3D anisotropic volumes. In: Frangi, A.F., Schnabel, J.A., Davatzikos, C., Alberola-López, C., Fichtinger, G. (eds.) MICCAI 2018. LNCS, vol. 11071, pp. 851–858. Springer, Cham (2018). https://doi.org/10.1007/978-3-030-00934-2_94

12. Men, K., Geng, H., Cheng, C., et al.: More accurate and efficient segmentation of organs-at-risk in radiotherapy with convolutional neural networks cascades. Med. Phys. **46**(1), 286–292 (2019)

13. Qi, K., et al.: X-Net: brain stroke lesion segmentation based on depthwise separable convolution and long-range dependencies. In: Shen, D., et al. (eds.) MICCAI 2019. LNCS, vol. 11766, pp. 247–255. Springer, Cham (2019). https://doi.org/10.1007/978-3-030-32248-9_28

14. Raudaschl, P.F., Zaffino, P., Sharp, G.C., et al.: Evaluation of segmentation methods on head and neck CT: auto-segmentation challenge 2015. Med. Phys. **44**(5), 2020–2036 (2017)

15. Ren, X., et al.: Interleaved 3D-CNNs for joint segmentation of small-volume structures in head and neck CT images. Med. Phys. **45**(5), 2063–2075 (2018)

16. Ronneberger, O., Fischer, P., Brox, T.: U-Net: convolutional networks for biomedical image segmentation. In: Navab, N., Hornegger, J., Wells, W.M., Frangi, A.F. (eds.) MICCAI 2015. LNCS, vol. 9351, pp. 234–241. Springer, Cham (2015). https://doi.org/10.1007/978-3-319-24574-4_28

17. Tang, H., Chen, X., Liu, Y., et al.: Clinically applicable deep learning framework for organs at risk delineation in CT images. Nat. Mach. Intell. **1**(10), 480–491 (2019)

18. Tang, H., Liu, X., Han, K., et al.: Spatial context-aware self-attention model for multi-organ segmentation. In: 2021 IEEE Winter Conference on Applications of Computer Vision (WACV), pp. 939–949. Virtual Conference (2021)

19. Torre, L.A., Bray, F., Siegel, R.L., Ferlay, J., Lortet-Tieulent, J., Jemal, A.: Global cancer statistics, 2012. Ophthalmology **65**(2), 87–108 (2015)

20. Wang, P., Chen, P., Yuan, Y., et al.: Understanding convolution for semantic segmentation. In: 2018 IEEE Winter Conference on Applications of Computer Vision (WACV), pp. 1451–1460. Lake Tahoe, NV, USA (2018)

21. Wang, Z., Wei, L., Wang, L., Gao, Y., Chen, W., Shen, D.: Hierarchical vertex regression-based segmentation of head and neck CT images for radiotherapy planning. IEEE Trans. Image. Process. **27**(2), 923–937 (2018)

22. Zhu, W., et al.: AnatomyNet: deep learning for fast and fully automated whole-volume segmentation of head and neck anatomy. Med. Phys. **46**(2), 576–589 (2019)

TumorCP: A Simple but Effective Object-Level Data Augmentation for Tumor Segmentation

Jiawei Yang[1], Yao Zhang[2,3], Yuan Liang[1], Yang Zhang[4], Lei He[1(✉)], and Zhiqiang He[4(✉)]

[1] Electrical and Computer Engineering, University of California, Los Angeles, USA
lhe@ee.ucla.edu
[2] Institute of Computing Technology, Chinese Academy of Sciences, Beijing, China
[3] University of Chinese Academy of Sciences, Beijing, China
[4] Lenovo Corporate Research and Development, Lenovo Ltd., Beijing, China
hezq@lenovo.com

Abstract. Deep learning models are notoriously data-hungry. Thus, there is an urging need for data-efficient techniques in medical image analysis, where well-annotated data are costly and time consuming to collect. Motivated by the recently revived "Copy-Paste" augmentation, we propose TumorCP, a simple but effective object-level data augmentation method tailored for tumor segmentation. TumorCP is online and stochastic, providing unlimited augmentation possibilities for tumors' subjects, locations, appearances, as well as morphologies. Experiments on kidney tumor segmentation task demonstrate that TumorCP surpasses the strong baseline by a remarkable margin of 7.12% on tumor Dice. Moreover, together with image-level data augmentation, it beats the current state-of-the-art by 2.32% on tumor Dice. Comprehensive ablation studies are performed to validate the effectiveness of TumorCP. Meanwhile, we show that TumorCP can lead to striking improvements in extremely low-data regimes. Evaluated with only 10% labeled data, TumorCP significantly boosts tumor Dice by **21.87%**. To the best of our knowledge, this is the very first work exploring and extending the "Copy-Paste" design in medical imaging domain. Code is available at: https://github.com/YaoZhang93/TumorCP.

Keywords: Data-efficiency · Tumor segmentation · Data augmentation

---

J. Yang and Y. Zhang—Equal contribution.
L. He and Z. He—Equal contribution as the corresponding authors.

Electronic supplementary material The online version of this chapter (https://doi.org/10.1007/978-3-030-87193-2_55) contains supplementary material, which is available to authorized users.

© Springer Nature Switzerland AG 2021
M. de Bruijne et al. (Eds.): MICCAI 2021, LNCS 12901, pp. 579–588, 2021.
https://doi.org/10.1007/978-3-030-87193-2_55

1   Introduction

Deep learning (DL) models work remarkably well over the past few years in computer vision tasks, including medical image analysis. Though DL models act like de facto standard, they are notoriously data-hungry, demanding more so than ever large and well-annotated datasets to achieve robust performance [16]. However, high-quality annotated datasets require intense labor and domain knowledge, which becomes more expensive in the medical domain.

To improve data-efficient learning, several successful approaches have been proposed from different perspectives, such as leveraging unlabeled data for semi-supervised self-training [1,15,16] or self-supervised pre-training [1,12,19], distilling priors from data as explicit constraints for model training [9,10], generating new data with the imaging of an anatomy of a different modality [8,11], or utilizing appropriate data augmentation methods to increase data diversity [2,5,14,15]. Some of them are designated for medical images. Particularly, Zhou et al. [19] designed a unified self-supervised learning framework, integrating multiple proxy tasks to exploit unlabeled medical data, and showed performance gains for downstream tasks. Xue et al. [17], and Shin et al. [13] used GANs to generate additional training data for histopathology image classification and brain tumor segmentation. The quality of the "realness" of synthesized training data dramatically affects model performance due to the risk of overfitting to fake data. Eaton et al. [3] studied Mix-up [18] augmentation for brain tumor segmentation. However, it requires a specific patch-level operation which involves complicated strategies, e.g. sampling of small patches to be mixed up.

Distinct from the trend of using increasingly sophisticated methods like GANs, we investigate "Copy-Paste", a straightforward augmentation technique [2,4] that has been recently revisited and made breakthroughs in natural image instance segmentation [5]. Copy-Paste augmentation avoids costly generation processes from representation space to pixel space by simply pasting the labeled instance onto new background images as additional training data. Despite its success in natural images, such method is largely unexplored in the medical image realm. Moreover, its effectiveness for medical tasks remains doubtable since the context information tends to be ignored in Copy-Paste. For instance, in the tumor segmentation, one would argue the importance of surrounding visual clues, i.e., context, for the emergence of a tumor. Besides, one would believe the inherent anatomical structures in medical image make the context indispensable for tumor segmentation. In this work, we also aim to fill the gap of understanding the role of *context* in medical domain by examining the effectiveness of Copy-Paste augmentation for tumor segmentation.

We propose TumorCP, a simple but effective object-level data augmentation method based on Copy-Paste for tumor segmentation tasks. Straightforwardly, TumorCP randomly chooses a tumor from a source image and paste it onto the organs in the target image after a series of spatial, contrast, and blurring augmentations. We use kidney tumor segmentation (KiTS19 dataset [6]) and a state-of-the-art model (nnUNet [7]) as the benchmark to evaluate the proposed method. We empirically show that though TumorCP inevitably generates artifacts after Copy-Paste, it consistently provides solid gains over all different settings in our experiments. Specifically, with only rigid spatial transformation and Copy-Paste within the same patient, TumorCP can surpass the baseline by 6.24% tumor Dice. Together with inter-patient Copy-Paste and other tumor-oriented augmentations, TumorCP further outperforms the baseline by 7.12% tumor Dice. Moreover, with image-level data augmentation (ImgDA), our best version beats state-of-the-art by 2.3%. Going one step further, we also study TumorCP for extremely low-data regime, where only 10% labeled data are exploited for training. Under this setting, TumorCP with ImgDA can improve the tumor Dice by **21.87%** compared with no-data-augmentation (noDA), which is unprecedented to our knowledge, convincingly demonstrating the effectiveness of TumorCP for data-efficiency learning.

The success of TumorCP is an empirical observation to support context-decoupled learning even in *medical domain*. We briefly discuss our understanding of the open question of the context's role and why TumorCP works in Sect. 2.2. We hope our work can provide some useful data points to our community and shed light on the importance of Copy-Paste augmentation, which is powerful but unfortunately nearly absent in the medical imaging field.

2 Method

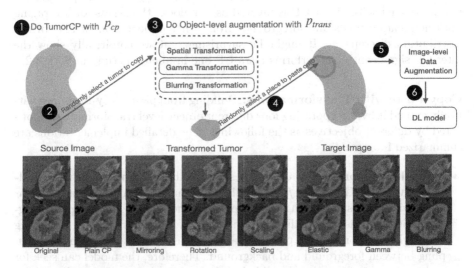

Fig. 1. Illustration of TumorCP's pipeline. A pair of source image and target image are sampled from the dataset. With probability of p_{cp}, TumorCP performs Copy-Paste once, following the step number of 2,3,4,5, and finally to 6; otherwise, it directly goes to step 6. In step 3, each of the transformation has its own probability (p_{trans}) to be invoked. The bottom illustrates two samples performing Copy-Paste with object-level data augmentation.

TumorCP is an online and stochastic augmentation process specified for tumor segmentation. Its implementation is easy and straightforward. As illustrated in Fig. 1, given a set of training samples $\mathcal{D}$, with the probability of $(1 - p_{cp})$, TumorCP does nothing; otherwise TumorCP samples a pair of images $(x_{src}, x_{tgt}) \sim \mathcal{D}$ and conducts Copy-Paste once. Let $\mathcal{O}_{src}$ be the set of tumor(s) on x_{src}, $\mathcal{V}_{tgt}$ be the set of volumetric coordinates of organ(s) on x_{tgt}, and $\mathcal{T}$ be the set of stochastic data transformations, each of which has a probability parameter called p_{trans}. To do once Copy-Paste, TumorCP first samples a tumor $o \sim \mathcal{O}_{src}$, a set of transformation(s) $\tau \sim \mathcal{T}$, and a target location $v \sim \mathcal{V}_{tgt}$, followed by centering $\tau(o)$ at v to replace the original data and annotation. To fully leverage the advantage of TumorCP, we carefully design two modes of Copy-Paste for tumors: intra-patient and inter-patient Copy-Paste. Meanwhile, we enhance Copy-Paste with several object-level transformation to obtain abundant augmentations.

2.1   TumorCP's augmentation

Intra-/Inter- Copy-Paste. In order to study the effect of inter-patient variance to TumorCP, we define two base settings: 1) intra-patient Copy-Paste (intra-CP) if the source and target images are identical, i.e., both from the same patient and 2) inter-patient Copy-Paste (inter-CP) if those are different. From the perspective of data distribution, the intra-CP is preferred as its intensity agreement with the data as a whole, but this limits data diversity. From the perspective of data diversity, the inter-CP is favored as it unlocks the access for leveraging both new backgrounds and foregrounds from other patients, but it also brings distribution discrepancy. It might be surprising that we empirically show the inter-CP significantly outperforms intra-CP one in ablation study in Sect. 3.2.

Copy-Paste with Transformations. Building from plain Copy-Paste, we naturally extend it by incorporating four different object-level transformations motivated by different objectives as the followings. The detailed implementations are summarized in appendix.

- **Spatial transformation decouples context and improves morphology diversity.** Given the *fixed* acquired CT images, tumors always appear along with their surrounding visual context. Though image-level spatial augmentation increases data diversity in terms of perspectives (e.g., mirroring and slight rotation), it still processes an image as a whole, remaining the coupling between foreground and background. Therefore, the model can seek for and tend to overfit to the plausible but de facto irrelevant surrounding clues. Note that plain Copy-Paste already addressed this problem by offering new background via the most basic spatial transformation—**shifting**. We further increase the morphology diversity by applying i) rigid transformation that includes scaling, rotation, and mirroring, and ii) elastic transformation that deforms tumors. Figure 1 demonstrates examples of transformed tumors.
- **Gamma transformation enhances contrast and improves intensity diversity.** Given a tumor, we apply gamma transformation to adjust its intensity distribution while retaining the whole intensity range. On the one

hand, the tumor intensity diversity is enhanced by randomly sample gamma parameter; on the other hand, the local contrast is enhanced by power-law non-linearity, facilitating tumor discrimination.

- **Blurring transformation improves texture diversity.** We use a Gaussian filter as the blurring transformation. Intuitively, a Gaussian filter with different sigma values can filter out the noise and smooth the tumor to some extent. Aggregating noise-perturbed low-level textures can indirectly increase the texture diversity to relatively high-level textures.

The whole pipeline can be incorporated together with image-level augmentation. It is worth mentioning that all the instance augmentation process is both **online** and **random**, bringing unlimited possibilities for tumors' locations and appearances within or across the subjects.

2.2 Intuitions on TumorCP's Effectiveness

As aforementioned, TumorCP has two goals: i) increase the data diversity, and ii) learn high-level and to abstract the *invariant* representation of tumor. Data diversity is increased as the new combinations of tumors, and their surroundings are generated with the augmentation. For learning high-level information, we discuss three properties of TumorCP to explain its effectiveness.

Eliminated Background Bias by Context-Invariant Prediction. As mentioned before, the semantic contexts are *fixed* for the acquired medical images. Convolutional Neural Network (CNN) inevitably convolutes surrounding visual contexts along with the objects themselves. This can bias the model towards plausible but indeed tumor-irrelevant clues, increasing the risk of overfitting. With both *random* and *online* spatial transformation, TumorCP offers access for tumor to preciously unattached zones and thus provides unlimited possibilities for tumors' surrounding contexts. It enforces the model's prediction to be invariant across different visual surroundings and eliminates background bias.

Improved Generalizability by Transformation-Invariant Prediction. The model should capture both high-level semantic information and low-level boundary information for successful segmentation. With both *random* and *online* Gamma & Blur transformations, TumorCP can generate diverse tumors in terms of size, shape, color and texture, which increase the intra-class disparity. It tasks the model to capture the golden semantics from the data. In other words, it enforces the model's prediction to be invariant across different data transformation (that potentially resembles real-world data) and improves generalizability.

Oversampling Behavior. Data imbalance is a widely experienced problem. Typical solutions usually re-weight loss function or re-sample training data according to the class distribution. In this work, the distribution of background, organ, and tumor is extremely imbalanced. From this perspective, TumorCP acts like a data re-sampler that significantly increases the volume of tumors in multiplication degree at a minor cost.

3 Experiments and Discussion

3.1 Experiment Settings

We evaluate TumorCP on KiTS19 [6], a publicly available dataset for kidney tumor segmentation. We randomly split the published 210 images into a training set with 168 images and a validation set with 42 images. As the limited computation resources, we majorly report ablation study results on the validation set if not specified. Note that this validation set is unaugmented and unseen i.e., neither used to tune hyper-parameters nor to monitor the training process. We use Sørensen-Dice Coefficient (Dice) score in all experiments, which measures the overlap of model's prediction y_{pred} and ground truth y_{true}, formulated as Dice $= |y_{true} \cap y_{pred}|/|y_{true} \cup y_{pred}|$. The average and standard deviation of the Dice score over all patients are reported.

We use publicly available[1] state-of-the-art nnUNet codebase for implementation, which includes data pre-processing, leading image-level augmentation pipelines, as well as top-performance models. It almost tops all biomedical image segmentation benchmarks [7]. This paper focuses on a general augmentation method for tumor segmentation, so the choices of datasets and running models are orthogonal to our goal. TumorCP can generalize to other segmentation models and tumor segmentation datasets at no cost.

All experiments are conducted on Nvidia V100 GPU with 500 epochs training of 3d_fullres nnUNet, instead of 1000 epochs by nnUNet's default. The batch size for training is 2. During training, each epoch takes 250 iterations, which means 250 batches of data are sampled and learned. Other settings in model training remain its default. We refer readers to [7] and the codebase link for more details.

3.2 Ablation Study

For simplicity and unification, we set the probability of TumorCP performing Copy-Paste as $p_{cp} = 0.8$ for all experiments.

Ablation on Intra-CP with Different Transformations. We first investigate TumorCP under intra-CP with various object-level transformations. In this ablation, no image-level augmentation is applied. All object-level transformations have a 0.5 probability of being invoked. For example, Intra-CP&Rigid means rigid transformation has a 0.5 probability to be conducted when Intra-CP is triggered. Table 1 presents the comparison on different methods. As

---

[1] https://github.com/MIC-DKFZ/nnUNet.

the first group of Table 1 demonstrates, all the models trained with TumorCP (shaded cells) consistently outperform the baseline model, no-data-augmentation (noDA). Specifically, the vanilla intra-CP itself can bring 1.09% Dice improvement over baseline; TumorCP with only rigid transformation can increase tumor Dice by 6.24%.

Table 1. Ablation study of TumorCP. The first group shows the results of TumorCP with different transformations in the intra-CP setting, while the second group shows the results TumorCP with intra-CP and inter-CP settings. The shaded rows denote our work.

Method	Mean dice ± std/improvement over baseline (%) ↑	
	Kidney	Tumor
noDA	96.62±2.41/baseline	72.59±26.97/baseline
Intra-CP	96.81±2.02/+0.19	73.68±26.99/+1.09
Intra-CP&Elastic	96.75±1.88/+0.13	73.95±28.20/+1.36
Intra-CP&Rigid	96.78±1.92/+0.16	**78.83±19.77/+6.24**
Intra-CP&Gamma	96.81±1.89/+0.19	76.32±23.97/+3.73
Intra-CP&Blur	**96.89±1.92/+0.27**	76.46±24.86/+3.87
Intra-CP	**96.81±2.02/+0.19**	73.68±26.99/+1.09
Inter-CP	96.73±2.03/+0.11	77.22±23.67/+4.63
Intra-&Inter-CP	96.78±1.98/+0.16	**77.44±23.46/+4.85**

Ablation on Intra-/inter-CP. Here we study the effect of intra-/inter-CP for the considerations in Sect. 2.1. The second group in Table 1 shows that inter-CP significantly outperforms intra-CP by 3.54% Tumor Dice, yielding a 4.63% improvement over the baseline model. Though surprised to some extent, this result meets our expectation as both the tumors' and the backgrounds' diversity from one patient are still limited compared to other patients. Copying others' tumors and pasting them onto current patients' cases is supposed to unlock more novel combinations and bring more data diversity. We also aggregate intra- and inter- CP by setting a 50% chance for each to sample data pairs from the dataset. The last line in Table 1 presents the result and is shown to the best entry among this ablation. It demonstrates the superiority of combining both intra- and inter-patient's context exchange.

Ablation on Compatibility. As the last step, we accumulate the composition of all object-level transformations and Intra-&Inter-CP to constitute TumorCP*. Previously we improve from noDA baseline. Here we also explore the compatibility between TumorCP and image-level augmentation. The image-level augmentation follows `nnUNetV2Trainer` default setting detailed here[2] [7]. Results in Table 2 shows that `TumorCP*` is compatible with image-level augmentations, and thus can act as a plug-in module in general augmentation pipeline. Together with image-level augmentation, `TumorCP*` can improve 7.12% from no image-level augmentation (noDA) baseline and 2.32% from image-level augmentation (ImgDA) baseline. It is worthy to mention that the ImgDA baseline currently still holds the state-of-the-art performance for KiTS Dataset, which means `TumorCP*` can further boost exisiting arts to higher performance. `TumorCP*` can generalize to other models and datasets at almost no cost.

`TumorCP` also improves organ segmentation. Though `TumorCP` is intended for better tumor segmentation, it also consistently improves kidney segmentation performance compared to its baselines. It also meets our intuitions for `TumorCP`, since from the perspective of kidney, tumors are the relative context and background to some extent, which resembles *"Eliminated background bias by context-invariant prediction"* but now for the kidney.

3.3 Towards Extremely Low-Data Regime

Finally, we demonstrate the potentials of `TumorCP` in extremely low-data regime via some additional ablations. Particularly, we randomly select 10% data from the training set same as before. Then, we train three models, `noDA`, `ImgDA` and `TumorCP*` + `ImgDA` on 10% data respectively, followed by the evaluation on the same validation set. Table 3 shows the results. Under this setting, our method can improve the noDA by **21.87%**, which, to the best of our knowledge, is

Table 2. Comparison of `TumorCP` and image-level augmentation. The shaded rows denote our work.

Method	Mean dice ± std/improvement over baseline (%) ↑	
	Kidney	Tumor
noDA	96.62±2.41/baseline	72.59±26.97/baseline
TumorCP*	**96.86±1.91/+0.24**	**79.71±22.56/+7.12**
ImgDA	97.06±1.48/baseline	82.43±21.29/baseline
TumorCP* + ImgDA	**97.15±1.43/+0.09**	**84.75±20.87/+2.32**

Table 3. Comparison of `TumorCP` and image-level augmentation for data-efficient segmentation. The shaded rows denote our work.

Method	Mean dice ± std/improvement over baseline (%) ↑	
	Kidney	Tumor
10%-data noDA	93.25±4.41/baseline	41.12±39.58/baseline
10%-data ImgDA	95.41±3.25/+2.16	54.34±31.59/+13.22
10%-data TumorCP* + ImgDA	**95.53±3.25/(+2.16/+2.28)**	**62.99±26.92/(+13.22/+21.87)**

---

[2] https://git.io/Jqvro.

unprecedented, convincingly demonstrating the effectiveness of TumorCP for data-efficiency learning. It breaks the trend of using sophisticated methods or strategies while achieving promising results in low-data regime of tumor segmentation.

4   Conclusion and Future Works

This key contribution of our work is the proposal and comprehensive study of TumorCP, a simple but effective object-level data augmentation for tumor segmentation. Extensive experiments confirm the remarkable effectiveness of our method. In addition to surpassing current art in kidney tumor segmentation by 2.31% in tumor Dice, we also demonstrate the potential of TumorCP for the extremely low-data regime. We prefer to call our TumorCP as a *new baseline*, as it does not involve any sophisticated techniques nor extensive hyper-parameter adjustment while achieving the new state-of-the-art. Besides, TumorCP does not directly handle the distribution mismatching in the inter-CP setting but still gets fabulous performance. Future works can easily extend TumorCP for other medical segmentation tasks without significant modifications, and are worth trying for further improving state-of-the-art accuracy.

References

1. Chen, T., Kornblith, S., Norouzi, M., Hinton, G.: A simple framework for contrastive learning of visual representations. In: International Conference on Machine Learning, pp. 1597–1607. PMLR (2020)
2. Dwibedi, D., Misra, I., Hebert, M.: Cut, paste and learn: surprisingly easy synthesis for instance detection. In: Proceedings of the IEEE International Conference on Computer Vision, pp. 1301–1310 (2017)
3. Eaton-Rosen, Z., Bragman, F., Ourselin, S., Cardoso, M.J.: Improving data augmentation for medical image segmentation (2018)
4. Fang, H.S., Sun, J., Wang, R., Gou, M., Li, Y.L., Lu, C.: Instaboost: boosting instance segmentation via probability map guided copy-pasting. In: Proceedings of the IEEE/CVF International Conference on Computer Vision, pp. 682–691 (2019)
5. Ghiasi, G., et al.: Simple copy-paste is a strong data augmentation method for instance segmentation. arXiv preprint arXiv:2012.07177 (2020)
6. Heller, N., et al.: The kits19 challenge data: 300 kidney tumor cases with clinical context, ct semantic segmentations, and surgical outcomes. arXiv preprint arXiv:1904.00445 (2019)
7. Isensee, F., Jaeger, P.F., Kohl, S.A., Petersen, J., Maier-Hein, K.H.: nnu-net: a self-configuring method for deep learning-based biomedical image segmentation. Nat. Methods **18**(2), 203–211 (2021)
8. Liang, Y., et al.: Oralviewer: 3d demonstration of dental surgeries for patient education with oral cavity reconstruction from a 2d panoramic x-ray. In: 26th International Conference on Intelligent User Interfaces, pp. 553–563 (2021)
9. Liang, Y., Song, W., Dym, J.P., Wang, K., He, L.: Comparenet: anatomical segmentation network with deep non-local label fusion. In: International Conference on Medical Image Computing and Computer-Assisted Intervention, pp. 292–300 (2019)

10. Liang, Y., Song, W., Yang, J., Qiu, L., Wang, K., He, L.: Atlas-aware convnet for accurate yet robust anatomical segmentation. In: Asian Conference on Machine Learning, pp. 113–128. PMLR (2020)
11. Liang, Y., Song, W., Yang, J., Qiu, L., Wang, K., He, L.: X2teeth: 3d teeth reconstruction from a single panoramic radiograph. In: International Conference on Medical Image Computing and Computer-Assisted Intervention, pp. 400–409 (2020)
12. Mitrovic, J., McWilliams, B., Walker, J., Buesing, L., Blundell, C.: Representation learning via invariant causal mechanisms. arXiv preprint arXiv:2010.07922 (2020)
13. Shin, H.C.: Medical image synthesis for data augmentation and anonymization using generative adversarial networks. In: Gooya, A., Goksel, O., Oguz, I., Burgos, N. (eds.) SASHIMI 2018. LNCS, vol. 11037, pp. 1–11. Springer, Cham (2018). https://doi.org/10.1007/978-3-030-00536-8_1
14. Shorten, C., Khoshgoftaar, T.M.: A survey on image data augmentation for deep learning. J. Big Data 6(1), 1–48 (2019)
15. Sohn, K., et al.: Fixmatch: simplifying semi-supervised learning with consistency and confidence. arXiv preprint arXiv:2001.07685 (2020)
16. Xie, Q., Luong, M.T., Hovy, E., Le, Q.V.: Self-training with noisy student improves imagenet classification. In: Proceedings of the IEEE/CVF Conference on Computer Vision and Pattern Recognition, pp. 10687–10698 (2020)
17. Xue, Y., et al.: Synthetic augmentation and feature-based filtering for improved cervical histopathology image classification. In: Shen, D., et al. (eds.) MICCAI 2019. LNCS, vol. 11764, pp. 387–396. Springer, Cham (2019). https://doi.org/10.1007/978-3-030-32239-7_43
18. Zhang, H., Cisse, M., Dauphin, Y.N., Lopez-Paz, D.: mixup: Beyond empirical risk minimization. arXiv preprint arXiv:1710.09412 (2017)
19. Zhou, Z., et al.: Models genesis: generic autodidactic models for 3D medical image analysis. In: Shen, D., et al. (eds.) MICCAI 2019. LNCS, vol. 11767, pp. 384–393. Springer, Cham (2019). https://doi.org/10.1007/978-3-030-32251-9_42

Modality-Aware Mutual Learning for Multi-modal Medical Image Segmentation

Yao Zhang[1,2], Jiawei Yang[3], Jiang Tian[4], Zhongchao Shi[4], Cheng Zhong[4],
Yang Zhang[5(✉)], and Zhiqiang He[1,5(✉)]

[1] Institute of Computing Technology, Chinese Academy of Sciences, Beijing, China
[2] University of Chinese Academy of Sciences, Beijing, China
[3] Electrical and Computer Engineering, University of California, Los Angeles, USA
[4] AI Lab, Lenovo Research, Beijing, China
[5] Lenovo Corporate Research and Development, Lenovo Ltd., Beijing, China
{zhangyang20,hezq@lenovo.com}

Abstract. Liver cancer is one of the most common cancers worldwide. Due to inconspicuous texture changes of liver tumor, contrast-enhanced computed tomography (CT) imaging is effective for the diagnosis of liver cancer. In this paper, we focus on improving automated liver tumor segmentation by integrating multi-modal CT images. To this end, we propose a novel mutual learning (**ML**) strategy for effective and robust multi-modal liver tumor segmentation. Different from existing multi-modal methods that fuse information from different modalities by a single model, with ML, an ensemble of modality-specific models learn collaboratively and teach each other to distill both the characteristics and the commonality between high-level representations of different modalities. The proposed ML not only enables the superiority for multi-modal learning but can also handle missing modalities by transferring knowledge from existing modalities to missing ones. Additionally, we present a modality-aware (**MA**) module, where the modality-specific models are interconnected and calibrated with attention weights for adaptive information exchange. The proposed modality-aware mutual learning (**MAML**) method achieves promising results for liver tumor segmentation on a large-scale clinical dataset. Moreover, we show the efficacy and robustness of MAML for handling missing modalities on both the liver tumor and public brain tumor (BRATS 2018) datasets. Our code is available at https://github.com/YaoZhang93/MAML.

1 Introduction

Liver cancer is one of the most common cancer diseases in the world [1]. CT images are the most commonly used imaging modality for the initial evaluation

---

Y. Zhang and Z. He—Equal contribution as the corresponding authors.
This work is done when Yao Zhang was an intern at AI Lab, Lenovo Research.

M. de Bruijne et al. (Eds.): MICCAI 2021, LNCS 12901, pp. 589–599, 2021.
https://doi.org/10.1007/978-3-030-87193-2_56

of liver cancer. The accurate measurements of liver tumor status from CT images, including tumor volume, shape, and location, can assist doctors in making hepatocellular carcinoma evaluation and surgical planning. However, a portion of textures of the liver tumor on CT volumes are inconspicuous and, therefore, can be easily neglected even by experienced radiologists. In clinical practice, radiologists usually enhance CT images by an injection protocol for clearly observing liver tumors. When the contrast agent goes through the liver within blood vessels, it yields a favorable contrast between liver tissues and abnormalities, including liver tumors. Contrast-enhanced CT imaging used in the dual-modality protocol is comprised of venous and arterial phases with intravenous contrast delay. Dual-phase images can make good complementary information for each other and thus can contribute to better diagnosis of liver tumor.

In recent years, deep learning has largely advanced the field of computer-aided diagnosis (CAD), especially medical image segmentation [15,17,30]. Fully Convolution Neural Networks (FCNs) go beyond the limitation of hand-crafted features and dramatically improve the performance of liver tumor segmentation with an encoder-decoder architecture [10,13,24–26]. There exists two major issues applying FCNs in multi-modal segmentation. One is how to integrate information from multi-modal medical images effectively. The other is how to deal with the scenario of missing modalities that is common in practice. We elaborate them in the followings.

Multi-modal information has been fused and applied for different purposes, e.g., brain segmentation [29], diagnosis [11], and 3D dental reconstruction [12,19], which is also extended to CT images. Most methods extend the single-modal method to a multi-stream model, where each stream is intended for a specific modality. The modality-specific features extracted by different streams are fused in subsequent modules. Notably, the input multi-modal images should be registered before feeding into the model. Based on the encoder-decoder architecture, the strategies for multi-modal feature fusion can be classified into four categories. The first one is an early-fusion strategy, where multi-modal images are integrated at the input and processed jointly along a single stream of network [8]. Second, instead of merging both phases at the input of the network, a middle-fusion strategy processes different modalities independently in the corresponding encoders, and these modalities share the same decoder for feature fusion and final segmentation [3]. Third, a late-fusion fashion makes each phase go through an independent stream of an encoder-decoder network, and the learned features are fused at the end of each stream [20]. At last, an ultimate one introduces hyper-connections between and within encoder-decoder networks to enable more effective information exchange between different modalities [29]. However, in these methods, the features from each modality are straightforwardly combined, and consequently, the diverse contribution of different modalities is neglected.

The strategies proposed to handle missing modalities include synthesizing missing modalities by a generative model [16] or learn a modality-invariant feature space [4,6]. However, synthesizing missing modalities requires heavy computations, and existing modality-invariant methods usually failed when most

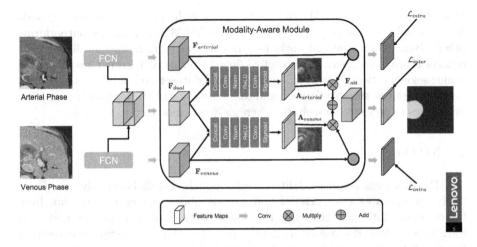

Fig. 1. Illustration of the framework. The input multi-modal CT images are first embedded by different modality-specific FCNs. Then a modality-aware module regresses attention maps, which reflect how to achieve an effective and interpretable fusion of the modality-specific features. The framework is trained by mutual learning strategy composed of intra- and join losses.

of the modalities are missing. Recent KD-Net [7] transfers knowledge from a multi-modal network from a mono-modal one by knowledge distillation. However, KD-Net relies on one student model for each missing modality and an additional teacher model to perform only one-way knowledge transfer to the student model, which brings extra computation cost and limits the multi-modal representation.

In this paper, we present a novel Modality-aware Mutual Learning (**MAML**) method for effective and robust multi-modal liver tumor segmentation. Specifically, we construct a set of modality-specific models to handle multi-modal data, where each model is intended for one modality. To enable more effective and interpretable information exchange across different modalities, we carefully design an Modality-aware (**MA**) module to adaptively aggregate the model-specific features in a learnable way. For each model, MA module produces weight maps to value the features pixel by pixel, and then the features are merged by a weighted aggregation for effective multi-modal segmentation. Moreover, inspired by [27] and [7], we design an novel Mutual Learning (**ML**) strategy. Different from [7], ML enables interactive knowledge transfer to improve the generalization ability of a model and avoid the use of superfluous teacher model. With ML, modality-specific models solve the task collaboratively. We achieve this by training the models through intra-modality and joint losses: the former encourages each model to learn discriminative modality-specific features, while the latter encourages each model to learn from each other to keep the commonality between high-level features for better incorporation of multi-modal information. To sufficiently leverage the deep learning method's power, we collect a large-scale clinical dataset with 654 CT volumes to evaluate the proposed method. Experimental results demonstrate that the proposed MAML significantly outperforms other advanced multi-modal works by a remarkable margin. Specifically, MAML reports a promising performance of 81.25% in terms of Dice per case for liver tumor segmentation. Moreover, on the clinical dataset and public BRATS 2018 dataset, we show the effectiveness and robustness of MAML for handling missing modalities in an extreme scenario where only one modality is available.

2   Method

MAML employs a set of modality-specific models to collaboratively and adaptively incorporates both arterial and venous phase images for accurate liver tumor segmentation. In this case, it consists of two modality-specific models to learn specific features in each modality and a MA module to explore correlated features between two modalities adaptively. Note that the proposed method can be easily extended for more modalities.

2.1   Modality-Specific Model

A modality-specific model is a common FCN for single-modal segmentation. As UNet [17] has been proven successful in medical image segmentation, MAML adopts the powerful nnUNet model [8], one of the state-of-the-art UNet-like framework for medical image segmentation, to achieve the feature extraction from raw CT images. The input of dual-phase CT volumes individually goes through each model, and the high-level semantic embeddings of specific phases from the last layer are obtained. It is worth to note that the high-level semantic embeddings share the same shape of the input image. The outputs of different modality-specific models are denoted as $\mathbf{F}_i \in \mathbb{R}^{C \times D \times H \times W}$, where $C = 32$ is the number of channels, D, H, W are the depth, height, and width, and $i \in \{AP, VP\}$. AP and VP are the abbreviations for arterial and venous phases respectively.

2.2   Modality-Aware Module

As illustrated in Fig. 1, we propose an MA module via an attention mechanism to adaptively measure the contribution of each phase. The attention model is widely used for various tasks, including semantic segmentation [5]. Several attention mechanisms have been proposed to enhance the representation of network [2,18, 21,23,28]. In this study, we explore the cross-modality attention mechanism to selectively highlight the target features embedded in a single modality to obtain more discriminative dual-modal features for liver tumor segmentation.

The outputs of modality-specific models are concatenated together along channels to generate $\mathbf{F}_{dual}$ by a followed convolution layer. Although $\mathbf{F}_{dual}$ encodes both arterial and venous information of liver tumor, it also inevitably introduces redundant noise from each modality for liver tumor segmentation.

Instead of obtaining straightforward segmentation from $\mathbf{F}_{dual}$, we propose MA via attention mechanism to adaptively measure each phase's contribution and visually interpret it.

MA module leverages $\mathbf{F}_{dual}$ and $\mathbf{F}_i$ as inputs and produces $\mathbf{F}_{att}$. Specifically, we first generate an attention map $\mathbf{A}_i$ for each $\mathbf{F}_i$, which indicates the significance of the features in $\mathbf{F}_{dual}$ for each specific phase. Given the $\mathbf{F}_i$ of each phase, we concatenate them with the $\mathbf{F}_{dual}$, and then produce the attention weights $\mathbf{A}_i$:

$$\mathbf{A}_i = \sigma(f_a([\mathbf{F}_{dual}; \mathbf{F}_i]; \theta_i)), i \in \{AP, VP\}, \tag{1}$$

where σ is a Sigmoid function, and θ represents the parameters learned by f_a, which consists of two cascaded convolutional layers. The first convolutional layer uses $3 \times 3 \times 3$ kernels, and the second convolutional layer applies $1 \times 1 \times 1$ kernels. Each convolutional layer is followed by an instance normalization [22], and a leaky rectified linear unit (Leaky ReLU). These convolutional operations are employed to model the correlation of the discriminative dual-modality information with respect to the features of each modality.

Then, we multiply the attention map $\mathbf{A}_i$ with the $\mathbf{F}_i$ in an element-wise manner. $\mathbf{F}_{att}$ is calculated by a weighted sum of each $\mathbf{F}_i$, defined as:

$$\mathbf{F}_{att} = \sum_{i \in \{AP, VP\}} \mathbf{A}_i * \mathbf{F}_i. \tag{2}$$

We apply the MA module for each phase to selectively emphasize their characteristics. During this process, the attention mechanism is used to generate a set of attention maps to indicate how much attention should be paid to the $\mathbf{F}_i$ for more discriminative $\mathbf{F}_{att}$. Furthermore, those attention maps provide a visual interpretation of the contribution of each phase for liver tumor segmentation, which is crucial in clinical practice.

2.3 Mutual Learning Strategy

The learning of the set of modality-specific models is formulated as a voxel-wise binary classification error minimization problem with respect to the ground-truth mask. We carefully design the ML strategy for multi-modal liver tumor segmentation. Concretely, each modality-specific model interacts as a teacher and a student mutually. Thus, the venous model not only draws clues for tumor segmentation from the venous phase but also learns from the arterial model and vice versa. To achieve this, we introduce an intra-phase loss and a joint one. The former encourages each stream to learn discriminative phase-specific features, while the latter encourages each stream to learn from each other to keep the commonality between high-level features for better incorporation of multi-modal information. Let $X = \{X_{venous}, X_{arterial}\}$ be the input venous and arterial volumes respectively, Y be the ground-truth annotations, and $W = \{W_{venous}, W_{arterial}\}$ be the weights in venous and arterial streams respectively.

Table 1. Results on multi-modal liver tumor segmentation. Best results are highlighted with bold.

Methods	Dice [%] ↑	ASSD [voxel] ↓
nnUNet [8]	78.76 ± 18.91	8.02 ± 20.21
OctopusNet [3]	78.89 ± 18.65	12.67 ± 42.43
MS+Ensemble	78.96 ± 19.37	5.88 ± 10.73
MS+MA	80.98 ± 18.58	5.38 ± 9.20
MAML	**81.25 ± 17.02**	**4.71 ± 6.13**

Table 2. Results on handling missing modalities for liver tumor segmentation. Best results are highlighted with bold.

Methods		Dice [%] ↑	ASSD [voxel] ↓
Arterial phase	nnUNet [8]	71.21 ± 25.87	9.51 ± 28.34
	MAML	**79.55 ± 19.06**	**6.38 ± 12.00**
Venous phase	nnUNet [8]	75.10 ± 20.65	9.26 ± 30.82
	MAML	**79.81 ± 18.42**	**6.35 ± 12.03**

The goal of the teacher-student training scheme is to minimize the following objective function

$$\mathcal{L} = \lambda \sum_{i \in \{AP, VP\}} \mathcal{L}_{intra}(Y|X_i; W_i) + \mathcal{L}_{joint}(Y|X; W), \tag{3}$$

where both intra-phase loss $\mathcal{L}_{intra}$ and joint loss $\mathcal{L}_{joint}$ are standard segmentation loss function, and λ is the weight factors that are empirically set as 0.5. We employ a combination of Cross-Entropy loss and Dice loss as the segmentation loss to reduce the effect of imbalanced data distribution of tumors.

The advantages of ML lie in the following three aspects: (1) it enables the model to be capable of dealing with both multi-modal segmentation and handling missing modalities without any modification, which is applicable and efficient in clinical practice; (2) each model for single modality can implicitly leverage dual-modality information by learning from the other models, which leads to better segmentation results even when other modalities are missing; (3) combined with characteristics and commonality of each modality, the collaboration of all model-specific models can make a better multi-modal segmentation.

3   Experiments and Results

Datasets and Evaluation Metrics. Experiments are conducted on contrast-enhanced CT volumes obtained from Chinese PLA General Hospital. We acquire

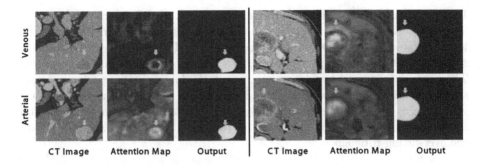

Fig. 2. Attention maps produced by Modality-Aware Module are able to capture enhanced part (left) as well as bleeding part and pseudo capsule (right) of the tumor.

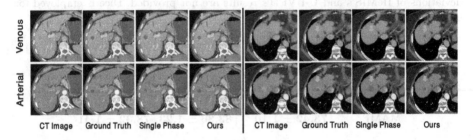

Fig. 3. Qualitative examples where our method detects the tumors while single-phase methods fail on arterial phase (left) or venous phase (right). The tumors are marked in red and highlighted with yellow arrows. (Color figure online)

654 contrast-enhanced CT volume[1] with arterial and venous phases. All CT volumes are obtained using SIEMENS scanners. The in-plane size of CT volumes is 512×512 with spacing ranges from 0.56 mm to 0.91 mm, and the number of slices ranges from 67 to 198 with spacing 1.5 mm. Three experienced clinicians from hepatobiliary surgery with extensive experience interpreting the CT images have been involved for validation. To effectively combine multi-modal CT images, we utilize a registration method [9] to obtain the spatial relation between the images of different phases. For data pre-processing, we truncate the raw intensity values within the range 0.5%–99.5% of the initial HU value and normalize each raw CT case to have zero mean and unit variance. BraTS 2018 dataset [14] contains MR scans from 285 patients with four modalities: T1, T2, T1 contrasted-enhanced (T1ce) and Flair. The goal of the dataset is to segment three sub-regions of brain tumors: whole tumor (WT), tumor core (TC), and enhancing tumor (ET). The metrics employed to quantitatively evaluate segmentation include Dice Similarity Coefficient (Dice) and Average Symmetric Surface Distance (ASSD) (Fig. 3).

---

[1] One volume corresponds to one phase from a patient.

Implementation Details. The framework is built with PyTorch on an Nvidia Tesla V100 GPU. The network is trained by the Adam optimizer with an initial learning rate of 0.0003. Due to the constraint of GPU memory, each volume is sliced into patches with a size of $128 \times 128 \times 128$ before feeding into the network. The network is trained for 600 epochs, about 150 h. No further post-processing strategies are applied as we only intend to evaluate the effectiveness of the network design. For data augmentation, we adopt on-the-fly random mirroring and rotation, deformation, and gamma correction for all training data to alleviate the over-fitting problem.

Table 3. Results on handling missing modalities for brain tumor segmentation. The results of HeMIS, U-HVED, and KD-Net are derived from [7], where the standard deviations of HeMIS's and U-HVED's results are not provided. Dice is employed for evaluation.

Methods	Enhanced tumor	Tumor core	Whole tumor
HeMIS [6]	60.8	58.5	58.5
U-HVED [4]	65.5	66.7	62.4
KD-Net [7]	71.67 ± 1.22	81.45 ± 1.25	76.98 ± 1.54
MAML	**73.42 ± 1.10**	**83.36 ± 1.23**	**78.32 ± 1.41**

Effectiveness of Multi-modal Modeling. To demonstrate the effectiveness of MAML, we make an ablation study for MA and ML respectively on the clinical dataset, where one fifth images is for testing, and the rest are for training. The baseline is a straightforward average of the outputs of modality-specific models, denoted as "MS+Ensemble". Then we apply MA to aggregate the modality-specific models adaptively, denoted as "MS+MA". Finally, we combine both MA and ML, denoted as "MAML". As shown in Table 1, MA outperforms the baseline in terms of both Dice and ASSD. Moreover, ML further boosts the performance with a remarkable margin. The experimental results demonstrate the effectiveness of MAML for multi-modal liver tumor segmentation. Then we compare MAML with recent advanced methods for multi-modal segmentation, nnUNet [8] and OctopusNet [3]. The former takes a concatenation of both phases as input while the latter individually encodes each phase and generate segmentation by one decoder. The results in Table 1 revealing the outstanding performance of MAML.

Interpretable Fusion. MA offers not only an effective fusion of different modalities, but also an interpretable one. We illustrate the interpretability by qualitatively visualizing the learned attention map. From Fig. 2 (left), we can observe that the venous attention map focuses on the edge of the tumor while the arterial attention map focuses on the body. Besides, a certain number of the tumors' surface and the adjacent liver is usually delineated with a pseudo capsule. In Fig. 2 (right), the venous attention map focuses on the pseudo capsule and the bleeding part inside the tumor. It proves that MA can capture the knowledge of medical imaging for an interpretable multi-modal liver tumor segmentation.

Handling Missing Modalities. A superiority of ML strategy is the capability of dealing with missing modalities in multi-modal segmentation. We consider an extreme scenario that only one modality is available. On the clinical dataset, the CT images with either arterial or venous phase are available at inference procedure. We set nnUNet, the counterpart of the modality-specific model in MAML, as a baseline and train it solely on arterial or venous phase. From Table 2, it is observed that MAML significantly outperforms the baseline. Besides, the performance gap between arterial and venous phases of MAML is significantly smaller than that of nnUNet, revealing the excellent ability of ML that transfers knowledge between modalities. We also compare MAML with methods specialized for dealing with missing modalities. Following [7], a 3-fold cross-validation on public BRATS 2018 dataset using only the T1ce modality as input. The results of KD-Net [7], U-HVED [4], and HeMIS [6], in terms of Dice, are directly taken from [7]. From Table 3, we observe that our method excels in the other three advanced methods, demonstrating the effectiveness of MAML for handling missing modalities. The limitation of the proposed framework in the current implementaiton is that it allows either for the full set of modalities or only one modality as input. We would like to enhance it for arbitrary number of missing modalities in the future work.

4   Conclusion

In this study, we propose MAML that enables effective and robust multi-modal segmentation. ML achieves an ensemble of modality-specific models collaboratively learning the complementary information. MA performs in an adaptive and explainable way for better multi-modal liver tumor segmentation. We illustrate that MAML can substantially improve the performance of multi-modal segmentation and effectively handle missing modalities, which is of great value in clinical practice.

References

1. Bray, F., Ferlay, J., Soerjomataram, I., Siegel, R.L., Torre, L.A., Jemal, A.: Global cancer statistics 2018: globocan estimates of incidence and mortality worldwide for 36 cancers in 185 countries. CA Cancer J. Clin. **68**(6), 394–424 (2018)
2. Chen, K., Bui, T., Fang, C., Wang, Z., Nevatia, R.: Amc: attention guided multi-modal correlation learning for image search. In: 2017 IEEE Conference on Computer Vision and Pattern Recognition (CVPR), pp. 6203–6211 (2017)

3. Chen, Y., Chen, J., Wei, D., Li, Y., Zheng, Y.: Octopusnet: a deep learning segmentation network for multi-modal medical images. In: International Workshop on Multiscale Multimodal Medical Imaging, pp. 17–25 (2019)

4. Dorent, R.P.R., Joutard, S.R.D., Modat, M., Ourselin, S., Vercauteren, T.: Heteromodal variational encoder-decoder for joint modality completion and segmentation. In: International Conference on Medical Image Computing and Computer-Assisted Intervention, pp. 74–82 (2019)

5. Fu, J., et al.: Dual attention network for scene segmentation. In: 2019 IEEE/CVF Conference on Computer Vision and Pattern Recognition (CVPR), pp. 3146–3154 (2019)

6. Havaei, M., Guizard, N., Chapados, N., Bengio, Y.: Hemis: hetero-modal image segmentation. In: International Conference on Medical Image Computing and Computer-Assisted Intervention, pp. 469–477 (2016)

7. Hu, M., et al.: Knowledge distillation from multi-modal to mono-modal segmentation networks. In: International Conference on Medical Image Computing and Computer-Assisted Intervention, pp. 772–781 (2020)

8. Isensee, F., Jaeger, P.F., Kohl, S.A.A., Petersen, J., Maier-Hein, K.H.: nnu-net: a self-configuring method for deep learning-based biomedical image segmentation. Nat. Methods 18(2), 203–211 (2021)

9. Klein, S., Staring, M., Murphy, K., Viergever, M., Pluim, J.: elastix: a toolbox for intensity-based medical image registration. IEEE Trans. Med. Imaging 29(1), 196–205 (2010)

10. Li, X., Chen, H., Qi, X., Dou, Q., Fu, C.W., Heng, P.A.: H-denseunet: hybrid densely connected unet for liver and tumor segmentation from ct volumes. IEEE Trans. Med. Imaging 37(12), 2663–2674 (2018)

11. Liang, Y., et al.: Oralcam: enabling self-examination and awareness of oral health using a smartphone camera. In: Proceedings of the 2020 CHI Conference on Human Factors in Computing Systems, pp. 1–13 (2020)

12. Liang, Y., Song, W., Yang, J., Qiu, L., Wang, K., He, L.: X2teeth: 3d teeth reconstruction from a single panoramic radiograph. In: International Conference on Medical Image Computing and Computer-Assisted Intervention, pp. 400–409 (2020)

13. Liu, S., et al.: 3d anisotropic hybrid network: transferring convolutional features from 2d images to 3d anisotropic volumes (2018)

14. Menze, B.H., et al.: The multimodal brain tumor image segmentation benchmark (brats). IEEE Trans. Med. Imaging 34(10), 1993–2024 (2015)

15. Milletari, F., Navab, N., Ahmadi, S.A.: V-net: fully convolutional neural networks for volumetric medical image segmentation. In: 2016 Fourth International Conference on 3D Vision (3DV), pp. 565–571 (2016)

16. Orbes-Arteaga, M., et al.: Simultaneous synthesis of flair and segmentation of white matter hypointensities from t1 mris. arXiv preprint arXiv:1808.06519 (2018)

17. Ronneberger, Olaf, Fischer, Philipp, Brox, Thomas: U-Net: convolutional networks for biomedical image segmentation. In: Navab, Nassir, Hornegger, Joachim, Wells, William M.., Frangi, Alejandro F.. (eds.) MICCAI 2015. LNCS, vol. 9351, pp. 234–241. Springer, Cham (2015). https://doi.org/10.1007/978-3-319-24574-4_28

18. Schlemper, J., et al.: Attention gated networks: learning to leverage salient regions in medical images. Med. Image Anal. 53, 197–207 (2019)

19. Song, W., Liang, Y., Yang, J., Wang, K., He, L.: Oral-3d: reconstructing the 3d structure of oral cavity from panoramic x-ray. In: Proceedings of the AAAI Conference on Artificial Intelligence, vol. 35, pp. 566–573 (2021)

20. Sun, C., et al.: Automatic segmentation of liver tumors from multiphase contrast-enhanced ct images based on fcns. Artif. Intell. Med. 83, 58–66 (2017)

21. Tian, J., Liu, L., Shi, Z., Xu, F.: Automatic couinaud segmentation from ct volumes on liver using glc-unet. In: International Workshop on Machine Learning in Medical Imaging, pp. 274–282 (2019)
22. Ulyanov, D., Vedaldi, A., Lempitsky, V.S.: Instance normalization: The missing ingredient for fast stylization. arXiv preprint arXiv:1607.08022 (2016)
23. Wang, G., et al.: Automatic segmentation of vestibular schwannoma from t2-weighted mri by deep spatial attention with hardness-weighted loss. In: International Conference on Medical Image Computing and Computer-Assisted Intervention, pp. 264–272 (2019)
24. Zhang, J., Xie, Y., Zhang, P., Chen, H., Xia, Y., Shen, C.: Light-weight hybrid convolutional network for liver tumor segmentation. In: IJCAI'19 Proceedings of the 28th International Joint Conference on Artificial Intelligence, pp. 4271–4277 (2019)
25. Zhang, Y., et al.: Sequentialsegnet: combination with sequential feature for multi-organ segmentation. In: 2018 24th International Conference on Pattern Recognition (ICPR), pp. 3947–3952 (2018)
26. Zhang, Y., Tian, J., Zhong, C., Zhang, Y., Shi, Z., He, Z.: Darn: deep attentive refinement network for liver tumor segmentation from 3d ct volume. In: 2020 25th International Conference on Pattern Recognition (ICPR), pp. 7796–7803 (2021)
27. Zhang, Y., Xiang, T., Hospedales, T.M., Lu, H.: Deep mutual learning. In: 2018 IEEE/CVF Conference on Computer Vision and Pattern Recognition, pp. 4320–4328 (2018)
28. Zhang, Z., Fu, H., Dai, H., Shen, J., Pang, Y., Shao, L.: Et-net: a generic edge-attention guidance network for medical image segmentation. In: International Conference on Medical Image Computing and Computer-Assisted Intervention, pp. 442–450 (2019)
29. Zhou, Y., et al.: Hyper-pairing network for multi-phase pancreatic ductal adenocarcinoma segmentation. In: International Conference on Medical Image Computing and Computer-Assisted Intervention, pp. 155–163 (2019)
30. Zhu, W., et al.: Anatomynet: deep learning for fast and fully automated whole-volume segmentation of head and neck anatomy. Med. Phys. **46**(2), 576–589 (2019)

Hybrid Graph Convolutional Neural Networks for Landmark-Based Anatomical Segmentation

Nicolás Gaggion(✉), Lucas Mansilla, Diego H. Milone, and Enzo Ferrante

Research Institute for Signals, Systems and Computational Intelligence,
sinc(i) CONICET, Universidad Nacional del Litoral, Santa Fe, Argentina
ngaggion@sinc.unl.edu.ar

Abstract. In this work we address the problem of landmark-based segmentation for anatomical structures. We propose HybridGNet, an encoder-decoder neural architecture which combines standard convolutions for image feature encoding, with graph convolutional neural networks to decode plausible representations of anatomical structures. We benchmark the proposed architecture considering other standard landmark and pixel-based models for anatomical segmentation in chest x-ray images, and found that HybridGNet is more robust to image occlusions. We also show that it can be used to construct landmark-based segmentations from pixel level annotations. Our experimental results suggest that Hybrid-Net produces accurate and anatomically plausible landmark-based segmentations, by naturally incorporating shape constraints within the decoding process via spectral convolutions.

Keywords: Landmark-based segmentation · Graph convolutional neural networks · Spectral convolutions

1 Introduction

Deep learning models based on convolutional neural networks have become the state-of-the-art for anatomical segmentation of biomedical images. The current practise is to employ standard convolutional neural networks (CNNs) trained to minimize a pixel level loss function, where dense segmentation masks are used as ground truth. Casting image segmentation as a pixel labeling problem is desirable in scenarios like lesion segmentation, where topology and location do not tend to be preserved across individuals. However, organs and anatomical structures usually present a characteristic topology which tends to be regular. Differently from dense segmentation masks, statistical shape models [15] and graph-based representations [4] provide a natural way to incorporate topological constraints

Electronic supplementary material The online version of this chapter (https://doi.org/10.1007/978-3-030-87193-2_57) contains supplementary material, which is available to authorized users.

© Springer Nature Switzerland AG 2021
M. de Bruijne et al. (Eds.): MICCAI 2021, LNCS 12901, pp. 600–610, 2021.
https://doi.org/10.1007/978-3-030-87193-2_57

by construction. Moreover, such shape representations make it easier to establish landmark correspondences among individuals, particularly important in the context of statistical shape analysis.

Since the early 1990s, variations of point distribution models (PDMs) have been proposed [8] to segment anatomical structures using landmarks. PDMs are flexible shape templates describing how the relative locations of important points can vary. Techniques based on PDMs, like active shape models (ASM) [8,31] and active appearance models (AAM) [7] became the defacto standard to deal with anatomical segmentation at the end of the century. Subsequently, during the next decade, the development of more powerful and robust image registration algorithms [32] positioned deformable template matching algorithms as the choice of option for anatomical segmentation and atlas construction [11,16,25]. More recently, with the advent of deep fully convolutional networks [27,28], great efforts were made to incorporate anatomical constraints into such models [17,20,24]. The richness of the image features learned by standard CNNs allowed them to achieve highly accurate results. However, most of these methods work directly on the pixel space, producing acceptable dense segmentations masks but without landmark annotations and connectivity structure. On the contrary, structured models like graphs appear as a natural way to represent landmarks, contours and surfaces. By defining the landmark position as a function on the graph nodes, and encoding the anatomical structure through its adjacency matrix, we can easily constrain the space of solutions and ensure topological correctness.

During the last years, the emerging field of geometric deep learning [5] extended the success of convolutional neural networks to non-Euclidean domains like graphs and meshes. While classical CNNs have been particularly successful when dealing with signals such as images or speech, more recent developments like spectral convolutions [6,9] and neural message passing [12] enabled the use of deep learning on graphs. Recently, graph generative models were proposed [19,26]. Of particular interest for this work is the convolutional mesh autoencoder proposed in [26]. The authors construct an encoder-decoder network using spectral graph convolutions, and train it in a variational setting using face meshes. By sampling the latent space, they are able to generate new expressive faces never seen during training. Inspired by this idea, we propose to exploit the generative power of convolutional graph autoencoders [10,26] to decode plausible anatomical segmentations from images.

Contributions. In this work, we revisit landmark-based segmentation in light of the latest developments on deep learning for Euclidean and non-Euclidean data. We aim at leveraging the best of both worlds, combining standard convolutions for image feature encoding, with generative models based on graph convolutional neural networks (GCNN) to decode plausible representations of anatomical structures. Under the hyphothesis that encoding connectivity information through the graph adjacency matrix will result in richer representations than standard landmark-based PDMs, we also compare our results with other statistical point distribution models which do not make explicit use of the graph con-

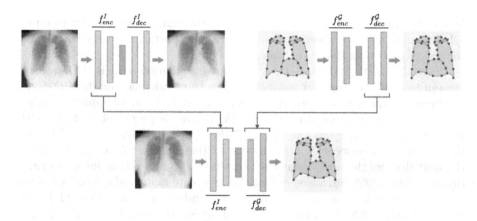

Fig. 1. The proposed HybridGNet (bottom) is an encoder-decoder architecture which combines standard convolutions for image feature encoding (blue), with graph spectral convolutions (green) to decode plausible anatomical graph-based representations. (Color figure online)

nectivity. Our contributions are 4-fold: (1) we propose HybridGNet, an encoder-decoder architecture which combines standard convolutions with GCNNs to extract graph representations directly from images; (2) we showcase the proposed architecture in the context of anatomical landmark-based segmentation of chest x-ray images and benchmark the results considering other landmark and pixel-based segmentation models; (3) we propose to use HybridGNet to create graph-based representations from dense anatomical masks without paired images and (4) we show that HybridGNet is more robust to image oclussions than state-of-the-art pixel-level segmentation methods.

2   Hybrid Graph Convolutional Neural Networks

Problem Setting. Let us have a dataset $\mathcal{D} = \{(I, \mathcal{G})_k\}_{0 < k < N}$, composed of N dimensional images I and their corresponding landmark-based segmentation represented as a graph $\mathcal{G} = < V, \mathbf{A}, \mathbf{X} >$. V is the set of nodes (landmarks), $\mathbf{A} \in \{0,1\}^{|V| \times |V|}$ is the adjacency matrix indicating the connectivity between pairs of nodes ($\mathbf{A}_{ij} = 1$ indicates an edge connecting vertices i and j, and $\mathbf{A}_{ij} = 0$ otherwise) and $\mathbf{X} \in \mathbb{R}^{|V| \times d}$ is a function (represented as a matrix) assigning a feature vector to every node. In our case, it assigns a d-dimensional spatial coordinate to every landmark. Without loss of generality, in this work we will showcase the proposed framework in 2D images (thus $\mathbf{X} \in \mathbb{R}^{|V| \times 2}$). However, note that extending our method to 3D images is straightforward, the only difference being that images I will be volumes and graphs $\mathcal{G}$, meshes.

In the context of landmark-based segmentation and point distribution models, it is common to have manual annotations with fixed number of points. Therefore, for all graphs, we assume that the set of nodes (landmarks) V and the connectivity matrices $\mathbf{A}$ are the same. The only difference among them is given by the spatial coordinates defined in $\mathbf{X}$. This assumption enables the use of spectral graph convolutions [9,26] to learn latent representations of anatomy.

Spectral Graph Convolutions. We have previously mentioned that extending discretized convolutions to graph structures is not straightforward. A myriad of graph convolution formulations have been recently proposed [5,12]. Here we adopt the localized spectral version since it has shown to be effective at learning powerful shape representations from graph structures with a fixed number of nodes [26]. Spectral convolutions are build using the eigendecomposition of the graph Laplacian matrix $\mathbf{L}$, exploiting the property that convolutions in the node domain are equivalent to multiplications in the graph spectral domain [30].

The graph Laplacian is defined as $\mathbf{L} = \mathbf{D} - \mathbf{A}$, where $\mathbf{D}$ is the diagonal degree matrix with $\mathbf{D}_{ii} = \sum_j \mathbf{A}_{ij}$, and $\mathbf{A}$ is the adjacency matrix. The Laplacian $\mathbf{L}$ can be decomposed as $\mathbf{L} = \mathbf{U}\mathbf{\Lambda}\mathbf{U}^T$, where $\mathbf{U} \in \mathbb{R}^{|V| x |V|} = [u_0, u_1, ..., u_{|V|-1}]$ is the matrix of eigenvectors (Fourier basis) and $\mathbf{\Lambda} = \mathrm{diag}(\lambda_0, \lambda_1 ... \lambda_{|V|-1})$ the diagonal matrix of eigenvalues (frequencies of the graph). By analogy with the classical Fourier transform for continuous or discrete signals, the graph Fourier transform of a function $\mathbf{X}$ defined on the graph domain is $\hat{\mathbf{X}} = \mathbf{U}^T\mathbf{X}$, while its inverse is given by $\mathbf{X} = \mathbf{U}\hat{\mathbf{X}}$. Based on this formulation, the spectral convolution between a signal $\mathbf{X}$ and a filter $\mathbf{g}_\phi = \mathrm{diag}(\phi)$ is defined as $\mathbf{g}_\phi * \mathbf{X} = \mathbf{g}_\phi(\mathbf{L})\mathbf{X} = \mathbf{g}_\phi(\mathbf{U}\mathbf{\Lambda}\mathbf{U}^T)\mathbf{X} = \mathbf{U}\mathbf{g}_\phi(\mathbf{\Lambda})\mathbf{U}^T\mathbf{X}$, where $\phi \in \mathbb{R}^n$ is a vector of coefficients parameterizing the filter. We follow the work of Defferrard et al. [9] and restrict the class of filters to polynomial filters $\mathbf{g}_\phi = \sum_{k=0}^K \phi_k \mathbf{\Lambda}^k$. Polynomial filters are strictly localized in the vertex domain (a K-order polynomial filter considers K-hop neighborhoods around the node) and reduce the computational complexity of the convolutional operator. Such filters can be well approximated by a truncated expansion in terms of Chebyshev polynomials computed recursively. Following [9,26] we adopt this approximation to implement the spectral convolutions. Note that a spectral convolutional layer will take feature matrices $\mathbf{X}^l$ as input, and produce filtered versions $\mathbf{X}^{l+1}$ akin to what standard convolutions do with images and feature maps.

Auto-Encoding Shape and Appearance. Autoencoders are neural networks designed to reconstruct their input. They follow an encoder-decoder scheme, where an encoder $z = f_{enc}(x)$ maps the input x to a lower dimensional latent code z, which is then processed by a decoder $f_{dec}(z)$ to reconstruct the original input. The bottleneck imposed by the low-dimensionality of the encoding z forces the model to retain useful information, learning powerful representations of the data distribution. The model is trained to minimize a reconstruction loss $\mathcal{L}_{rec}(x, f_{dec}(f_{enc}(x)))$ between the input and the output reconstruction. To constrain the distribution of the latent space z, we add a variational loss term to the objective function, resulting in a variational autoencoder (VAE) [18]. We assume that the latent codes z are sampled from a distribution $Q(z)$ for which we will impose a unit multivariate Gaussian prior. In practise, during training, this results in the latent codes z being sampled from a distribution $\mathcal{N}(\mu, \sigma)$ via the reparametrization trick [18], where μ, σ are deterministic parameters generated by the encoder $f_{enc}(x)$. Given a sample z, we can generate (reconstruct) the corresponding data point by using the decoder $f_{dec}(x)$. This model is trained by minimizing a loss function defined as:

$$\mathcal{L}_{vae} = \mathcal{L}_{rec}(x, f_{dec}(z)) + w\text{KL}(\mathcal{N}(0, 1)||Q(z|x)), \tag{1}$$

where the first term is the reconstruction loss, and the second term imposes a unit Gaussian prior $\mathcal{N}(0,1)$ via the KL divergence loss. Depending on the type of data x that will be processed, here we will implement f_{enc} and f_{dec} using standard convolutional layers (to encode image appearence) or spectral convolutions (to encode graph structures).

Hybrid Graph Convolutional Neural Networks (HybridGNet). To construct the HybridGNet, we pre-train two independent VAEs. The first one, defined by the corresponding encoder f_{enc}^I and decoder f_{dec}^I, is trained to reconstruct images I. The second one, defined by f_{enc}^G and f_{dec}^G, is trained using graphs. In both cases, the bottleneck latent representation is modelled by fully connected neurons and have the same size. The encoder/decoder blocks are implemented using standard convolutional layers in the first case, and spectral convolutions in the later case. As depicted in Fig. 1, once both models are trained, we decouple the encoders and decoders, keeping only the image encoder f_{enc}^I and graph decoder f_{dec}^G. The HybridGNet is thus constructed by connecting these two networks as $f_{\text{HybridGNet}}(x) = f_{dec}^G(f_{enc}^I(x))$. The coupled model is initialized with the pre-trained weights, and re-trained until convergence using the landmarks MSE as reconstruction loss, and the KL divergence term for regularization. Thus, HybridGNet takes images I as input and produce graphs G as output, combining standard with spectral convolutions in a single model.

Dual Models. Based on the HybridGNet model, we implemented two more architectures incorporating a second decoder which outputs dense segmentation masks. The first one is the so called HybridGNet *Dual* model, which has two decoders: the graph f_{dec}^G and an extra f_{dec}^I whose output is a dense probabilistic segmentation. The model is trained to minimize a reconstruction loss $\mathcal{L}_{dual} = \mathcal{L}_{rec} + \mathcal{L}_{seg}$, combining the landmark localization error ($\mathcal{L}_{rec}$) with the dense segmentation error ($\mathcal{L}_{seg}$). The second model, entitled HybridGNet *Dual SC* incorporates skip connections (implemented as sums) between corresponding blocks from the image encoder f_{enc}^I and decoder f_{dec}^I. Even though in this paper we do not focus on dense segmentation models, these models were implemented to evaluate the effect of incorporating dense masks into the learning process. Since the dataset only contains contours, we derive the corresponding dense masks for training by filling the organ contours, assigning different labels to every organ.

Baseline Models. Our work builds on the hypothesis that encoding connectivity information through graph structures will result in richer representations than standard landmark-based point distribution models. To assess the validity of this hypothesis, we construct standard point distribution models (like those used in ASM) from the graph representations. Differently from the graph structures which incorporate connectivity information, here the landmarks are treated as independent points. For a given graph $\mathcal{G} =< V, \mathbf{A}, \mathbf{X} >$, we construct a vectorized representation S by concatenating the rows of $\mathbf{X}$ in a single vector as $S = (\mathbf{X}_{0,0}, \mathbf{X}_{0,1}, \mathbf{X}_{1,0}, \mathbf{X}_{1,1}, ... \mathbf{X}_{|V|-1,0}, \mathbf{X}_{|V|-1,1})$.

We implement two baseline models using the vectorized representation. In the first one, similar to [3,23], we simply use principal component analysis (PCA) to turn the vectorized representations S into lower-dimensional embeddings. We then optimize the pre-trained image encoder f_{enc} to produce the coefficients of the PCA modes, reconstructing the final landmark-based segmentation as a linear combination of the modes. In the second model, we train a fully connected VAE, f^S, to reconstruct the vectorized representations S. Following the same methodology used to construct the HybridGNet, we keep only the decoder f^S_{dec} and connect it with the pre-trained image encoder f^I_{enc}. These baselines are also trained end-to-end computing the loss on the final location of the reconstructed landmarks, and using the pre-trained weights for initialization.

We also include a third baseline model for comparison, which uses a multi-atlas segmentation (MAS) approach [1,2]. Given a test image for which we want to predict the landmarks, we take the 5 most similar images (in terms of mutual information) from the training set. We then perform pairwise non-rigid registration (with affine initialization) using Simple Elastix [21]. The final landmark-based segmentation is obtained by simply averaging the position of the corresponding transferred landmarks to the target space.

Detailed Architectures and Training Details. The VAE image encoder f^I_{enc} consists of 5 residual blocks [14], alternated with 4 max pooling operations. The decoder f^I_{dec} is a mirrored version of f^I_{enc}. The graph autoencoder consists on 4 layers of Chebyshev convolutions for the encoder $f^{\mathcal{G}}_{enc}$ and 5 layers for the decoder $f^{\mathcal{G}}_{dec}$, with 16 feature maps and a polynomial order of 6. In both models the latent space was fixed to 64 features and implemented as a fully connected layer, see Tables 1–4 in Sup. Mat. for more details. The autoencoders were trained using MSE as $\mathcal{L}_{rec}$ (for both the image and landmark reconstruction). The HybridGNet was initalized with the pre-trained weights and refined for 2000 epochs using the MSE to compute the landmark reconstruction error. The dual models also incorporated the $\mathcal{L}_{seg}$ term based on the average between soft Dice [22] and Cross Entropy. All models used the KL divergence for regularization (with $w = 1e-5$, choosen by grid search) and were trained using Adam optimizer. All the models were implemented in PyTorch 1.7.1 and PyTorch Geometric 1.6.3. Our code is available at https://github.com/ngaggion/HybridGNet.

3   Experiments and Discussion

We evaluated the proposed approaches using the Japanese Society of Radiological Technology (JSRT) Database [29], which consists on 247 high resolution X-Ray images, with expert annotations for lung, hearth and clavicles [13]. Annotations consist on 166 landmarks for every image. We constructed the graph adjacency matrix following a connectivity pattern as shown in Fig. 2. This matrix

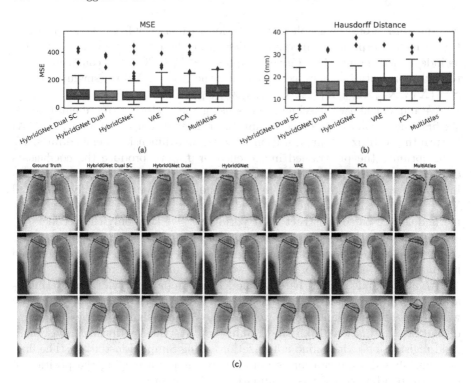

Fig. 2. Quantitative and qualitative results. The difference between the means of the HybridGNet variants and the baselines for MSE (a) and HD (b) are statistically significant according to a Wilcoxon paired test (see the Sup. Mat. for details). We also include qualitative results (c) reflecting the improvement in anatomically plausibility obtained when using the HybridGNets (particularly in the clavicles which are the most challenging).

was shared across all images. The dataset was divided in 3 folds, using 70% of the images for training, 10% for validation and 20% for testing. We showcase the potential of the proposed HybridGNet architecture in three different experiments.

Experiment 1: Anatomical Landmark-Based Segmentation. We evaluated the proposed models using landmark specific metrics, including the mean square error (MSE, in pixel space) over the vectorized landmark location and the contour Hausdorff distance (HD, in millimeters). Figure 2 shows that the Hybrid models outperform the baselines in terms of both MSE and HD. However, incorporating the dual path with dense mask reconstruction does not seem to make a difference.

Experiment 2: Generating Landmark-Based Representations from Dense Segmentations. The proposed HybridGNet offers a natural way to recover landmark based representations from dense segmentation masks. This

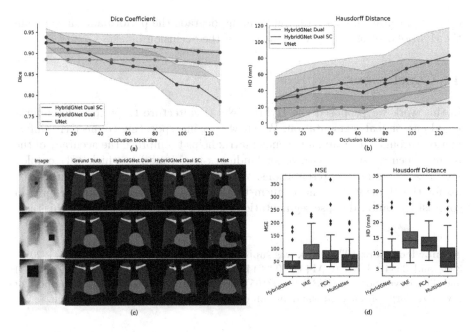

Fig. 3. Figures (a), (b) and (c) show quantitative and qualitative results for the Experiment 3 (oclussion study). We can see that the HybridGNet Dual models are much more robust to missing parts than a standard UNet model. Figure (d) includes quantitative results for Experiment 2, showing that HybridGNet is more accurate at recovering landmark contours from dense segmentation masks.

could be useful in scenarios where dense segmentation masks are available, but we require matched landmarks for statistical shape analysis. We trained our HybridGNet using dense segmentation masks as input and compared the resulting model with the other approaches trained in the same conditions. Figure 3d shows that proposed HybridGNet outperformed the three baselines, confirming that it can be used to construct statistical shape models with landmark correspondences from pixel level annotations (see Sup. Mat. for visual results).

Experiment 3: Robustness Study. We assessed the robustness of the proposed model to image occlusions which may be due to anonymization reasons. We created artificial occlusions by overlapping random black boxes of different size on the test images. In order to highlight the advantages of the HybridGNet over standard dense segmentation models, we trained a UNet architecture (same as the HybridGNet Dual SC but without the graph decoder). Figure 3 shows the results when comparing the dense UNet with our HybridGNet variants. Since we cannot compute the landmark MSE for the dense segmentations, we compared the models evaluating the Dice coefficient and Hausdorff distance on the dense masks obtained from the UNet and the convolutional decoder of the dual models. We can see that not only the UNet, but also the dual models which incorporate dense segmentation masks during training, degrade the performance faster than the HybridGNet as we increase the size of the occlusion block.

4   Conclusions

In this work we proposed the HybridGNet architecture to perform landmark-based anatomical segmentation. We show that incorporating connectivity information through the graph adjacency matrix helps to improve the accuracy of the results when compared with other landmark-based models which only employ vectorized landmark representations. We showcased different application scenarios for the HybridGNet and confirm that it is robust to image occlusions, in contrast to standard dense segmentation methods which tend to fail in this task.

Acknowledgment. This work was supported by grants from ANPCyT (PICT 2018-3907 and 3384), UNL (CAI+D 50220140100-084LI, 50620190100-145LI and 115LI) and The Royal Society (IES/R2/202165). We gratefully acknowledge the support of NVIDIA Corporation with the donation of the Titan Xp used for this research.

References

1. Alvén, J., Kahl, F., Landgren, M., Larsson, V., Ulén, J.: Shape-aware multi-atlas segmentation. In: 2016 23rd International Conference on Pattern Recognition (Icpr), pp. 1101–1106. IEEE (2016)
2. Alvén, J., Kahl, F., Landgren, M., Larsson, V., Ulén, J., Enqvist, O.: Shape-aware label fusion for multi-atlas frameworks. Pattern Recogn. Lett. **124**, 109–117 (2019)
3. Bhalodia, R., Elhabian, S.Y., Kavan, L., Whitaker, R.T.: DeepSSM: a deep learning framework for statistical shape modeling from raw images. In: Reuter, M., Wachinger, C., Lombaert, H., Paniagua, B., Lüthi, M., Egger, B. (eds.) ShapeMI 2018. LNCS, vol. 11167, pp. 244–257. Springer, Cham (2018). https://doi.org/10.1007/978-3-030-04747-4_23
4. Boussaid, H., Kokkinos, I., Paragios, N.: Discriminative learning of deformable contour models. In: 2014 IEEE 11th International Symposium on Biomedical Imaging (ISBI), pp. 624–628. IEEE (2014)
5. Bronstein, M.M., Bruna, J., LeCun, Y., Szlam, A., Vandergheynst, P.: Geometric deep learning: going beyond euclidean data. IEEE Signal Process. Mag. **34**(4), 18–42 (2017)
6. Bruna, J., Zaremba, W., Szlam, A., LeCun, Y.: Spectral networks and locally connected networks on graphs. arXiv preprint arXiv:1312.6203 (2013)
7. Cootes, T.F., Edwards, G.J., Taylor, C.J.: Active appearance models. In: Burkhardt, H., Neumann, B. (eds.) ECCV 1998. LNCS, vol. 1407, pp. 484–498. Springer, Heidelberg (1998). https://doi.org/10.1007/BFb0054760
8. Cootes, T.F., Taylor, C.J., Cooper, D.H., Graham, J.: Training models of shape from sets of examples. In: BMVC92, pp. 9–18. Springer, Heidelberg (1992). https://doi.org/10.1007/978-1-4471-3201-1_2
9. Defferrard, M., Bresson, X., Vandergheynst, P.: Convolutional neural networks on graphs with fast localized spectral filtering. arXiv preprint arXiv:1606.09375 (2016)

10. Foti, S., Foti, S., et al.: Intraoperative liver surface completion with graph convolutional VAE. In: Sudre, C.H., et al. (eds.) UNSURE/GRAIL -2020. LNCS, vol. 12443, pp. 198–207. Springer, Cham (2020). https://doi.org/10.1007/978-3-030-60365-6_19

11. Frangi, A.F., Niessen, W.J., Rueckert, D., Schnabel, J.A.: Automatic 3D ASM construction via atlas-based landmarking and volumetric elastic registration. In: Insana, M.F., Leahy, R.M. (eds.) IPMI 2001. LNCS, vol. 2082, pp. 78–91. Springer, Heidelberg (2001). https://doi.org/10.1007/3-540-45729-1_7

12. Gilmer, J., Schoenholz, S.S., Riley, P.F., Vinyals, O., Dahl, G.E.: Neural message passing for quantum chemistry. In: International Conference on Machine Learning, pp. 1263–1272. PMLR (2017)

13. van Ginneken, B., Stegmann, M., Loog, M.: Segmentation of anatomical structures in chest radiographs using supervised methods: a comparative study on a public database. Med. Image Anal. **10**(1), 19–40 (2006)

14. He, K., Zhang, X., Ren, S., Sun, J.: Deep residual learning for image recognition (2015)

15. Heimann, T., Meinzer, H.P.: Statistical shape models for 3d medical image segmentation: a review. Med. Image Anal. **13**(4), 543–563 (2009)

16. Heitz, G., Rohlfing, T., Maurer Jr, C.R.: Automatic generation of shape models using nonrigid registration with a single segmented template mesh. In: VMV, pp. 73–80 (2004)

17. Jurdia, R.E., Petitjean, C., Honeine, P., Cheplygina, V., Abdallah, F.: High-level prior-based loss functions for medical image segmentation: A survey. arXiv preprint arXiv:2011.08018 (2020)

18. Kingma, D.P., Welling, M.: Auto-encoding variational bayes. arXiv preprint arXiv:1312.6114 (2013)

19. Kipf, T.N., Welling, M.: Variational graph auto-encoders. arXiv preprint arXiv:1611.07308 (2016)

20. Larrazabal, A.J., Martinez, C., Ferrante, E.: Anatomical priors for image segmentation via post-processing with denoising autoencoders. In: Shen, D., et al. (eds.) MICCAI 2019. LNCS, vol. 11769, pp. 585–593. Springer, Cham (2019). https://doi.org/10.1007/978-3-030-32226-7_65

21. Marstal, K., Berendsen, F., Staring, M., Klein, S.: Simpleelastix: a user-friendly, multi-lingual library for medical image registration. In: Proceedings of the IEEE Conference on Computer Vision and Pattern Recognition Workshops, pp. 134–142 (2016)

22. Milletari, F., Navab, N., Ahmadi, S.A.: V-net: fully convolutional neural networks for volumetric medical image segmentation. In: 2016 Fourth International Conference on 3D Vision (3DV), pp. 565–571. IEEE (2016)

23. Milletari, F., Rothberg, A., Jia, J., Sofka, M.: Integrating statistical prior knowledge into convolutional neural networks. In: Descoteaux, M., Maier-Hein, L., Franz, A., Jannin, P., Collins, D.L., Duchesne, S. (eds.) MICCAI 2017. LNCS, vol. 10433, pp. 161–168. Springer, Cham (2017). https://doi.org/10.1007/978-3-319-66182-7_19

24. Oktay, O., et al.: Anatomically constrained neural networks (acnns): application to cardiac image enhancement and segmentation. IEEE Trans. Med. Imaging **37**(2), 384–395 (2017)

25. Paulsen, R., Larsen, R., Nielsen, C., Laugesen, S., Ersbøll, B.: Building and testing a statistical shape model of the human ear canal. In: Dohi, T., Kikinis, R. (eds.) MICCAI 2002. LNCS, vol. 2489, pp. 373–380. Springer, Heidelberg (2002). https://doi.org/10.1007/3-540-45787-9_47

26. Ranjan, A., Bolkart, T., Sanyal, S., Black, M.J.: Generating 3d faces using convolutional mesh autoencoders. In: Proceedings of the European Conference on Computer Vision (ECCV), pp. 704–720 (2018)
27. Ronneberger, O., Fischer, P., Brox, T.: U-Net: convolutional networks for biomedical image segmentation. In: Navab, N., Hornegger, J., Wells, W.M., Frangi, A.F. (eds.) MICCAI 2015. LNCS, vol. 9351, pp. 234–241. Springer, Cham (2015). https://doi.org/10.1007/978-3-319-24574-4_28
28. Shakeri, M., et al.: Sub-cortical brain structure segmentation using f-cnn's. In: 2016 IEEE 13th International Symposium on Biomedical Imaging (ISBI), pp. 269–272. IEEE (2016)
29. 1 Shiraishi, J., et al.: Development of a digital image database for chest radiographs with and without a lung nodule: receiver operating characteristic analysis of radiologists' detection of pulmonary nodules. Am. J. Roentgenol. **174**(1), 71–74 (2000)
30. Shuman, D.I., Narang, S.K., Frossard, P., Ortega, A., Vandergheynst, P.: The emerging field of signal processing on graphs: extending high-dimensional data analysis to networks and other irregular domains. IEEE Signal Process. Mag. **30**(3), 83–98 (2013)
31. Sozou, P.D., Cootes, T.F., Taylor, C.J., Di Mauro, E., Lanitis, A.: Non-linear point distribution modelling using a multi-layer perceptron. Image Vision Comput. **15**(6), 457–463 (1997)
32. Zitova, B., Flusser, J.: Image registration methods: a survey. Image Vision Comput. **21**(11), 977–1000 (2003)

RibSeg Dataset and Strong Point Cloud Baselines for Rib Segmentation from CT Scans

Jiancheng Yang[1,2], Shixuan Gu[1], Donglai Wei[3], Hanspeter Pfister[3], and Bingbing Ni[1(✉)]

[1] Shanghai Jiao Tong University, Shanghai, China
nibingbing@sjtu.edu.cn
[2] Dianei Technology, Shanghai, China
[3] Harvard University, Cambridge, MA, USA

Abstract. Manual rib inspections in computed tomography (CT) scans are clinically critical but labor-intensive, as 24 ribs are typically elongated and oblique in 3D volumes. Automatic rib segmentation methods can speed up the process through rib measurement and visualization. However, prior arts mostly use in-house labeled datasets that are publicly unavailable and work on dense 3D volumes that are computationally inefficient. To address these issues, we develop a labeled rib segmentation benchmark, named *RibSeg*, including 490 CT scans (11,719 individual ribs) from a public dataset. For ground truth generation, we used existing morphology-based algorithms and manually refined its results. Then, considering the sparsity of ribs in 3D volumes, we thresholded and sampled sparse voxels from the input and designed a point cloud-based baseline method for rib segmentation. The proposed method achieves state-of-the-art segmentation performance (Dice $\approx$ 95%) with significant efficiency (10–40$\times$ faster than prior arts). The RibSeg dataset, code, and model in PyTorch are available at https://github.com/M3DV/RibSeg.

Keywords: Rib segmentation · Rib centerline · Medical image dataset · Point clouds · Computed tomography

1 Introduction

Detection and diagnosis of rib-related diseases, *e.g.*, rib fracture and bone lesions, is essential and common in clinical practice and forensics. For instance, rib fracture detection can identify chest trauma severity that accounts for 10%–15% of all traumatic injuries [17], and bone metastases are common in solid tumours [3]. Chest computed tomography (CT) is a primary choice for examining chest trauma thanks to its advantage in revealing occult fractures and fracture-related complications [6]. However, understanding ribs in CT imaging

---

J. Yang and S. Gu—Contributed equally.

M. de Bruijne et al. (Eds.): MICCAI 2021, LNCS 12901, pp. 611–621, 2021.
https://doi.org/10.1007/978-3-030-87193-2_58

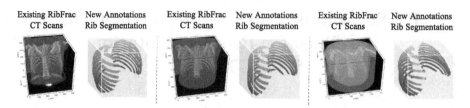

Fig. 1. Illustration of RibSeg Dataset: 490 CT scans from the existing RibFrac dataset [7] and new annotations for rib segmentation. Colors in rib segmentation figures denotes the axial depth.

is challenging: 24 ribs in human bodies are typically elongated and oblique in 3D volumes, with complex geometries across numerous CT sections; in other words, a large number of CT slices must be evaluated sequentially, rib-by-rib and side-by-side, which is tedious and labor-intensive for clinicians. Besides, fractures/lesions could be inconspicuous when reading 2D slices. These challenges urge the development and application of rib visualization tools, *e.g.*, rib unfolding [15], whose core technique is (automatic) rib segmentation and centerline extraction.

A few prior arts have addressed rib segmentation [8,16,18] before the era of deep learning. Rib tracing is a popular method, while it is highly sensitive to initially detected seed points and vulnerable to local ambiguities. Supervised deep learning-based segmentation [9] from CT volumes is robust as it learns hierarchical visual features from raw voxels; However, this study does not consider the sparsity and elongated geometry of ribs in 3D volumes. There are also studies focusing on rib centerline extraction instead of full rib segmentation, *e.g.*, rib tracing [14] and deformable template matching [21]. This study focuses on rib segmentation, where rib centerlines could be extracted with geometric post-processing algorithms.

Although researchers have made progress in rib segmentation and centerline extraction, there is no public dataset in this topic, making it difficult to benchmark existing methods and develop downstream applications (*e.g.*, rib fracture detection). Besides, existing methods work on the dense 3D volumes instead of the sparse rib voxels, which are computationally inefficient: around $5s$ to $20s$ to segment ribs [9] or $40s$ to extract rib centerlines [21]. To address these issues, we first develop a large-scale CT dataset, named *RibSeg* for rib segmentation (Fig. 1). The CT scans come from the public RibFrac dataset [7] consisting of 660 chest-abdomen CT scans for rib fracture segmentation, detection, and classification. As ribs are relatively recognizable compared to other anatomical structures, we use hand-crafted 3D image processing algorithms to generate segmentation and then manually check and refine the labels. This procedure, with little annotation effort, generates visually satisfactory rib segmentation labels for 490 CT scans. Moreover, considering the sparsity of ribs in 3D volumes ($<0.5\%$ voxels) and high HU values (>200) of bone structures in CT scans, we propose an efficient point cloud-based model to segment ribs, which is an order of magnitude faster than previous methods. The proposed method converts dense CT volume into sparse point clouds via thresholding and random downsampling, and produces high-quality and robust rib segmentation (Dice ≈95%). We can further extract rib centerlines from the predicted rib segmentation with post-processing.

The RibSeg dataset could be used for the development of downstream rib-related applications. Besides, considering the differences from standard medical image datasets [1,23] with pixel/voxel grids, the elongated shapes and oblique poses of ribs enable the RibSeg dataset to serve as a benchmark for curvilinear structures and geometric deep learning (*e.g.*, point clouds).

Contributions. 1) The first public benchmark for rib segmentation, which enables downstream applications and method comparison. 2) A novel point-based perspective on modeling 3D medical images beyond voxel grids. 3) A point cloud-based rib segmentation baseline with high efficiency and accuracy.

2   Materials and Methods

2.1   RibSeg Dataset

Dataset Overview. The RibSeg dataset uses the public computed tomography (CT) scans from RibFrac dataset [7], an open dataset with 660 chest-abdomen CT scans for rib fracture segmentation, detection, and classification. The CT scans are saved in NIFTI (.nii) format with volume sizes of $N \times 512 \times 512$, where 512×512 is the size of CT slices, and N is the number of CT slices (typically 300–400). Most cases are confirmed with complete rib cages and manually annotated with at least one rib fracture by radiologists.

As ribs are relatively recognizable compared to other anatomical structures, we use a semi-automatic approach (see details in the following section) to generate rib segmentation, with hand-crafted morphology-based image processing algorithms, as well as manual checking and refinement. Though computationally intensive, this approach produces visually satisfactory labels with few annotation efforts. Finally, there are 490 qualified CT cases in the RibSeg dataset with 11,648 individual ribs in total, where each case has segmentation labels of 24 (or 22 in some cases) ribs. We also provide the rib centerline ground truth extracted from the rib segmentation labels. Note that the rib segmentation and centerline ground truth are imperfect, as the annotations are generated with algorithms. Besides, only voxels with higher HU than a threshold (200) are included, making the rib segmentation annotations in the RibSeg dataset hollow. However, we manually check the ground truth labels for rib segmentation and centerlines to ensure that the included 490 datasets are high-quality enough to develop downstream applications. The data split of the RibSeg dataset is summarized in Table 1: training set (320 cases, to train the deep learning system), development set (a.k.a validation set, 50 cases, to tune hyperparameters of the deep learning system), and test set (120 cases, to evaluate the model). The RibSeg training, development, and test set are from those of the RibFrac dataset respectively, enabling the development of downstream applications (*e.g.*, rib fracture detection) in the MICCAI 2020 RibFrac challenge[1].

Table 1. Overview of RibSeg dataest.

Subset	No. of CT scans	No. of individual ribs
Training/Development/Test	320/50/120	7,670/1,187/2,862

Semi-Automatic Annotation and Quality Control. We describe the primary steps of the annotation procedure as follows:

Rib Segmentation. For each volume, we first filter out non-target voxels by thresholding and removing regions outside the bodies. Considering the geometric differences between ribs and vertebra, we separate the ribs from vertebra using morphology-based image processing algorithms(*e.g.*, dilation, erosion). In some cases, the segmentation result contains parts of the clavicle and scapula. Therefore, we manually locate those non-target voxels and remove them according to the coordinates of their connected components.

Centerline Extraction. Based on the rib segmentation, we extract the centerline by implementing the following procedure on each rib (connected component): randomly select two points at both ends of the cylinder dilated from the rib, calculate the shortest path between the points[2] [19], and smoothen the path to obtain centerline. This procedure produces high-quality centerlines even from coarse rib segmentation. At the end of extraction, we label both centerlines and rib segmentation in the order of top to bottom and left to right.

Manual Checking and Refinement. The abnormal cases, along with the pursuit of high annotation quality, motivate us to perform laborious checking and refinement after both rib segmentation and centerline extraction stages. For instance, a few cases miss floating ribs after segmentation, which reduces the connected components in annotation to 22 or less. Hence we have to check and refine the annotation case by case manually. To recover and annotate missed ribs, we turn back to the previous stage to ensure segmentation completeness by modifying the corresponding components.

2.2   Rib Segmentation from a Viewpoint of Point Clouds

The key insight of the proposed method is that simple algorithm (*i.e.*, thresholding in this study) can produce the candidate voxels for bone structures. Thus, we can avoid heavy computation on dense voxels with sparse point clouds instead. Besides, the point cloud methods use geometric information directly, reducing

---

[1] https://ribfrac.grand-challenge.org/.

[2] https://github.com/pangyuteng/simple-centerline-extraction.

the texture bias of pixel/voxel-based CNNs [4]. The point cloud viewpoint has the potential to generalize to other anatomical structures whose coarse prediction could be obtained cheaply.

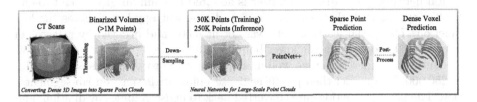

Fig. 2. Rib Segmentation from a Viewpoint of Point Clouds. The CT volumes are first binarized to obtain candidate bone voxels as inputs, then a point cloud neural network (*e.g.*, PointNet++ [13]) is used to segment each point in downsampled input point clouds. Note that the downsampling scale is different during training (30K points) and inference (250K points).

Deep Learning for Point Cloud Analysis. Deep learning for point cloud analysis [5] has been an emerging field in computer vision thanks to the popularity of 3D sensors (*e.g.*, LiDAR and RGB-D cameras) and computer graphics applications (*e.g.*, games, VR/AR/MR), PointNet [12] and DeepSet [26] pioneer this direction, where a symmetric function (*e.g.*, shared FC) is used for learning high-level representation before aggregation (*e.g.*, pooling). Following studies introduces sophisticated feature aggregation based on spatial graphs [10,13] or attention [24]. However, only a few studies have applied deep shape analysis in medical imaging scenarios [20,22,25].

Model Pipeline. Considering the sparsity of ribs in 3D volumes (<0.5% voxels) and high HU values of bones in CT scans, we design a point cloud-based model to segment ribs on binarized sparse voxels. As depicted in Fig. 2, we first set a threshold of 200 HU (Hounsfield Unit) to filter out the non-bone voxels roughly. The resultant binarized volumes are randomly downsampled and converted to point sets for ease of computation before forwarding to the network.

Our point cloud-based model is expected to infer dense predictions from large-scale point sets, which has to address the memory issue. Hence a custom PointNet++ [13], with its adjustable memory footprint, is adopted as backbone. Capable of learning local features with increasing contextual scales, PointNet++ has shown compelling robustness on sparse 3D point cloud segmentation tasks. Through set abstraction, geometric features of ribs can be extracted from binarized sparse voxels facilitating the rib segmentation task. For post-processing, the model output point prediction is converted back to volumes (voxel prediction) by morphology-based image processing algorithms.

Model Training and Inference. During the training stage, batches are down-sampled to 30,000 points per volume considering the trade-off between batch size and input size. We apply online data augmentations, including scaling, translation, and jittering, to all downsampled point sets before forwarding them to the neural network. The Adam optimizer is adopted to train all models end-to-end for 250 epochs with the batch size of 8 and cross-entropy loss (CE) as the loss function. The initial learning rate was set at 0.001 and decayed by a factor of 0.5 every 20 epochs with the lower bound of 10^{-5}.

During the inference stage, volumes are converted to point sets with the size of 250,000. The model then produces dense point predictions on rib segmentation. The point predictions are converted back to dense volumes by dilation, and we obtain the voxel predictions of rib segmentation by taking the intersections between the dilated volumes and the binarized volumes.

Model Evaluation. As the model only outputs sparse point predictions, we post-process the point predictions back to volumes (dense voxel predictions). Both point-wise and voxel-wise segmentation performance is evaluated,

$$Dice^{(L)} = 2 \cdot |y_{(L)} \cdot \hat{y}_{(L)}|/(|y_{(L)}| + |\hat{y}_{(L)}|), \ L \in \{P, V\}, \tag{1}$$

where $Dice^{(P,V)}$ indicates the sparse point-wise and dense voxel-wise Dice.

Apart from segmentation performance, we also report the missing ratio of individual ribs to evaluate the clinical applicability. Specifically, a missing of an individual rib i is counted if $recall_i < 0.5$, and then the missing ratio can be calculated with ease. As the segmentation of first and twelfth rib pairs tend to be more difficult, we calculate and report the missing ratio of all/first/intermediate/twelfth rib pairs, as depicted in Table 2.

3   Results

3.1   Quantitative Analysis

For model accuracy comparison, we first implement a 3D UNet [2] taking patches of CT volumes as input with the same setting of FracNet [7]. Moreover, we train two models with and without data augmentation, respectively. The models are evaluated with two input (point sets) sizes: 30K (input size in the training stage) and 250K. As point-wise Dice is only a proxy metric, we focus on voxel-wise Dice, as it is fair for any methods in rib segmentation. As depicted in Table 2, all point-based methods significantly outperform voxel-based 3D UNet; Besides, as point-based methods take whole volumes as inputs, it is more efficient than voxel-based method. The methods with data augmentation are at least 2% higher than the methods without data augmentation, and methods with large-scale input enjoy 0.9%–1.3% higher values, as dense point prediction leads to rich details in voxel prediction. When it comes to the missing ratio of ribs, the method with data augmentation and a large input size performs best. The comparison results show that training-time data augmentation and inference-time large input volume size

Table 2. Quantitative metrics on RibSeg test set, including Dice over sparse points ($Dice^{(P)}$), Dice over dense voxels after post-processing ($Dice^{(v)}$), ratio of missed all/first/intermediate/twelfth rib pairs (A/F/I/T) at recall> 0.5, and the model forward time in second. Post-processing time is not included as it heavily depends on the implementation.

Methods	$Dice^{(P)}$	$Dice^{(V)}$	Missed Ribs (A/F/I/T)	Forward (s)
Voxel-based 3D UNet [2,7]	-	86.3%	4.6%/7.9%/2.3%/24.6%	30.63
PN++ [13] (30K)	92.3%	91.0%	1.6%/2.9%/0.7%/10.4%	**0.32**
PN++ [13] (250K)	91.5%	92.3%	0.9%/3.3%/0.3%/4.7%	1.12
PN++ [13] (30K) + aug	**94.9%**	94.3%	1.1%/0.8%/0.4%/9.0%	**0.32**
PN++ [13] (250K) + aug	94.6%	**95.2%**	**0.6%/0.4%/0.2%/5.2%**	1.12

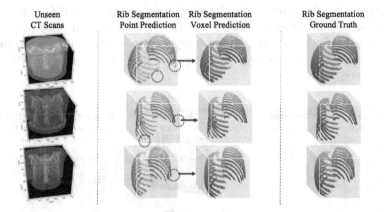

Fig. 3. Visualization of Predicted Rib Segmentation.. Red circles denote imperfect (sparse) point prediction. (Color figure online)

can improve the result. These quantitative metrics also indicate the potentials of our method in clinical applications.

In terms of the run-time, point-based methods have a clear advantage. While methods with a 250K-point input size have a little bit higher time consumption (0.8s), it is acceptable in consideration of its performance boost. The inference time was measured with the implementation of PyTorch 1.7.1 [11] and Python 3.7, on a machine with a single NVIDIA Tesla P100 GPU with Intel(R) Xeon(R) CPU @ 2.20 GHz and 150 G memory. As a reference, prior art [9] based on 3D networks takes a model forward time of 5–20 s, with a Dice value of 84% at best. The previous work [21] takes 40 s to extract the rib centerline. However, a direct comparison of metrics and speed is unfair since these results were measured with different infrastructures on different datasets. Note that, post-processing time is not included as it heavily depends on the implementation (*e.g.*, programming languages, parallel computing).

3.2   Qualitative Analysis

Visualization on Predicted Rib Segmentation. Figure 3 visualizes the point-level and voxel-level rib segmentation prediction. As depicted in Fig. 3, point predictions of the first 3 CT scans are visually acceptable, which can smoothly produce voxel-level segmentation predictions. The results are promising, and it is even hard to tell the visual difference between predictions and ground truths.

As depicted in Fig. 3, the point prediction is imperfect even with high segmentation Dice. After manually-tuned post-processing, these imperfect predictions could be fixed to some degree. For instance, the point prediction of the second CT scan contains a certain part of the scapula that can be nicely filtered during post-processing. However, the point prediction of the third CT scan suffers a missing on the first pair of ribs, which can not be ignored. The post-process is not able to fix it and produces incomplete voxel segmentation. Despite the small missing part, the rib cages in the predictions on rib segmentation are still visually acceptable.

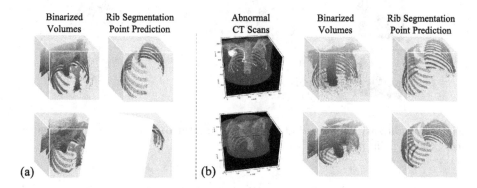

Fig. 4. Robustness Test on Extreme Cases. (a) Point prediction on unseen incomplete rib cages. (b) Point prediction on unseen abnormal CT scans.

Robustness Test on Extreme Cases. To evaluate the robustness of the proposed point cloud-based model, three kinds of extreme cases are selected for inference: CT scans of incomplete rib cages, CT scans of serious spinal pathology, and CT scans containing metal objects inseparable from ribs (*e.g.*, pacemaker). Incomplete CT scans are rather common in clinical cases, which makes the tests practically critical. As depicted in Fig. 4(a), we randomly select unseen CT scans and take the upper half of their binarized volumes for inference. Delightfully, the point predictions are visually qualified for clinical applications.

For further robustness evaluation, we test on a case of serious spinal pathology and a case containing a pacemaker of high density. Regarding the case with a pacemaker, it is inseparable from ribs, which makes the segmentation extremely

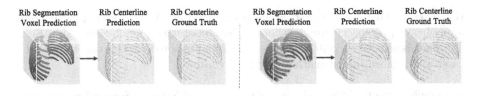

| Rib Segmentation Voxel Prediction | Rib Centerline Prediction | Rib Centerline Ground Truth | Rib Segmentation Voxel Prediction | Rib Centerline Prediction | Rib Centerline Ground Truth |

Fig. 5. Post-Processing Rib Centerlines from predicted rib segmentation.

laborious to obtain. Hence we save the trouble of manual segmentation by setting it as an abnormal case when building the RibSeg Dataset. As depicted in Fig. 4(b), our prediction on the case of spinal pathology looks complete as if it is nicely segmented. While the prediction on the other case contains a small number of voxels belong to the pacemaker, the whole rib cage is well segmented. Considering that all cases selected for the robustness test are unseen and geometrically difficult to segment, such promising prediction results may confirm the strong robustness of the proposed method as well as its potentials to be clinically applicable.

Post-Processing Rib Centerlines. With the high-quality rib segmentation, as illustrated in Fig. 5, the rib centerlines could be obtained by post-processing (*i.e.*, shortest path between points, same as the procedure with ground truth in Sect. 2.1). Although not end-to-end, the rib centerline predictions are visually acceptable for most cases. However, the post-processing algorithms for rib centerlines are sensitive to rib fractures and other abnormal cases. Considering the high clinical importance of rib centerlines, it urges a more robust method for rib centerline extraction with rib segmentation and centerline labels provided by the RibSeg dataset.

4   Conclusion and Further Work

We built the RibSeg dataset, which is the first open dataset for rib segmentation. On this dataset, we benchmarked a point cloud-based method with high performance and significant efficiency. The proposed method shows potentials to be clinically applicable, enhancing the efficiency and performance of downstream tasks, such as the diagnosis of rib fractures and bone lesions. Besides the clinical application, the RibSeg dataset could also serve as an interesting benchmark for curvilinear structures and geometric deep learning (*e.g.*, point clouds), considering the special geometry of rib structures.

There are several limitations in this study. The annotations in this paper are generated with hand-crafted morphological algorithms, and then manually checked by a junior radiologist with 3D Slicers. While such pipeline reduces the annotation cost, it cannot handle cases when the initial automatic method fails. Thus, we only managed to annotate the segmentation for 490 cases out of the 660 cases in the RibFrac dataset. Also, for the centerline extraction task that is essential for rib-related applications, we take a two-stage approach and apply heavy post-processing method to the first-stage segmentation result. Such approach is sensitive to rib fractures and segmentation errors for other abnormal cases and a more robust method will be favorable.

Acknowledgment. This work was supported by the National Science Foundation of China (U20B2072, 61976137).

References

1. Antonelli, M., Reinke, A., Bakas, S., et al.: The medical segmentation decathlon. arXiv preprint arXiv:2106.05735 (2021)
2. Çiçek, Ö., Abdulkadir, A., Lienkamp, S.S., Brox, T., Ronneberger, O.: 3D U-Net: learning dense volumetric segmentation from sparse annotation. In: Ourselin, S., Joskowicz, L., Sabuncu, M.R., Unal, G., Wells, W. (eds.) MICCAI 2016. LNCS, vol. 9901, pp. 424–432. Springer, Cham (2016). https://doi.org/10.1007/978-3-319-46723-8_49
3. Coleman, R., Body, J.J., Aapro, M., Hadji, P., Herrstedt, J., Group, E.G.W., et al.: Bone health in cancer patients: Esmo clinical practice guidelines. Ann. Oncology **25**, iii124-iii137 (2014)
4. Geirhos, R., Rubisch, P., Michaelis, C., Bethge, M., Wichmann, F.A., Brendel, W.: Imagenet-trained cnns are biased towards texture; increasing shape bias improves accuracy and robustness. In: ICLR (2019)
5. Guo, Y., Wang, H., Hu, Q., Liu, H., Liu, L., Bennamoun, M.: Deep learning for 3d point clouds: a survey. IEEE Trans. Pattern Anal. Mach. Intell. (2020)
6. Jin, L., Ge, X., Lu, F., Sun, Y., et al.: Low-dose CT examination for rib fracture evaluation: a pilot study. Medicine 97(30) (2018)
7. Jin, L., et al.: Deep-learning-assisted detection and segmentation of rib fractures from CT scans: development and validation of fracnet. EBioMedicine **62**, 103106 (2020)
8. Klinder, T., et al.: Automated model-based rib cage segmentation and labeling in CT images. In: Ayache, N., Ourselin, S., Maeder, A. (eds.) MICCAI 2007. LNCS, vol. 4792, pp. 195–202. Springer, Heidelberg (2007). https://doi.org/10.1007/978-3-540-75759-7_24
9. Lenga, M., Klinder, T., Bürger, C., von Berg, J., Franz, A., Lorenz, C.: Deep learning based rib centerline extraction and labeling. In: Vrtovec, T., Yao, J., Zheng, G., Pozo, J.M. (eds.) MSKI 2018. LNCS, vol. 11404, pp. 99–113. Springer, Cham (2019). https://doi.org/10.1007/978-3-030-11166-3_9
10. Liu, J., Ni, B., Li, C., Yang, J., Tian, Q.: Dynamic points agglomeration for hierarchical point sets learning. In: ICCV, pp. 7546–7555 (2019)
11. Paszke, A., et al.: Automatic differentiation in pytorch (2017)
12. Qi, C.R., Su, H., Mo, K., Guibas, L.J.: Pointnet: deep learning on point sets for 3d classification and segmentation. In: CVPR, pp. 652–660 (2017)
13. Qi, C.R., Yi, L., Su, H., Guibas, L.J.: Pointnet++: deep hierarchical feature learning on point sets in a metric space. In: NIPS (2017)
14. Ramakrishnan, S., Alvino, C., Grady, L., Kiraly, A.: Automatic three-dimensional rib centerline extraction from CT scans for enhanced visualization and anatomical context. In: Medical Imaging 2011: Image Processing, vol. 7962, p. 79622X (2011)

15. Ringl, H., Lazar, M., Töpker, M., Woitek, R., Prosch, H., Asenbaum, U., Balassy, C., Toth, D., Weber, M., Hajdu, S., et al.: The ribs unfolded-a CT visualization algorithm for fast detection of rib fractures: effect on sensitivity and specificity in trauma patients. Eur. Radiol. **25**(7), 1865–1874 (2015)

16. Shen, H., Liang, L., Shao, M., Qing, S.: Tracing based segmentation for the labeling of individual rib structures in chest CT volume data. In: Barillot, C., Haynor, D.R., Hellier, P. (eds.) MICCAI 2004. LNCS, vol. 3217, pp. 967–974. Springer, Heidelberg (2004). https://doi.org/10.1007/978-3-540-30136-3_117

17. Sirmali, M., et al.: A comprehensive analysis of traumatic rib fractures: morbidity, mortality and management. Eur. J. Cardiothorac. Surg. **24**(1), 133–138 (2003)

18. Staal, J., van Ginneken, B., Viergever, M.A.: Automatic rib segmentation and labeling in computed tomography scans using a general framework for detection, recognition and segmentation of objects in volumetric data. Med. Image Anal. **11**(1), 35–46 (2007)

19. Teng, P.y., Bagci, A.M., Alperin, N.: Automated prescription of an optimal imaging plane for measurement of cerebral blood flow by phase contrast magnetic resonance imaging. IEEE Trans. Biomed. Eng. **58**(9), 2566–2573 (2011)

20. Wickramasinghe, U., Remelli, E., Knott, G., Fua, P.: Voxel2Mesh: 3D mesh model generation from volumetric data. In: Martel, A.L., Abolmaesumi, P., Stoyanov, D., Mateus, D., Zuluaga, M.A., Zhou, S.K., Racoceanu, D., Joskowicz, L. (eds.) MICCAI 2020. LNCS, vol. 12264, pp. 299–308. Springer, Cham (2020). https://doi.org/10.1007/978-3-030-59719-1_30

21. Wu, D., et al.: A learning based deformable template matching method for automatic rib centerline extraction and labeling in ct images. In: CVPR, pp. 980–987. IEEE (2012)

22. Yang, J., Fang, R., Ni, B., Li, Y., Xu, Y., Li, L.: Probabilistic radiomics: ambiguous diagnosis with controllable shape analysis. In: Shen, D., Liu, T., Peters, T.M., Staib, L.H., Essert, C., Zhou, S., Yap, P.-T., Khan, A. (eds.) MICCAI 2019. LNCS, vol. 11769, pp. 658–666. Springer, Cham (2019). https://doi.org/10.1007/978-3-030-32226-7_73

23. Yang, J., Shi, R., Ni, B.: Medmnist classification decathlon: a lightweight automl benchmark for medical image analysis. In: ISBI (2021)

24. Yang, J., Zhang, Q., Ni, B., Li, L., Liu, J., Zhou, M., Tian, Q.: Modeling point clouds with self-attention and gumbel subset sampling. In: CVPR, pp. 3323–3332 (2019)

25. Yang, X., Xia, D., Kin, T., Igarashi, T.: Intra: 3d intracranial aneurysm dataset for deep learning. In: CVPR, pp. 2656–2666 (2020)

26. Zaheer, M., Kottur, S., Ravanbakhsh, S., Póczos, B., Salakhutdinov, R.R., Smola, A.J.: Deep sets. In: NIPS (2017)

Hierarchical Self-supervised Learning for Medical Image Segmentation Based on Multi-domain Data Aggregation

Hao Zheng$^{(\boxtimes)}$, Jun Han, Hongxiao Wang, Lin Yang, Zhuo Zhao, Chaoli Wang, and Danny Z. Chen

Department of Computer Science and Engineering, University of Notre Dame, Notre Dame, IN 46556, USA
hzheng3@nd.edu

Abstract. A large labeled dataset is a key to the success of supervised deep learning, but for medical image segmentation, it is highly challenging to obtain sufficient annotated images for model training. In many scenarios, unannotated images are abundant and easy to acquire. Self-supervised learning (SSL) has shown great potentials in exploiting raw data information and representation learning. In this paper, we propose Hierarchical Self-Supervised Learning (HSSL), a new self-supervised framework that boosts medical image segmentation by making good use of unannotated data. Unlike the current literature on task-specific self-supervised pretraining followed by supervised fine-tuning, we utilize SSL to learn task-agnostic knowledge from heterogeneous data for various medical image segmentation tasks. Specifically, we first aggregate a dataset from several medical challenges, then pre-train the network in a self-supervised manner, and finally fine-tune on labeled data. We develop a new loss function by combining contrastive loss and classification loss, and pre-train an encoder-decoder architecture for segmentation tasks. Our extensive experiments show that multi-domain joint pre-training benefits downstream segmentation tasks and outperforms single-domain pre-training significantly. Compared to learning from scratch, our method yields better performance on various tasks (e.g., +0.69% to +18.60% in Dice with 5% of annotated data). With limited amounts of training data, our method can substantially bridge the performance gap with respect to denser annotations (e.g., 10% vs. 100% annotations).

Keywords: Self-supervised learning · Image segmentation · Multi-domain

Electronic supplementary material The online version of this chapter (https://doi.org/10.1007/978-3-030-87193-2_59) contains supplementary material, which is available to authorized users.

M. de Bruijne et al. (Eds.): MICCAI 2021, LNCS 12901, pp. 622–632, 2021.
https://doi.org/10.1007/978-3-030-87193-2_59

1   Introduction

Although supervised deep learning has achieved great success on medical image segmentation [15,17,24,33], it heavily relies on sufficient good-quality manual annotations which are usually hard to obtain due to expensive acquisition, data privacy, etc. Public medical image datasets are normally smaller than the generic image datasets (see Fig. 1(a)), and may hinder improving segmentation performance. Deficiency of annotated data has driven studies to explore alternative solutions. Transfer learning fine-tunes models pre-trained on ImageNet for target tasks [12,35,36], but it could be impractical and inefficient due to the predefined model architectures [18] and is not as good as transferred from medical images due to image characteristics differences [36]. Semi-supervised learning utilizes unlimited amounts of unlabeled data to boost performance, but it usually assumes that the labeled data sufficiently covers the data distribution, and needs to address consequent non-trivial challenges such as adversarial learning [19,31] and noisy labels [30,34]. Active learning selects the most representative samples for annotation [28,32,35] but focuses on saving manual effort and does not utilize unannotated data. Considering these limitations and the fact that considerable unlabeled medical images are easy to acquire and free to use, we seek to answer the question: *Can we improve segmentation performance with limited training data by directly exploiting raw data information and representation learning?*

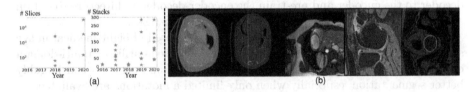

Fig. 1. (a) The number of images for each medical image segmentation challenge every year since 2016 at MICCAI (left: 2D images; right: 3D stacks). (b) Diverse medical image and mask examples: spleen, liver & tumours, cardiovascular structures, knee bones & cartilages, and prostate.

Recently, self-supervised learning (SSL) approaches, which initialize models by constructing and training surrogate tasks with unlabeled data, attracted much attention due to soaring performance on representation learning [8–10,14,16,20,21,23] and downstream tasks [4,5,22,26,36,37]. It was shown that the learned representation by *contrastive learning*, a variant of SSL, gradually approaches the effectiveness of representations learned through strong supervision, even under circumstances when only limited data or a small-scale dataset is available [6,11]. However, three key factors of contrastive learning have not been well explored for medical segmentation tasks: (1) A medical image dataset is often insufficiently large due to the intrusive nature of some imaging techniques or expensive annotations (e.g., 3D(+T) images), which suppresses self-supervised pre-training and hinders representation learning using a single dataset. (2) The contrastive strategy considers only congenetic image pairs generated by different transformations used in data augmentation, which suppresses the model from learning task-agnostic representations from heterogeneous data collected from different sources (see Fig. 1(b)). (3) Most studies focused on extracting high-level representations by pre-training the encoder while neglecting to learn low-level features explicitly and initialize the decoder, which hinders the performance of dense prediction tasks such as semantic segmentation.

To address these challenges, in this paper, we propose a new *hierarchical self-supervised learning* (HSSL) framework to pre-train on heterogeneous unannotated data and obtain an initialization beneficial for training multiple downstream medical image segmentation tasks with limited annotations. First, we investigate available public challenge datasets on medical image segmentation and propose to aggregate a multi-domain (modalities, organs, or facilities) dataset. In this way, our collected dataset is considerably larger than a task-specific dataset and the pretext model is forced to learn task-agnostic knowledge (e.g., texture, intensity distribution, etc.). Second, we construct pretext tasks at multiple abstraction levels to learn hierarchical features and explicitly force the model to learn richer semantic features for segmentation tasks on medical images. Specifically, our HSSL utilizes contrasting and classification strategies to supervise image-, task-, and group-level pretext tasks. We also extract multi-level features from the network encoding path to bridge the gap between low-level texture and high-level semantic representations. Third, we attach a lightweight decoder to the encoder and pre-train the encoder-decoder architecture to obtain a suitable initialization for downstream segmentation tasks.

We experiment on our aggregated dataset composed of eight medical image segmentation tasks and show that our HSSL is effective in utilizing multi-domain data to initialize model parameters for target tasks and achieves considerably better segmentation, especially when only limited annotations are available.

2   Methodology

We discuss the necessity and feasibility of aggregating multi-domain image data and show how to construct such a dataset in Sect. 2.1, and then introduce our hierarchical self-supervised learning pretext tasks (shown in Fig. 2) in Sect. 2.2. After pre-training, we fine-tune the trained encoder-decoder network on downstream segmentation tasks with limited annotations.

2.1   Multi-Domain Data Aggregation

Necessity. As shown in Fig. 1(a), most publicly available medical image segmentation datasets are of relatively small sizes. Yet, recent progresses on contrastive learning empirically showed that training on a larger dataset often learns better representations and brings larger performance improvement in downstream tasks [6,7,11]. Similarly, a larger dataset is beneficial for supervised classification tasks and unsupervised image reconstruction tasks, because such a dataset tends to be more diverse and better cover the true image space distribution.

Feasibility. First, there are quite a few medical image dataset archives (e.g., TCIA) and public challenges (e.g., Grand Challenge). Typical imaging modalities (CT, MRI, X-ray, etc.) of multiple regions-of-interest (ROIs, organs, structures, etc.) are covered. Second, common/similar textures or intensity distributions are shared among different datasets (see Fig. 1(b)), and their raw images may cover the same physical regions (e.g., abdominal CT for the spleen dataset and liver dataset). Therefore, an aggregated multi-domain dataset can (1) enlarge the data size of a shared image space and (2) force the model to distinguish different contents from the raw images. In this way, task-agnostic knowledge is extracted.

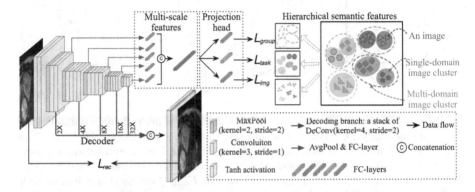

Fig. 2. An overview of our proposed hierarchical self-supervised learning (HSSL) framework (best viewed in color). The backbone encoder builds a pyramid of multi-scale features from the input image, forming a rich latent vector. Then it is stratified to represent hierarchical semantic features of the aggregated multi-domain data, supervised by different pretext tasks in the hierarchy. Besides, an auxiliary reconstruction pretext task helps initialize the decoder. (Color figure online)

Dataset Aggregation. To ensure the effectiveness of multi-domain data aggregation, three principles should be considered. (1) Representativeness: The datasets considered for aggregation should cover a moderate range of medical imaging techniques/modalities. (2) Relevance: The datasets considered should not drastically differ in content/appearance. Otherwise, it is easy for the model to distinguish them and a less common feature space is shared among them. (3) Diversity: The datasets considered should benefit a range of applications. In this work, we focus on CT and MRI of various ROIs (i.e., heart, liver, prostate, pancreas, knee, and spleen). The details of aggregated dataset are shown in Table 1.

2.2   Hierarchical Self-supervised Learning (HSSL)

Having aggregated multiple datasets, $\mathcal{D} = \{D_1, D_2, \ldots, D_N\}$, where D_i is a dataset for a certain segmentation task. A straightforward method to use $\mathcal{D}$ is to directly extend some known pretext tasks (e.g., SimCLR [6]) and conduct joint pre-training. However, such pretext tasks only explicitly force the model to learn a global representation and are not tailored for the target segmentation tasks. Hence, taking imaging techniques and prior knowledge (e.g., appearance, ROIs) into account, we propose to extract richer semantic features from hierarchical abstract levels and devise the network for target segmentation tasks.

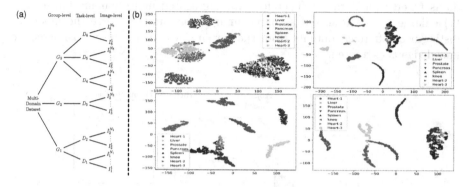

Fig. 3. (a) An example of the hierarchical structure of a multi-domain dataset. Each chosen dataset/task D_i forms a domain consisting of a set of images $\{I_i^k\}_{k=1}^{N_i}$, where N_i is the total number of images in D_i. Multiple tasks form a multi-domain cluster called a *group* (G_j). (b) t-SNE projection of extracted features (best viewed in color). Top-left: F_{VGG-19}; top-right: F_{image}; bottom-left: F_{task} (forming single-domain task-level clusters as in Table 1); bottom-right: F_{group} (forming multi-domain group-level clusters as in Table 1). (Color figure online)

We formulate three hierarchical levels (see Fig. 3(a)). (1) *Image-level*: Each image I is a learning subject; we want to extract distinguishable features of I w.r.t. another image, regardless of which dataset it originally comes from or what ROIs it contains. Specifically, we follow the state-of-the-art SimCLR [6] and build positive and negative pairs with various data augmentations. (2) *Task-level*: Each D_i is originally imaged for a specific purpose (e.g., CT for spleen). Generally, images belonging to a same dataset are similar inherently. As shown in Fig. 3(b), images of different modalities and ROIs are easier to distinguish. For abdominal CTs of spleen and liver, although the images are similar, their contents are different. Thus, each task's dataset forms a single domain of certain ROI and image types. (3) *Group-level*: Despite the differences among different segmentation tasks, the contents of images may show a different degree of similarity. For example, in the physical space, liver CT scans have overlapping with spleen CT scans; cardiac MRIs scanned for different purposes (e.g., diverse cardiovascular structures) contain the same ROI (i.e., the heart) regardless of the image size and contrast. In this way, we categorize multiple domains of images into a group, which forms a multi-domain cluster in the feature space. Assigned with both task-level and group-level labels, each image constitutes a tuple (I, y^t, y^g), where t and g are task-class and group-class, respectively (see Table 1).

Further, to better aggregate low- and high-level features from the encoder, we compress multi-scale feature vectors from the feature pyramid and concatenate them together, and then attach three different projection heads to automatically extract hierarchical representations (see Fig. 2).

Image-Level Loss. Given an input image I, the contrastive loss is formulated as: $l(\tilde{I}, \hat{I}) = -\log \frac{\exp\{sim(\tilde{z},\hat{z})/\tau\}}{\exp\{sim(\tilde{z},\hat{z})/\tau\} + \sum_{\bar{I} \in \Lambda^-} \exp\{sim(\tilde{z},\bar{z})/\tau\}}$, where $\tilde{z} = P_l(E(\tilde{I}))$, $\hat{z} = P_l(E(\hat{I}))$, $\bar{z} = P_l(E(\bar{I}))$, $P_l(\cdot)$ is the image-level projection head, $E(\cdot)$ is the encoder, $\tilde{I}$ and $\hat{I}$ are two different augmentations of image I (i.e., $\tilde{I} = \tilde{t}(I)$ and $\hat{I} = \hat{t}(I)$), $\bar{I} \in \Lambda^-$ consisting of all negative samples of I, and $\tilde{t}, \hat{t} \in \mathcal{T}$ are two augmentations. The augmentations $\mathcal{T}$ include random cropping, resizing, blurring, and adding noise. $sim(\cdot, \cdot)$ is cosine similarity, and τ is a temperature scaling parameter. Given our multi-domain dataset $\mathcal{D}$, the image-level loss is defined as: $\mathcal{L}_{img} = \frac{1}{|\Lambda^+|} \sum_{\forall(\tilde{I},\hat{I}) \in \Lambda^+} [l(\tilde{I},\hat{I}) + l(\hat{I},\tilde{I})]$, where Λ^+ is a set of all similar pairs sampled from $\mathcal{D}$.

Table 1. Details of our data obtained from public sources. The left two columns: their task-classes and group-classes based on our multi-domain data aggregation principles.

Task ID	Group ID	ROI-Type	Segmentation class	# of slices	Source
1	1	Heart-MRI	1: left atrium	1262	LASC [27]
2	2	Liver-CT	1: liver, 2: tumor	4342	LiTS [3]
3	3	Prostate-MRI	1: central gland, 2: peripheral zone	483	MSD [25]
4	2	Pancreas-CT	1: Pancreas, 2: tumor	8607	MSD [25]
5	2	Spleen-CT	1: spleen	1466	MSD [25]
6	4	Knee-MRI	1: femur bone, 2: tibia bone, 3: femur cartilage, 4: tibia cartilage	8187	Knee [29]
7	1	Heart-MRI	1: left ventricle, 2: right ventricle, 3: myocardium	1891	ACDC [2]
8	1	Heart-MRI		3120	M& Ms [1]

Task-Level Loss and Group-Level Loss. Given task-class and group-class, we formulate task- and group-level pretext tasks as classification tasks. The training objectives are: $\mathcal{L}_{task} = -\sum_{c=1}^{T} y_c^t \log(p_c^t)$; $\mathcal{L}_{group} = -\sum_{c=1}^{G} y_c^g \log(p_c^g)$, where $p_c^t = P_t(E(I))$, $p_c^g = P_g(E(I))$, $P_t(\cdot)$ (or $P_g(\cdot)$) is the task-level (or group-level) projection head, $E(\cdot)$ is the encoder, y_c^t (or y_c^g) is the task-class (or group-class) of input image I, and T (or G) is the number of classes of tasks (or groups).

We visualize some sample learned features in Fig. 3(b), in which the hierarchical layout is as expected, implying that our model is capable of extracting richer semantic features at different abstract levels of the input images.

Decoder Initialization. A decoder is also indispensable for semantic segmentation tasks. We devise a multi-scale decoder and formulate a reconstruction pretext task. The loss is defined as: $\mathcal{L}_{rec} = \frac{1}{|\mathcal{D}|} \sum_{I \in \mathcal{D}} ||S(E(I)) - I||_2$, where $E(\cdot)$ is the encoder, $S(\cdot)$ is the decoder, and $|| \cdot ||_2$ is the L_2 norm.

In summary, we combine the hierarchical self-supervised losses at all the levels and the auxiliary reconstruction loss to jointly optimize the model: $\mathcal{L}_{total} = \lambda_1 \mathcal{L}_{img} + \lambda_2 \mathcal{L}_{task} + \lambda_3 \mathcal{L}_{group} + \lambda_4 \mathcal{L}_{rec}$, where $\lambda_i (i = 1, 2, 3, 4)$ are the weights to balance loss terms. For simplicity, we let $\lambda_1 = \lambda_2 = \lambda_3 = 1/3, \lambda_4 = 50$.

Segmentation. Once trained, the encoder-decoder can be fine-tuned for downstream multi-domain segmentation tasks. For a give task D_i, we acquire some annotations (e.g., 10%) and optimize the network with cross-entropy loss.

3   Experiments and Results

Datasets and Experimental Setup. We employ multiple MRI and CT image sets from 8 different data sources with distribution shift, and sample 2D slices from each stack (see Table 1 for a summary of their sample numbers and downstream tasks). Each dataset is split into X_{tr}, X_{val}, and X_{te} in the ratios of $7:1:2$. We use all images for the pre-training stage and then fine-tune the pre-trained network with labeled images from X_{tr}. We experiment with different

Table 2. Quantitative results on Task-1 (heart), Task-3 (prostate), and Task-5 (spleen). Dice scores for each class are listed and the average scores are in parentheses. TFS: training from scratch. Same network architecture is used for fair comparison in all the experiments. Our HSSL achieves the best performance in most settings (in bold).

Task-#	Anno	TFS	Rotation [9]	In-painting [23]	MoCo [11]	SimCLR [6]	HSSL (Ours)
1	5%	71.56	72.83	65.40	75.97	73.45	**81.46**
	10%	79.64	**82.31**	81.99	79.07	81.19	81.79
	100%	85.81	87.43	86.56	87.19	87.06	**87.65**
3	5%	20.65; 47.56 (34.10)	28.74; 67.11 (47.93)	20.13; 52.16 (36.14)	29.55; 64.95 (47.25)	39.67; 68.35 (**54.01**)	35.30; 70.08 (52.69)
	10%	40.10; 66.95 (53.53)	44.15; 70.63 (57.39)	33.81;67.14 (50.48)	40.16; 67.98 (54.07)	46.04; 70.39 (58.22)	46.97; 72.21 (**59.59**)
	100%	50.19; 76.74 (63.47)	55.21; 78.21 (66.71)	53.19 77.97 (65.59)	56.31; 77.59 (66.95)	56.53; 77.86 (67.20)	58.80; 78.35 (**68.58**)
5	5%	48.75	56.74	47.86	54.91	63.40	**67.35**
	10%	67.44	74.68	71.30	68.22	78.25	**80.95**
	100%	85.88	86.96	85.96	85.75	87.76	**88.45**

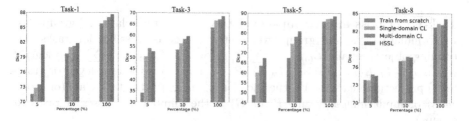

Fig. 4. Quantitative results of TFS *vs.* single-domain CL *vs.* multi-domain CL *vs.* HSSL for Task-1/-3/-5/-8 with different ratios (5%, 10%, 100%) of labeled data, respectively.

amounts of training data X_{tr}^s, where $s \in \{5\%, 10\%, 100\%\}$ denotes the ratio of $\frac{X_{tr}^s}{X_{tr}}$. The segmentation accuracy is measured by the Dice-Sørensen Coefficient.

Implementation Details. For self-supervised pre-training, we use ResNet-34 [13] as the base encoder network, FC layers to obtain latent vectors, and sequential DeConv layers to reconstruct images. Detailed structures can be found in Supplementary Material. The model is optimized using Adam with a linear learning rate scaling for $1k$ epochs (initial learning rate $= 3e^{-4}$). For segmentation tasks, we optimize the network using Adam with the "poly" learning rate policy, $L_r \times \left(1 - \frac{epoch}{\#epoch}\right)^{0.9}$, where the initial learning rate $L_r = 5e^{-4}$ and $\#epoch = 10k$. Random cropping and rotation are applied for augmentation. In all the experiments, the mini-batch size is 30 and input image size is 192×192.

Main Results. Our approach contributes to the "pre-training + fine-tuning" scheme in two aspects: hierarchical self-supervised learning (HSSL) and multi-domain data aggregation. **(1)** *Effectiveness of HSSL*. We compare with state-

Table 3. Quantitative results of different models on Task-1/-3/-5 with 5%, 10%, and 50% annotated data, respectively. Our HSSL achieves the best performance in most the settings (highest scores in bold).

Method	Param. (M)	Task-1 (heart)			Task-3 (prostate)			Task-5 (spleen)		
		5%	10%	50%	5%	10%	50%	5%	10%	50%
UNet [24]	39.40	75.43	77.72	86.75	38.19	49.44	62.61	54.71	62.81	81.48
UNet3+ [15]	26.97	78.48	78.81	**87.52**	42.06	50.94	63.50	60.05	64.83	82.74
HSSL (Ours)	22.07	**81.46**	**81.79**	87.02	**52.69**	**59.59**	**66.64**	**67.35**	**80.95**	**85.86**

Table 4. Ablation study of loss functions on Task-1 & Task-5 w/ 5% anno. data.

L_{rec}	L_{img}	L_{task}	L_{group}	Task-1	Task-5
✓				65.71	46.13
	✓			73.45	63.40
✓	✓			77.26	65.01
✓	✓	✓		79.32	66.67
✓	✓	✓	✓	81.46	67.35

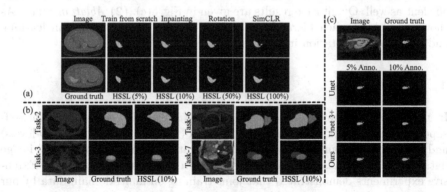

Fig. 5. Qualitative comparison (best viewed in color). (a) Top: results of different methods on Task-5 (10% annotated data); Bottom: results of our HSSL with different ratios of annotated data. (b) Results of Task-2/-3/-6/-7 (10% annotated data). (c) Results of different models on Task-1 trained with 5% and 10% annotated data, respectively.

of-the-art pretext task training methods [6,9,11,23] on seven downstream segmentation tasks, and quantitative results of three representative tasks are summarized in Table 2. First, our method surpasses training from scratch (TFS) substantially, showing the effectiveness of better model initialization. More can be found in Supplementary Material. Second, our approach outperforms known SSL-based methods in almost all the settings, indicating a better capability to extract features for segmentation tasks. Third, our HSSL can more effectively boost performance, especially when extremely limited annotations are available (e.g., +18.60% with 5% annotated data on Task-3), implying potential applicability when abundant images are acquired but few are labeled. Fourth, with more annotations, our method can further improve accuracy and achieve state-of-the-art performance (e.g., +1.84% to +2.57% with 100% annotated data over TFS). Qualitative results are given in Fig. 5 and Supplementary Material. (2) *Effectiveness of Multi-Domain Data Aggregation.* We conduct pre-training on single-domain and aggregated multi-domain data, and compare the segmentation performances. "Single-domain CL" and "Multi-domain CL" are all based on the state-of-the-art SimCLR [6]. As sketched in Fig. 4, one can see that multi-domain data aggregation consistently outperforms (sometimes significantly) single-domain pre-training (e.g., with 10% annotated data on Task-5, multi-domain CL and HSSL outperform single-domain CL by 3.74% and 6.41%, respectively). This suggests that more data varieties can provide complementary information and help improve the overall performance.

Discussions.   (1) *Comparison with State-of-the-Art Models.* As shown in Table 3, our method outperforms the state-of-the-art UNet3+ [15] significantly in almost all the settings. Further, with limited annotated data (e.g., 5%), our method bridges the performance gap significantly w.r.t. the results obtained by training with more annotated data. Also, our model is most lightweight, and thus efficient as well. Qualitative results are given in Fig. 5(c). (2) *Ablation Study.* As shown in Table 4, each hierarchical loss contributes to representation learning and leads to segmentation improvement.

4   Conclusions

In this paper, we proposed *hierarchical self-supervised learning*, a novel self-supervised framework that learns hierarchical (image-, task-, and group-levels) and multi-scale semantic features from aggregated multi-domain medical image data. A decoder is also initialized for downstream segmentation tasks. Extensive experiments demonstrated that joint training on multi-domain data by our method outperforms training from scratch and conventional pre-training strategies, especially in limited annotation scenarios.

Acknowledgement. This research was supported in part by the U.S. National Science Foundation through grants IIS-1455886, CCF-1617735, CNS-1629914, and IIS-1955395.

References

1. Multi-centre, multi-vendor & multi-disease cardiac image segmentation challenge (M&Ms). https://www.ub.edu/mnms/. Accessed 01 July 2021
2. Bernard, O., et al.: Deep learning techniques for automatic MRI cardiac multi-structures segmentation and diagnosis: is the problem solved? IEEE Trans. Med. Imaging **37**(11), 2514–2525 (2018)
3. Bilic, P., et al.: The liver tumor segmentation benchmark (LiTS). arXiv preprint arXiv:1901.04056 (2019)
4. Chaitanya, K., Erdil, E., Karani, N., Konukoglu, E.: Contrastive learning of global and local features for medical image segmentation with limited annotations. In: NeurIPS, pp. 12546–12558 (2020)
5. Chen, L., Bentley, P., Mori, K., Misawa, K., Fujiwara, M., Rueckert, D.: Self-supervised learning for medical image analysis using image context restoration. Med. Image Anal. **58**, 101539 (2019)
6. Chen, T., Kornblith, S., Norouzi, M., Hinton, G.: A simple framework for contrastive learning of visual representations. In: ICML, pp. 1597–1607 (2020)
7. Chen, T., Kornblith, S., Swersky, K., Norouzi, M., Hinton, G.E.: Big self-supervised models are strong semi-supervised learners. In: NeurIPS, pp. 22243–22255 (2020)
8. Doersch, C., Gupta, A., Efros, A.A.: Unsupervised visual representation learning by context prediction. In: ICCV, pp. 1422–1430 (2015)
9. Gidaris, S., Singh, P., Komodakis, N.: Unsupervised representation learning by predicting image rotations. In: ICLR (2018)
10. Grill, J.B., et al.: Bootstrap your own latent: a new approach to self-supervised learning. In: NeurIPS, pp. 21271–21284 (2020)
11. He, K., Fan, H., Wu, Y., Xie, S., Girshick, R.: Momentum contrast for unsupervised visual representation learning. In: CVPR, pp. 9729–9738 (2020)
12. He, K., Girshick, R., Dollár, P.: Rethinking ImageNet pre-training. In: ICCV, pp. 4918–4927 (2019)
13. He, K., Zhang, X., Ren, S., Sun, J.: Deep residual learning for image recognition. In: CVPR, pp. 770–778 (2016)
14. Hjelm, R.D., et al.: Learning deep representations by mutual information estimation and maximization. In: ICLR (2019)
15. Huang, H., et al.: UNet 3+: a full-scale connected UNet for medical image segmentation. In: Proceedings of the IEEE International Conference on Acoustics, Speech and Signal Processing, pp. 1055–1059 (2020)
16. Larsson, G., Maire, M., Shakhnarovich, G.: Learning representations for automatic colorization. In: ECCV, pp. 577–593 (2016)
17. Liang, P., Chen, J., Zheng, H., Yang, L., Zhang, Y., Chen, D.Z.: Cascade decoder: a universal decoding method for biomedical image segmentation. In: Proceedings of the IEEE International Symposium on Biomedical Imaging, pp. 339–342 (2019)
18. Liu, S., Xu, D., Zhou, S.K., Grbic, S., Cai, W., Comaniciu, D.: Anisotropic hybrid network for cross-dimension transferable feature learning in 3D medical images. In: Deep Learning and Convolutional Neural Networks for Medical Imaging and Clinical Informatics, pp. 199–216 (2019)
19. Madani, A., Moradi, M., Karargyris, A., Syeda-Mahmood, T.: Semi-supervised learning with generative adversarial networks for chest X-ray classification with ability of data domain adaptation. In: ISBI, pp. 1038–1042 (2018)
20. Noroozi, M., Favaro, P.: Unsupervised learning of visual representations by solving jigsaw puzzles. In: ECCV, pp. 69–84 (2016)

21. Oord, A.v.d., Li, Y., Vinyals, O.: Representation learning with contrastive predictive coding. arXiv preprint arXiv:1807.03748 (2018)

22. Ouyang, C., Biffi, C., Chen, C., Kart, T., Qiu, H., Rueckert, D.: Self-supervision with superpixels: Training few-shot medical image segmentation without annotation. In: ECCV, pp. 762–780 (2020)

23. Pathak, D., Krahenbuhl, P., Donahue, J., Darrell, T., Efros, A.A.: Context encoders: Feature learning by inpainting. In: CVPR, pp. 2536–2544 (2016)

24. Ronneberger, O., Fischer, P., Brox, T.: U-Net: convolutional networks for biomedical image segmentation. In: Navab, N., Hornegger, J., Wells, W.M., Frangi, A.F. (eds.) MICCAI 2015. LNCS, vol. 9351, pp. 234–241. Springer, Cham (2015). https://doi.org/10.1007/978-3-319-24574-4_28

25. Simpson, A.L., et al.: A large annotated medical image dataset for the development and evaluation of segmentation algorithms. arXiv preprint arXiv:1902.09063 (2019)

26. Tao, X., Li, Y., Zhou, W., Ma, K., Zheng, Y.: Revisiting rubik's cube: self-supervised learning with volume-wise transformation for 3D medical image segmentation. In: MICCAI, pp. 238–248 (2020)

27. Tobon-Gomez, C., Geers, A.J., Peters, J., Weese, J., Pinto, K., Karim, R., Ammar, M., Daoudi, A., Margeta, J., Sandoval, Z., et al.: Benchmark for algorithms segmenting the left atrium from 3D CT and MRI datasets. IEEE Trans. Med. Imaging **34**(7), 1460–1473 (2015)

28. Yang, L., Zhang, Y., Chen, J., Zhang, S., Chen, D.Z.: Suggestive annotation: a deep active learning framework for biomedical image segmentation. In: MICCAI, pp. 399–407 (2017)

29. Yin, Y., Zhang, X., Williams, R., Wu, X., Anderson, D.D., Sonka, M.: LOGISMOS—layered optimal graph image segmentation of multiple objects and surfaces: cartilage segmentation in the knee joint. IEEE Trans. Med. Imaging **29**(12), 2023–2037 (2010)

30. Yu, L., Wang, S., Li, X., Fu, C.W., Heng, P.A.: Uncertainty-aware self-ensembling model for semi-supervised 3D left atrium segmentation. In: MICCAI, pp. 605–613 (2019)

31. Zhang, Y., Yang, L., Chen, J., Fredericksen, M., Hughes, D.P., Chen, D.Z.: Deep adversarial networks for biomedical image segmentation utilizing unannotated images. In: MICCAI, pp. 408–416 (2017)

32. Zheng, H., et al.: Biomedical image segmentation via representative annotation. In: AAAI, pp. 5901–5908 (2019)

33. Zheng, H., et al.: HFA-Net: 3D cardiovascular image segmentation with asymmetrical pooling and content-aware fusion. In: MICCAI, pp. 759–767 (2019)

34. Zheng, H., Zhang, Y., Yang, L., Wang, C., Chen, D.Z.: An annotation sparsification strategy for 3D medical image segmentation via representative selection and self-training. In: AAAI, pp. 6925–6932 (2020)

35. Zhou, Z., Shin, J., Zhang, L., Gurudu, S., Gotway, M., Liang, J.: Fine-tuning convolutional neural networks for biomedical image analysis: Actively and incrementally. In: CVPR, pp. 7340–7351 (2017)

36. Zhou, Z., et al.: Models genesis: Generic autodidactic models for 3D medical image analysis. In: MICCAI, pp. 384–393 (2019)

37. Zhuang, X., Li, Y., Hu, Y., Ma, K., Yang, Y., Zheng, Y.: Self-supervised feature learning for 3D medical images by playing a Rubik's cube. In: MICCAI, pp. 420–428 (2019)

CCBANet: Cascading Context and Balancing Attention for Polyp Segmentation

Tan-Cong Nguyen[1,2,5], Tien-Phat Nguyen[1,5], Gia-Han Diep[1,4,5],
Anh-Huy Tran-Dinh[2,5], Tam V. Nguyen[3], and Minh-Triet Tran[1,4,5]([✉])

[1] University of Science, VNU-HCM, Ho Chi Minh City, Vietnam
`tmtriet@fit.hcmus.edu.vn`
[2] University of Social Sciences and Humanities, VNU-HCM,
Ho Chi Minh City, Vietnam
[3] University of Dayton, Dayton, USA
[4] John von Neumann Institute, VNU-HCM,
Ho Chi Minh City, Vietnam
[5] Vietnam National University, Ho Chi Minh City, Vietnam

Abstract. Polyps detection plays an important role in colonoscopy, cancer diagnosis, and early treatment. Many efforts have been made to improve the encoder-decoder framework using the global feature with an attention mechanism to enhance local features, helping to effectively segment diversity polyps. However, using only global information derived from the last encoder block leads to the loss of regional information from intermediate layers. Furthermore, defining the boundaries of some polyps is challenging because there is visual interference between the benign region and the polyps at the border. To address these problems, we propose two novel modules: the Cascading Context module (CCM) and the Attention Balance module (BAM), aiming to build an effective polyp segmentation model. Specifically, CCM combines the extracted regional information of the current layer and the lower layer, then pours it into the upper layer - fusing regional and global information analogous to a waterfall pattern. The BAM uses the prediction output of the adjacent lower layer as a guide map to implement the attention mechanism for the three regions separately: the background, polyp, and boundary curve. BAM enhances local context information when deriving features from the encoder block. Our proposed approach is evaluated on three benchmark datasets with six evaluation metrics for segmentation quality and gives competitive results compared to other advanced methods, for both accuracy and efficiency. Code is available at https://github.com/ntcongvn/CCBANet.

Keywords: Polyp segmentation · Semantic segmentation · Colonoscopy

---

T.-C. Nguyen, T.-P. Nguyen and G.-H. Diep—Equal contribution.

© Springer Nature Switzerland AG 2021
M. de Bruijne et al. (Eds.): MICCAI 2021, LNCS 12901, pp. 633–643, 2021.
https://doi.org/10.1007/978-3-030-87193-2_60

1   Introduction

Colorectal Cancer (CRC) is one of the most common causes of human mortality in the world, as it is responsible for 9.4% of worldwide cancer deaths, nearly 1 million cases in 2020 [12]. Therefore, early detection and accurate diagnosis are crucial for effective treatment with a low mortality rate. Colonoscopy becomes the most popular procedure for examining an abnormal region, polyps or adenoma in the colon, and preventing itfrom becoming cancerous. However, owing to numerous reasons, colonoscopy remains high miss rate detection. Hence, an automatic diagnosis system could play an auxiliary role by providing credible predictions and helping physicians not to neglect cancer signals.

Deep neural networks have shown their outstanding performance over most image processing tasks. Especially for biomedical image segmentation, deep convolutional networks based on the encoder-decoder structure has succeeded in localizing adenoma regions in both accuracy and efficiency aspect. The encoder constructed from several convolutional blocks aims to enlarge the receptive field by downscale the feature map over each layer. Reversely, the decoder is gradually upsampling, propagating context information to upper layers, and then producing the final prediction mask in the original resolution. There are also relative skip connections between the two branches to reduce information lost during encoding and achieve effective training.

The architecture is first introduced by the popular UNet [11] framework and becomes a fundamental design for almost state-of-the-art segmentation methods later. UNet++ [17] concentrates on eliminating the gap of semantic meaning between encoder and decoder layers by replacing the original skip connections with the nested and dense layers. ResUNet [16] uses the residual units instead to improve the learning performance of the original convolutional blocks. Based on the success of this improvement, ResUNet++ [6] effectively combines novel modules to strengthen the encoding process, the skip operation and synthesize better context information. U2-Net [10] defined a nested Unet model trying to capture both local and global features at each layer, guiding them with a residual loss from the lowest to the highest level feature.

PraNet [2] constraints the model to focus on the polyp boundary also. With the help from high-level feature aggregation and boundary attention block, PraNet achieves outstanding results compared to all the above methods on both accuracy and efficiency. The higher level the feature map is, the more general and global information that it contains. From this point of view, ACSNet [15] utilizes both local and global context features from the encoder branch to provide guided signals for each decoding step. However, using only global context information derived from the last encoder block on top of the decoder branch may lead to ignoring some useful information of the intermediate encoder layers. In addition, most current attention mechanism approaches focus only on the part of images such as background regional [2] or boundary curve [15].

Hence, we propose two innovative modules: Cascading Context Module (CCM) and Balancing Attention Module (BAM). The CCM helps to combine regional and global contextual information for each layer according to a waterfall

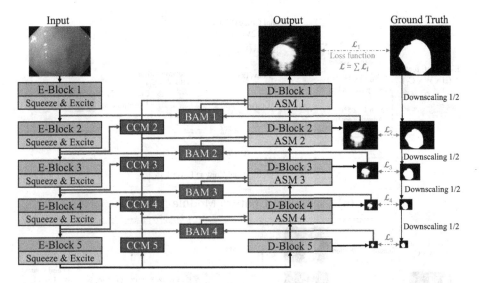

Fig. 1. Overview of the proposed CCBANet

pattern. The BAM implements the attention mechanism for the three regions separately, including the background, the polyp, and boundary site. These two module types can be integrated into the encoder-decoder framework to build an efficient semantic segmentation model.

Our main contributions are as follows. (1) A Cascading Context Module (CCM) similar to a waterfall model to extract the local and global features. (2) A Balancing Attention Module (BAM) to add attention to foreground, background, and boundary regions separately. Our proposed method is detailed in Sect. 2. Besides, we also provide extensive experiments and comparisons with other advanced methods (Sect. 3) on three benchmark datasets, two of which are seenable testing domain (Kvasir-SEG, CVC-ClinicDB) to test the learning ability and CVC-EndoSceneStill (different training and testing domain) to test the generalization capability of our model. We achieve the best Dice score compared to other methods: 92.59%, 95.43%, 85.79%, respectively, with an average inference speed of 39 frames-per-second, which is suitable for real-time testing.

2   Method

An overview of our proposed model is shown in Fig. 1, which is based on an Encoder-Decoder architecture and consists of five layers. Each layer contains two main blocks, an E-Block and a D-Block, to encode and decode the colonoscopy images. The output of each E-Block is passed through a Squeeze-and-Excite block [5] to re-calibrate the response of the feature by a channel-wise weighting mechanism. The model also integrates two types of modules, Cascading Context Module (CCM) and Balancing Attention Module (BAM), which can help to

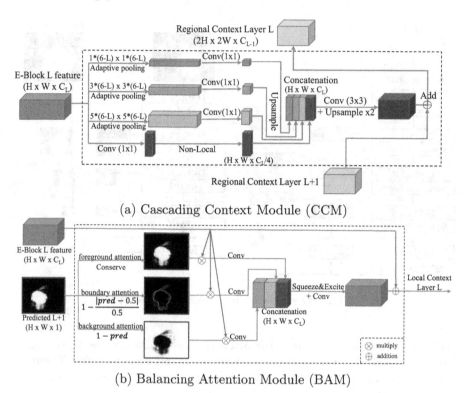

(a) Cascading Context Module (CCM)

(b) Balancing Attention Module (BAM)

Fig. 2. Cascading Context Module and Balancing Attention Module

extract regional and local context information. At each layer L, CCM receives the output feature of E-Block and extracts regional context information, then combines it with the previous one from the lower layer. Concurrently, BAM plays the role of enhancing local context information for the feature derived from E-Block by obtaining the predictive output from the previous D-Block and uses it as a guide map to separately apply attention mechanism for three regions of interest (background, foreground, and boundary). The previous D-Block's context information and output feature are synthesized through the Adaptive Select Module (ASM) before moving to the next one. We follow [15] to implement the ASM. Throughout the decoding process, each D-Block provides segmentation output with the corresponding resolution. The highest block output mask is the final prediction of the model, while others are used as guide maps for the BAM modules and supervised by the resized ground-truth, respectively.

Cascading Context Module (CCM). Motivated by using the context information from [8,9,15], we design the CCM module to capture global context as well as the regional context derived from E-Blocks of intermediate layers, thereby providing the ability to enable multi-scale receptive fields. As shown in

Fig. 2a, our CCM receives the feature from the E-Block and extracts context information with four processing branches. One of which implements a non-local operation [14] to capture the long-range dependence. The other three use adaptive average pooling with a kernel size of $1 * (6 - L)$, $3 * (6 - L)$ and $5 * (6 - L)$, respectively (for the Lth layer), later perform channel reduction by a quarter with a Conv 1×1, and upsample the features to the same spatial size. Finally, the features from the four branches are concatenated altogether, extracted for the final representative feature of this layer, and then upsampled to add pixel-wised with the regional context feature from the previous CCM layer. This mechanism helps to combine region context from many different layers. By the cascading connection of the CCMs, the regional context also contains the global context captured by the deepest CCM on top of the encoder branch.

Balancing Attention Module (BAM). While PraNet [2] only applies attention to the regional background and ACSNet [15] applies only to the boundary site, we design the BAM is more general. The BAM module implements separately and balances the attention mechanism for three regions: background, foreground (polyp), and boundary site. As shown in Fig. 2b, BAM receives the feature map from the E-Block and predictive map of the previous D-Block as input. With the obtained predictive map, the BAM calculates three attention maps: conserves the original predictive map for the foreground attention map, computes $1 - \frac{|pred - 0.5|}{0.5}$ to obtain the boundary attention map [15] and $1 - pred$ for the background attention map [2]. Then, the attention maps are channel-wise multiplied with the decoding feature, which helps to determine which part of the encoding feature needs attention for each attention map. The obtained attention features are then fed to the Conv layers to reduce the channels by one-third and then concatenated altogether. This final feature is passed through a Squeeze-and-Excite block [5] and a Conv layer to obtain the BAM output. Due to using the Squeeze-and-Excite block in the last step, BAM can rebalance the attention strategy for all three separate regionals by weighting between channels.

Novelty. *Cascading Context Module (CCM)* refers to two new ideas. First, CCM proposes the concept of regional context information at the intermediate layers, which has not been mentioned in previous studies. Global context information can be useful for segmenting large polyps, but regional context information is more useful for small polyps. Here, the global context contains the features of the entire image; however, the small polyps occupy only a small region in the image. Second, CCM proposed a new way to effectively combine the global context and the regional context information. The current approaches using only global context information derived from the last encoder block may ignore some useful information of the intermediate encoder layers. In contrast, the CCM module can capture the global context and the regional context derived from E-Blocks of all deeper layers. In this way, the context information of regions in the image is weighted more precisely.

Attention Balance Module (BAM) performs as a general attention mechanism. The current approaches focus only on a regional semantic concept for the attention mechanism by using only either background or boundary. In contrast, BAM uses the background and boundary and extends the attention mechanism on the third region of interest that is the foreground. These three regions cover all the semantic space of the entire image. Therefore directing attention to individual regions will help strengthen the discriminative regions leading to a more accurate segmentation. BAM proposed a new approach that has not been mentioned in previous studies to implement separately and balances the attention mechanism for all three regions to learn a better feature representation.

Loss Function. We use BCE-Dice Loss which is generally the default for training segmentation models. The BCE-Dice loss is applied to the result predicted by D-Block. The downsample ground truth respectively for each layer in the decoder branch calculates the residual loss by summing the losses in all layers. We assumed that the method of calculating the residual loss by upsampling the lower-resolution prediction to the ground truth size might gain unnecessary additional loss to our objective function [2,10,15].

Hyperparameter Setting. We use ResNet34 [3] as backbone model for the CCBANet. We adopt the Adam optimization algorithm with a momentum beta1 of 0.9, a momentum beta2 of 0.999, and a weight decay of 1e–5 to optimize the overall parameters of the model. A learning rate scheduler is applied to adjust the initial learning rate, and further decreases it. We follow [15], the learning rate scheduler is defined by $lr = init\_lr \times (1 - \frac{epoch}{nEpoch})^{power}$, where $init\_lr = 0.001$, $nEpoch = 200$ and $power = 0.9$.

3   Experiments

We evaluate and compare our proposed method on three benchmark datasets: Kvasir-SEG [7], CVC-ClinicDB [1] and CVC-EndoSceneStill [13]. Kvasir-SEG is the largest dataset with 1,000 various-sized images acquired and annotated by experienced doctors. CVC-ClinicDB comprises 612 colonoscopy images captured from 31 colorectal sequences of respective 31 different polyps. All frames have the same resolution of 384×288. Meanwhile, CVC-EndoSceneStill contains 912 samples in total, which is the combination of CVC-ClinicDB and CVC-ColonDB (constructed from 300 500×574 images).

We follow the same training strategy as in [2,6]. For Kavaser-SEG and CVC-ClinicDB datasets, we randomly split into training, validation, and testing sets with the ratio of 8 : 1 : 1. For the CVC-EndoSceneStill dataset, the sets are pre-divided. We refer to the setting of [15], the images of Kvasir-SEG and CVC-ClinicDB are resized to 320 × 320 then randomly cropped to 224 × 224 as input with a batch size of 14. For CVC-EndoSceneStill, we resize the images uniformly to 384 × 288 and then randomly crop to 256 × 256 as input with a batch size of 9. For ACSNET, we use a batch size of 4 and U2-NET we use a batch size of 30 for Kvasir-SEG and CVC-ClinicDB, and 22 for CVC-EndoSceneStill. For a deeper insight into the performance of the model, we use six widely-used metrics for semantic segmentation, as shown in Table 1a.

To evaluate the effectiveness of the proposed method, we compare CCBANet with eight advanced methods including U-Net [11] and a modified version of Unet'[1], U-Net++ [17], Residual U-Net [16], ResUnet++ [6], PraNet[2] [2], ACSNet [15], and U2Net [10]. Since the experiment scenarios of these methods are different, for the ease of comparison, we rerun the experiments for all methods on the same training, validation, and test sets. We also apply the same data augmentation, and all experiments are done on the same hardware environment using an NVIDIA Tesla P100 PCIe 16 GB GPU.

3.1   Experiment Results

Table 1a shows the quantitative results of our method compared to the others. Our proposed approach still achieves competitive results on all three datasets (best Dice score, best or second-best results on all metrics except Recall for CVC-EndoSceneStill). We also tested the ACSNET with a batch size of 15 on Kvasir-SEG; it converges faster than itself with smaller batch size, e.g., 4, albeit the score results do not vary much. Besides testing the learning ability of our model using the two Kvasir-SEG and CVC-ClinicDB (CVC-612) where the distribution of the train and test sets are close, we double-tested our method on CVC-EndoSceneStill, to check the ability to generalize the dataset of our model, with another setting (image size of 288 × 288 and batch size of 5) as shown in the last line of Table 1a that gives better result than the previously described setting.

For the qualitative result, Fig. 3 illustrates some example results from Kvasir-SEG, a seenable dataset: our model seems to cover the region of interest more effectively, which suggests its ability to help to reduce the rate of missed polyp segmentation if acted as an aiding system for diagnosis especially when the inference speed of our model is 39 frames per second on average - suitable for real-time prediction.

Table 1. Comparison results on three datasets and ablation study on Kvasir-SEG

(a) Comparison results on Kvasir-SEG, CVC-ClinicDB (CVC-612) and CVC-EndoSceneStill datasets. The **best**, _second best_, third best results are highlighted.

Dataset	Method	Dice↑	IoU (Jaccard)↑	Recall↑	Precision↑	Accuracy↑	F2↑
Kvasir-SEG	U-Net[c] [11]	79.94	69.35	81.51	82.91	82.17	81.79
	U-Net'[a,c]	84.57	77.21	88.09	86.49	88.78	85.00
	U-Net++[c] [17]	88.10	81.68	_91.09_	89.30	91.82	_90.73_
	Residual U-Net[c] [16]	72.50	59.84	72.42	79.35	73.12	73.71
	ResUnet++[c] [6]	83.48	75.74	87.69	83.67	88.40	86.86
	PraNet[b] [2]	_89.84_	_83.81_	_92.14_	91.12	96.53	_91.93_
	ACSNet[d] [15]	_91.38_	_84.12_	90.05	_92.74_	_97.04_	90.58
	U2-Net[d] [10]	86.88	76.80	84.02	89.94	95.58	85.14
	CCBANet[d] (Our)	**92.59**	**86.21**	**92.21**	**92.98**	**97.43**	**92.36**
CVC-ClinicDB (CVC-612)	U-Net[c] [11]	87.62	79.47	87.32	89.99	87.36	87.84
	U-Net'[a,c]	90.38	83.94	90.46	91.45	90.49	90.28
	U-Net++[c] [17]	88.77	81.35	89.08	90.39	89.13	89.34
	Residual U-Net[c] [16]	86.73	78.17	87.44	88.20	87.48	87.59
	ResUnet++[c] [6]	87.93	81.06	88.23	90.40	88.30	88.66
	PraNet[b] [2]	_94.59_	_90.26_	**95.00**	94.50	**99.23**	_94.90_
	ACSNet[d] [15]	_94.27_	_89.15_	_92.86_	_95.72_	_99.03_	93.42
	U2-Net[d] [10]	92.88	86.70	89.65	**96.34**	98.82	90.91
	CCBANet[d] (Our)	**95.43**	**91.26**	_94.79_	_96.08_	_99.22_	**95.05**
CVC-EndoSceneStill	U-Net[c] [11]	65.87	54.08	76.75	69.39	76.75	75.16
	U-Net'[a,c]	75.53	67.20	_84.90_	76.02	84.91	78.06
	U-Net++[c] [17]	75.51	67.57	_86.87_	74.14	86.88	_83.99_
	Residual U-Net[c] [16]	59.98	47.26	68.60	65.80	68.60	68.02
	ResUnet++[c] [6]	51.09	42.74	78.27	47.57	78.28	69.32
	PraNet[b] [2]	_83.62_	**76.55**	**88.33**	87.18	96.60	**88.10**
	ACSNet[d] [15]	_84.78_	_73.58_	79.37	_90.97_	_97.37_	81.45
	U2-Net[d] [10]	62.42	45.37	46.97	_93.03_	_94.77_	52.13
	CCBANet[d] (Our)	**85.79**	_75.12_	79.29	**93.45**	**97.57**	_81.77_
	Our (288x288)x5bs	86.70	76.52	81.89	93.45	97.68	83.97

(b) Ablation study for CCBANet on the Kvasir-SEG dataset.

Settings	Dice↑	IoU (Jaccard)↑	Recall↑	Precision↑	Accuracy↑	F2↑
Backbone	89.66	81.27	90.25	89.09	96.37	90.02
Backbone+CCM	91.64	84.56	90.09	93.23	97.13	90.70
Backbone+BAM(fg)	91.69	84.65	89.68	93.79	97.16	90.47
Backbone+BAM(bg)	92.15	85.44	90.52	93.84	97.31	91.17
Backbone+BAM(bo)	91.97	85.13	90.54	93.44	97.24	91.11
Backbone+BAM	92.31	85.71	92.52	92.09	97.31	92.43
Backbone+CCM+BAM	93.04	86.98	92.80	93.28	97.58	92.90

[a] U-Net': Unet with parametric relu [4] and reflection padding, source code from jeffwen
[b] PraNet is trained with image size of 352 × 352, batch size of 16 without random crop as we tested that random crop significantly reduce the performance of this model.
[c] trained with imgage size of 256 × 256, batch size of 16
[d] trained with imgage size of 224 × 224, batch size of 14 for Kvasir-SEG, CVC-ClinicDB and imgage size of 256 × 256, batch size of 9 for EndoScene. ACSNET is trained with batch size of 4. U2-Net is trained with batch size of 30, 30, 22 respectively.

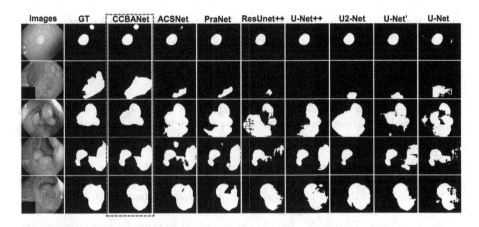

Fig. 3. Qualitative results of different methods

3.2 Ablation Study

In order to evaluate the effectiveness of the proposed method, we perform the ablation experiment with four different variations of CCBANet; the results are presented in Table 1b. We obtain the backbone of the CCBANet by removing all the CCM and BAM modules. The experiments in this table use the input size of 256×256. The backbone of CCBANet has a Dice score of 89.66%. When adding CCM and BAM alternately, the performance is improved, boosting Dice by 1.98% and 2.65%, respectively. This proves that the attention mechanism enhances local information better than using global information in the decoding process. In BAM settings, using only the background shows better results than the foreground and boundary. When combining CCM and BAM modules, the Dice score is increased by 3.38% and reaches 93.03%.

4   Conclusion

In this paper, we propose using the Cascading Context Module (CCM) to combine the global context and the regional context information, which provides the ability to enable multi-scale receptive fields to help enhance the consistency of pixel-wise segmentation. In addition, we also present the Balancing Attention Module (BAM), which implements separately and balances the attention mechanism for all three regionals: background, foreground, and boundary curve. The CCM and the BAM are flexible modules, meaning that these modules can be integrated into any encoder-decoder model to further improve the accuracy.

Acknowledgements. This research is funded by Vietnam National University HoChiMinh City (VNU-HCM) under grant number DS2020-42-01.

Gia-Han Diep was funded by Vingroup Joint Stock Company and supported by the Domestic Master/ PhD Scholarship Programme of Vingroup Innovation

Foundation (VINIF), Vingroup Big Data Institute (VINBIGDATA), code VINIF. 2020.ThS.JVN.04.

References

1. Bernal, J., Sánchez, F.J., Fernández-Esparrach, G., Gil, D., Rodríguez, C., Vilariño, F.: Wm-dova maps for accurate polyp highlighting in colonoscopy: Validation vs saliency maps from physicians. Comput Med. Imaging Graph. **43**, 99–111 (2015)
2. Fan, D.P., et al.: Pranet: parallel reverse attention network for polyp segmentation. In: Martel, A.L., Aet al. (eds.) Medical Image Computing and Computer Assisted Intervention - MICCAI 2020, pp. 263–273. Springer, Cham (2020). https://doi. org/10.1007/978-3-030-59725-2_26
3. He, K., Zhang, X., Ren, S., Sun, J.: Deep residual learning for image recognition. In: 2016 IEEE Conference on Computer Vision and Pattern Recognition (CVPR), pp. 770–778 (2016). https://doi.org/10.1109/CVPR.2016.90
4. He, K., Zhang, X., Ren, S., Sun, J.: Delving deep into rectifiers: surpassing human-level performance on imagenet classification. CoRR abs/1502.01852 (2015), http:// arxiv.org/abs/1502.01852
5. Hu, J., Shen, L., Sun, G.: Squeeze-and-excitation networks. In: 2018 IEEE/CVF Conference on Computer Vision and Pattern Recognition, pp. 7132–7141 (2018). https://doi.org/10.1109/CVPR.2018.00745
6. Jha, D., et al.: Resunet++: an advanced architecture for medical image segmentation. In: 2019 IEEE International Symposium on Multimedia (ISM), pp. 225–2255 (2019). https://doi.org/10.1109/ISM46123.2019.00049
7. Nguyen, D.V., Tran, H.T.T., Thang, T.C.: A delay-aware adaptation framework for cloud gaming under the computation constraint of user devices. In: Ro, Y.M., et al. (eds.) MMM 2020. LNCS, vol. 11962, pp. 27–38. Springer, Cham (2020). https://doi.org/10.1007/978-3-030-37734-2_3
8. Li, H., Xiong, P., An, J., Wang, L.: Pyramid attention network for semantic segmentation (2018)
9. Lin, C.Y., Chiu, Y.C., Ng, H.F., Shih, T.K., Lin, K.H.: Global-and-local context network for semantic segmentation of street view images. Sensors **20**(10) (2020). https://doi.org/10.3390/s20102907, https://www.mdpi.com/1424-8220/ 20/10/2907
10. Qin, X., Zhang, Z., Huang, C., Dehghan, M., Zaiane, O.R., Jagersand, M.: U2-net: going deeper with nested u-structure for salient object detection. Pattern Recogn. **106**, 107404 (2020). https://doi.org/10.1016/j.patcog.2020.107404, http://dx.doi. org/10.1016/j.patcog.2020.107404
11. Ronneberger, O., Fischer, P., Brox, T.: U-net: convolutional networks for biomedical image segmentation. In: Navab, N., Hornegger, J., Wells, W.M., Frangi, A.F. (eds.) Medical Image Computing and Computer-Assisted Intervention - MICCAI 2015, pp. 234–241. Springer, Cham (2015). https://doi.org/10.1007/978-3-319-24574-4_28
12. Sung, H., et al.: Global cancer statistics 2020: globocan estimates of incidence and mortality worldwide for 36 cancers in 185 countries. CA Cancer J. Clin. (2021)
13. Vázquez, D., et al.: A benchmark for endoluminal scene segmentation of colonoscopy images. J. Healthcare Eng. **2017** (2017)

14. Wang, X., Girshick, R., Gupta, A., He, K.: Non-local neural networks. In: 2018 IEEE/CVF Conference on Computer Vision and Pattern Recognition, pp. 7794–7803 (2018). https://doi.org/10.1109/CVPR.2018.00813

15. Zhang, R., Li, G., Li, Z., Cui, S., Qian, D., Yu, Y.: Adaptive context selection for polyp segmentation. In: Martel, A.L., et al. (eds.) Medical Image Computing and Computer Assisted Intervention - MICCAI 2020, pp. 253–262. Springer, Cham (2020). https://doi.org/10.1007/978-3-030-59725-2_25

16. Zhang, Z., Liu, Q., Wang, Y.: Road extraction by deep residual u-net. IEEE Geosci. Remote Sens. Lett. **15**(5), 749–753 (2018)

17. Zhou, Z., Rahman Siddiquee, M.M., Tajbakhsh, N., Liang, J.: UNet++: a nested u-net architecture for medical image segmentation. In: Stoyanov, D., et al. (eds.) DLMIA/ML-CDS -2018. LNCS, vol. 11045, pp. 3–11. Springer, Cham (2018). https://doi.org/10.1007/978-3-030-00889-5_1

Point-Unet: A Context-Aware Point-Based Neural Network for Volumetric Segmentation

Ngoc-Vuong Ho[1]([✉]), Tan Nguyen[1], Gia-Han Diep[3], Ngan Le[4],
and Binh-Son Hua[1,2]

[1] VinAI Research, Hanoi, Vietnam
{v.vuonghn,v.tannh10,v.sonhb}@vinai.io
[2] VinUniversity, Hanoi, Vietnam
[3] University of Science, VNU-HCM, Ho Chi Minh, Vietnam
han.diep@ict.jvn.edu.vn
[4] Department of Computer Science and Computer Engineering,
University of Arkansas, Fayetteville 72701, USA
thile@uark.edu

Abstract. Medical image analysis using deep learning has recently been prevalent, showing great performance for various downstream tasks including medical image segmentation and its sibling, volumetric image segmentation. Particularly, a typical volumetric segmentation network strongly relies on a voxel grid representation which treats volumetric data as a stack of individual voxel 'slices', which allows learning to segment a voxel grid to be as straightforward as extending existing image-based segmentation networks to the 3D domain. However, using a voxel grid representation requires a large memory footprint, expensive test-time and limiting the scalability of the solutions. In this paper, we propose *Point-Unet*, a novel method that incorporates the efficiency of deep learning with 3D point clouds into volumetric segmentation. Our key idea is to first predict the regions of interest in the volume by learning an attentional probability map, which is then used for sampling the volume into a sparse point cloud that is subsequently segmented using a point-based neural network. We have conducted the experiments on the medical volumetric segmentation task with both a small-scale dataset Pancreas and large-scale datasets BraTS18, BraTS19, and BraTS20 challenges. A comprehensive benchmark on different metrics has shown that our context-aware Point-Unet robustly outperforms the SOTA voxel-based networks at both accuracies, memory usage during training, and time consumption during testing. Our code is available at https://github.com/VinAIResearch/Point-Unet.

Keywords: Volumetric segmentation · Medical image segmentation · Medical representation · Point cloud

Electronic supplementary material The online version of this chapter (https://doi.org/10.1007/978-3-030-87193-2_61) contains supplementary material, which is available to authorized users.

M. de Bruijne et al. (Eds.): MICCAI 2021, LNCS 12901, pp. 644–655, 2021.
https://doi.org/10.1007/978-3-030-87193-2_61

1   Introduction

Medical image segmentation has played an important role in medical analysis and is widely developed for many clinical applications. Although deep learning can achieve accuracy close to human performance for many computer vision tasks on 2D images, it is still challenging and limited for applying to medical imaging tasks such as volumetric segmentation. Existing voxel-based neural networks for volumetric segmentation have prohibitive memory requirements: nnNet [12] uses a volume patch size of $160 \times 192 \times 128$, which requires a GPU with 32 GB of memory for training to achieve the state-of-the-art performance [1]. To mitigate high memory usage, some previous work resort to workarounds such as using smaller grid size (e.g., 25^3 and 19^3 in DeepMedic [22]) for computation, resulting in degraded performance.

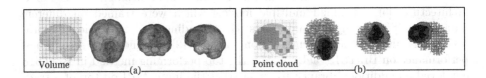

Fig. 1. (a): a 2D voxel grid and a segmentation rendered by volume rendering; (b): a PC from the grid and the point-based segmentation results.

In this work, we propose to leverage the 3D PC representation for the problem of medical volumetric segmentation as inspired by recent success in 3D point cloud (PC) analysis [8,20,28,29,37]. Having a PC representation is advantageous in that we can have fine-grained control of the segmentation quality, i.e., to sample the volume and focus points at the important areas. PCs are also suitable for capturing global features that are challenging and costly to have with a regular voxel grid. A summary of the difference between PC and voxel grid on an MRI image is shown in Fig. 1.

Our so-called *Point-Unet* is a point-based volumetric segmentation framework with three main modules: the saliency attention, the context-aware sampling, and the point-based segmentation module. The saliency attention module takes a volume as input and predicts an attentional probability map that guides the context-aware point sampling in the subsequent module to transform the volume into a PC. The point-based segmentation module then processes the PCs and outputs the segmentation, which is finally fused back to the volume to obtain the final segmentation results.

In summary, our main contributions in this work are: (1) Point-Unet, a new perspective and formulation to solve medical volumetric segmentation using a PC representation; (2) A saliency proposal network to extract an attentional probability map which emphasizes the regions of interests in a volume; (3) An efficient context-aware point sampling mechanism for capturing better local dependencies within regions of interest while maintaining global relations; (4) A comprehensive benchmark that demonstrates the advantage of our point-based method over other SOTA voxel-based 3D networks at both accuracies, memory usage during training, and inference time.

2 Related Work

Volumetric Segmentation. Deep learning-based techniques, especially CNNs, have shown excellent performance in the volumetric medical segmentation. Early methods include the standard Unet [30], Vnet [24], and then DeepMedic [22], which improves robustness with multi-scale segmentation. Recently, by utilizing hard negative mining to achieve the final voxel-level classification, [16] improved the patch-based CNNs performance. KD-Net [7] fused information from different modalities through knowledge distillation. Instead of picking the best model architecture, [5] ensembled multiple models which were trained on different datasets or different hyper-parameters. By extending U-Net with leaky ReLU activations and instance normalization, nnNet [12] obtained the second-best performance on BraTS18. aeUnet [25], the top-performing method in BraTS18, employed an additional branch to reconstruct the input MRI on top of a traditional encoder-decoder 3D CNN architecture. The top performance of BraTS19 is [14], which is a two-stage cascaded U-Net. The first stage had a U-Net architecture. In the second stage, the output of the first stage was concatenated to the original input and fed to a similar encoder-decoder to obtain the final segmentation.

Point Cloud Segmentation. In 3D deep learning, the semantic segmentation task can be solved by directly analyzing PCs data. Many point-based techniques have been recently developed for PC semantic segmentation [8,18,28]. PointNet [28] used MLPs to learn the representation of each point, whereas the global representation was extracted by applying a symmetric function like max pooling on the per-point features. PointNet++ [29] was then developed to address the lack of local features by using a hierarchy of PointNet itself to capture local geometric features in a local point neighborhood. PointCNN [20] used a $\mathcal{X}$-transformation to learn features from unstructured PCs. In order to extract richer edge features [18] proposed SuperPoint Graph where each superpoint was embedded in a PointNet and then refined by RNNs. Inspired by the idea of the attention mechanism, [36] proposed a graph-based convolution with attention to capture the structural features of PCs while avoiding feature contamination between objects. Recently, RandLA-Net [8] has achieved SOTA performance on semantic segmentation of large point clouds by leveraging random sampling at inference. In this work, we aim to bring the efficiency of deep learning with point clouds into volumetric segmentation for medical 3D data.

3   Proposed Point-Unet

Our proposed Point-Unet for volumetric segmentation contains three modules i.e., to saliency attention module, context-aware sampling and point-based segmentation module. The overall architecture is given in Fig. 2.

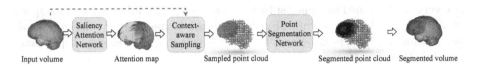

Fig. 2. Point-Unet takes a volume as input and consists of 3 modules: saliency attention network, context-aware sampling and point segmenation network.

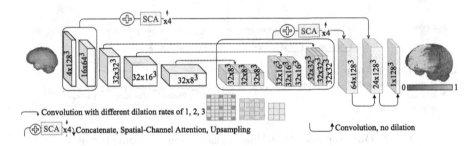

Fig. 3. Our proposed saliency attention network.

3.1   Saliency Attention

Our saliency attention network is leveraged by [3,41], and designed as contextual pyramid to capture multi-scale with multi-receptive-field at high-level features. The network is illustrated in Fig. 3 and contains two high-level layers and two low-level layers. At the high-level features, we adopt atrous convolution with different dilation rates set to 1, 2, and 3 to capture multi receptive field context information. The feature maps from different atrous layers are then combined by concatenation while the smaller ones are upsampled to the largest one. Then, we combine them by cross channel concatenation and channel-wise attention (SCA) [3] as the output of the high-level feature extraction. At the low-level features, we apply SCA [3] to combine two low-level features maps after upsampling the smaller ones. The high-level feature is then upsampled and combined with the low-level feature to form a feature map at original resolution.

3.2 Context-Aware Sampling

Random sampling (RS) used in the original RandLA-Net [8] has been successfully applied into 3D shapes, but it is not a good sampling technique for medical volumetric data because of following reasons: (i) there is no mechanism in RS to handle intra-imbalance; (ii) topological structure is important in medical analysis but there is no attention mechanism in RS to focus on the object boundary which are very weak in medical images; (iii) RS samples points all over the data space, it may skip small objects while objects of interest in medical are relatively small; (iv) volumetric data is large and RS requires running inference multiple times which is time consuming.

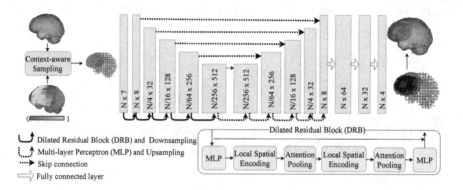

Fig. 4. Our proposed Point-Unet segmentation on volumetric data.

Our context-aware sampling is designed to take all such limitations into account. Here the main conceptual differences are that our sampling is a single pass and our sampling only samples i.e., it is necessary just to sample the volume and perform the inference once. Our context-aware sampling is as follows. Firstly, we sample the points identified by the attentional probability map obtained by the saliency attention module. On the salient region where the probabilities are higher, we densely sample points to better learn contextual local representation. In the non-salient region, we apply random sampling to keep global relations. By doing that, our context-aware sampling can capture better local dependencies within regions of interest while maintaining global relations.

3.3 Point-Based Segmentation

Given a volumetric data, we first sample a PC by Subsect. 3.2. Our point-based segmentation departs from the original design of RandLA-Net [8] in that we introduce a context-aware sampling technique for effectively sampling PCs from a volume. We also redesign RandLA-Net under Unet architecture [30] together with an appropriate loss to better fit the task of medical volumetric segmentation.

Table 1. Comparison on BraTS18. The **best**, second best and *third best* are highlighted.

BraTS18	Offline validation set			Online validation set			
Methods	Dice score ↑		HD95 ↓	Methods	Dice score ↑		HD95 ↓
	ET/WT/TC	AVG	AVG		ET/WT/TC	AVG	AVG
3DUnet [15]	66.82/81.19/77.58	75.20	9.32	3DUnet [35]	72.05/84.24/76.41	77.56	17.83
3DUNet [11]	68.43/*89.91*/86.77	81.70	**5.33**	KD-Net [7]	71.67/81.45/76.98	76.70	–
h-Dense [19]	70.69/89.51/82.76	80.99	6.41	DenseNet [39]	*80.00*/90.00/82.00	84.00	–
DMF [2]	*76.35*/89.09/82.70	82.71	–	aeUnet [25]^b	**81.45**/*90.42*/**85.96**	**85.94**	5.52
aeUnet [25]	75.31/85.69/81.98	80.99	8.64	aeUnet [25]^a	72.60/85.02/77.33	78.32	18.58
S3D [4]	73.95/88.81/*84.42*	*82.39*	*5.40*	S3D [4]	74.93/89.35/83.09	82.56	*5.63*
nnNet [12]	*76.65*/81.57/84.21	80.81	12.59	nnNet [12]^b	79.59/**90.80**/*84.32*	*84.90*	**5.36**
KaoNet [16]	73.50/*90.20*/81.30	81.67	*5.92*	nnNet [12]^a	75.01/82.23/81.84	79.69	7.14
RandLA [8]	70.05/88.13/80.32	79.50	6.36	RandLA [8]	73.05/87.30/76.94	79.10	5.79
Ours	**80.76/90.55/87.09**	**86.13**	6.01	**Ours**	*80.97/90.50/84.11*	*85.19*	6.30

^{a} Reproduce the results on the network trained with 100 epochs.
^{b} The results reported in the paper.

The architecture of the proposed point-based segmentation network is illustrated in Fig. 4. Our network takes N points as its input $\{p_i\}_{i=1}^{N}$, each point $p_i = \{x^i, f^i\}$ where x^i is a tuple of coordinates in sagittal, coronal, transverse planes and f^i is a tuple of point features including point intensity in T_1, T_2, T_{ce}, and *Flair* modalities.

The input PC is first processed through encoder path, which contains sequences Dilated Residual Block (DRB) and downsampling. Each DRB includes multiple units of multi-layer perceptrons (MLP), local spatial encoding and attentive pooling stacks and the DRBs are connected through skip-connections as proposed in RandLA-Net [8]. The output of the encoder path is then processed through the decoder path, which consists of a sequence of MLP and upsampling. The network also makes use of the skip connection while upsampling to learn feature representation at multi-scale. Finally, the decoder output is passed through three fully connected layers with drop-out regularization.

Far apart from RandLA-Net[8] and most existing segmentation networks, which use the cross-entropy (CE) loss, we utilize Generalized Dice Loss (GDL) [32] for training. GDL [32] has been proven to be efficient at dealing with data imbalance problems that often occur in medical image segmentation. GDL is computed as $GDL = 1 - 2\frac{\sum_{l=1} w_l \sum_n r_{ln} p_{ln}}{\sum_{l=1} w_l \sum_n (r_{ln}+p_{ln})}$ where $w_l = \frac{1}{\sum_n^N r_{ln}}$ is used to provide invariance to different label set properties. R is gold standard with value at voxel n^{th} denoted as r_n. P the predicted segmentation map with value at voxel n^{th} denoted as p_n. For class l, the groundtruth and predicted labels are r_{ln} and p_{ln}.

4   Experimental Results

We evaluate our method and compare to the state-of-the-art methods on two datasets: Pancreas and BraTS. Pancreas [31] contains 82 abdominal contrast

Table 2. Comparison on BraTS19. The **best**, second best and *third best* are highlighted.

BraTS19 Methods	Offline validation set ET/WT/TC	AVG	AVG	Online validation set Methods	ET/WT/TC	ET/WT/TC	AVG
	Dice score ↑		HD95 ↓		Dice score ↑	HD95 ↓	
3DUnet [15]	67.74/80.17/78.92	75.61	11.69	3DUnet [35]	64.26/79.65/72.07	72.00	23.68
nnNet [12]	79.46/81.13/*87.08*	82.67	6.75	nnNet [12]a	70.42/81.53/78.22	76.72	8.30
aeUnet [25]	80.55/86.26/85.78	84.19	10.94	aeUnet [25]a	64.81/83.02/74.48	74.10	21.75
HNF [13]	*80.96*/*91.12*/86.40	*86.16*	–	HNF [13]	**81.16**/**91.12**/84.52	85.60	3.81
N3D [34]	83.0/**91.60**/**88.80**	87.35	**3.58**	Synth [6]	76.65/*89.65*/79.01	81.77	5.75
				2stage [14]b	*79.67*/90.80/**85.89**	85.45	**3.74**
				3Unet [34]	73.70/89.40/80.70	81.27	5.84
				3DSe [26]	80.00/89.40/*83.40*	*84.27*	*4.91*
				Bag-trick [42]c	70.20/88.30/80.00	79.50	4.93
RandLA [8]	76.68/89.01/84.81	83.50	4.45	RandLA [8]	70.77/86.95/74.27	70.77	7.09
Ours	**85.67**/91.18/**90.10**	**88.98**	4.92	**Ours**	79.01/87.63/79.70	82.11	10.39

a Reproduce the results on the network trained with 100 epochs.
b The first place of BraTS19. We choose the Ensemble of 5-fold.
c The second place on BraTS19.

Table 3. Comparison on BraTS20. The **best**, second best and *third best* are highlighted.

BraTS20 Methods	Offline validation set ET/WT/TC	AVG	AVG	Online validation set Methods	ET/WT/TC	AVG	AVG
	Dice score ↑		HD95 ↓		Dice score ↑		HD95 ↓
3DUnet [15]	66.92/82.86/72.98	74.25	30.19	3DUNet [35]	67.66/87.35/79.30	78.10	21.16
nnNet [12]	*73.64*/80.99/*81.60*	*78.74*	*14.33*	nnNet [12]a	68.69/81.34/78.06	78.03	24.30
aeUnet [25]	*71.31*/*84.72*/*79.02*	*78.35*	15.43	nnUNet [10]c	*77.67*/**90.60**/**84.26**	**84.18**	15.30
				aeNet [25]a	64.00/83.16/74.66	73.95	33.91
				Cascade [21]4	**78.81**/*89.92*/*82.06*	*83.60*	*12.00*
				KiUNet [33]	73.21/87.60/73.92	78.24	**8.38**
RandLA [8]	67.40/*87.74*/76.85	77.33	**7.03**	RandLA [8]	66.31/88.01/77.03	77.17	16.65
Ours	**76.43**/**89.67**/**82.97**	**83.02**	8.26	**Ours**	**78.98**/*89.71*/*82.75*	**83.81**	11.73

a Reproduce the results on the network trained with 100 epochs.
b We choose the model with as similar batch size as ours.
c We choose single model with 190 epoches, stage 1. The best model at Brats2020.

enhanced 3D CT scans. The CT scans have resolutions of 512×512 pixels with varying pixel sizes and slice thickness between 1.5–2.5 mm. BraTS [23] consists of a large-scale brain tumor dataset. The training set includes 285/335/369 patients and validation set contains 66/125/125 patients in BraTS18/BraTS19/BraTS20. Each image is registered to a common space, sampled to an isotropic with skull-stripped and has a dimension of $240 \times 240 \times 155$. For point sampling in our training and inference, we sample 180,000 points in Pancreas, and 350,000 points in BraTS.

Evaluation Setup: For each dataset, we experiment on both offline validation set and online validation set. The evaluation on the offline validation set is conducted locally by partitioning the training set into training subset (80%) and evaluation subset (20%). We train the network on TensorFlow empowered by Tensorpack [9] to speed up the training. We use Momentum optimizer with momentum value 0.9, learning rate 0.01 with decay, and batch size 2. The model is trained on an NVIDIA Tesla V100 32 GB GPU for 100 epochs. For BraTS, we compare our method with SOTA voxel-based results, and with RandLA-Net [8], the SOTA point-based segmentation method. Note that our evaluation is done on volume as our point-based segmentation results can be transferred directly to the volume without further processing thanks to our sampling scheme.

Evaluation Results: The evaluation on Brats is given in Tables 1, 2 and 3 for BraTS 2018, BraTS 2019, and BraTS 2020, respectively. Whereas the evaluation on Pancreas is given in Table 4. On Brats, there are two groups of methods corresponding to SOTA voxel-based and point-based reported in each table. While our Point-Unet achives either better or competitive performance compared to SOTA voxel-based methods, it is better than SOTA point-based method (RandLA-Net) at both Dice score and HD95. For nnNet [12] and aeNet [25], we train and reproduce the resutls. It shows that the reproduced results are always lower than the ones reported which were postprocessed. Without postprocessing, our results outperforms both nnNet [12] and aeNet [25]. In other words, our Point-Unet obtains SOTA performance on both offline validation set and online validation set without postprocessing. Not only on the large-scale dataset such as Brats, our Point-Unet also obtains the SOTA performance on small-scale dataset such as Pancreas as shown Table 4.

Please also refer to the supplementary material for full comparisons on Pancreas, an ablation study of our network, and other implementation details.

Table 4. Dice score comparison on Pancreas dataset.

Method	Average ↑	Method	Average ↑
Oktay et al. [27]	83.10 ± 3.80	Yu et al. [40]	84.50 ± 4.97
Zhu et al. [43]	84.59 ± 4.86	**Ours**	**85.68** ±5.96

Performance Analysis: Figure 5(a) shows the comparison between our Point-Unet against RandLA-Net [8] in terms of the number of iterations performed during inference. The experiment is conducted on BraTS20 offline validation set. The performance of RandLA-Net with RS strategy highly depend on the number of iterations. It reaches the best performance when RS covers the entire volume, which requires up to eight iterations. By using context-aware sampling, our PC covers regions of interest in just a single iteration while outperforming RandLA-Net. Figure 5(b) provides the memory requirement during training with batch size set to 1 on different input volume patch sizes. We also measure inference time by three runs and then take the averages. In general, for voxel-based networks, a smaller patch size requires less memory during training but it takes more time at inference. By contrast, point-based networks including our Point-Unet and RandLA-Net [8] require much less memory to handle the entire volume while keeping the inference time plausible.

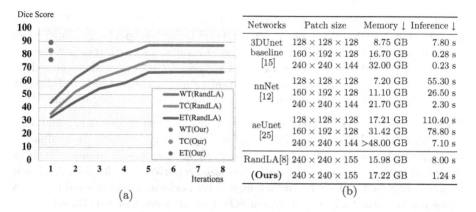

Fig. 5. Performance analysis. (a) With a single inference, our Point-Unet outperforms RandLA-Net, which requires multiple iterations at inference. (b) Memory requirement for training with batch size 1 and inference time with difference volume patch sizes.

5 Conclusion

In this work, we introduced Point-Unet, a point-based framework for volumetric segmentation. We tested our framework on the problem of brain tumor and pancreas segmentation and showed that our point-based neural segmentation is robust, scalable, and more accurate than existing voxel-based segmentation methods. Future investigations might aim for better techniques for volume-point sampling and label reconstruction. Techniques for segmentation boundary adaptive sampling [17] and attention-based convolution [38] are also potential extensions for performance improvement.

References

1. Brügger, R., Baumgartner, C.F., Konukoglu, E.: A partially reversible U-Net for memory-efficient volumetric image segmentation. In: Shen, D., et al. (eds.) MIC-CAI 2019. LNCS, vol. 11766, pp. 429–437. Springer, Cham (2019). https://doi.org/10.1007/978-3-030-32248-9_48

2. Chen, C., Liu, X., Ding, M., Zheng, J., Li, J.: 3D dilated multi-fiber network for real-time brain tumor segmentation in MRI. In: Shen, D., et al. (eds.) MICCAI 2019. LNCS, vol. 11766, pp. 184–192. Springer, Cham (2019). https://doi.org/10.1007/978-3-030-32248-9_21

3. Chen, L., et al.: SCA-CNN: spatial and channel-wise attention in convolutional networks for image captioning. In: CVPR (2017)

4. Chen, W., Liu, B., Peng, S., Sun, J., Qiao, X.: S3D-UNet: separable 3D U-Net for brain tumor segmentation. In: BrainLes 2018. LNCS, vol. 11384, pp. 358–368. Springer, Cham (2019). https://doi.org/10.1007/978-3-030-11726-9_32

5. Feng, X., Tustison, N.J., Patel, S.H., Meyer, C.H.: Brain tumor segmentation using an ensemble of 3D U-Nets and overall survival prediction using radiomic features. Front. Comput. Neurosci. **14**, 25 (2020)

6. Hamghalam, M., Lei, B., Wang, T.: Brain tumor synthetic segmentation in 3D multimodal MRI scans. In: Crimi, A., Bakas, S. (eds.) BrainLes 2019. LNCS, vol. 11992, pp. 153–162. Springer, Cham (2020). https://doi.org/10.1007/978-3-030-46640-4_15

7. Hu, M., et al.: Knowledge distillation from multi-modal to mono-modal segmentation networks. In: Martel, A.L., et al. (eds.) MICCAI 2020. LNCS, vol. 12261, pp. 772–781. Springer, Cham (2020). https://doi.org/10.1007/978-3-030-59710-8_75

8. Hu, Q., et al.: RandLA-Net: efficient semantic segmentation of large-scale point clouds. In: CVPR (2020)

9. Huf, P., Carminati, J.: TensorPack: a maple-based software package for the manipulation of algebraic expressions of tensors in general relativity. J. Phys: Conf. Ser. **633**, 012021 (2015)

10. Isensee, F., Jaeger, P.F., Full, P.M., Vollmuth, P., Maier-Hein, K.H.: nnU-Net for brain tumor segmentation (2020)

11. Isensee, F., Kickingereder, P., Wick, W., Bendszus, M., Maier-Hein, K.H.: Brain tumor segmentation and radiomics survival prediction: contribution to the BRATS 2017 challenge. In: Crimi, A., Bakas, S., Kuijf, H., Menze, B., Reyes, M. (eds.) BrainLes 2017. LNCS, vol. 10670, pp. 287–297. Springer, Cham (2018). https://doi.org/10.1007/978-3-319-75238-9_25

12. Isensee, F., Kickingereder, P., Wick, W., Bendszus, M., Maier-Hein, K.H.: No new-net. In: Crimi, A., Bakas, S., Kuijf, H., Keyvan, F., Reyes, M., van Walsum, T. (eds.) BrainLes 2018. LNCS, vol. 11384, pp. 234–244. Springer, Cham (2019). https://doi.org/10.1007/978-3-030-11726-9_21

13. Jia, H., Xia, Y., Cai, W., Huang, H.: Learning High-resolution and efficient non-local features for brain glioma segmentation in MR images. In: Martel, A.L., et al. (eds.) MICCAI 2020. LNCS, vol. 12264, pp. 480–490. Springer, Cham (2020). https://doi.org/10.1007/978-3-030-59719-1_47

14. Jiang, Z., Ding, C., Liu, M., Tao, D.: Two-stage cascaded U-Net: 1st place solution to BraTS challenge 2019 segmentation task. In: Crimi, A., Bakas, S. (eds.) BrainLes 2019. LNCS, vol. 11992, pp. 231–241. Springer, Cham (2020). https://doi.org/10.1007/978-3-030-46640-4_22

15. Kamnitsas, K., et al.: Ensembles of multiple models and architectures for robust brain tumour segmentation. In: Crimi, A., Bakas, S., Kuijf, H., Menze, B., Reyes, M. (eds.) BrainLes 2017. LNCS, vol. 10670, pp. 450–462. Springer, Cham (2018). https://doi.org/10.1007/978-3-319-75238-9_38

16. Kao, P.-Y., Ngo, T., Zhang, A., Chen, J.W., Manjunath, B.S.: Brain tumor segmentation and tractographic feature extraction from structural mr images for overall survival prediction. In: Crimi, A., Bakas, S., Kuijf, H., Keyvan, F., Reyes, M., van Walsum, T. (eds.) BrainLes 2018. LNCS, vol. 11384, pp. 128–141. Springer, Cham (2019). https://doi.org/10.1007/978-3-030-11726-9_12

17. Kirillov, A., Wu, Y., He, K., Girshick, R.: PointRend: image segmentation as rendering. In: CVPR (2020)

18. Landrieu, L., Simonovsky, M.: Large-scale point cloud semantic segmentation with superpoint graphs. In: CVPR (2018)

19. Li, X., Chen, H., Qi, X., Dou, Q., Fu, C.W., Heng, P.A.: H-DenseUNet: hybrid densely connected UNet for liver and tumor segmentation from CT volumes. IEEE TMI **37**(12), 2663–2674 (2018)

20. Li, Y., Bu, R., Sun, M., Chen, B.: PointCNN: convolution on x-transformed points. In: NIPS (2018)

21. Lyu, C., Shu, H.: A two-stage cascade model with variational autoencoders and attention gates for MRI brain tumor segmentation (2020)

22. McKinley, R., Meier, R., Wiest, R.: Ensembles of densely-connected CNNs with label-uncertainty for brain tumor segmentation. In: Crimi, A., Bakas, S., Kuijf, H., Keyvan, F., Reyes, M., van Walsum, T. (eds.) BrainLes 2018. LNCS, vol. 11384, pp. 456–465. Springer, Cham (2019). https://doi.org/10.1007/978-3-030-11726-9_40

23. Menze, B.H., Jakab, A., Bauer, S., et al.: The multimodal brain tumor image segmentation benchmark (BRATS). IEEE TMI **34**(10), 1993–2024 (2015)

24. Milletari, F., Navab, N., Ahmadi, S.: V-Net: fully convolutional neural networks for volumetric medical image segmentation. CoRR (2016)

25. Myronenko, A.: 3D MRI brain tumor segmentation using autoencoder regularization. In: Crimi, A., Bakas, S., Kuijf, H., Keyvan, F., Reyes, M., van Walsum, T. (eds.) BrainLes 2018. LNCS, vol. 11384, pp. 311–320. Springer, Cham (2019). https://doi.org/10.1007/978-3-030-11726-9_28

26. Myronenko, A., Hatamizadeh, A.: Robust semantic segmentation of brain tumor regions from 3D MRIs. In: Crimi, A., Bakas, S. (eds.) BrainLes 2019. LNCS, vol. 11993, pp. 82–89. Springer, Cham (2020). https://doi.org/10.1007/978-3-030-46643-5_8

27. Oktay, O., Schlemper, J., Folgoc, L.L., et al.: Attention U-Net: learning where to look for the pancreas. arXiv preprint arXiv:1804.03999 (2018)

28. Qi, C.R., Su, H., Mo, K., Guibas, L.J.: PointNet: deep learning on point sets for 3D classification and segmentation. In: CVPR (2017)

29. Qi, C.R., Yi, L., Su, H., Guibas, L.J.: PointNet++: deep hierarchical feature learning on point sets in a metric space. In: NIPS (2017)

30. Ronneberger, O., Fischer, P., Brox, T.: U-Net: convolutional networks for biomedical image segmentation. arXiv e-prints arXiv:1505.04597 (May 2015)

31. Roth, H.R., Farag, A., et al.: Data from pancreas-CT (2016)

32. Sudre, C.H., Li, W., Vercauteren, T., Ourselin, S., Jorge Cardoso, M.: Generalised dice overlap as a deep learning loss function for highly unbalanced segmentations. In: DLMIA/ML-CDS@MICCAI (2017)

33. Valanarasu, J., Sindagi, V.A., Hacihaliloglu, I., Patel, V.M.: KiU-Net: overcomplete convolutional architectures for biomedical image and volumetric segmentation (2020)

34. Wang, F., Jiang, R., Zheng, L., Meng, C., Biswal, B.: 3D U-Net based brain tumor segmentation and survival days prediction. In: Crimi, A., Bakas, S. (eds.) BrainLes 2019. LNCS, vol. 11992, pp. 131–141. Springer, Cham (2020). https://doi.org/10.1007/978-3-030-46640-4_13

35. Wang, G., Li, W., Ourselin, S., Vercauteren, T.: Automatic brain tumor segmentation using cascaded anisotropic convolutional neural networks. In: Crimi, A., Bakas, S., Kuijf, H., Menze, B., Reyes, M. (eds.) BrainLes 2017. LNCS, vol. 10670, pp. 178–190. Springer, Cham (2018). https://doi.org/10.1007/978-3-319-75238-9_16

36. Wang, L., Huang, Y., Hou, Y., Zhang, S., Shan, J.: Graph attention convolution for point cloud semantic segmentation. In: CVPR (2019)

37. Wang, Y., Sun, Y., Liu, Z., Sarma, S.E., Bronstein, M.M., Solomon, J.M.: Dynamic graph CNN for learning on point clouds. ACM Trans. Graph. **38**, 1–12 (2019)

38. Woo, S., Park, J., Lee, J.-Y., Kweon, I.S.: CBAM: convolutional block attention module. In: Ferrari, V., Hebert, M., Sminchisescu, C., Weiss, Y. (eds.) ECCV 2018. LNCS, vol. 11211, pp. 3–19. Springer, Cham (2018). https://doi.org/10.1007/978-3-030-01234-2_1

39. Bangalore Yogananda, C.G., et al.: A fully automated deep learning network for brain tumor segmentation. Tomography **6**(2), 186–193 (2020)

40. Yu, Q., Xie, L., Wang, Y., Zhou, Y., Fishman, E.K., Yuille, A.L.: Recurrent saliency transformation network: incorporating multi-stage visual cues for small organ segmentation. In: CVPR (2018)
41. Zhao, T., Wu, X.: Pyramid feature attention network for saliency detection. In: CVPR (2019)
42. Zhao, Y.-X., Zhang, Y.-M., Liu, C.-L.: Bag of tricks for 3D MRI brain tumor segmentation. In: Crimi, A., Bakas, S. (eds.) BrainLes 2019. LNCS, vol. 11992, pp. 210–220. Springer, Cham (2020). https://doi.org/10.1007/978-3-030-46640-4_20
43. Zhu, Z., Xia, Y., Shen, W., Fishman, E., Yuille, A.: A 3D coarse-to-fine framework for volumetric medical image segmentation. In: 3DV, pp. 682–690. IEEE (2018)

TUN-Det: A Novel Network for Thyroid Ultrasound Nodule Detection

Atefeh Shahroudnejad[1,2], Xuebin Qin[1,2(✉)], Sharanya Balachandran[1,2],
Masood Dehghan[1,2], Dornoosh Zonoobi[2], Jacob Jaremko[1,2], Jeevesh Kapur[2,3],
Martin Jagersand[1], Michelle Noga[1], and Kumaradevan Punithakumar[1]

[1] University of Alberta, Edmonton, Canada
xuebin@ualberta.ca
[2] Medo.ai, Alberta, Canada
[3] National University Hospital, Singapore, Singapore

Abstract. This paper presents a novel one-stage detection model, TUN-Det, for thyroid nodule detection from ultrasound scans. The main contributions are (i) introducing Residual U-blocks (RSU) to build the backbone of our TUN-Det, and (ii) a newly designed multi-head architecture comprised of three parallel RSU variants to replace the plain convolution layers of both the classification and regression heads. Residual blocks enable each stage of the backbone to extract both local and global features, which plays an important role in detection of nodules with different sizes and appearances. The multi-head design embeds the ensemble strategy into one end-to-end module to improve the accuracy and robustness by fusing multiple outputs generated by diversified sub-modules. Experimental results conducted on 1268 thyroid nodules from 700 patients, show that our newly proposed RSU backbone and the multi-head architecture for classification and regression heads greatly improve the detection accuracy against the baseline model. Our TUN-Det also achieves very competitive results against the state-of-the-art models on overall Average Precision (AP) metric and outperforms them in terms of AP_{35} and AP_{50}, which indicates its promising performance in clinical applications. The code is available at: https://github.com/Medo-ai/TUN-Det.

Keywords: Thyroid nodule detection · Deep convolutional networks · Ultrasound image · Multi-scale features · Multi-head architecture

1 Introduction

Ultrasound (US) is the primary diagnostic tool for both the detection and characterization of thyroid nodules. As part of clinical workflow in thyroid sonography, thyroid nodules are measured and their sizes are monitored over time as significant growth could be a sign of thyroid cancer. Hence, finding Region of Interest (ROI) of nodules for further processing becomes the preliminary step

---

A. Shahroudnejad and X. Qin—Equal contribution.

© Springer Nature Switzerland AG 2021
M. de Bruijne et al. (Eds.): MICCAI 2021, LNCS 12901, pp. 656–667, 2021.
https://doi.org/10.1007/978-3-030-87193-2_62

of the Computer-Aided Diagnosis (CAD) systems. In traditional CAD systems, the ROIs are manually defined by experts, which is time-consuming and highly relies on the experience of the radiologists and sonographers. Therefore, automatic thyroid nodule detection, which predicts the bounding boxes of thyroid nodules, from ultrasound images could play a very important role in computer aided thyroid cancer diagnosis [11,33].

Thyroid nodule detection in ultrasound images is an important yet challenging task in both medical image analysis and computer vision fields [4,18,26,29]. In the past decades, many traditional object detection approaches have been proposed [7,34,35,40], such as BING [5], EdgeBox [39] and Selective Search [32]. However, due to the large variations of the targets, there is still significant room for the improvements of traditional object detection approaches in terms of accuracy and robustness. In recent years, object detection has achieved great improvements by introducing machine learning and deep learning techniques. These methods can be mainly categorized into three groups: (i) two-stage models: such as RCNN [10], Fast-RCNN [9], Faster-RCNN [24], SPP-Net [12], R-FCN [6], Cascaded-RCNN [3] and so on; (ii) one-stage models: such as OverFeat [25], YOLO (v1, v2, v3, v4, v5) [1,2,21–23], SSD [19], RetinaNet [16] and so on; (iii) anchor-free models, such as CornerNet [15], CenterNet [8], ExtremeNet [38], RepPoints [37], FoveaBox [14] and FCOS [31]. As we know, the two-stage models are originally more accurate but less efficient than one-stage models. However, with the development of new losses, e.g. focal loss [16] and training strategies, one-stage models are now able to achieve comparable performance against two-stage models while requires less time costs. The anchor-free models relies on the object center or key points, which are relatively less accessible in ultrasound images.

Almost all of the above detection models are originally designed for object detection from natural images, which have different characteristics than ultrasound images. Particularly, ultrasound images have variable spatial resolution, heavy speckle noise, and multiple acoustic artifacts, which make the detection task challenging. In addition, thyroid nodules have diverse sizes, shapes and appearances. Sometimes, thyroid nodules are very similar to the thyroid tissue and are not defined by clear boundaries (e.g. ill-defined nodule). Some nodules are heterogeneous due to diffuse thyroid disease, which makes these nodules difficult to differentiate from each other and their backgrounds. In addition, the occasional occurrence of multiple thyroid nodules within the same image, and large thyroid nodules with complex interior textures, which could be considered internal nodules, further increase the difficulty of the nodule detection task. These characteristics lead to high inter-observer variability among human readers, and analogous challenges for machine learning tools, which often lead to inaccurate or unreliable nodule detection.

To address the above issues, multi-scale features are very important. Therefore, we propose a novel one-stage thyroid nodule detection model, called *TUN-Det*, whose backbone is built upon the ReSidual U-blocks (RSU) [20], which is able to extract richer multi-scale features from feature maps with different resolutions. In addition, we design a multi-head architecture for both the nodule

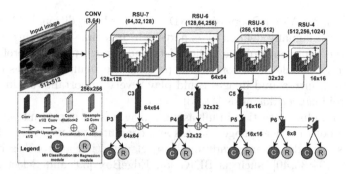

Fig. 1. Architecture of the proposed TUN-Det.

bounding boxes classification and regression in our TUN-Det to predict more reliable results. Each multi-head module is comprised of three different heads, which are variants of the RSU block and arranged in parallel. Each multi-head module outputs three separate outputs, which are supervised by losses computed independently in the training process. In the inference step, multi-head outputs are combined to achieve better detection performance. The Weighted Boxes Fusion (WBF) algorithm [28] is introduced to fuse the outputs of each multi-head module. In summary, our contributions are threefold: (i) a novel one-stage thyroid nodule detection network, TUN-Det, built upon the Residual U-blocks [20]; (ii) a novel multi-head architecture for both bounding boxes classification and regression heads, in which the ensemble strategy is embedded; (iii) Very competitive performance against the state-of-the-art models on our newly built thyroid nodule detection dataset.

2   Proposed Method

2.1   TUN-Det Architecture

Feature Pyramid Network (FPN) is one of the most popular architecture in object detection. Because the FPN architecture is able to efficiently extract high-level and low-level features from deeper and shallow layers, respectively. As we know, multi-scale features play very important roles in object detection. High-level features are responsible for predicting the classification scores while low-level features are used to guarantee the bounding boxes' regression accuracy. The FPN architectures usually take existing image classification networks, such VGG [27], ResNet [13] and so on, as their backbones. However, each stage of these backbones is only able to capture single-scale features because image classification backbones are designed to perceive only high-level semantic meaning while paying less attention to the low-level or multi-scale features[20]. To capture more multi-scale features from different stages, we build our TUN-Det upon the Residual U-blocks (RSU), which was first proposed in salient object detection U^2-Net [20]. Our proposed TUN-Det is also a one-stage FPN similar to RetinaNet [16].

Figure 1 illustrates the overall architecture of our newly proposed TUN-Det for thyroid nodule detection. As we can see, the backbone of our TUN-Det consists of five stages. The first stage is a plain convolution layer with stride of two, which is used to reduce the feature maps resolution. The second to the fifth stages are RSU-7, RSU-6, RSU-5 and RSU-4, respectively. There is a maxpooling operation between the neighboring stages. Compared with other plain convolution, the RSUs are able to capture both local and global information from feature maps with arbitrary resolutions[20]. Therefore, richer multi-scale features $\{C_3, C_4, C_5\}$ can be extracted by the backbone built upon these blocks for supporting the nodule detection. Then, an FPN [16] is applied on top of the backbone's features $\{C_3, C_4, C_5\}$ to create multi-scale pyramid features $\{P_3, P_4, P_5, P_6, P_7\}$, which will be used for bounding boxes regression and classification.

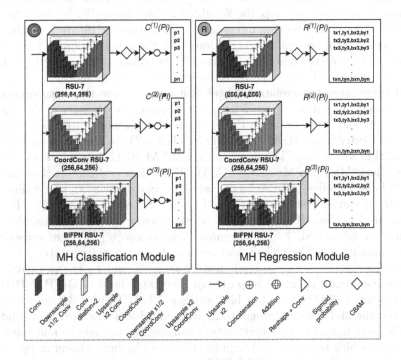

Fig. 2. Multi-head classification and regression module.

2.2 Multi-head Classification and Regression Module

After obtaining the multi-scale pyramid features $\{P_3, P_4, P_5, P_6, P_7\}$, the most important step is regressing the bounding boxes' coordinates and predicting their probabilities of being nodules. These two processes are usually implemented by a regression module $BBOX_i = R(P_i)$ and a classification module $CLAS_i = C(P_i)$, respectively. The regression outputs $\{BBOX_3, BBOX_4, \ldots, BBOX_7\}$ and the classification outputs $\{CLAS_3, CLAS_4, \ldots, CLAS_7\}$ from different features are then fused to achieve the final detection results by conducting non-maximum suppression (NMS).

To further reduce the False Positives (FP) and False Negatives (FN) in the detection results, multi-model ensemble strategy is usually considered. However, this approach is not preferable in real-world applications due to high computational and time costs. Hence, we design a multi-head (three-head) architecture for both classification and regression modules to address this issue. Particularly, each classification and regression module consists of three parallel-configured heads, $\{C^{(1)}, C^{(2)}, C^{(3)}\}$, and $\{R^{(1)}, R^{(2)}, R^{(3)}\}$, respectively. Given a feature map P_i, three classification outputs, $\{C^{(1)}(P_i), C^{(2)}(P_i), C^{(3)}(P_i)\}$, and three regression outputs, $\{R^{(1)}(P_i), R^{(2)}(P_i), R^{(3)}(P_i)\}$, will be produced. In the training process, their losses will be computed separately and summed to supervise the model training. In the inference step, the Weighted Boxes Fusion (WBF) algorithm [28] is used to fuse the regression and classification outputs of different heads. This design embeds the ensemble strategy into both the classification and regression module to improve the detection accuracy while avoiding training multiple models, which is a standard procedure in common ensemble methods.

In this paper, the architectures of $R^{(i)}$ and $C^{(i)}$ are the same except for the last convolution layer (see Fig. 2). To increase the diversity of the prediction results and hence reducing the variance, three variants of RSU-7 (**CBAM RSU-7, CoordConv RSU-7** and **BiFPN RSU-7**) are developed to construct the multi-head modules. The first head is **CBAM RSU-7**, in which a Convolutional Block Attention Module (CBAM) [36] block is added after the standard RSU-7 block to refine features by channel (M_c) and spatial (M_s) attention. The formulation can be described as $F_c = M_c(F_{in}) \otimes F_{in}$ and $F_s = M_s(F_c) \otimes F_c$. The second head is **CoordConv RSU-7**, which replaces the plain convolution layers in the original RSU-7 by Coordinate Convolution [17] layers to encode geometric information. CoordConv can be described as $conv(concat(F_{in}, F_i, F_j))$, where $F_{in} \in \mathbb{R}^{(h \times w \times c)}$ is an input feature map, F_i and F_j are extra row and column coordinate channels respectively. The third head is **BiFPN RSU-7**, which expands RSU-7 by adding bi-directional FPN (BiFPN) [30] layer between the encoding and decoding stages to improve multi-scale feature representation. BiFPN layer has a ∩-shape architecture consisted of bottom-up and top-down pathways, which helps to learn high-level features by fusing them in two directions. Here, we use four-stage BiFPN layer to avoid complexity and reduce the number of trainable parameters.

2.3 Supervision

As shown in Fig. 1, our newly proposed TUN-Det has five groups of classification and regression outputs. Therefore, the total loss is the summation of these five groups of outputs: $\mathcal{L} = \sum_{i=1}^{5} \alpha_i \mathcal{L}_i$, where α_i is the weight of each group (all α are set to 1.0 here). For every anchor, each group produces three classification outputs $\{C^{(1)}, C^{(2)}, C^{(3)}\}$ and three regression outputs $\{R^{(1)}, R^{(2)}, R^{(3)}\}$. Therefore, the loss of each group can be defined as

$$\mathcal{L}_i = \sum_{j=1}^{3} \lambda_i^{C^{(j)}} \mathcal{L}_i^{C^{(j)}} + \sum_{j=1}^{3} \lambda_i^{R^{(j)}} \mathcal{L}_i^{R^{(j)}}, \tag{1}$$

where $\mathcal{L}_i^{C^{(j)}}$ and $\mathcal{L}_i^{R^{(j)}}$ are the corresponding losses for classification and regression outputs respectively. $\lambda_i^{C^{(j)}}$ and $\lambda_i^{R^{(j)}}$ are their corresponding weights to determine the importance of each output. We set all the λ weights to 1.0 in our experiments. $\mathcal{L}_i^{C^{(j)}}$ is the focal loss [16] for classification. It can be defined as follows:

$$\mathcal{L}_i^{C^{(i)}} = \text{Focal}(p_t) = \alpha_t(1 - p_t)^{\gamma} \times BCE(p_c, y_c),$$

$$p_t = \begin{cases} p_c & \text{if } y_c = 1 \\ 1 - p_c & \text{otherwise} \end{cases}, \quad \alpha_t = \begin{cases} \alpha & \text{if } y_c = 1 \\ 1 - \alpha & \text{otherwise,} \end{cases} \tag{2}$$

where p_c and y_c are predicted and target classes respectively. α and γ are focal weighting factor and focusing parameters that are set to 0.25 and 2.0, respectively. $\mathcal{L}_i^{R^{(j)}}$ is the Smooth-L1 loss [9] for regression, which is defined as:

$$\mathcal{L}_i^{R^{(j)}} = \text{Smooth-L1}(p_r, y_r) = \begin{cases} 0.5(\sigma x)^2 & \text{if } |x| < \frac{1}{\sigma^2} \\ |x| - \frac{0.5}{\sigma^2} & \text{otherwise,} \end{cases}, \quad x = p_r - y_r \tag{3}$$

where p_r and y_r are predicted and ground truth bounding boxes respectively. σ defines where the regression loss changes from L2 to L1 loss. It is set to 3.0 in our experiments.

3 Experimental Results

3.1 Datasets and Evaluation Metrics

To validate the performance of our newly proposed TUN-Det on ultrasound thyroid nodule detection task, we build a new thyroid nodule detection dataset. The dataset was retrospectively collected from 700 patients aged between 18–82 years who presented at 12 different imaging centers for a thyroid ultrasound examination. Our retrospective study was approved by the health research ethics boards of the participating centers. There are a total of 3941 ultrasound images, which are extracted from 1924 transverse (TRX) and 2017 sagittal (SAG) scans. These images are split into three subsets for training (2534), validation (565) and testing (842) with 3554, 981, and 1268 labeled nodule bounding boxes, respectively. There is no common patient in the training, validation and testing sets. All nodule bounding boxes are manually labeled by 5 experienced sonographers (with ≥8 years of experience in thyroid sonography) and validated by 3 radiologists. To evaluate the performance of our TUN-Det against other models, Average Precision (AP) [16] is used as the evaluation metric. The validation set is only used to select the model weights in the training process. All the performance evaluation conducted in this paper is based on the testing set.

3.2   Implementation Details

Our proposed TUN-Det is implemented in Tensorflow 1.14 and Keras. The input images are resized to 512×512 and the batch size is set to 1. The model parameters are initialized by Xavier and Adam optimizer with default parameters is used to train the model. Both our training and testing process are conducted on a 12-core, 24-thread PC with an AMD Ryzen Threadripper 2920x 4.3 GHz CPU (128 GB RAM) with an NVIDIA GTX 1080Ti GPU (11GB memory). The model converges after 200 epochs and takes 20 h in total. The average inference time per image (512×512) is 94 ms.

Table 1. Ablation on different backbones and heads configurations. AP_{35}, AP_{50}, AP_{75} are average precision at the fixed 35%, 50%, 75% IoU thresholds, respectively. AP is the average of AP computed over ten different IoU thresholds from 50% to 95% [AP_{50}, AP_{55}, $\cdots$, AP_{95}].

Model	AP	AP_{35}	AP_{50}	AP_{75}
RetinaNet w/ ResNet-50 backbone (baseline) [16]	39.50	74.03	69.07	41.39
w/ RSU backbone	40.73	79.56	74.81	41.62
w/ RSU + CBAM-RSU heads	42.63	80.92	75.49	**45.58**
w/ RSU + CoordConv-RSU heads	41.85	79.62	75.24	43.55
w/ RSU + BiFPN-RSU heads	41.70	80.11	74.20	43.54
w/ RSU + CoordConv-CBAM-BiFPN MH (Our TUN-Det)	**42.75**	**81.22**	**75.66**	45.53

3.3   Ablation Study

To validate the effectiveness of our proposed architecture, ablation studies are conducted on different configurations and the results are summarized in Table 1. The first two rows show the comparison between the original RetinaNet and the RetinaNet-like detection model with our newly developed backbones built upon the RSU-blocks. As we can see, our new adaptation greatly improves the performance against the original RetinaNet. The bottom part of the table illustrates the ablation studies on different configurations of classification and regression modules. It can be observed that our multi-head classification and regression modules, CoordConv-CBAM-BiFPN, shows better performance against other configurations in terms of the AP, AP_{35} and AP_{50}.

3.4   Comparisons Against State-of-the-Arts

Quantitative Comparisons. To evaluate the performance of our newly proposed TUN-Det, we compare our model against six typical state-of-the-art detection models including (i) Faster-RCNN [24] as a two-stage model; (ii) RetinaNet

[16], SSD [19], YOLO-v4 [2] and YOLO-v5 [1] as one stage models; and (iii) FCOS [31] as an anchor-free model. As shown in Table 2, our TUN-Det greatly improves the AP, AP_{35}, AP_{50}, and AP_{75} against Faster-RCNN, RetinaNet, SSD, YOLOV4 and FCOS. Compared with YOLO-v5, our TUN-Det achieves better performance in terms of AP_{35} and Although our model is inferior in terms of AP_{75}, it is doing a better job in terms of FN (i.e. our Average Recall at 75%, AR_{75}, is 45.5 vs. 40.3 in YOLO-v5), which is a priority in the context of thyroid nodule detection to not missing any nodules. Having low Recall with high Precision is unacceptable as it would miss many cancers. Regarding AP, it is usually reported to show the average performance. However, in practice we seek a threshold for achieving final detection results in real-world clinical applications. According to the experiments, our model achieves the best performance under different IoU thresholds (e.g. 35%, 50%), which means our model is more applicable to clinical workflow.

Table 2. Comparisons against the state-of-the-arts.

Model	Backbone	AP	AP_{35}	AP_{50}	AP_{75}
Faster-RCNN [24]	VGG16	0.91	42.13	29.65	2.58
SSD [19]	VGG16	19.05	40.10	36.55	18.10
FCOS [31]	ResNet-50	33.15	62.74	58.67	32.44
RetinaNet [16]	ResNet-50	39.50	74.03	69.07	41.39
YOLO-v4 [2]	CSPDarknet-53	40.43	78.21	72.48	42.04
YOLO-v5 [1]	CSPNet	**45.19**	78.71	74.74	**50.90**
TUN-Det (ours)	RSU	42.75	**81.22**	**75.66**	45.53

Qualitative Comparisons. Figure 3 shows the qualitative comparison of our TUN-Det with other SOTA models on sampled sagittal scans (first two rows) and transverse scans (last two rows). Each column shows the result of one method. The ground truth is shown with green and detection result is shown in red. Figure 3 (1st row) shows that TUN-Det can correctly detect the challenging case of a non-homogeneous large hypo-echoic nodule, while all other methods fail. The 2nd row illustrate that TUN-Det performs well in detecting nodules with ill-defined boundaries, while others miss them. The 3rd and 4th rows highlight that our TUN-Det successfully excludes the false positive and false negative nodules. The last column of Fig. 3 signifies that our TUN-Det produces the most accurate nodule detection results.

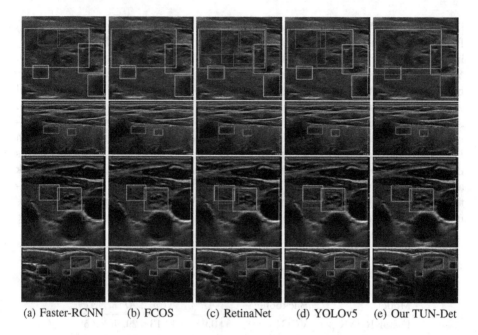

| (a) Faster-RCNN | (b) FCOS | (c) RetinaNet | (d) YOLOv5 | (e) Our TUN-Det |

Fig. 3. Qualitative comparison of ground truth (green) and detection results (red) for different methods. Each column shows the result of one method. (Color figure online)

4   Conclusion and Discussion

This paper proposes a novel detection network, TUN-Det. The novel backbone, built upon the RSU blocks, of our TUN-Det greatly improves the detection accuracy by extracting richer multi-scale features from feature maps with different resolutions. The newly proposed multi-head architecture for both classification and regression heads further improves the nodule detection performance by fusing outputs from diversified sub-modules. Experimental results show that our TUN-Det achieves very competitive performance against existing detection models on overall AP and outperforms other models in terms of AP_{35} and AP_{50}, which indicates its promising performance in practical applications. We believe that this architecture is also promising for other detection tasks on ultrasound images. In the near future, we will focus on improving the detection consistency between neighboring slices of 2D sweeps and exploring new representations for describing nodules merging and splitting in 3D space.

References

1. Ultralytics/yolov5. https://github.com/ultralytics/yolov5. Accessed Oct 2020
2. Bochkovskiy, A., Wang, C.Y., Liao, H.Y.M.: YOLOv4: optimal speed and accuracy of object detection. arXiv preprint arXiv:2004.10934 (2020)

3. Cai, Z., Vasconcelos, N.: Cascade R-CNN: delving into high quality object detection. In: Proceedings IEEE Conference on Computer Vision and Pattern Recognition, pp. 6154–62 (2018)
4. Chen, J., You, H., Li, K.: A review of thyroid gland segmentation and thyroid nodule segmentation methods for medical ultrasound images. Comput. Methods Programs Biomed. **185**, 105329 (2020)
5. Cheng, M.M., Zhang, Z., Lin, W.Y., Torr, P.: Bing: Binarized normed gradients for objectness estimation at 300fps. In: Proceedings of the IEEE Conference on Computer Vision and Pattern Recognition, pp. 3286–3293 (2014)
6. Dai, J., Li, Y., He, K., Sun, J.: R-FCN: Object detection via region-based fully convolutional networks. In: Advances in Neural Information Processing Systems, pp. 379–387 (2016)
7. Dalal, N., Triggs, B.: Histograms of oriented gradients for human detection. In: Proceedings of the IEEE Conference on Computer Vision and Pattern Recognition, pp. 886–893 (2005)
8. Duan, K., Bai, S., Xie, L., Qi, H., Huang, Q., Tian, Q.: CenterNet: keypoint triplets for object detection. In: Proceedings of the IEEE international Conference on Computer Vision, pp. 6569–6578 (2019)
9. Girshick, R.: Fast R-CNN. In: Proceedings of the IEEE International Conference on Computer Vision, pp. 1440–1448 (2015)
10. Girshick, R., Donahue, J., Darrell, T., Malik, J.: Rich feature hierarchies for accurate object detection and semantic segmentation. In: Proceedings of the IEEE Conference on Computer Vision and Pattern Recognition, pp. 580–587 (2014)
11. Haugen, B.R., et al.: 2015 American Thyroid Association management guidelines for adult patients with thyroid nodules and differentiated thyroid cancer. Thyroid **26**(1), 1–133 (2016)
12. He, K., Zhang, X., Ren, S., Sun, J.: Spatial pyramid pooling in deep convolutional networks for visual recognition. IEEE Trans. Pattern Anal. Mach. Intell. **37**(9), 1904–1916 (2015)
13. He, K., Zhang, X., Ren, S., Sun, J.: Deep residual learning for image recognition. In: Proceedings of the IEEE Conference on Computer Vision and Pattern Recognition, pp. 770–778. IEEE Computer Society (2016)
14. Kong, T., Sun, F., Liu, H., Jiang, Y., Li, L., Shi, J.: FoveaBox: beyond anchor-based object detection. IEEE Trans. Image Process. **29**, 7389–7398 (2020)
15. Law, H., Deng, J.: CornerNet: detecting objects as paired keypoints. In: Ferrari, V., Hebert, M., Sminchisescu, C., Weiss, Y. (eds.) Computer Vision – ECCV 2018. LNCS, vol. 11218, pp. 765–781. Springer, Cham (2018). https://doi.org/10.1007/978-3-030-01264-9_45
16. Lin, T.Y., Goyal, P., Girshick, R., He, K., Dollár, P.: Focal loss for dense object detection. In: Proceedings of the IEEE International Conference on Computer Vision, pp. 2980–2988 (2017)
17. Liu, R., et al.: An intriguing failing of convolutional neural networks and the CoordConv solution. arXiv preprint arXiv:1807.03247 (2018)
18. Liu, T., et al.: Automated detection and classification of thyroid nodules in ultrasound images using clinical-knowledge-guided convolutional neural networks. Med. Image Anal. **58**, 101555 (2019)
19. Liu, W., et al.: SSD: single shot multibox detector. In: Leibe, B., Matas, J., Sebe, N., Welling, M. (eds.) ECCV 2016. LNCS, vol. 9905, pp. 21–37. Springer, Cham (2016). https://doi.org/10.1007/978-3-319-46448-0_2

20. Qin, X., Zhang, Z., Huang, C., Dehghan, M., Zaiane, O., Jagersand, M.: U2-net: going deeper with nested U-structure for salient object detection, vol. 106, p. 107404 (2020)
21. Redmon, J., Divvala, S., Girshick, R., Farhadi, A.: You only look once: unified, real-time object detection. In: Proceedings of the IEEE Conference on Computer Vision and Pattern Recognition, pp. 779–788 (2016)
22. Redmon, J., Farhadi, A.: YOLO9000: better, faster, stronger. In: Proceedings of the IEEE Conference on Computer Vision and Pattern Recognition, pp. 7263–7271 (2017)
23. Redmon, J., Farhadi, A.: YOLOv3: an incremental improvement. arXiv preprint arXiv:1804.02767 (2018)
24. Ren, S., He, K., Girshick, R., Sun, J.: Faster R-CNN: towards real-time object detection with region proposal networks. In: Advances in Neural Information Processing Systems, pp. 91–99 (2015)
25. Sermanet, P., Eigen, D., Zhang, X., Mathieu, M., Fergus, R., LeCun, Y.: OverFeat: integrated recognition, localization and detection using convolutional networks. arXiv preprint arXiv:1312.6229 (2013)
26. Sharifi, Y., Bakhshali, M.A., Dehghani, T., DanaiAshgzari, M., Sargolzaei, M., Eslami, S.: Deep learning on ultrasound images of thyroid nodules. Biocybern. Biomed. Eng. (2021)
27. Simonyan, K., Zisserman, A.: Very deep convolutional networks for large-scale image recognition. arXiv preprint arXiv:1409.1556 (2014)
28. Solovyev, R., Wang, W., Gabruseva, T.: Weighted boxes fusion: ensembling boxes from different object detection models. Image Vis. Comput. **107**, 1–6 (2021)
29. Song, W., et al.: Multitask cascade convolution neural networks for automatic thyroid nodule detection and recognition. IEEE J. Biomed. Health Inform. **23**(3), 1215–1224 (2018)
30. Tan, M., Pang, R., Le, Q.V.: EfficientDet: scalable and efficient object detection. In: Proceedings IEEE Conference on Computer Vision and Pattern Recognition, pp. 10781–10790 (2020)
31. Tian, Z., Shen, C., Chen, H., He, T.: FCOS: fully convolutional one-stage object detection. In: Proceedings of the IEEE International Conference on Computer Vision, pp. 9627–9636 (2019)
32. Uijlings, J.R., Van De Sande, K.E., Gevers, T., Smeulders, A.W.: Selective search for object recognition. Int. J. Comput. Vis. **104**(2), 154–171 (2013)
33. Vaccarella, S., Franceschi, S., Bray, F., Wild, C.P., Plummer, M., Dal Maso, L., et al.: Worldwide thyroid-cancer epidemic? The increasing impact of overdiagnosis. N. Engl. J. Med. **375**(7), 614–617 (2016)
34. Viola, P., Jones, M.: Rapid object detection using a boosted cascade of simple features. In: Proceedings of the IEEE Conference on Computer Vision and Pattern Recognition, vol. 1, p. I. IEEE (2001)
35. Viola, P., Jones, M.J.: Robust real-time face detection. Int. J. Comput. Vis. **57**(2), 137–154 (2004)
36. Woo, S., Park, J., Lee, J.-Y., Kweon, I.S.: CBAM: convolutional block attention module. In: Ferrari, V., Hebert, M., Sminchisescu, C., Weiss, Y. (eds.) ECCV 2018. LNCS, vol. 11211, pp. 3–19. Springer, Cham (2018). https://doi.org/10.1007/978-3-030-01234-2_1
37. Yang, Z., Liu, S., Hu, H., Wang, L., Lin, S.: RepPoints: point set representation for object detection. In: Proceedings of the IEEE International Conference on Computer Vision, pp. 9657–9666 (2019)

38. Zhou, X., Zhuo, J., Krahenbuhl, P.: Bottom-up object detection by grouping extreme and center points. In: Proceedings of the IEEE Conference on Computer Vision and Pattern Recognition, pp. 850–859 (2019)
39. Zitnick, C.L., Dollár, P.: Edge boxes: locating object proposals from edges. In: Fleet, D., Pajdla, T., Schiele, B., Tuytelaars, T. (eds.) ECCV 2014. LNCS, vol. 8693, pp. 391–405. Springer, Cham (2014). https://doi.org/10.1007/978-3-319-10602-1_26
40. Zou, Z., Shi, Z., Guo, Y., Ye, J.: Object detection in 20 years: a survey. arXiv preprint arXiv:1905.05055 (2019)

Distilling Effective Supervision for Robust Medical Image Segmentation with Noisy Labels

Jialin Shi[1(✉)] and Ji Wu[1,2]

[1] Department of Electronic Engineering, Tsinghua University, Beijing, China
`shi-jl16@mails.tsinghua.edu.cn`, `wuji_ee@mail.tsinghua.edu.cn`
[2] Institute for Precision Medicine, Tsinghua University, Beijing, China

Abstract. Despite the success of deep learning methods in medical image segmentation tasks, the human-level performance relies on massive training data with high-quality annotations, which are expensive and time-consuming to collect. The fact is that there exist low-quality annotations with label noise, which leads to suboptimal performance of learned models. Two prominent directions for segmentation learning with noisy labels include pixel-wise noise robust training and image-level noise robust training. In this work, we propose a novel framework to address segmenting with noisy labels by distilling effective supervision information from both pixel and image levels. In particular, we explicitly estimate the uncertainty of every pixel as pixel-wise noise estimation, and propose pixel-wise robust learning by using both the original labels and pseudo labels. Furthermore, we present an image-level robust learning method to accommodate more information as the complements to pixel-level learning. We conduct extensive experiments on both simulated and real-world noisy datasets. The results demonstrate the advantageous performance of our method compared to state-of-the-art baselines for medical image segmentation with noisy labels.

Keywords: 3D segmentation · Noisy labels · Robust learning

1 Introduction

Image segmentation plays an important role in biomedical image analysis. With rapid advances in deep learning, many models based on deep neural networks (DNNs) have achieved promising segmentation performance [1]. The success relies on massive training data with high-quality manual annotations, which are expensive and time-consuming to collect. Especially for medical images, the

Electronic supplementary material The online version of this chapter (https://doi.org/10.1007/978-3-030-87193-2_63) contains supplementary material, which is available to authorized users.

M. de Bruijne et al. (Eds.): MICCAI 2021, LNCS 12901, pp. 668–677, 2021.
https://doi.org/10.1007/978-3-030-87193-2_63

annotations heavily rely on expert knowledge. The fact is that there exist low-quality annotations with label noise. Many studies have shown that label noise can significantly affect the accuracy of the learned models [2]. In this work, we address the following problem: how to distill more effective information on noisy labeled datasets for the medical segmentation tasks?

Many efforts have been made to improve the robustness of a deep classification model from noisy labels, including loss correction based on label transition matrix [3–5], reweighting samples [6,7], selecting small-loss instances [8,9], etc. Although effective on image classification tasks, these methods cannot be straightforwardly applied to the segmentation tasks [10].

There are some deep learning solutions for medical segmentation with noisy labels. Previous works can be categorized into two groups. Firstly, some methods are proposed to against label noise using pixel-wise noise estimation and learning. For example, [11] proposed to learn spatially adaptive weight maps and adjusted the contribution of each pixel based on meta-reweighting framework. [10] proposed to train three networks simultaneously and each pair of networks selected reliable pixels to guide the third network by extending the co-teaching method. [12] employed the idea of disagreement strategy to develop label-noise-robust method, which updated the models only on the pixel-wise predictions of the two models differed. The second group of methods concentrates on image-level noise estimation and learning. For example, [13] introduced a label quality evaluation strategy to measure the quality of image-level annotations and then re-weighted the loss to tune the network. To conclude, most existing methods either focus on pixel-wise noise estimation or image-level quality evaluation for medical image segmentation.

However, when evaluating the label noise degree of a segmentation task, we not only judge whether image-level labels are noisy, but also pay attention to which pixels in the image have pixel-wise noisy labels. There are two types of noise for medical image segmentation tasks: pixel-wise noise and image-level noise. Despite the individual advances in pixel-wise and image-level learning, their connection has been underexplored. In this paper, we propose a novel two-phase framework PINT (Pixel-wise and Image-level Noise Tolerant learning) for medical image segmentation with noisy labels, which distills effective supervision information from both pixel and image levels.

Concretely, we first propose a novel pixel-wise noise estimation method and corresponding robust learning strategy for the first phase. The intuition is that the predictions under different perturbations for the same input would agree on the relative clean labels. Based on agreement maximization principle, our method relabels the noisy pixels and further explicitly estimates the uncertainty of every pixel as pixel-wise noise estimation. With the guidance of the estimated pixel-wise uncertainty, we propose pixel-wise noise tolerant learning by using both the original pixel-wise labels and generated pseudo labels. Secondly, we propose image-level noise tolerant learning for the second phase. For pixel-wise noise-tolerant learning, the pixels with high uncertainty tends to be noisy. However, there are also some clean pixels which show high uncertainty when they lie in the boundaries. If only pixel-wise robust learning is considered, the network

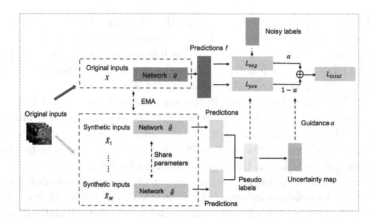

Fig. 1. Illustration of proposed pixel-wise noise tolerant learning framework. We generate multiple mini-batches of synthetic inputs $\{\hat{X}_m\}_{m=1}^{M}$ with different perturbations. The synthetic mini-batch images go through the network $\tilde{\theta}$ to get their predictions. We regard the mean predictions as the pseudo labels and choose the predictive entropy as the metric to estimate uncertainty. The rectified total loss is calculated with L_{seg} and L_{pse} guided by factor α based on uncertainty map. The image-level noise tolerant learning has the similar pipeline.

will inevitably neglect these useful pixels. We extend pixel-wise robust learning to image-level robust learning to address this problem. Based on the pixel-wise uncertainty, we calculate the image-level uncertainty as the image-level noise estimation. We design the image-level robust learning strategy according to the original image-level labels and pseudo labels. Our image-level method could distill more effective information as the complement to pixel-level learning. Last, to show that our method improves the robustness of deep learning on noisy labels, we conduct extensive experiments on simulated and real-world noisy datasets. Experimental results demonstrate the effectiveness of our method.

2   Method

2.1   Pixel-Wise Robust Learning

Pixel-Wise Noise Estimation. In this section, we apply the agreement maximization principle to tackle the problem of noisy labels. The motivation is that the predictions under different perturbations for the same input would agree on the relatively clean pixel-wise labels, and it is unlikely for these predictions to agree on relatively incorrect pixel-wise labels. Inspired by this, we propose our pixel-wise robust learning. Figure 1 shows the pixel-wise noise tolerant learning framework. We study the segmentation tasks with noisy labels for 3D medical images. To satisfy the limitations of GPU memory, we follow the inspiration of mean-teacher model [14]. We formulate the proposed PINT approach with two deep neural networks. The main network is parameterized by θ and the auxiliary network is parameterized by $\widetilde{\theta}$, which is computed as the exponential moving average (EMA) [14] of the θ. At training step t, $\widetilde{\theta}$ is updated with $\widetilde{\theta}_t = \gamma\widetilde{\theta}_{t-1} + (1 - \gamma)\theta_t$, where γ is a smoothing coefficient.

For each mini-batch of training data, we generate synthetic inputs $\{\hat{X}_m\}_{m=1}^M$ on the same images with different perturbations. Formally, we consider a mini-batch data (X, Y) sampled from the training set, where $X = \{x_1, \cdots, x_K\}$ are K samples, and $Y = \{y_1, \cdots, y_K\}$ are the corresponding noisy labels. In our study, we choose Gaussian noises as the perturbations. Afterwards, we perform M stochastic forward passes on the auxiliary network $\widetilde{\theta}$ and obtain a set of probability vector $\{p_m\}_{m=1}^M$ for each pixel in the input. In this way, we choose the mean prediction as the pseudo label of v-th pixel: $\hat{p}_v = \frac{1}{M}\sum_m p_m^v$, where p_m^v is the probability of the m-th auxiliary network for v-th pixel. Inspired by the uncertainty estimation in Bayesian networks [15], we choose the entropy as the metric to estimate the uncertainty. When a pixel-wise label tends to be clean, it is likely to have a peaky prediction probability distribution, which means a small entropy and a small uncertainty. Conversely, if a pixel-wise label tends to be noisy, it is likely to have a flat probability distribution, which means a large entropy and a high uncertainty. As a result, we regard the uncertainty of every pixel as pixel-wise noise estimation:

$$u_v = \mathcal{E}[-\hat{p}_v log\hat{p}_v] \tag{1}$$

where u_v is the uncertainty of v-th pixel and $\mathcal{E}$ is the expectation operator. The relationship between label noise and uncertainty is verified in Experiments 3.2.

Pixel-Wise Loss. We propose pixel-wise noise tolerant learning. Considering that the pseudo labels obtained by predictions also contain noisy pixels and the original labels also have useful information, we train our segmentation network leveraging both the original pixel-wise labels and pesudo pixel-wise labels. For the v-th pixel, the loss is formulated by:

$$L_v = \alpha_v L_v^{seg} + (1 - \alpha_v)L_v^{pse} \tag{2}$$

where L_v^{seg} is the pixel-wise loss between the prediction of main network f_v and original noisy label y_v; L_v^{seg} adopts the cross-entropy loss and is formulated by: $L_v^{seg} = \mathcal{L}_{ce}(f_v, y_v) = \mathcal{E}[-y_v log f_v]$. L_v^{pse} is the pixel-wise loss between the prediction f_v and pseudo label $\hat{y}_v$. $\hat{y}_v$ is equal to $\hat{p}_v$ for soft label and is the one-hot version of $\hat{p}_v$ for hard label. L_v^{pse} is designed as pixel-level mean squared error (MSE) and is formulated by: $L_v^{pse} = \mathcal{L}_{mse}(f_v, \hat{y}_v) = \mathcal{E}[||f_v - \hat{y}_v||^2]$. α_v is the weight factor which controls the importance of L_v^{seg} and L_v^{pse}. Instead of manually setting a fixed value, we provide automatic factor α_v based on pixel-wise uncertainty u_v. We introduce α_v as $\exp(-u_v)$. If the uncertainty has received one large value, this pixel-wise label is prone to be noisy. This factor α_v tends to zero, which drives the model to neglect original label and focus on the pseudo label. In contrast, when the value of uncertainty is small, this pixel-wise label is likely to be reliable. The factor α_v tends to one and the model will focus on the original label. The rectified pixel-wise total loss could be written as:

$$L_v^{total} = \mathcal{E}[\exp(-u_v)L_v^{seg} + (1 - \exp(-u_v))L_v^{pse}] \tag{3}$$

2.2   Image-Level Robust Learning

Image-Level Noise Estimation. For our 3D volume, we regard every slice-level data as image-level data. Based on the estimated pixel uncertainty, the image-level uncertainty can be summarized as: $U_i = \frac{1}{N_i} \sum_v u_v$, where U_i is the uncertainty of i-th image (i-th slice); v denotes the pixel and N_i denotes the number of pixels in the given image. In this case, the image with small uncertainty tends to provide more information even if some pixels involved have noisy labels. The pipeline is similar to pixel-wise framework and the differences lie in the noise estimation method and corresponding robust total loss construction.

Image-Level Loss. For image-level robust learning, we train our segmentation network leveraging both the original image-level labels and pseudo image-level labels. For the i-th image, the loss is formulated by:

$$L_i = \alpha_i L_i^{seg} + (1 - \alpha_i) L_i^{pse} \tag{4}$$

where L_i^{seg} is the image-level cross-entropy loss between the prediction f_i and original noisy label y_i; L_i^{pse} is the image-level MSE loss between the prediction f_i and pseudo label $\hat{y}_i$; Image-level pseudo label $\hat{y}_i$ is composed of pixel-level $\hat{y}_v$. α_i is the automatic weight factor to control the importance of L_i^{seg} and L_i^{pse}. Similarity, we provide automatic factor α_i as $\exp(-U_i)$ based on image-level uncertainty U_i. The rectified image-level total loss is expressed as:

$$L_i^{total} = \mathcal{E}[\exp(-U_i) L_i^{seg} + (1 - \exp(-U_i)) L_i^{pse}] \tag{5}$$

Our PINT framework has two phases for training with noisy labels. In the first phase, we apply the pixel-wise noise tolerant learning. Based on the guidance of the estimated pixel-wise uncertainty, we can filter out the unreliable pixels and preserve only the reliable pixels. In this way, we distill effective information for learning. However, for segmentation tasks, there are also some clean pixels have high uncertainty when they lie in the marginal areas. Thus, we adopt the image-level noise tolerant learning for the second phase. Based on the estimated image-level uncertainty, we can learn from the images with relative more information. That is, image-level learning enables us to investigate the easily neglected hard pixels based on the whole images. Image-level robust learning can be regarded as the complement to pixel-level robust learning.

3   Experiments and Results

3.1   Datasets and Implementation Details

Datasets. For synthetic noisy labels, we use the publicly available Left Atrial (LA) Segmentation dataset. We refer the readers to the Challenge [20] for more details. LA dataset provides 100 3D MR image scans and segmentation masks for training and testing. We split the 100 scans into 80 scans for training and 20 scans for testing. We randomly crop $112 \times 112 \times 80$ sub-volumes as the inputs. All data are pre-processed by zero-mean and unit-variance intensity normalization.

Table 1. Segmentation performance comparison on simulated LA noisy dataset with varying noise rates (25%, 50% and 75%). The average values (±std) over 3 repetitions are reported. The arrows indicate which direction is better.

Method	25%		50%		75%	
	Dice (%)↑	ASD ↓	Dice (%)↑	ASD ↓	Dice (%)↑	ASD ↓
V-net[16]	86.34 ± 0.59	2.72 ± 0.36	82.55 ± 0.26	3.35 ± 0.01	72.76 ± 1.00	5.48 ± 0.06
Reweighting [11]	87.31 ± 0.28	2.46 ± 0.35	83.24 ± 0.70	3.20 ± 0.17	73.02 ± 0.32	5.30 ± 0.12
Tri-network [10]	87.92 ± 0.44	2.37 ± 0.27	84.79 ± 0.44	2.83 ± 0.16	73.88 ± 0.46	5.22 ± 0.11
Pick-and-learn [13]	88.47 ± 0.30	1.92 ± 0.24	85.09 ± 0.56	2.73 ± 0.20	73.30 ± 0.27	5.11 ± 0.08
PNT	88.29 ± 0.43	1.82 ± 0.11	86.16 ± 0.69	2.43 ± 0.05	74.92 ± 0.19	5.16 ± 0.01
INT	89.24 ± 0.21	1.75 ± 0.21	85.78 ± 0.55	2.56 ± 0.12	74.42 ± 0.23	5.20 ± 0.08
PINT	**90.49 ± 0.39**	**1.60 ± 0.06**	**89.04 ± 0.71**	**1.92 ± 0.17**	**76.25 ± 0.44**	**4.56 ± 0.18**

For real-world dataset, we have collected CT scans with 30 patients (average 72 slices/patient). The dataset is used to delineate the Clinical Target Volume (CTV) of cervical cancer for radiotherapy. Ground truths are defined as the reference segmentations generated by two radiation oncologists via consensus. Noisy labels are provided by the less experienced operators. 20 patients are randomly selected as training images and the remaining 10 patients are selected as testing images. We resize the images to $256 \times 256 \times 64$ for inputs.

Implementation Details. The framework is implemented with PyTorch, using a GTX 1080Ti GPU. We employ V-net [16] as the backbone network and add two dropout layers after the L-5 and R-1 stage layers with dropout rate 0.5 [17]. We set the EMA decay γ as 0.99 referring to the work [14] and set batch size as 4. We use the SGD optimizer to update the network parameters (weight decay $= 0.0001$, momentum $= 0.9$). Gaussian noises are generated from a normal distribution. For the uncertainty estimation, we set $M = 4$ for all experiments to balance the uncertainty estimation quality and training efficiency. The effect of hyper-parameters M is shown in supplementary materials. Code will be made publicly available upon acceptance.

For the first phase, we apply the pixel-wise noise tolerant learning for 6000 iterations. At this time, the performance difference between different iterations is small enough in our experiments. The learning rate is initially set to 0.01 and is divided by 10 every 2500 iterations. For the second phase, we apply the image-level noise tolerant learning. When trained on noisy labels, deep models have been verified to first fit the training data with clean labels and then memorize the examples with false labels. Following the promising works [18,19], we adopt "high learning rate" and "early-stopping" strategies to prevent the network from memorizing the noisy labels. In our experiments, we set a high learning rate as lr $= 0.01$ and the small number of iterations as 2000. All hyper-parameters are empirically determined based on the validation performance of LA dataset.

3.2 Results

Experiments on LA Dataset. We conduct experiments on LA dataset with simulated noisy labels. We randomly select 25%, 50% and 75% training samples

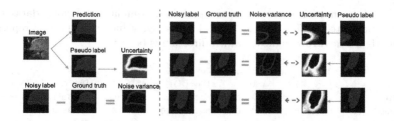

Fig. 2. Illustration of noise variance and pixel-wise uncertainty. The white color means higher uncertainty. The pixels with high uncertainty usually lie in the noise areas or marginal areas. (Color figure online)

and further randomly erode/dilate the contours with 5–18 pixels to simulate the non-expert noisy labels. We train our framework with non-expert noisy annotations and evaluate the model by the Dice coefficient score and the average surface distance (ASD [voxel]) between the predictions and the accurate ground truth annotations [17]. We compare our PINT framework with multiple baseline frameworks. 1) V-net [16]: which uses a cross-entropy loss to directly train the network on the noisy training data; 2) Reweighting framework [11]: a pixel-wise noise tolerant strategy based on the meta-reweight framework; 3) Tri-network [10]: a pixel-wise noise tolerant method based on tri- network extended by co-teaching method. 4) Pick-and-learn framework [13]: an image-level noise tolerant strategy based on image-level quality estimation. We use PNT to represent our PINT framework with only pixel-wise robust learning and INT to represent our PINT framework with only image-level robust learning. Our PINT framework contains two-phase pixel-wise and image-level noise tolerant learning.

Table 1 illustrates the experimental results on the testing data. For clean-annotated dataset, the V-net has the upper bound of average Dice 91.14% and average ASD 1.52 voxels. (1) We can observe that as the noise percentage increase (from clean labels to 25%, 50% and 75% noise rate), the segmentation performance of baseline V-net decreases sharply. In this case, the trained model tends to overfit to the label noise. When adopting noise-robust strategy, the segmentation network begins to recover its performance. (2) For pixel-wise noise robust learning, we compare Reweighting method [11] and our PNT with only pixel-wise distillation. Our method gains 2.92% improvement of Dice for 50% noise rate (83.24% vs 86.16%). For image-level noise robust learning, we compare Pick-and-learn [13] and our INT with only image-level distillation. Our method achieves 1.12% average gains of Dice for 75% noise rate (73.30% vs 74.42%). These results verify that our pixel-wise and image-level noise robust learning are effective. (3) We can observe that our PINT outperforms other baselines by a large margin. Moreover, comparing to PNT and INT methods, our PINT with both pixel-wise and image-level learning shows better performance, which verifies that our PINT can distill more effective supervision information.

Label Noise and Uncertainty. To investigate the relationship between pixel-wise uncertainty estimation and noisy labels, we illustrates the results of ran-

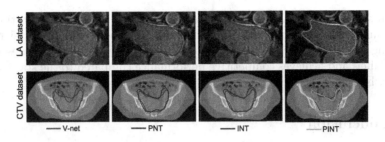

Fig. 3. Qualitative results of segmentation with noisy labels on simulated LA dataset and real-world CTV dataset. The ground truths, the predictions by V-net, the predictions by PNT, the predictions by INT and the predictions by PINT are colored with red, purple, green, blue and yellow, respectively. (Color figure online)

domly selected samples on synthetic noisy LA dataset with 50% noise rate in Fig. 2. The discrepancy between ground-truth and noisy label is approximated as the noise variance. We can observe that the noise usually exists in the areas with high uncertainty (shown in white color on the left). Inspired by this, we provide our pixel-wise noise estimation based on pixel-wise uncertainty awareness. Apart from noisy labels, pseudo labels also suffer from the noise effect. The best way for training robust model is to use both original noisy labels and pseudo labels. Furthermore, multiple examples are shown on the right. We observe that there are some clean pixels show high uncertainty when they lie in the boundaries. If only pixel-wise robust learning is considered, the network will neglect these useful pixels. Therefore, we propose image-level robust learning to learn from the whole images for distilling more effective information.

Visualization. As shown in Fig. 3, we provide the qualitative results of the simulated noisy LA segmentation dataset and real-world noisy CTV dataset. For noisy LA segmentation, we show some random selected examples with 50% noise rate. Compared to the baselines, our PINT with both pixel-wise and image-level robust learning yields more reasonable segmentation predictions.

Experiments on Real-World Dataset. We explore the effectiveness of our approach on a real CTV dataset with noisy labels. Due to the lack of professional medical knowledge, the non-expert annotators often generate noisy annotations. The results are shown in Table 2. 'No noise' means we train the segmentation network with clean labels. The other methods including V-net, Re-weighting,

Table 2. Segmentation performance comparison on real-world CTV dataset. The arrows indicate which direction is better and the average values (±std) over 3 repetitions are reported.

Method	No noise [16]	V-net [16]	Re-weighting [11]	Pick-and-learn [13]	PNT	INT	PINT
Dice(%)↑	77.26 ± 0.53	68.26 ± 0.21	69.31 ± 0.43	70.79 ± 0.31	73.57 ± 0.37	72.08 ± 0.56	**75.31 ± 0.15**
ASD [voxel]↓	1.38 ± 0.03	2.25 ± 0.02	2.05 ± 0.06	2.11 ± 0.07	1.85 ± 0.04	1.92 ± 0.08	**1.76 ± 0.13**

Pick-and-learn, PNT, INT and PINT are the same with LA segmentation. All the results show that our PINT with both pixel-wise and image-level robust learning can successfully recognize the clinical target volumes in the presence of noisy labels and achieves competitive performance compared to the state-of-the-art methods.

4   Conclusion

In this paper, we propose a novel framework PINT, which distills effective supervision information from both pixel and image levels for medical image segmentation with noisy labels. We explicitly estimate the uncertainty of every pixel as pixel-wise noise estimation, and propose pixel-wise robust learning by using both the original labels and pseudo labels. Furthermore, we present the image-level robust learning method to accommodate more informative locations as the complements to pixel-level learning. As a result, we achieve the competitive performance on the synthetic noisy dataset and real-world noisy dataset. In the future, we will continue to investigate the joint estimation and learning of pixel and image levels for medical segmentation tasks with noisy labels.

References

1. Litjens, G., et al.: A survey on deep learning in medical image analysis. Med. Image Anal. **42**, 60–88 (2017)
2. Karimi, D., et al.: Deep learning with noisy labels: exploring techniques and remedies in medical image analysis. Med. Image Anal. **65**, 101759 (2020)
3. Patrini, G., et al.: Making deep neural networks robust to label noise: a loss correction approach. In Proceedings of the IEEE Conference on Computer Vision and Pattern Recognition, pp. 1944–1952, (2017)
4. Hendrycks, D., Mazeika, M., Wilson, D., et al.: Using trusted data to train deep networks on labels corrupted by severe noise. In: Advances in Neural Information Processing Systems, pp. 10456–10465 (2018)
5. Wang, Z., Hu, G., Hu, Q.: Training noise-robust deep neural networks via meta-learning. In Proceedings of the IEEE/CVF Conference on Computer Vision and Pattern Recognition, pp. 4524–4533 (2020)
6. Ren, M., Zeng, W., Yang, B., et al.: Learning to reweight examples for robust deep learning. In: International Conference on Machine Learning (2018)
7. Shu, J., Xie, Q., Yi, L., et al.:Meta-weight-net: learning an explicit mapping for sample weighting. In: Advances in Neural Information Processing Systems, pp. 1919–1930 (2019)
8. Han, B., Yao, Q., Yu, X., et al.: Co-teaching: Robust training of deep neural networks with extremely noisy labels. In: Advances in Neural Information Processing Systems (2018)
9. Yu, X., Han, B., Yao, J., et al.: How does disagreement help generalization against label corruption? In: International Conference on Machine Learning, pp. 7164–7173 (2019)

10. Zhang, T., Yu, L., Hu, N., Lv, S., Gu, S.: Robust medical image segmentation from non-expert annotations with tri-network. In: Martel, A.L., et al. (eds.) MICCAI 2020. LNCS, vol. 12264, pp. 249–258. Springer, Cham (2020). https://doi.org/10.1007/978-3-030-59719-1_25
11. Mirikharaji, Z., Yan, Y., Hamarneh, G.: Learning to segment skin lesions from noisy annotations. In: Domain Adaptation and Representation Transfer and Medical Image Learning with Less Labels and Imperfect Data, pp. 207–215 (2019)
12. Min, S., Chen, X., Zha, Z., et al.: A two-stream mutual attention network for semi-supervised biomedical segmentation with noisy labels. Proc. AAAI Conf. Artif. Intell. **33**(01), 4578–4585 (2019)
13. Zhu, H., Shi, J., Wu, J.: Pick-and-learn: automatic quality evaluation for noisy-labeled image segmentation. In: International Conference on Medical Image Computing and Computer-Assisted Intervention, pp. 576–584 (2019)
14. Tarvainen, A., Valpola, H.: Mean teachers are better role models: Weight-averaged consistency targets improve semi-supervised deep learning results. In: Advances in Neural Information Processing Systems, pp. 1195–1204 (2017)
15. Kendall, A., Gal, Y.: What uncertainties do we need in Bayesian deep learning for computer vision? In: Advances in Neural Information Processing Systems (2017)
16. Milletari, F., Navab, N., Ahmadi, S. A.: V-net: fully convolutional neural networks for volumetric medical image segmentation. In: 2016 4th International Conference on 3D Vision (3DV), pp. 565–571 (2016)
17. Ma, J., Wei, Z., Zhang, Y., et al.: How distance transform maps boost segmentation CNNs: an empirical study. In: Medical Imaging with Deep Learning, pp. 479–492 (2020)
18. Tanaka, D., Ikami, D., Yamasaki, T., et al.: Joint optimization framework for learning with noisy labels. In: Proceedings of the IEEE Conference on Computer Vision and Pattern Recognition, pp. 5552–5560 (2018)
19. Liu, S., Niles-Weed, J., Razavian, N., et al.: Early-learning regularization prevents memorization of noisy labels. In: Advances in Neural Information Processing Systems (2020)
20. MICCAI 2018 left atrial segmentation. http://atriaseg2018.cardiacatlas.org/

On the Relationship Between Calibrated Predictors and Unbiased Volume Estimation

Teodora Popordanoska[✉], Jeroen Bertels, Dirk Vandermeulen, Frederik Maes, and Matthew B. Blaschko

Center for Processing Speech and Images, Department of ESAT,
KU Leuven, Leuven, Belgium
`teodora.popordanoska@kuleuven.be`

Abstract. Machine learning driven medical image segmentation has become standard in medical image analysis. However, deep learning models are prone to overconfident predictions. This has lead to a renewed focus on *calibrated predictions* in the medical imaging and broader machine learning communities. Calibrated predictions are estimates of the probability of a label that correspond to the true expected value of the label conditioned on the confidence. Such calibrated predictions have utility in a range of medical imaging applications, including surgical planning under uncertainty and active learning systems. At the same time it is often an accurate *volume measurement* that is of real importance for many medical applications. This work investigates the relationship between model calibration and volume estimation. We demonstrate both mathematically and empirically that if the predictor is calibrated *per image*, we can obtain the correct volume by taking an expectation of the probability scores per pixel/voxel of the image. Furthermore, we show that linear combinations of calibrated classifiers preserve volume estimation, but do not preserve calibration. Therefore, we conclude that having a calibrated predictor is a sufficient, but not necessary condition for obtaining an unbiased estimate of the volume. We validate our theoretical findings empirically on a collection of 18 different (calibrated) training strategies on the tasks of glioma volume estimation on BraTS 2018, and ischemic stroke lesion volume estimation on ISLES 2018 datasets.

Keywords: Calibration · Uncertainty · Volume · Segmentation

1 Introduction

In recent years the segmentation performance of CNNs improved dramatically. Despite these improvements, the adoption of automated segmentation systems

Electronic supplementary material The online version of this chapter (https://doi.org/10.1007/978-3-030-87193-2_64) contains supplementary material, which is available to authorized users.

© Springer Nature Switzerland AG 2021
M. de Bruijne et al. (Eds.): MICCAI 2021, LNCS 12901, pp. 678–688, 2021.
https://doi.org/10.1007/978-3-030-87193-2_64

into clinical routine is rather slow. Having calibrated model predictions would foster this relationship by providing the clinician with valuable information on failure detection or when manual intervention is needed [15]. Reasoning under uncertainty is a central part of surgical planning [11], and uncertainty estimates are central to many active learning frameworks [28]. With this in mind, current research orients towards the design of automated segmentation systems that provide a measure of confidence alongside its predictions.

The empirical findings that modern CNNs are poorly calibrated [13] stimulated a variety of research to improve model calibration, including deep ensembles [18], Bayesian NNs [22] and MC dropout [10]. The ability to deliver realistic segmentations would take calibration even further [3,16]. Remarkably, a large part of ongoing research stands orthogonal by design. Choosing a loss function that is consistent with a certain target metric [9] often means the outputs cannot be interpreted as voxel-wise probabilities [4], let alone that the predictions would be calibrated. Nonetheless, post-hoc calibration might come to the rescue and has proven to be competitive to training-time calibration [24].

While the automated segmentation is an important task in its own right, it is mostly used as an intermediary for calculating certain biomarkers. In that respect, volume is by far the most important biomarker in medical imaging. For example, tumor volume is a basic and specific response predictor in radiotherapy [7] and the volume of an acute stroke lesion can be used to decide on the type of endovascular treatment [12]. When the CNN outputs voxel-wise probabilities, they can be propagated to produce volumetric uncertainty [8]. However, similar to its effect on calibration, the loss function determines how to calculate volume, and thus when the former is applicable [4].

This work bridges the gap between model calibration and volume estimation. There will be theoretical grounds that calibration error bounds volume error. This relationship is confirmed in an empirical validation on BraTS 2018 [1,2,20] and ISLES 2018 [27]. Furthermore, there is a clear empirical correlation between calibration error and volume error, and between object size and volume error. As a result, this work acknowledges and encourages research towards calibrated systems. Such systems not only provide additional robustness; they will also produce correct volume estimates.

2 The Relationship Between Calibration and Volume Bias

In this section, we demonstrate that calibrated uncertainty estimates are intimately related to unbiased volume estimates. We develop novel mathematical results showing that calibration error upper bounds the absolute value of the volume bias (Proposition 1), which implies that as the calibration error goes to zero, the resulting function has unbiased volume estimates. We further show that unbiased volume estimates do not imply that the classifier is calibrated, and that solely enforcing an unbiased classifier does not result in a calibrated classifier (Proposition 2 and Corollary 2). This motivates our subsequent experimental study where we measure the empirical relationship between calibration error and volume bias in Sect. 3.

Definition 1 (Volume bias [4]). *Let f be a function that predicts from an image/tomography x the probability of each pixel/voxel $\{y_i\}_{i=1}^P$ belonging to a given class. The volume bias of f is:*

$$\mathrm{Bias}(f) := \mathbb{E}_{(x,y)\sim P}\left[f(x) - y\right]. \tag{1}$$

Definition 2 (Calibration error [17,21,26]). *The calibration error of f : $\mathcal{X} \to [0,1]$ is:*

$$\mathrm{CE}(f) = \mathbb{E}_{(x,y)\sim P}\left[\left|\mathbb{E}_{(x,y)\sim P}\left[[y = 1] \mid f(x)\right] - f(x)\right|\right] \tag{2}$$

In plain English: A classifier is calibrated if its confidence score is equal to the probability of the prediction being correct. Note that this definition is specific to a binary classification setting. The extension of binary calibration methods to multiple classes is usually done by reducing the problem of multiclass classification to K one-vs.-all binary problems [13]. The so-called *marginal CE* [17, Definition 2.4]) is then measured as an average of the per-class CEs.

Proposition 1. *The absolute value of dataset (respectively volume) bias is upper bounded by dataset (respectively volume) calibration error:*

$$\mathrm{CE}(f) \geq |\mathrm{Bias}(f)|. \tag{3}$$

Proof.

$$|\mathrm{Bias}(f)| = \left|\mathbb{E}_{(x,y)\sim P}\left[y - f(x)\right]\right| \tag{4}$$

$$= \left|\underbrace{\mathbb{E}\left[y - \mathbb{E}\left[[y = 1] \mid f(x)\right]\right]}_{=0} + \mathbb{E}_{(x,y)\sim P}\left[\mathbb{E}\left[[y = 1] \mid f(x)\right] - f(x)\right]\right| \tag{5}$$

$$\leq \underbrace{\mathbb{E}_{(x,y)\sim P}\left[\left|\mathbb{E}\left[[y = 1] \mid f(x)\right] - f(x)\right|\right]}_{=\mathrm{CE}(f)}. \tag{6}$$

Focusing on the first term of the right hand side of (5),

$$\mathbb{E}\left[y - \mathbb{E}[[y = 1] \mid f(x)]\right] = \mathbb{E}[y] - \underbrace{\mathbb{E}[\mathbb{E}\left[[y = 1] \mid f(x)\right]]}_{=\mathbb{E}[y]}. \tag{7}$$

In the second term of the r.h.s., we may take the expectation with respect to $f(x)$ in place of x as f is a deterministic function. This term is therefore also equal to $\mathbb{E}[y]$ by the law of total expectation. Finally, the inequality in (6) is obtained due to the convexity of the absolute value and by application of Jensen's inequality. □

We note that Proposition 1 holds for all problem settings and class distributions. In multiclass settings, for each individual class we can obtain a bound by measuring the bias and the CE per class.

Corollary 1. $CE(f) = 0$ *implies that* f *yields unbiased volume estimates.*

Proposition 2. $Bias(f) = 0$ *does not imply that* $CE(f) = 0$.

Proof. Consider the following example. Let the dataset consist of 100 positive and 200 negative points. Let f_1 and f_2 be binary classifiers that rank 1/4 of the negative points with a score of zero, and the remaining negative points with a score of 0.25. Let f_1 rank the first half of the positive points with a score of one and the second half with a score of 0.25, and f_2 vice-versa. In this case, $CE(f_1) = 0$ and $CE(f_2) = 0$, and therefore by Corollary 1, f_1 and f_2 are both unbiased estimates of the volume. Let f_3 be a classifier that performs a linear combination (e.g. an average) of the scores of f_1 and f_2. We note that a convex combination of unbiased estimators is unbiased [19] and therefore f_3 is unbiased. Even though f_3 has a perfect accuracy and $Bias(f_3) = 0$, the scores are no longer calibrated, i.e., $CE(f_3) \neq 0$. □

Corollary 2. *There exists no multiplicative bound of the form* $CE(f) \leq \gamma |Bias(f)|$ *for some finite* $\gamma > 0$ *(cf. [9, Definition 2.2]).*

Thus we see that control over $CE(f)$ minimizes an upper bound on $|Bias(f)|$, but the converse is not true. There are several implications of these theoretical results for the design of medical image analysis systems: (i) Optimizing calibration error per-subject is an attractive method to simultaneously control the bias of the volume estimate, but we need to empirically validate if bias and CE are correlated in practice as we only know a priori that one bounds the other; and (ii) optimizing volume bias alone (e.g. by empirical risk minimization) does not automatically give us the additional benefits of calibrated uncertainty estimates, and does not even provide a multiplicative bound on how poor the calibration error could be. We consequently empirically evaluate the relationship between bias and calibration error in the remainder of this work.

3   Empirical Setup

The empirical validation of our theoretical results will be performed by analyzing two segmentation tasks, each requiring a different distribution in the predicted confidences, with multiple different models, each trained with respect to a different loss function and subject to different post-hoc calibration strategies.

Tasks. The data from two publicly available medical datasets is used and two segmentation tasks are formulated as follows: (i) Whole tumor segmentation using the BraTS 2018 [1,2,20] (BR18) dataset. BR18 contains 285 multi-modal MR volumes with accompanying manual tumor delineations. Due to a rather low inter/intra-rater variability [20] the voxel-wise confidences will be distributed towards the high-confidence ranges; (ii) Ischemic core segmentation using the

ISLES 2018 [27] (IS18) dataset. IS18 contains data from 94 CT perfusion scans with manual delineations of the ischemic infractions on co-registered DWI MR imaging. The identification of the ischemic infarction on CT perfusion data is generally considered non-trivial [6]. This means that a rather high intra/inter-rater variability is to be expected, which in turn will distribute the voxel-wise confidences across the entire range. For both datasets there was a five-fold split of the data identical to [4,24].

Models. For the two former tasks the pre-trained and publicly available models from [24] are used. They investigated the effects of the loss function in combination with a multitude of different post-hoc calibration strategies on the Dice score and model calibration, but without any consideration to volume estimates or volume bias. Their base model shares a U-Net [23] CNN architecture similar to [14]. The three loss functions for the initial training were: (i) cross-entropy (CrE); (ii) soft-Dice (SD); and (iii) a combination of CrE pre-training with SD fine-tuning (CrE-SD). In addition, these base models were calibrated using different post-hoc calibration strategies: (i) Platt scaling and its variants (auxiliary network and fine-tuning); and (ii) two Monte Carlo (MC) dropout methods (MC-Dropout and MC-Center) with different positioning of the dropout layers. For further details on the exact training procedures the reader is referred to [24]. Nevertheless, it is important to note that the initial training and the post-hoc calibration was done on the training sets, and thus the predictions on the validation sets may be aggregated for further testing.

Bias and ECE. The Bias is calculated by direct implementation of Definition 1, i.e. the expectation of the probability scores per voxel. The CE from Definition 2 is a theoretical quantity that in practice is approximated by a binned estimator of the expected calibration error (ECE) (Eq. (3) from [24]). The ECE ranges from 0 to 1, with lower values representing better calibration. We use 20 bins for binning the CNN outputs. Following the example of [15,24], we only consider voxels within the skull-stripped brain/lesion and report the mean per-volume Bias and ECE, as being more clinically relevant versions opposed to their dataset-level variants.

Code. The source code is available at https://github.com/tpopordanoska/calibration_and_bias.

4   Results and Discussion

Figure 1 shows scatter plots of ECE and Bias for the base model and a calibrated model with the fine-tuning strategy for BR18 and IS18. We can confirm that, as predicted by Proposition 1, all the points (volumes measured in ml) lie in-between the lines ECE = ± Bias. We note further that the correlation between ECE and Bias is strictly higher for a calibrated model compared to the base model. There are many volumes for which ECE = | Bias | holds, which supports the findings in [15] where per-volume calibration tends to be off, either resulting in a complete under- or over-estimation. This is also visible in Fig. 3, when the calibration curve lies below or above the unity line ECE = | Bias |.

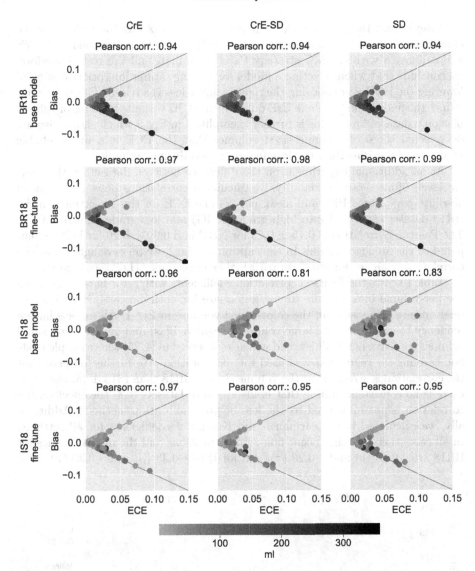

Fig. 1. Scatter plots on BR18 (top two rows) and IS18 (bottom two rows), color-coded by tumor/lesion size (in ml). Every point in the plot represents an image. The Pearson correlation between per-volume ECE and absolute per-volume bias is shown above the plots. Note that the Bias is calculated in voxels, while the color-coding has the units converted to ml.

The ECE and Bias for all 18 models are visually presented in a scatter plot in Fig. 2. Analogously to [24], we find that for the BR18 data only the models trained with SD are on the Pareto front. However, contrary to their result that MC methods are Pareto dominated if optimizing the Dice score is of interest, we observe that the MC-Decoder calibration strategy is Pareto-efficient for the clinical application of measuring volume. For IS18, the model trained with CrE and calibrated with the MC strategy Pareto dominates all the rest. Therefore, we conclude that when selecting a model for volume estimation, prioritization of lowering the ECE over choosing the right training loss is advised. This somewhat refines the findings in [4] where CrE outperformed SD variants in terms of volume bias on a dataset level. This is further exemplified in Fig. 3 which shows visually on selected slices of different-sized volumes that the ECE is a more reliable predictor of the Bias than the loss function used during training.

As an additional experiment on the difference between datasets with different levels of inter-rater variability, we calculated correlations between the mean absolute per-volume Bias and mean per-volume ECE for the 18 settings on the BR18 dataset separated into high grade (HGG) and low grade glioma (LGG). The Pearson correlation is 0.15 ± 0.23 for HGG and 0.39 ± 0.20 for LGG. Compared to the analogous result in the caption of Fig. 2, we observe again that the correlation is higher for the data with higher uncertainty (LGG and IS18).

Table 1 shows the Pearson correlation coefficients with error bars [5] between the per-volume bias and the tumor/lesion size for both datasets. Negative correlations are also visible in the color-coded volumes in Fig. 1. All models have a tendency to underestimate large volumes (negative bias) and overestimate small volumes (positive bias). This trend was also observed in [4,25] where simple post-hoc training-set regression was used for recalibration. We further observe that the correlation is stronger for the volumes in the BR18 than IS18 data. However, there is no consistent finding that holds for both datasets (e.g. the effect on the correlation is not influenced by the loss or post-calibration method). Additionally, we calculated the Spearman ρ and Kendall τ coefficients for all settings. In all cases there are significant non-zero correlations and the median values for BR18 are -0.43 (ρ) and -0.29 (τ), and for IS18 -0.43 (ρ) and -0.30 (τ).

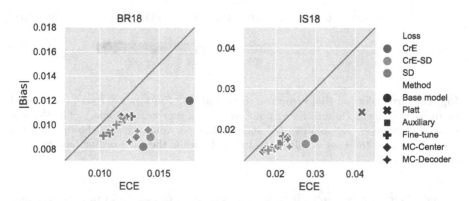

Fig. 2. Scatter plots of ECE versus |Bias| for all combinations of loss functions and calibration methods. The Pearson correlation between |Bias| (computed as mean of absolute per-volume biases) and ECE (mean per-volume ECE) is 0.30 ± 0.21 for BR18 and 0.91 ± 0.04 for IS18.

Table 1. Pearson correlation coefficients between per-volume bias and volume size for BR18 and IS18.

Loss→	CrE	CrE-SD	SD	CrE	CrE-SD	SD
Method ↓		BR18			IS18	
Base model	−0.73 ± 0.03	−0.53 ± 0.04	−0.48 ± 0.05	−0.39 ± 0.09	−0.23 ± 0.10	−0.20 ± 0.10
Platt	−0.58 ± 0.04	−0.67 ± 0.03	−0.62 ± 0.04	−0.50 ± 0.08	−0.39 ± 0.09	−0.48 ± 0.08
Auxiliary	−0.57 ± 0.04	−0.65 ± 0.03	−0.61 ± 0.04	−0.32 ± 0.09	−0.31 ± 0.09	−0.43 ± 0.08
Fine-tune	−0.59 ± 0.04	−0.62 ± 0.04	−0.55 ± 0.04	−0.37 ± 0.09	−0.36 ± 0.09	−0.49 ± 0.08
MC-Decoder	−0.55 ± 0.04	−0.47 ± 0.05	−0.42 ± 0.05	−0.23 ± 0.10	−0.27 ± 0.10	−0.36 ± 0.09
MC-Center	−0.53 ± 0.04	−0.49 ± 0.04	−0.49 ± 0.04	−0.27 ± 0.10	−0.24 ± 0.10	−0.39 ± 0.09

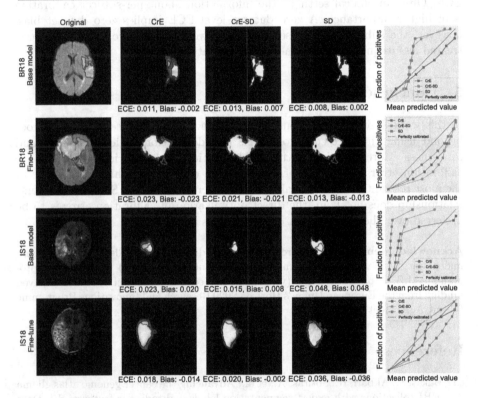

Fig. 3. Qualitative examples of the predictions from a base model and a calibrated model with fine-tuning on BR18 (top two rows) and IS18 (bottom two rows). The red line represents the delineation of the ground truth. The predicted delineations after thresholding at 0.5 are overlayed in blue. The ECE and the bias (shown below the plot) are calculated for the selected slice. The last column shows the calibration curves for the volume. The images are chosen to be representative of different sizes of tumors/lesions and the middle slice of the volume is shown.

Limitations. We showed theoretically and empirically that the CE is an upper bound for the Bias, and that the Bias does not provide a bound on CE. Nonetheless, we did not properly characterize when CE may be strictly larger than this quantity and whether there is a meaningful interpretation of CE $-|$ Bias $|$. Furthermore, we proved that a convex combination of calibrated classifiers does not preserve calibration (although it preserves unbiasedness), but we did not explore ways to combine calibrated models to obtain a calibrated predictor.

Designing a calibrated estimator is a non-trivial task and the field of confidence calibration is an active area of research. However, investing resources to find out which calibration strategy works best for the problem at hand is of high importance, especially in medical applications.

Finally, we wish to emphasize the distinction between dataset and per-volume ECE. Often in clinical settings, the information about per-subject calibration is of higher importance. A zero dataset level ECE implies zero dataset bias, however, the subject-level biases may very well be non-zero. This subject is treated less frequently in the calibration literature.

5   Conclusions

In this work, it was shown that the importance of confidence calibration goes beyond using the predicted voxel-wise confidences for clinical guidance and robustness. More specifically, there is theoretical and empirical evidence that calibration error bounds the error of volume estimation, which still is one of the most relevant biomarkers calculated further downstream. Since the converse relationship does not hold, the direct optimization of calibration error is to be preferred over the optimization of volume bias alone.

Acknowledgments. This research received funding from the Flemish Government under the "Onderzoeksprogramma Artificiële Intelligentie (AI) Vlaanderen" programme. J.B. is part of NEXIS (www.nexis-project.eu), a project that has received funding from the European Union's Horizon 2020 Research and Innovations Programme (Grant Agreement #780026).

References

1. Bakas, S., Akbari, H., Sotiras, A., et al.: Advancing the cancer genome atlas glioma MRI collections with expert segmentation labels and radiomic features. Sci. Data **4**, 1–13 (2017)
2. Bakas, S., Reyes, M., Jakab, A., et al.: Identifying the best machine learning algorithms for brain tumor segmentation, progression assessment, and overall survival prediction in the BRATS challenge (2018)
3. Baumgartner, C.F., et al.: PHiSeg: capturing uncertainty in medical image segmentation. In: Shen, D., et al. (eds.) MICCAI 2019. LNCS, vol. 11765, pp. 119–127. Springer, Cham (2019). https://doi.org/10.1007/978-3-030-32245-8_14
4. Bertels, J., Robben, D., Vandermeulen, D., Suetens, P.: Theoretical analysis and experimental validation of volume bias of soft dice optimized segmentation maps in the context of inherent uncertainty. Med. Image Anal. **67**, 101833 (2021)
5. Bowley, A.L.: The standard deviation of the correlation coefficient. J. Am. Stat. Assoc. **23**(161), 31–34 (1928)

6. Demeestere, J., Garcia-Esperon, C., Garcia-Bermejo, P., et al.: Evaluation of hyper-acute infarct volume using ASPECTS and brain CT perfusion core volume. Neurology **88**(24), 2248–2253 (2017)

7. Dubben, H.H., Thames, H.D., Beck-Bornholdt, H.P.: Tumor volume: a basic and specific response predictor in radiotherapy. Radiother. Oncol. **47**(2), 167–174 (1998)

8. Eaton-Rosen, Z., Bragman, F., Bisdas, S., Ourselin, S., Cardoso, M.J.: Towards safe deep learning: accurately quantifying biomarker uncertainty in neural network predictions. In: Frangi, A.F., Schnabel, J.A., Davatzikos, C., Alberola-López, C., Fichtinger, G. (eds.) MICCAI 2018. LNCS, vol. 11070, pp. 691–699. Springer, Cham (2018). https://doi.org/10.1007/978-3-030-00928-1_78

9. Eelbode, T., Bertels, J., Berman, M., et al.: Optimization for medical image segmentation: Theory and practice when evaluating with Dice score or Jaccard index. IEEE Trans. Med. Imaging **39**(11), 3679–3690 (2020)

10. Gal, Y., Ghahramani, Z.: Dropout as a Bayesian approximation: representing model uncertainty in deep learning. In: Balcan, M.F., Weinberger, K.Q. (eds.) Proceedings of the 33rd International Conference on Machine Learning, pp. 1050–1059 (2016)

11. Gillmann, C., Maack, R.G., Post, T., Wischgoll, T., Hagen, H.: An uncertainty-aware workflow for keyhole surgery planning using hierarchical image semantics. Vis. Inf. **2**(1), 26–36 (2018)

12. Goyal, M., Menon, B.K., Zwam, W.H.V., et al.: Endovascular thrombectomy after large-vessel Ischaemic stroke: a meta-analysis of individual patient data from five randomised trials. The Lancet **387**(10029), 1723–1731 (2016)

13. Guo, C., Pleiss, G., Sun, Y., Weinberger, K.Q.: On calibration of modern neural networks. In: Proceedings of the 34th International Conference on Machine Learning, vol. 70, pp. 1321–1330 (2017)

14. Isensee, F., Kickingereder, P., Wick, W., Bendszus, M., Maier-Hein, K.H.: No new-net. In: Crimi, A., Bakas, S., Kuijf, H., Keyvan, F., Reyes, M., van Walsum, T. (eds.) BrainLes 2018. LNCS, vol. 11384, pp. 234–244. Springer, Cham (2019). https://doi.org/10.1007/978-3-030-11726-9_21

15. Jungo, A., Balsiger, F., Reyes, M.: Analyzing the quality and challenges of uncertainty estimations for brain tumor segmentation. Front. Neurosci. **14**, 282 (2020)

16. Kohl, S.A., Romera-Paredes, B., Meyer, C., et al.: A probabilistic U-net for segmentation of ambiguous images. In: Advances in Neural Information Processing Systems, pp. 6965–6975 (2018)

17. Kumar, A., Liang, P.S., Ma, T.: Verified uncertainty calibration. In: Wallach, H., Larochelle, H., Beygelzimer, A., d'Alché-Buc, F., Fox, E., Garnett, R. (eds.) Advances in Neural Information Processing Systems 32, pp. 3792–3803 (2019)

18. Lakshminarayanan, B., Pritzel, A., Blundell, C.: Simple and scalable predictive uncertainty estimation using deep ensembles. In: Advances in Neural Information Processing Systems, pp. 6403–6414 (2017)

19. Lee, A.J.: U-Statistics: Theory and Practice. Taylor & Francis (1990)

20. Menze, B.H., Jakab, A., Bauer, S., et al.: The multimodal brain tumor image segmentation benchmark (BRATS). IEEE Trans. Med. Imaging **34**, 1993–2024 (2015)

21. Naeini, M.P., Cooper, G.F., Hauskrecht, M.: Obtaining well calibrated probabilities using Bayesian binning. In: Proceedings of the 29th AAAI Conference on Artificial Intelligence, pp. 2901–2907 (2015)

22. Neal, R.M.: Bayesian Learning for Neural Networks. Springer, New York (2012). https://doi.org/10.1007/978-1-4612-0745-0

23. Ronneberger, O., Fischer, P., Brox, T.: U-Net: convolutional networks for biomedical image segmentation. In: Navab, N., Hornegger, J., Wells, W.M., Frangi, A.F. (eds.) MICCAI 2015. LNCS, vol. 9351, pp. 234–241. Springer, Cham (2015). https://doi.org/10.1007/978-3-319-24574-4_28
24. Rousseau, A.J., Becker, T., Bertels, J., Blaschko, M., Valkenborg, D.: Post training uncertainty calibration of deep networks for medical image segmentation. In: ISBI (2021)
25. Tilborghs, S., Maes, F.: Left ventricular parameter regression from deep feature maps of a jointly trained segmentation CNN. In: Pop, M., et al. (eds.) STACOM 2019. LNCS, vol. 12009, pp. 395–404. Springer, Cham (2020). https://doi.org/10.1007/978-3-030-39074-7_41
26. Wenger, J., Kjellström, H., Triebel, R.: Non-parametric calibration for classification. In: International Conference on Artificial Intelligence and Statistics, pp. 178–190 (2020)
27. Winzeck, S., Hakim, A., McKinley, R., et al.: ISLES 2016 and 2017-benchmarking ischemic stroke lesion outcome prediction based on multispectral MRI. Front. Neurol. 9, 679 (2018)
28. Wu, J., Ruan, S., Lian, C., et al.: Active learning with noise modeling for medical image annotation. In: ISBI, pp. 298–301 (2018)

High-Resolution Segmentation of Lumbar Vertebrae from Conventional Thick Slice MRI

Federico Turella[1]([✉]), Gustav Bredell[1], Alexander Okupnik[1],
Sebastiano Caprara[2,3][ID], Dimitri Graf[2][ID], Reto Sutter[2][ID],
and Ender Konukoglu[1]

[1] Computer Vision Lab, ETH Zürich, Zürich, Switzerland
`fturella@student.ethz.ch`
[2] Institute for Biomechanics, ETH Zürich, Zürich, Switzerland
[3] Universitätsklinik Balgrist, Zürich, Switzerland
`https://github.com/FedeTure/ReconNet`

Abstract. Imaging plays a crucial role in treatment planning for lumbar spine related problems. Magnetic Resonance Imaging (MRI) in particular holds great potential for visualizing soft tissue and, to a lesser extent, the bones, thus enabling the construction of detailed patient-specific 3D anatomical models. One challenge in MRI of the lumbar spine is that the images are acquired with thick slices to shorten acquisitions in order to minimize patient discomfort as well as motion artifacts. In this work we investigate whether detailed 3D segmentation of the vertebrae can be obtained from thick-slice acquisitions. To this end, we extend a state-of-the-art segmentation algorithm with a simple segmentation reconstruction network, which aims to recover fine-scale shape details from segmentations obtained from thick-slice images. The overall method is evaluated on a paired dataset of MRI and corresponding Computed Tomography (CT) images for a number of subjects. Fine scale segmentations obtained from CT are compared with those reconstructed from thick-slice MRI. Results demonstrate that detailed 3D segmentations can be recovered to a great extent from thick-slice MRI acquisitions for the vertebral bodies and processes in the lumbar spine.

Keywords: Segmentation · High-resolution · MRI

1 Introduction

Patient specific anatomic modeling holds the promise to improve treatment planning as well as our understanding of the anatomical variation in the population.

Partially supported by Hochschulmedizin Zürich through the Surgent project.

Electronic supplementary material The online version of this chapter (https://doi.org/10.1007/978-3-030-87193-2_65) contains supplementary material, which is available to authorized users.

© Springer Nature Switzerland AG 2021
M. de Bruijne et al. (Eds.): MICCAI 2021, LNCS 12901, pp. 689–698, 2021.
https://doi.org/10.1007/978-3-030-87193-2_65

One region of particular interest for patient-specific modeling is the lumbar spine due to the high prevalence of lower-back pain in today's society and its crucial role in healthy mobility. Accurate patient specific anatomical models of the lumbar spine would allow a better understanding of a patient's current condition and assessment of possible treatment options through personalized biomechanical simulations. Especially for lumbar spine fusion, which is a frequent surgical intervention, personalized modeling could be crucial for forecasting the influence of different stages of fusion. Vertebrae anatomy is the first step towards constructing a useful personalized anatomical model of the lumbar spine.

Computed Tomography has been the modality of choice to construct accurate models of the lumbar vertebrae over the last decades due to its excellent properties for imaging bone structures at high resolution and contrast. Moreover, following the developments in deep learning and availability of large annotated CT datasets [10,16,17], accurate automatic algorithms became available for segmenting vertebrae from CT images [8,16]. Even though the segmentation performance in CT is impressive and allows constructing accurate patient specific lumbar spine models [2], CT may not be the best modality for lumbar spine investigation for two reasons. First, CT exposes patients to ionizing radiation increasing the possibility of inducing cancer, and second, CT does not visualize soft tissue with high contrast, which is important for building anatomical models with more details including muscles and nerve roots.

Magnetic Resonance Imaging (MRI) does not share the mentioned shortcomings with CT and can be preferred in practice for lumbar spine inspection, however, it has its own shortcomings. Firstly, MRI acquisitions are time intensive, leading to a trade-off between image quality and acquisition time. At the operating point used in clinical practice, images are often acquired with high in-plane resolution but thick slices. Secondly, MRI visualizes soft tissue nicely, however, it does not have the same contrast quality for bone structures. Thirdly, we have observed that even in thin sliced MRIs that were manually segmented, obvious bone structures were missing (Supplement Material, section 5.2). This seems to especially be the case for the processes of vertebrae. These three factors make extracting detailed anatomical models of vertebrae from MRI challenging. Recent work has shown that the second problem, i.e., contrast, can be addressed with deep learning methods [13] to segment the low-contrast bone structures in clinical quality MRI. What thus remains open is the thick slice nature of MRI in addition to the absence of fine bone structures. Increasing the through-plane resolution of the final segmentations in addition to augmenting fine bone details, would ultimately allow extracting high quality patient specific vertebrae models from MRI.

There are various alternatives that can be used for increasing through plane resolution of vertebrae segmentation. Arguably the most obvious ones are (i) super-resolving the underlying MRI using a state-of-the-art DL method, e.g., [9] and (ii) predicting high through plane resolution segmentations from the input images by constraining output shapes by using for instance [12]. While these approaches are conceivable, they are not viable in practice. They would require a large number of thin slice MRIs for training, and such publicly available datasets do not exist to the best of our knowledge for the lumbar region.

In addition, by relying on thin slice MRI these methods would not be able to address the missing fine bone structures. An alternative is to utilize synthesis techniques that can utilize available high resolution CT images, which are in abundance compared to MRI, without requiring paired MRI and CT datasets. A Cycle-GAN-based alternative can convert a thick slice MRI into a high resolution CT and the segmentation can be directly performed on the CT, as it has been done on different anatomy [5,6,11]. While promising, this unsupervised approach can introduce artifacts during image translation (Supplement material, section 5.1), and the segmentation quality of the synthesized images can be low as demonstrated even in easier synthesis tasks [7]. One last alternative, which is the direction taken here, is to directly work on the segmentation space where CT and MRI datasets can be combined directly without considering differences in modalities.

We propose a segmentation reconstruction network, referred to as Recon-Net, that transforms low-quality segmentations, extracted from thick slice MRIs, to high-quality segmentations, as extracted from thin-slice high resolution CT images. The ReconNet is trained on the segmentations of widely available CT lumbar spine datasets. In addition, we propose to use a Variational Autoencoder (VAE) as a post-processing step to improve the segmentation quality after reconstruction. We evaluate the proposed segmentation reconstruction using a dataset with paired MRI and CT. Segmentations extracted from thick-slice MRI before and after reconstruction are compared with those extracted from high resolution CT images. In addition, we investigate the usability of the reconstructed segmentations in downstream tasks such as registration and compare them to CT-based segmentation.

2 Methods

The goal of the proposed segmentation reconstruction network is to extract a high-quality segmentation of the lumbar spine based on the segmentation of thick slice MRI. To obtain the segmentation of the vertebrae in the thick slice MRI we employ deep learning methods from literature [8,13]. These segmentations, which differ from the actual shapes of the vertebrae due to the low through plane resolution and missing fine bone structures, are used as input for ReconNet to produce a high-quality segmentation, which is subsequently refined by a VAE. This process can be seen in Fig. 1. In this section, the parts of the pipeline are described in more detail.

2.1 Segmentation

Current models segmenting lumbar spine in CT images utilize 3D models due to the high resolution in all planes as well as large data availability. For thick sliced MRI, however, 2D segmentation is preferred due to the very coarse through plane resolution and the lack of large publicly available datasets. SpineSparseNet [13] for instance, first employs a 3D graph convolutional segmentation network for a coarse grained segmentation and then a 2D network for refinement and accurate

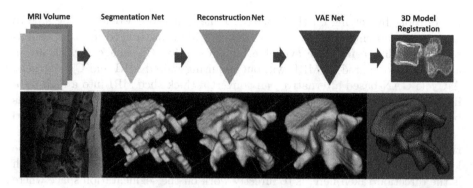

Fig. 1. Illustration of how our reconstruction method fits into a full model extraction pipeline consisting of segmentation, reconstruction, VAE enhancement and a possible downstream task of 3D Model registration presented in Sect. 3. For each step of the pipeline, a representative snapshot is shown, documenting the transformation from thick sliced MRI volume to 3D Model.

segmentation. In the same spirit, we adjusted another method referred to as iterative fully convolutional neural networks (IFCNN) [8] to a 2D version to compare which segmentation approach works the best. IFCNN was adopted to a 2D version by instead of having a sliding $128 \times 128 \times 128$ patch to have a 128×64 sliding window on the central slice of the volume due to the most structures being visible on this slice. Once the sliding window has detected a vertebra it is fixed and windows at the same location are extracted from all the other slices. 2D segmentation is then performed on all of the extracted windows before moving on to the next vertebra. In addition, no classification module was employed while omitting the labeling error as well as completeness classification error in the loss function. We refer the reader to the original paper for more details on training and inference [8].

2.2 Reconstruction

The extracted segmentation will still be of coarse nature due to the thick slices of the MRI and missing fine bone structures. Here we introduce ReconNet for segmentation reconstruction and the VAE approach used for segmentation refinement.

Reconstruction Network. ReconNet is trained in a supervised manner to take as input low-quality segmentations and transforms the segmentations to high-quality. To this end, a dataset of CT images with corresponding ground truth masks (T) are used. At each iteration during training, a $128\,\mathrm{mm}^3$ patch of the high-quality segmentation is eroded for two iterations with an element with square connectivity equal to one in order to simulate the mismatch in the visibility of bone structures between MRI and CT. Then, the high-quality mask is transformed to a low-quality segmentation by only selecting at random between

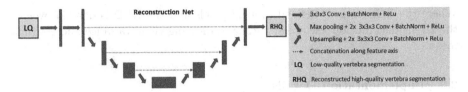

Fig. 2. Schematic illustration of the ReconNet architecture. The input and output are 3D volumes, as well as each feature channel across the network.

17–21 sagittal slices with a constant spacing between them (T_s), thus simulating a thick sliced MRI acquisition with a spacing of around 4 mm between sagittal slices. This spacing is similar to the range found in clinical MRI of the lumbar spine. Subsequently, a bicubic upsampling (R) to 1 mm^3 is done based on the selected slices, which is then used as the simulated low quality segmentation. In addition to selecting only a limited number of slices to generate the low-quality segmentation for input, one could also further augment the generated mask to compensate for segmentation errors. The explored augmentations (A) consist of random eroding and dilating of the segmentation. Each slice has a probability α to be augmented, and if so, half the time it gets eroded and the other half dilated for a number of iterations extracted randomly from the $[1, \beta]$ interval. The higher α and β, the heavier the modification will be, the more the model will trust its prior knowledge about vertebra shape rather than the input mask. The network can then be trained in an end-to-end fashion using the generated low-quality segmentation as input, while using the original high-quality segmentation as ground truth. False positives (FP) and false negatives (FN) are used for the loss function along with a weighting mask (W), giving higher weight to vertebra borders, to optimize for the parameters of the reconstruction network F_θ:

$$L = argmin_\theta W * [(1-T) * F_\theta(A_{\alpha,\beta}(R(T_s))) + \lambda * (1 - F_\theta(A_{\alpha,\beta}(R(T_s)))) * T] \quad (1)$$

whereas W is defined as: $\zeta + \eta * exp(-D^2/\omega^2)$. With $\zeta = 1$, $\eta = 15$, $\omega = 4$ and D defined as distance to a vertebra border pixel.

Variational Autoencoder Reconstruction. The reconstruction of the low-quality segmentation might not always be perfect as the reconstruction network might focus too much on certain structures of the segmentation that are not anatomically feasible. To ensure that the reconstructed vertebra is anatomically feasible one could train a VAE and push the reconstructed vertebra through the VAE to ensure that the final reconstruction lies within the probability distribution of real vertebrae. The VAE is trained by encoding high-quality 3D CT ground truths to a latent space of 100 distributions. The loss function used for the VAE network with parameters theta (V_θ) consists out of a binary cross-entropy (BCE) and Kullback-Leibler divergence (KLD) term:

$$L = argmin_\theta - (T * log(V_\theta) + (1-T) * log(1-V_\theta)) - 0.5 * (1 + log(\Sigma) - M^2 - \Sigma) \quad (2)$$

Where the BCE term is defined over all pixels and the KLD is defined over all the latent distribution variables, with Σ indicating their variance and M their mean.

2.3   Implementation Details

The networks were implemented in PyTorch [14] and auxiliary functions in Python. The 2D IFCNN network was implemented as 2D U-Net [15], trained for 250 epochs using an Adam optimizer with a learning rate of $1*10^{-3}$ and a batch size of 32. The reconstruction net is a 3D U-Net [3], it was trained for 120 epochs with a learning rate of $2*10^{-4}$ and batch size of 4. Its structure is shown in Fig. 2 and the code can be found in the project repository [18]. The scaling factor λ in the loss function is set to 0.1 for the first 20 epochs to avoid the constantly empty prediction fail-case, and after that it is set to 1. The VAE network is a 3D U-Net with a latent space of 100 distributions, it was trained for 140 epochs with a learning rate of $2*10^{-5}$ and a batch size of 4. The training took place on the in-house GPU cluster mainly consisting of GeForce GTX TITAN X with 12 GB memory.

3   Results

3.1   Thick Slice MRI Segmentation

Thick Slice MRI Dataset. A local hospital provided thick slice MRIs of 30 patients collected during clinical routine. Five patients had either scoliosis or degenerate discs, whereas the remaining 25 were largely healthy. The visible lumbar spine vertebrae in all the volumes were segmented manually and reviewed by radiologists. The sagittal plane has a resolution of either 512×512 or 418×418 and a corresponding pixel spacing between 0.59 mm and 0.71 mm. The number of sagittal slices per volume ranges from 14 to 19, with a pixel spacing between 4.4 mm and 4.95 mm. We used both T1 and T2 modalities for a total of 60 volumes, 54 of which were used as training set and 6 for the testing set. Note that both modalities of the same subject were assigned to the same set.

Segmentation Results. Both the 2D IFCNN network and SpineSparseNet were trained with the training set of the thick slice MRI dataset and the segmentation results for the test set are shown in Table 1. For training the 2D IFCNN, the volumes were resampled to 1 mm^2 in the sagittal plane. In addition, the slices were separated and formed a training set of 804 2D sagittal slices. For 2D IFCNN the images were cropped in the middle of the highest visible vertebra and for SpineSparseNet such that only the T12 to L5 vertebrae were visible. The test scans were resampled to 1 mm^2 in the sagittal plane as well, and the Dice's Similarity Coefficient (DSC) was calculated on the 3D mask of the vertebrae with respect to ground truth. For all lumbar spine vertebrae, the DSC of the

Table 1. Quantitative results of the semantic segmentation performance. Reported is the mean ± standard deviation of the DSC (%). The results refer to a test set of 6 MRIs, which includes 2 modalities (T1 and T2) for 3 subjects.

Method	L1	L2	L3	L4	L5
2D iterative FCN	87.79 ± 2.08%	86.12 ± 1.73%	87.01 ± 2.48%	85.97 ± 1.85%	86.24 ± 1.66%
SpineSparseNet	84.82 ± 1.64%	83.91 ± 0.85%	85.18 ± 1.19%	84.41 ± 1.45%	83.47 ± 1.90%

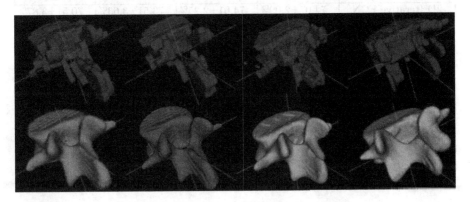

Fig. 3. Qualitative results showing L2 to L5 reconstructed vertebrae, from left to right. The top row images are the masks produced by the 2D IFCNN on thick sliced MRI. The bottom row shows the output of the VAE network, which was fed with the reconstruction of ReconNet Augmented. The samples are representative and come from different patients of the test set.

2D IFCNN is, on average, 2–3% higher. The DSC of SpineSparseNet is significantly lower than what is stated in the original paper. However, the number of subjects in our dataset is an order of magnitude smaller, which indicates that SpineParseNet relies on more data in order to reach its peak performance.

3.2   Reconstruction

Training and Test Datasets. For training the different segmentation reconstruction network options the ground truths from 121 scans of a large publicly available dataset of CT lumbar spine was used [10,16,17]. The masks were resampled to 1 mm³ and the lumbar vertebrae (L1–L5) extracted in 128 × 128 × 128 patches. In total 106 lumbar spine volumes were used for training and 15 were used as the validation set. Extracting the individual vertebrae lead to 74 vertebra patches for validation and 519 for training.

To test the reconstruction results an independent dataset was used that contained paired thick slice MRI and CT acquisitions for 20 patients [1]. The segmentations for the dataset were obtained by manual labeling by radiologists and experts. Due to the high manual labour cost only 11 patients were selected based on the number of vertebrae visible in the volume and CT image quality. Out of

Table 2. The upper table shows quantitative results of the reconstruction performance. Reported is the mean ± standard deviation of the DSC (%) on the test dataset [1]. The VAE scores are obtained by feeding the VAE the vertebrae reconstructed by ReconNet Augmented. The bottom table shows quantitative results of the 3D model registration. Reported are the mean of the DSC(%) and Hausdorff distance (HD) calculated using 3D models obtained from the ground truth masks.

Reconstruction	L2		L3		L4		L5	
2D Iterative FCN	74.07±2.09%		74.01±2.25%		73.51±3.63%		70.51±3.87%	
ReconNet	84.78±1.31%		84.56±1.61%		83.39±2.81%		81.91±2.38%	
ReconNet Augmented	85.33±1.64%		85.55±2.03%		84.07±2.55%		83.25±2.50%	
VAE	86.01±1.42%		86.14±1.68%		85.31±2.17%		83.96±3.10%	
3D Iterative FCN	91.37±1.46%		91.78±2.09%		91.04±2.48%		88.35±4.17%	
Registration	**L2**		**L3**		**L4**		**L5**	
	DSC	HD	DSC	HD	DSC	HD	DSC	HD
2D Iterative FCN	79.55	14.68	79.44	15.39	79.53	15.10	76.40	19.38
ReconNet Augmented	82.36	8.73	83.10	10.67	81.11	11.21	80.89	14.54
VAE	82.58	8.93	83.10	10.90	81.48	10.95	81.33	13.87
3D Iterative FCN	85.30	7.75	85.52	8.56	84.21	9.37	83.23	12.59

the used scans, 8 volumes have fully visible L2 to L5, and 3 have fully-visible L3 to L5. Moreover, 2 of the 11 patients have scoliosis.

Reconstruction Results. Table 2 shows the comparison between different reconstruction approaches by reporting the mean and standard deviation of the DSCs (%) calculated by comparing the 3D vertebrae reconstructed from MRI with the 3D masks created by manual segmentation based on CT scans. Note that for all the methods the reconstructed vertebrae were registered to the ground truth to compensate for the position change of the subjects in between different acquisition techniques. The 2D IFCNN entry refers to the linear upsampling of the thick slice MRI segmentation to 1 mm³. The ReconNet entries report the values for the reconstruction network with different levels of augmentation: in the plain case none of the slices was eroded or dilated, in the augmented case 50% of slices were either eroded or dilated with a number of [1,8] iterations. For the VAE entry, the vertebrae reconstructed by ReconNet Augmented were further refined by the VAE network. The 3D IFCNN entry represents the values obtained by segmenting the CT scans using the approach outlined by Lessmann et al. [8,13] (segmentation path only). The network was trained using 88 scans from the same training set as the reconstruction networks. This approach is used to compare the quality of our reconstruction network with state-of-the-art CT segmentation performance. For the VAE case, Fig. 3 shows some examples of reconstructed vertebrae. After vertebrae assembly, we observe an average voxel overlap of $0.39 \pm 0.45\%$ per patient.

3.3 Registration

To investigate the usefulness of the reconstructed vertebrae for downstream tasks we extracted the anatomical 3D meshes which could be used to build a finite element (FE) model of the lumbar spine. Previously developed Statistical Shape Models (SSMs) of the lumbar vertebrae were non-rigidly registered on the reconstructed vertebrae using a point-set to image registration method described in Clogenson et al. [4]. The resulting 3D models were compared to the meshes obtained with the ground truth data to analyze the registration performance. The non-rigid registration of the SSMs was automatically performed following the steps described in Caprara et al. [2] using the results of the different reconstruction strategies. The resulting metrics are shows in Table 2.

4   Conclusion

We show that leveraging abundant CT datasets of lumbar spine can be used to train a reconstruction network and VAE to transform low quality segmentations based on thick sliced MRI to high quality. With this work we closed the DSC gap between segmentations based on CT versus that of thick sliced MRI from 18% to 5%. In addition, we show that the reconstructed high quality segmentations can be successfully used in the downstream task of registration leading to 3D models that can be used for biomechanical simulation.

References

1. Cai, Y., Osman, S., Sharma, M., Landis, M., Li, S.: Multi-modality vertebra recognition in arbitrary views using 3D deformable hierarchical model. IEEE Trans. Med. Imaging **34**(8), 1676–1693 (2015)
2. Caprara, S., Carrillo, F., Snedeker, J.G., Farshad, M., Senteler, M.: Automated pipeline to generate anatomically accurate patient-specific biomechanical models of healthy and pathological FSUs. Front. Bioeng. Biotechnol. **9** (2021)
3. Çiçek, Ö., Abdulkadir, A., Lienkamp, S.S., Brox, T., Ronneberger, O.: 3D U-Net: learning dense volumetric segmentation from sparse annotation. In: Ourselin, S., Joskowicz, L., Sabuncu, M.R., Unal, G., Wells, W. (eds.) MICCAI 2016. LNCS, vol. 9901, pp. 424–432. Springer, Cham (2016). https://doi.org/10.1007/978-3-319-46723-8_49
4. Clogenson, M., et al.: A statistical shape model of the human second cervical vertebra. Int. J. Comput. Assist. Radiol. Surg. **10**(7), 1097–1107 (2014). https://doi.org/10.1007/s11548-014-1121-x
5. Ge, Y., et al.: Unpaired MR to CT synthesis with explicit structural constrained adversarial learning. In: 2019 IEEE 16th International Symposium on Biomedical Imaging (ISBI 2019), pp. 1096–1099 (2019). https://doi.org/10.1109/ISBI.2019.8759529
6. Hiasa, Y., et al.: Cross-modality image synthesis from unpaired data using Cycle-GAN. In: Gooya, A., Goksel, O., Oguz, I., Burgos, N. (eds.) SASHIMI 2018. LNCS, vol. 11037, pp. 31–41. Springer, Cham (2018). https://doi.org/10.1007/978-3-030-00536-8_4

7. Iglesias, J.E., Konukoglu, E., Zikic, D., Glocker, B., Van Leemput, K., Fischl, B.: Is synthesizing MRI contrast useful for inter-modality analysis? In: Mori, K., Sakuma, I., Sato, Y., Barillot, C., Navab, N. (eds.) MICCAI 2013. LNCS, vol. 8149, pp. 631–638. Springer, Heidelberg (2013). https://doi.org/10.1007/978-3-642-40811-3_79

8. Lessmann, N., Van Ginneken, B., De Jong, P.A., Išgum, I.: Iterative fully convolutional neural networks for automatic vertebra segmentation and identification. Med. Image Anal. **53**, 142–155 (2019)

9. Lyu, Q., et al.: Multi-contrast super-resolution MRI through a progressive network. IEEE Trans. Med. Imaging **39**(9), 2738–2749 (2020). https://doi.org/10.1109/TMI.2020.2974858

10. Löffler, M.T., et al.: A vertebral segmentation dataset with fracture grading. Radiol. Artif. Intell. **2**(4), e190138 (2020). https://doi.org/10.1148/ryai.2020190138

11. Nie, D., et al.: Medical image synthesis with context-aware generative adversarial networks. In: Descoteaux, M., Maier-Hein, L., Franz, A., Jannin, P., Collins, D.L., Duchesne, S. (eds.) MICCAI 2017. LNCS, vol. 10435, pp. 417–425. Springer, Cham (2017). https://doi.org/10.1007/978-3-319-66179-7_48

12. Oktay, O., et al.: Anatomically constrained neural networks (ACNNs): application to cardiac image enhancement and segmentation. IEEE Trans. Med. Imaging **37**(2), 384–395 (2017)

13. Pang, S., et al.: SpineParseNet: Spine parsing for volumetric MR image by a two-stage segmentation framework with semantic image representation. IEEE Trans. Med. Imaging **40**(1), 262–273 (2021). https://doi.org/10.1109/TMI.2020.3025087

14. Paszke, A., et al.: Automatic differentiation in PyTorch (2017)

15. Ronneberger, O., Fischer, P., Brox, T.: U-Net: convolutional networks for biomedical image segmentation. In: Navab, N., Hornegger, J., Wells, W.M., Frangi, A.F. (eds.) MICCAI 2015. LNCS, vol. 9351, pp. 234–241. Springer, Cham (2015). https://doi.org/10.1007/978-3-319-24574-4_28

16. Sekuboyina, A., et al.: VerSe: a vertebrae labelling and segmentation benchmark for multi-detector CT images. Elsevier (2020, under review)

17. Sekuboyina, A., Rempfler, M., Valentinitsch, A., Menze, B.H., Kirschke, J.S.: Labeling vertebrae with two-dimensional reformations of multidetector CT images: an adversarial approach for incorporating prior knowledge of spine anatomy. Radiol. Artif. Intell. **2**(2), e190074 (2020). https://doi.org/10.1148/ryai.2020190074

18. Turella, F.: High-resolution segmentation of lumbar vertebrae from conventional thick-slice MRI code (2021). https://github.com/FedeTure/ReconNet

Shallow Attention Network for Polyp Segmentation

Jun Wei[1,2], Yiwen Hu[1,2,5], Ruimao Zhang[1,2], Zhen Li[1,2(✉)], S. Kevin Zhou[1,3,4], and Shuguang Cui[1,2]

[1] School of Science and Engineering, The Chinese University of Hong Kong (Shenzhen), Shenzhen, China
lizhen@cuhk.edu.cn
[2] Shenzhen Research Institute of Big Data, Shenzhen, China
[3] School of Biomedical Engineering and Suzhou Institute for Advanced Research, University of Science and Technology of China, Suzhou, China
[4] Institute of Computing Technology, Chinese Academy of Sciences, Beijing, China
[5] Institute of Urology, The Third Affiliated Hospital of Shenzhen University (Luohu Hospital Group), Shenzhen, China

Abstract. Accurate polyp segmentation is of great importance for colorectal cancer diagnosis. However, even with a powerful deep neural network, there still exists three big challenges that impede the development of polyp segmentation. (i) Samples collected under different conditions show inconsistent colors, causing the feature distribution gap and overfitting issue; (ii) Due to repeated feature downsampling, small polyps are easily degraded; (iii) Foreground and background pixels are imbalanced, leading to a biased training. To address the above issues, we propose the **Sh**allow **A**ttention **Net**work (**SANet**) for polyp segmentation. Specifically, to eliminate the effects of color, we design the color exchange operation to decouple the image contents and colors, and force the model to focus more on the target shape and structure. Furthermore, to enhance the segmentation quality of small polyps, we propose the shallow attention module to filter out the background noise of shallow features. Thanks to the high resolution of shallow features, small polyps can be preserved correctly. In addition, to ease the severe pixel imbalance for small polyps, we propose a probability correction strategy (PCS) during the inference phase. Note that even though PCS is not involved in the training phase, it can still work well on a biased model and consistently improve the segmentation performance. Quantitative and qualitative experimental results on five challenging benchmarks confirm that our proposed SANet outperforms previous state-of-the-art methods by a large margin and achieves a speed about **72FPS**.

Keywords: Polyp segmentation · Colonoscopy · Colorectal Cancer

---

J. Wei, Y. Hu—Equal contributions.

© Springer Nature Switzerland AG 2021
M. de Bruijne et al. (Eds.): MICCAI 2021, LNCS 12901, pp. 699–708, 2021.
https://doi.org/10.1007/978-3-030-87193-2_66

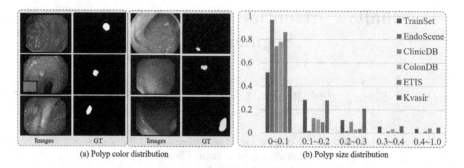

(a) Polyp color distribution   (b) Polyp size distribution

Fig. 1. (a) Polyp samples with different colors. (b) The histogram of polyp size. The horizontal axis shows the proportion of the polyp area to the image area. The vertical axis shows the proportion of polyp samples of a specific size to the total samples. One training set and five testing sets are shown in different colors. (Color figure online)

1   Introduction

Colorectal Cancer (CRC) has become a serious threat for human health, causing the fourth highest cancer death rate worldwide [7]. Polyps in the intestinal mucosa are considered the harbinger of CRC, which are easily transformed into malignant lesions [9]. Therefore, early diagnosis and treatment of polyps are of great significance. Fortunately, with the assistance of computer technology, a lot of automatic polyp segmentation models [1,4–6,13,15,17,22,23,25] have been developed and achieved remarkable progress.

However, polyp segmentation has always been a challenging task due to the inconsistent color distribution and small size of the targets. As shown in Fig. 1(a), polyp samples collected under the same conditions usually correspond to the same color, while those collected under different conditions appear in different colors. Thus, a strong correlation between color and polyp segmentation is implicitly contained in the dataset, which is harmful to the model training and will cause the model to overfit the color. Besides, most of polyp areas are very small. As shown in Fig. 1(b), the vast majority of polyps have an area which is less than 0.1 of the total image area. This brings two difficulties for existing segmentation models. First, small polyps are prone to get lost and hard to restore because of the repeated feature downsampling. Second, for images with small polyps, there exists a large imbalance between foreground and background pixels, leading to a biased model and a poor performance.

To break the correlation between color and polyp segmentation, we propose the color exchange (CE) operation. Specifically, for each input image, we randomly select another one and transfer its color to the input image. By multiple CE operations, each input image could appear in different colors, thus the connection between color and polyp segmentation is reduced and the model will not overfit to the fake causality. Besides, small polyp segmentation relies more on shallow features. Because they have higher resolutions and contain richer details, compared with deep ones [19]. Unfortunately, shallow features are too noisy to

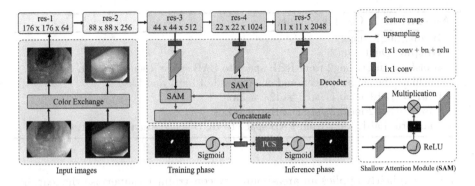

Fig. 2. An overview of the proposed model. (Color figure online)

be used directly. Thus, we propose the shallow attention module (SAM), which could help to remove the background noise, using clearer deep features. Furthermore, to ease the imbalance between foreground and background, we propose the probability correction strategy (PCS) during inference, which can adjust the biased prediction to the correct one with simple post-processing.

In summary, our contributions are four-folds: (1) We propose the color exchange operation to decouple the contents and colors, which reduces the overfitting issue; (2) We design the novel SANet to focus more on shallow features with richer details, improving the small polyp segmentation; (3) We introduce the probability correction strategy (PCS) for inference to balance the biased predictions, which can consistently improve model performance with negligible computation cost; (4) Extensive experiments demonstrate that the proposed model achieves state-of-the-art performance on five widely used public benchmarks.

2   Related Work

Polyp segmentation has went through two periods of development. Earlier models mainly rely on hand-crafted features (e.g., color and texture) [13,17,24], which can hardly capture the global context information and are not robust to complex scenarios. Recently, fully convolutional network (FCN [12]) has been applied to polyp segmentation and made great progress. For example, U-Net [15] is a famous structure for medical image segmentation, which consists of a contracting path to capture context and an expanding path to restore the precise detail. SegNet [20] adopts a similar structure to U-Net, but utilizes the max pooling indices to restore features in the upsampling operation. U-Net++ [25] and ResUNet++ [11] further improve original U-Net by dense connection and better pretrained backbone, achieving promising segmentation performance.

Though the body part of polyps could be well handled, boundaries are ignored. To enhance the boundary segmentation, Psi-Net [14] proposes to combine both body and boundary features in the segmentation model. Furthermore,

---

Algorithm 1: Color Exchange

Input: Img1, Img2
Output: Out1, Out2

1 transform Img1, Img2 from RGB space to LAB space, then get Lab1, Lab2
2 calculate the channel mean and channel std of Lab1 and Lab2
3 Lab1 = (Lab1-mean1)/std1*std2+mean2
4 Lab2 = (Lab2-mean2)/std2*std1+mean1
5 transform Lab1, Lab2 from LAB space to RGB space, then get Out1, Out2

SFA [6] explicitly applies an area-boundary constraint to supervise the learning of both polyp regions and boundaries. PraNet [5] proposes the reverse attention to firstly locate the polyp areas and then refine object boundaries implicitly. Though the above models have made great progress, polyp segmentation still faces big challenges because of the limited data. Thus, we propose the SANet to further improve the polyp segmentation, as shown in the Fig. 1.

3   Method

Figure 2 depicts the concrete architecture of the proposed SANet, where Res2Net [8] is adopted as the encoder backbone. According to the feature scale, Res2Net could be divided into five blocks, which have different receptive fields. Considering Wu *et al.* [21] have shown that low-level features bring too much computational cost with limited performance gains, thus we use the features of last three blocks $\{f_i | i \in (3, 4, 5)\}$ for the following experiments.

3.1   Color Exchange

As show in Fig. 1(a), polyp samples collected under different conditions show very different color distributions. However, these colors are far less important than shapes and structures for polyp segmentation, which actually leads to the overfitting issue due to the limited dataset. To avoid this effect, we propose the color exchange (CE) operation to explicitly decouple the image content and color. Specifically, for each input image, we randomly pick another one from the dataset and transfer its color to the input image. Algorithm 1 shows the specific steps. We calculate the mean and standard deviation of the colors in the LAB space and exchange these statistics between images. As show in Fig. 2, after color exchange, we could get the new input image with exactly the same content but a different color. Exchanging with different auxiliary images, the same input image could show a variety of colors but correspond to the same ground truth, so the model will focus more on the image contents and will not be affected by the color distribution, which could largely alleviate the influence of color distribution on model training. It is worth noting that color exchange is performed only during training. For inference, we directly use the original images, thus it will not bring any time overhead during testing.

3.2 Shallow Attention Module

As shown in Fig. 1(b), small polyps make up the majority of both training and testing datasets, facing the serious information loss during the repeated down-sampling in CNNs. To avoid this limitation, shallow features f_s deserve more attention due to their high resolutions. Namely, f_s have clear object boundaries, which are important for the accurate polyp segmentation. However, due to the limitation of receptive field, these features are submerged by background noise and hard to be used directly. In contrast, deep features f_d are coarse in boundaries but have clean background. Therefore, we propose the shallow attention module (SAM) to filter out background noise of f_s with the assistance of f_d, as shown in Fig. 2. Specifically, SAM involves in both f_s and f_d. Different from self-attention, SAM makes use of the complementarity between different features. In SAM, f_d will be firstly upsampled into the same size with f_s and then regarded as the attention maps for f_s. Finally, these maps will be multiplied with f_s to help suppress the background noise. The whole process is shown in following equations:

$$Att = \sigma(Up(f_d)), \tag{1}$$
$$f_s = Att \otimes f_s, \tag{2}$$

where $Up(\cdot)$ represents the upsampling operation. $\sigma(\cdot)$ is the ReLU function and $\otimes$ is the element-wise multiplication. After SAM, shallow features f_s will become much cleaner and provide important cues for the segmentation of small polyps. Furthermore, SAM plays an important role in balancing features of different blocks. Instead of aggregating features from all levels with the same weight, SAM could dynamically assign weight to different features according to their contributions.

3.3 Probability Correction Strategy

Probability correction strategy (PCS) is an effective way to improve final predictions, especially for small polyps where foreground and background pixels are extremely unbalanced. This imbalance brings in difficulties for the model training and leads to a biased model. Namely, negative samples (background pixels) are dominant in the training process, which leads to the tendency of the model to give lower confidence to positive samples (foreground pixels). To enhance the predictions of the positive samples, we propose to explicitly correct the predicted probability through logit reweighting. Specifically, for each image, we extract the predicted features before the Sigmoid function as the target (*i.e.*, logit) to be corrected. We count the proportion ($rate^p$) of positive samples (*i.e.*, $logit > 0$) in the image, which is relatively small. At the same time, due to the model bias, the logit of the positive sample also is small. Thus, we propose to normalize the logit of positive samples with its proportion, as shown in Eq. 3.

$$logit_{ij}^{p_{norm}} = logit_{ij}^p/rate^p \tag{3}$$

where i, j represent the coordinate of the sample. Similarly, the logits of negative (*i.e.*, $logit < 0$) samples could also be normalized, as shown in Eq. 4.

$$logit_{ij}^{n_{norm}} = logit_{ij}^n / rate^n \tag{4}$$

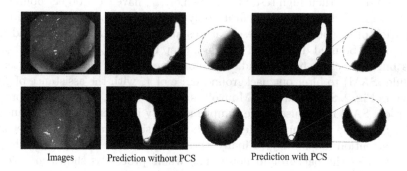

Images   Prediction without PCS   Prediction with PCS

Fig. 3. Visual comparison between the predictions with PCS and without PCS.

After normalization, the effect of sample number has been attenuated and the segmentation results will be more accurate. It is worth nothing that PCS is only applied in the inference phase with very little computation cost. Figure 3 visualizes some predictions. Obviously, predictions with PCS have clearer boundaries.

3.4 Loss Function

We use binary cross entropy (BCE) loss and Dice loss for supervision, as shown in Eq. 5, where P, G are the prediction and ground truth, respectively. λ_1, λ_2 are the weighting coefficients, which are set to 1 for simplification.

$$Loss = \lambda_1 BCE(P, G) + \lambda_2 Dice(P, G) \tag{5}$$

4   Experiments

4.1   Datasets and Training Settings

To evaluate the performance of the proposed SANet, five polyp segmentation datasets are adopted, including Kvasir [10], CVC-ClinicDB [2], CVC-ColonDB [3], EndoScene [18] and ETIS [16]. To keep the fairness of the experiments, we follow [5] advice and take exactly the same training and testing dataset division. Besides, six state-of-the-art methods are used for comparison, namely U-Net [15], U-Net++ [25], ResUNet [23], ResUNet++ [11], SFA [6] and PraNet [5]. We use Pytorch to implement our model. All input images are uniformly resized to 352×352. For data augmentation, we adopt the random flip, random rotation and multi-scale training. The whole network is trained in an end-to-end way, using stochastic gradient descent (SGD). Initial learning rate and batch size are set to 0.04 and 64, respectively. We train the entire model for 128 epoches.

Table 1. Performance comparison with different polyp segmentation models. The highest and second highest scores are highlighted in red and blue colors, respectively.

Methods	Kvasir		ClinicDB		ColonDB		EndoScene		ETIS	
	mDice	mIoU	mDice	mIoU	mDice	mIoU	mDice	mIoU	mDice	mIoU
U-Net	0.818	0.746	0.823	0.750	0.512	0.444	0.710	0.627	0.398	0.335
U-Net++	0.821	0.743	0.794	0.729	0.483	0.410	0.707	0.624	0.401	0.344
ResUNet	0.791	–	0.779	–	–	–	–	–	–	–
ResUNet++	0.813	0.793	0.796	0.796	–	–	–	–	–	–
SFA	0.723	0.611	0.700	0.607	0.469	0.347	0.467	0.329	0.297	0.217
PraNet	0.898	0.840	0.899	0.849	0.712	0.640	0.871	0.797	0.628	0.567
SANet(Ours)	0.904	0.847	0.916	0.859	0.753	0.670	0.888	0.815	0.750	0.654

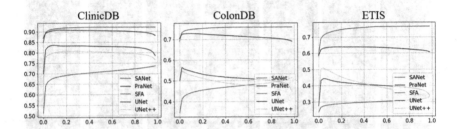

Fig. 4. Dice curves under different thresholds on three polyp datasets.

4.2 Quantitative Comparison

To prove the effectiveness of the proposed SANet, six state-of-the-art models are used for comparison, as shown in Table 1. SANet achieves the best scores across five datasets on both mIoU and mDice, demonstrating the superior performance of the proposed model. In addition, Fig. 4 shows the dice values of above models under different thresholds. From these curves, we could observe that SANet consistently outperforms other models, which proves its good capability for polyp segmentation. Furthermore, SANet achieves a speed about 72FPS on RTX 2080Ti GPU, faster than the 64FPS of previous PraNet [5].

4.3 Visual Comparison

Figure 5 visualizes some predictions of different models. Compared with other counterparts, our method could not only clearly highlight the polyp regions but also suppress the background noise. Even for the challenging scenarios, our model could handle well and generate accurate segmentation mask.

4.4 Ablation Study

To investigate the importance of each component in SANet, both ColonDB and Kvasir datasets are used for controlled experiments. As shown in Table 2, all the modules or strategies are necessary for the final predictions. Combining all the proposed methods, our model achieves the new state-of-the-art performance.

Table 2. Ablation study for SANet on the ColonDB and Kvasir datasets.

Settings	ColonDB		Kvasir	
	mDice	mIoU	mDice	mIoU
backbone	0.676	0.608	0.853	0.780
backbone+SAM	0.728	0.645	0.882	0.821
backbone+SAM+CE	0.745	0.662	0.896	0.838
backbone+SAM+CE+PCS	0.753	0.670	0.904	0.847

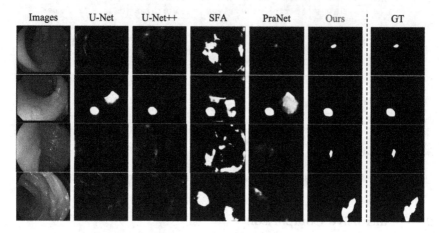

Fig. 5. Visual comparison between the proposed method and four state-of-the-art ones.

5 Conclusion

Because of the limited dataset, polyp segmentation model is easy to corrupt due to overfitting. In this paper, we attempt to alleviate this problem from two aspects. For the false color causality, we propose to decouple the image color and content by color exchange. For the difficult small polyp segmentation, we design the shallow attention to reduce the data noise. All of these can reduce the interference of irrelevant factors on the model. In the future, we will combine more prior knowledge to design more robust features to remove the interference of independent factors.

Acknowledgement. The work was supported in part by Key Area R&D Program of Guangdong Province with grant No. 2018B030338001, by the National Key R&D Program of China with grant No. 2018YFB1800800, by Shenzhen Outstanding Talents Training Fund, by Guangdong Research Project No. 2017ZT07X152, by NSFC-Youth 61902335, by Guangdong Regional Joint Fund-Key Projects 2019B1515120039, by The National Natural Science Foundation Fund of China (61931024), by helix0n biotechnology company Fund and CCF-Tencent Open Fund.

References

1. Akbari, M., et al.: Polyp segmentation in colonoscopy images using fully convolutional network. In: 2018 40th Annual International Conference of the IEEE Engineering in Medicine and Biology Society (EMBC), pp. 69–72 (2018)
2. Bernal, J., Sánchez, F.J., Fernández-Esparrach, G., Gil, D., Rodríguez, C., Vilariño, F.: WM-DOVA maps for accurate polyp highlighting in colonoscopy: validation vs. saliency maps from physicians. Comput. Med. Imaging Graph. **43**, 99–111 (2015)
3. Bernal, J., Sánchez, J., Vilarino, F.: Towards automatic polyp detection with a polyp appearance model. Pattern Recogn. **45**(9), 3166–3182 (2012)
4. Brandao, P., et al.: Fully convolutional neural networks for polyp segmentation in colonoscopy. In: Medical Imaging 2017: Computer-Aided Diagnosis, vol. 10134, p. 101340F (2017)
5. Fan, D.-P., et al.: PraNet: parallel reverse attention network for polyp segmentation. In: Martel, A.L., et al. (eds.) MICCAI 2020. LNCS, vol. 12266, pp. 263–273. Springer, Cham (2020). https://doi.org/10.1007/978-3-030-59725-2_26
6. Fang, Y., Chen, C., Yuan, Y., Tong, K.: Selective feature aggregation network with area-boundary constraints for polyp segmentation. In: Shen, D., et al. (eds.) MICCAI 2019. LNCS, vol. 11764, pp. 302–310. Springer, Cham (2019). https://doi.org/10.1007/978-3-030-32239-7_34
7. Favoriti, P., Carbone, G., Greco, M., Pirozzi, F., Pirozzi, R.E.M., Corcione, F.: Worldwide burden of colorectal cancer: a review. Updates Surg. **68**(1), 7–11 (2016). https://doi.org/10.1007/s13304-016-0359-y
8. Gao, S., Cheng, M., Zhao, K., Zhang, X., Yang, M., Torr, P.H.S.: Res2Net: a new multi-scale backbone architecture. IEEE Trans. Pattern Anal. Mach. Intell. **43**(2), 652–662 (2021)
9. Granados-Romero, J.J., et al.: Colorectal cancer: a review. Int. J. Res. Med. Sci. **5**(11), 4667–4676 (2017)
10. Jha, D., et al.: Kvasir-SEG: a segmented polyp dataset. In: Ro, Y.M., et al. (eds.) MMM 2020. LNCS, vol. 11962, pp. 451–462. Springer, Cham (2020). https://doi.org/10.1007/978-3-030-37734-2_37
11. Jha, D., et al.: ResUNet++: an advanced architecture for medical image segmentation. In: 2019 IEEE International Symposium on Multimedia (ISM), pp. 225–2255. IEEE (2019)
12. Long, J., Shelhamer, E., Darrell, T.: Fully convolutional networks for semantic segmentation. In: Proceedings of the IEEE Conference on Computer Vision and Pattern Recognition, pp. 3431–3440 (2015)
13. Mamonov, A.V., Figueiredo, I.N., Figueiredo, P.N., Tsai, Y.H.R.: Automated polyp detection in colon capsule endoscopy. IEEE Trans. Med. Imaging **33**(7), 1488–1502 (2014)
14. Murugesan, B., Sarveswaran, K., Shankaranarayana, S.M., Ram, K., Joseph, J., Sivaprakasam, M.: Psi-Net: shape and boundary aware joint multi-task deep network for medical image segmentation. In: 2019 41st Annual International Conference of the IEEE Engineering in Medicine and Biology Society (EMBC), pp. 7223–7226 (2019)
15. Ronneberger, O., Fischer, P., Brox, T.: U-Net: convolutional networks for biomedical image segmentation. In: Navab, N., Hornegger, J., Wells, W.M., Frangi, A.F. (eds.) MICCAI 2015. LNCS, vol. 9351, pp. 234–241. Springer, Cham (2015). https://doi.org/10.1007/978-3-319-24574-4_28

16. Silva, J., Histace, A., Romain, O., Dray, X., Granado, B.: Toward embedded detection of polyps in WCE images for early diagnosis of colorectal cancer. Int. J. Comput. Assist. Radiol. Surg. **9**(2), 283–293 (2014)
17. Tajbakhsh, N., Gurudu, S.R., Liang, J.: Automated polyp detection in colonoscopy videos using shape and context information. IEEE Trans. Med. Imaging **35**(2), 630–644 (2015)
18. Vázquez, D., et al.: A benchmark for endoluminal scene segmentation of colonoscopy images. J. Healthcare Eng. **2017**, article ID 4037190, 9 p. (2017). https://doi.org/10.1155/2017/4037190
19. Wei, J., Wang, S., Huang, Q.: F^3Net: fusion, feedback and focus for salient object detection. In: Proceedings of the AAAI Conference on Artificial Intelligence, vol. 34, pp. 12321–12328 (2020)
20. Wickstrøm, K., Kampffmeyer, M., Jenssen, R.: Uncertainty and interpretability in convolutional neural networks for semantic segmentation of colorectal polyps. Med. Image Anal. **60**, 101619101619 (2020)
21. Wu, Z., Su, L., Huang, Q.: Cascaded partial decoder for fast and accurate salient object detection. In: CVPR, June 2019
22. Zhang, R., Li, G., Li, Z., Cui, S., Qian, D., Yu, Y.: Adaptive context selection for polyp segmentation. In: Martel, A.L., et al. (eds.) MICCAI 2020. LNCS, vol. 12266, pp. 253–262. Springer, Cham (2020). https://doi.org/10.1007/978-3-030-59725-2_25
23. Zhang, Z., Liu, Q., Wang, Y.: Road extraction by deep residual U-Net. IEEE Geosci. Remote Sens. Lett. **15**(5), 749–753 (2018)
24. Zhou, S., et al.: A review of deep learning in medical imaging: image traits, technology trends, case studies with progress highlights, and future promises. Proc. IEEE **109**, 820–838 (2020)
25. Zhou, Z., Rahman Siddiquee, M.M., Tajbakhsh, N., Liang, J.: UNet++: a nested U-Net architecture for medical image segmentation. In: Stoyanov, D., et al. (eds.) DLMIA/ML-CDS -2018. LNCS, vol. 11045, pp. 3–11. Springer, Cham (2018). https://doi.org/10.1007/978-3-030-00889-5_1

A Line to Align: Deep Dynamic Time Warping for Retinal OCT Segmentation

Heiko Maier[1,3(✉)], Shahrooz Faghihroohi[1], and Nassir Navab[1,2]

[1] Computer Aided Medical Procedures, Technical University of Munich, Munich, Germany
heiko.maier@tum.de
[2] Computer Aided Medical Procedures, Johns Hopkins University, Baltimore, USA
[3] Deutsches Herzzentrum München, Munich, Germany

Abstract. In order to scan for or monitor retinal diseases, OCT is a useful diagnostic tool that allows to take high-resolution images of the retinal layers. For the aim of fully automated, semantic segmentation of OCT images, both graph based models and deep neural networks have been used so far. Here, we propose to interpret the semantic segmentation of 2D OCT images as a sequence alignment task. Splitting the image into its constituent OCT scanning lines (A-Modes), we align an anatomically justified sequence of labels to these pixel sequences, using dynamic time warping. Combining this dynamic programming approach with learned convolutional filters allows us to leverage the feature extraction capabilities of deep neural networks, while at the same time enforcing explicit guarantees in terms of the anatomical order of layers through the dynamic programming. We investigate both the solitary training of the feature extraction stage, as well as an end-to-end learning of the alignment. The latter makes use of a recently proposed, relaxed formulation of dynamic time warping, that allows us to backpropagate through the dynamic program to enable end-to-end training of the network. Complementing these approaches, a local consistency criterion for the alignment task is investigated, that allows to improve consistency in the alignment of neighbouring A-Modes. We compare this approach to two state of the art methods, showing favourable results.

Keywords: Image segmentation · Ophthalmology · OCT

H.M. was supported by TUM International Graduate School of Science and Engineering (IGSSE). H.M. and S.F. were supported by the ICL-TUM Joint Academy of Doctoral Studies (JADS) program. N.N. was partially supported by U.S. National Institutes of Health under grant number 1R01EB025883-01A1.

Electronic supplementary material The online version of this chapter (https://doi.org/10.1007/978-3-030-87193-2_67) contains supplementary material, which is available to authorized users.

M. de Bruijne et al. (Eds.): MICCAI 2021, LNCS 12901, pp. 709–719, 2021.
https://doi.org/10.1007/978-3-030-87193-2_67

1   Introduction

Optical coherence tomography (OCT) is a non-invasive imaging modality that can provide high-resolution, cross-sectional images of the human retina. It allows for detection and monitoring of retinal pathologies and assessment of associated, clinically relevant parameters. An exemplary disease detectable in the OCT is diabetic macular edema, which manifests as fluid accumulations in the retinal layers [17]. Manually obtaining parameters like layer thickness and fluid occurrences requires expert knowledge and experience, and is a time-consuming, labour-intensive task, especially when performed on larger sets of data like for study cohorts. Thus, there is a need for the (fully) automated segmentation of retinal OCT images and the assessment of the associated clinical parameters.

2   State of the Art

2.1   Automated Retinal Layer Segmentation

Many methods have been proposed for the automated segmentation of retinal layers in OCT images. Early works often employed handcrafted features, like intensity gradients, as input to graph-based algorithms [4,13]. With deep neural networks becoming prevalent in image segmentation tasks, a variety of works has applied them to retinal OCT images. Most of them are based on U-Net [18] -like fully convolutional architectures, employing an end-to-end trainable encoder-decoder architecture with skip connections. RelayNet [19] is such a fully convolutional architecture and capable of segmenting both the retinal layer boundaries and fluid pockets. Approaches like adding dilated convolutions for increased receptive field size [16] or attention mechanisms and multiscale inputs and outputs [14] to the fully convolutional architecture further improved the results.

Although these end-to-end trained architectures offer strong segmentation performance, they cannot directly take into account prior knowledge like the anatomical order of retinal layers, let alone guarantee to satisfy such constraints. To integrate such prior knowledge, one group of works has focused on adding graph-based post processing on top of trained networks. For example, extracting layer border probabilities using a CNN [7], RNN [11] or FCN [12] and adding a graph-based shortest path extraction as post processing have been proposed. Other works have focused on using multiple networks that are consecutively applied to each other's output, correcting for earlier mistakes or ensuring constraint adherence. Passing classified image patches that do not adhere to given topological constraints to a second network for refined prediction lead to more plausible segmentations on macular OCTs [10]. Similarly, replacing the second network with one that regresses (non-negative) layer thicknesses helped in guaranteeing predictions that follow the anatomical order of retinal layers [9]. Modified loss functions, that penalize violations of a desired topology, can also help in encouraging networks to make topologically sound predictions, albeit without guaranteeing them [2,23]. Instead of viewing the task as patch-based or full-image segmentation, an OCT can also be seen as the set of its constituting A-Modes. These are the individual depth-profiles acquired at each position during an OCT recording [1]. From this point of view, reformulating OCT segmentation as language processing and viewing the topological order as a grammar that has to be learned by GRU or LSTM networks has also been proposed [22].

Despite the multitude of approaches, one observation can be made: On the one hand, there are approaches that can guarantee topological ordering, but are then not fully end-to-end trainable. On the other hand, there are approaches that are fully end-to-end trainable, but cannot guarantee to adhere to topological constraints. In this work, we propose to utilize a relaxed variant of the well known dynamic time warping algorithm for the task at hand. It allows for end-to-end learning of a segmentation that follows the topological constraint of retinal layer order. The trained model can then either be used as-is, strongly adhering to the constraints, or with a small modification, guaranteeing them.

2.2 Dynamic Time Warping and Relaxed Formulation

Dynamic time warping (DTW) is a widely known distance measure for sequences, which is based on finding an optimal, cost minimizing alignment using dynamic programming. For readers unfamiliar with DTW, we point to external material [20]. In the following, we refer to the DTW local distance function as $d(i,j)$, the distance matrix as $\theta(i,j) = [d(a_i, b_j)]$ and the binary alignment matrix as Y.

In case of sequences for which the local distance function $d(i,j)$ is hard to define, one could instead try and learn it supervisedly, e.g. using gradient descent. The standard formulation of DTW hinders such approaches, as the DTW loss is not differentiable everywhere (w.r.t. the distance function) [15]. To be able to use DTW as a loss for learning tasks, replacing the discontinuous min operator used in the dynamic program with a relaxed soft-min operator was proposed for learning shapelets [21] and to perform averaging under the DTW distance [5]. In doing so, the previously binary alignment matrix becomes a soft alignment matrix with values between 0 and 1. Using this relaxation, it is also possible to derive gradients w.r.t. the alignment matrix itself, allowing to build differentiable neural network layers that predict soft alignment matrices [15].

DTW was once before applied to retinal layer segmentation, but only in a semi-automated fashion and without any form of learning [6]. As there, DTW was used to align multiple A-Modes to each other, it differs even more from our approach where we align single A-Modes to a symbolic reference sequence.

3   Proposed Approach

We investigate two approaches. The first one pretrains a CNN on image patches in a classification scenario and uses this CNN as local distance function for the alignment task. The second makes use of the relaxed DTW formulation from [15] to learn segmentation and alignment in an end-to-end fashion.

3.1 Segmentation as Sequence Alignment

To reformulate retinal OCT segmentation as a sequence alignment task, we propose to split each image into its constituting A-Modes (i.e. pixel columns). Each A-Mode serves as the first sequence in the alignment task. The second sequence (from here on referred to as the topological reference sequence, TRS) represents the anatomically given succession of retinal layers. This includes the seven retinal layers, as well as the space above and below the retina. It is also interspersed by fluid, with each fluid entry representing the possibility that fluid can or can not occur in this place. An example of one such sequence would be $[A, L_1, F, L_2, F, L_3, F, L_4, F, L_5, F, L_6, F, L_7, B]$, where A stands for the space above the retina, B for the space below the retina, L_1 to L_7 for the respective retinal layers and F for fluid pockets. We can then perform the segmentation of the respective A-Mode by aligning these two sequences such that

1. each pixel is assigned to exactly one entry in the TRS
2. each entry in the TRS is assigned to 0 or more pixels, implying that intermediate TRS entries can be skipped by the alignment
3. the first pixel is matched onto entry A and the last pixel onto B.

In the second requirement, the zero pixel case is necessary when a layer does not appear at all (thickness of zero) and to ignore fluid entries when no fluid is present. A visualization of our overall approach is given in Fig. 1. We refer to different TRSes by the notation LF_i, where i is the number of repetitions of each retinal Layer. The above mentioned sequence would then be an LF_1 sequence, while an LF_2 sequence would contain one repetition of each layer $(...L_3, F, L_3, F, ...)$. Due to the repetitions, LF_i with higher i allow for a higher numeber of fluid occurences within one layer.

3.2 First Approach: Pretrained CNN Features

For our first approach, a CNN serves as the learnable distance function, calculating the distance between one pixel of the A-Mode and an entry from the TRS. For each A-Mode, we extract a stripe with a width of 30 pixels, which was found to work well for the same dataset in [22]. We do this in order to provide the network with more information than just the pixels in the A-Mode. The entries in the TRS are represented by integers (0 to 6: layers; 7: fluid; 8: space above and below the retina), forming our 9 classes. The architecture of the CNN is depicted in Fig. 2. For dtw, we use the dtw-python package [8], extending it by some new, asymmetric step patterns for our purposes. Training was done using Tensorflow (version 2.4.1).

Training. The stripes are cropped into overlapping, rectangular patches and the CNN is trained to predict the class of the central pixel of each patch, minimizing a categorical cross entropy loss over the nine classes. To account for the imbalance of the different classes in the dataset, class weighting is used, where each class is weighted inversely to the number of times it appears in the training dataset.

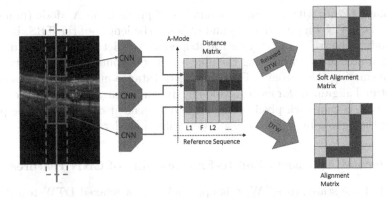

Fig. 1. Overview of the approach. Around an A-Mode (red line), a neighborhood stripe (red dotted rectangle) is extracted. The stripe is cropped into patches that are sequentially processed by the same CNN. For each patch, the CNN fills one row of the distance matrix. Then, DTW is used to derive the optimal alignment from this distance matrix. In the case of standard DTW, the CNN is pretrained and the resulting alignment matrix is binary. In the case of the relaxed DTW formulation, the alignment matrix is soft, and the CNN weights can be trained end-to-end on the given task. (Color figure online)

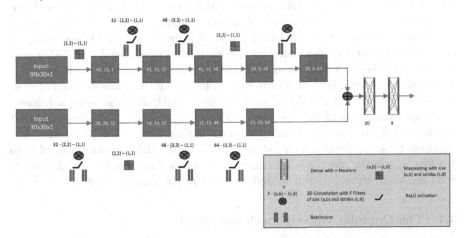

Fig. 2. Architecture of our CNN. The upper path has a large receptive field of low spatial resolution due to an immediate maxpooling layer. The lower path has a smaller field of view, yet a higher spatial resolution. Both paths are concatenated after a set of convolutional operations and fed into common dense layers. A larger version of this figure can be found in our supplementary materials.

Alignment. For the alignment, we combine the pretrained CNN with a standard DTW. The local distance between a pixel p_i and a TRS element r_j is calculated as $d(p_i, r_j) = 1.0 - softmax(c_p(r_j))$, where $softmax(c_p(r_j))$ is the softmax score for the class c_p that corresponds to entry r_j. Our distance matrix is of shape $[P, Q]$, with P being the number of pixels in an A-Mode (minus the height of the CNN input window) and Q being the length of the TRS. By vertically sliding the trained CNN over the stripes, we predict one row of the distance matrix at a time by rearranging the nine softmax outputs according to the TRS and subtracting them from 1. From the filled distance matrix θ, we then obtain the optimal alignment for each A-Mode by dynamic programming. The outcome of this step is that each pixel is aligned to one element of the TRS. The pixel's class is then given by the class associated with that element.

3.3   Second Approach: End-to-End Learning of CNN Features

Instead of the standard DTW, this approach uses a relaxed DTW to calculate a soft alignment matrix and backpropagate through it for end-to-end learning. We used the python implementation of the relaxed DTW by the authors of [15], integrated it into a keras layer and ported it to cython for speed improvements.

Training. Depending on the chosen TRS, there can be multiple optimal alignment matrices. In an LF_2 sequence, it is e.g. ambiguous whether a pixel belonging to Layer 3 should be assigned to the first or the second L_3 entry in the TRS when there's no fluid present within the layer. Due to this ambiguity, we decided not to construct alignment matrices as the network's learning target. Instead, we evaluate the soft alignment matrix by summing up all values from entries belonging to the same class along each row. This process extracts the underlying segmentation from ambiguous alignment matrices. The resulting vector is then interpreted as logits, and the loss is calculated as $crossentropy(logits, groundtruth) + \frac{1}{2} \cdot diceloss$ (with dice loss being a differentiable approximation of the dice score, as used e.g. in [19]). As we assumed that the alignment process would make use of similar features as the patch-wise classification task, we decided to initialize the CNN with the weights obtained from our first approach, effectively finetuning the classification weights for alignment. For this, we froze all but the dense layers before starting the training process.

3.4   The Cumulative Neighborhood Matrix

Segmentation results should be smooth between neighbouring A-Modes, as jumps between layers are not to be expected. We want to encourage alignments that follow this consistency prior. This means we prefer alignments that not only achieve low alignment cost on the current A-Mode, but would also result in low costs when applied to neighbouring A-Modes. We achieve this by adding weighted versions of the distance matrices θ_n from neighbouring A-Modes to the distance matrix of the current one. The resulting distance matrix is termed a cumulative neighborhood matrix.

4   Experiments

We use the publicly available DUKE OCT dataset [3] for DME patients. It contains 10 subjects with fluid pockets due to diabetic macular edema. For each subject, two experts manually annotated 11 B-Scans for the 7 retinal layers as well as for fluid. We split the dataset into subjects 1 to 6 for training, 7 and 8 for validation and 9 and 10 for testing. As both approaches perform the segmentations along A-Modes and work on the same dataset with the same train/validation/test splits, we compare our results to the ones reported in [22], who also evaluated a RelayNet on exactly this experimental setup. In preliminary experiments, we tested different TRSes with a pretrained CNN and a standard DTW and found the resulting dice scores to not differ significantly. We thus chose to use an LF_2 TRS for the final experiments, together with an asymmetric step pattern that allows to skip up to one full layer (i.e. 6 TRS entries).

Hyperparameters. We used a batch size of 128, dropout of 0.25 in the penultimate Dense Layer and an Adam optimizer (learning rate: 0.0001). The stride between two stripes was 10 px in horizontal direction (i.e. they overlapped); the stride for patches in vertical direction was also 10 px. Training lasted 30 Epochs. The second approach used the same hyperparameters except for a batch size of 64. The model with the lowest validation loss was selected as the final model. For the weights for constructing the neighborhood matrix, we optimize performance on the validation dataset by grid search, using the closest and second closest stripe on both sides as neighbours with weights between 0.0 and 0.4 in steps of 0.1. Resulting values were 0.2 and 0.1 for the direct and second closest neighbor.

Results. We compare the performance of four different versions for our model: Ours (pretr.only) is the pretrained CNN from approach 1 together with a standard DTW. Ours (soft) is the end-to-end finetuned CNN, using the soft-DTW for alignment. Ours (Soft-Hard) has been finetuned using the relaxed DTW, but for alignment uses the standard DTW, thus guaranteeing adherence to topological order. Ours (Soft-Hard + C) further applies the cumulative neighborhood matrix in order to smooth the predictions made by the DTW. The results of the comparison are given in Tables 1 and 2; for a visual impression see Fig. 3.

Comparing our own models, it can be seen that Ours (pretr. only) performs reasonable, but still lacks performance, giving the lowest dice score among our models. Finetuning the CNN with the relaxed DTW improves the mean dice score. The (Soft-Hard) variant even outperforms the end-to-end trained model by a small margin. Finally, adding the cumulative neighborhood matrix both smoothes the predictions (see supplementary material for visual impressions) and further increases the mean dice score. Comparing our models to the Language model and RelayNet, all but (pretr. only) give increased mean dice scores. Interestingly, these mean improvements are, despite other small fluctuations, coming nearly exclusively from improvements in the dice score of the fluid class,

Table 1. Comparison of our results with results for a language-model based approach and RelayNet (results for both latter ones taken from [22]).

		ILM	NFL-IPL	INL	OPL	ONL-ISM	ISE	OS-RPE	Fluid	Mean
DICE	Ours (Soft-Hard+C)	0.85	0.89	0.75	0.74	0.90	0.90	0.87	0.56	0.81
	Ours (Soft-Hard)	0.85	0.89	0.75	0.73	0.90	0.90	0.86	0.52	0.80
	Ours (Soft)	0.85	0.89	0.75	0.73	0.90	0.90	0.86	0.48	0.79
	Ours (pretr. only)	0.80	0.87	0.73	0.70	0.88	0.88	0.79	0.49	0.77
	Language	0.85	0.89	0.75	0.75	0.89	0.90	0.87	0.39	0.78
	RelayNet	0.78	0.85	0.70	0.71	0.87	0.88	0.84	0.30	0.75

Table 2. Standard Deviations of the Dice Scores obtained from our model over the 22 test B-Modes. Methods same as in Table 1 (abbreviated for spacing reasons).

		ILM	NFL-IPL	INL	OPL	ONL-ISM	ISE	OS-RPE	Fluid	Mean
DICE	Ours (SH+C)	0.045	0.045	0.072	0.071	0.034	0.021	0.037	0.322	0.065
	Ours (SH)	0.047	0.047	0.075	0.070	0.036	0.021	0.036	0.322	0.066
	Ours (Soft)	0.042	0.046	0.077	0.064	0.032	0.022	0.037	0.326	0.065
	Ours (PO)	0.046	0.053	0.075	0.071	0.033	0.022	0.038	0.332	0.076

where all our models significantly outperform our competitors. This is a welcome improvement in view of the connection between fluid pockets and pathology.

Evaluating the std deviation of dice sores over the 22 B-Modes in the test set (Table 2) shows our methods have standard deviations that are on average, over all classes, between 0.065 and 0.076, with the largest contributions coming from the fluid class (between 0.322 and 0.332). The latter makes sense as fluid comprises one of the smallest classes and is often concentrated in only one region in the B-Mode. If the network does not recognize this region as containing fluid, the dice score for fluid in one single B-Mode can easily become 0, leading to a higher std deviation than for the retinal layer classes.

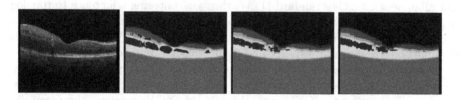

Fig. 3. An exemplary segmentation result (from left to right: OCT image, groundtruth annotations, our segmentation (Soft-Hard), our segmentation (Soft). More images can be found in the supplementary material.

We also evaluated the topological correctness of our approaches. A topology violation was defined as the number of times the method assigned a pixel to a retinal layer (N-1) after it had already assigned at least one pixel as being of retinal layer (N) (because anatomically, layers are monotonically increasing). Means and std deviations of these violations over 4 training runs were: using only a CNN (no DTW at all): (27570 ± 2504); Soft-DTW: (28 ± 17); pretr. only and soft-hard DTW (with and without cumulative dist. matrix): (0 ± 0). This clearly shows that using soft DTW strongly diminishes the violations a simple CNN alone produces and using a "hard" DTW completely prevents violations.

5   Discussion and Conclusion

In this paper, we presented the idea of formulating OCT image segmentation as finding the optimal alignment between A-Modes and a symbolic reference sequence. We used DTW to find the optimal alignment, evaluating the standard formulation together with pretrained CNN filters and a recently proposed relaxed formulation allowing for end-to-end finetuning. The proposed method could successfully segment both retinal layers and fluid jointly in a public dataset and compared favourable to two other baselines on the same dataset and subject splits. The (soft) approach is fully end-to-end learned, while the (soft-hard) one is not end-to-end trained, but guarantees topological constraints.

In the future, we want to investigate how our method compares to a larger set of approaches; its performance on different datasets and possible combinations with other approaches. Especially, we want to investigate the combination of the relaxed DTW formulation with fully convolutional architectures, in order to create the distance matrix in one pass instead of sliding a CNN over each A-Mode. This would allow to formulate smoothness of neighbouring predictions as an additional loss term, as we would get predictions not only for the centering A-Mode, but also for its neighbours in a single forward pass. Thereby, it could allow to incorporate the neighborhood matrix into the end-to-end pipeline.

References

1. Aumann, S., Donner, S., Fischer, J., Müller, F.: Optical Coherence Tomography (OCT): principle and technical realization. In: Bille, J.F. (ed.) High Resolution Imaging in Microscopy and Ophthalmology, pp. 59–85. Springer, Cham (2019). https://doi.org/10.1007/978-3-030-16638-0_3
2. BenTaieb, A., Hamarneh, G.: Topology aware fully convolutional networks for histology gland segmentation. In: Ourselin, S., Joskowicz, L., Sabuncu, M.R., Unal, G., Wells, W. (eds.) MICCAI 2016. LNCS, vol. 9901, pp. 460–468. Springer, Cham (2016). https://doi.org/10.1007/978-3-319-46723-8_53
3. Chiu, S.J., Allingham, M.J., Mettu, P.S., Cousins, S.W., Izatt, J.A., Farsiu, S.: Kernel regression based segmentation of optical coherence tomography images with diabetic macular edema. Biomed. Opt. Express 6(4), 1172–1194 (2015). https://doi.org/10.1364/BOE.6.001172
4. Chiu, S.J., Li, X.T., Nicholas, P., Toth, C.A., Izatt, J.A., Farsiu, S.: Automatic segmentation of seven retinal layers in sdoct images congruent with expert manual segmentation. Opt. Express 18(18), 19413–19428 (2010)

5. Cuturi, M., Blondel, M.: Soft-DTW: a differentiable loss function for time-series. In: Precup, D., Teh, Y.W. (eds.) Proceedings of the 34th International Conference on Machine Learning. Proceedings of Machine Learning Research, vol. 70, pp. 894–903. PMLR, International Convention Centre, Sydney, Australia, 06–11 Aug 2017

6. Duan, W., et al.: A generative model for oct retinal layer segmentation by group-wise curve alignment. IEEE Access **6**, 25130–25141 (2018)

7. Fang, L., Cunefare, D., Wang, C., Guymer, R.H., Li, S., Farsiu, S.: Automatic segmentation of nine retinal layer boundaries in oct images of non-exudative amd patients using deep learning and graph search. Biomed. Opt. Express **8**(5), 2732–2744 (2017)

8. Giorgino, T.: Computing and visualizing dynamic time warping alignments in R: the DTW package. J. Stat. Softw. **31**, 1–24 (2009)

9. He, Y., et al.: Topology guaranteed segmentation of the human retina from oct using convolutional neural networks. arXiv preprint arXiv:1803.05120v1 (2018)

10. He, Y., et al.: Towards topological correct segmentation of macular oct from cas-caded fcns. In: Cardoso, M.J., et al. (eds.) Fetal, Infant and Ophthalmic Medical Image Analysis, pp. 202–209. Springer International Publishing, Cham (2017)

11. Kugelman, J., et al.: Automatic choroidal segmentation in oct images using super-vised deep learning methods. Sci. Rep. **9**(1), 1–13 (2019)

12. Kugelman, J., Alonso-Caneiro, D., Read, S.A., Vincent, S.J., Collins, M.J.: Auto-matic segmentation of oct retinal boundaries using recurrent neural networks and graph search. Biomed. Opt. Express **9**(11), 5759–5777 (2018)

13. Lee, K., Niemeijer, M., Garvin, M.K., Kwon, Y.H., Sonka, M., Abramoff, M.D.: Segmentation of the optic disc in 3-d oct scans of the optic nerve head. IEEE Trans. Med. Imaging **29**(1), 159–168 (2010). https://doi.org/10.1109/TMI.2009.2031324

14. Liu, W., Sun, Y., Ji, Q.: Mdan-unet: multi-scale and dual attention enhanced nested u-net architecture for segmentation of optical coherence tomography images. Algorithms **13**(3), 60 (2020)

15. Mensch, A., Blondel, M.: Differentiable dynamic programming for structured pre-diction and attention. In: Dy, J., Krause, A. (eds.) Proceedings of the 35th International Conference on Machine Learning. Proceedings of Machine Learning Research, vol. 80, pp. 3462–3471. PMLR, Stockholmsmässan, Stockholm Sweden (10–15 Jul 2018)

16. Guru Pradeep Reddy, T., et al.: Retinal-layer segmentation using dilated convolu-tions. In: Chaudhuri, B.B., Nakagawa, M., Khanna, P., Kumar, S. (eds.) Proceed-ings of 3rd International Conference on Computer Vision and Image Processing. AISC, vol. 1022, pp. 279–292. Springer, Singapore (2020). https://doi.org/10.1007/978-981-32-9088-4_24

17. Romero-Aroca, P., Baget-Bernaldiz, M., Pareja-Rios, A., Lopez-Galvez, M., Navarro-Gil, R., Verges, R.: Diabetic macular edema pathophysiology: Vasogenic versus inflammatory. J. Diabetes Res. **2016**, 2156273 (2016). https://doi.org/10.1155/2016/2156273

18. Ronneberger, O., Fischer, P., Brox, T.: U-net: Convolutional networks for biomed-ical image segmentation. arxiv 2015. arXiv preprint arXiv:1505.04597 (2015)

19. Roy, A.G., et al.: Relaynet: retinal layer and fluid segmentation of macular optical coherence tomography using fully convolutional networks. Biomed. Opt. Express **8**(8), 3627–3642 (2017)

20. Sakoe, H., Chiba, S.: Dynamic programming algorithm optimization for spoken word recognition. IEEE Trans. Acoust. Speech Signal Process. **26**(1), 43–49 (1978). https://doi.org/10.1109/TASSP.1978.1163055

21. Shah, M., Grabocka, J., Schilling, N., Wistuba, M., Schmidt-Thieme, L.: Learning dtw-shapelets for time-series classification. In: Proceedings of the 3rd IKDD Conference on Data Science, 2016. CODS '16, Association for Computing Machinery, New York, NY, USA (2016). https://doi.org/10.1145/2888451.2888456, https://doi.org/10.1145/2888451.2888456

22. Tran, A., Weiss, J., Albarqouni, S., Faghi Roohi, S., Navab, N.: Retinal layer segmentation reformulated as OCT language processing. In: Martel, A.L., et al. (eds.) MICCAI 2020. LNCS, vol. 12265, pp. 694–703. Springer, Cham (2020). https://doi.org/10.1007/978-3-030-59722-1_67

23. Wei, H., Peng, P.: The segmentation of retinal layer and fluid in sd-oct images using mutex dice loss based fully convolutional networks. IEEE Access **8**, 60929–60939 (2020)

Learnable Oriented-Derivative Network for Polyp Segmentation

Mengjun Cheng[1], Zishang Kong[1], Guoli Song[2], Yonghong Tian[3],
Yongsheng Liang[4], and Jie Chen[1,2(✉)]

[1] School of Electronic and Computer Engineering, Peking University, Beijing, China
chenj@pcl.ac.cn
[2] Peng Cheng Laboratory, Shenzhen, China
[3] School of Electronics Engineering and Computer Science, Peking University,
Beijing, China
[4] School of Electronic and Information Engineering, Harbin Institute of Technology,
Harbin, China

Abstract. Gastrointestinal polyps are the main cause of colorectal cancer. Given the polyp variations in terms of size, color, texture and poor optical conditions brought by endoscopy, polyp segmentation is still a challenging problem. In this paper, we propose a Learnable Oriented-Derivative Network (LOD-Net) to refine the accuracy of boundary predictions for polyp segmentation. Specifically, it firstly calculates eight oriented derivatives at each pixel for a polyp. It then selects those pixels with large oriented-derivative values to constitute a candidate border region of a polyp. It finally refines boundary prediction by fusing border region features and also those high-level semantic features calculated by a backbone network. Extensive experiments and ablation studies show that the proposed LOD-Net achieves superior performance compared to the state-of-the-art methods by a significant margin on publicly available datasets, including *CVC-ClinicDB, CVC-ColonDB, Kvasir, ETIS, and EndoScene*. For examples, for the dataset *Kvasir*, we achieve an mIoU of 88.5% vs. 82.9% by PraNet; for the dataset *ETIS*, we achieve an mIoU of 88.4% vs. 72.7% by PraNet. The code is available at https://github.com/midsdsy/LOD-Net.

Keywords: Endoscopy · Polyp segmentation · Learnable oriented derivative

1 Introduction

Colorectal cancer(CRC) is the third most common type of cancer worldwide[24]. Colonoscopy examinations are effective for early diagnosis and prevention of CRC[11]. It allows gastroenterologists to diagnose a wide range of diseases and abscesses, like polyps and so on. However, polyp diagnose highly depends on the skills and experiences of endoscopists. Several research groups developed

© Springer Nature Switzerland AG 2021
M. de Bruijne et al. (Eds.): MICCAI 2021, LNCS 12901, pp. 720–730, 2021.
https://doi.org/10.1007/978-3-030-87193-2_68

computer-aided systems, which use texture and morphological features to differentiate the polyp regions from folding regions on the colon wall. Accurate segmentation of polyps helps gastroenterologists to estimate the shape and size of polyps quickly for the diagnose of illness. Nevertheless, as shown in Fig. 1(a), how to bulid a polyp segmentation model is quite challenging because of low contrast between polyps and their surrounding mucosa.

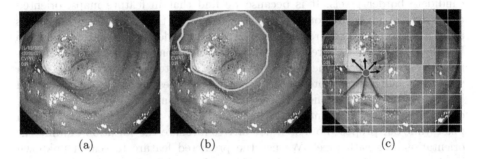

(a)   (b)   (c)

Fig. 1. Illustration of our proposed pixel-based learnable oriented-derivative network; (a) low contrast between a polyp and its surrounding mucosa; (b) ground truth contour (yellow curve); (c) learnable derivatives orientations to search border region (yellow pixels). Here, arrows refer to learned derivative orientations; oriented derivatives in red color are employed as the reference value for candidate border pixel, and oriented derivatives in black color are ignored because these oriented-derivative magnitude are small. In our case, we calculate eight oriented derivatives for each pixel. Best viewed in color. (Color figure online)

Among various polyp segmentation methods, the early algorithms [10,14] which rely on extracting low-level hand-crafted features does not work well on complex clinical situations. Convolutional Neural Networks (CNN) recently significantly improves the performance on computer-vision tasks including classification [18], detection [22] and segmentation [12,23], especially in natural scene. Likewise, several deep learning based methods are proposed for polyp segmentation[8,15]. Fully Convolutional Network(FCN) architecture is widely used for colonoscopy image segmentation [1]. Since the success of U-Net [23] applied in biomedical image segmentation, U-Net++ [28] and ResUNet++ [16] are developed for better performance. Mask R-CNN [12] is one of the popular deep-learning models for instance segmentation which detects targets in an image and predicts mask for each detected target. It is also extended for polyp segmentation [17]. However, most existing CNN-based methods suffer from inaccurate performance for lesion-boundary regions during segmentation. Some researchers focus on boundary information learning in their proposed models. Several works enhance boundary context information by fusing high-level semantic features and low-level detailed features to refine segmentation results [5,26]. SFA [9] and PraNet [8] explicitly and implicitly consider the dependency between areas and boundaries to obtain good result, respectively. All these methods attempt to address inaccurate boundary prediction by enhancing features with boundary information.

In this paper, we propose a Learnable Oriented-Derivative Network (LOD-Net) to refine the accuracy of boundary prediction during polyp segmentation. Specifically, we firstly learn eight derivative orientations for each pixel in a polyp proposal detected by a backbone network, and then calculate derivatives along these orientations. Those pixels with large oriented derivatives are employed as candidate border pixels. It is because we find that in feature maps, oriented derivatives of pixels in boundary region are larger than those of other pixels away from the boundary region. Further fusing border region features and the high-level semantic features calculated by a backbone network, our LOD-Net achieves superior performance compared to the state-of-the-art methods by a significant margin. Our proposed oriented-derivatives feature is different with [4] who learns a direction field. Specially, the direction field [4] predicts a unique direction vector that points from the boundary to the central area for each pixel while our proposed oriented derivatives measure the variation on different orientations for each pixel. We use the predicted feature to search unknown border region and refine the feature of boundary while the predicted direction field is used for rectifying the initial feature map.

We summarize our main contributions as follows.

- We propose a Learnable Oriented-derivative Network (LOD-Net) to effectively utilize the representation capacity of oriented derivative for border region searching.

- Our proposed LOD-Net outperforms the state-of-the-art methods on polyp segmentation benchmarks over publicly available datasets, including Kvasir,CVC-ClinicDB, CVC-ColonDB, ETIS-LaribPolypDB, and EndoScene.

2 Methods

2.1 Overview

Our framework is illustrated in Fig. 2. To address the polyp segmentation problem, we propose learnable derivative representation to model the probability of each pixel locating in border region (Sect. 2.2). After acquiring pixel-based oriented-derivative representation in instance level, we adopt an adaptive thresholding policy to search a possible border region and a fusion policy to enhance feature representation (Sect. 2.3).

2.2 Learnable Oriented-Derivative Representation

As mentioned above, oriented derivatives of pixels in boundary region are larger than those of other pixels. The larger the derivative is, the more the orientation of derivative is perpendicular to contour, which is our expectant orientation. However traditional methods ignore oriented derivative other than orientation of gradients, whose representation capability is inferior.

Inspired by this, we relax the directional limitation from the calculation of gradients to an arbitrary direction. By taking more directions into consideration, we assess the boundary pixels in a comprehensive way. According to the experience of 8-adjacent images [20], we simply use 8-adjacent orientations at each pixel to approximately represent their directions.

Oriented Derivative. We use Δ to denote 2D preset offsets as shown in Fig. 3, which means the initial derivative orientation from current position(x_0, y_0) to its neighbouring points on eight orientations, and $|\Delta| = 8$. We then calculate the positions of its neighbouring points and sample value of them. The oriented derivatives are defined as the difference of values between current point and sampled points after normalization, which is expressed as:

$$OD^i(x_0, y_0) = \frac{1}{D}(X(x_0, y_0) - X(x_0 + \Delta_i^x, y_0 + \Delta_i^y)) \tag{1}$$

$$\Delta = \{(\Delta_i^x, \Delta_i^y)|i = 0, 1, ..., 7\} = \{(-1, -1), (-1, 0), ..., (0, 1), (1, 1)\} \tag{2}$$

where Δ^x and Δ^y are preset offsets on x-axis and y-axis. i is the index of orientation on Δ. X is input feature maps and $X(x, y)$ is a sample operator on (x, y). Normalized parameter D is the euclidean distance between current point and sampled point.

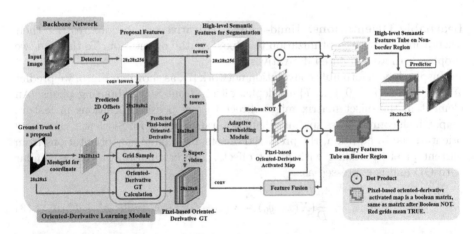

Fig. 2. Illustration of our proposed LOD-Net during training. We predict 2D offsets Φ from proposal features to calculate positions and values of sampled shifted points points through grid sample operator. We then calculate pixel-based oriented derivatives (OD) ground truth (GT) in Oriented Derivative Learning Module, which is shown in Fig. 3. After that, the network uses an Adaptive Thresholding Module to search a possible border region, which is shown in Fig. 4. We finally combine different features in border region and non-border region to make final segmentation prediction. Best viewed in color. (Color figure online)

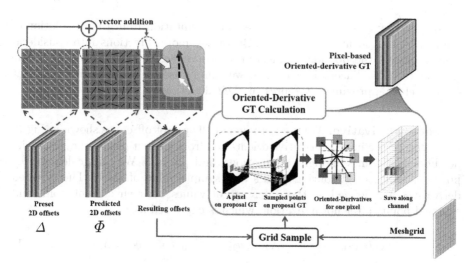

Fig. 3. Illustration of Oriented-Derivative (OD) Learning Module. By adding preset and predicted offsets, we calculate positions of sampled points on proposal ground truth. For the module of oriented-derivative GT calculation, it calculates 8 oriented derivatives based on sampled points on proposal GT for each pixel and outputs the pixel-based oriented-derivatives GT for supervision. For more details, please refer to Eq. (3). Best viewed in color. (Color figure online)

Learnable Orientations. Hand-crafted derivative orientations restrict their orientation derivative optimization to a fixed spatial orientation. To this end, we propose learnable oriented derivatives.

Inspired by deformable convolution network [7], as shown in Fig. 3, we predict $\Phi = \{(\Phi_i^x, \Phi_i^y)|i = 0, 1, ..., 7\}$ to depict offsets from 8-neighbouring pixels. An element in the offset matrix might direct to any possible positions in feature map, which compensates spatial calculation range of OD for 8-orientation to any orientations. In addition, the predicted offsets also define new orientations from current pixel by adding with preset offsets. The calculation process is similar with OD and is expressed as:

$$LOD^i(x_0, y_0) = \frac{1}{D}(X(x_0, y_0) - X(x_0 + \Delta_i^x + \Phi_i^x, y_0 + \Delta_i^y + \Phi_i^y)) \quad (3)$$

where $(x_0 + \Delta_i^x + \Phi_i^x, y_0 + \Delta_i^y + \Phi_i^y)$ denotes the sampled shifted position which can be an arbitrary location, $X(x, y)$ is a bilinear interpolation opertor on (x, y). With the learned orientations, LOD representation selects eight adaptable orientations for each pixel and improves the representation capacity of oriented derivatives. It avoids the situation that any preset orientation of a pixel is closely parallel to tangent line of object contour which leads to a small magnitude of oriented derivative and weakens the real response value of the pixel.

Oriented Derivative Sensitive Loss. Smooth L1 loss is used for pixel-based oriented-derivative representation learning:

$$L_{OD} = \frac{1}{M} \sum_{i=0}^{h} \sum_{j=0}^{w} Smooth_{L_1}(o_{i,j}, o_{i,j}^*) \qquad (4)$$

where h and w are height and width of feature maps. $o_{i,j}$ is the predicted oriented derivative, $o_{i,j}^*$ is ground truth and $M = h * w$. Besides the proposed oriented derivative sensitive loss L_{OD}, other losses for training are the same with that in Mask R-CNN.

2.3 Border Region Searching

As shown in Fig. 2, based on pixel-based oriented-derivative representation which models the possibility of each pixel locating in border region, we employ an adaptive thresholding policy to select border points among all candidate points. Then by fusing boundary features from OD representation on border region and high-level semantic features, our LOD-Net enhances the feature representation capacity on border region and improves model performance.

Channel-Wise Sampling. For the adaptive thresholding policy, we use a direction-based sampling method. As shown in Fig. 4, we first choose top k points with large oriented-derivative response values on each channel, which preserves partial oriented-derivative value of points instead of preserving all information. We add them through channel-wise operator to get a pixel-based OD activated features of instances, whose pixels value only represents its salient derivatives on selected orientations. The top k points with large response value are selected and considered as possible border region, which is referred to pixel-based oriented-derivative activated map.

Boundary-Aware Mask Scoring. We use a fusion to enhance features representation for final mask segmentation. As shown in Fig. 2, for border points, we acquire their boundary features through a pixel multiplication between intermediate features and OD activated map. The intermediate feature is acquired by fusing pixel-based OD representation and high-level semantic features. For non-border points, we use dot product between high-level semantic features and the opposite of OD activated map. Final prediction is based on the combination of features of border points and non-border points.

3 Experiments

3.1 Datasets and Evaluation Metrics

Five publicly available datasets are used in our experiments including CVC-ColonDB [3], Kvasir [21], ETIS [24], CVC-ClinicDB [2] and testing set of

EndoScene[25]. Following the training settings in [8,15], our training sets contain 1450 images which are composed of 900 images from Kvasir and 550 images from CVC-ClinicDB, and the testing sets contain 380 images from CVC-ColonDB, 100 images from Kvasir, 196 images from ETIS, 62 images from CVC-ClinicDB and 60 images from EndoScene. By testing our model on two seen datasets, and three unseen datasets, we evaluate the learning ability and generalization capability of our method. For polyp segmentation, we adopt mDice and mIoU as evaluation metrics following [8].

3.2   Implementation Details

Our network is developed based on Mask R-CNN [12] and implemented on Detectron2 [27] architecture. We use ResNet [13] and FPN [19] as our backbone and the weights pretrained on ImageNet are used to initialize it. Our models are trained for 10k iterations using a stochastic gradient descent (SGD) optimizer, with a base learning rate of 0.005 and a batch size of 8 images on four GeForce RTX 2080Ti GPUs. The learning rate is reduced by a factor of 0.1 at iteration 6k and 8k, respectively. Weight decay and momentum are set as 0.0001 and 0.9, respectively. Input images are resized following the settings of Detectron2. Specially, motionblur, randomfilp augmentation and multi-scale training are used during training for polyp data.

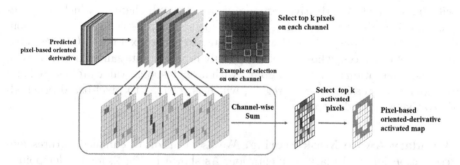

Fig. 4. Illustration of the Adaptive Thresholding Module. We select top k pixels with large response value on each channel and add them together to gain a pixel-based oriented-derivative activated map. Best viewed in color. (Color figure online)

3.3   Experiment Results

Compare with SOTA Methods. As shown in Table 1, our network shows a superior performance on five datasets over state-of-the-art methods. In particular, LOD-Net acheives 5.6% improvement over the PraNet [8] on seen dataset Kvasir and a comparable result on CVC-ClinicDB dataset. It also acheives 15.7% improvement over the PraNet on unseen dataset ETIS in the metric of mIoU.

Table 1. Comparison with SOTA methods for polyp segmentation.

Model	ClinicDB		Kvasir		ETIS		ColonDB		EndoScene	
	mDice	mIoU	mDice	mIoU	mDice	mIoU	mDice	mIoU	mDice	mIoU
U-Net++ [28]	78.6	64.7	73.5	58.1	45.1	29.1	31.6	18.8	64.0	47.0
SFA [9]	74.0	58.8	66.1	49.3	71.8	56.0	56.0	38.9	84.4	73.0
PraNet [8]	95.1	90.7	90.6	82.9	84.2	72.7	67.1	50.5	95.1	90.6
HarDNet-MSEG [15]	**95.3**	**91.1**	90.5	82.7	83.9	72.3	58.2	41.0	94.4	89.5
baseline [12]	92.5	86.0	92.4	85.9	89.7	81.4	58.6	41.5	94.0	88.7
Ours	92.6	86.2	**93.9**	**88.5**	**93.8**	**88.4**	**70.0**	**53.9**	**95.7**	**91.7**

The results of other methods are calculated from official result maps supported online or generated by official model weights. We use the open source code of mmsegmentation [6] to calculate the mDice and mIoU in a fixed threshold of 0.5. The total number of our model parameters is 15.28M, which is trivially more than that of Mask R-CNN (14.76M) [12] and less than that of PraNet (30.49M) [8] and HarDNet (17.42M) [15].

3.4   Ablation Study

Oriented-Derivative vs Learnable Oriented-Derivative. We use a simple implementation version of ResNet-50[18] for ablation study experiment and record average results of four repeated experiments. Table 2 shows that our proposed method with learned orientations receives a better performance with 3.53% improvement on seen cvc-ClinicDB dataset and 1.44% improvement on unseen ETIS dataset for mIoU due to its ability of learning adaptive orientations.

Table 2. Ablation study for OD and LOD.

Model	CVC-ClinicDB (seen)		ETIS (unseen)	
	mDice	mIoU	mDice	mIoU
OD with hand-crafted orientations	90.42	79.27	86.31	75.93
LOD with learned orientations	**90.59**	**82.80**	**87.26**	**77.37**

Different Adaptive Thresholding Policies. For the adaptive thresholding policy, we consider three policies which are pixel-based average, pixel-based maximum and direction-based method. The first two policies refer to using an average or maximum value along channels to represent pixel features. As shown in Table 3, direction-based sample policy is more suitable.

Table 3. Ablation study for different adaptive thresholding policies.

Policy	CVC-ClinicDB (seen)		ETIS (unseen)	
	mDice	mIoU	mDice	mIoU
Point-based average	87.98	78.55	85.58	74.80
Point-based maximum	88.31	79.08	86.98	77.17
Direction-based	**90.59**	**82.80**	**87.26**	**77.37**

4 Conclusion

We address the issue of coarse boundaries and imprecise localization caused by low contrast between polys and their surroundings in polyp segmentation task. We propose a novel learnable oriented-derivative Network(LOD-Net), which learns pixel-based oriented-derivatives to search border region. LOD-Net gains a fine boundary features fused by oriented-derivative representation and high-level semantic features for accurate localization and boundary prediction. Our experiments demonstrate that our method achieves remarkable and stable improvements on several publicly available datasets.

Acknowledgment. This work is supported by the Nature Science Foundation of China (No. 61972217, 62081360152, 62006133, 32071459), Guangdong Basic and Applied Basic Research Foundation (No. 2019B1515120049) and Guangdong Science and Technology Department (No. 2020B1111340056).

References

1. Akbari, M., et al.: Polyp segmentation in colonoscopy images using fully convolutional network. In: 2018 40th Annual International Conference of the IEEE Engineering in Medicine and Biology Society (EMBC), pp. 69–72. IEEE (2018)
2. Bernal, J., Sánchez, F.J., Fernández-Esparrach, G., Gil, D., Rodríguez, C., Vilariño, F.: Wm-dova maps for accurate polyp highlighting in colonoscopy: Validation vs. saliency maps from physicians. Computerized Med. Imaging Graph. **43**, 99–111 (2015)
3. Bernal, J., Sánchez, J., Vilarino, F.: Towards automatic polyp detection with a polyp appearance model. Pattern Recogn. **45**(9), 3166–3182 (2012)
4. Cheng, F., et al.: Learning directional feature maps for cardiac MRI segmentation. In: Martel, A.L., Abolmaesumi, P., Stoyanov, D., Mateus, D., Zuluaga, M.A., Zhou, S.K., Racoceanu, D., Joskowicz, L. (eds.) MICCAI 2020. LNCS, vol. 12264, pp. 108–117. Springer, Cham (2020). https://doi.org/10.1007/978-3-030-59719-1_11
5. Cheng, T., Wang, X., Huang, L., Liu, W.: Boundary-preserving mask R-CNN. In: Vedaldi, A., Bischof, H., Brox, T., Frahm, J.-M. (eds.) ECCV 2020. LNCS, vol. 12359, pp. 660–676. Springer, Cham (2020). https://doi.org/10.1007/978-3-030-58568-6_39
6. Contributors, M.: MMSegmentation: openmmlab semantic segmentation toolbox and benchmark (2020). https://github.com/open-mmlab/mmsegmentation

7. Dai, J., et al.: Deformable convolutional networks. In: 2017 IEEE International Conference on Computer Vision (ICCV), pp. 764–773. IEEE Computer Society (2017)

8. Fan, D.-P., et al.: PraNet: parallel reverse attention network for polyp segmentation. In: Martel, A.L., Abolmaesumi, P., Stoyanov, D., Mateus, D., Zuluaga, M.A., Zhou, S.K., Racoceanu, D., Joskowicz, L. (eds.) MICCAI 2020. LNCS, vol. 12266, pp. 263–273. Springer, Cham (2020). https://doi.org/10.1007/978-3-030-59725-2_26

9. Fang, Y., Chen, C., Yuan, Y., Tong, K.: Selective feature aggregation network with area-boundary constraints for polyp segmentation. In: Shen, D., et al. (eds.) MICCAI 2019. LNCS, vol. 11764, pp. 302–310. Springer, Cham (2019). https://doi.org/10.1007/978-3-030-32239-7_34

10. Ganster, H., Pinz, P., Rohrer, R., Wildling, E., Binder, M., Kittler, H.: Automated melanoma recognition. IEEE Trans. Med. Imaging **20**(3), 233–239 (2001). https://doi.org/10.1109/42.918473

11. Haggar, F.A., Boushey, R.P.: Colorectal cancer epidemiology: incidence, mortality, survival, and risk factors. Clinics Colon Rectal Surgery **22**(4), 191 (2009)

12. He, K., Gkioxari, G., Dollar, P., Girshick, R.: Mask r-cnn. IEEE Trans. Pattern Anal. Mach. Intell. **42**(2), 386–397 (2018)

13. He, K., Zhang, X., Ren, S., Sun, J.: Deep residual learning for image recognition. In: 2016 IEEE Conference on Computer Vision and Pattern Recognition (CVPR), pp. 770–778. IEEE Computer Society (2016)

14. He, Y., Xie, F.: Automatic skin lesion segmentation based on texture analysis and supervised learning. In: Lee, K.M., Matsushita, Y., Rehg, J.M., Hu, Z. (eds.) ACCV 2012. LNCS, vol. 7725, pp. 330–341. Springer, Heidelberg (2013). https://doi.org/10.1007/978-3-642-37444-9_26

15. Huang, C.H., Wu, H.Y., Lin, Y.L.: Hardnet-mseg: a simple encoder-decoder polyp segmentation neural network that achieves over 0.9 mean dice and 86 fps. arXiv preprint arXiv:2101.07172 (2021)

16. Jha, D., et al.: Resunet++: an advanced architecture for medical image segmentation. In: 2019 IEEE International Symposium on Multimedia (ISM), pp. 225–2255. IEEE (2019)

17. Kang, J., Gwak, J.: Ensemble of instance segmentation models for polyp segmentation in colonoscopy images. IEEE Access **7**, 26440–26447 (2019)

18. Krizhevsky, A., Sutskever, I., Hinton, G.E.: Imagenet classification with deep convolutional neural networks. In: Advances in Neural Information Processing Systems, pp. 1097–1105 (2012)

19. Lin, T.Y., Dollar, P., Girshick, R., He, K., Hariharan, B., Belongie, S.: Feature pyramid networks for object detection. In: 2017 IEEE Conference on Computer Vision and Pattern Recognition (CVPR), pp. 936–944. IEEE Computer Society (2017)

20. Ojala, T., Pietikainen, M., Maenpaa, T.: Multiresolution gray-scale and rotation invariant texture classification with local binary patterns. IEEE Trans. Pattern Anal. Mach. Intell. **24**(7), 971–987 (2002)

21. Pogorelov, K., et al.: Kvasir: a multi-class image dataset for computer aided gastrointestinal disease detection. In: Proceedings of the 8th ACM on Multimedia Systems Conference, pp. 164–169 (2017)

22. Ren, S., He, K., Girshick, R., Sun, J.: Faster r-cnn: towards real-time object detection with region proposal networks. In: Advances in Neural Information Processing Systems, pp. 91–99 (2015)

23. Ronneberger, O., Fischer, P., Brox, T.: U-Net: convolutional networks for biomedical image segmentation. In: Navab, N., Hornegger, J., Wells, W.M., Frangi, A.F. (eds.) MICCAI 2015. LNCS, vol. 9351, pp. 234–241. Springer, Cham (2015). https://doi.org/10.1007/978-3-319-24574-4_28

24. Siegel, R.L., et al.: Colorectal cancer statistics. CA: a Cancer J. Clinicians **70**(3), 145–164 (2020)

25. Vázquez, D., et al.: A benchmark for endoluminal scene segmentation of colonoscopy images. Journal of healthcare engineering (2017)

26. Wang, R., Chen, S., Ji, C., Fan, J., Li, Y.: Boundary-aware context neural network for medical image segmentation. arXiv preprint arXiv:2005.00966 (2020)

27. Wu, Y., Kirillov, A., Massa, F., Lo, W.Y., Girshick, R.: Detectron2 (2019). https://github.com/facebookresearch/detectron2

28. Zhou, Z., Rahman Siddiquee, M.M., Tajbakhsh, N., Liang, J.: UNet++: a nested U-Net architecture for medical image segmentation. In: Stoyanov, D., et al. (eds.) DLMIA/ML-CDS -2018. LNCS, vol. 11045, pp. 3–11. Springer, Cham (2018). https://doi.org/10.1007/978-3-030-00889-5_1

LambdaUNet: 2.5D Stroke Lesion Segmentation of Diffusion-Weighted MR Images

Yanglan Ou[1]([✉]), Ye Yuan[2], Xiaolei Huang[1], Kelvin Wong[3], John Volpi[4], James Z. Wang[1], and Stephen T. C. Wong[3]

[1] The Pennsylvania State University, University Park, PA, USA
yanglanou@psu.edu
[2] Carnegie Mellon University, Pittsburgh, PA, USA
[3] TT and WF Chao Center for BRAIN & Houston Methodist Cancer Center, Houston Methodist Hospital, Houston, TX, USA
[4] Eddy Scurlock Comprehensive Stroke Center, Department of Neurology, Houston Methodist Hospital, Houston, TX, USA

Abstract. Diffusion-weighted (DW) magnetic resonance imaging is essential for the diagnosis and treatment of ischemic stroke. DW images (DWIs) are usually acquired in multi-slice settings where lesion areas in two consecutive 2D slices are highly discontinuous due to large slice thickness and sometimes even slice gaps. Therefore, although DWIs contain rich 3D information, they cannot be treated as regular 3D or 2D images. Instead, DWIs are somewhere in-between (or 2.5D) due to the volumetric nature but inter-slice discontinuities. Thus, it is not ideal to apply most existing segmentation methods as they are designed for either 2D or 3D images. To tackle this problem, we propose a new neural network architecture tailored for segmenting highly discontinuous 2.5D data such as DWIs. Our network, termed LambdaUNet, extends UNet by replacing convolutional layers with our proposed Lambda+ layers. In particular, Lambda+ layers transform both intra-slice and inter-slice context around a pixel into linear functions, called lambdas, which are then applied to the pixel to produce informative 2.5D features. LambdaUNet is simple yet effective in combining sparse inter-slice information from adjacent slices while also capturing dense contextual features within a single slice. Experiments on a unique clinical dataset demonstrate that LambdaUNet outperforms existing 3D/2D image segmentation methods including recent variants of UNet. Code for LambdaUNet is available. (URL: https://github.com/YanglanOu/LambdaUNet.)

Keywords: Stroke · Lesion segmentation · Inter- and intra-slice context · 2.5-dimensional images

Electronic supplementary material The online version of this chapter (https://doi.org/10.1007/978-3-030-87193-2_69) contains supplementary material, which is available to authorized users.

M. de Bruijne et al. (Eds.): MICCAI 2021, LNCS 12901, pp. 731–741, 2021.
https://doi.org/10.1007/978-3-030-87193-2_69

1   Introduction

In the United States, stroke is the second leading cause of death and the third leading cause of disability [9]. About 795,000 people in the US have a stroke each year [12]. A stroke happens when some brain cells suddenly die or are damaged due to lack of oxygen when blood flow to parts of the brain is lost or reduced due to blockage or rupture of an artery [14]. Locating the lesion areas where brain tissue is prevented from getting oxygen and nutrients is essential for accurate evaluation and timely treatment. Diffusion-weighted imaging (DWI) is a commonly performed magnetic resonance imaging (MRI) sequence for evaluating acute ischemic stroke and is sensitive in detecting small and early infarcts [11].

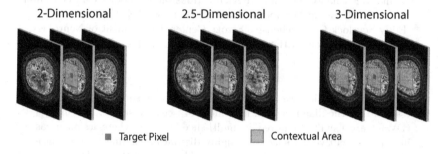

Fig. 1. Comparison of 2D, 2.5D, and 3D feature extraction methods. When extracting features for a target pixel, our 2.5D method restricts the context area in adjacent slices to focus on the most relevant pixels to reduce noise and improve generalization.

Segmenting stroke lesions on DWIs manually is time-consuming and subjective [10]. With the advancement of deep learning, numerous automatic segmentation methods based on deep neural networks (DNNs) have emerged to detect stroke lesions. Some of them perform segmentation on each 2D slice individually [2,4], while others treat DWIs as 3D data and apply 3D segmentation networks [19]. Beyond methods for lesion segmentation in DWIs, there have been many successful methods for general medical image segmentation. For instance, UNet [16] has shown the advantage of skip-connections on biomedical image segmentation. Based on UNet, Oktay *et al.* proposed Attention UNet by adding attention gates that filter the features propagated through the skip connections in U-Net [13]; Chen *et al.* proposed TransUNet, as they find that transformers make strong encoders for medical image segmentation [3]. Çiçek [5] extend UNet to 3D field for volumetric segmentation. Wang *et al.* proposed volumetric attention combined with Mask-RCNN to address the GPU memory limitation of 3D U-net. Zhang *et al.* [19] proposed a 3D fully convolutional and densely connected convolutional network which is derived from the powerful DenseNet [8].

Although previous medical image segmentation methods work well for 2D or 3D data by design, they are not well suited for DWIs, which have contextual characteristics between 2D and 3D. We term such data type as 2.5D [18].[1] Different from 2D data, DWIs contain 3D volumetric information by having multiple DWI slices. However, unlike typical 3D medical images that are isotropic or near isotropic in all three dimensions, DWIs are highly anisotropic with slice dimension at least five times more than in-plane dimensions. Therefore, neighboring slices can have abrupt changes around the same area which is especially problematic for early infarcts that are small and do not extend beyond a few slices. Due to the 2.5D characteristics of DWIs, if we apply 2D segmentation methods to DWIs, we lose valuable 3D contextual information from neighboring slices (Fig. 1 (left)). On the other hand, if we apply a traditional 3D CNN-based segmentation method, due to the high discontinuity between slices, many irrelevant features from neighboring slices are processed by the network (Fig. 1 (right)), which adds substantial noise to the learning process and also makes the network prone to over-fitting.

In this work, our goal is to design a segmentation network tailored for images with 2.5D characteristics like DWIs. To this end, we propose LambdaUNet which adopts the UNet [16] structure but replaces convolutional layers with our proposed Lambda+ layers which can capture both dense intra-slice features and sparse inter-slice features effectively. Lambda+ layers are inspired by the Lambda layers [1] which transform both global and local context around a pixel into linear functions, called lambdas, and produce features by applying these lambdas to the pixel. Although Lambda layers have shown strong performance for 2D image classification, they are not suitable for 2.5D DWIs because they are designed for 2D data and cannot capture sparse inter-slice features. Our proposed Lambda+ layers are designed specifically for 2.5D DWI data, where they consider both the intra-slice and inter-slice contexts of each pixel. Here the inter-slice context of a pixel consists of pixels at the same 2D location but in neighboring slices (Fig. 1 (middle)). Note that, unlike many 3D feature extraction methods, Lambda+ layers do not consider pixels in neighboring slices that are at different 2D locations, because these pixels are less likely to contain relevant features and we suppress them to reduce noise and prevent over-fitting. Lambda+ layers transform the inter-slice context into a different linear function–inter-slice lambda–which complements other intra-slice Lambdas to derive sparse inter-slice features. As illustrated in Fig. 1, the key design of Lambda+ layers is that they treat intra-slice and inter-slice features differently by using a dense intra-slice context and a sparse inter-slice context, which suits well the 2.5D DWI data.

Existing works in 2.5D segmentation [7,17,20] also recognize the anisotropy challenge of CT scans. However, they simply combine 3D and 2D convolutions without explicitly considering the anisotropy. To our knowledge, the proposed LambdaUNet is the first 2.5D segmentation model that is designed specifically for 2.5D data like DWIs and treats intra-slice and inter-slice pixels differently. Exten-

---

[1] Note that our definition of 2.5D is different from that in computer vision, where 2.5D means the 2D retinal projections of 3D environments.

sive experiments on a large annotated clinical DWI dataset of stroke patients show that `LambdaUNet` significantly outperforms previous art in terms of segmentation accuracy.

2   Methods

Denote a DWI volume as $I \in \mathbb{R}^{T \times H \times W \times C}$, where T is the number of DWI slices, H and W are the spatial dimensions (in pixels) of each 2D slice, respectively, and C is the number of DWI channels. The DWI volumes are preprocessed by skullstripping to remove non-brain tissues in all the DWI channels.

Our goal is to predict the segmentation map $O \in \mathbb{R}^{T \times H \times W}$ of stroke lesions. The spatial resolution within each slice is 1 mm between adjacent pixels while the inter-slice resolution is 6 mm between slices. We can observe that the inter-slice resolution of DWIs is much lower than the intra-slice resolution, which leads to the high discontinuity between adjacent slices—the main characteristic of 2.5D data like DWIs. As discussed in Sect. 1, both 3D and 2D segmentation models are not ideal for DWIs, because common 3D models are likely to overfit irrelevant features in neighboring slices, while 2D models completely disregard 3D contextual information. This motivates us to propose the `LambdaUNet`, a 2.5D segmentation model specifically designed for DWIs. Below, we will first provide an overview of `LambdaUNet` and then elaborate on how its `Lambda+` layers effectively capture 2.5D contextual features.

LambdaUNet. The main structure of our `LambdaUNet` follows the UNet [16] for its strong ability to preserve both high-level semantic features and low-level details. The key difference of `LambdaUNet` from the original UNet is that we replace convolutional layers in the UNet encoder with our proposed `Lambda+` layers (detailed in Sect. 2.1), which can extract both dense intra-slice features and sparse inter-slice features effectively. Since all layers except `Lambda+` layers in `LambdaUNet` are identical with those in UNet, they require 2D features as input; we address this by merging the slice dimension T with the batch dimension to reshape 3D features into 2D features for non-`Lambda+` layers, while `Lambda+` layers undo this reshaping to recover the slice dimension and regenerate a 3D input that is used to extract both intra- and inter-slice features. The final output of `LambdaUNet` is the lesion segmentation mask $O \in \mathbb{R}^{T \times H \times W}$. The Binary Cross-Entropy (BCE) loss is used to train `LambdaUNet` for the pixel-wise binary classification task.

2.1   Lambda+ Layers

`Lambda+` layers are an enhanced version of Lambda layers [1], which transform context around a pixel into linear functions, called *lambdas*, and mimic the attention operation by applying lambdas to the pixel to produce features. Different from attention, the lambdas can encode positional information as we will elaborate later, which affords them a stronger ability to model spatial relations.

`Lambda+` layers extend Lambda layers, which are designed for 2D data, by adding inter-slice lambdas with a restricted context region to effectively extract features from 2.5D data such as DWIs.

The input to a `Lambda+` layer is a 3D feature map $\boldsymbol{X} \in \mathbb{R}^{|n| \times |c|}$, where $|c|$ is the number of channels and n is the linearized pixel index into both spatial (height H and width W) and slice (T) dimensions of the feature map, *i.e.*, n iterates over all pixels $\mathcal{P}$ inside the 3D volume, and $|n|$ equals the total number of pixels $|\mathcal{P}|$. Besides input $\boldsymbol{X}$, we also have context $\boldsymbol{C} \in \mathbb{R}^{|m| \times |c|}$ where $\boldsymbol{C} = \boldsymbol{X}$ (same as self-attention) and m also iterates over all pixels $\mathcal{P}$ in the 3D volume. Importantly, when extracting features for each pixel n, we restrict the region of context pixels m to a 2.5D area $\mathcal{A}(n) \subset \mathcal{P}$. As shown in Fig. 2 (a), the 2.5D context area consists of the entire slice where pixel n is in, as well as pixels with the same 2D location in adjacent T slices where T is the inter-slice kernel size.

Similar to attention, `Lambda+` layer computes queries $\boldsymbol{Q} = \boldsymbol{X} \boldsymbol{W}_Q \in \mathbb{R}^{|n| \times |k|}$, keys $\boldsymbol{K} = \boldsymbol{C} \boldsymbol{W}_K \in \mathbb{R}^{|m| \times |k| \times |u|}$, and values $\boldsymbol{V} = \boldsymbol{C} \boldsymbol{W}_V \in \mathbb{R}^{|m| \times |v| \times |u|}$, where $\boldsymbol{W}_Q \in \mathbb{R}^{|c| \times |k|}$, $\boldsymbol{W}_K \in \mathbb{R}^{|c| \times |k| \times |u|}$ and $\boldsymbol{W}_V \in \mathbb{R}^{|c| \times |v| \times |u|}$ are learnable projection matrices, $|k|$ and $|v|$ are the dimensions of queries (keys) and values, and $|u|$ is an additional dimension to increase model capacity. We normalize the keys across pixels using softmax: $\bar{\boldsymbol{K}} = \text{softmax}(\boldsymbol{K})$. We denote $\boldsymbol{q}_n \in \mathbb{R}^{|k|}$ as the n-th query in $\boldsymbol{Q}$ for a pixel n. We also denote $\bar{\boldsymbol{K}}_m \in \mathbb{R}^{|k| \times |u|}$ and $\boldsymbol{V}_m \in \mathbb{R}^{|v| \times |u|}$ as the m-th key and value in $\boldsymbol{K}$ and $\boldsymbol{V}$ for a context pixel m.

For a target pixel $n \in \mathcal{P}$ inside a slice t, a `lambda+` layer computes three types of lambdas (linear functions) as illustrated in Fig. 2: (1) a **global lambda** that encodes global context within slice t, (2) a **local lambda** that summarizes the local context around pixel n in slice t, and (3) an **inter-slice lambda** that captures inter-slice features from adjacent slices.

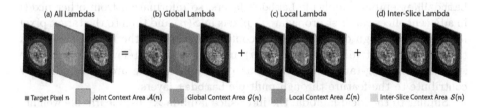

■ Target Pixel n ▨ Joint Context Area $\mathcal{A}(n)$ ▨ Global Context Area $\mathcal{G}(n)$ ▨ Local Context Area $\mathcal{L}(n)$ ▨ Inter-Slice Context Area $\mathcal{S}(n)$

Fig. 2. Context areas of the global lambda, local lambda, and inter-slice lambda.

Global Lambda. As shown in Fig. 2(b), the global lambda aims to encode the global context within slice t where the target pixel n is in, so the context area $\mathcal{G}(n)$ of the global lambda includes all pixels within slice t. For each context pixel $m \in \mathcal{G}(n)$, its contribution to the global lambda is computed as:

$$\boldsymbol{\mu}_m^{\text{G}} = \bar{\boldsymbol{K}}_m \boldsymbol{V}_m^T, \quad m \in \mathcal{G}(n). \tag{1}$$

The global lambda $\boldsymbol{\lambda}_n^{\mathcal{G}}$ is the sum of the contributions from each pixel $m \in \mathcal{G}(n)$:

$$\boldsymbol{\lambda}_n^{\mathcal{G}} = \sum_{m \in \mathcal{G}(n)} \boldsymbol{\mu}_m^{\mathcal{G}} = \sum_{m \in \mathcal{G}(n)} \bar{\boldsymbol{K}}_m \boldsymbol{V}_m^T \in \mathbb{R}^{|k| \times |v|} . \tag{2}$$

Note that $\boldsymbol{\lambda}_n^{\mathcal{G}}$ is invariant for all n within the same slice as $\mathcal{G}(n)$ is the same.

Local Lambda. The local lambda encodes the context of a local $R \times R$ area $\mathcal{L}(n)$ centered around the target pixel n in slice t (see Fig. 2(c)). Compared with the global lambda, besides the difference in context areas, the local lambda uses *learnable* relative-position-dependent weights $\boldsymbol{E}_{nm} \in \mathbb{R}^{|k| \times |u|}$ to encode the position-aware contribution of a context pixel m to the local lambda:

$$\boldsymbol{\mu}_{nm}^{\mathrm{L}} = \boldsymbol{E}_{nm} \boldsymbol{V}_m^T, \quad m \in \mathcal{L}(n) . \tag{3}$$

Note that the weights $\boldsymbol{E}_{nm}$ are shared for any pairs of pixels (n, m) with the same relative position between n and m. The local lambda $\boldsymbol{\lambda}^{\mathrm{L}}$ is obtained by:

$$\boldsymbol{\lambda}_n^{\mathrm{L}} = \sum_{m \in \mathcal{L}(n)} \boldsymbol{\mu}_{nm}^{\mathrm{L}} = \sum_{m \in \mathcal{L}(n)} \boldsymbol{E}_{nm} \boldsymbol{V}_m^T \in \mathbb{R}^{|k| \times |v|} . \tag{4}$$

Inter-Slice Lambda. The inter-slice lambda defines a context area $\mathcal{S}(n)$ including pixels in adjacent slices sharing the same 2D location with the target pixel n, as shown in Fig. 2(d). As discussed before, we use this restricted context area for extracting inter-slice features due to the high discontinuity between slices for 2.5D data like DWIs. Although one context pixel per adjacent slice seems very restrictive, one pixel of a feature map at coarse (downsampled) 2D scales in `LambdaUNet` corresponds to a large area in the original scale. Furthermore, `LambdaUNet` employs multiple `Lambda+` layers, so information from other pixels in adjacent slices can first propagate to pixels in $\mathcal{S}(n)$ and then to the target pixel n. Thus, our design of the restricted context area makes the network focus on the most-informative pixels inside $\mathcal{S}(n)$ and suppress less-relevant pixels, while still allowing long-range interactions as pixels outside the area can indirectly contribute to the feature through multiple `Lambda+` layers.

Similar to the local lambda, the inter-slice lambda $\boldsymbol{\lambda}_n^{\mathrm{S}}$ uses learnable weights $\boldsymbol{F}_{nm} \in \mathbb{R}^{|k| \times |u|}$ to encode position-aware contribution of context pixels:

$$\boldsymbol{\mu}_{nm}^{\mathrm{S}} = \boldsymbol{F}_{nm} \boldsymbol{V}_m^T, \quad m \in \mathcal{S}(n) , \tag{5}$$

$$\boldsymbol{\lambda}_n^{\mathrm{S}} = \sum_{m \in \mathcal{S}(n)} \boldsymbol{\mu}_{nm}^{\mathrm{S}} = \sum_{m \in \mathcal{S}(n)} \boldsymbol{F}_{nm} \boldsymbol{V}_m^T \in \mathbb{R}^{|k| \times |v|} . \tag{6}$$

Applying Lambdas. After computing the global lambda $\boldsymbol{\lambda}_n^{\mathcal{G}}$, local lambda $\boldsymbol{\lambda}_n^{\mathrm{L}}$, and inter-slice lambda $\boldsymbol{\lambda}_n^{\mathrm{S}}$, we are ready to apply them to the query $\boldsymbol{q}_n$ of the target pixel n. The output feature $\boldsymbol{y}_n$ for the target pixel n is:

$$\boldsymbol{y}_n = \boldsymbol{q}_n^T \left(\boldsymbol{\lambda}_n^{\mathcal{G}} + \boldsymbol{\lambda}_n^{\mathrm{L}} + \boldsymbol{\lambda}_n^{\mathrm{S}} \right) \in \mathbb{R}^{|v|} . \tag{7}$$

The final output of Lambda+ layer is a 3D feature map $Y \in \mathbb{R}^{|n| \times |v|}$ formed by the output features y_n of all pixels $n \in \mathcal{P}$. Although the above procedure for computing lambdas is for a single pixel n, we can easily parallelize the computation for all pixels using standard convolution operations, which makes Lambda+ layers computationally efficient. We refer readers to the pseudocode in the supplementary materials for detailed implementation.

3   Experiments

The primary focus of our experiments is to answer the following questions: (1) Does LambdaUNet predict lesion segmentation maps more accurately than baselines? (2) Is our 2.5D Lambda+ layer more effective than the 2D or 3D Lambda layer? (3) Based on qualitative results, does LambdaUNet has clinical significance?

Dataset. The clinical data we use to evaluate our model is provided by an urban academic hospital. We sampled 99 acute ischemic stroke cases with large ($n = 42$) and small ($n = 57$) infarct size. The data has an equal distribution of samples from stroke with the left or right middle cerebral artery (MCA), posterior cerebral artery (PCA), and anterior cerebral artery (ACA) origins. The cases contain a mix of 1.5T and 3.0T scans. Certain cases even have a mix of MCA and ACA. The ischemic infarcts are manually segmented by three experts based on diffusion-weighted imaging (DWI) (b = $1000 \, \text{s/mm}^2$) and the calculated exponential apparent diffusion map (eADC) using MRIcro v1.4. We use the eADC and DWI images from ischemic stroke patients to form the two channels of input DWIs I. We use 67 of the 99 fully labeled cases for training and the remaining 32 fully labeled cases for validation and testing. More specifically, we split the 32 cases into three folds of roughly the same size. Two of the three folds are used for validation and one remaining fold is used for testing. Each of the three folds is used for testing once, and the average result is reported as the final testing result. The 32 cases used for testing were carefully chosen to make sure the stroke size, location, and type are nicely balanced in the testing set.

Implementation Details. Our implementation is using the PyTorch [15] and the Lightning [6] frameworks. All experiments are conducted using four NVIDIA Quadro RTX 6000 GPUs with 24 GB memory. For Lambda+ layers, both the inter-slice kernel size T and the local kernel size R are set to 3. We train the model for 100 epochs using the RMSprop optimizer; an initial learning rate of 1e-4 is used for 20 epochs and then the learning rate is linearly reduced to 0. We randomly select 12 DWI sequence segments of 8 slices to form a mini-batch during training. The whole training process takes about 4h to finish. The training converges after 40 epochs. For testing, we select the model that gives the highest dice score for validation data.

Baselines and Metrics. We compare our method against well-known and recent 2D segmentation methods, U-Net [16], AttnUNet [13], and TransUNet [3], as well as one 3D segmentation method: 3D UNet [5]. All the baseline methods are reproduced based on their open-sourced code with careful hyperparameter tuning. Besides, we also report the results of two variants of LambdaUNet to further evaluate the effectiveness of the proposed 2.5D lambda+ layer. We use four common evaluation metrics—dice score coefficient (DSC), recall, precision, and F_1 score—for stroke lesion segmentation to provide quantitative comparisons.

3.1 Results

In the first group of Table 1, we show the slice-level accuracy of all baselines on our stroke lesion dataset. One can observe that the proposed LambdaUNet has significant improvements over baselines, e.g., performance gains range from 3.06% to 8.31% for average DSC. The improvement suggests that our Lambda+ layers are more suitable for feature extraction of 2.5D DWI data. In the second group, we compare LambdaUNet with its 2D and 3D variants. LambdaUNet2D directly removes the inter-slice lambda from the LambdaUNet while LambdaUNet3D uses a 3D local context area $\mathcal{L}(n)$ instead of the inter-slice lambda. As indicated in Table 1, both variants perform worse than LambdaUNet in terms of DSC and the F_1 score. This demonstrates the effectiveness of the 2.5D design of the proposed Lambda+ layers. Although LambdaUNet does not achieve the highest precision or recall over the baselines and variants, it can maintain a good balance between recall and precision, which sometimes cancel each other out (e.g., AttnUNet and 3D UNet). This is further confirmed by the superior F_1 score of LambdaUNet.

Table 1. Segmentation performance comparison between different models.

Method	2D/3D	DSC	Recall/Precision	F_1 Score
UNet [16]	2D	82.15	80.28/86.29	81.61
AttnUNet [13]	2D	81.83	77.45/86.74	80.82
TransUNet [3]	2D	83.45	83.24/87.15	84.48
3D UNet [5]	3D	78.20	**83.54**/78.39	78.21
(Ours) LambdaUNet-2D	2D	84.03	82.27/87.10	84.19
(Ours) LambdaUNet-3D	3D	84.76	79.92/**89.86**	84.09
(Ours) LambdaUNet	2.5D	**86.51**	81.76/89.39	**84.84**

Figure 3 visualizes the predicted segmentation masks on five consecutive slices for one stroke case. We can see that the masks produced by our LambdaUNet (last column) are the closest to the ground truth than the baselines. For instance, in slice 3 (S3), the baselines either miss some details (UNet, AttnUNet, TransUNet) indicated by the white areas or generate some false positive predictions (3D UNet) denoted by the red areas, while our LambdaUNet captures the irregular shape of lesions well. S4 and S5 also show that LambdaUNet performs the best on difficult small lesions. More results are provided in the supplementary materials.

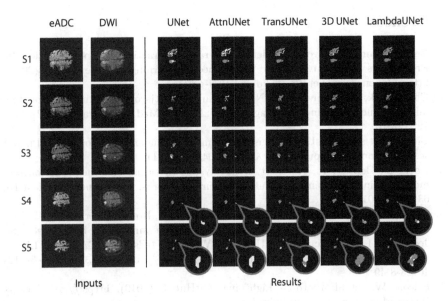

Fig. 3. Qualitative results on five consecutive slices of one ischemic stroke clinical case. Green indicates the correct predictions. White areas are false negative while red areas are false positive. Red circles show a close-up view of the lesion areas. (Color figure online)

3.2 Discussion

Our LambdaUNet not only shows advantages on both quantitative and qualitative measurements, the way it extracts features is more like clinicians. As clinicians consider all adjacent slices but only focus on the most informative areas, our Lambda+ layers capture intra- and inter-slice features and automatically suppress irrelevant 3D interference. Lesion areas of acute stroke are an important end-point for clinical trials, as proper treatment relies on measuring the infarction core volume and estimating salvageable tissue. Therefore, an accurate and reproducible DWI-suited segmentation model like LambdaUNet will be of high interest in clinical practice.

4 Conclusion

We defined DWIs as 2.5D data for their dense intra-slice resolution and sparse inter-slice resolution. Based on the 2.5D characteristics, we proposed a segmentation network LambdaUNet, which includes a new 2.5D feature extractor, termed Lambda+ layers. Lambda+ layers effectively capture features in 2.5D data by using dense intra-slice and sparse inter-slice context areas. This design allows the network to focus on informative features while suppressing less relevant features to reduce noise and improve generalization. Experiments on the clinical stroke dataset verify that our LambdaUNet outperforms state-of-the-art segmentation methods and shows strong potential in clinical practice.

References

1. Bello, I.: LambdaNetworks: modeling long-range interactions without attention. In: Proceedings of the International Conference on Learning Representations (2021). https://openreview.net/forum?id=xTJEN-ggl1b

2. Charoensuk, W., Covavisaruch, N., Lerdlum, S., Likitjaroen, Y.: Acute stroke brain infarct segmentation in DWI images. Int. J. Pharma Med. Biological Sci. **4**(2), 115–122 (2015)

3. Chen, J., et al.: TransUNet: transformers make strong encoders for medical image segmentation. arXiv preprint arXiv:2102.04306 (2021)

4. Chen, L., Bentley, P., Rueckert, D.: Fully automatic acute ischemic lesion segmentation in DWI using convolutional neural networks. NeuroImage: Clinical **15**, 633–643 (2017)

5. Çiçek, Ö., Abdulkadir, A., Lienkamp, S.S., Brox, T., Ronneberger, O.: 3D U-Net: learning dense volumetric segmentation from sparse annotation. In: Ourselin, S., Joskowicz, L., Sabuncu, M.R., Unal, G., Wells, W. (eds.) MICCAI 2016. LNCS, vol. 9901, pp. 424–432. Springer, Cham (2016). https://doi.org/10.1007/978-3-319-46723-8_49

6. Falcon, W., et al: PyTorch Lightning. GitHub 3 (2019). https://github.com/PyTorchLightning/pytorch-lightning

7. Han, L., Chen, Y., Li, J., Zhong, B., Lei, Y., Sun, M.: Liver segmentation with 2.5D perpendicular UNets. Comput. Electr. Eng. **91**, 107118 (2021)

8. Huang, G., Liu, Z., Van Der Maaten, L., Weinberger, K.Q.: Densely connected convolutional networks. In: Proceedings of the IEEE Conference on Computer Vision and Pattern Recognition, pp. 4700–4708 (2017)

9. Johnson, W., Onuma, O., Owolabi, M., Sachdev, S.: Stroke: a global response is needed. Bull. World Health Organ. **94**(9), 634 (2016)

10. Kanchana, R., Menaka, R.: Ischemic stroke lesion detection, characterization and classification in CT images with optimal features selection. Biomed. Eng. Lett. **10**, 333–344 (2020)

11. Lansberg, M.G.: Diffusion-weighted MRI in Acute Stroke. Utrecht University Dissertation (2002)

12. Mozaffarian, D., Benjamin, E.J., Go, A.S., Arnett, D.K., Blaha, M.J., Cushman, M., Das, S.R., De Ferranti, S., Després, J.P., Fullerton, H.J., et al.: Heart disease and stroke statistics–2016 update: a report from the American Heart Association. Circulation **133**(4), e38–e360 (2016)

13. Oktay, O., et al.: Attention U-Net: Learning where to look for the pancreas. arXiv preprint arXiv:1804.03999 (2018)

14. Owolabi, M.O., et al.: The burden of stroke in Africa: a glance at the present and a glimpse into the future. Cardiovascular J. Africa **26**(2 H3Africa Suppl), S27 (2015)

15. Paszke, A., et al.: Pytorch: An imperative style, high-performance deep learning library. arXiv preprint arXiv:1912.01703 (2019)

16. Ronneberger, O., Fischer, P., Brox, T.: U-Net: convolutional networks for biomedical image segmentation. In: Navab, N., Hornegger, J., Wells, W.M., Frangi, A.F. (eds.) MICCAI 2015. LNCS, vol. 9351, pp. 234–241. Springer, Cham (2015). https://doi.org/10.1007/978-3-319-24574-4_28

17. Zhang, C., Hua, Q., Chu, Y., Wang, P.: Liver tumor segmentation using 2.5D UV-Net with multi-scale convolution. Comput. Biology Med. **133**, 104424 (2021)

18. Zhang, H., Valcarcel, A.M., Bakshi, R., Chu, R., Bagnato, F., Shinohara, R.T., Hett, K., Oguz, I.: Multiple sclerosis lesion segmentation with tiramisu and 2.5D stacked slices. In: Shen, D., Liu, T., Peters, T.M., Staib, L.H., Essert, C., Zhou, S., Yap, P.-T., Khan, A. (eds.) MICCAI 2019. LNCS, vol. 11766, pp. 338–346. Springer, Cham (2019). https://doi.org/10.1007/978-3-030-32248-9_38

19. Zhang, R., Zhao, L., Lou, W., Abrigo, J.M., Mok, V.C., Chu, W.C., Wang, D., Shi, L.: Automatic segmentation of acute ischemic stroke from DWI using 3-D fully convolutional DenseNets. IEEE Trans. Med. Imaging **37**(9), 2149–2160 (2018)

20. Zhou, D., et al.: Eso-Net: a novel 2.5D segmentation network with the multistructure response filter for the cancerous esophagus. IEEE Access **8**, 155548–155562 (2020)

Correction to: Interactive Segmentation via Deep Learning and B-Spline Explicit Active Surfaces

Helena Williams, João Pedrosa, Laura Cattani, Susanne Housmans,
Tom Vercauteren, Jan Deprest, and Jan D'hooge

Correction to:
Chapter "Interactive Segmentation via Deep Learning
and B-Spline Explicit Active Surfaces" in: M. de Bruijne et al.
(Eds.): *Medical Image Computing and Computer Assisted*
Intervention – MICCAI 2021, **LNCS 12901,**
https://doi.org/10.1007/978-3-030-87193-2_30

The original version of this chapter was revised. The figure 2 with missing information was corrected.

The updated version of this chapter can be found at
https://doi.org/10.1007/978-3-030-87193-2_30

© Springer Nature Switzerland AG 2022
M. de Bruijne et al. (Eds.): MICCAI 2021, LNCS 12901, p. C1, 2022.
https://doi.org/10.1007/978-3-030-87193-2_70

Author Index

Printed in the United States
by Baker & Taylor Publisher Services

Printed in the United States
by Baker & Taylor Publisher Services